Theory of Elasticity

ENGINEERING SOCIETIES MONOGRAPHS

Bakhmeteff: Hydraulics of Open Channels
Bleich: Buckling Strength of Metal Structures
Crandall: Engineering Analysis
Elevatorski: Hydraulic Energy Dissipators
Ippen: Estuary for Coastline Hydrodynamics
Leontovich: Frames and Arches
Nadai: Theory of Flow and Fracture of Solids
Timoshenko and Gere: Theory of Elastic Stability
Timoshenko and Goodier: Theory of Elasticity
Timoshenko and Woinowsky-Krieger: Theory of Plates and Shells

Five national engineering societies, the American Society of Civil Engineers, the American Institute of Mining, Metallurgical, and Petroleum Engineers, the American Society of Mechanical Engineers, the Institute of Electrical and Electronics Engineers, and the American Institute of Chemical Engineers, have an arrangement with the McGraw-Hill Book Company for the production of a series of selected books adjudged to possess usefulness for engineers and industry.

The purposes of this arrangement are: to provide monographs of high technical quality within the field of engineering; to rescue from obscurity important technical manuscripts which might not be published commercially because of too limited sale without special introduction; to develop manuscripts to fill gaps in existing literature; to collect into one volume scattered information of especial timeliness on a given subject.

The societies assume no responsibility for any statements made in these books. Each book before publication has, however, been examined by one or more representatives of the societies competent to express an opinion on the merits of the manuscript.

Theory of Elasticity

THIRD EDITION

S. P. Timoshenko

PROFESSOR EMERITUS OF ENGINEERING MECHANICS

J. N. Goodier

PROFESSOR OF APPLIED MECHANICS
STANFORD UNIVERSITY

INTERNATIONAL STUDENT EDITION

McGRAW-HILL BOOK COMPANY

Auckland Bogotá Guatemala Hamburg Lisbon
London Madrid Mexico New Delhi Panama Paris San Juan
São Paulo Singapore Sydney Tokyo

Theory of Elasticity
INTERNATIONAL EDITION 1970

30 29 28 27 26 25 24 23
20 09 08 07 06 05 04 03
BJE

Library of Congress Catalog Card Number 69-13617

When ordering this title use ISBN 0-07-085805-5

Printed in Singapore

Preface to the Third Edition

In the revision of this book for a third edition, the primary intention and plan of the first edition have been preserved—to provide for engineers, in as simple a form as the subject allows, the essential fundamental knowledge of the theory of elasticity together with a compilation of solutions of special problems that are important in engineering practice and design. The numerous footnote references indicate how the several topics may be pursued further. Since these are now readily supplemented by means of *Applied Mechanics Reviews*, new footnotes have been added sparingly with this in mind. Small print again indicates sections that can be omitted from a first reading.

The whole text has been reexamined, and many minor improvements have been made throughout by elimination and rearrangement as well as addition.

The major additions reflect developments and extensions of interest and practical applicability that have occurred since the appearance of the second edition in 1951. End effects and eigensolutions associated with the principle of Saint-Venant are treated in Chapters 3 and 4. In view of the rapid growth of the applications of dislocational elastic solutions in materials science, these discontinuous displacement solutions have been given more explicit treatment as edge dislocations and screw dislocations in Chapters 4, 8, 9, and 12. An introduction to the moiré method with a practical illustration has been added to Chapter 5. The treatment of strain energy and variational principles has been recast in three-dimensional form and embodied in Chapter 8, which now provides a basis for new sections on thermoelasticity in Chapter 13. The discussion of the use of complex potentials for two-dimensional problems has been extended by a group of new articles based on the now well-known methods of Muskhelishvili. Moreover, the approach is somewhat different, in that advantage has been taken of solutions previously developed in order to deal with analytic functions only. Further solutions for the elliptic hole, important in current fracture mechanics (cracks), are given explicit treatment. The discussion of axisymmetric stress in Chapter 12 has been simplified; and new sections have been added that replace the approximate analysis by a more exact one for the cut ring, as one turn of a helical spring. In view of its greatly increased applications, as in nuclear energy equipment, Chapter 13, on thermal stress, has been extended by inclusion of a thermoelastic reciprocal theorem and several useful results obtained from it; and by an introduction to thermal stress concentrations due to disturbance of heat flow by cavities and inclusions has also been added. In addition, treatment of two-dimensional problems has been supplemented by the two final articles, the last bringing the two-dimensional thermoelastic problems into connection with the complex potentials and Muskhelishvili procedures of Chapter 6. In Chapter 14, on wave propagation, a rearrangement gives prominence to the basic three-dimensional theory. A solution for explosive pressure in a spherical cavity has been added. The Appendix on numerical finite difference methods includes an example of the use of a digital computer to cope with a large number of unknowns.

Some of these changes offer simplifications of analysis arrived at in the experience of giving courses at Stanford University over the past twenty years. Many valuable suggestions, corrections, and even completely formulated problems with solutions, have come from numerous students and correspondents, to whom a blanket but most cordial acknowledgment is both unavoidable and inadequate.

Almost all the "Problems" are from examinations set and given at Stanford University. The reader may see roughly from these what parts of the book correspond to a course sequence occupying somewhat less than three hours per week for the academic year.

J. N. Goodier

Preface to the Second Edition

The many developments and clarifications in the theory of elasticity and its applications which have occurred since the first edition was written are reflected in numerous additions and emendations in the present edition. The arrangement of the book remains the same for the most part.

The treatments of the photoelastic method, two-dimensional problems in curvilinear coordinates, and thermal stress have been rewritten and enlarged into separate new chapters which present many methods and solutions not given in the former edition. An appendix on the method of finite differences and its applications, including the relaxation method, has been added. New articles and paragraphs incorporated in the other chapters deal with the theory of the strain gauge rosette, gravity stresses, Saint-Venant's principle, the components of rotation, the reciprocal theorem, general solutions, the approximate character of the plane stress solutions, center of twist and center of shear, torsional stress concentration at fillets, the approximate treatment of slender (*e.g.*, solid airfoil) sections in torsion and bending, and the circular cylinder with a band of pressure.

Problems for the student have been added covering the text as far as the end of the chapter on torsion.

It is a pleasure to make grateful acknowledgment of the many helpful suggestions which have been contributed by readers of the book.

S. P. Timoshenko

J. N. Goodier

Preface to the First Edition

During recent years the theory of elasticity has found considerable application in the solution of engineering problems. There are many cases in which the elementary methods of strength of materials are inadequate to furnish satisfactory information regarding stress distribution in engineering structures, and recourse must be made to the more powerful methods of the theory of elasticity. The elementary theory is insufficient to give information regarding local stresses near the loads and near the supports of beams. It fails also in the cases when the stress distribution in bodies, all the dimensions of which are of the same order, has to be investigated. The stresses in rollers and in balls of bearings can be found only by using the methods of the theory of elasticity. The elementary theory gives no means of investigating stresses in regions of sharp variation in cross section of beams or shafts. It is known that at reentrant corners a high stress concentration occurs and as a result of this cracks are likely to start at such corners, especially if the structure is submitted to a reversal of stresses. The majority of fractures of machine parts in service can be attributed to such cracks.

During recent years considerable progress has been made in solving such practically important problems. In cases where a rigorous solution cannot be readily obtained, approximate methods have been developed. In some cases solutions have been obtained by using experimental methods. As an example of this the photoelastic method of solving two-dimensional problems of elasticity may be mentioned. The photoelastic equipment may be found now at universities and also in many industrial research laboratories. The results of photoelastic experiments have proved especially useful in studying various cases of stress concentration at points of sharp variation of cross-sectional dimensions and at sharp fillets of reentrant corners. Without any doubt these results have considerably influenced the modern design of machine parts and helped in many cases to improve the construction by eliminating weak spots from which cracks may start.

Another example of the successful application of experiments in the solution of elasticity problems is the soap-film method for determining stresses in torsion and bending of prismatical bars. The difficult problems of the solution of partial differential equations with given boundary conditions are replaced in this case by measurements of slopes and deflections of a properly stretched and loaded soap film. The experiments show that in this way not only a visual picture of the stress distribution but also the necessary information regarding magnitude of stresses can be obtained with an accuracy sufficient for practical application.

Again, the electrical analogy which gives a means of investigating torsional stresses in shafts of variable diameter at the fillets and grooves is interesting. The analogy between the problem of bending of plates and the two-dimensional problem of elasticity has also been successfully applied in the solution of important engineering problems.

In the preparation of this book the intention was to give to engineers, in a simple form, the necessary fundamental knowledge of the theory of elasticity. It was also intended to bring together solutions of special problems which may be of practical importance and to describe approximate and experimental methods of the solution of elasticity problems.

Having in mind practical applications of the theory of elasticity, matters of more theoretical interest and those which have not at present any direct applications in engineering have been omitted in favor of the discussion of specific cases. Only by studying such cases with all the details and by comparing the results of exact investigations with the approximate solutions usually given in the elementary books on strength of materials can a designer acquire a thorough understanding of stress distribution in engineering structures, and learn to use, to his advantage, the more rigorous methods of stress analysis.

In the discussion of special problems in most cases the method of direct determination of stresses and the use of the compatibility equations in terms of stress components has been applied. This method is more familiar to engineers who are usually interested in the magnitude of stresses. By a suitable introduction of stress functions this method is also often simpler than that in which equations of equilibrium in terms of displacements are used.

In many cases the energy method of solution of elasticity problems has been used. In this way the integration of differential equations is replaced by the investigation of minimum conditions of certain integrals. Using Ritz's method this problem of variational calculus is reduced to a simple problem of finding a minimum of a function. In this manner useful approximate solutions can be obtained in many practically important cases.

To simplify the presentation, the book begins with the discussion of two-dimensional problems and only later, when the reader has familiarized himself with the various methods used in the solution of problems of the theory of elasticity, are three-dimensional problems discussed. The portions of the book that, although of practical importance, are such that they can be omitted during the first reading are put in small type. The reader may return to the study of such problems after finishing with the most essential portions of the book.

The mathematical derivations are put in an elementary form and usually do not require more mathematical knowledge than is given in engineering schools. In the cases of more complicated problems all necessary explanations and intermediate calculations are given so that the reader can follow without difficulty through all the derivations. Only in a few cases are final results given without complete derivations. Then the necessary references to the papers in which the derivations can be found are always given.

In numerous footnotes references to papers and books on the theory of elasticity which may be of practical importance are given. These references may be of interest to engineers who wish to study some special problems in more detail. They give also a picture of the modern development of the theory of elasticity and may be of some use to graduate students who are planning to take their work in this field.

In the preparation of the book the contents of a previous book ("Theory of Elasticity," vol. I, St. Petersburg, Russia, 1914) on the same subject, which represented a course of lectures on the theory of elasticity given in several Russian engineering schools, were used to a large extent.

The author was assisted in his work by Dr. L. H. Donnell and Dr. J. N. Goodier, who read over the complete manuscript and to whom he is indebted for many corrections and suggestions. The author takes this opportunity to thank also Prof. G. H. MacCullough, Dr. E. E. Weibel, Prof. M. Sadowsky, and Mr. D. H. Young, who assisted in the final preparation of the book by reading some portions of the manuscript. He is indebted also to Mr. L. S. Veenstra for the preparation of drawings and to Mrs. E. D. Webster for the typing of the manuscript.

S. P. Timoshenko

Contents

Chapter 5 | Photoelastic and Moiré Experimental Methods 150

Chapter 6 | Two-dimensional Problems in Curvilinear Coordinates 168

Notation

x, y, z Rectangular coordinates.
r, θ Polar coordinates.
ξ, η Orthogonal curvilinear coordinates; sometimes rectangular coordinates.
R, ψ, θ Spherical coordinates.
N Outward normal to the surface of a body.
l, m, n Direction cosines of the outward normal.
A Cross-sectional area.
I_x, I_y Moments of inertia of a cross section with respect to x and y axes.
I_p Polar moment of inertia of a cross section.
g Gravitational acceleration.
ρ Density.
q Intensity of a continuously distributed load.
p Pressure.
X, Y, Z Components of a body force per unit volume.
$\bar{X}, \bar{Y}, \bar{Z}$ Components of a distributed surface force per unit area.
M Bending moment.
M_t Torque.
$\sigma_x, \sigma_y, \sigma_z$ Normal components of stress parallel to x, y, and z axes.
σ_n Normal component of stress parallel to n.
σ_r, σ_θ Radial and tangential normal stresses in polar coordinates.
σ_ξ, σ_η Normal stress components in curvilinear coordinates.
$\sigma_r, \sigma_\theta, \sigma_z$ Normal stress components in cylindrical coordinates.
$\Theta = \sigma_x + \sigma_y + \sigma_z = \sigma_r + \sigma_\theta + \sigma_z.$
τ Shearing stress.
$\tau_{xy}, \tau_{xz}, \tau_{yz}$ Shearing-stress components in rectangular coordinates.
$\tau_{r\theta}$ Shearing stress in polar coordinates.
$\tau_{\xi\eta}$ Shearing stress in curvilinear coordinates.
$\tau_{r\theta}, \tau_{\theta z}, \tau_{rz}$ Shearing-stress components in cylindrical coordinates.
S Total stress on a plane. Surface tension.
u, v, w Components of displacements.
ϵ Unit elongation.
$\epsilon_x, \epsilon_y, \epsilon_z$ Unit elongations in x, y, and z directions.
$\epsilon_r, \epsilon_\theta$ Radial and tangential unit elongations in polar coordinates.

$e = \epsilon_x + \epsilon_y + \epsilon_z$	Volume expansion.
γ	Unit shear.
$\gamma_{xy}, \gamma_{xz}, \gamma_{yz}$	Shearing-strain components in rectangular coordinates.
$\gamma_{r\theta}, \gamma_{\theta z}, \gamma_{rz}$	Shearing-strain components in cylindrical coordinates.
E	Modulus of elasticity in tension and compression.
G	Modulus of elasticity in shear. Modulus of rigidity.
ν	Poisson's ratio.
$\mu = G, \lambda = \dfrac{\nu E}{(1+\nu)(1-2\nu)}$	Lamé's constants.
ϕ	Stress function.
$\phi(z), \psi(z), \chi(z)$	Complex potentials; functions of the complex variable $z = x + iy$.
$\bar{z}$	The conjugate complex variable $x - iy$.
C	Torsional rigidity.
θ	Angle of twist per unit length.
$F = 2G\theta$	Used in torsional problems.
V	Strain energy.
V_0	Strain energy per unit volume.
t	Time.
T	Certain interval of time. Temperature.
α	Coefficient of thermal expansion. Angle.
c_1, c_2	Wave velocities.

chapter 1

Introduction

1 | Elasticity

Almost all engineering materials possess to a certain extent the property of *elasticity*. If the external forces producing *deformation* do not exceed a certain limit, the deformation disappears with the removal of the forces. Throughout this book it will be assumed that the bodies undergoing the action of external forces are *perfectly elastic*, i.e., that they resume their initial form completely after removal of the forces.

Atomic structure will not be considered here. It will be assumed that the matter of an elastic body is *homogeneous* and continuously distributed over its volume so that the smallest element cut from the body possesses the same specific physical properties as the body. To simplify the discussion it will also be assumed that for the most part the body is *isotropic*, i.e., that the elastic properties are the same in all directions.

Structural materials do not satisfy the above assumptions completely. Such an important material as steel, for instance, when studied with a microscope, is seen to consist of crystals of various kinds and various orientations. The material is very far from being homogeneous, but experience shows that solutions of the theory of elasticity based on the assumptions of homogeneity and isotropy can be applied to steel structures with very great accuracy. The explanation of this is that the crystals are very small; usually there are millions of them in one cubic inch of steel. While the elastic properties of a single crystal may be very different in different directions, the crystals are ordinarily distributed at random and the elastic properties of larger pieces of metal represent averages of properties of the crystals. So long as the geometrical dimensions defining the form of a body are very large in comparison with the dimensions of a single crystal the assumption of homogeneity can be used with great accuracy, and if the crystals are orientated at random the material can be treated as isotropic.

When, due to certain technological processes such as rolling, a certain orientation of the crystals in a metal prevails, the elastic properties of the metal become different in different directions and the condition of *anisotropy* must be considered. We have such a condition, for instance, in cold-rolled copper.

2 | Stress

Figure 1 indicates a body in equilibrium. Under the action of external forces $P_1, \ldots, P_7$, internal forces will be produced between the parts of the body. To study the magnitude of these forces at any point O, let us imagine the body divided into two parts A and B by a cross section mm through this point. Considering one of these parts, for instance, A, it can be stated that it is in equilibrium under the action of external forces $P_1, \ldots, P_7$ and the internal forces distributed over the cross section mm and representing the actions of the material of the part B on the material of the part A. It will be assumed that these forces are continuously distributed over the area mm in the same way that hydrostatic pressure or wind pressure is continuously distributed over the sur face on which it acts. The magnitudes of such forces are usually defined by their *intensity*, i.e., by the amount of force per unit area of the surface on which they act. In discussing internal forces this intensity is called *stress*.

In the simplest case of a prismatical bar submitted to tension by forces uniformly distributed over the ends (Fig. 2), the internal forces are also

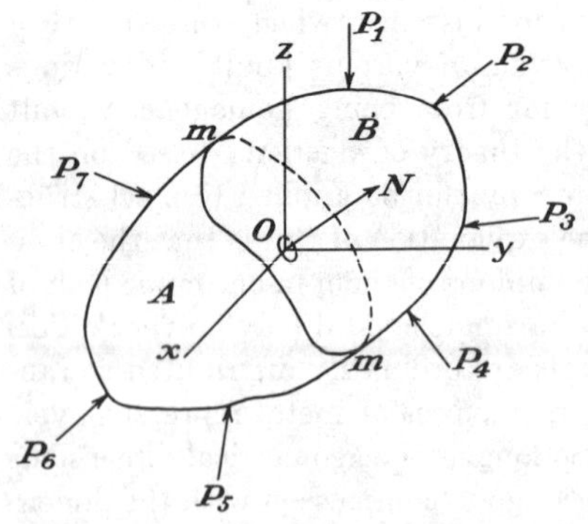

Fig. 1

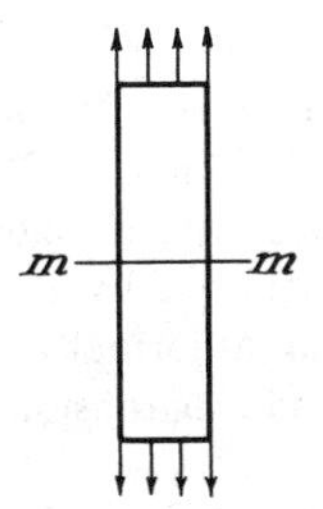

Fig. 2

uniformly distributed over any cross section mm. Hence the intensity of this distribution, i.e., the stress, can be obtained by dividing the total tensile force P by the cross-sectional area A.

In the case just considered the stress was uniformly distributed over the cross section. In the general case of Fig. 1 the stress is not uniformly distributed over mm. To obtain the magnitude of stress acting on a small area δA, cut out from the cross section mm at any point O, we observe that the forces acting across this elemental area, due to the action of material of the part B on the material of the part A, can be reduced to a resultant δP. If we now continuously contract the elemental area δA, the limiting value of the ratio $\delta P/\delta A$ gives us the magnitude of the stress acting on the cross section mm at the point O. The limiting direction of the resultant δP is the direction of the stress. In the general case the direction of stress is inclined to the area δA on which it acts and we can resolve it into two components: a *normal stress* perpendicular to the area and a *shearing stress* acting in the plane of the area δA.

3 | Notation for Forces and Stresses

There are two kinds of external forces which may act on bodies. Forces distributed over the surface of the body, such as the pressure of one body on another or hydrostatic pressure, are called *surface forces*. Forces distributed over the volume of a body, such as gravitational forces, magnetic forces, or in the case of a body in motion, inertia forces, are called *body forces*. The surface force per unit area we shall resolve into three components parallel to cartesian coordinate axes x, y, z, and use for these components the notation $\bar{X}$, $\bar{Y}$, $\bar{Z}$. We shall also resolve the body force per unit volume into three components and denote these components by X, Y, Z.

We shall use the letter σ for normal stress and the letter τ for shearing stress. To indicate the direction of the plane on which the stress is acting, subscripts to these letters are used. We take a very small cubic element at a point P (Fig. 1), with sides parallel to the coordinate axes. The notations for the components of stress acting on the sides of this element and the directions taken as positive are as indicated in Fig. 3. For the sides of the element perpendicular to the y axis, for instance, the normal components of stress acting on these sides are denoted by σ_y. The subscript y indicates that the stress is acting on a plane normal to the y axis. The normal stress is taken positive when it produces tension and negative when it produces compression.

The shearing stress is resolved into two components parallel to the coordinate axes. Two subscript letters are used in this case, the first indicating the direction of the normal to the plane under consideration and the second indicating the direction of the component of the stress. For instance, if we again consider the sides perpendicular to the y axis, the component in the x direction is denoted by τ_{yx} and that in the z direction by τ_{yz}. The positive directions of the components of shearing stress on any side of the cubic element are taken as the positive directions of the coordinate axes if a tensile stress on the same side would have the positive direction of the corresponding axis. If the tensile stress has a direction opposite to the positive axis, the positive directions of the shearing-stress components should be reversed. Following this rule, the positive directions of all the components of stress acting on the right side of the cubic element (Fig. 3) coincide with the positive directions of the coordinate axes. The positive directions are all reversed if we are considering the left side of this element.

4 | Components of Stress

From the discussion of the previous article, we see that for each pair of parallel sides of a cubic element, such as in Fig. 3, one symbol is needed

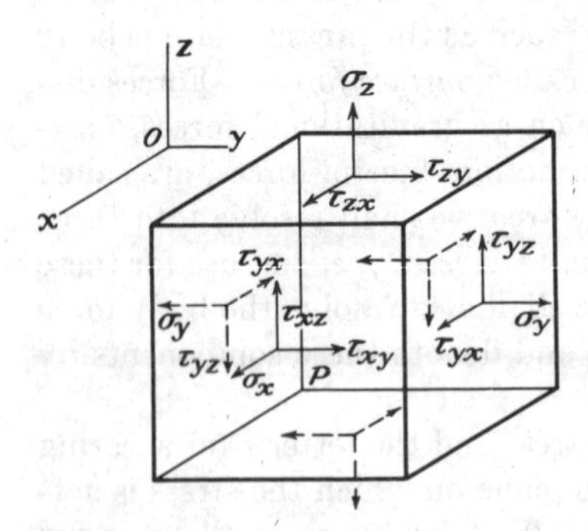

Fig. 3

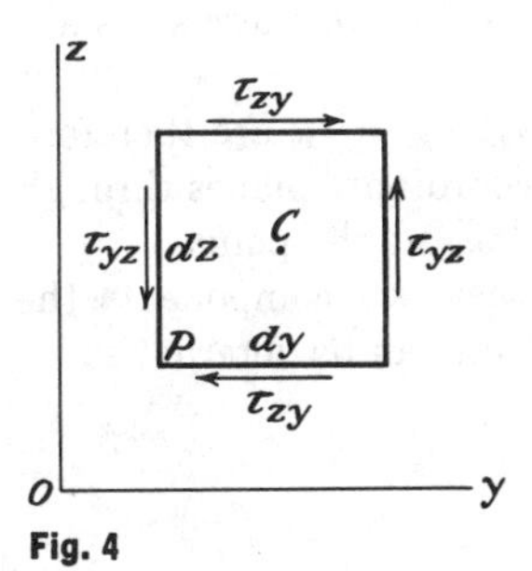

Fig. 4

to denote the normal component of stress and two more symbols to denote the two components of shearing stress. To describe the stresses acting on the six sides of the element three symbols σ_x, σ_y, σ_z are necessary for normal stresses; and six symbols τ_{xy}, τ_{yx}, τ_{xz}, τ_{zx}, τ_{yz}, τ_{zy}, for shearing stresses. By a simple consideration of the equilibrium of the element the number of symbols for shearing stresses can be reduced to three.

If we take the moments of the forces acting on the element about a line through the midpoint C and parallel to the x axis, for instance, only the surface stresses shown in Fig. 4 need be considered. Body forces, such as the weight of the element, can be neglected in this instance because in reducing the dimensions of the element the body forces acting on it diminish as the cube of the linear dimensions, whereas the surface forces diminish as the square of the linear dimensions. Hence, for a very small element, body forces are small quantities of higher order than surface forces and can be omitted in calculating the moments. Similarly, moments due to nonuniformity of distribution of normal forces are of higher order than those due to the shearing forces and vanish in the limit. Also the forces on each side can be considered to be the area of the side times the stress at the middle. Denoting the dimensions of the small element in Fig. 4 by dx, dy, dz, the equation of equilibrium of this element, taking moments of forces about C, is then

$$\tau_{zy}\, dx\, dy\, dz = \tau_{yz}\, dx\, dy\, dz$$

The two other equations can be obtained in the same manner. From these equations we find

$$\tau_{xy} = \tau_{yx} \qquad \tau_{zx} = \tau_{xz} \qquad \tau_{zy} = \tau_{yz} \tag{1}$$

Hence for two perpendicular sides of a cubic element the components of shearing stress perpendicular to the line of intersection of these sides are equal.[1]

The six quantities σ_x, σ_y, σ_z, $\tau_{xy} = \tau_{yx}$, $\tau_{xz} = \tau_{zx}$, $\tau_{yz} = \tau_{zy}$ are therefore sufficient to describe the stresses acting on the coordinate planes through a point; these will be called the *components of stress* at the point.

It will be shown later (Art. 74) that with these six components the stress on any inclined plane through the same point can be determined.

5 | Components of Strain

In discussing the deformation of an elastic body it will be assumed that there are enough constraints to prevent the body from moving as a rigid body so that no displacements of particles of the body are possible without a deformation of it.

In this book, only small deformations such as commonly occur in engineering structures will be considered. The small displacements of particles of a deformed body will first be resolved into components u, v, w parallel to the coordinate axes x, y, z, respectively. It will be assumed that these components are very small quantities varying continuously over the volume of the body. Consider a small element $dx\, dy\, dz$ of an elastic body (Fig. 5). If the body undergoes a deformation and u, v, w are the components of the displacement of the point P, the displacement in the x direction of an adjacent point A on the x axis is, to the first order in dx,

$$u + \frac{\partial u}{\partial x}\, dx$$

due to the increase $(\partial u/\partial x)\, dx$ of the function u with increase of the coordinate x. The increase in length of the element PA due to deformation is therefore $(\partial u/\partial x)\, dx$. Hence the *unit elongation* at point P in the x direction is $\partial u/\partial x$. In the same manner it can be shown that the

[1] There are exceptions, especially when stress is induced by electric and magnetic fields (see Prob. 2, p. 14).

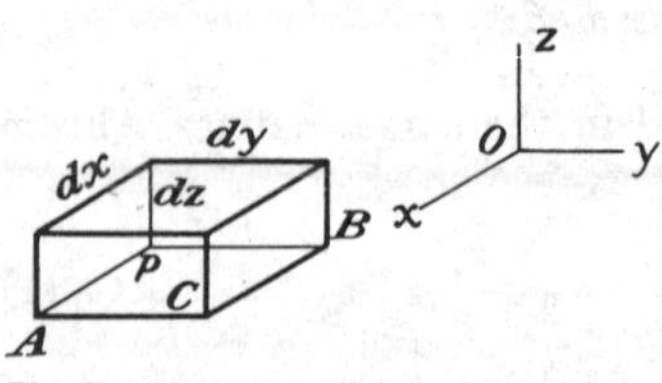

Fig. 5

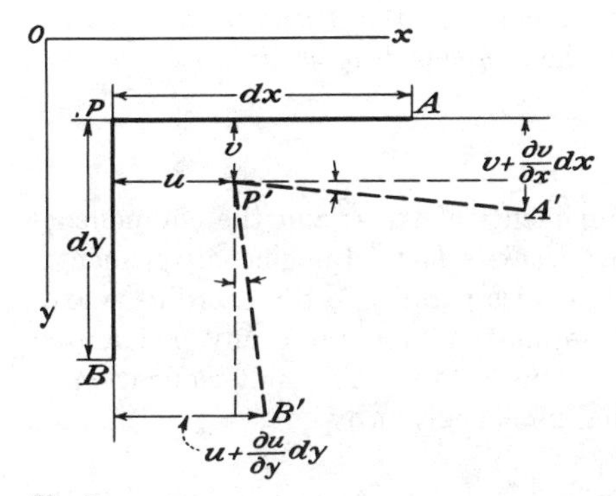

Fig. 6

unit elongations in the y and z directions are given by the derivatives $\partial v/\partial y$ and $\partial w/\partial z$.

Let us consider now the distortion of the angle between the elements PA and PB, Fig. 6. If u and v are the displacements of the point P in the x and y directions, the displacement of the point A in the y direction and of the point B in the x direction are $v + (\partial v/\partial x)\,dx$ and $u + (\partial u/\partial y)\,dy$, respectively. Owing to these displacements the new direction $P'A'$ of the element PA is inclined to the initial direction by the small angle indicated in the figure, equal to $\partial v/\partial x$. In the same manner the direction $P'B'$ is inclined to PB by the small angle $\partial u/\partial y$. From this it will be seen that the initially right angle APB between the two elements PA and PB is diminished by the angle $\partial v/\partial x + \partial u/\partial y$. This is the *shearing strain* between the planes xz and yz. The shearing strains between the planes xy and xz and the planes yx and yz can be obtained in the same manner.

We shall use the letter ϵ for unit elongation and the letter γ for unit shearing strain. To indicate the directions of strain we shall use the same subscripts to these letters as for the stress components. Then from the above discussion

$$
\begin{aligned}
\epsilon_x &= \frac{\partial u}{\partial x} & \epsilon_y &= \frac{\partial v}{\partial y} & \epsilon_z &= \frac{\partial w}{\partial z} \\
\gamma_{xy} &= \frac{\partial u}{\partial y} + \frac{\partial v}{\partial x} & \gamma_{xz} &= \frac{\partial u}{\partial z} + \frac{\partial w}{\partial x} & \gamma_{yz} &= \frac{\partial v}{\partial z} + \frac{\partial w}{\partial y}
\end{aligned}
\tag{2}
$$

It will be shown later that, having the three unit elongations in three perpendicular directions and three unit shear strains related to the same directions, the elongation in *any* direction and the distortion of the angle between *any* two directions can be calculated (see Art. 81). The six quantities $\epsilon_x, \ldots, \gamma_{yz}$ are called the *components of strain*.

6 | Hooke's Law

Linear relations between the components of stress and the components of strain are known generally as *Hooke's law*. Imagine an elemental rectangular parallelepiped with the sides parallel to the coordinate axes and submitted to the action of normal stress σ_x uniformly distributed over two opposite sides, as in the tensile test. The unit elongation of the element up to the proportional limit is given by

$$\epsilon_x = \frac{\sigma_x}{E} \qquad (a)$$

in which E is the *modulus of elasticity in tension*. Materials used in engineering structures have moduli which are very large in comparison with allowable stresses, and the unit elongation (*a*) is a very small quantity. In the case of structural steel, for instance, it is usually smaller than 0.001.

This extension of the element in the x direction is accompanied by lateral strain components (contractions)

$$\epsilon_y = -\nu \frac{\sigma_x}{E} \qquad \epsilon_z = -\nu \frac{\sigma_x}{E} \qquad (b)$$

in which ν is a constant called *Poisson's ratio*. For many materials Poisson's ratio can be taken equal to 0.25. For structural steel it is usually taken equal to 0.3.

Equations (*a*) and (*b*) can be also used for simple compression. The modulus of elasticity and Poisson's ratio in compression are the same as in tension.

If the above element is submitted simultaneously to the action of normal stresses $\sigma_x, \sigma_y, \sigma_z$, uniformly distributed over the sides, the resultant components of strain can be obtained from Eqs. (*a*) and (*b*). If we superpose the strain components produced by each of the three stresses, we obtain the equations

$$\begin{aligned} \epsilon_x &= \frac{1}{E}[\sigma_x - \nu(\sigma_y + \sigma_z)] \\ \epsilon_y &= \frac{1}{E}[\sigma_y - \nu(\sigma_x + \sigma_z)] \\ \epsilon_z &= \frac{1}{E}[\sigma_z - \nu(\sigma_x + \sigma_y)] \end{aligned} \qquad (3)$$

which have been found consistent with very numerous test measurements.

In our further discussion we shall often use this *method of superposition* in calculating total deformations and stresses produced by several forces. It is legitimate as long as the deformations are small and the corresponding small displacements do not affect substantially the action of the external forces. In such cases we neglect small changes in dimensions of deformed bodies and also small displacements of points of application of external forces and base our calculations on initial dimensions and initial shape of the body. The resultant displacements will then be obtained by superposition in the form of linear functions of external forces, as in deriving Eqs. (3).

There are, however, exceptional cases in which small deformations cannot be neglected but must be taken into consideration. As an example of this kind, the case of the simultaneous action on a thin bar of axial and lateral forces may be mentioned. Axial forces alone produce simple tension or compression, but they may have a substantial effect on the bending of the bar if they are acting simultaneously with lateral forces. In calculating the deformation of bars under such conditions, the effect of the deflection on the moment of the external forces must be considered, even though the deflections are very small.[1] Then the total deflection is no longer a linear function of the forces and cannot be obtained by simple superposition.

In Eqs. (3), the relations between elongations and stresses are completely defined by two physical constants E and ν. The same constants can also be used to define the relation between shearing strain and shearing stress.

Let us consider the particular case of deformation of the rectangular parallelepiped in which $\sigma_z = \sigma$, $\sigma_y = -\sigma$, and $\sigma_x = 0$. Cutting out an element *abcd* by planes parallel to the x axis and at 45° to the y and z axes (Fig. 7), it may be seen from Fig. 7*b*, by summing up the forces along and perpendicular to *bc*, that the normal stress on the sides of this element is

[1] Several examples of this kind can be found in S. Timoshenko, "Strength of Materials," 3d ed., vol. 2, chap. 2, D. Van Nostrand Company, Inc., Princeton, N.J., 1956.

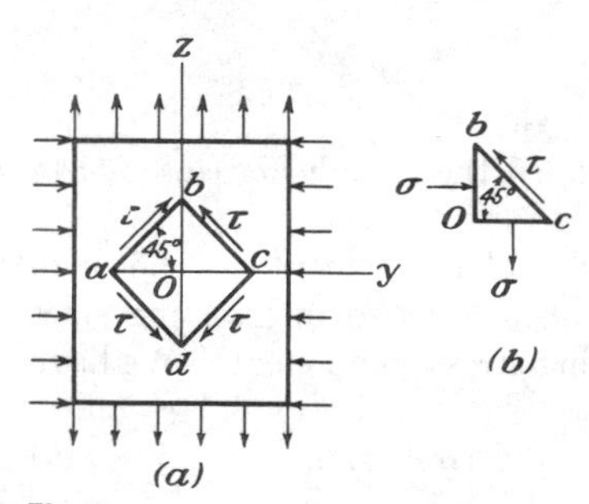

Fig. 7

zero and the shearing stress on the sides is

$$\tau = \tfrac{1}{2}(\sigma_z - \sigma_y) = \sigma \tag{c}$$

Such a condition of stress is called *pure shear*. The elongation of the vertical element Ob is equal to the shortening of the horizontal elements Oa and Oc, and neglecting a small quantity of the second order we conclude that the lengths ab and bc of the element do not change during deformation. The angle between the sides ab and bc changes, and the corresponding magnitude of shearing strain γ may be found from the triangle Obc. After deformation, we have

$$\frac{Oc}{Ob} = \tan\left(\frac{\pi}{4} - \frac{\gamma}{2}\right) = \frac{1 + \epsilon_y}{1 + \epsilon_z}$$

Substituting, from Eqs. (3),

$$\epsilon_z = \frac{1}{E}(\sigma_z - \nu\sigma_y) = \frac{(1 + \nu)\sigma}{E}$$

$$\epsilon_y = -\frac{(1 + \nu)\sigma}{E}$$

and noting that for small γ

$$\tan\left(\frac{\pi}{4} - \frac{\gamma}{2}\right) = \frac{\tan\frac{\pi}{4} - \tan\frac{\gamma}{2}}{1 + \tan\frac{\pi}{4}\tan\frac{\gamma}{2}} = \frac{1 - \frac{\gamma}{2}}{1 + \frac{\gamma}{2}}$$

we find

$$\gamma = \frac{2(1 + \nu)\sigma}{E} = \frac{2(1 + \nu)\tau}{E} \tag{4}$$

Thus the relation between shearing strain and shearing stress is defined by the constants E and ν. Often the notation

$$G = \frac{E}{2(1 + \nu)} \tag{5}$$

is used. Then Eq. (4) becomes

$$\gamma = \frac{\tau}{G}$$

The constant G, defined by Eq. (5), is called the *modulus of elasticity in shear*, or the *modulus of rigidity*.

If shearing stresses act on all the sides of an element, as shown in Fig. 3, the distortion of the angle between any two intersecting sides depends only on the corresponding shearing-stress component. We have

$$\gamma_{xy} = \frac{1}{G}\tau_{xy} \qquad \gamma_{yz} = \frac{1}{G}\tau_{yz} \qquad \gamma_{zx} = \frac{1}{G}\tau_{zx} \tag{6}$$

The elongations (3) and the distortions (6) are independent of each other. The general case of strain, produced by three normal and three shearing components of stress, can be obtained by superposition: on the three elongations given by Eqs. (3) are superposed three shearing strains given by Eqs. (6).

Equations (3) and (6) give the components of strain as functions of the components of stress. Sometimes the components of stress expressed as functions of the components of strain are needed. These can be obtained as follows. Adding Eqs. (3) together and using the notations

$$\begin{aligned} e &= \epsilon_x + \epsilon_y + \epsilon_z \\ \Theta &= \sigma_x + \sigma_y + \sigma_z \end{aligned} \tag{7}$$

we obtain the following relation between the volume expansion e and the sum of normal stresses:

$$e = \frac{1 - 2\nu}{E} \Theta \tag{8}$$

In the case of a uniform hydrostatic pressure of the amount p we have

$$\sigma_x = \sigma_y = \sigma_z = -p$$

and Eq. (8) gives

$$e = -\frac{3(1 - 2\nu)p}{E}$$

which represents the relation between unit volume expansion e and hydrostatic pressure p.

The quantity $E/3(1 - 2\nu)$ is called the *modulus of volume expansion.*

Using notations (7) and solving Eqs. (3) for σ_x, σ_y, σ_z, we find

$$\begin{aligned} \sigma_x &= \frac{\nu E}{(1 + \nu)(1 - 2\nu)} e + \frac{E}{1 + \nu} \epsilon_x \\ \sigma_y &= \frac{\nu E}{(1 + \nu)(1 - 2\nu)} e + \frac{E}{1 + \nu} \epsilon_y \\ \sigma_z &= \frac{\nu E}{(1 + \nu)(1 - 2\nu)} e + \frac{E}{1 + \nu} \epsilon_z \end{aligned} \tag{9}$$

or using the notation

$$\lambda = \frac{\nu E}{(1 + \nu)(1 - 2\nu)} \tag{10}$$

and Eq. (5), these become

$$\begin{aligned} \sigma_x &= \lambda e + 2G\epsilon_x \\ \sigma_y &= \lambda e + 2G\epsilon_y \\ \sigma_z &= \lambda e + 2G\epsilon_z \end{aligned} \tag{11}$$

7 | Index Notation

The notation already introduced for components of force, stress, displacement, and strain is one that has become well established in many countries, particularly for engineering purposes. It will be used throughout this book. For the concise representation of general equations and the theorems derived from them, however, the alternative *index notation* is advantageous and is often encountered. The displacement components for instance are written u_1, u_2, u_3, or collectively as u_i, with the understanding that the *index* i can be 1, 2, or 3. The coordinates themselves are written x_1, x_2, x_3, or simply x_i, instead of x, y, z.

In Fig. 3 nine stress components appear. They can be arranged as in the table or array on the *left* below.

$$\begin{matrix} \sigma_x & \tau_{xy} & \tau_{xz} \\ \tau_{yx} & \sigma_y & \tau_{yz} \\ \tau_{zx} & \tau_{zy} & \sigma_z \end{matrix} \qquad \begin{matrix} \tau_{xx} & \tau_{xy} & \tau_{xz} \\ \tau_{yx} & \tau_{yy} & \tau_{yz} \\ \tau_{zx} & \tau_{zy} & \tau_{zz} \end{matrix} \qquad \begin{matrix} \tau_{11} & \tau_{12} & \tau_{13} \\ \tau_{21} & \tau_{22} & \tau_{23} \\ \tau_{31} & \tau_{32} & \tau_{33} \end{matrix} \qquad (a)$$

Writing τ_{xx} instead of σ_x, τ_{yy} for σ_y, and τ_{zz} for σ_z, we have the middle array above. Here the first suffix indicates the direction of the normal to the side of the element on which the component acts and the second suffix indicates the axis to which the stress component arrow is parallel. In the array on the right, above, the suffixes are changed to the corresponding numerical indices. To write the nine components collectively we need now two indices i and j, each being 1, 2, 3 independently. Then all nine components are comprised in

$$\tau_{ij} \quad \text{with } i, j = 1, 2, \text{ or } 3 \qquad (b)$$

The relations (1) which reduced the nine components to six distinct numbers (but we still have nine entries in the array), can now be expressed as

$$\tau_{ji} = \tau_{ij} \qquad i \neq j \qquad (c)$$

If we permit $i = j$ we have merely three identities such as $\tau_{11} = \tau_{11}$.

In place of the strain-displacement relations (2) we can take nine strain components ϵ_{ij} (with $\epsilon_{ji} = \epsilon_{ij}$, as the definition of shearing strain requires) according to the relations

$$\epsilon_{ij} = \frac{1}{2}\left(\frac{\partial u_i}{\partial x_j} + \frac{\partial u_j}{\partial x_i}\right) \qquad (d)$$

Taking $i = j = 1$ this reproduces the first of (2) in the form of the first of the three relations

$$\epsilon_{11} = \frac{\partial u_1}{\partial x_1} \qquad \epsilon_{22} = \frac{\partial u_2}{\partial x_2} \qquad \epsilon_{33} = \frac{\partial u_3}{\partial x_3} \qquad (e)$$

Taking $i = 1, j = 2$ we get from (d) the first of the three relations

$$\epsilon_{12} = \frac{1}{2}\left(\frac{\partial u_1}{\partial x_2} + \frac{\partial u_2}{\partial x_1}\right) \qquad \epsilon_{23} = \frac{1}{2}\left(\frac{\partial u_2}{\partial x_3} + \frac{\partial u_3}{\partial x_2}\right) \qquad \epsilon_{31} = \frac{1}{2}\left(\frac{\partial u_3}{\partial x_1} + \frac{\partial u_1}{\partial x_3}\right) \qquad (f)$$

We observe that $2\epsilon_{12}$, $2\epsilon_{13}$, $2\epsilon_{23}$ are the same as γ_{xy}, γ_{xz}, γ_{yz} in (2). Thus ϵ_{12} is *half* the reduction of the original right angle between line elements dx_1, dx_2, at x_1, x_2, x_3.

To express the sum of the three terms appearing in the first of (7) we may write

$$\epsilon_{11} + \epsilon_{22} + \epsilon_{33} \qquad \text{or} \qquad \sum_{i=1,2,3} \epsilon_{ii} \qquad (g)$$

But in this notation it is customary to suppress the summation symbol, and write simply

ϵ_{ii}. The summation is implied by the *repeated index*. This is known as the *summation convention*. Thus, in the stress components,

$$\tau_{ii} = \tau_{11} + \tau_{22} + \tau_{33} \tag{h}$$

Use of j (or any other literal index we may introduce) instead of i does not change the meaning. For this reason such a repeated index is often referred to as a "dummy" index.

The six stress components are expressed in terms of the six strain components by (11) together with (6). To put these together under the index notation we require the array

$$\begin{matrix} 1 & 0 & 0 \\ 0 & 1 & 0 \\ 0 & 0 & 1 \end{matrix}$$

This is written δ_{ij}. Evidently this symbol means zero when $i \neq j$, and it means unity when $i = j = 1$ or 2 or 3. It is referred to as the "Kronecker delta." The six relations obtainable from

$$\tau_{ij} = \lambda\delta_{ij}\epsilon_{kk} + 2G\epsilon_{ij} \qquad i, j, k = 1 \text{ or } 2 \text{ or } 3 \tag{j}$$

reproduce the six relations (11) with (6). The symbol ϵ_{kk} means, of course, a sum like τ_{ii} in (h). But the reader will see here the necessity of using a dummy index k distinct from i and j. For instance, to reproduce the first of (11) we take $i = 1$, $j = 1$, and find from (j)

$$\begin{aligned} \tau_{11} &= \lambda\delta_{11}\epsilon_{kk} + 2G\epsilon_{11} \\ &= \lambda\epsilon_{kk} + 2G\epsilon_{11} \end{aligned} \tag{k}$$

and ϵ_{kk} means the same thing as e, by (7).

Differentiation with respect to coordinates, as for instance in (d), is often expressed more concisely by the use of commas. Thus (d) may be written as

$$\epsilon_{ij} = \tfrac{1}{2}(u_{i,j} + u_{j,i}) \tag{l}$$

Writing 3τ for the sum in (h), τ is the mean of the three normal stress components. The stress τ_{ij} can be regarded as a superposition of the two stress states

$$\begin{matrix} \tau & 0 & 0 \\ 0 & \tau & 0 \\ 0 & 0 & \tau \end{matrix} \qquad \text{and} \qquad \begin{matrix} \tau_{11} - \tau & \tau_{12} & \tau_{13} \\ \tau_{21} & \tau_{22} - \tau & \tau_{23} \\ \tau_{31} & \tau_{32} & \tau_{33} - \tau \end{matrix} \tag{m}$$

The first, often called simply the *mean stress*,[1] can be represented by $\tau\delta_{ij}$. The second, called the *deviatoric* stress, or stress *deviator*, can be represented by τ_{ij}' where

$$\tau_{ij}' = \tau_{ij} - \tau\delta_{ij} \tag{n}$$

Similarly we can separate the strain ϵ_{ij} into a mean strain $\epsilon_{ii}/3$ or $e/3$, and a *deviatoric strain* ϵ_{ij}' where

$$\epsilon_{ij}' = \epsilon_{ij} - \tfrac{1}{3}e\delta_{ij} \tag{o}$$

The six equations expressing Hooke's law are equivalent to

$$\tau_{ij}' = 2G\epsilon_{ij}' \qquad \text{with } 3\tau = (3\lambda + 2G)e \tag{p}$$

It is a simple exercise to deduce these from Eqs. (j), or conversely to begin with (p) and recover (j).

[1] If $\tau = -p$, $p > 0$, it is a hydrostatic pressure p.

The form (p) is particularly convenient as an ingredient of the theory of plasticity or the theory of viscoelasticity. The constant $3\lambda + 2G$ is often written $3K$. Then K is the *modulus of volume expansion*, already introduced on p. 11.

PROBLEMS

1. Show that Eqs. (1) continue to hold if the element of Fig. 4 is in motion and has an angular acceleration like a rigid body.
2. Suppose an elastic material contains a large number of evenly distributed small magnetized particles, so that a magnetic field exerts on any element $dx\,dy\,dz$ a moment $\mu\,dx\,dy\,dz$ about an axis parallel to the x axis. What modification will be needed in Eqs. (1)?
3. Give some reasons why the formulas (2) wili be valid for *small* strains only.
4. An elastic layer is sandwiched between two perfectly rigid plates, to which it is bonded. The layer is compressed between the plates, the direct stress being σ_z. Supposing that the attachment to the plates prevents lateral strain ϵ_x, ϵ_y completely, find the apparent Young's modulus (that is, σ_z/ϵ_z) in terms of E and ν. Show that it is many times E if the material of the layer has a Poisson's ratio only slightly less than 0.5, e.g., rubber.
5. Prove that Eq. (8) follows from Eqs. (11), (10), and (5).

chapter 2

Plane Stress and Plane Strain

8 | Plane Stress

If a thin plate is loaded by forces applied at the boundary, parallel to the plane of the plate and distributed uniformly over the thickness (Fig. 8), the stress components σ_z, τ_{xz}, τ_{yz} are zero on both faces of the plate, and it may be assumed, tentatively, that they are zero also within the plate. The state of stress is then specified by σ_x, σ_y, τ_{xy} only, and is called *plane stress*. It may also be assumed tentatively[1] that these three components are independent of z, i.e., they do not vary through the thickness. They are then functions of x and y only.

9 | Plane Strain

A similar simplification is possible at the other extreme when the dimension of the body in the z direction is very large. If a long cylindrical or prismatical body is loaded by forces that are perpendicular to the longitudinal elements and do not vary along the length, it may be assumed that all cross sections are in the same condition. It is simplest to suppose at first that the end sections are confined between fixed smooth rigid planes, so that displacement in the axial direction is prevented. The effect of removing these will be examined later. Since there is no axial displacement at the ends and, by symmetry, at the midsection, it may be assumed that the same holds at every cross section.

There are many important problems of this kind, for instance, a retain-

[1] The assumptions made here are examined critically in Art. 98. Variation of stress does occur, but in a sufficiently thin plate it can be ignored, like the meniscus on the column of fluid in the capillary tube of a thermometer.

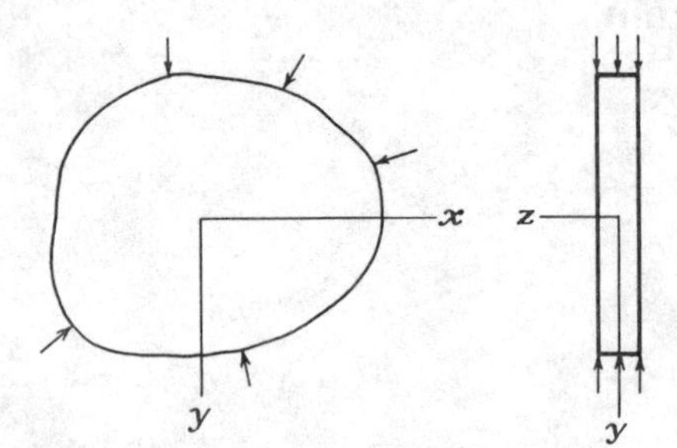

Fig. 8

ing wall with lateral pressure (Fig. 9), a culvert or tunnel (Fig. 10), a cylindrical tube with internal pressure, a cylindrical roller compressed by forces in a diametral plane as in a roller bearing (Fig. 11). In each case, of course, the loading must not vary along the length. Since conditions are the same at all cross sections, it is sufficient to consider only a slice between two sections unit distance apart. The components u and v of the displacement are functions of x and y but are independent of the longitudinal coordinate z. Since the longitudinal displacement w is zero, Eqs. (2) give

$$\gamma_{yz} = \frac{\partial v}{\partial z} + \frac{\partial w}{\partial y} = 0$$

$$\gamma_{xz} = \frac{\partial u}{\partial z} + \frac{\partial w}{\partial x} = 0 \qquad (a)$$

$$\epsilon_z = \frac{\partial w}{\partial z} = 0$$

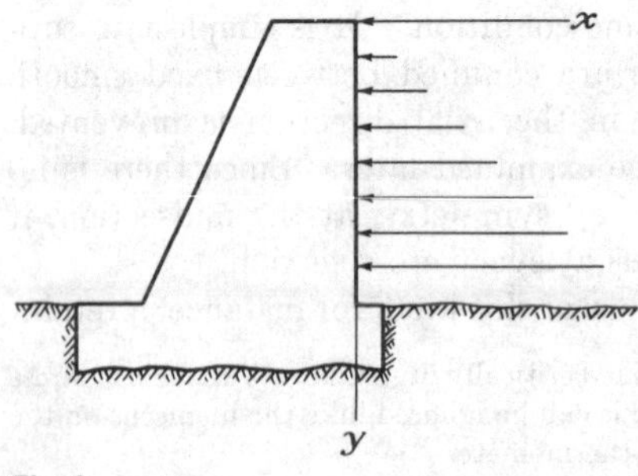

Fig. 9

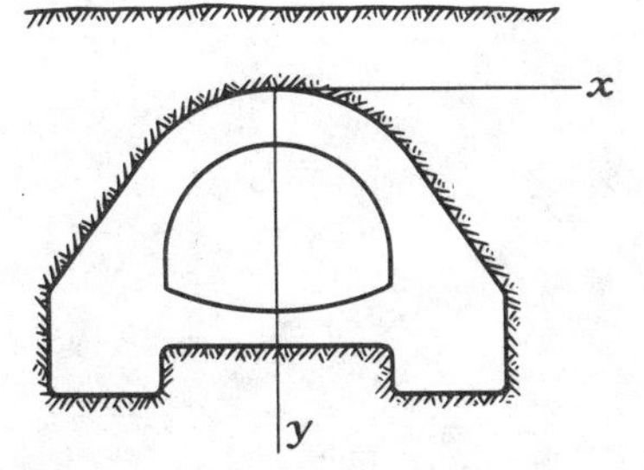

Fig. 10

The longitudinal normal stress σ_z can be found in terms of σ_x and σ_y by means of Hooke's law, Eqs. (3). Since $\epsilon_z = 0$ we find

$$\sigma_z - \nu(\sigma_x + \sigma_y) = 0$$

or

$$\sigma_z = \nu(\sigma_x + \sigma_y) \tag{b}$$

These normal stresses act over the cross sections, including the ends, where they represent forces required to maintain the plane strain and provided by the fixed smooth rigid planes.

By Eqs. (*a*) and (6), the stress components τ_{xz} and τ_{yz} are zero, and, by Eq. (*b*), σ_z can be found from σ_x and σ_y. Thus the plane strain problem, like the plane stress problem, reduces to the determination of σ_x, σ_y, and τ_{xy} as functions of x and y only.

10 | Stress at a Point

Knowing the stress components σ_x, σ_y, τ_{xy} at any point of a plate in a condition of plane stress or plane strain, the stress acting on any plane through this point perpendicular to the plate and inclined to the x and y axes can be calculated from the equations of statics. Let P be a point of the stressed plate and suppose the stress components σ_x, σ_y, τ_{xy} are

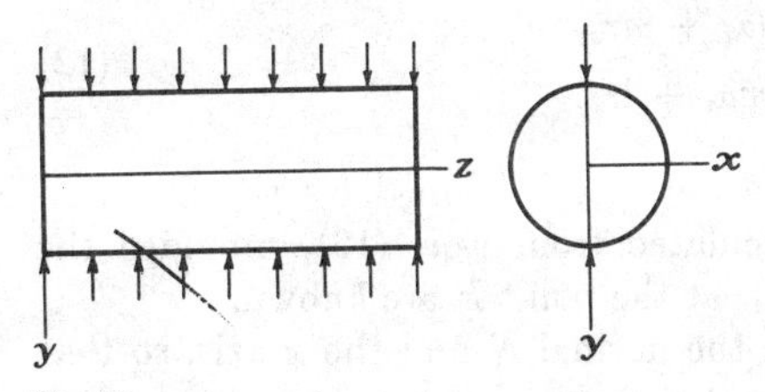

Fig. 11

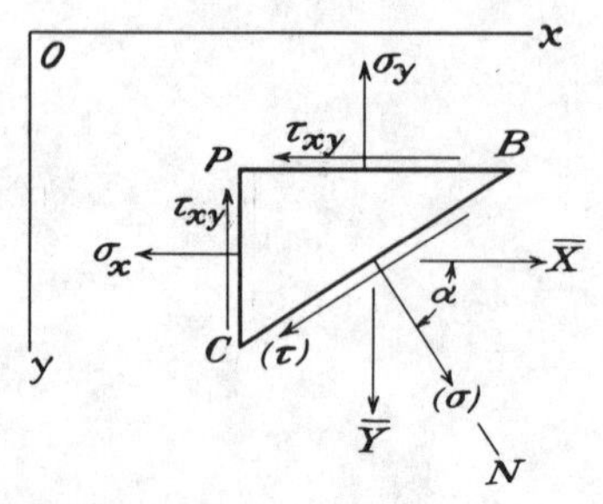

Fig. 12

known (Fig. 12). We take a plane BC parallel to the z axis, at a small distance from P, so that this plane together with the coordinate planes cuts out from the plate a very small triangular prism PBC. Since the stresses vary continuously over the volume of the body, the stress acting on the plane BC will approach the stress on the parallel plane through P as the element is made smaller.

In discussing the conditions of equilibrium of the small triangular prism, the body force can be neglected as a small quantity of a higher order. Likewise, if the element is very small, we can neglect the variation of the stresses over the sides and assume that the stresses are uniformly distributed. The forces acting on the triangular prism can therefore be determined by multiplying the stress components by the areas of the sides. Let N be the direction of the normal to the plane BC and denote the cosines of the angles between the normal N and the axes x and y by

$$\cos Nx = l \qquad \cos Ny = m$$

Then, if A denotes the area of the side BC of the element, the areas of the other two sides are Al and Am.

If we denote by $\bar{X}$ and $\bar{Y}$ the components of stress acting on the side BC, the equations of equilibrium of the prismatical element give

$$\begin{aligned} \bar{X} &= l\sigma_x + m\tau_{xy} \\ \bar{Y} &= m\sigma_y + l\tau_{xy} \end{aligned} \tag{12}$$

Thus the components of stress on any plane defined by the direction cosines l and m can easily be calculated from Eqs. (12), provided the three components of stress σ_x, σ_y, τ_{xy} at the point P are known.

Letting α be the angle between the normal N and the x axis, so that $l = \cos\alpha$ and $m = \sin\alpha$, the normal and shearing components of stress on the plane BC are (from Eqs. 12)

$$\begin{aligned} \sigma &= \bar{X} \cos \alpha + \bar{Y} \sin \alpha = \sigma_x \cos^2 \alpha + \sigma_y \sin^2 \alpha + 2\tau_{xy} \sin \alpha \cos \alpha \\ \tau &= \bar{Y} \cos \alpha - \bar{X} \sin \alpha = \tau_{xy}(\cos^2 \alpha - \sin^2 \alpha) \\ &\qquad + (\sigma_y - \sigma_x) \sin \alpha \cos \alpha \end{aligned} \tag{13}$$

It may be seen that the angle α can be chosen in such a manner that the shearing stress τ becomes equal to zero. For this case we have

$$\tau_{xy}(\cos^2 \alpha - \sin^2 \alpha) + (\sigma_y - \sigma_x) \sin \alpha \cos \alpha = 0$$

or

$$\frac{\tau_{xy}}{\sigma_x - \sigma_y} = \frac{\sin \alpha \cos \alpha}{\cos^2 \alpha - \sin^2 \alpha} = \frac{1}{2} \tan 2\alpha \tag{14}$$

From this equation, two perpendicular directions can be found for which the shearing stress is zero. These directions are called *principal directions* and the corresponding normal stresses *principal stresses*.

If the principal directions are taken as the x and y axes, τ_{xy} is zero and Eqs. (13) are simplified to

$$\begin{aligned} \sigma &= \sigma_x \cos^2 \alpha + \sigma_y \sin^2 \alpha \\ \tau &= \tfrac{1}{2}(\sigma_y - \sigma_x) \sin 2\alpha \end{aligned} \tag{13'}$$

The variation of the stress components σ and τ, as we vary the angle α, can be easily represented graphically by making a diagram in which σ and τ are taken as coordinates.[1] For each plane there will correspond a point on this diagram, the coordinates of which represent the values of σ and τ for this plane. Figure 13 represents such a diagram. For the planes perpendicular to the principal directions we obtain points A and B with

[1] This graphical method is due to O. Mohr, *Zivilingenieur*, 1882, p. 113. See also his "Technische Mechanik," 2d ed., 1914.

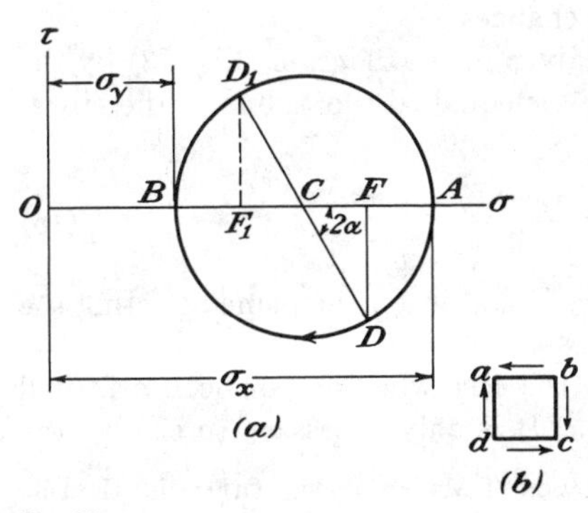

Fig. 13

abscissas σ_x and σ_y, respectively. Now it can be proved that the stress components for any plane BC with an angle α (Fig. 12) will be represented by coordinates of a point on the circle having AB as a diameter. To find this point it is only necessary to measure from the point A in the same direction as α is measured in Fig. 12 an arc subtending an angle equal to 2α. If D is the point obtained in this manner, then, from the figure,

$$OF = OC + CF = \frac{\sigma_x + \sigma_y}{2} + \frac{\sigma_x - \sigma_y}{2} \cos 2\alpha = \sigma_x \cos^2 \alpha + \sigma_y \sin^2 \alpha$$

$$DF = CD \sin 2\alpha = \tfrac{1}{2}(\sigma_x - \sigma_y) \sin 2\alpha$$

Comparing with Eqs. (13′), it is seen that the coordinates of point D give the numerical values of stress components on the plane BC at the angle α. To bring into coincidence the sign of the shearing component, we take τ positive in the upward direction (Fig. 13) and consider shearing stresses as positive when they give a couple in the clockwise direction, as on the sides bc and ad of the element $abcd$ (Fig. 13b). Shearing stresses of opposite direction, as on the sides ab and dc of the element, are considered as negative.[1]

As the plane BC rotates about an axis perpendicular to the xy plane (Fig. 12) in the clockwise direction, and α varies from 0 to $\pi/2$, the point D in Fig. 13 moves from A to B, so that the lower half of the circle determines the stress variation for all values of α within these limits. The upper half of the circle gives stresses for $\pi/2 \leq \alpha \leq \pi$.

Prolonging the radius CD to the point D_1 (Fig. 13), i.e., taking the angle $\pi + 2\alpha$, instead of 2α, the stresses on the plane perpendicular to BC (Fig. 12) are obtained. This shows that the shearing stresses on two perpendicular planes are numerically equal, as previously proved. As for normal stresses, we see from the figure that $OF_1 + OF = 2OC$, that is, the sum of the normal stresses over two perpendicular cross sections remains constant when the angle α changes.

The maximum shearing stress is given in the diagram (Fig. 13) by the maximum ordinate of the circle, i.e., is equal to the radius of the circle. Hence

$$\tau_{\max} = \frac{\sigma_x - \sigma_y}{2} \tag{15}$$

It acts on the plane for which $\alpha = \pi/4$, that is, on the plane bisecting the angle between the two principal stresses.

The diagram can also be used in the case when one or both principal stresses are negative (compression). It is only necessary to change the

[1] This rule is used only in the construction of Mohr's circle. Otherwise the rule given on p. 5 holds.

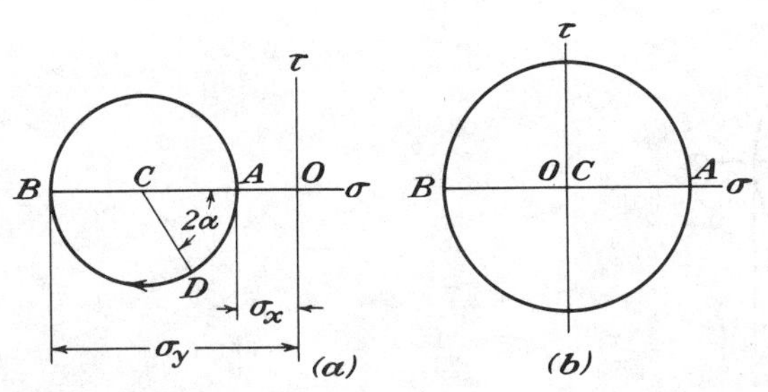

Fig. 14

sign of the abscissa for compressive stress. In this manner Fig. 14*a* represents the case when both principal stresses are negative and Fig. 14*b* the case of pure shear.

From Figs. 13 and 14 it is seen that the stress at a point can be resolved into two parts: one, biaxial tension or compression, the two components being equal and of magnitude given by the abscissa of the center of the circle; and the other, pure shear, the magnitude of which is given by the radius of the circle. When several plane stress distributions are superposed, the uniform tensions or compressions can be added together algebraically. The pure shears must be added together by taking into account the directions of the planes on which they are acting. It can be shown that if we superpose two systems of pure shear whose planes of maximum shear make an angle of β with each other, the resulting system will be another case of pure shear. For example, Fig. 15 represents the determination of stress on any plane defined by

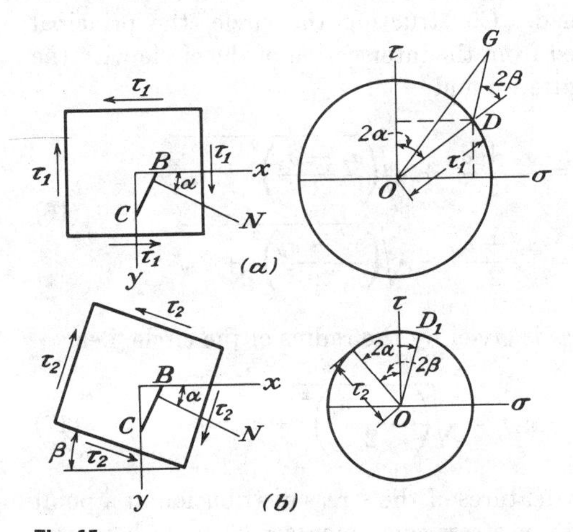

Fig. 15

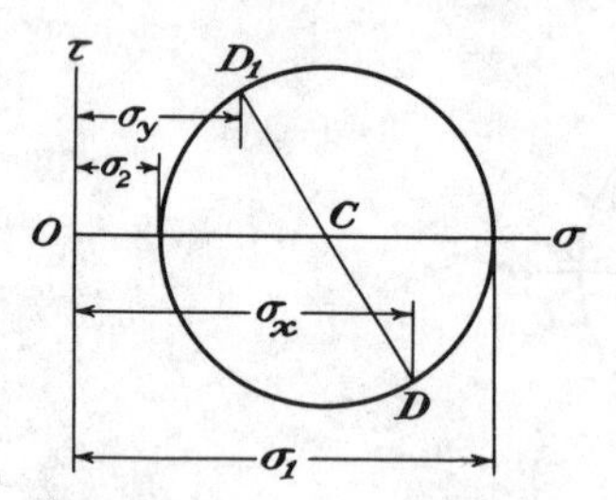

Fig. 16

α, produced by two pure shears of magnitude τ_1 and τ_2 acting one on the planes xz and yz (Fig. 15*a*) and the other on the planes inclined to xz and yz by the angle β (Fig. 15*b*). In Fig. 15*a* the coordinates of point D represent the shear and normal stress on plane CB produced by the first system, whereas the coordinate of D_1 (Fig. 15*b*) gives the stresses on this plane for the second system. Adding OD and OD_1 geometrically we obtain OG, the resultant stress on the plane due to both systems, the coordinates of G giving us the shear and normal stress. Note that the magnitude of OG does not depend upon α. Hence, as the result of the superposition of two shears, we obtain a Mohr circle for pure shear, the magnitude of which is given by OG, the planes of maximum shear being inclined to the xz and yz planes by an angle equal to half the angle GOD.

A diagram, such as is shown in Fig. 13, can also be used for determining principal stresses if the stress components σ_x, σ_y, τ_{xy} for any two perpendicular planes (Fig. 12) are known. We begin in such a case with the plotting of the two points D and D_1, representing stress conditions on the two coordinate planes (Fig. 16). In this manner the diameter DD_1 of the circle is obtained. Constructing the circle, the principal stresses σ_1 and σ_2 are obtained from the intersection of the circle with the abscissa axis. From the figure we find

$$\begin{aligned} \sigma_1 &= OC + CD = \frac{\sigma_x + \sigma_y}{2} + \sqrt{\left(\frac{\sigma_x - \sigma_y}{2}\right)^2 + \tau_{xy}^2} \\ \sigma_2 &= OC - CD = \frac{\sigma_x + \sigma_y}{2} - \sqrt{\left(\frac{\sigma_x - \sigma_y}{2}\right)^2 + \tau_{xy}^2} \end{aligned} \tag{16}$$

The maximum shearing stress is given by the radius of the circle, i.e.,

$$\tau_{\max} = \frac{1}{2}(\sigma_1 - \sigma_2) = \sqrt{\left(\frac{\sigma_x - \sigma_y}{2}\right)^2 + \tau_{xy}^2} \tag{17}$$

In this manner, all necessary features of the stress distribution at a point can be obtained if only the three stress components σ_x, σ_y, τ_{xy} are known.

11 | Strain at a Point

When the strain components ϵ_x, ϵ_y, γ_{xy} at a point are known, the unit elongation for any direction and the decrease of a right angle—the shearing strain—of any orientation at the point can be found. A line element PQ (Fig. 17*a*) between the points (x,y), $(x + dx,\ y + dy)$ is translated, stretched (or contracted), and rotated into the line element $P'Q'$ when the deformation occurs. The displacement components of P are u, v, and those of Q are

$$u + \frac{\partial u}{\partial x}\,dx + \frac{\partial u}{\partial y}\,dy \qquad v + \frac{\partial v}{\partial x}\,dx + \frac{\partial v}{\partial y}\,dy$$

If $P'Q'$ in Fig. 17*a* is now translated so that P' is brought back to P, it is in the position PQ'' of Fig. 17*b*, and QR, RQ'' represent the components of the displacement of Q relative to P. Thus

$$QR = \frac{\partial u}{\partial x}\,dx + \frac{\partial u}{\partial y}\,dy \qquad RQ'' = \frac{\partial v}{\partial x}\,dx + \frac{\partial v}{\partial y}\,dy \tag{a}$$

The components of this relative displacement QS, SQ'', normal to PQ'' and along PQ'', can be found from these as

$$QS = -QR \sin\theta + RQ'' \cos\theta \qquad SQ'' = QR\cos\theta + RQ''\sin\theta \tag{b}$$

ignoring the small angle QPS in comparison with θ. Since the short line QS may be identified with an arc of a circle with center P, SQ'' gives the stretch of PQ. The unit elongation of $P'Q'$, denoted by ϵ_θ, is SQ''/PQ. Using (*b*) and (*a*) we have

$$\begin{aligned}\epsilon_\theta &= \cos\theta\left(\frac{\partial u}{\partial x}\frac{dx}{ds} + \frac{\partial u}{\partial y}\frac{dy}{ds}\right) + \sin\theta\left(\frac{\partial v}{\partial x}\frac{dx}{ds} + \frac{\partial v}{\partial y}\frac{dy}{ds}\right)\\ &= \frac{\partial u}{\partial x}\cos^2\theta + \left(\frac{\partial u}{\partial y} + \frac{\partial v}{\partial x}\right)\sin\theta\cos\theta + \frac{\partial v}{\partial y}\sin^2\theta\end{aligned}$$

or
$$\epsilon_\theta = \epsilon_x \cos^2\theta + \gamma_{xy}\sin\theta\cos\theta + \epsilon_y\sin^2\theta \tag{c}$$

which gives the unit elongation for any direction θ.

The angle ψ_θ through which PQ is rotated is QS/PQ. Thus from (*b*) and (*a*),

$$\psi_\theta = -\sin\theta\left(\frac{\partial u}{\partial x}\frac{dx}{ds} + \frac{\partial u}{\partial y}\frac{dy}{ds}\right) + \cos\theta\left(\frac{\partial v}{\partial x}\frac{dx}{ds} + \frac{\partial v}{\partial y}\frac{dy}{ds}\right)$$

or
$$\psi_\theta = \frac{\partial v}{\partial x}\cos^2\theta + \left(\frac{\partial v}{\partial y} - \frac{\partial u}{\partial x}\right)\sin\theta\cos\theta - \frac{\partial u}{\partial y}\sin^2\theta \tag{d}$$

The line element PT at right angles to PQ makes an angle $\theta + (\pi/2)$ with the x direction, and its rotation $\psi_{\theta+\pi/2}$ is therefore given by (*d*) when $\theta + (\pi/2)$ is substituted for θ. Since $\cos[\theta + (\pi/2)] = -\sin\theta$,

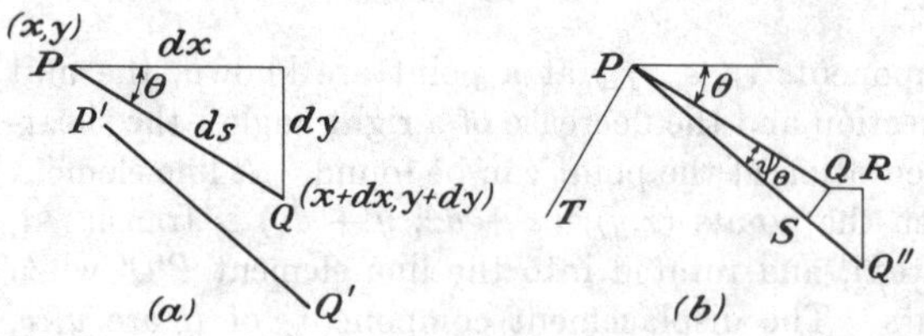

Fig. 17

$\sin[\theta + (\pi/2)] = \cos\theta$, we find

$$\psi_{\theta+\pi/2} = \frac{\partial v}{\partial x}\sin^2\theta - \left(\frac{\partial v}{\partial y} - \frac{\partial u}{\partial x}\right)\sin\theta\cos\theta - \frac{\partial u}{\partial y}\cos^2\theta \qquad (e)$$

The shear strain γ_θ for the directions PQ, PT is $\psi_\theta - \psi_{\theta+\pi/2}$, so

$$\gamma_\theta = \left(\frac{\partial v}{\partial x} + \frac{\partial u}{\partial y}\right)(\cos^2\theta - \sin^2\theta) + \left(\frac{\partial v}{\partial y} - \frac{\partial u}{\partial x}\right)2\sin\theta\cos\theta$$

or

$$\tfrac{1}{2}\gamma_\theta = \tfrac{1}{2}\gamma_{xy}(\cos^2\theta - \sin^2\theta) + (\epsilon_y - \epsilon_x)\sin\theta\cos\theta \qquad (f)$$

Comparing (*c*) and (*f*) with (13), we observe that they may be obtained from (13) by replacing σ by ϵ_θ, τ by $\gamma_\theta/2$, σ_x by ϵ_x, σ_y by ϵ_y, τ_{xy} by $\gamma_{xy}/2$, and α by θ. Consequently, for each deduction made from (13) as to σ and τ, there is a corresponding deduction from (*c*) and (*f*) as to ϵ_θ and $\gamma_\theta/2$. There are thus two values of θ, differing by 90°, for which γ_θ is zero. They are given by

$$\frac{\gamma_{xy}}{\epsilon_x - \epsilon_y} = \tan 2\theta$$

The corresponding strains ϵ_θ are *principal strains*. A Mohr circle diagram analogous to Fig. 13 or 16 may be drawn, the ordinates representing $\gamma_\theta/2$ and the abscissas ϵ_θ. The principal strains ϵ_1, ϵ_2 will be the algebraically greatest and least values of ϵ_θ as a function of θ. The greatest value of $\gamma_\theta/2$ will be represented by the radius of the circle. Thus the greatest shearing strain $\gamma_{\theta\,\max}$ is given by

$$\gamma_{\theta\,\max} = \epsilon_1 - \epsilon_2$$

12 | Measurement of Surface Strains

The strains, or unit elongations, on a surface are usually most conveniently measured by means of electric-resistance strain gauges.[1] The

[1] A detailed account of this method is given in M. Hetényi (ed.), "Handbook of Experimental Stress Analysis," Chaps. 5 and 9, John Wiley & Sons, Inc., New York, 1950.

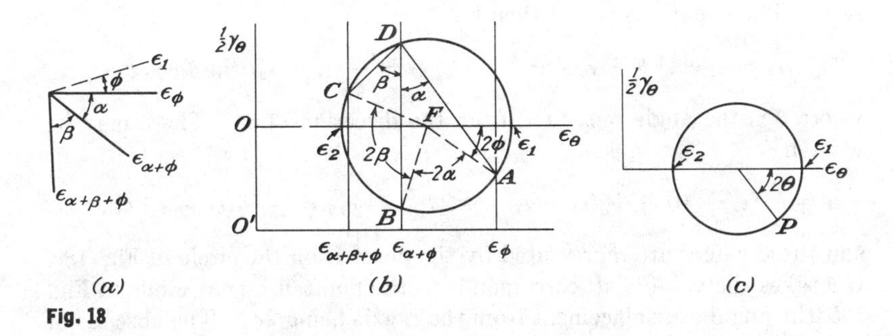

Fig. 18

simplest form of such a gauge is a short length of wire insulated from and glued to the surface. When stretching occurs the resistance of the wire is increased, and the strain can thus be measured electrically. The effect is usually magnified by looping the wires backward and forward several times, to form several gauge lengths connected in series. The wire is glued between two tabs of paper, and the assembly is glued to the surface.

The use of these gauges is simple when the principal directions are known. One gauge is placed along each principal direction and direct measurements of ϵ_1, ϵ_2 are obtained. The principal stresses σ_1, σ_2 may then be calculated from Hooke's law, Eqs. (3), with $\sigma_x = \sigma_1$, $\sigma_y = \sigma_2$, $\sigma_z = 0$, the last holding on the assumption that there is no stress acting on the surface to which the gauges are attached. Then

$$(1 - \nu^2)\sigma_1 = E(\epsilon_1 + \nu\epsilon_2) \qquad (1 - \nu^2)\sigma_2 = E(\epsilon_2 + \nu\epsilon_1)$$

When the principal directions are not known in advance, three measurements are needed. Thus the state of strain is completely determined if ϵ_x, ϵ_y, γ_{xy} can be measured. But since the strain gauges measure extensions, and not shearing strain directly, it is convenient to measure the unit elongations in three directions at the point. Such a set of gauges is called a "strain rosette." The Mohr circle can be drawn by the simple construction[1] given in Art. 13, and the principal strains can then be read off. The three gauges are represented by the three full lines in Fig. 18*a*. The broken line represents the (unknown) direction of the larger principal strain ϵ_1, from which the direction of the first gauge is obtained by a clockwise rotation ϕ.

If the x and y directions for Eqs. (c) and (f) of Art. 11 had been taken as the principal directions, ϵ_x would be ϵ_1, ϵ_y would be ϵ_2, and γ_{xy} would be

[1] Glenn Murphy, *J. Appl. Mech.*, vol. 12, p. A-209, 1945; N. J. Hoff, *ibid.*

zero. The equations would then be

$$\epsilon_\theta = \epsilon_1 \cos^2 \theta + \epsilon_2 \sin^2 \theta \qquad \tfrac{1}{2}\gamma_\theta = -(\epsilon_1 - \epsilon_2) \sin \theta \cos \theta$$

where θ is the angle measured from the direction of ϵ_1. These may be written

$$\epsilon_\theta = \tfrac{1}{2}(\epsilon_1 + \epsilon_2) + \tfrac{1}{2}(\epsilon_1 - \epsilon_2) \cos 2\theta \qquad \tfrac{1}{2}\gamma_\theta = -\tfrac{1}{2}(\epsilon_1 - \epsilon_2) \sin 2\theta$$

and these values are represented by the point P on the circle in Fig. 18c. If θ takes the value ϕ, P corresponds to the point A on the circle in Fig. 18b, the angular displacement from the ϵ_θ axis being 2ϕ. The abscissa of this point is ϵ_ϕ, which is known. If θ takes the value $\phi + \alpha$, P moves to B, through a further angle $AFB = 2\alpha$, and the abscissa is the known value $\epsilon_{\alpha+\phi}$. If θ takes the value $\phi + \alpha + \beta$, P moves on to C, through a further angle $BFC = 2\beta$, and the abscissa is $\epsilon_{\alpha+\beta+\phi}$.

The problem is to draw the circle when these three abscissas and the two angles α, β are known.

13 | Construction of Mohr Strain Circle for Strain Rosette

A temporary horizontal ϵ axis is drawn horizontally from any origin O', Fig. 18b, and the three measured strains ϵ_ϕ, $\epsilon_{\alpha+\phi}$, $\epsilon_{\alpha+\beta+\phi}$ are laid off along it. Verticals are drawn through these points. Selecting any point D on the vertical through $\epsilon_{\alpha+\phi}$, lines DA, DC are drawn at angles α and β to the vertical at D as shown, to meet the other two verticals at A and C. The circle drawn through D, A, and C is the required circle. Its center F is determined by the intersection of the perpendicular bisectors of CD, DA. The points representing the three gauge directions are A, B, and C. The angle AFB, being twice the angle ADB at the circumference, is 2α, and BFC is 2β. Thus A, B, C are at the required angular intervals round the circle and have the required abscissas. The ϵ_θ axis can now be drawn as OF, and the distances from O to the intersections with the circle give ϵ_1, ϵ_2. The angle 2ϕ is the angle of FA below this axis.

14 | Differential Equations of Equilibrium

We now consider the equilibrium of a small rectangular block of edges h, k, and unity (Fig. 19). The stresses acting on the faces 1, 2, 3, 4, and their positive directions are indicated in the figure. On account of the variation of stress throughout the material, the value of, for instance, σ_x is not quite the same for face 1 as for face 3. The symbols σ_x, σ_y, τ_{xy} refer to the point x, y, the midpoint of the rectangle in Fig. 19. The values at the midpoints of the faces are denoted by $(\sigma_x)_1$, $(\sigma_x)_2$, etc. Since

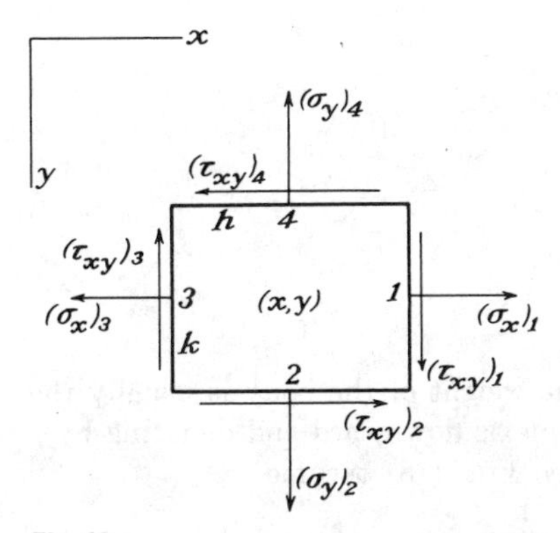

Fig. 19

the faces are very small, the corresponding forces are obtained by multiplying these values by the areas of the faces on which they act.[1]

The body force on the block, which was neglected as a small quantity of higher order in considering the equilibrium of the triangular prism of Fig. 12, must be taken into consideration, because it is of the same order of magnitude as the terms owing to the variations of the stress components that are now under consideration. If X, Y denote the components of body force per unit volume, the equation of equilibrium for forces in the x direction is

$$(\sigma_x)_1 k - (\sigma_x)_3 k + (\tau_{xy})_2 h - (\tau_{xy})_4 h + Xhk = 0$$

or, dividing by hk,

$$\frac{(\sigma_x)_1 - (\sigma_x)_3}{h} + \frac{(\tau_{xy})_2 - (\tau_{xy})_4}{k} + X = 0$$

If now the block is taken smaller and smaller, that is, $h \to 0$, $k \to 0$, the limit of $[(\sigma_x)_1 - (\sigma_x)_3]/h$ is $\partial\sigma_x/\partial x$ by the definition of such a derivative. Similarly $[(\tau_{xy})_2 - (\tau_{xy})_4]/k$ becomes $\partial\tau_{xy}/\partial y$. The equation of equilibrium for forces in the y direction is obtained in the same manner. Thus

$$\begin{aligned} \frac{\partial\sigma_x}{\partial x} + \frac{\partial\tau_{xy}}{\partial y} + X = 0 \\ \frac{\partial\sigma_y}{\partial y} + \frac{\partial\tau_{xy}}{\partial x} + Y = 0 \end{aligned} \tag{18}$$

These are the differential equations of equilibrium for two-dimensional problems.

[1] More precise considerations would introduce terms of higher order that vanish in the final limiting process.

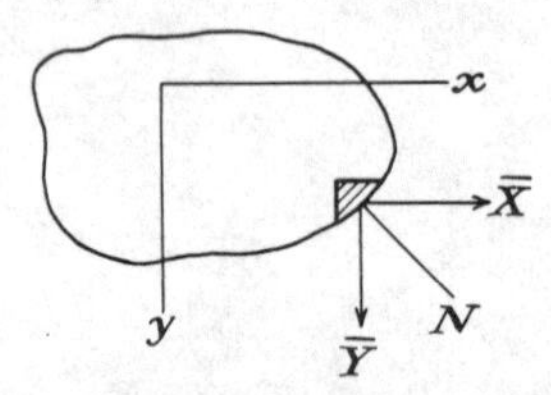

Fig. 20

In many practical applications the weight of the body is usually the only body force. Then, taking the y axis downward and denoting by ρ the mass per unit volume of the body, Eqs. (18) become

$$
\begin{aligned}
\frac{\partial \sigma_x}{\partial x} + \frac{\partial \tau_{xy}}{\partial y} &= 0 \\
\frac{\partial \sigma_y}{\partial y} + \frac{\partial \tau_{xy}}{\partial x} + \rho g &= 0
\end{aligned}
\tag{19}
$$

15 | Boundary Conditions

Equations (18) or (19) must be satisfied at all points throughout the volume of the body. The stress components vary over the volume of the plate; and when we arrive at the boundary they must be such as to be in equilibrium with the external forces on the boundary of the plate, so that external forces may be regarded as a continuation of the internal stress distribution. These conditions of equilibrium at the boundary can be obtained from Eqs. (12). Taking the small triangular prism PBC (Fig. 12), so that the side BC coincides with the boundary of the plate, as shown in Fig. 20, and denoting by $\bar{X}$ and $\bar{Y}$ the components of the surface forces per unit area at this point of the boundary, we have

$$
\begin{aligned}
\bar{X} &= l\sigma_x + m\tau_{xy} \\
\bar{Y} &= m\sigma_y + l\tau_{xy}
\end{aligned}
\tag{20}
$$

in which l and m are the direction cosines of the normal N to the boundary.

In the particular case of a rectangular plate, the coordinate axes are usually taken parallel to the sides of the plate and the boundary conditions (20) can be simplified. Taking, for instance, a side of the plate parallel to the x axis we have for this part of the boundary the normal N parallel to the y axis; hence $l = 0$ and $m = \pm 1$. Equations (20) then become

$$\bar{X} = \pm\tau_{xy} \qquad \bar{Y} = \pm\sigma_y$$

Here the positive sign should be taken if the normal N has the positive direction of the y axis and the negative sign for the opposite direction of N. It is seen from this that at the boundary the stress components become equal to the components of the surface forces per unit area of the boundary.

16 | Compatibility Equations

It is a fundamental problem of the theory of elasticity to determine the state of stress in a body submitted to the action of given forces. In a two-dimensional problem it is necessary to solve the differential equations of equilibrium (18), and the solution must be such as to satisfy the boundary conditions (20). These equations, derived by application of the equations of statics and containing three stress components σ_x, σ_y, τ_{xy}, are not sufficient for the determination of these components. The problem is a statically indeterminate one, and in order to obtain the solution the elastic deformation of the body must also be considered.

The mathematical formulation of the condition for compatibility of stress distribution with the existence of continuous functions u, v, w defining the deformation will be obtained from Eqs. (2). For two-dimensional problems we consider three strain components, namely,

$$\epsilon_x = \frac{\partial u}{\partial x} \qquad \epsilon_y = \frac{\partial v}{\partial y} \qquad \gamma_{xy} = \frac{\partial u}{\partial y} + \frac{\partial v}{\partial x} \tag{a}$$

These three strain components are expressed by two functions u and v; hence, they cannot be taken arbitrarily, and there exists a certain relation between the strain components that can easily be obtained from (*a*). Differentiating the first of the Eqs. (*a*) twice with respect to y, the second twice with respect to x, and the third once with respect to x and once with respect to y, we find

$$\frac{\partial^2 \epsilon_x}{\partial y^2} + \frac{\partial^2 \epsilon_y}{\partial x^2} = \frac{\partial^2 \gamma_{xy}}{\partial x \, \partial y} \tag{21}$$

This differential relation, called the *condition of compatibility*, must be satisfied by the strain components to secure the existence of functions u and v connected with the strain components by Eqs. (*a*). By using Hooke's law, [Eqs. (3)], the condition (21) can be transformed into a relation between the components of stress.

In the case of plane stress distribution (Art. 8), Eqs. (3) reduce to

$$\epsilon_x = \frac{1}{E}(\sigma_x - \nu\sigma_y) \qquad \epsilon_y = \frac{1}{E}(\sigma_y - \nu\sigma_x) \tag{22}$$

$$\gamma_{xy} = \frac{1}{G}\tau_{xy} = \frac{2(1+\nu)}{E}\tau_{xy} \tag{23}$$

Substituting in Eq. (21), we find

$$\frac{\partial^2}{\partial y^2}(\sigma_x - \nu\sigma_y) + \frac{\partial^2}{\partial x^2}(\sigma_y - \nu\sigma_x) = 2(1 + \nu)\frac{\partial^2 \tau_{xy}}{\partial x\, \partial y} \tag{b}$$

This equation can be written in a different form by using the equations of equilibrium. For the case when the weight of the body is the only body force, differentiating the first of Eqs. (19) with respect to x and the second with respect to y and adding them, we find

$$2\frac{\partial^2 \tau_{xy}}{\partial x\, \partial y} = -\frac{\partial^2 \sigma_x}{\partial x^2} - \frac{\partial^2 \sigma_y}{\partial y^2}$$

Substituting in Eq. (*b*), the compatibility equation in terms of stress components becomes

$$\left(\frac{\partial^2}{\partial x^2} + \frac{\partial^2}{\partial y^2}\right)(\sigma_x + \sigma_y) = 0 \tag{24}$$

Proceeding in the same manner with the general equations of equilibrium (18), we find

$$\left(\frac{\partial^2}{\partial x^2} + \frac{\partial^2}{\partial y^2}\right)(\sigma_x + \sigma_y) = -(1 + \nu)\left(\frac{\partial X}{\partial x} + \frac{\partial Y}{\partial y}\right) \tag{25}$$

In the case of plane strain (Art. 9), we have

$$\sigma_z = \nu(\sigma_x + \sigma_y)$$

and from Hooke's law (Eqs. 3), we find

$$\begin{aligned} \epsilon_x &= \frac{1}{E}[(1 - \nu^2)\sigma_x - \nu(1 + \nu)\sigma_y] \\ \epsilon_y &= \frac{1}{E}[(1 - \nu^2)\sigma_y - \nu(1 + \nu)\sigma_x] \end{aligned} \tag{26}$$

$$\gamma_{xy} = \frac{2(1 + \nu)}{E}\tau_{xy} \tag{27}$$

Substituting in Eq. (21) and using, as before, the equations of equilibrium (19), we find that the compatibility equation (24) holds also for plane strain. For the general case of body forces we obtain from Eqs. (21) and (18) the compatibility equation in the following form:

$$\left(\frac{\partial^2}{\partial x^2} + \frac{\partial^2}{\partial y^2}\right)(\sigma_x + \sigma_y) = -\frac{1}{1 - \nu}\left(\frac{\partial X}{\partial x} + \frac{\partial Y}{\partial y}\right) \tag{28}$$

The equations of equilibrium (18) or (19) together with the boundary conditions (20) and one of the above compatibility equations give us a system of equations that is usually sufficient for the complete determination of the stress distribution in a two-dimensional problem.[1] The particular cases in which certain additional considerations are necessary will be discussed later (page 133). It is interesting to note that in the case of constant body forces the equations determining stress distribution do not contain the elastic constants of the material. Hence, the stress distribution is the same for all isotropic materials, provided the equations are sufficient for the complete determination of the stresses. The conclusion is of practical importance: we shall see later that in the case of transparent materials, such as glass or xylonite, it is possible to determine stresses by an optical method using polarized light (page 150). From the above discussion it is evident that experimental results obtained with a transparent material in most cases can be applied immediately to any other material, such as steel.

It should also be noted that in the case of constant body forces the compatibility equation (24) holds both for the case of plane stress and for the case of plane strain. The stress distribution is hence the same in these two cases, provided the shape of the boundary and the external forces are the same.[2]

17 | Stress Function

It has been shown that a solution of two-dimensional problems reduces to the integration of the differential equations of equilibrium together with the compatibility equation and the boundary conditions. If we begin with the case when the weight of the body is the only body force, the equations to be satisfied are [see Eqs. (19) and (24)]

$$\frac{\partial \sigma_x}{\partial x} + \frac{\partial \tau_{xy}}{\partial y} = 0 \qquad\qquad (a)$$

$$\frac{\partial \sigma_y}{\partial y} + \frac{\partial \tau_{xy}}{\partial x} + \rho g = 0$$

$$\left(\frac{\partial^2}{\partial x^2} + \frac{\partial^2}{\partial y^2}\right)(\sigma_x + \sigma_y) = 0 \qquad\qquad (b)$$

To these equations the boundary conditions (20) should be added. The usual method of solving these equations is by introducing a new function,

[1] In plane stress there are compatibility conditions other than (21) that are in fact violated by our assumptions. It is shown in Art. 131 that in spite of this the method of the present chapter gives good approximations for thin plates.

[2] This statement may require modification when the plate or cylinder has holes, for then the problem can be correctly solved only by considering the displacements as well as the stresses. See Art. 43.

called the *stress function.*[1] As is easily checked, Eqs. (*a*) are satisfied by taking any function ϕ of x and y and putting the following expressions for the stress components:

$$\sigma_x = \frac{\partial^2 \phi}{\partial y^2} - \rho g y \qquad \sigma_y = \frac{\partial^2 \phi}{\partial x^2} - \rho g y \qquad \tau_{xy} = -\frac{\partial^2 \phi}{\partial x \, \partial y} \tag{29}$$

In this manner we can get a variety of solutions of the equations of equilibrium (*a*). The true solution of the problem is that which satisfies also the compatibility equation (*b*). Substituting expressions (29) for the stress components into Eq. (*b*), we find that the stress function ϕ must satisfy the equation

$$\frac{\partial^4 \phi}{\partial x^4} + 2\frac{\partial^4 \phi}{\partial x^2 \, \partial y^2} + \frac{\partial^4 \phi}{\partial y^4} = 0 \tag{30}$$

Thus, the solution of a two-dimensional problem, when the weight of the body is the only body force, reduces to finding a solution of Eq. (30) that satisfies the boundary conditions (20) of the problem. In the following chapters, this method of solution will be applied to several examples of practical interest.

Let us now consider a more general case of body forces and assume that these forces have a potential. Then the components X and Y in Eqs. (18) are given by the equations

$$X = -\frac{\partial V}{\partial x} \qquad Y = -\frac{\partial V}{\partial y} \tag{c}$$

in which V is the potential function. Equations (18) become

$$\frac{\partial}{\partial x}(\sigma_x - V) + \frac{\partial \tau_{xy}}{\partial y} = 0$$

$$\frac{\partial}{\partial y}(\sigma_y - V) + \frac{\partial \tau_{xy}}{\partial x} = 0$$

These equations are of the same form as Eqs. (*a*) and can be satisfied by taking

$$\sigma_x - V = \frac{\partial^2 \phi}{\partial y^2} \qquad \sigma_y - V = \frac{\partial^2 \phi}{\partial x^2} \qquad \tau_{xy} = -\frac{\partial^2 \phi}{\partial x \, \partial y} \tag{31}$$

in which ϕ is the stress function. Substituting expressions (31) in the compatibility equation (25) for plane stress distribution, we find

$$\frac{\partial^4 \phi}{\partial x^4} + 2\frac{\partial^4 \phi}{\partial x^2 \, \partial y^2} + \frac{\partial^4 \phi}{\partial y^4} = -(1 - \nu)\left(\frac{\partial^2 V}{\partial x^2} + \frac{\partial^2 V}{\partial y^2}\right) \tag{32}$$

An analogous equation can be obtained for the case of plane strain.

[1] This function was introduced in the solution of two-dimensional problems by G. B. Airy, *Brit. Assoc. Advan. Sci. Rept.*, 1862, and is sometimes called the *Airy stress function.*

When the body force is simply the weight, the potential V is $-\rho g y$. In this case the right-hand side of Eq. (32) reduces to zero. By taking the solution $\phi = 0$ of (32), or of (30), we find the stress distribution from (31), or (29),

$$\sigma_x = -\rho g y \qquad \sigma_y = -\rho g y \qquad \tau_{xy} = 0 \tag{d}$$

as a possible state of stress due to gravity. This is a state of hydrostatic pressure $\rho g y$ in two dimensions, with zero stress at $y = 0$. It can exist in a plate or cylinder of any shape provided the corresponding boundary forces are applied. Considering a boundary element as in Fig. 12, Eqs. (13) show that there must be a normal *pressure* $\rho g y$ on the boundary, and zero shear stress. If the plate or cylinder is to be supported in some other manner we have to superpose a boundary normal *tension* $\rho g y$ and the new supporting forces. The two together will be in equilibrium, and the determination of their effects is a problem of boundary forces only, without body forces.[1]

PROBLEMS

1. Show that Eqs. (12) remain valid when the element of Fig. 12 has acceleration.
2. Find graphically the principal strains and their directions from rosette measurements

$$\epsilon_\phi = 2 \times 10^{-3} \qquad \epsilon_{\alpha+\phi} = 1.35 \times 10^{-3} \qquad \epsilon_{\alpha+\beta+\phi} = 0.95 \times 10^{-3} \text{ in. per in.}$$

   where $\alpha = \beta = 45°$.
3. Show that the line elements at the point x, y that have the maximum and minimum rotation are those in the two perpendicular directions θ determined by

$$\tan 2\theta = \frac{\partial v/\partial y - \partial u/\partial x}{\partial v/\partial x + \partial u/\partial y}$$

4. The stresses in a rotating disk (of unit thickness) can be regarded as due to centrifugal force as body force in a stationary disk. Show that this body force is derivable from the potential $V = -\frac{1}{2}\rho\omega^2(x^2 + y^2)$, where ρ is the density and ω the angular velocity of rotation (about the origin).
5. A disk with its axis horizontal has the gravity stress represented by Eqs. (*d*) of Art. 16. Make a sketch showing the boundary forces that support its weight. Show by another sketch the auxiliary problem of boundary forces that must be solved when the weight is entirely supported by the reaction of a horizontal surface on which the disk stands.
6. A cylinder with its axis horizontal has the gravity stress represented by Eqs. (*d*) of Art. 16. Its ends are confined between smooth fixed rigid planes that maintain the condition of *plane strain*. Sketch the forces acting on its surface, including the ends.
7. Using the stress-strain relations, and Eqs. (*a*) of Art. 15 in the equations of equilibrium (18), show that in the absence of body forces the displacements in problems of plane stress must satisfy

$$\frac{\partial^2 u}{\partial x^2} + \frac{\partial^2 u}{\partial y^2} + \frac{1+\nu}{1-\nu}\frac{\partial}{\partial x}\left(\frac{\partial u}{\partial x} + \frac{\partial v}{\partial y}\right) = 0$$

   and a companion equation.

[1] This problem, and the general case of a potential V such that the right-hand side of Eq. (32) vanishes, have been discussed by M. Biot, *J. Appl. Mech.*, 1935, p. A-41.

8. The figure represents a "tooth" on a plate in a state of plane stress in the plane of the paper. The faces of the tooth (the two straight lines) are free from force. On the supposition that the stress components are all finite and continuous throughout the region, prove that there is no stress at all at the apex of the tooth.

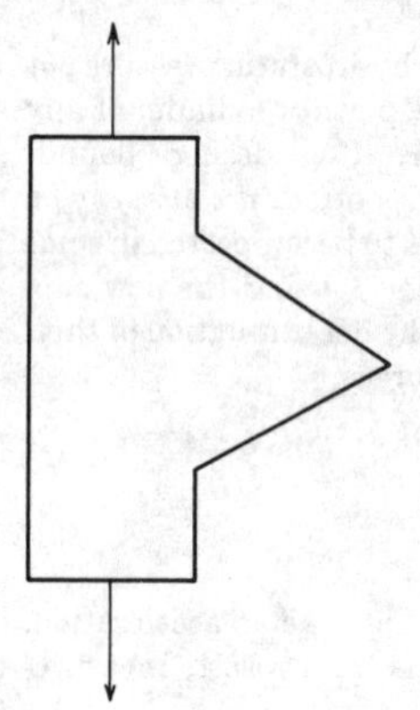

chapter 3

Two-dimensional Problems in Rectangular Coordinates

18 | Solution by Polynomials

It has been shown that the solution of two-dimensional problems, when body forces are absent or are constant, is reduced to the integration of the differential equation

$$\frac{\partial^4 \phi}{\partial x^4} + 2\frac{\partial^4 \phi}{\partial x^2 \partial y^2} + \frac{\partial^4 \phi}{\partial y^4} = 0 \tag{a}$$

having regard to boundary conditions (20). In the case of long rectangular strips, solutions of Eq. (*a*) in the form of polynomials are of interest. By taking polynomials of various degrees, and suitably adjusting their coefficients, a number of practically important problems can be solved.[1]

Beginning with a polynomial of the second degree

$$\phi_2 = \frac{a_2}{2}x^2 + b_2xy + \frac{c_2}{2}y^2 \tag{b}$$

which evidently satisfies Eq. (*a*), we find from Eqs. (29), putting $\rho g = 0$,

$$\sigma_x = \frac{\partial^2 \phi_2}{\partial y^2} = c_2 \qquad \sigma_y = \frac{\partial^2 \phi_2}{\partial x^2} = a_2 \qquad \tau_{xy} = -\frac{\partial^2 \phi_2}{\partial x\, \partial y} = -b_2$$

All three stress components are constant throughout the body, i.e., the stress function (*b*) represents a combination of uniform tensions or compressions[2] in two perpendicular directions and a uniform shear. The

[1] A. Mesnager, *Compt. Rend.*, vol. 132, p. 1475, 1901. See also A. Timpe, *Z. Math. Physik*, vol. 52, p. 348, 1905.

[2] The arrows in Fig. 21 are all drawn in the standard sense, as defined in Art. 3. The numbers a_2, $-b_2$, c_2 attached to them may be positive or negative. Thus all possibilities can be covered without changing the directions of the arrows. In Fig. 22, however, the arrows show directly the intended directions of the applied forces.

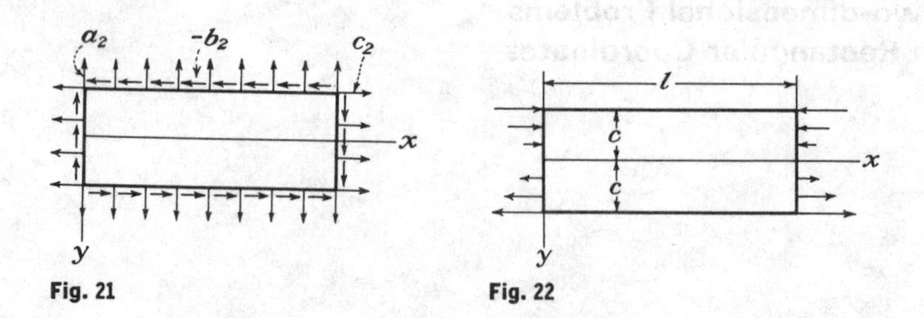

Fig. 21

Fig. 22

forces on the boundaries must equal the stresses at these points as discussed on page 28; in the case of a rectangular plate with sides parallel to the coordinate axes, these forces are shown in Fig. 21.

Let us consider now a stress function in the form of a polynomial of the third degree:

$$\phi_3 = \frac{a_3}{3(2)} x^3 + \frac{b_3}{2} x^2 y + \frac{c_3}{2} xy^2 + \frac{d_3}{3(2)} y^3 \tag{c}$$

This also satisfies Eq. (*a*). Using Eqs. (29) and putting $\rho g = 0$, we find

$$\sigma_x = \frac{\partial^2 \phi_3}{\partial y^2} = c_3 x + d_3 y$$

$$\sigma_y = \frac{\partial^2 \phi_3}{\partial x^2} = a_3 x + b_3 y$$

$$\tau_{xy} = -\frac{\partial^2 \phi_3}{\partial x \, \partial y} = -b_3 x - c_3 y$$

For a rectangular plate, taken as in Fig. 22, assuming all coefficients except d_3 equal to zero, we obtain pure bending. If only coefficient a_3 is different from zero, we obtain pure bending by normal stresses applied to the sides $y = \pm c$ of the plate. If coefficient b_3 or c_3 is taken different from zero, we obtain not only normal but also shearing stresses acting on the sides of the plate. Figure 23 represents, for instance, the case in

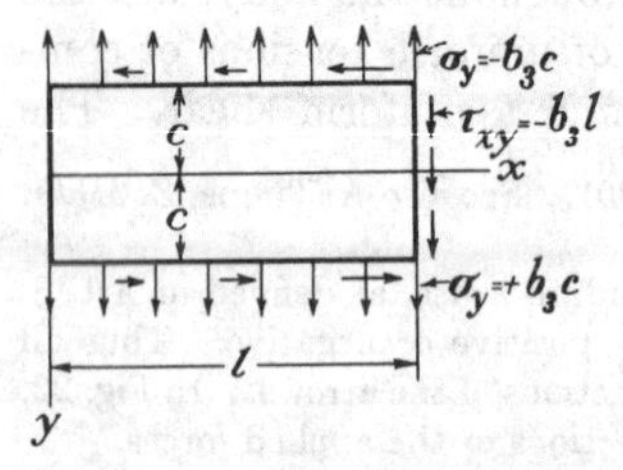

Fig. 23

which all coefficients except b_3 in function (c) are equal to zero. Along the sides $y = \pm c$ we have uniformly distributed tensile and compressive stresses, respectively, and shearing stresses proportional to x. On the side $x = l$ we have only the constant shearing stress $-b_3 l$, and there are no stresses acting on the side $x = 0$. An analogous stress distribution is obtained if coefficient c_3 is taken different from zero.

In taking the stress function in the form of polynomials of the second and third degrees we are completely free in choosing the magnitudes of the coefficients, since Eq. (a) is satisfied whatever values they may have. In the case of polynomials of higher degrees Eq. (a) is satisfied only if certain relations between the coefficients are satisfied. Taking, for instance, the stress function in the form of a polynomial of the fourth degree,

$$\phi_4 = \frac{a_4}{4(3)} x^4 + \frac{b_4}{3(2)} x^3 y + \frac{c_4}{2} x^2 y^2 + \frac{d_4}{3(2)} x y^3 + \frac{e_4}{4(3)} y^4 \tag{d}$$

and substituting it into Eq. (a), we find that the equation is satisfied only if

$$e_4 = -(2c_4 + a_4)$$

The stress components in this case are

$$\sigma_x = \frac{\partial^2 \phi_4}{\partial y^2} = c_4 x^2 + d_4 xy - (2c_4 + a_4) y^2$$

$$\sigma_y = \frac{\partial^2 \phi_4}{\partial x^2} = a_4 x^2 + b_4 xy + c_4 y^2$$

$$\tau_{xy} = -\frac{\partial^2 \phi_4}{\partial x \, \partial y} = -\frac{b_4}{2} x^2 - 2c_4 xy - \frac{d_4}{2} y^2$$

Coefficients $a_4, \ldots, d_4$ in these expressions are arbitrary, and by suitably adjusting them we obtain various conditions of loading of a rectangular plate. For instance, taking all coefficients except d_4 equal to zero, we find

$$\sigma_x = d_4 xy \qquad \sigma_y = 0 \qquad \tau_{xy} = -\frac{d_4}{2} y^2 \tag{e}$$

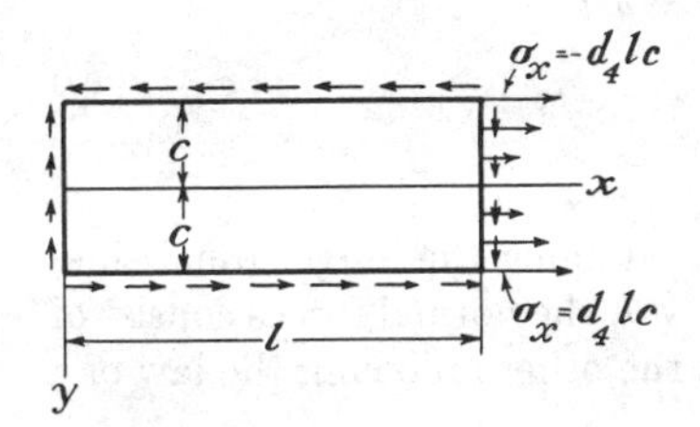

Fig. 24

Assuming d_4 positive, the forces acting on the rectangular plate shown in Fig. 24 and producing the stresses (e) are as given. On the longitudinal sides, $y = \pm c$ are uniformly distributed shearing forces; on the ends, shearing forces are distributed according to a parabolic law. The shearing forces acting on the boundary of the plate reduce to the couple[1]

$$M = \frac{d_4 c^2 l}{2} 2c - \frac{1}{3} \frac{d_4 c^2}{2} 2cl = \frac{2}{3} d_4 c^3 l$$

This couple balances the couple produced by the normal forces along the side $x = l$ of the plate.

Let us consider a stress function in the form of a polynomial of the fifth degree.

$$\phi_5 = \frac{a_5}{5(4)} x^5 + \frac{b_5}{4(3)} x^4 y + \frac{c_5}{3(2)} x^3 y^2 + \frac{d_5}{3(2)} x^2 y^3 + \frac{e_5}{4(3)} x y^4 + \frac{f_5}{5(4)} y^5 \tag{f}$$

Substituting in Eq. (a) we find that this equation is satisfied if

$$\begin{aligned} e_5 &= -(2c_5 + 3a_5) \\ f_5 &= -\tfrac{1}{3}(b_5 + 2d_5) \end{aligned}$$

The corresponding stress components are:

$$\sigma_x = \frac{\partial^2 \phi_5}{\partial y^2} = \frac{c_5}{3} x^3 + d_5 x^2 y - (2c_5 + 3a_5) x y^2 - \frac{1}{3}(b_5 + 2d_5) y^3$$

$$\sigma_y = \frac{\partial^2 \phi_5}{\partial x^2} = a_5 x^3 + b_5 x^2 y + c_5 x y^2 + \frac{d_5}{3} y^3$$

$$\tau_{xy} = -\frac{\partial^2 \phi_5}{\partial x \, \partial y} = -\frac{1}{3} b_5 x^3 - c_5 x^2 y - d_5 x y^2 + \frac{1}{3}(2c_5 + 3a_5) y^3$$

Again coefficients $a_5, \ldots, d_5$ are arbitrary, and in adjusting them we obtain solutions for various loading conditions of a plate. Taking, for instance, all coefficients, except d_5, equal to zero we find

$$\begin{aligned} \sigma_x &= d_5(x^2 y - \tfrac{2}{3} y^3) \\ \sigma_y &= \tfrac{1}{3} d_5 y^3 \\ \tau_{xy} &= -d_5 x y^2 \end{aligned} \tag{g}$$

The normal forces are uniformly distributed along the longitudinal sides of the plate (Fig. 25a). Along the side $x = l$, the normal forces consist of two parts, one following a linear law and the other following the law of a

[1] The thickness of the plate is taken equal to unity.

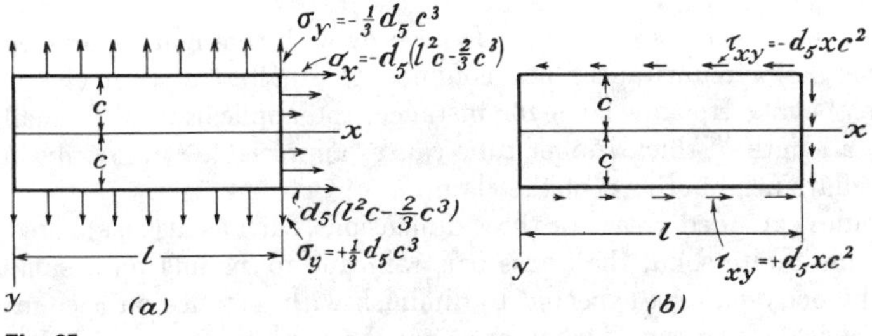

Fig. 25

cubic parabola. The shearing forces are proportional to x on the longitudinal sides of the plate and follow a parabolic law along the side $x = l$. The distribution of these stresses is shown in Fig. 25*b*.

Since Eq. (*a*) is a linear differential equation, a sum of several solutions of this equation is also a solution. We can superpose the elementary solutions considered in this article to arrive at new solutions of practical interest. Several examples of the application of this method of superposition will be considered.

19 | End Effects. Saint-Venant's Principle

In the previous article several solutions for rectangular plates were obtained from very simple forms of the stress function ϕ. In each case the boundary forces must be distributed exactly as the solution itself requires. In the case of pure bending, for instance (Fig. 22), the loading on the ends must consist of normal traction (σ_x, at $x = 0$ or $x = l$) proportional to y. If the couples on the ends are applied in any other manner, the solution given in Art. 18 is no longer correct. Another solution must be found if the changed boundary conditions on the ends are to be exactly satisfied. Many such solutions have been obtained (some are referred to later) not only for rectangular regions but for prismatic, cylindrical, and tapered shapes. These show that a change in the distribution of the load on an end, without change of the resultant, alters the stress significantly only near the end. In such cases then, simple solutions such as those of the present chapter can give sufficiently accurate results except near the ends.

The change of distribution of the load is equivalent to the superposition of a system of forces statically equivalent to zero force and zero couple. The expectation that such a system, applied to a small part of the surface of the body, would give rise to localized stress and strain only, was enunciated by Saint-Venant[1] in 1855 and came to be

[1] B. de Saint-Venant, "Mémoires des Savants Etrangers," vol. 14, 1855.

known as *Saint-Venant's principle.* It accords with common experience in a variety of circumstances not confined to small strains in elastic materials obeying Hooke's law—for instance, the application of a small clamp to a length of thick rubber tube causes appreciable strain only in the immediate neighborhood of the clamp.

For bodies extended in two or three dimensions, such as disks, spheres, or the semi-infinite solid, the stress or strain due to loading on a small part of the body may be expected to diminish with distance on account of "geometrical divergence," whether or not the resultant is zero. It has been shown[1] that vanishing of the resultant is not an adequate criterion for the degree of localization.

20 | Determination of Displacements

When the components of stress are found from the previous equations, the components of strain can be obtained by using Hooke's law, Eqs. (3) and (6). Then the displacements u and v can be obtained from the equations

$$\frac{\partial u}{\partial x} = \epsilon_x \qquad \frac{\partial v}{\partial y} = \epsilon_y \qquad \frac{\partial u}{\partial y} + \frac{\partial v}{\partial x} = \gamma_{xy} \tag{a}$$

The integration of these equations in each particular case does not present any difficulty, and we shall have several examples of their application. It may be seen at once that the strain components (*a*) remain unchanged if we add to u and v the linear functions

$$u_1 = a + by \qquad v_1 = c - bx \tag{b}$$

in which a, b, and c are constants. This means that the displacements are not entirely determined by the stresses and strains. A displacement like that of a rigid body can be superposed on the displacements due to the internal strains. The constants a and c in Eqs. (*b*) represent a translatory motion of the body and the constant b is a small angle of rotation of the rigid body about the z axis.

It has been shown (see page 31) that in the case of constant body forces the stress distribution is the same for plane stress distribution or plane strain. The displacements are different for these two problems, however, since in the case of plane stress distribution the components of strain, entering into Eqs. (*a*), are given by equations

$$\epsilon_x = \frac{1}{E}(\sigma_x - \nu\sigma_y) \qquad \epsilon_y = \frac{1}{E}(\sigma_y - \nu\sigma_x) \qquad \gamma_{xy} = \frac{1}{G}\tau_{xy}$$

[1] R. von Mises, *Bull. Am. Math. Soc.*, vol. 51, p. 555, 1945; E. Sternberg, *Quart. Appl. Math.*, vol. 11, p. 393, 1954; E. Sternberg and W. T. Koiter, *J. Appl. Mech.*, vol. 25, pp. 575–581, 1958.

and in the case of plane strain the strain components are:

$$\epsilon_x = \frac{1}{E}[\sigma_x - \nu(\sigma_y + \sigma_z)] = \frac{1}{E}[(1 - \nu^2)\sigma_x - \nu(1 + \nu)\sigma_y]$$

$$\epsilon_y = \frac{1}{E}[\sigma_y - \nu(\sigma_x + \sigma_z)] = \frac{1}{E}[(1 - \nu^2)\sigma_y - \nu(1 + \nu)\sigma_x]$$

$$\gamma_{xy} = \frac{1}{G}\tau_{xy}$$

It is easily verified that these equations can be obtained from the preceding set for plane stress by replacing F in the latter by $E/(1 - \nu^2)$, and ν by $\nu/(1 - \nu)$. These substitutions leave G, which is $E/2(1 + \nu)$, unchanged. The integration of Eqs. (*a*) will be shown later in discussing particular problems.

21 | Bending of a Cantilever Loaded at the End

Consider a cantilever having a narrow rectangular cross section of unit width bent by a force P applied at the end (Fig. 26). The upper and lower edges are free from load, and shearing forces, having a resultant P, are distributed along the end $x = 0$. These conditions can be satisfied by a proper combination of pure shear with the stresses (*e*) of Art. 18 represented in Fig. 24. Superposing the pure shear $\tau_{xy} = -b_2$ on the stresses (*e*), we find

$$\sigma_x = d_4 xy \qquad \sigma_y = 0$$
$$\tau_{xy} = -b_2 - \frac{d_4}{2} y^2 \tag{a}$$

To have the longitudinal sides $y = \pm c$ free from forces we must have

$$(\tau_{xy})_{y=\pm c} = -b_2 - \frac{d_4}{2} c^2 = 0$$

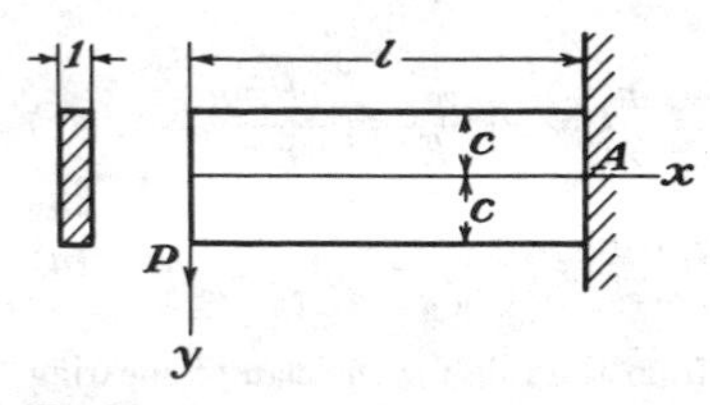

Fig. 26

from which

$$d_4 = -\frac{2b_2}{c^2}$$

To satisfy the condition on the loaded end the sum of the shearing forces distributed over this end must be equal to P. Hence[1]

$$-\int_{-c}^{c} \tau_{xy}\, dy = \int_{-c}^{c} \left(b_2 - \frac{b_2}{c^2} y^2\right) dy = P$$

from which

$$b_2 = \frac{3}{4}\frac{P}{c}$$

Substituting these values of d_4 and b_2 in Eqs. (a) we find

$$\sigma_x = -\frac{3}{2}\frac{P}{c^3} xy \qquad \sigma_y = 0$$

$$\tau_{xy} = -\frac{3P}{4c}\left(1 - \frac{y^2}{c^2}\right)$$

Noting that $\frac{2}{3}c^3$ is the moment of inertia I of the cross section of the cantilever, we have

$$\sigma_x = -\frac{Pxy}{I} \qquad \sigma_y = 0$$
$$\tau_{xy} = -\frac{P}{I}\frac{1}{2}(c^2 - y^2) \tag{b}$$

This coincides completely with the elementary solution as given in books on the strength of materials. It should be noted that this solution represents an exact solution only if the shearing forces on the ends are distributed according to the same parabolic law as the shearing stress τ_{xy} and the intensity of the normal forces at the built-in end is proportional to y. If the forces at the ends are distributed in any other manner, the stress distribution (b) is not a correct solution for the ends of the cantilever, but, by virtue of Saint-Venant's principle, it can be considered satisfactory for cross sections at a considerable distance from the ends.

Let us consider now the displacement corresponding to the stresses (b). Applying Hooke's law we find

$$\epsilon_x = \frac{\partial u}{\partial x} = \frac{\sigma_x}{E} = -\frac{Pxy}{EI} \qquad \epsilon_y = \frac{\partial v}{\partial y} = -\frac{\nu\sigma_x}{E} = \frac{\nu Pxy}{EI} \tag{c}$$

$$\gamma_{xy} = \frac{\partial u}{\partial y} + \frac{\partial v}{\partial x} = \frac{\tau_{xy}}{G} = -\frac{P}{2IG}(c^2 - y^2) \tag{d}$$

[1] The minus sign before the integral follows from the rule for the sign of shearing stresses. Stress τ_{xy} on the end $x = 0$ is positive if it is upward (see p. 4).

The procedure for obtaining the components u and v of the displacement consists in integrating Eqs. (c) and (d). By integration of Eqs. (c) we find

$$u = -\frac{Px^2y}{2EI} + f(y) \qquad v = \frac{\nu Pxy^2}{2EI} + f_1(x)$$

in which $f(y)$ and $f_1(x)$ are as yet unknown functions of y only and x only. Substituting these values of u and v in Eq. (d) we find

$$-\frac{Px^2}{2EI} + \frac{df(y)}{dy} + \frac{\nu Py^2}{2EI} + \frac{df_1(x)}{dx} = -\frac{P}{2IG}(c^2 - y^2)$$

In this equation some terms are functions of x only, some are functions of y only, and one is independent of both x and y. Denoting these groups by $F(x)$, $G(y)$, K, we have

$$F(x) = -\frac{Px^2}{2EI} + \frac{df_1(x)}{dx} \qquad G(y) = \frac{df(y)}{dy} + \frac{\nu Py^2}{2EI} - \frac{Py^2}{2IG}$$

$$K = -\frac{Pc^2}{2IG}$$

and the equation may be written

$$F(x) + G(y) = K$$

Such an equation means that $F(x)$ must be some constant d and $G(y)$ some constant e. Otherwise $F(x)$ and $G(y)$ would vary with x and y, respectively, and by varying x alone, or y alone, he equality would be violated. Thus

$$e + d = -\frac{Pc^2}{2IG} \tag{e}$$

and $$\frac{df_1(x)}{dx} = \frac{Px^2}{2EI} + d \qquad \frac{df(y)}{dy} = -\frac{\nu Py^2}{2EI} + \frac{Py^2}{2IG} + e$$

Functions $f(y)$ and $f_1(x)$ are then

$$f(y) = -\frac{\nu Py^3}{6EI} + \frac{Py^3}{6IG} + ey + g$$

$$f_1(x) = \frac{Px^3}{6EI} + dx + h$$

Substituting in the expressions for u and v we find

$$\begin{aligned} u &= -\frac{Px^2y}{2EI} - \frac{\nu Py^3}{6EI} + \frac{Py^3}{6IG} + ey + g \\ v &= \frac{\nu Pxy^2}{2EI} + \frac{Px^3}{6EI} + dx + h \end{aligned} \tag{g}$$

The constants d, e, g, h may now be determined from Eq. (e) and from the three conditions of constraint that are necessary to prevent the beam from moving as a rigid body in the xy plane. Assume that the point A, the centroid of the end cross section, is fixed. Then u and v are zero for $x = l$, $y = 0$, and we find from Eqs. (g)

$$g = 0 \qquad h = -\frac{Pl^3}{6EI} - dl$$

The deflection curve is obtained by substituting $y = 0$ into the second of Eqs. (g). Then

$$(v)_{y=0} = \frac{Px^3}{6EI} - \frac{Pl^3}{6EI} - d(l - x) \tag{h}$$

For determining the constant d in this equation, we must use the third condition of constraint, eliminating the possibility of rotation of the beam in the xy plane about the fixed point A. This constraint can be realized in various ways. Let us consider two cases: (1) When an element of the axis of the beam is fixed at the end A. Then the condition of constraint is

$$\left(\frac{\partial v}{\partial x}\right)_{\substack{x=l\\y=0}} = 0 \tag{k}$$

(2) When a vertical element of the cross section at the point A is fixed. Then the condition of constraint is

$$\left(\frac{\partial u}{\partial y}\right)_{\substack{x=l\\y=0}} = 0 \tag{l}$$

In the first case we obtain from Eq. (h)

$$d = -\frac{Pl^2}{2EI}$$

and from Eq. (e) we find

$$e = \frac{Pl^2}{2EI} - \frac{Pc^2}{2IG}$$

Substituting all the constants in Eqs. (g), we find

$$\begin{aligned} u &= -\frac{Px^2y}{2EI} - \frac{\nu Py^3}{6EI} + \frac{Py^3}{6IG} + \left(\frac{Pl^2}{2EI} - \frac{Pc^2}{2IG}\right)y \\ v &= \frac{\nu Pxy^2}{2EI} + \frac{Px^3}{6EI} - \frac{Pl^2x}{2EI} + \frac{Pl^3}{3EI} \end{aligned} \tag{m}$$

The equation of the deflection curve is

$$(v)_{y=0} = \frac{Px^3}{6EI} - \frac{Pl^2x}{2EI} + \frac{Pl^3}{3EI} \tag{n}$$

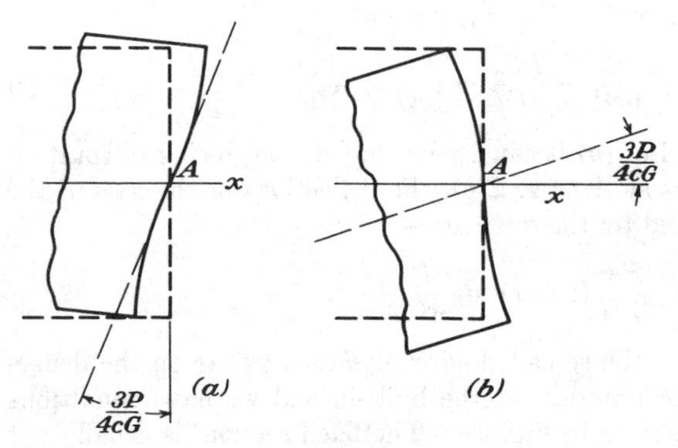

Fig. 27

which gives for the deflection at the loaded end ($x = 0$) the value $Pl^3/3EI$. This coincides with the value usually derived in elementary books on the strength of materials.

To illustrate the distortion of cross sections produced by shearing stresses, let us consider the displacement u at the fixed end ($x = l$). For this end we have from Eqs. (m),

$$
\begin{aligned}
(u)_{x=l} &= -\frac{\nu P y^3}{6EI} + \frac{P y^3}{6IG} - \frac{Pc^2 y}{2IG} \\
\left(\frac{\partial u}{\partial y}\right)_{x=l} &= -\frac{\nu P y^2}{2EI} + \frac{P y^2}{2IG} - \frac{Pc^2}{2IG} \\
\left(\frac{\partial u}{\partial y}\right)_{\substack{x=l\\ y=0}} &= -\frac{Pc^2}{2IG} = -\frac{3}{4}\frac{P}{cG}
\end{aligned}
\qquad (o)
$$

The shape of the cross section after distortion is as shown in Fig. 27a. Owing to the shearing stress $\tau_{xy} = -3P/4c$ at the point A, an element of the cross section at A rotates in the xy plane about the point A through an angle $3P/4cG$ in the clockwise direction.

If a vertical element of the cross section is fixed at A (Fig. 27b), instead of a horizontal element of the axis, we find from condition (l) and the first of Eqs. (g)

$$
e = \frac{Pl^2}{2EI}
$$

and from Eq. (e) we find

$$
d = -\frac{Pl^2}{2EI} - \frac{Pc^2}{2IG}
$$

Substituting in the second of Eqs. (g) we find

$$(v)_{y=0} = \frac{Px^3}{6EI} - \frac{Pl^2x}{2EI} + \frac{Pl^3}{3EI} + \frac{Pc^2}{2IG}(l - x) \tag{r}$$

Comparing this with Eq. (n) it can be concluded that, owing to rotation of the end of the axis at A (Fig. 27b), the deflections of the axis of the cantilever are increased by the quantity

$$\frac{Pc^2}{2IG}(l - x) = \frac{3P}{4cG}(l - x)$$

This is an estimate[1] of the so-called *effect of shearing force* on the deflection of the beam. In practice, at the built-in end we have conditions different from those shown in Fig. 27. The fixed section[2] is usually not free to distort and the distribution of forces at this end is different from that given by Eqs. (b). However, solution (b) is satisfactory for comparatively long cantilevers at considerable distances from the terminals.

22 | Bending of a Beam by Uniform Load

Let a beam of narrow rectangular cross section of unit width, supported at the ends, be bent by a uniformly distributed load of intensity q, as shown in Fig. 28. The conditions at the upper and lower edges of the beam are:

$$(\tau_{xy})_{y=\pm c} = 0 \qquad (\sigma_y)_{y=+c} = 0 \qquad (\sigma_y)_{y=-c} = -q \tag{a}$$

The conditions at the ends $x = \pm l$ are

$$\int_{-c}^{c} \tau_{xy}\, dy = \mp ql \qquad \int_{-c}^{c} \sigma_x\, dy = 0 \qquad \int_{-c}^{c} \sigma_x y\, dy = 0 \tag{b}$$

The last two of Eqs. (b) state that there is no longitudinal force and no bending couple applied at the ends of the beam. All the conditions (a)

[1] Others are indicated in Prob. 3, p. 63, and in the text on p. 49.

[2] The effect of elasticity in the support itself is examined experimentally and analytically by W. J. O'Donnell, *J. Appl. Mech.*, vol. 27, pp. 461–464, 1960.

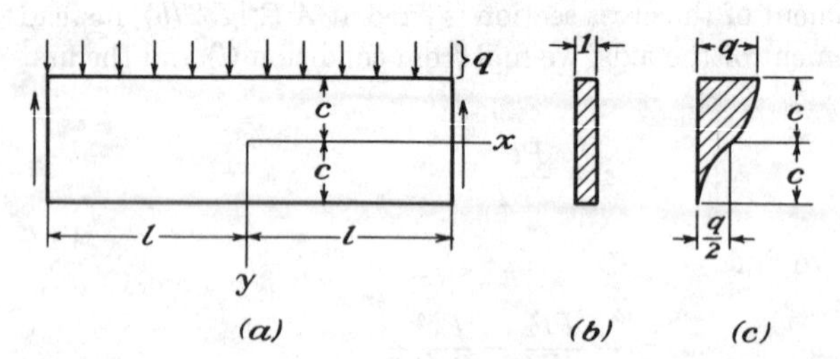

Fig. 28

and (*b*) can be satisfied by combining certain solutions in the form of polynomials as obtained in Art. 18. We begin with solution (*g*), illustrated by Fig. 25. To remove the tensile stresses along the side $y = c$ and the shearing stresses along the sides $y = \pm c$, we superpose a simple compression $\sigma_y = a_2$ from solution (*b*), Art. 18, and the stresses $\sigma_y = b_3 y$ and $\tau_{xy} = -b_3 x$ in Fig. 23. In this manner we find

$$\begin{aligned} \sigma_x &= d_5(x^2y - \tfrac{2}{3}y^3) \\ \sigma_y &= \tfrac{1}{3}d_5y^3 + b_3y + a_2 \\ \tau_{xy} &= -d_5xy^2 - b_3x \end{aligned} \tag{c}$$

From the conditions (*a*) we find

$$\begin{aligned} -d_5c^2 - b_3 &= 0 \\ \tfrac{1}{3}d_5c^3 + b_3c + a_2 &= 0 \\ -\tfrac{1}{3}d_5c^3 - b_3c + a_2 &= -q \end{aligned}$$

from which

$$a_2 = -\frac{q}{2} \qquad b_3 = \frac{3}{4}\frac{q}{c} \qquad d_5 = -\frac{3}{4}\frac{q}{c^3}$$

Substituting in Eqs. (*c*) and noting that $2c^3/3$ is equal to the moment of inertia I of the rectangular cross-sectional area of unit width, we find

$$\begin{aligned} \sigma_x &= -\frac{3}{4}\frac{b}{c^3}\left(x^2y - \frac{2}{3}y^3\right) = -\frac{q}{2I}\left(x^2y - \frac{2}{3}y^3\right) \\ \sigma_y &= -\frac{3q}{4c^3}\left(\frac{1}{3}y^3 - c^2y + \frac{2}{3}c^3\right) = -\frac{q}{2I}\left(\frac{1}{3}y^3 - c^2y + \frac{2}{3}c^3\right) \\ \tau_{xy} &= -\frac{3q}{4c^3}(c^2 - y^2)x = -\frac{q}{2I}(c^2 - y^2)x \end{aligned} \tag{d}$$

It can easily be checked that these stress components satisfy not only conditions (*a*) on the longitudinal sides but also the first two conditions (*b*) at the ends. To make the couples at the ends of the beam vanish, we superpose on solution (*d*) a pure bending, $\sigma_x = d_3y$, $\sigma_y = \tau_{xy} = 0$, shown in Fig. 22, and determine the constant d_3 from the condition at $x = \pm l$

$$\int_{-c}^{c} \sigma_x y\, dy = \int_{-c}^{c} \left[-\frac{3}{4}\frac{q}{c^3}\left(l^2y - \frac{2}{3}y^3\right) + d_3y\right] y\, dy = 0$$

from which

$$d_3 = \frac{3}{4}\frac{q}{c}\left(\frac{l^2}{c^2} - \frac{2}{5}\right)$$

Hence, finally,

$$\sigma_x = -\frac{3}{4}\frac{q}{c^3}\left(x^2y - \frac{2}{3}y^3\right) + \frac{3}{4}\frac{q}{c}\left(\frac{l^2}{c^2} - \frac{2}{5}\right)y$$
$$= \frac{q}{2I}(l^2 - x^2)y + \frac{q}{2I}\left(\frac{2}{3}y^3 - \frac{2}{5}c^2y\right) \tag{33}$$

The first term in this expression represents the stresses given by the usual elementary theory of bending, and the second term gives the necessary correction. This correction does not depend on x and is small in comparison with the maximum bending stress, provided the span of the beam is large in comparison with its depth. For such beams the elementary theory of bending gives a sufficiently accurate value for the stresses σ_x. It should be noted that expression (33) is an exact solution only if at the ends $x = \pm l$ the normal forces are distributed according to the law

$$\bar{X} = \pm\frac{3}{4}\frac{q}{c^3}\left(\frac{2}{3}y^3 - \frac{2}{5}c^2y\right)$$

i.e., if the normal forces at the ends are the same as σ_x for $x = \pm l$ from Eq. (33). These forces have zero resultant force and zero resultant couple. Hence, from Saint-Venant's principle we can conclude that their effects on the stresses at considerable distances from the ends, say at distances larger than the depth of the beam, can be neglected. Solution (33) at such points is therefore accurate enough when no such forces $\bar{X}$ are applied.

The discrepancy between the exact solution (33) and the approximate solution, given by the first term of (33), appears because in deriving the approximate solution it is assumed that the longitudinal fibers of the beam are in a condition of simple tension. From solution (*d*) it can be seen that there are compressive stresses σ_y between the fibers. These stresses are responsible for the correction represented by the second term of solution (33). The distribution of the compressive stresses σ_y over the depth of the beam is shown in Fig. 28*c*. The distribution of shearing stress τ_{xy}, given by the third of Eqs. (*d*), over a cross section of the beam coincides with that given by the usual elementary theory.

When the beam is loaded by its own weight instead of the distributed load q, the solution must be modified by putting $q = 2\rho gc$ in (33) and the last two of Eqs. (*d*) and adding the stresses

$$\sigma_x = 0 \qquad \sigma_y = \rho g(c - y) \qquad \tau_{xy} = 0 \tag{e}$$

For the stress distribution, (*e*) can be obtained from Eqs. (29) by taking

$$\phi = \tfrac{1}{6}\rho g(y^3 + 3cx^2)$$

and therefore represents a possible state of stress due to weight and boundary forces. On the upper edge $y = -c$ we have $\sigma_y = 2\rho gc$, and on the lower edge $y = c$, $\sigma_y = 0$.

Thus, when the stresses (*e*) are added to the previous solution, with $q = 2\rho g c$, the stress on both horizontal edges is zero, and the load on the beam consists only of its own weight.

The displacements u and v can be calculated by the method indicated in the previous article. Assuming that at the centroid of the middle cross section ($x = 0$, $y = 0$) the horizontal displacement is zero and the vertical displacement is equal to the deflection δ, we find, using solutions (*d*) and (33),

$$u = \frac{q}{2EI}\left[\left(l^2x - \frac{x^3}{3}\right)y + x\left(\frac{2}{3}y^3 - \frac{2}{5}c^2y\right) + \nu x\left(\frac{1}{3}y^3 - c^2y + \frac{2}{3}c^3\right)\right]$$

$$v = -\frac{q}{2EI}\left\{\frac{y^4}{12} - \frac{c^2y^2}{2} + \frac{2}{3}c^3y + \nu\left[(l^2 - x^2)\frac{y^2}{2} + \frac{y^4}{6} - \frac{1}{5}c^2y^2\right]\right\}$$
$$- \frac{q}{2EI}\left[\frac{l^2x^2}{2} - \frac{x^4}{12} - \frac{1}{5}c^2x^2 + \left(1 + \frac{1}{2}\nu\right)c^2x^2\right] + \delta$$

It can be seen from the expression for u that the neutral surface of the beam is not at the centerline. Owing to the compressive stress

$$(\sigma_y)_{y=0} = -\frac{q}{2}$$

the centerline has a tensile strain $\nu q/2E$, and we find

$$(u)_{y=0} = \frac{\nu q x}{2E}$$

From the expression for v we find the equation of the deflection curve,

$$(v)_{y=0} = \delta - \frac{q}{2EI}\left[\frac{l^2x^2}{2} - \frac{x^4}{12} - \frac{1}{5}c^2x^2 + \left(1 + \frac{1}{2}\nu\right)c^2x^2\right] \qquad (f)$$

Assuming that the deflection is zero at the ends ($x = \pm l$) of the centerline, we find

$$\delta = \frac{5}{24}\frac{ql^4}{EI}\left[1 + \frac{12}{5}\frac{c^2}{l^2}\left(\frac{4}{5} + \frac{\nu}{2}\right)\right] \qquad (34)$$

The factor before the brackets is the deflection that is derived by the elementary analysis, assuming that cross sections of the beam remain plane during bending. The second term in the brackets represents the correction usually called the *effect of shearing force.*

By differentiating Eq. (*f*) for the deflection curve twice with respect to x, we find the following expression for the curvature:

$$\left(\frac{d^2v}{dx^2}\right)_{y=0} = \frac{q}{EI}\left[\frac{l^2 - x^2}{2} + c^2\left(\frac{4}{5} + \frac{\nu}{2}\right)\right] \qquad (35)$$

It will be seen that the curvature is not exactly proportional to the bending moment[1] $q(l^2 - x^2)/2$. The additional term in the brackets represents the necessary correction to the usual elementary formula. A more general investigation of the curvature of beams shows[2] that the correction term given in expression (35) can also be used for any case of continuously varying intensity of load. The effect of shearing force on the deflection in the case of a concentrated load will be discussed later (page 122).

An elementary derivation of the effect of the shearing force on the curvature of the deflection curve of beams was given by Rankine[3] in England and by Grashof[4] in Germany. Taking the maximum shearing strain at the neutral axis of a rectangular beam of unit width as $\frac{3}{2}(Q/2cG)$, where Q is the shearing force, the corresponding increase in curvature is given by the derivative of the above shearing strain with respect to x, which gives $\frac{3}{2}(q/2cG)$. The corrected expression for the curvature by elementary analysis then becomes

$$\frac{q}{EI}\frac{l^2 - x^2}{2} + \frac{3}{2}\frac{q}{2cG} = \frac{q}{EI}\left[\frac{l^2 - x^2}{2} + c^2(1 + \nu)\right]$$

Comparing this with expression (35), it is seen that the elementary solution gives an exaggerated value[5] for the correction.

The correction term in expression (35) for the curvature cannot be attributed to the shearing force alone. It is produced partially by the compressive stresses σ_y. These stresses are not uniformly distributed over the depth of the beam. The lateral expansion in the x direction produced by these stresses diminishes from the top to the bottom of the beam, and in this way a reversed curvature (convex upwards) is produced. This curvature together with the effect of shearing force accounts for the correction term in Eq. (35).

23 | Other Cases of Continuously Loaded Beams

By increasing the degree of polynomials representing solutions of the two-dimensional problem (Art. 18), we may obtain solutions of bending problems with various types of continuously varying load.[6] By taking, for instance, a solution in the form of a polynomial of the sixth degree and combining it with the previous solutions of Art. 18, we may obtain the stresses in a vertical cantilever loaded by hydrostatic pressure, as shown in Fig. 29. In this manner it can be shown that all conditions on the longitudinal sides of the cantilever are satisfied by the following

[1] This was pointed out first by K. Pearson, *Quart. J. Math.*, vol. 24, p. 63, 1889.

[2] See paper by T. v. Kármán, *Abhandl. Aerodynam. Inst., Tech. Hochschule, Aachen*, vol. 7, p. 3, 1927.

[3] Rankine, "Applied Mechanics," 14th ed., p. 344, 1895.

[4] Grashof, "Elastizität und Festigkeit," 2d ed., 1878.

[5] A better approximation is given by elementary strain-energy considerations. See S. Timoshenko, "Strength of Materials," 3d ed., vol. 1, p. 318.

[6] See paper by Timpe, *loc. cit.*; W. R. Osgood, *J. Res. Nat. Bur. Std.*, ser. B, vol. 28, p. 159, 1942.

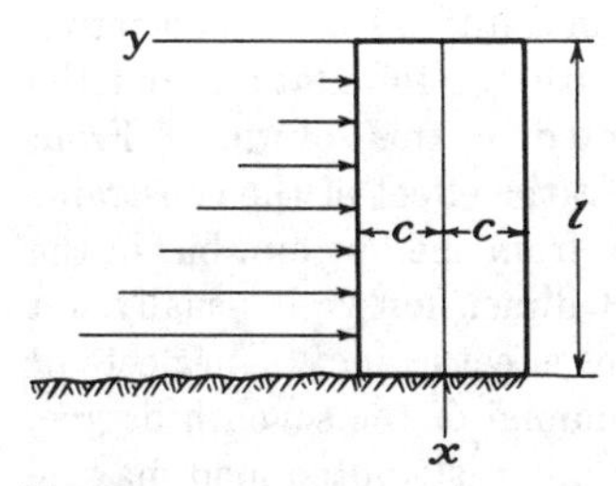

Fig. 29

system of stresses:

$$\sigma_x = \frac{qx^3y}{4c^3} + \frac{q}{4c^3}\left(-2xy^3 + \frac{6}{5}c^2xy\right)$$

$$\sigma_y = -\frac{qx}{2} + qx\left(\frac{y^3}{4c^3} - \frac{3y}{4c}\right) \tag{a}$$

$$\tau_{xy} = \frac{3qx^2}{8c^3}(c^2 - y^2) - \frac{q}{8c^3}(c^4 - y^4) + \frac{q}{4c^3}\frac{3}{5}c^2(c^2 - y^2)$$

Here q is the weight of unit volume of the fluid, so that the intensity of the load at a depth x is qx. The shearing force and the bending moment at the same depth are $qx^2/2$ and $qx^3/6$, respectively. It is evident that the first terms in the expressions for σ_x and τ_{xy} are the values of the stresses calculated by the usual elementary formulas.

On the top end of the beam ($x = 0$) the normal stress is zero. The shearing stress is

$$\tau_{xy} = -\frac{q}{8c^3}(c^4 - y^4) + \frac{q}{4c^3}\frac{3}{5}c^2(c^2 - y^2)$$

Although this is not zero, it is small all over the cross section and the resultant is zero, so that the condition approaches that of an end free from external forces.

By adding to σ_x in Eqs. (*a*) the term $-q_1x$, in which q_1 is the weight of unit volume of the material of the cantilever, the effect of the weight of the beam on the stress distribution is taken into account. It has been proposed[1] to use the solution obtained in this way for calculating the stresses in masonry dams of rectangular cross section. It should be noted that this solution does not satisfy the conditions at the bottom of the dam. Solution (*a*) is exact if, at the bottom, forces are acting which are

[1] M. Levy, *Compt. Rend.*, vol. 126, p. 1235, 1898.

distributed in the same manner as σ_x and τ_{xy} in solution (*a*). In an actual case the bottom of the dam is connected with the foundation, and the conditions are different from those represented by this solution. From Saint-Venant's principle it can be stated that the effect of the constraint at the bottom is negligible at large distances from the bottom, but in the case of a masonry dam the cross-sectional dimension $2c$ is usually not small in comparison with the height l and this effect cannot be neglected.[1]

By taking for the stress function a polynomial of the seventh degree, the stresses in a beam loaded by a parabolically distributed load may be obtained. In Chap. 6 (page 178) it is shown how, by use of the complex variable, the polynomial stress function of any degree may be written down at once.

In the general case of a continuous distribution of load q (Fig. 30) the stresses at any cross section at a considerable distance from the ends, say at a distance larger than the depth of the beam, can be approximately calculated from the following equations:[2]

$$
\begin{aligned}
\sigma_x &= \frac{My}{I} + q\left(\frac{y^3}{2c^3} - \frac{3}{10}\frac{y}{c}\right) \\
\sigma_y &= -\frac{q}{2} + q\left(\frac{3y}{4c} - \frac{y^3}{4c^3}\right) \\
\tau_{xy} &= \frac{Q}{2I}(c^2 - y^2)
\end{aligned}
\tag{36}
$$

in which M and Q are the bending moment and shearing forces calculated in the usual

[1] The problem of stresses in masonry dams is of great practical interest and has been discussed by various authors. See K. Pearson, On Some Disregarded Points in the Stability of Masonry Dams, *Drapers' Co. Research Mem.*, 1904; K. Pearson and C. Pollard, An Experimental Study of the Stresses in Masonry Dams, *Drapers' Co. Research Mem.*, 1907. See also papers by L. F. Richardson, *Trans. Roy. Soc. (London)*, ser. A, vol. 210, p. 307, 1910; and S. D. Carothers, *Proc. Roy. Soc. Edinburgh*, vol. 33, p. 292, 1913. I. Muller, *Publ. Lab. Photoélasticité*, Zürich, 1930. Fillunger, *Oesterr. Wochschr. Öffentl. Baudienst*, 1913, No. 35. K. Wolf, *Sitzber. Akad. Wiss. Wien*, vol. 123, 1914.

[2] F. Seewald, *Abhandl. Aerodynam. Inst., Tech. Hochschule, Aachen*, vol. 7, p. 11, 1927. Concerning further development of such approximations see B. E. Gatewood and R. Dale, *J. Appl. Mech.*, vol. 29, 1962, pp. 747–749.

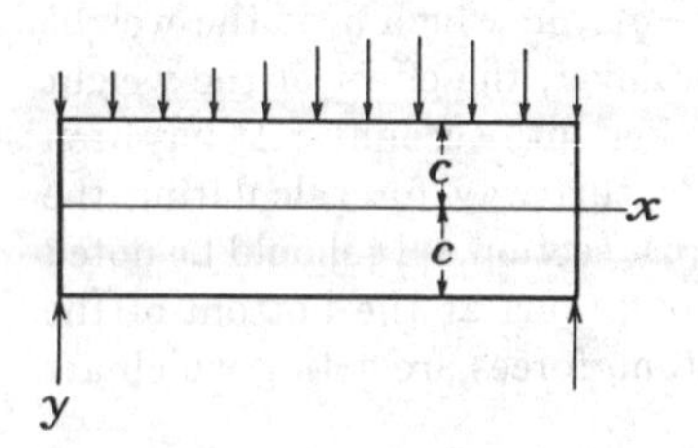

Fig. 30

way and q is the intensity of load at the cross section under consideration. These equations agree with those previously obtained for a uniformly loaded beam (see Art. 22).

If the load of intensity q, in the downward direction, is distributed along the lower edge ($y = +c$) of the beam, the expressions for the stresses are obtained from Eqs. (36) by superposing a uniform tensile stress, $\sigma_y = q$, and

$$\sigma_x = \frac{My}{I} + q\left(\frac{y^3}{2c^3} - \frac{3}{10}\frac{y}{c}\right)$$

$$\sigma_y = \frac{q}{2} + q\left(\frac{3y}{4c} - \frac{y^3}{4c^3}\right) \tag{36'}$$

$$\tau_{xy} = \frac{Q}{2I}(c^2 - y^2)$$

24 | Solution of the Two-dimensional Problem in the Form of a Fourier Series[1]

It has been shown that if the load is continuously distributed along the length of a rectangular beam of narrow cross section, a stress function in the form of a polynomial may be used in certain simple cases. A much greater degree of generality is attained by taking the function as a Fourier series (in x). Each component of load on the upper and lower edges can then have the generality possible in such series. For instance, it may have discontinuities.

The equation for the stress function,

$$\frac{\partial^4\phi}{\partial x^4} + 2\frac{\partial^4\phi}{\partial x^2\,\partial y^2} + \frac{\partial^4\phi}{\partial y^4} = 0 \tag{a}$$

may be satisfied by taking the function ϕ in the form

$$\phi = \sin\frac{m\pi x}{l} f(y) \tag{b}$$

in which m is an integer and $f(y)$ a function of y only. Substituting (b) into Eq. (a) and using the notation $m\pi/l = \alpha$, we find the following equation for determining $f(y)$:

$$\alpha^4 f(y) - 2\alpha^2 f''(y) + f^{IV}(y) = 0 \tag{c}$$

The general integral of this linear differential equation with constant coefficients is

$$f(y) = C_1\cosh\alpha y + C_2\sinh\alpha y + C_3 y\cosh\alpha y + C_4 y\sinh\alpha y$$

The stress function then is

$$\phi = \sin\alpha x(C_1\cosh\alpha y + C_2\sinh\alpha y + C_3 y\cosh\alpha y + C_4 y\sinh\alpha y) \tag{d}$$

[1] Perhaps the earliest investigation of Fourier solutions, and still one of the most thorough, is given by E. Mathieu, "Théorie de l'Elasticité des Corps Solides," seconde partie, chap. 10, pp. 140–178, Gauthier-Villars, Paris, 1890. Single Fourier series in x and y are superposed to solve problems of finite rectangles. Convergence in the determination of the Fourier coefficients from an infinite set of simultaneous algebraic equations is examined.

and the corresponding stress components are

$$
\begin{aligned}
\sigma_x &= \frac{\partial^2 \phi}{\partial y^2} = \sin \alpha x[C_1\alpha^2 \cosh \alpha y + C_2\alpha^2 \sinh \alpha y + C_3\alpha(2 \sinh \alpha y \\
&\qquad + \alpha y \cosh \alpha y) + C_4\alpha(2 \cosh \alpha y + \alpha y \sinh \alpha y)] \\
\sigma_y &= \frac{\partial^2 \phi}{\partial x^2} = -\alpha^2 \sin \alpha x(C_1 \cosh \alpha y + C_2 \sinh \alpha y + C_3 y \cosh \alpha y + C_4 y \sinh \alpha y) \qquad (e) \\
\tau_{xy} &= -\frac{\partial^2 \phi}{\partial x\, \partial y} = -\alpha \cos \alpha x[C_1\alpha \sinh \alpha y + C_2\alpha \cosh \alpha y + C_3(\cosh \alpha y \\
&\qquad + \alpha y \sinh \alpha y) + C_4(\sinh \alpha y + \alpha y \cosh \alpha y)]
\end{aligned}
$$

Let us consider a particular case of a rectangular beam supported at the ends and subjected along the upper and lower edges to the action of continuously distributed vertical forces of the intensity $A \sin \alpha x$ and $B \sin \alpha x$, respectively. Figure 31 shows the case when $\alpha = 4\pi/l$ and indicates also the positive values of A and B. The stress distribution for this case can be obtained from solution (e). The constants of integration $C_1, \ldots, C_4$ may be determined from the conditions on the upper and lower edges of the beam, $y = \pm c$. These conditions are:

For $y = +c$,

$$\tau_{xy} = 0 \qquad \sigma_y = -B \sin \alpha x$$

For $y = -c$, $\qquad (f)$

$$\tau_{xy} = 0 \qquad \sigma_y = -A \sin \alpha x$$

Substituting these values in the third of Eqs. (e), we find

$$C_1\alpha \sinh \alpha c + C_2\alpha \cosh \alpha c + C_3(\cosh \alpha c + \alpha c \sinh \alpha c) + C_4(\sinh \alpha c + \alpha c \cosh \alpha c) = 0$$

$$-C_1\alpha \sinh \alpha c + C_2\alpha \cosh \alpha c + C_3(\cosh \alpha c + \alpha c \sinh \alpha c) - C_4(\sinh \alpha c + \alpha c \cosh \alpha c) = 0$$

from which

$$
\begin{aligned}
C_3 &= -C_2 \frac{\alpha \cosh \alpha c}{\cosh \alpha c + \alpha c \sinh \alpha c} \\
C_4 &= -C_1 \frac{\alpha \sinh \alpha c}{\sinh \alpha c + \alpha c \cosh \alpha c}
\end{aligned} \qquad (g)
$$

Using the conditions on the sides $y = \pm c$ in the second of Eqs. (e), we find

$$\alpha^2(C_1 \cosh \alpha c + C_2 \sinh \alpha c + C_3 c \cosh \alpha c + C_4 c \sinh \alpha c) = B$$

$$\alpha^2(C_1 \cosh \alpha c - C_2 \sinh \alpha c - C_3 c \cosh \alpha c + C_4 c \sinh \alpha c) = A$$

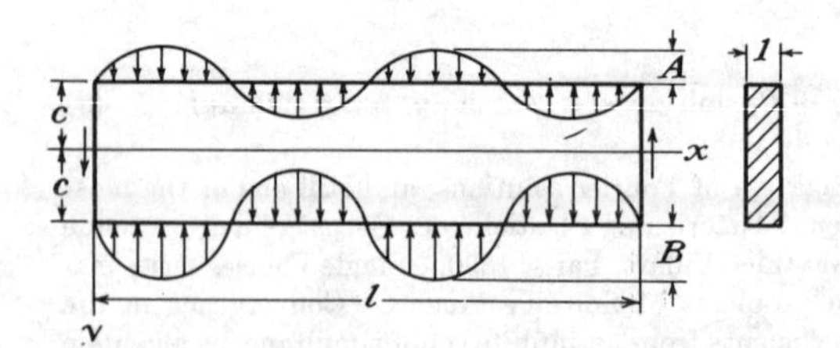

Fig. 31

By adding and subtracting these equations and using Eqs. (g), we find

$$
\begin{aligned}
C_1 &= \frac{A+B}{\alpha^2}\frac{\sinh \alpha c + \alpha c \cosh \alpha c}{\sinh 2\alpha c + 2\alpha c} \\
C_2 &= -\frac{A-B}{\alpha^2}\frac{\cosh \alpha c + \alpha c \sinh \alpha c}{\sinh 2\alpha c - 2\alpha c} \\
C_3 &= \frac{A-B}{\alpha^2}\frac{\alpha \cosh \alpha c}{\sinh 2\alpha c - 2\alpha c} \\
C_4 &= -\frac{A+B}{\alpha^2}\frac{\alpha \sinh \alpha c}{\sinh 2\alpha c + 2\alpha c}
\end{aligned}
\tag{h}
$$

Substituting in Eqs. (e), we find the following expressions for the stress components:

$$
\begin{aligned}
\sigma_x &= (A+B)\frac{(\alpha c \cosh \alpha c - \sinh \alpha c)\cosh \alpha y - \alpha y \sinh \alpha y \sinh \alpha c}{\sinh 2\alpha c + 2\alpha c}\sin \alpha x \\
&\quad - (A-B)\frac{(\alpha c \sinh \alpha c - \cosh \alpha c)\sinh \alpha y - \alpha y \cosh \alpha y \cosh \alpha c}{\sinh 2\alpha c - 2\alpha c}\sin \alpha x \\
\sigma_y &= -(A+B)\frac{(\alpha c \cosh \alpha c + \sinh \alpha c)\cosh \alpha y - \alpha y \sinh \alpha y \sinh \alpha c}{\sinh 2\alpha c + 2\alpha c}\sin \alpha x \\
&\quad + (A-B)\frac{(\alpha c \sinh \alpha c + \cosh \alpha c)\sinh \alpha y - \alpha y \cosh \alpha y \cosh \alpha c}{\sinh 2\alpha c - 2\alpha c}\sin \alpha x \\
\tau_{xy} &= -(A+B)\frac{\alpha c \cosh \alpha c \sinh \alpha y - \alpha y \cosh \alpha y \sinh \alpha c}{\sinh 2\alpha c + 2\alpha c}\cos \alpha x \\
&\quad + (A-B)\frac{\alpha c \sinh \alpha c \cosh \alpha y - \alpha y \sinh \alpha y \cosh \alpha c}{\sinh 2\alpha c - 2\alpha c}\cos \alpha x
\end{aligned}
\tag{k}
$$

These stresses satisfy the conditions shown in Fig. 31 along the sides $y = \pm c$. At the ends of the beam $x = 0$ and $x = l$, the stresses σ_x are zero and only shearing stress τ_{xy} is present. This stress is represented by two terms [see Eqs. (k)]. The first term, proportional to $A + B$, represents stresses which, for the upper and lower halves of the end cross section, are of the same magnitude but of opposite sign. The resultant of these stresses over the end is zero. The second term, proportional to $A - B$, has resultants at the ends of the beam that maintain equilibrium with the loads applied to the longitudinal sides ($y = \pm c$).

If these loads are the same for both sides, coefficient A is equal to B, and the reactive forces at the ends vanish. Let us consider this particular case more in detail, assuming that the length of the beam is large in comparison with its depth. From the second of Eqs. (k) the normal stresses σ_y over the middle plane $y = 0$ of the beam are

$$\sigma_y = -2A\frac{\alpha c \cosh \alpha c + \sinh \alpha c}{\sinh 2\alpha c + 2\alpha c}\sin \alpha x \tag{l}$$

For long beams αc, equal to $m\pi c/l$, is small, provided the number of half-waves m is not large. Then, substituting in (l),

$$\sinh \alpha c = \alpha c + \frac{(\alpha c)^3}{6} + \frac{(\alpha c)^5}{120} + \cdots \qquad \cosh \alpha c = 1 + \frac{(\alpha c)^2}{2} + \frac{(\alpha c)^4}{24} + \cdots$$

and neglecting small quantities of higher order than $(\alpha c)^4$, we find

$$\sigma_y = -A \sin \alpha x \left(1 - \frac{(\alpha c)^4}{24}\right)$$

Hence for small values of αc the distribution of stresses over the middle plane is practically the same as on both horizontal edges ($y = \pm c$) of the beam. It can be

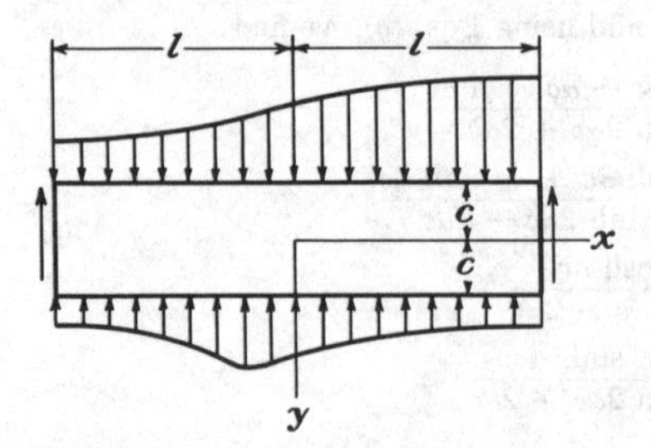

Fig. 32

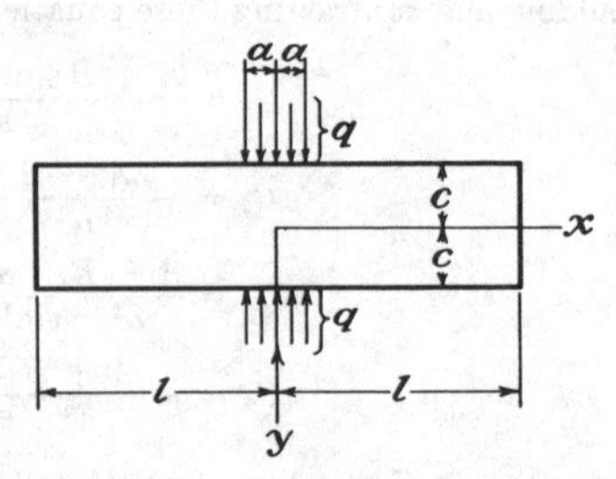

Fig. 33

concluded that pressures are transmitted through a beam or plate without any substantial change, provided the variation of these pressures along the sides is not rapid.

The shearing stresses τ_{xy} for this case are very small. On the upper and lower halves of the end cross sections they add up to the small resultants necessary to balance the small difference between the pressures on the horizontal edges ($y = \pm c$) and the middle plane ($y = 0$).

In the most general case the distribution of vertical loading along the upper and lower edges of a beam (Fig. 32) can be represented by the following series:[1]

For the upper edge,

$$q_u = A_0 + \sum_{m=1}^{\infty} A_m \sin \frac{m\pi x}{l} + \sum_{m=1}^{\infty} A_m' \cos \frac{m\pi x}{l}$$

For the lower edge, $\qquad (m)$

$$q_l = B_0 + \sum_{m=1}^{\infty} B_m \sin \frac{m\pi x}{l} + \sum_{m=1}^{\infty} B_m' \cos \frac{m\pi x}{l}$$

The constant terms A_0 and B_0 represent a uniform loading of the beam, which was discussed in Art. 22. Stresses produced by terms containing $\sin (m\pi x/l)$ are obtained by summing up solutions (k). The stresses produced by terms containing $\cos (m\pi x/l)$ are easily obtained from (k) by exchanging $\sin \alpha x$ for $\cos \alpha x$ and vice versa, and by changing the sign of τ_{xy}.

To illustrate the application[2] of this general method of stress calculation in rectangular plates, let us consider the case shown in Fig. 33. For this case of symmetrical loading, the terms with $\sin (m\pi x/l)$ vanish from expressions (m) and the coefficients A_0 and A_m' are obtained in the usual manner:

$$A_0 = B_0 = \frac{qa}{l} \qquad A_m' = B_m' = \frac{1}{l}\int_{-a}^{a} q \cos \frac{m\pi x}{l}\, dx = \frac{2q \sin (m\pi a/l)}{m\pi} \qquad (n)$$

The terms A_0 and B_0 represent a uniform compression in the y direction equal to qa/l. The stresses produced by the trigonometric terms are obtained by using solutions (k), exchanging $\sin \alpha x$ for $\cos \alpha x$ in this solution, and changing the sign of τ_{xy}.

[1] For Fourier series see Osgood, "Advanced Calculus," 1928; or Byerly, "Fourier Series and Spherical Harmonics," 1902; or Churchill, "Fourier Series and Boundary Value Problems," 1963.

[2] Several examples are worked out by M. C. Ribière, *Compt. Rend.*, vol. 126, pp. 402–404 and 1190–1192, 1898, and by F. Bleich, *Bauingenieur*, vol. 4, p. 255, 1923.

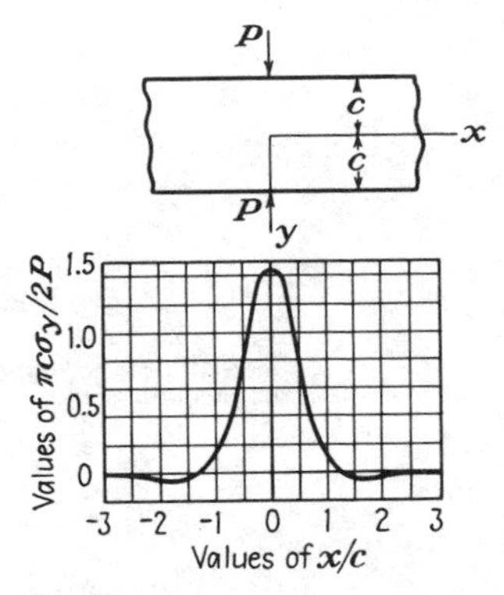

Fig. 34

Let us consider the middle plane $y = 0$, on which there is only the normal stress σ_y. By using the second of Eqs. (k) we find

$$\sigma_y = -\frac{qa}{l} - \frac{4q}{\pi}\sum_{m=1}^{\infty}\frac{\sin(m\pi a/l)}{m}\frac{(m\pi c/l)\cosh(m\pi c/l) + \sinh(m\pi c/l)}{\sinh(2m\pi c/l) + 2(m\pi c/l)}\cos\frac{m\pi x}{l}$$

This stress was evaluated by Filon[1] for an infinitely long strip when the dimension a is very small (i.e., concentrated force $P = 2qa$). The results of this calculation are shown in Fig. 34. It will be seen that σ_y diminishes very rapidly with x. At a value $x/c = 1.35$, it becomes zero and is then replaced by tension. Filon discusses also the case shown in Fig. 35 when the forces P are displaced one with respect to the other. The distribution of shearing stresses over the cross section nn in this case is of practical interest and is shown in Fig. 36. It may be seen that for small values of the ratio b/c this distribution does not resemble the parabolic distribution given by the elementary

[1] L. N. G. Filon, *Trans. Roy. Soc. (London)*, ser. A, vol. 201, p. 67, 1903. The same problem was discussed also by A. Timpe, *Z. Math. Physik*, vol. 55, p. 149, 1907; G. Mesmer, Vergleichende spannungsoptische Untersuchungen . . . , dissertation, Göttingen, 1929; F. Seewald, *loc. cit.*, and H. Bay, *Ingenieur-Arch.*, vol. 3, p. 435, 1932. An approximate solution of the same problem was given by M. Pigeaud, *Compt. Rend.*, vol. 161, p. 673, 1915. The investigation of the problem in the case of a rectangular plate of finite length was made by J. N. Goodier, *J. Appl. Mech.*, vol. 54, no. 18, p. 173, 1932.

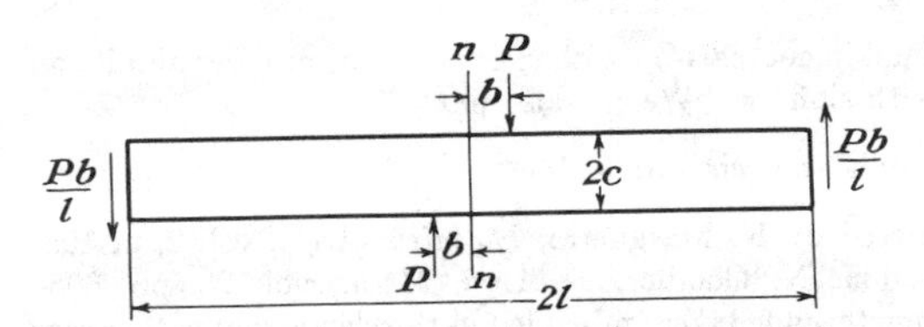

Fig. 35

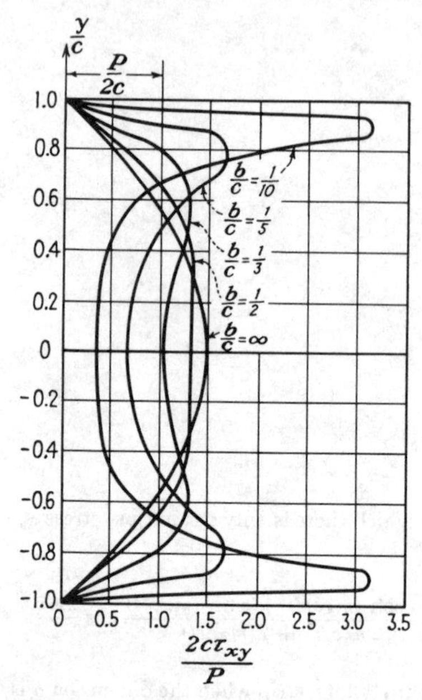

Fig. 36

theory, and that there are very large stresses at the top and bottom of the beam while the middle portion of the beam is practically free from shearing stresses.

In the problem of Fig. 34 there will by symmetry be no shear stress and no vertical displacement at the middle line $y = 0$. The upper half therefore corresponds to an elastic layer resting on a rigid smooth base.[1]

Let us consider now another extreme case when the depth of the plate $2c$ is large in comparison with the length $2l$ (Fig. 37). We shall use this case to show that the distribution of stresses over cross sections rapidly approaches uniformity as the distance from the point of application of the forces P increases. By using the second of Eqs. (k) with $\cos \alpha x$ instead of $\sin \alpha x$ and expressions (n) for coefficients A_m', equal to B_m', we find

$$\sigma_y = -\frac{qa}{l} - \frac{4q}{\pi} \sum_{m=1}^{\infty} \frac{\sin \alpha a}{m} \frac{(\alpha c \cosh \alpha c + \sinh \alpha c) \cosh \alpha y - \alpha y \sinh \alpha y \sinh \alpha c}{\sinh 2\alpha c + 2\alpha c} \cos \alpha x \tag{p}$$

in which $qa = P/2$. If l is small in comparison with c, αc is a large number and it can be neglected in comparison with $\sinh \alpha c$. We can also put

$$\sinh \alpha c = \cosh \alpha c = \tfrac{1}{2} e^{\alpha c}$$

[1] The rough base is considered by K. Marguerre, *Ingenieur-Arch.*, vol. 2, p. 108, 1931; G. R. Abrahamson and J. N. Goodier, *J. Appl. Mech.*, vol. 28, pp. 608–610, 1961; and a flexible but inextensible layer embedded in the elastic material, a case of interest in soil mechanics, by M. A. Biot, *Phys.*, vol. 6, p. 367, 1935.

For cross sections at a large distance from the middle of the plate we can write $\sinh \alpha y = \cosh \alpha y = \frac{1}{2}e^{\alpha y}$. Substituting these in Eq. (p), we find

$$\sigma_y = -\frac{qa}{l} - \frac{4q}{\pi} \sum_{m=1}^{\infty} \frac{\sin \alpha a}{2m} [(\alpha c + 1)e^{\alpha(y-c)} - \alpha y e^{\alpha(y-c)}] \cos \alpha x$$

$$= -\frac{qa}{l} - \frac{4q}{\pi} \sum_{m=1}^{\infty} \frac{\sin (m\pi a/l)}{2m} \left[\frac{m\pi}{l}(c - y) + 1\right] e^{(m\pi/l)(y-c)} \cos \frac{m\pi x}{l}$$

If $c - y$ is not very small, say $c - y > l/2$, this series converges very rapidly and it is only necessary to take a few terms in calculating σ_y. Then we can take

$$\sin \frac{m\pi a}{l} = \frac{m\pi a}{l}$$

and putting $2aq = P$, we find

$$\sigma_y = -\frac{P}{2l} - \frac{P}{l} \sum_{m=1}^{\infty} \left[\frac{m\pi}{l}(c - y) + 1\right] e^{(m\pi/l)(y-c)} \cos \frac{m\pi x}{l}$$

For $y = c - l$, for instance,

$$\sigma_y = -\frac{P}{2l} - \frac{P}{l}\left(\frac{\pi + 1}{e^{\pi}} \cos \frac{\pi x}{l} + \frac{2\pi + 1}{e^{2\pi}} \cos \frac{2\pi x}{l} + \frac{3\pi + 1}{e^{3\pi}} \cos \frac{3\pi x}{l} + \cdots\right)$$

The first three terms of the series are sufficient to give good accuracy and the stress distribution is as shown in Fig. 38b. In the same figure the stress distributions for $c - y = l/2$ and $c - y = 2l$ are also given.[1] It is evident that at a distance from the end equal to the width of the strip the stress distribution is practically uniform, which confirms the conclusion usually made on the basis of Saint-Venant's principle.

For a long strip such as in Fig. 37 the σ_x stresses will be transmitted through the width $2l$ of the plate with little change, provided the rate of variation along the edge is

[1] F. Bleich, *loc. cit.* These results are confirmed by the more complete analysis and measurements of P. Theocaris: (1) *J. Appl. Mech.*, vol. 26, pp. 401–406, 1959; (2) *Intern. J. Engr. Sci.*, vol. 2, pp. 1–19, 1964.

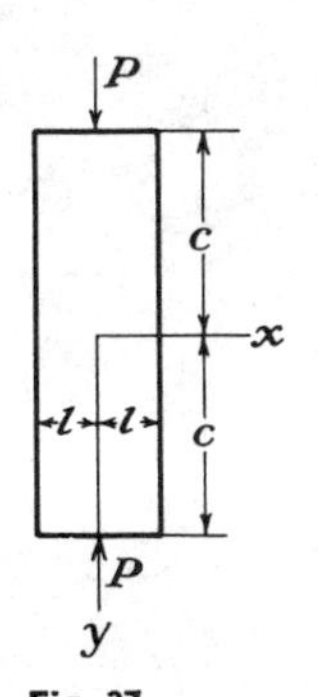

Fig. 37

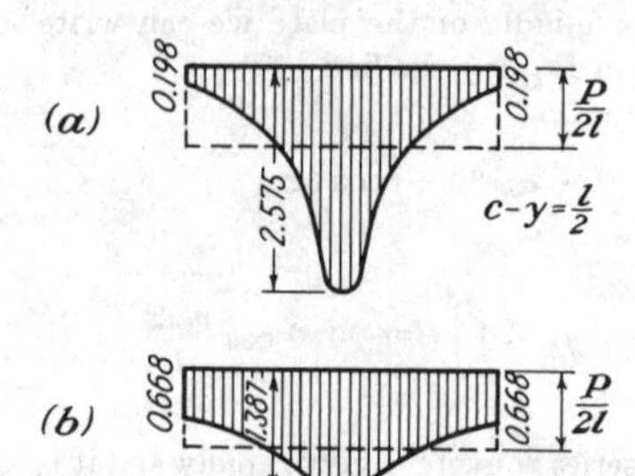

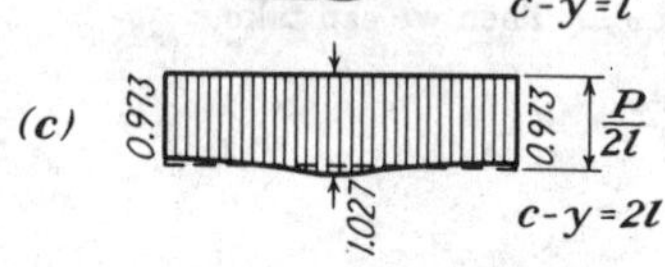

Fig. 38

not too rapid. The stresses of the present solution will, however, require some correction on this account, especially near the ends, $y = \pm c$. A solution of the problem of Fig. 37 with $c = 2l$, by a different method,[1] yields a practically uniform compressive stress over the middle horizontal section, in agreement with Fig. 38c. The stresses in the vicinity of the points of application of the loads P will be discussed later (see page 97).

25 | Other Applications of Fourier Series. Gravity Loading

The problems considered in Art. 24 concerned a single "span" l or $2l$. The solutions, however, can equally well be regarded as representing periodic states of stress in long strips parallel to the x axis, since a Fourier series represents a periodic function. A continuous beam consisting of a sequence of equal spans similarly loaded will have such a periodic stress distribution if the end conditions are appropriate. If, as in certain reinforced-concrete bunker constructions, the beam is essentially a wall supported at points whose distance apart is comparable with the depth (Fig. 39), useful

[1] J. N. Goodier, *Trans. ASME*, vol. 54, p. 173, 1932.

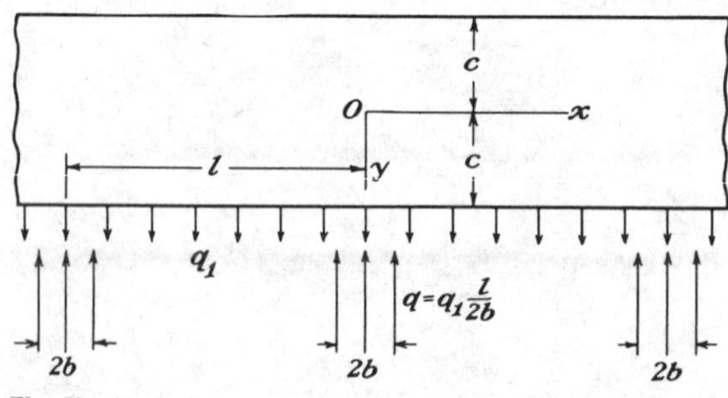

Fig. 39

results can be obtained by the present method.[1] The elementary beam theory is not adequate. A uniformly distributed load q_1 on the lower edge, supported by upward reactions uniformly distributed in widths $2b$ at intervals l, presents a special case covered by Eqs. (m) of Art. 24. If the load q_1 is applied on the upper edge it is merely necessary to add the stress distribution due to equal and opposite uniformly distributed pressures q_1 on both upper and lower edges.

If the load is the weight of the beam itself the resulting body-force problem may at once be reduced to an edge-load problem. The simple stress distribution

$$\sigma_x = 0 \qquad \sigma_y = -\rho g(y + c) \qquad \tau_{xy} = 0$$

satisfies the equations of equilibrium and compatibility (19) and (24). It clearly represents support by uniformly distributed pressure $2\rho gc$ on the lower edge in Fig. 39. The condition that σ_y is zero at the lower edge, except at the supports (of width $2b$), is satisfied by adding this stress distribution to that represented by Fig. 39 when q_1 is replaced by $2\rho gc$, and the stress is due to q and q_1 without body force.

26 | End Effects. Eigensolutions

The Fourier series form of stress function considered in Arts. 24 and 25 is appropriate for given loading (or displacement) on two opposite edges. For given conditions on all four sides of a rectangular region, it is not sufficiently general. A stress function that is a Fourier series in y instead of x can be added. This leads to the method of "crosswise superposition" of single series developed by Mathieu (see footnote on p. 53).

A different kind of stress function can be used for the investigation of end loading when the edges have zero load (or displacement, or other homogeneous conditions). Consider, for instance, in relation to free edges $y = \pm c$, the function

$$\phi = Ce^{-\gamma x/c}\left(\kappa \cos \frac{\gamma y}{c} + \frac{\gamma y}{c} \sin \frac{\gamma y}{c}\right) \tag{a}$$

which satisfies the differential equation (30) for arbitrary values of the constants C, γ, κ. The conditions $\sigma_y = 0$, $\tau_{xy} = 0$, on $y = \pm c$ are satisfied if we make

$$\phi = 0 \qquad \frac{\partial \phi}{\partial y} = 0 \qquad \text{on } y = \pm c \tag{b}$$

since these imply

$$\frac{\partial^2 \phi}{\partial x^2} = 0 \qquad \frac{\partial^2 \phi}{\partial x\, \partial y} = 0 \qquad \text{on } y = \pm c$$

But the conditions (b) also imply zero resultant force and couple on any section x = constant. The symmetry about the x axis means that only the force in the x direction need be examined. For that we have

$$\int_{-c}^{c} \sigma_x \, dy = \int_{-c}^{c} \frac{\partial^2 \phi}{\partial y^2} dy = \left[\frac{\partial \phi}{\partial y}\right]_{y=-c}^{y=c} = 0$$

Thus the load on each end of the strip is self-equilibrating.

Since the function (a) is even in y, it is sufficient to apply the conditions (b) at

[1] Problems of this kind are discussed, with references, in the book "Die Statik im Eisenbetonbau," by K. Beyer, 2d ed., p. 723, 1934; see also H. Craemer, *Ingenieur-Arch.*, vol. 7, p. 325, 1936.

$y = c$ only. The result is

$$\kappa \cos \gamma + \gamma \sin \gamma = 0 \qquad \gamma \cos \gamma + (1 - \kappa) \sin \gamma = 0 \tag{c}$$

Then, by elimination of κ,

$$\sin 2\gamma + 2\gamma = 0 \tag{d}$$

provided $\cos \gamma \neq 0$. The roots of (*d*)—other than the obvious root $\gamma = 0$, which is of no interest—are complex. They occur in conjugate complex pairs, and if γ is a root, so is $-\gamma$. Those having positive real part give stress functions of the type (*a*), which diminish as x increases and are therefore suitable for problems of self-equilibrating load on the end ($x = 0$) of a strip ($x > 0$). In order of increasing real part, the first two nonzero roots[1] are

$$\gamma_2 = 2.1061 + 1.1254i \qquad \gamma_4 = 5.3563 + 1.5516i \tag{e}$$

Even suffixes are used because we have considered only an even function of y in (*a*). If instead we consider the odd function

$$\kappa' \sin \frac{\gamma y}{c} \cos \frac{\gamma y}{c} \tag{f}$$

the equation (*d*) is replaced by

$$\sin 2\gamma - 2\gamma = 0 \tag{g}$$

The first two nonzero roots[1] are

$$\gamma_3 = 3.7488 + 1.3843i \qquad \gamma_5 = 6.9500 + 1.6761i \tag{h}$$

and to find corresponding values of κ' we have, in place of the first of (*c*),

$$\kappa' \sin \gamma + \gamma \cos \gamma = 0 \tag{i}$$

Returning to the symmetrical case represented by (*a*), the substitution of a chosen root γ, for instance, γ_2 from (*e*) and its associated value of κ from the first (or second) of (*c*), yields a complex form of stress function, regarding the coefficient C as unity for the moment. Since this satisfies the differential equation (30), its real and imaginary parts individually satisfy this equation and may be used as real stress functions. Each may then be given its own real coefficient. The real part of γ implies an exponential factor governing the rate of attenuation as x increases. The lowest such rate occurs in the functions corresponding to γ_2; and from (*e*), the exponential factor is

$$e^{-2.1061x/c}$$

This provides a measure of the attenuation, anticipated qualitatively by Saint-Venant's principle, provided that the infinite set of "eigenfunctions" considered here is able to represent any self-equilibrating end load we may have. Although this is the case, the actual determination of coefficients can lead to laborious computations. To avoid these, approximation functions of a simpler kind have been tabulated[2] and used in a series of papers.

[1] From J. Fadle, *Ingenieur-Arch.*, vol. 11, 1941, p. 125. Functions of the kind considered here were introduced independently by Fadle, and by Papcovicz (1940). For references to these and later papers see (1) J. P. Benthem, *Quart. J. Mech. Appl. Math.*, vol. 16, 1963, pp. 413–429; (2) G. Horvay and J. S. Born, *J. Appl. Mech.*, vol. 24, 1957, pp. 261–268; (3) J. N. Goodier and P. G. Hodge, "Elasticity and Plasticity," p. 20, John Wiley & Sons, Inc., New York, 1958; (4) M. W. Johnson, Jr. and R. W. Little, *Quart. Appl. Math.*, vol. 22, pp. 335–344, 1965.

[2] See Horvay and Born, *op. cit.*

Instead of prescribed loading, the end conditions may prescribe displacements. In certain cases the stress will then have singularities at the corners $x = 0$, $y = \pm c$, and it becomes important to distinguish the character of the singular terms[1] and, if possible, to represent them in closed form so that the series part of the solution is asked to represent only a nonsingular part. An example occurs in the problem of the strip with one end clamped to zero displacements, and loaded in tension, which has been solved in this manner.[2] The problem of the compound tension strip, having elastic constants in the part $x > 0$ different from those in the part $x < 0$, has also been investigated.[3]

PROBLEMS

1. Investigate what problem of plane stress is solved by the stress function

$$\phi = \frac{3F}{4c}\left(xy - \frac{xy^3}{3c^2}\right) + \frac{P}{2}y^2$$

2. Investigate what problem is solved by

$$\phi = -\frac{F}{d^3}xy^2(3d - 2y)$$

applied to the region included in $y = 0$, $y = d$, $x = 0$, on the side x positive.

3. Show that

$$\phi = \frac{q}{8c^3}\left[x^2\left(y^3 - 3c^2y + 2c^3\right) - \frac{1}{5}y^3\left(y^2 - 2c^2\right)\right]$$

is a stress function, and find what problem it solves when applied to the region included in $y = \pm c$, $x = 0$, on the side x positive.

4. The stress function

$$\phi = s\left(\frac{1}{4}xy - \frac{xy^2}{4c} - \frac{xy^3}{4c^2} + \frac{ly^2}{4c} + \frac{ly^3}{4c^2}\right)$$

is proposed as giving the solution for a cantilever ($y = \pm c$, $0 < x < l$) loaded by uniform shear along the lower edge, the upper edge and the end $x = l$ being free from load. In what respects is this solution imperfect? Compare the expressions for the stresses with those obtainable from elementary tension and bending formulas.

5. In the cantilever problem of Fig. 26, the support conditions at $x = l$ are given as

At $x = l$, $y = 0$: $\quad u = v = 0$

At $x = l$, $y = \pm c$: $\quad u = 0$

Show that the deflection is now

$$(v)_{\substack{x=0\\y=0}} = \frac{Pl^3}{3EI}\left[1 + \frac{1}{2}(4 + 5\nu)\frac{c^2}{l^2}\right]$$

[1] This requires separate consideration of the corner region as in Chap. 4, Art. 42.

[2] See Benthem, *op. cit.*

[3] K. T. S. Iyengar and R. S. Alwar, *Z. Angew. Math. Phys.*, vol. 14, pp. 344–352, 1963; and *Z. Angew. Math. Mech.*, vol. 43, pp. 249–258, 1963.

Sketch the deformed shape of the supported end ($x = l$), and indicate on the sketch how this mode of support could be realized (hinges? rollers bearing on fixed planes?).

6. The beam of Fig. 28 is loaded by its own weight instead of the load q on the upper edge. Find expressions for the displacement components u and v. Find also an expression for the change of the (originally unit) thickness.

7. The cantilever of Fig. 26, instead of having a narrow rectangular cross section, has a wide rectangular cross section, and is maintained in plane strain by suitable forces along the vertical sides. The load is P per unit width on the end.

Justify the statement that the stresses σ_x, σ_y, τ_{xy} are the same as those found in Art. 21. Find an expression for the stress σ_z, and sketch its distribution along the sides of the cantilever. Write down expressions for the displacement components u and v when a horizontal element of the axis is fixed at $x = l$.

8. Show that if V is a plane harmonic function, i.e., it satisfies the Laplace equation

$$\frac{\partial^2 V}{\partial x^2} + \frac{\partial^2 V}{\partial y^2} = 0$$

then the functions xV, yV, $(x^2 + y^2)V$ will each satisfy Eq. (*a*) of Art. 18, and so can be used as stress functions.

9. Show that

$$(Ae^{\alpha y} + Be^{-\alpha y} + Cye^{\alpha y} + Dye^{-\alpha y}) \sin \alpha x$$

is a stress function.

Derive series expressions for the stresses in a semi-infinite plate, $y > 0$, with normal pressure on the straight edge ($y = 0$) having the distribution

$$\sum_{m=1}^{\infty} b_m \sin \frac{m\pi x}{l}$$

Show that the stress σ_x at a point on the edge is a compression equal to the applied pressure at that point. Assume that the stress tends to disappear as y becomes large.

10. Show that (*a*) the stresses given by Eqs. (*e*) of Art. 24 and (*b*) the stresses in Prob. 9 satisfy Eq. (*b*) of Art. 17.

chapter 4

Two-dimensional Problems in Polar Coordinates

27 | General Equations in Polar Coordinates

In discussing stresses in circular rings and disks, curved bars of narrow rectangular cross section with a circular axis, etc., it is advantageous to use polar coordinates. The position of a point in the middle plane of a plate is then defined by the distance from the origin O (Fig. 40) and by the angle θ between r and a certain axis Ox fixed in the plane.

Let us now consider the equilibrium of a small element 1234 cut out from the plate by the radial sections 04, 02, normal to the plate, and by two cylindrical surfaces 3, 1, normal to the plate. The normal stress component in the radial direction is denoted by σ_r, the normal component in the circumferential direction by σ_θ, and the shearing-stress component by $\tau_{r\theta}$, each symbol representing stress at the point r, θ, which is the midpoint P of the element. On account of the variation of stress the values at the midpoints of the sides 1, 2, 3, 4 are not quite the same

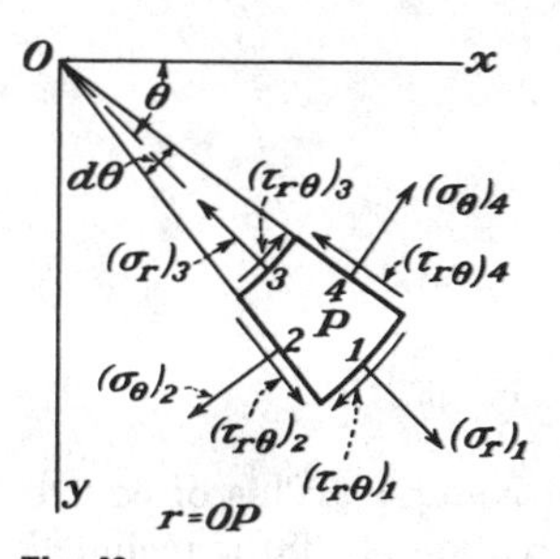

Fig. 40

as the values σ_r, σ_θ, $\tau_{r\theta}$, and are denoted by $(\sigma_r)_1$, etc., in Fig. 40. The radii of the sides 3, 1 are denoted by r_3, r_1. The radial force on the side 1 is $(\sigma_r)_1 r_1\, d\theta$ which may be written $(\sigma_r r)_1\, d\theta$, and similarly the radial force on side 3 is $-(\sigma_r r)_3\, d\theta$. The normal force on side 2 has a component along the radius through P of $-(\sigma_\theta)_2(r_1 - r_3) \sin (d\theta/2)$, which may be replaced by $-(\sigma_\theta)_2\, dr\, (d\theta/2)$. The corresponding component from side 4 is $-(\sigma_\theta)_4\, dr\, (d\theta/2)$. The shearing forces on sides 2 and 4 give $[(\tau_{r\theta})_2 - (\tau_{r\theta})_4]\, dr$.

Summing up forces in the radial direction, including body force R per unit volume in the radial direction, we obtain the equation of equilibrium

$$(\sigma_r r)_1\, d\theta - (\sigma_r r)_3\, d\theta - (\sigma_\theta)_2\, dr\, \frac{d\theta}{2} - (\sigma_\theta)_4\, dr\, \frac{d\theta}{2} + [(\tau_{r\theta})_2 - (\tau_{r\theta})_4]\, dr + Rr\, d\theta\, dr = 0$$

Dividing by $dr\, d\theta$ this becomes

$$\frac{(\sigma_r r)_1 - (\sigma_r r)_3}{dr} - \frac{1}{2}[(\sigma_\theta)_2 + (\sigma_\theta)_4] + \frac{(\tau_{r\theta})_2 - (\tau_{r\theta})_4}{d\theta} + Rr = 0$$

If the dimensions of the element are now taken smaller and smaller, to the limit zero, the first term of this equation is in the limit $\partial(\sigma_r r)/\partial r$. The second becomes σ_θ, and the third $\partial\tau_{r\theta}/\partial\theta$. The equation of equilibrium in the tangential direction may be derived in the same manner. The two equations take the final form

$$\begin{aligned} \frac{\partial\sigma_r}{\partial r} + \frac{1}{r}\frac{\partial\tau_{r\theta}}{\partial\theta} + \frac{\sigma_r - \sigma_\theta}{r} + R &= 0 \\ \frac{1}{r}\frac{\partial\sigma_\theta}{\partial\theta} + \frac{\partial\tau_{r\theta}}{\partial r} + \frac{2\tau_{r\theta}}{r} + S &= 0 \end{aligned} \tag{37}$$

where S is the component of body force (per unit volume) in the tangential direction (θ-increasing).

These equations take the place of Eqs. (18) when we solve two-dimensional problems by means of polar coordinates. When the body force is zero they are satisfied by putting

$$\begin{aligned} \sigma_r &= \frac{1}{r}\frac{\partial\phi}{\partial r} + \frac{1}{r^2}\frac{\partial^2\phi}{\partial\theta^2} \\ \sigma_\theta &= \frac{\partial^2\phi}{\partial r^2} \\ \tau_{r\theta} &= \frac{1}{r^2}\frac{\partial\phi}{\partial\theta} - \frac{1}{r}\frac{\partial^2\phi}{\partial r\,\partial\theta} = -\frac{\partial}{\partial r}\left(\frac{1}{r}\frac{\partial\phi}{\partial\theta}\right) \end{aligned} \tag{38}$$

where ϕ is the stress function as a function of r and θ. This of course may be verified by direct substitution. A derivation of (38) is included in what follows.

Instead of deriving (37) and observing that when $R = S = 0$ they are satisfied by (38), we can consider the stress distribution in question as first given in xy components σ_x, σ_y, τ_{xy}, as in Chap. 3. We can then obtain from these the polar components σ_r, σ_θ, $\tau_{r\theta}$. From (13) we have (identifying α with θ)

$$\begin{aligned}\sigma_r &= \sigma_x \cos^2\theta + \sigma_y \sin^2\theta + 2\tau_{xy}\sin\theta\cos\theta \\ \sigma_\theta &= \sigma_x \sin^2\theta + \sigma_y \cos^2\theta - 2\tau_{xy}\sin\theta\cos\theta \\ \tau_{r\theta} &= (\sigma_y - \sigma_x)\sin\theta\cos\theta + \tau_{xy}(\cos^2\theta - \sin^2\theta)\end{aligned} \tag{a}$$

We can similarly express σ_x, σ_y, τ_{xy} in terms of σ_r, σ_θ, $\tau_{r\theta}$ by the relations (see Prob. 1, page 144)

$$\begin{aligned}\sigma_x &= \sigma_r \cos^2\theta + \sigma_\theta \sin^2\theta - 2\tau_{r\theta}\sin\theta\cos\theta \\ \sigma_y &= \sigma_r \sin^2\theta + \sigma_\theta \cos^2\theta + 2\tau_{r\theta}\sin\theta\cos\theta \\ \tau_{xy} &= (\sigma_r - \sigma_\theta)\sin\theta\cos\theta + \tau_{r\theta}(\cos^2\theta - \sin^2\theta)\end{aligned} \tag{b}$$

To obtain (38) we consider next the relations between derivatives in the two coordinate systems. First we have

$$r^2 = x^2 + y^2 \qquad \theta = \arctan\frac{y}{x}$$

which yield

$$\frac{\partial r}{\partial x} = \frac{x}{r} = \cos\theta \qquad \frac{\partial r}{\partial y} = \frac{y}{r} = \sin\theta$$

$$\frac{\partial \theta}{\partial x} = -\frac{y}{r^2} = -\frac{\sin\theta}{r} \qquad \frac{\partial \theta}{\partial y} = \frac{x}{r^2} = \frac{\cos\theta}{r}$$

Thus for any function $f(x,y)$, in polar coordinates $f(r\cos\theta, r\sin\theta)$, we have

$$\frac{\partial f}{\partial x} = \frac{\partial f}{\partial r}\frac{\partial r}{\partial x} + \frac{\partial f}{\partial \theta}\frac{\partial \theta}{\partial x} = \cos\theta\frac{\partial f}{\partial r} - \frac{\sin\theta}{r}\frac{\partial f}{\partial \theta} \tag{c}$$

To get $\partial^2 f/\partial x^2$ we repeat the operation indicated in the last member of (c). Then

$$\begin{aligned}\frac{\partial^2 f}{\partial x^2} &= \left(\cos\theta\frac{\partial}{\partial r} - \frac{\sin\theta}{r}\frac{\partial}{\partial\theta}\right)\left(\cos\theta\frac{\partial f}{\partial r} - \frac{\sin\theta}{r}\frac{\partial f}{\partial\theta}\right) \\ &= \cos^2\theta\frac{\partial^2 f}{\partial r^2} - \cos\theta\sin\theta\frac{\partial}{\partial r}\left(\frac{1}{r}\frac{\partial f}{\partial\theta}\right) \\ &\qquad - \frac{\sin\theta}{r}\frac{\partial}{\partial\theta}\left(\cos\theta\frac{\partial f}{\partial r}\right) + \frac{\sin\theta}{r^2}\frac{\partial}{\partial\theta}\left(\sin\theta\frac{\partial f}{\partial\theta}\right)\end{aligned}$$

With a little rearrangement, this takes the form

$$\frac{\partial^2 f}{\partial x^2} = \cos^2\theta\frac{\partial^2 f}{\partial r^2} + \sin^2\theta\left(\frac{1}{r}\frac{\partial f}{\partial r} + \frac{1}{r^2}\frac{\partial^2 f}{\partial\theta^2}\right) - 2\sin\theta\cos\theta\frac{\partial}{\partial r}\left(\frac{1}{r}\frac{\partial f}{\partial\theta}\right) \tag{d}$$

Similarly, we find

$$\frac{\partial^2 f}{\partial y^2} = \sin^2\theta \frac{\partial^2 f}{\partial r^2} + \cos^2\theta\left(\frac{1}{r}\frac{\partial f}{\partial r} + \frac{1}{r^2}\frac{\partial^2 f}{\partial \theta^2}\right) + 2\sin\theta\cos\theta\frac{\partial}{\partial r}\left(\frac{1}{r}\frac{\partial f}{\partial \theta}\right) \quad (e)$$

$$-\frac{\partial^2 f}{\partial x\,\partial y} = \sin\theta\cos\theta\left(\frac{1}{r}\frac{\partial f}{\partial r} + \frac{1}{r^2}\frac{\partial^2 f}{\partial \theta^2} - \frac{\partial^2 f}{\partial r^2}\right) - (\cos^2\theta - \sin^2\theta)\frac{\partial}{\partial r}\left(\frac{1}{r}\frac{\partial f}{\partial \theta}\right) \quad (f)$$

When we take for f the stress-function $\phi(x,y)$ as in (29)—but with $\rho g = 0$—the derivatives on the left-hand sides of (*d*), (*e*), and (*f*) become σ_y, σ_x, τ_{xy}, respectively. The expressions on the right-hand sides of (*d*), (*e*), and (*f*) can therefore be substituted for these stress components in the right-hand sides of (*a*). It is easily verified that the results reduce to (38).

To convert the differential equation (*a*), page 35, to polar form, we first add (*d*) and (*e*) above to obtain

$$\left(\frac{\partial^2}{\partial x^2} + \frac{\partial^2}{\partial y^2}\right) f = \left(\frac{\partial^2}{\partial r^2} + \frac{1}{r}\frac{\partial}{\partial r} + \frac{1}{r^2}\frac{\partial^2}{\partial \theta^2}\right) f \quad (g)$$

showing that the operator on the right is the polar equivalent of the laplacian operator on the left. Next, we find by addition of the first two of equations (*b*)

$$\sigma_x + \sigma_y = \sigma_r + \sigma_\theta \quad (h)$$

For zero body force we have, as on page 30,

$$\left(\frac{\partial^2}{\partial x^2} + \frac{\partial^2}{\partial y^2}\right)(\sigma_x + \sigma_y) = 0 \quad (i)$$

In view of (*i*), (*h*), and (*g*), this becomes

$$\left(\frac{\partial^2}{\partial r^2} + \frac{1}{r}\frac{\partial}{\partial r} + \frac{1}{r^2}\frac{\partial^2}{\partial \theta^2}\right)\left(\frac{\partial^2 \phi}{\partial r^2} + \frac{1}{r}\frac{\partial \phi}{\partial r} + \frac{1}{r^2}\frac{\partial^2 \phi}{\partial \theta^2}\right) = 0 \quad (39)$$

From various solutions of this partial differential equation we obtain solutions of two-dimensional problems in polar coordinates for various boundary conditions. Several examples of such problems will be discussed in this chapter.

28 | Stress Distribution Symmetrical about an Axis

When the stress function depends on r only, the equation of compatibility (39) becomes

$$\left(\frac{d^2}{dr^2} + \frac{1}{r}\frac{d}{dr}\right)\left(\frac{d^2\phi}{dr^2} + \frac{1}{r}\frac{d\phi}{dr}\right) = \frac{d^4\phi}{dr^4} + \frac{2}{r}\frac{d^3\phi}{dr^3} - \frac{1}{r^2}\frac{d^2\phi}{dr^2} + \frac{1}{r^3}\frac{d\phi}{dr} = 0 \quad (40)$$

This is an ordinary differential equation, which can be reduced to a linear differential equation with constant coefficients by introducing a new variable t such that $r = e^t$. In this manner the general solution of Eq. (40) can be easily obtained. This solution has four constants of integration, which must be determined from the boundary conditions. By substitution it can be checked that

$$\phi = A \log r + Br^2 \log r + Cr^2 + D \tag{41}$$

is the general solution. The solutions of a group of problems of symmetrical stress distribution[1] with no body forces can be obtained from this. The corresponding stress components from Eqs. (38) are

$$\begin{aligned}
\sigma_r &= \frac{1}{r}\frac{\partial \phi}{\partial r} = \frac{A}{r^2} + B(1 + 2 \log r) + 2C \\
\sigma_\theta &= \frac{\partial^2 \phi}{\partial r^2} = -\frac{A}{r^2} + B(3 + 2 \log r) + 2C \\
\tau_{r\theta} &= 0
\end{aligned} \tag{42}$$

If there is no hole at the origin of coordinates, constants A and B vanish, since otherwise the stress components (42) become infinite when $r = 0$. Hence, for a plate without a hole at the origin and with no body forces, only one case of stress distribution symmetrical with respect to the axis may exist, namely, that when $\sigma_r = \sigma_\theta =$ constant and the plate is in a condition of uniform tension or uniform compression in all directions in its plane.

If there is a hole at the origin, other solutions than uniform tension or compression can be derived from expressions (42). Taking B as zero,[2] for instance, Eqs. (42) become

$$\begin{aligned}
\sigma_r &= \frac{A}{r^2} + 2C \\
\sigma_\theta &= -\frac{A}{r^2} + 2C
\end{aligned} \tag{43}$$

This solution may be adapted to represent the stress distribution in a hollow cylinder submitted to uniform pressure on the inner and outer surfaces[3] (Fig. 41). Let a and b denote the inner and outer radii of the cylinder, and p_i and p_o the uniform internal and external pressures. Then the boundary conditions are

$$(\sigma_r)_{r=a} = -p_i \qquad (\sigma_r)_{r=b} = -p_o \tag{a}$$

[1] The stress function independent of θ does not give all stress distributions independent of θ. The function of the form $A\theta$ as in (q) on p. 126 illustrates this.

[2] Proof that B must be zero requires consideration of displacements. See p. 78.

[3] The solution of this problem is due to Lamé, "Leçons sur la théorie . . . de l'élasticité," Gauthier-Villars, Paris, 1852.

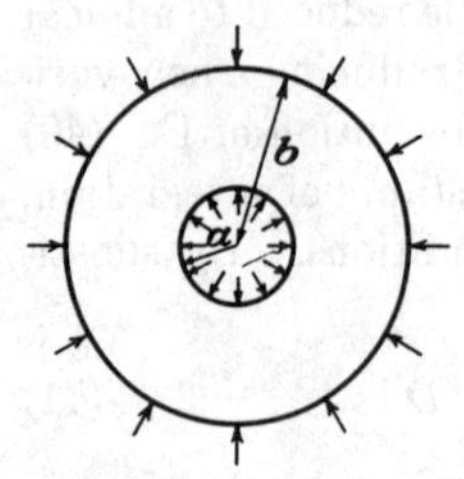

Fig. 41

Substituting in the first of Eqs. (43), we obtain the following equations to determine A and C:

$$\frac{A}{a^2} + 2C = -p_i$$

$$\frac{A}{b^2} + 2C = -p_o$$

from which

$$A = \frac{a^2b^2(p_o - p_i)}{b^2 - a^2}$$

$$2C = \frac{p_ia^2 - p_ob^2}{b^2 - a^2}$$

Substituting these in Eqs. (43) the following expressions for the stress components are obtained:

$$\sigma_r = \frac{a^2b^2(p_o - p_i)}{b^2 - a^2}\frac{1}{r^2} + \frac{p_ia^2 - p_ob^2}{b^2 - a^2}$$

$$\sigma_\theta = -\frac{a^2b^2(p_o - p_i)}{b^2 - a^2}\frac{1}{r^2} + \frac{p_ia^2 - p_ob^2}{b^2 - a^2} \tag{44}$$

The radial displacement u is easily found since here $\epsilon_\theta = u/r$, and for plane stress

$$E\epsilon_\theta = \sigma_\theta - \nu\sigma_r$$

It is interesting to note that the sum $\sigma_r + \sigma_\theta$ is constant through the thickness of the wall of the cylinder. Hence the stresses σ_r and σ_θ produce a uniform extension or contraction in the direction of the axis of the cylinder, and cross sections perpendicular to this axis remain plane. Hence the deformation produced by the stresses (44) in an element of the cylinder cut out by two adjacent cross sections does not interfere with the deformation of the neighboring elements, and it is justifiable to consider the element in the condition of plane stress as we did in the above discussion.

In the particular case when $p_o = 0$ and the cylinder is submitted to internal pressure only, Eqs. (44) give

$$\sigma_r = \frac{a^2 p_i}{b^2 - a^2}\left(1 - \frac{b^2}{r^2}\right)$$
$$\sigma_\theta = \frac{a^2 p_i}{b^2 - a^2}\left(1 + \frac{b^2}{r^2}\right) \tag{45}$$

These equations show that σ_r is always a compressive stress and σ_θ a tensile stress. The latter is greatest at the inner surface of the cylinder, where

$$(\sigma_\theta)_{\max} = \frac{p_i(a^2 + b^2)}{b^2 - a^2} \tag{46}$$

$(\sigma_\theta)_{\max}$ is always numerically greater than the internal pressure and approaches this quantity as b increases, so that it can never be reduced below p_i, however much material is added on the outside. Various applications of Eqs. (45) and (46) in machine design are usually discussed in elementary books on the strength of materials.[1]

The corresponding problem for a cylinder with an eccentric bore was solved by G. B. Jeffery.[2] If the radius of the bore is a and that of the external surface b, and if the distance between their centers is e, the maximum stress, when the cylinder is under an internal pressure p_i, is the tangential stress at the internal surface at the thinnest part, if $e < \frac{1}{2}a$, and is of the magnitude

$$\sigma = p_i \left[\frac{2b^2(b^2 + a^2 - 2ae - e^2)}{(a^2 + b^2)(b^2 - a^2 - 2ae - e^2)} - 1 \right]$$

If $e = 0$, this coincides with Eq. (46).

29 | Pure Bending of Curved Bars

Let us consider a curved bar with a constant narrow rectangular cross section[3] and a circular axis bent in the plane of curvature by couples M applied at the ends (Fig. 42). The bending moment in this case is constant along the length of the bar and it is natural to expect that the stress distribution is the same in all radial cross sections, and that the solution of the problem can therefore be obtained by using expression (41).

[1] See, for instance, S. Timoshenko, "Strength of Materials," 3d ed., vol. 2, chap. 6, D. Van Nostrand Company, Inc., Princeton, N.J., 1956.

[2] *Trans. Roy. Soc. (London)*, ser. A, vol. 221, p. 265, 1921. See also *Brit. Assoc. Advan. Sci. Rept.*, 1921. A complete solution by a different method is given in Art. 66 of the present book.

[3] From the general discussion of the two-dimensional problem, Art. 16, it follows that the solution obtained below for the stress holds also for plane strain.

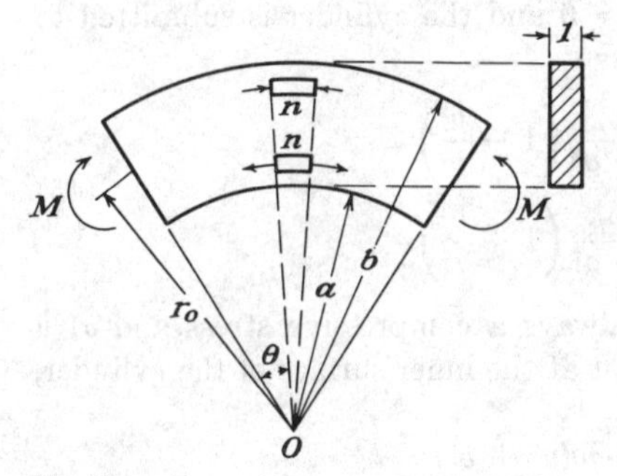

Fig. 42

Denoting by a and b the inner and the outer radii of the boundary and taking the width of the rectangular cross section as unity, the boundary conditions are

$$\begin{array}{lll} (1) & \sigma_r = 0 & \text{for } r = a \text{ and } r = b \\ (2) & \int_a^b \sigma_\theta \, dr = 0 \qquad \int_a^b \sigma_\theta r \, dr = -M & (a) \\ (3) & \tau_{r\theta} = 0 & \text{at the boundary} \end{array}$$

Condition (1) means that the convex and concave boundaries of the bar are free from normal forces; condition (2) indicates that the normal stresses at the ends give rise to the couple M only, and condition (3) indicates that there are no tangential forces applied at the boundary. Using the first of Eqs. (42) with (1) of the boundary conditions (a) we obtain

$$\begin{aligned} \frac{A}{a^2} + B(1 + 2 \log a) + 2C &= 0 \\ \frac{A}{b^2} + B(1 + 2 \log b) + 2C &= 0 \end{aligned} \qquad (b)$$

Condition (2) in (a) is now necessarily satisfied. The use of a stress function guarantees equilibrium. A nonzero force-resultant on each end would violate equilibrium. To have the bending couple equal to M, the condition

$$\int_a^b \sigma_\theta r \, dr = \int_a^b \frac{\partial^2 \phi}{\partial r^2} r \, dr = -M \qquad (d)$$

must be fulfilled. We have

$$\int_a^b \frac{\partial^2 \phi}{\partial r^2} r \, dr = \left| \frac{\partial \phi}{\partial r} r \right|_a^b - \int_a^b \frac{\partial \phi}{\partial r} dr = \left| \frac{\partial \phi}{\partial r} r \right|_a^b - |\phi|_a^b$$

and noting that on account of (*b*),

$$\left| \frac{\partial \phi}{\partial r} r \right|_a^b = 0$$

we find from (*d*),

$$|\phi|_a^b = M$$

or substituting expression (41) for ϕ,

$$A \log \frac{b}{a} + B(b^2 \log b - a^2 \log a) + C(b^2 - a^2) = M \qquad (e)$$

This equation, together with the two Eqs. (*b*), completely determines the constants A, B, C, and we find

$$\begin{aligned} A &= -\frac{4M}{N} a^2 b^2 \log \frac{b}{a} \qquad B = -\frac{2M}{N}(b^2 - a^2) \\ C &= \frac{M}{N}[b^2 - a^2 + 2(b^2 \log b - a^2 \log a)] \end{aligned} \qquad (f)$$

where for simplicity we have put

$$N = (b^2 - a^2)^2 - 4a^2b^2 \left(\log \frac{b}{a}\right)^2 \qquad (g)$$

Substituting the values (*f*) of the constants into the expressions (42) for the stress components, we find

$$\begin{aligned} \sigma_r &= -\frac{4M}{N}\left(\frac{a^2b^2}{r^2} \log \frac{b}{a} + b^2 \log \frac{r}{b} + a^2 \log \frac{a}{r}\right) \\ \sigma_\theta &= -\frac{4M}{N}\left(-\frac{a^2b^2}{r^2} \log \frac{b}{a} + b^2 \log \frac{r}{b} + a^2 \log \frac{a}{r} + b^2 - a^2\right) \\ \tau_{r\theta} &= 0 \end{aligned} \qquad (47)$$

This gives the stress distribution satisfying all the boundary conditions[1] (*a*) for pure bending and represents the exact solution of the problem, provided the distribution of the normal forces at the ends is that given by the second of Eqs. (47). If the forces giving the bending couple M are distributed over the ends of the bar in some other manner, the stress distribution at the ends will be different from that of the solution (47). But, as Saint-Venant's principle suggests, the deviations from solution (47) may be negligible away from the ends, say at distances greater than the depth of the bar. This is illustrated by Fig. 102.

[1] This solution is due to H. Golovin, *Trans. Inst. Tech.*, St. Petersburg, 1881. The paper, published in Russian, remained unknown in other countries, and the same problem was solved later by M. C. Ribière (*Compt. Rend.*, vol. 108, 1889, and vol. 132, 1901) and by L. Prandtl. See A. Föppl, "Vorlesungen uber Technische Mechanik," vol. 5, p. 72, 1907; also A. Timpe, *Z. Math. Physik*, vol. 52, p. 348, 1905.

It is of practical interest to compare solution (47) with the elementary solutions usually given in books on the strength of materials. If the depth of the bar, $b - a$, is small in comparison with the radius of the central axis, $(b + a)/2$, the same stress distribution as that for straight bars is usually assumed. If this depth is not small, it is usual in practice to assume that cross sections of the bar remain plane during the bending, from which it can be shown that the distribution of the normal stresses σ_θ over any cross sections follows a hyperbolic law.[1] In all cases the maximum[2] and minimum values of the stress σ_θ can be presented in the form

$$\sigma_\theta = m \frac{M}{a^2} \tag{h}$$

The following table gives the values of the numerical factor m calculated by the two elementary methods, referred to above, and by the exact formula (47).[3] It can be seen from this table that the elementary solution based on the hypothesis of plane cross sections gives very accurate results.

Coefficient m of Eq. (h)

$\frac{b}{a}$	Linear stress distribution	Hyperbolic stress distribution		Exact solution	
1.3	±66.67	+72.98,	−61.27	+73.05,	−61.35
2	± 6.000	+ 7.725,	− 4.863	+ 7.755,	− 4.917
3	± 1.500	+ 2.285,	− 1.095	+ 2.292,	− 1.130

It will be shown later that, in the case of pure bending, the cross sections actually do remain plane, and the discrepancy between the elementary and the exact solutions comes from the fact that in the elementary solution the stress component σ_r is neglected and it is assumed that longitudinal fibers of the bent bar are in simple tension or compression.

[1] This approximate theory was developed by H. Résal, *Ann. Mines*, p. 617, 1862, and by E. Winkler, *Zivilingenieur*, vol. 4, p. 232, 1858; see also his book "Die Lehre von der Elastizität und Festigkeit," chap. 15, Prag, 1867. Further development of the theory was made by F. Grashof, "Elastizität und Festigkeit," p. 251, 1878, and by K. Pearson, "History of the Theory of Elasticity," vol. 2, pt. 1, p. 422, 1893.

[2] The greatest value of σ_θ in (47) always occurs at the inside ($r = a$). A proof is given by J. E. Brock, *J. Appl. Mech.*, vol. 31, p. 559, 1964.

[3] The results are taken from the doctorate thesis, University of Michigan, 1931, of V. Billevicz.

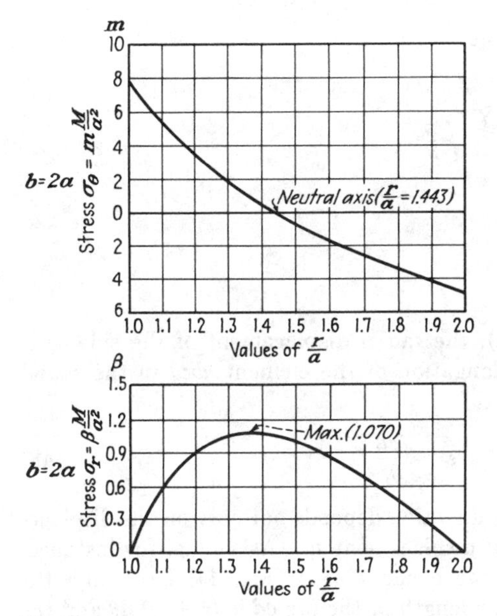

Fig. 43

From the first of Eqs. (47) it can be shown that the stress σ_r is always positive for the direction of bending shown in Fig. 42. The same can be concluded at once from the direction of stresses σ_θ acting on the elements $n - n$ in Fig. 42. The corresponding tangential forces give resultants in the radial direction tending to separate longitudinal fibers and producing tensile stress in the radial direction. This stress increases toward the neutral surface and becomes a maximum near this surface. This maximum is always much smaller than $(\sigma_\theta)_{\max}$. For instance, for $b/a = 1.3$, $(\sigma_r)_{\max} = 0.060(\sigma_\theta)_{\max}$; for $b/a = 2$, $(\sigma_r)_{\max} = 0.138(\sigma_\theta)_{\max}$; for $b/a = 3$, $(\sigma_r)_{\max} = 0.193(\sigma_\theta)_{\max}$. In Fig. 43 the distribution of σ_θ and σ_r for $b/a = 2$ is given. From this figure we see that the point of maximum stress σ_r is somewhat displaced from the neutral axis in the direction of the center of curvature.

30 | Strain Components in Polar Coordinates

In considering the displacement in polar coordinates let us denote by u and v the components of the displacement in the radial and tangential directions, respectively. If u is the radial displacement of the side ad

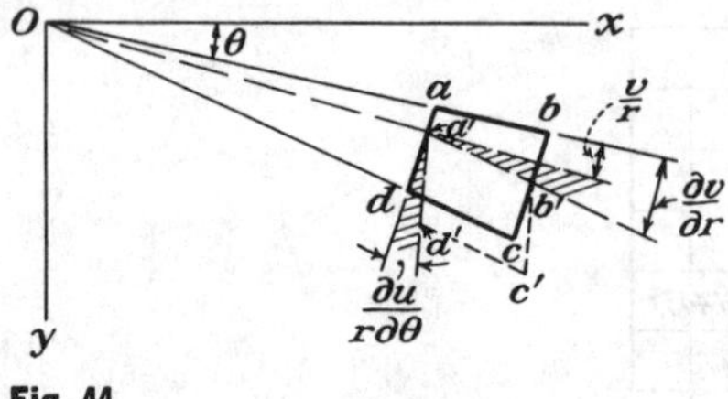

Fig. 44

of the element *abcd* (Fig. 44), the radial displacement of the side *bc* is $u + (\partial u/\partial r)\,dr$. The unit elongation of the element *abcd* in the radial direction is then

$$\epsilon_r = \frac{\partial u}{\partial r} \tag{48}$$

The strain in the tangential direction depends not only on the displacement v but also on the radial displacement u. Assuming, for instance, that the points a and d of the element *abcd* (Fig. 44) have only the radial displacement u, the new length of the arc *ad* is $(r + u)\,d\theta$ and the tangential strain is therefore

$$\frac{(r + u)\,d\theta - r\,d\theta}{r\,d\theta} = \frac{u}{r}$$

The difference in the tangential displacement of the sides *ab* and *cd* of the element *abcd* is $(\partial v/\partial\theta)\,d\theta$, and the tangential strain due to the displacement v is accordingly $\partial v/r\,\partial\theta$. The total tangential strain is thus[1]

$$\epsilon_\theta = \frac{u}{r} + \frac{\partial v}{r\,\partial\theta} \tag{49}$$

Considering now the shearing strain, let $a'b'c'd'$ be the position of the element *abcd* after deformation (Fig. 44). The angle between the direction *ad* and $a'd'$ is due to the radial displacement u and is equal to $\partial u/r\,\partial\theta$. In the same manner, the angle between $a'b'$ and *ab* is equal to $\partial v/\partial r$. It should be noted that only part of this angle (shaded in the figure) contributes to the shearing strain and the other part, equal to v/r, represents the angular displacement due to rotation of the element *abcd* as a rigid body about the axis through O. Hence the total change in the angle *dab*, representing the shearing strain, is

$$\gamma_{r\theta} = \frac{\partial u}{r\,\partial\theta} + \frac{\partial v}{\partial r} - \frac{v}{r} \tag{50}$$

[1] The symbol ϵ_θ was used with a different meaning in Art. 11.

Substituting now the expressions for the strain components (48), (49), (50) into the equations of Hooke's law for plane stress,

$$
\begin{aligned}
\epsilon_r &= \frac{1}{E}(\sigma_r - \nu\sigma_\theta) \\
\epsilon_\theta &= \frac{1}{E}(\sigma_\theta - \nu\sigma_r) \\
\gamma_{r\theta} &= \frac{1}{G}\tau_{r\theta}
\end{aligned}
\tag{51}
$$

we can obtain sufficient equations for determining u and v.

31 | Displacements for Symmetrical Stress Distributions

Substituting in the first of Eqs. (51) the stress components from Eqs. 42, we find

$$\frac{\partial u}{\partial r} = \frac{1}{E}\left[\frac{(1+\nu)A}{r^2} + 2(1-\nu)B\log r + (1-3\nu)B + 2(1-\nu)C\right]$$

By integration we obtain

$$u = \frac{1}{E}\left[-\frac{(1+\nu)A}{r} + 2(1-\nu)Br\log r - B(1+\nu)r + 2C(1-\nu)r\right] + f(\theta) \tag{a}$$

in which $f(\theta)$ is a function of θ only. From the second of Eqs. (51), we find, by using Eq. 49,

$$\frac{\partial v}{\partial \theta} = \frac{4Br}{E} - f(\theta)$$

from which, by integration

$$v = \frac{4Br\theta}{E} - \int f(\theta)\,d\theta + f_1(r) \tag{b}$$

where $f_1(r)$ is a function of r only. Substituting (*a*) and (*b*) in Eq. (50) and noting that $\gamma_{r\theta}$ is zero since $\tau_{r\theta}$ is zero, we find

$$\frac{1}{r}\frac{\partial f(\theta)}{\partial \theta} + \frac{\partial f_1(r)}{\partial r} + \frac{1}{r}\int f(\theta)\,d\theta - \frac{1}{r}f_1(r) = 0 \tag{c}$$

from which

$$f_1(r) = Fr \qquad f(\theta) = H\sin\theta + K\cos\theta \tag{d}$$

where F, H, and K are constants to be determined from the conditions of constraint of the curved bar or ring. Substituting expressions (*d*) into

Eqs. (*a*) and (*b*), we find the following expressions for the displacements:[1]

$$u = \frac{1}{E}\left[-\frac{(1+\nu)A}{r} + 2(1-\nu)Br\log r - B(1+\nu)r + 2C(1-\nu)r\right] + H\sin\theta + K\cos\theta \tag{52}$$

$$v = \frac{4Br\theta}{E} + Fr + H\cos\theta - K\sin\theta$$

in which the values of constants A, B, and C for each particular case should be substituted. Consider, for instance, pure bending. Taking the centroid of the cross section from which θ is measured (Fig. 42) and also an element of the radius at this point, as rigidly fixed, the conditions of constraint are

$$u = 0 \qquad v = 0 \qquad \frac{\partial v}{\partial r} = 0 \qquad \text{for } \theta = 0 \text{ and } r = r_0 = \frac{a+b}{2}$$

Applying these to expressions (52), we obtain the following equations for calculating the constants of integration F, H, and K:

$$\frac{1}{E}\left[-\frac{(1+\nu)A}{r_0} + 2(1-\nu)Br_0\log r_0 - B(1+\nu)r_0 + 2C(1-\nu)r_0\right] + K = 0$$

$$Fr_0 + H = 0$$

$$F = 0$$

From this it follows that $F = H = 0$, and for the displacement v we obtain

$$v = \frac{4Br\theta}{E} - K\sin\theta \tag{53}$$

This means that the displacement of any cross section consists of a translatory displacement $-K\sin\theta$, the same for all points in the cross section, and of a rotation of the cross section by the angle $4B\theta/E$ about the center of curvature O (Fig. 42). We see that cross sections remain plane in pure bending as is usually assumed in the elementary theory of the bending of curved bars.

In discussing the symmetrical stress distribution in a full ring (page 69) the constant B in the general solution (42) was taken as zero, and in this manner we arrived at a solution of Lamé's problem. Now, after obtaining expressions (52) for displacements, we see what is implied by taking B as zero. B contributes to the displacement v the term $4Br\theta/E$. This term is not *single-valued;* it changes when we increase θ by 2π, that is,

[1] Equation (*c*) is satisfied only when $\int f(\theta)\,d\theta$ is taken from (*d*) without an additive constant.

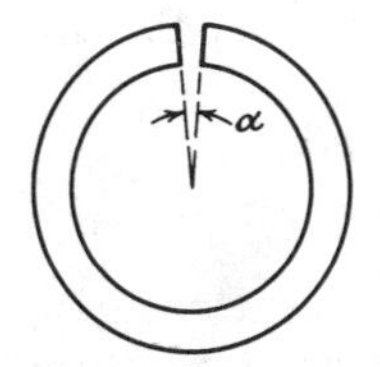

Fig. 45

if we return to a given point after making a complete circle round the ring. Such a *many-valued* expression for a displacement is physically impossible in a full ring, and so, for this case, we must take $B = 0$ in the general solution (42).

A full ring is an example of a multiply-connected body, that is, a body such that some sections can be cut clear across without dividing the body into two parts. In determining the stresses in such bodies, the boundary conditions referring to the stresses are not sufficient to determine completely the stress distribution, and additional equations, representing the conditions that the displacements should be single-valued, must be considered (see Arts. 34 and 43).

The physical meaning of many-valued solutions can be explained by considering *initial stresses* in a multiply-connected body. If a portion of the ring between two adjacent cross sections is cut out (Fig. 45) and the ends of the ring are joined again by welding or other means, a ring with initial stresses is obtained, that is, there are stresses in the ring when external forces are absent. If α is the small angle measuring the portion of the ring that was cut out, the tangential displacement necessary to bring the ends of the ring together is

$$v = \alpha r \tag{e}$$

The same displacement, obtained from Eq. (53) by putting $\theta = 2\pi$, is

$$v = 2\pi \frac{4Br}{E} \tag{f}$$

From (*e*) and (*f*) we find

$$B = \frac{\alpha E}{8\pi} \tag{g}$$

The constant B, entering into the many-valued term for the displacement (53), has now a definite value depending on the way in which the initial stresses were produced in the ring. Substituting (*g*) into Eqs. (*f*) of Art. 29, we find that the bending moment necessary to bring the ends of the ring together (Fig. 45) is

$$M = -\frac{\alpha E}{8\pi} \frac{(b^2 - a^2)^2 - 4a^2b^2[\log (b/a)]^2}{2(b^2 - a^2)} \tag{h}$$

The initial stresses in the ring can easily be calculated from this by using the solution (47) for pure bending.

32 | Rotating Disks

The stress distribution in rotating circular disks is of great practical importance.[1] If the thickness of the disk is small in comparison with its radius, the variation of radial and tangential stresses over the thickness can be neglected[2] and the problem can be easily solved.[3] If the thickness of the disk is constant Eqs. (37) can be applied, and it is only necessary to put the body force equal to the inertia force.[4] Then

$$R = \rho\omega^2 r \qquad S = 0 \tag{a}$$

where ρ is the mass per unit volume of the material of the disk and ω the angular velocity of the disk. Because of the symmetry, $\tau_{r\theta}$ vanishes and σ_r, σ_θ are independent of θ. The second of Eqs. (37) is identically satisfied. The first can be written in the form

$$\frac{d}{dr}(r\sigma_r) - \sigma_\theta + \rho\omega^2 r^2 = 0 \tag{b}$$

The strain components in the case of symmetry are, from Eqs. (48) and (49)

$$\epsilon_r = \frac{du}{dr} \qquad \epsilon_\theta = \frac{u}{r} \tag{c}$$

We can solve the first two stress-strain relations (51) as equations for the stress components to obtain

$$\sigma_r = \frac{E}{1 - \nu^2}(\epsilon_r + \nu\epsilon_\theta) \qquad \sigma_\theta = \frac{E}{1 - \nu^2}(\epsilon_\theta + \nu\epsilon_r)$$

and then, using (c),

$$\sigma_r = \frac{E}{1 - \nu^2}\left(\frac{du}{dr} + \nu\frac{u}{r}\right) \qquad \sigma_\theta = \frac{E}{1 - \nu^2}\left(\frac{u}{r} + \nu\frac{du}{dr}\right) \tag{d}$$

[1] A complete discussion of this problem, and an extensive bibliography of the subject can be found in the book K. Löffler, "Die Berechnung von Rotierenden Scheiben und Schalen," Springer-Verlag OHG, Göttingen, Germany, 1961.

[2] An exact solution of the problem for a disk having the shape of a flat ellipsoid of revolution was obtained by C. Chree, see *Proc. Roy. Soc. (London)*, vol. 58, p. 39, 1895. It shows that the difference between the maximum and the minimum stress at the axis of revolution is only 5 percent of the maximum stress in a uniform disk with thickness one-eighth of its diameter.

[3] A more detailed discussion of the problem will be given later (see Art. 134).

[4] The weight of the disk is neglected.

When these are substituted in (*b*), we find that u must satisfy

$$r^2 \frac{d^2u}{dr^2} + r \frac{du}{dr} - u = -\frac{1-\nu^2}{E} \rho\omega^2 r^3 \tag{e}$$

The general solution of this equation is

$$u = \frac{1}{E}\left[(1-\nu)Cr - (1+\nu)C_1 \frac{1}{r} - \frac{1-\nu^2}{8} \rho\omega^2 r^3\right] \tag{f}$$

where C and C_1 are arbitrary constants. The corresponding stress components are now found from (*d*) as

$$\begin{aligned} \sigma_r &= C + C_1 \frac{1}{r^2} - \frac{3+\nu}{8} \rho\omega^2 r^2 \\ \sigma_\theta &= C - C_1 \frac{1}{r^2} - \frac{1+3\nu}{8} \rho\omega^2 r^2 \end{aligned} \tag{g}$$

The integration constants C and C_1 are determined from the boundary conditions.

For a *solid disk* we must take $C_1 = 0$ to have $u = 0$ at the center. The constant C is determined from the condition at the periphery ($r = b$) of the disk. If there are no forces applied there, we have

$$(\sigma_r)_{r=b} = C - \frac{3+\nu}{8} \rho\omega^2 b^2 = 0$$

from which

$$C = \frac{3+\nu}{8} \rho\omega^2 b^2$$

The stress components, from Eqs. (*g*), are now

$$\begin{aligned} \sigma_r &= \frac{3+\nu}{8} \rho\omega^2 (b^2 - r^2) \\ \sigma_\theta &= \frac{3+\nu}{8} \rho\omega^2 b^2 - \frac{1+3\nu}{8} \rho\omega^2 r^2 \end{aligned} \tag{54}$$

These stresses are greatest at the center[1] of the disk, where

$$\sigma_r = \sigma_\theta = \frac{3+\nu}{8} \rho\omega^2 b^2 \tag{55}$$

In the case of a disk with a *circular hole* of radius a at the center, the constants of integration in Eqs. (*g*) are obtained from the conditions at the inner and outer boundaries. If there are no forces acting on these boundaries, we have

$$(\sigma_r)_{r=a} = 0 \qquad (\sigma_r)_{r=b} = 0 \tag{h}$$

[1] It can be seen from the definitions of σ_r, σ_θ that when they are independent of θ they must be equal at the center.

from which we find that

$$C = \frac{3+\nu}{8}\rho\omega^2(b^2 + a^2) \qquad C_1 = -\frac{3+\nu}{8}\rho\omega^2 a^2 b^2$$

Substituting in Eqs. (g),

$$\begin{aligned} \sigma_r &= \frac{3+\nu}{8}\rho\omega^2\left(b^2 + a^2 - \frac{a^2b^2}{r^2} - r^2\right) \\ \sigma_\theta &= \frac{3+\nu}{8}\rho\omega^2\left(b^2 + a^2 + \frac{a^2b^2}{r^2} - \frac{1+3\nu}{3+\nu}r^2\right) \end{aligned} \tag{56}$$

We find the maximum radial stress at $r = \sqrt{ab}$, where

$$(\sigma_r)_{\max} = \frac{3+\nu}{8}\rho\omega^2(b-a)^2 \tag{57}$$

The maximum tangential stress is at the inner boundary, where

$$(\sigma_\theta)_{\max} = \frac{3+\nu}{4}\rho\omega^2\left(b^2 + \frac{1-\nu}{3+\nu}a^2\right) \tag{58}$$

It will be seen that this stress is larger than $(\sigma_r)_{\max}$.

When the radius a of the hole approaches zero, the maximum tangential stress approaches a value twice as great as that for a solid disk (55); i.e., by making a small circular hole at the center[1] of a solid rotating disk we double the maximum stress. Further instances of this phenomenon of *stress concentration at a hole* will be discussed later (see pages 90–96).

Assuming that the stresses do not vary over the thickness of the disk, the method of analysis developed above for disks of constant thickness can be extended also to *disks of variable thickness.* If h is the thickness of the disk, varying with radius r, the equation of equilibrium of such an element as shown in Fig. 40 is

$$\frac{d}{dr}(hr\sigma_r) - h\sigma_\theta + h\rho\omega^2 r^2 = 0 \tag{k}$$

Using (d) to express the stress components in terms of u, Eq. (k) becomes

$$r^2\frac{d^2u}{dr^2} + r\frac{du}{dr} - u + \frac{r}{h}\frac{dh}{dr}\left(r\frac{du}{dr} + \nu u\right) = -\frac{1-\nu^2}{E}\rho\omega^2 r^3 \tag{l}$$

This is a differential equation for u when h is given as a function of r. It is easily integrated for the case

$$h = Hr^n \tag{m}$$

[1] For eccentric holes see Ta-Cheng Ku, *J. Appl. Mech.*, vol. 27, pp. 359–360, 1960, and the references given there.

in which H and n are constants. The general solution takes the form

$$u = mr^{n+3} + Ar^{\alpha} + Br^{\beta}$$

where
$$m = \frac{(1 - \nu^2)\rho\omega^2}{E[8 + (3 + \nu)n]}$$

and α, β are the roots of the quadratic

$$x^2 + nx + n\nu - 1 = 0$$

A and B are arbitrary constants.

A good approximation to the actual shapes of rotating disks can be obtained by dividing the disk into parts and fitting approximately to each part a curve of the type (m).[1] The case of a conical disk has been discussed by several authors.[2] Very often the calculations are made by dividing the disk into parts and considering each part as a disk of constant thickness.[3]

33 | Bending of a Curved Bar by a Force at the End[4]

We begin with the simple case shown in Fig. 46. A bar of a narrow rectangular cross section and with a circular axis is constrained at the lower end and bent by a force P applied at the upper end in the radial direction. The bending moment at any cross section mn is proportional to $\sin\theta$, and the normal stress σ_θ, according to elementary theory of the bending of curved bars, is proportional to the bending moment. Assuming that

[1] M. Grübler, *VDI*, vol. 50, p. 535, 1906.

[2] A. Fischer, *Z. Oesterr. Ing. Arch. Vereins*, vol. 74, p. 46, 1922; H. M. Martin, *Eng.*, vol. 115, p. 1, 1923; B. Hodkinson, *Eng.*, vol. 116, p. 274, 1923; K. E. Bisshopp, *J. Appl. Mech.*, vol. 11, p. A-1, 1944.

[3] This method was developed by M. Donath; see his book, "Die Berechnung Rotierender Scheiben und Ringe nach einem neuen Verfahren," Berlin, 1929; also the book by K. Löffler, *op. cit.*

[4] H. Golovin, *op. cit.*

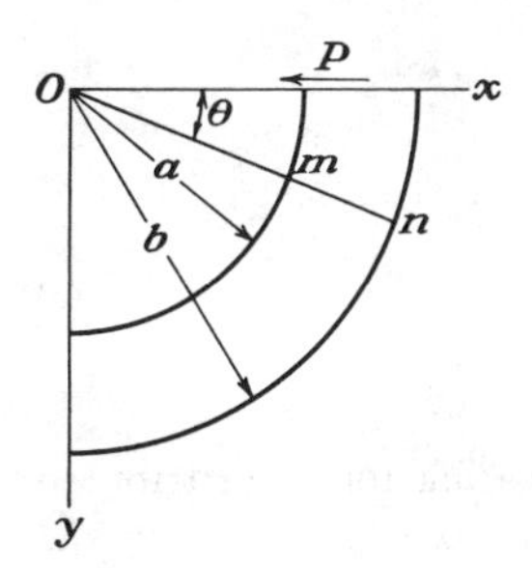

Fig. 46

this holds also for the exact solution, an assumption which the results will justify, we find from the second of Eqs. (38) that the stress function ϕ, satisfying the equation

$$\left(\frac{\partial^2}{\partial r^2} + \frac{1}{r}\frac{\partial}{\partial r} + \frac{1}{r^2}\frac{\partial^2}{\partial \theta^2}\right)\left(\frac{\partial^2 \phi}{\partial r^2} + \frac{1}{r}\frac{\partial \phi}{\partial r} + \frac{1}{r^2}\frac{\partial^2 \phi}{\partial \theta^2}\right) = 0 \tag{a}$$

should be proportional to sin θ. Taking

$$\phi = f(r) \sin \theta \tag{b}$$

and substituting in Eq. (*a*), we find that $f(r)$ must satisfy the following ordinary differential equation:

$$\left(\frac{d^2}{dr^2} + \frac{1}{r}\frac{d}{dr} - \frac{1}{r^2}\right)\left(\frac{d^2 f}{dr^2} + \frac{1}{r}\frac{df}{dr} - \frac{f}{r^2}\right) = 0 \tag{c}$$

This equation can be transformed into a linear differential equation with constant coefficients (see page 69), and its general solution is

$$f(r) = Ar^3 + B\frac{1}{r} + Cr + Dr \log r \tag{d}$$

in which A, B, C, and D are constants of integration, which are determined from the boundary conditions. Substituting solution (*d*) in expression (*b*) for the stress function, and using the general formulas (38), we find the following expressions for the stress components:

$$\begin{aligned}
\sigma_r &= \frac{1}{r}\frac{\partial \phi}{\partial r} + \frac{1}{r^2}\frac{\partial^2 \phi}{\partial \theta^2} = \left(2Ar - \frac{2B}{r^3} + \frac{D}{r}\right)\sin \theta \\
\sigma_\theta &= \frac{\partial^2 \phi}{\partial r^2} = \left(6Ar + \frac{2B}{r^3} + \frac{D}{r}\right)\sin \theta \\
\tau_{r\theta} &= -\frac{\partial}{\partial r}\left(\frac{1}{r}\frac{\partial \phi}{\partial \theta}\right) = -\left(2Ar - \frac{2B}{r^3} + \frac{D}{r}\right)\cos \theta
\end{aligned} \tag{59}$$

From the conditions that the outer and inner boundaries of the curved bar (Fig. 46) are free from external forces, we require that

$$\sigma_r = \tau_{r\theta} = 0 \qquad \text{for } r = a \text{ and } r = b$$

or, from Eqs. (59),

$$\begin{aligned}
2Aa - \frac{2B}{a^3} + \frac{D}{a} &= 0 \\
2Ab - \frac{2B}{b^3} + \frac{D}{b} &= 0
\end{aligned} \tag{e}$$

The final condition is that the sum of the shearing forces distributed over the upper end of the bar should equal the force P. Taking the width of the cross section as unity—or P as the load per unit thickness of the plate—we obtain for $\theta = 0$,

$$\int_a^b \tau_{r\theta}\, dr = -\int_a^b \frac{\partial}{\partial r}\left(\frac{1}{r}\frac{\partial \phi}{\partial \theta}\right) dr = \left| \frac{1}{r}\frac{\partial \phi}{\partial \theta} \right|_b^a$$

$$= \left| Ar^2 + \frac{B}{r^2} + C + D \log r \right|_b^a = P$$

or

$$-A(b^2 - a^2) + B\frac{(b^2 - a^2)}{a^2 b^2} - D \log \frac{b}{a} = P \tag{f}$$

From Eqs. (*e*) and (*f*) we find

$$A = \frac{P}{2N} \qquad B = -\frac{Pa^2b^2}{2N} \qquad D = -\frac{P}{N}(a^2 + b^2) \tag{g}$$

in which

$$N = a^2 - b^2 + (a^2 + b^2) \log \frac{b}{a}$$

Substituting the values (*g*) of the constants of integration in Eqs. (59), we obtain the expressions for the stress components. For the upper end of the bar, $\theta = 0$, we find

$$\sigma_\theta = 0$$

$$\tau_{r\theta} = -\frac{P}{N}\left[r + \frac{a^2b^2}{r^3} - \frac{1}{r}(a^2 + b^2)\right] \tag{h}$$

For the lower end, $\theta = \pi/2$,

$$\tau_{r\theta} = 0$$

$$\sigma_\theta = \frac{P}{N}\left[3r - \frac{a^2b^2}{r^3} - (a^2 + b^2)\frac{1}{r}\right] \tag{k}$$

The expressions (59) constitute an exact solution of the problem only when the forces at the ends of the curved bar are distributed in the manner given by Eqs. (*h*) and (*k*). For any other distribution of forces the stress distribution near the ends will be different from that given by solution (59), but at larger distances this solution may still be valid, by Saint-Venant's principle. Calculations show that the simple theory, based on the assumption that cross sections remain plane during bending, again gives very satisfactory results.

In Fig. 47 the distribution of the shearing stress $\tau_{r\theta}$ over the cross section $\theta = 0$ (for the cases $b = 3a$, $2a$, and $1.3a$) is shown. The abscissas are the radial distances from the inner boundary ($r = a$). The ordinates represent numerical factors with which we multiply the average shearing stress $P/(b - a)$ to get the shearing stress at the point in question. A value 1.5 for this factor gives the maximum shearing stress as calculated from the parabolic distribution for rectangular straight beams. From

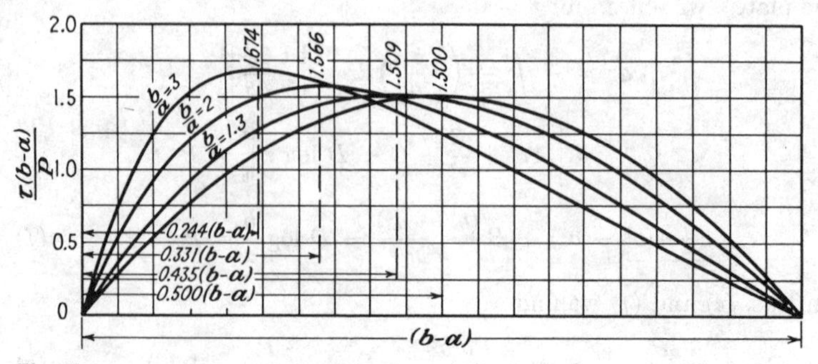

Fig. 47

the figures it may be seen that the distribution of shearing stresses approaches the parabolic distribution when the depth of the cross section is small. For such proportions as are usual in arches and vaults the parabolic distribution of shearing stress, as in straight rectangular bars, can be assumed with sufficient accuracy.

Let us consider now the displacements produced by the force P (Fig. 46). By using Eqs. (48) to (51), and substituting for the stress components the expressions (59), we find

$$\frac{\partial u}{\partial r} = \frac{\sin\theta}{E}\left[2Ar(1-3\nu) - \frac{2B}{r^3}(1+\nu) + \frac{D}{r}(1-\nu)\right]$$

$$\frac{\partial v}{\partial \theta} = r\epsilon_\theta - u \tag{l}$$

$$\gamma_{r\theta} = \frac{\partial u}{r\,\partial\theta} + \frac{\partial v}{\partial r} - \frac{v}{r}$$

From the first of these equations we obtain by integration

$$u = \frac{\sin\theta}{E}\left[Ar^2(1-3\nu) + \frac{B}{r^2}(1+\nu) + D(1-\nu)\log r\right] + f(\theta) \tag{m}$$

where $f(\theta)$ is a function of θ only. Substituting (m) in the second of Eqs. (l) together with the expression for ϵ_θ and integrating, we find

$$v = -\frac{\cos\theta}{E}\left[Ar^2(5+\nu) + \frac{B}{r^2}(1+\nu) - D\log r(1-\nu) + D(1-\nu)\right] - \int f(\theta)\,d\theta + F(r) \tag{n}$$

in which $F(r)$ is a function of r only. Substituting now (m) and (n) in the third of Eqs. (l) we arrive at the equation

$$\int f(\theta)\,d\theta + f'(\theta) + rF'(r) - F(r) = -\frac{4D\cos\theta}{E}$$

This equation is satisfied by putting

$$F(r) = Hr \qquad f(\theta) = -\frac{2D}{E}\theta\cos\theta + K\sin\theta + L\cos\theta \tag{p}$$

in which H, K, and L are arbitrary constants, to be determined from the conditions of constraint. The components of displacements, from (m) and (n), are then

$$\begin{aligned} u &= -\frac{2D}{E}\theta\cos\theta + \frac{\sin\theta}{E}\left[D(1-\nu)\log r + A(1-3\nu)r^2 + \frac{B(1+\nu)}{r^2}\right] + K\sin\theta + L\cos\theta \\ v &= \frac{2D}{E}\theta\sin\theta - \frac{\cos\theta}{E}\left[A(5+\nu)r^2 + \frac{B(1+\nu)}{r^2} - D(1-\nu)\log r\right] \\ &\quad + \frac{D(1+\nu)}{E}\cos\theta + K\cos\theta - L\sin\theta + Hr \end{aligned} \tag{q}$$

The radial deflection of the upper end of the bar is obtained by putting $\theta = 0$ in the expression for u, which gives

$$(u)_{\theta=0} = L \tag{r}$$

The constant L is obtained from the condition at the built-in end (Fig. 46). For $\theta = \pi/2$ we have $v = 0$; $\partial v/\partial r = 0$, hence, from the second of Eqs. (q),

$$H = 0 \qquad L = \frac{D\pi}{E} \tag{s}$$

The deflection of the upper end is, therefore, using (g),

$$(u)_{\theta=0} = \frac{D\pi}{E} = -\frac{P\pi(a^2+b^2)}{E[(a^2-b^2)+(a^2+b^2)\log(b/a)]} \tag{60}$$

The application of this formula will be given later. When b approaches a, and the depth of the curved bar, $h = b - a$, is small in comparison with a, we can use the expression

$$\log\frac{b}{a} = \log\left(1+\frac{h}{a}\right) = \frac{h}{a} - \frac{1}{2}\frac{h^2}{a^2} + \frac{1}{3}\frac{h^3}{a^3} - \cdots$$

Substituting in (60) and neglecting small terms of higher order, we obtain

$$(u)_{\theta=0} = -\frac{3\pi a^3 P}{Eh^3}$$

which coincides with the elementary formula for this case.[1]

[1] See S. Timoshenko, "Strength of Materials," vol. 1, Art. 80, 1955.

By taking the stress function in the form

$$\phi = f(r) \cos \theta$$

and proceeding as above, we get a solution for the case when a vertical force and a couple are applied to the upper end of the bar (Fig. 46). Subtracting from this solution the stresses produced by the couple (see Art. 29), the stresses due to a vertical force applied at the upper end of the bar remain. Having the solutions for a horizontal and for a vertical load, the solution for any inclined force can be obtained by superposition.

In the above discussion it was always assumed that Eqs. (*e*) are satisfied and that the circular boundaries of the bar are free from forces. By taking the expressions in (*e*) different from zero, we obtain the case when normal and tangential forces proportional to $\sin \theta$ and $\cos \theta$ are distributed over circular boundaries of the bar. Combining such solutions with the solutions previously obtained for pure bending and for bending by a force applied at the end we can approach the condition of loading of a vault covered with sand or soil.[1]

34 | Edge Dislocation

In Art. 33 the displacement components (*q*) were derived from the stress-components (59). The constants A, B, D were given by (*g*) for the problem illustrated in Fig. 46.

The application of this solution to the quarter ring was a matter of choice, not necessity. The same solution can be applied to a nearly complete ring, Fig. 48*a* or *b*. We can also interpret it for imposed *displacements* instead of imposed forces.

Considering the displacements (*q*) of Art. 33, we observe that the first term in the expression for u can give rise to a *discontinuity*. In Fig. 48*b*, a fine radial saw cut has been made in the originally complete ring, at $\theta = 0$. The lower face of the cut has $\theta = 0$. The upper face has $\theta = 2\pi - \epsilon$, where ϵ is infinitesimal. If u in (*q*) is evaluated for

[1] Several examples of this kind were discussed by Golovin, *loc. cit.*, and Ribière, *loc. cit.*

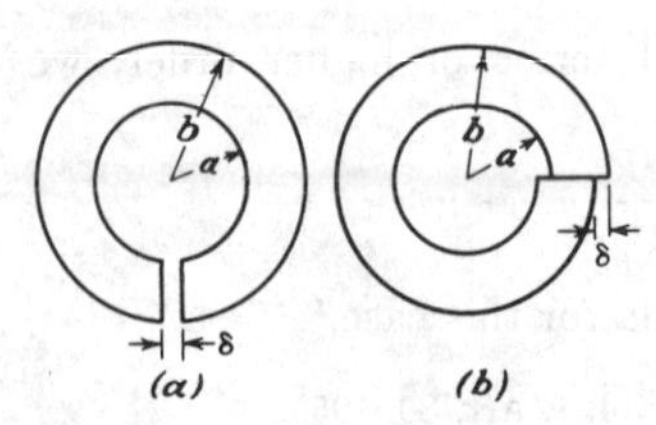

Fig. 48

these two values of θ, the results differ by an amount δ. Thus

$$\delta = (u)_{\theta=2\pi-\epsilon} - (u)_{\theta=0} \tag{a}$$

From (q) we then have

$$\delta = -\frac{2D}{E}2\pi \tag{b}$$

This relative displacement of the two faces of the cut is indicated by δ in Fig. 48*b*. The forces P required to effect it are found from the last of (g), Art. 33, with D given by (b) above. If the two faces are welded together after the relative displacement δ has been imposed, each applies the required force P to the other, as action and reaction. The ring is in a self-strained state, called an "edge dislocation." The corresponding plane strain state is the basis of the explanation of plastic deformation in metal crystals.[1]

Figure 48*a* shows a ring with a parallel gap of thickness δ. If a thin saw cut is first made, then relative displacements are imposed to open up the gap, the discontinuity of displacement now occurs in v, not u. It can be obtained from the solution of Art. 33 by taking the right-hand face of the cut at $\theta = -\pi/2$, the left-hand face at $\theta = 3\pi/2$. We then have (since v is in the direction θ-increasing)

$$\delta = (v)_{\theta=-\pi/2} - (v)_{\theta=3\pi/2} \tag{c}$$

Using the second of Eqs. (q) in Art. 33 we now find

$$\delta = \frac{2D}{E}\left(-\frac{\pi}{2}\right)\sin\left(-\frac{\pi}{2}\right) - \frac{2D}{E}\frac{3\pi}{2}\sin\frac{3\pi}{2} = \frac{4\pi D}{E} \tag{d}$$

The fact that the values of δ in (b) and (d) differ only in sign means that the stresses in the two cases will also differ only in sign. P is found from the third of (g) in Art. 33, then A and B follow from the first two. This correspondence is predictable from the fact that if the cuts of Fig. 48*a* and *b* are both made, the quadrant is cut free. The relative displacement δ of Fig. 48*a*, and a relative displacement $-\delta$ in Fig. 48*b*, can be effected simultaneously by sliding the quadrant to the right by the amount δ. No stress is induced by this, and therefore the two dislocations must have equal and opposite stresses when existing separately. This is an instance of a general[2] "theorem of equivalent cuts."

[1] G. I. Taylor, *Proc. Roy. Soc.* (London), ser. A, vol. 134, pp. 362–387, 1934. Or see, for instance, A. H. Cottrell, "Dislocations and Plastic Flow in Crystals," chap. 2, 1956.

[2] The demonstration used here was given by J. N. Goodier, *Proc. 5th Intern. Congr. Appl. Mech.*, pp. 129–133, 1938. The theorem is due to V. Volterra, who gave a general theory in *Ann. Ecole. Norm.* (*Paris*), ser. 3, vol. 24, pp. 401–517, 1907. See also A. E. H. Love, "Mathematical Theory of Elasticity," 4th ed., p. 221, Cambridge University Press, New York, 1927; A. Timpe, *Z. Math. Physik*, *loc. cit.*

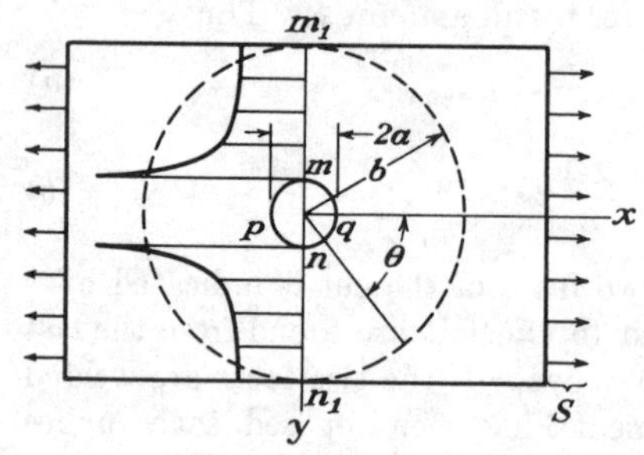

Fig. 49

35 | The Effect of Circular Holes on Stress Distributions in Plates

Figure 49 represents a plate submitted to a uniform tension of magnitude S in the x direction. If a small circular hole is made in the middle of the plate, the stress distribution in the neighborhood of the hole will be changed, but we can conclude from Saint-Venant's principle that the change is negligible at distances which are large compared with a, the radius of the hole.

Consider the portion of the plate within a concentric circle of radius b, large in comparison with a. The stresses at the radius b are effectively the same as in the plate without the hole and are therefore given by

$$\begin{aligned}(\sigma_r)_{r=b} &= S\cos^2\theta = \tfrac{1}{2}S(1+\cos 2\theta)\\ (\tau_{r\theta})_{r=b} &= -\tfrac{1}{2}S\sin 2\theta\end{aligned}\qquad (a)$$

These forces, acting around the outside of the ring having the inner and outer radii $r = a$ and $r = b$, give a stress distribution within the ring which we may regard as consisting of two parts. The first is due to the constant component $\frac{1}{2}S$ of the normal forces. The stresses it produces can be calculated by means of Eqs. (44). The remaining part, consisting of the normal forces $\frac{1}{2}S \cos 2\theta$, together with the shearing forces $-\frac{1}{2}S \sin 2\theta$, produces stresses that may be derived from a stress function of the form

$$\phi = f(r)\cos 2\theta \qquad (b)$$

Substituting this into the compatibility equation

$$\left(\frac{\partial^2}{\partial r^2}+\frac{1}{r}\frac{\partial}{\partial r}+\frac{1}{r^2}\frac{\partial^2}{\partial\theta^2}\right)\left(\frac{\partial^2\phi}{\partial r^2}+\frac{1}{r}\frac{\partial\phi}{\partial r}+\frac{1}{r^2}\frac{\partial^2\phi}{\partial\theta^2}\right)=0$$

we find the following ordinary differential equation to determine $f(r)$:

$$\left(\frac{d^2}{dr^2}+\frac{1}{r}\frac{d}{dr}-\frac{4}{r^2}\right)\left(\frac{d^2f}{dr^2}+\frac{1}{r}\frac{df}{dr}-\frac{4f}{r^2}\right)=0$$

The general solution is

$$f(r) = Ar^2 + Br^4 + C\frac{1}{r^2} + D$$

The stress function is therefore

$$\phi = \left(Ar^2 + Br^4 + C\frac{1}{r^2} + D\right)\cos 2\theta \tag{c}$$

and the corresponding stress components, from Eqs. (38), are

$$\begin{aligned} \sigma_r &= \frac{1}{r}\frac{\partial\phi}{\partial r} + \frac{1}{r^2}\frac{\partial^2\phi}{\partial\theta^2} = -\left(2A + \frac{6C}{r^4} + \frac{4D}{r^2}\right)\cos 2\theta \\ \sigma_\theta &= \frac{\partial^2\phi}{\partial r^2} = \left(2A + 12Br^2 + \frac{6C}{r^4}\right)\cos 2\theta \\ \tau_{r\theta} &= -\frac{\partial}{\partial r}\left(\frac{1}{r}\frac{\partial\phi}{\partial\theta}\right) = \left(2A + 6Br^2 - \frac{6C}{r^4} - \frac{2D}{r^2}\right)\sin 2\theta \end{aligned} \tag{d}$$

The constants of integration are now to be determined from conditions (*a*) for the outer boundary and from the condition that the edge of the hole is free from external forces. These conditions give

$$\begin{aligned} 2A + \frac{6C}{b^4} + \frac{4D}{b^2} &= -\frac{1}{2}S \\ 2A + \frac{6C}{a^4} + \frac{4D}{a^2} &= 0 \\ 2A + 6Bb^2 - \frac{6C}{b^4} - \frac{2D}{b^2} &= -\frac{1}{2}S \\ 2A + 6Ba^2 - \frac{6C}{a^4} - \frac{2D}{a^2} &= 0 \end{aligned}$$

Solving these equations and putting $a/b = 0$, that is, assuming an infinitely large plate, we obtain

$$A = -\frac{S}{4} \qquad B = 0 \qquad C = -\frac{a^4}{4}S \qquad D = \frac{a^2}{2}S$$

Substituting these values of constants into Eqs. (*d*) and adding the stresses produced by the uniform tension $\frac{1}{2}S$ on the outer boundary calculated from Eqs. (44) we find[1]

$$\begin{aligned} \sigma_r &= \frac{S}{2}\left(1 - \frac{a^2}{r^2}\right) + \frac{S}{2}\left(1 + \frac{3a^4}{r^4} - \frac{4a^2}{r^2}\right)\cos 2\theta \\ \sigma_\theta &= \frac{S}{2}\left(1 + \frac{a^2}{r^2}\right) - \frac{S}{2}\left(1 + \frac{3a^4}{r^4}\right)\cos 2\theta \\ \tau_{r\theta} &= -\frac{S}{2}\left(1 - \frac{3a^4}{r^4} + \frac{2a^2}{r^2}\right)\sin 2\theta \end{aligned} \tag{61}$$

[1] This solution was obtained by G. Kirsch; see *VDI*, vol. 42, 1898. It has been well confirmed many times by strain measurements and by the *photoelastic method* (see Chap. 5 and the books cited on p. 151).

The displacements u, v (but for a rigid body movement) can be found from these, using Eqs. (48) to (51). This is left as an exercise for the reader (Prob. 6, page 144). They are free from discontinuities.

If r is very large, σ_r and $\tau_{r\theta}$ approach the values given in Eqs. (a). At the edge of the hole, $r = a$ and we find

$$\sigma_r = \tau_{r\theta} = 0 \qquad \sigma_\theta = S - 2S \cos 2\theta$$

It can be seen that σ_θ is greatest when $\theta = \pi/2$ or $\theta = 3\pi/2$, that is, at the ends m and n of the diameter perpendicular to the direction of the tension (Fig. 49). At these points $(\sigma_\theta)_{max} = 3S$. This is the maximum tensile stress and is three times the uniform stress S, applied at the ends of the plate.

At the points p and q, θ is equal to π and 0 and we find

$$\sigma_\theta = -S$$

so that there is a compression stress in the tangential direction at these points.

For the cross section of the plate through the center of the hole and perpendicular to the x axis, $\theta = \pi/2$, and from Eqs. (61),

$$\tau_{r\theta} = 0 \qquad \sigma_\theta = \frac{S}{2}\left(2 + \frac{a^2}{r^2} + 3\frac{a^4}{r^4}\right)$$

It is evident that the effect of the hole is of a *localized character*. As r increases, the stress σ_θ rapidly approaches the value S. The distribution of this stress is shown in Fig. 49 by the shaded area. The localized character of the stresses around the hole justifies the application of the solution (e), derived for an infinitely large plate, to a plate of finite width. If the width of the plate is not less than four diameters of the hole, the error of the solution (61) in calculating $(\sigma_\theta)_{max}$ does not exceed 6 percent.[1]

Having the solution (d) for tension for compression in one direction, the solution for tension or compression in two perpendicular directions can easily be obtained by superposition. By taking, for instance, tensile stresses in two perpendicular directions equal to S, we find at the boundary of the hole a tensile stress $\sigma_\theta = 2S$ (see page 82). By taking a tensile stress S in the x direction (Fig. 50) and a compressive stress $-S$ in the y direction, we obtain the case of pure shear. The tangential stresses

[1] See S. Timoshenko, *Bull. Polytech. Inst., Kiew,* 1907. We must take S equal to the load divided by the gross area of the plate.

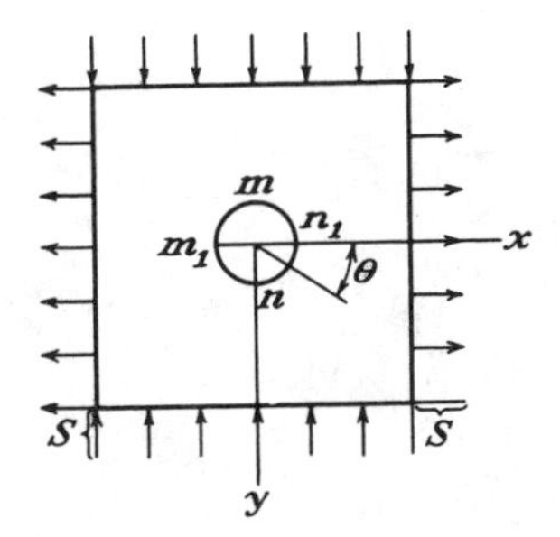

Fig. 50

at the boundary of the hole are, from Eqs. (61),

$$\sigma_\theta = S - 2S \cos 2\theta - [S - 2S \cos (2\theta - \pi)]$$

For $\theta = \pi/2$ or $\theta = 3\pi/2$, that is, at the points n and m, we find $\sigma_\theta = 4S$. For $\theta = 0$ or $\theta = \pi$, that is, at n_1 and m_1, $\sigma_\theta = -4S$. Hence, for a large plate under pure shear, the maximum tangential stress at the boundary of the hole is four times the applied pure shear stress.

The high *stress concentration* found at the edge of a hole is of great practical importance. As an example, holes in ships' decks may be mentioned. When the hull of a ship is bent, tension or compression is produced in the decks and there is a high stress concentration at the holes. Under the cycles of stress produced by waves, fatigue of the metal at the overstressed portions may result finally in fatigue cracks.[1]

It is often necessary to reduce the stress concentration at holes, such as access holes in airplane wings and fuselages. This can be done by adding a bead[2] or reinforcing ring.[3] The analytical problem has been solved by extending the method employed for the hole, and the results have been compared with strain-gauge measurements.[3]

The case of a circular hole near the straight boundary of a semi-infinite plate under tension parallel to this boundary (Fig. 51) was

[1] See the Introduction, and Bibliography, in Thein Wah (ed.), "A Guide for the Analysis of Ship Structures," Office of Technical Services, U.S. Dept. of Commerce, Washington, D.C., 1960.

[2] See S. Timoshenko, *J. Franklin Inst.*, vol. 197, p. 505, 1924; also S. Timoshenko, "Strength of Materials," 3d ed., vol. 2, p. 305, D. Van Nostrand Company, Inc., Princeton, N.J., 1956.

[3] S. Levy, A. E. McPherson, and F. C. Smith, *J. Appl. Mech.*, vol. 15, p. 160, 1948. References to prior work may be found in this paper. For references up to 1955 see J. N. Goodier and P. G. Hodge, "Elasticity and Plasticity," 1958, p. 11.

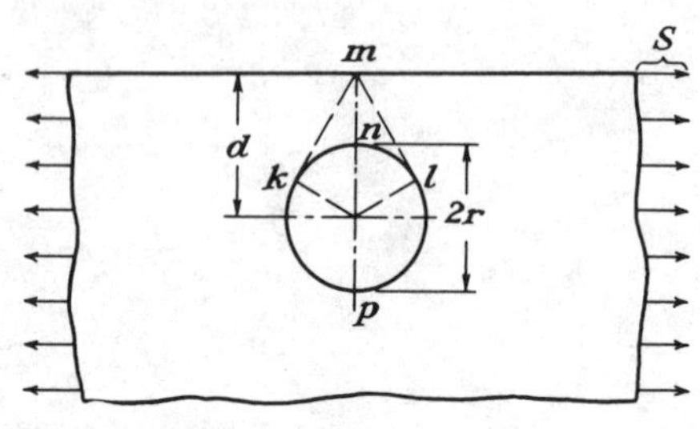

Fig. 51

analyzed by G. B. Jeffery.[1] A corrected result and a comparison with photoelastic tests (see Chap. 5) were given later by R. D. Mindlin.[2] The stress at the hole, at the point n nearest the edge, becomes a very large multiple of the undisturbed tensile stress when mn is small[3] compared with np.

Jeffery also investigated the case of a uniform normal pressure p_i acting on the boundary of the hole. This is a special case of the eccentric-bore problem described on page 71. If the hole is very far from the straight edge the stresses at the boundary of the hole, from Eqs. (45), are

$$\sigma_\theta = p_i \qquad \sigma_r = -p_i$$

If the hole is near the straight edge, the tangential stresses are no longer constant along the boundary of the hole. The maximum tangential stress is at the points k and l and is given by the formula

$$(\sigma_\theta)_{\max} = p_i \frac{d^2 + r^2}{d^2 - r^2} \tag{62}$$

This stress should be compared with the tensile stress at the point m on the straight edge of the plate, given by the formula

$$\sigma_\theta = \frac{4p_i r^2}{d^2 - r^2} \tag{63}$$

For $d = r\sqrt{3}$, the two expressions have the same magnitude. If d is greater than this, the maximum stress is at the circular boundary, and if it is less, the maximum stress is at the point m.

The case of a plate of finite width with a circular hole on the axis of

[1] *Loc. cit.*
[2] *Proc. Soc. Exptl. Stress Anal.*, vol. 5, p. 56, 1948.
[3] See also W. T. Koiter, *Quart. Appl. Math.*, vol. 15, p. 303, 1957.

symmetry (Fig. 52) was discussed by R. C. J. Howland.[1] He found, for instance, that when $2r = \frac{1}{2}d$, $\sigma_\theta = 4.3S$ at the point n and

$$\sigma_\theta = 0.75S$$

at the point m.

The method used in this article for analyzing stresses round a small circular hole can be applied when the plate is subjected to pure bending.[2] Many specific cases for both tension and bending have been worked out.[1] These include one hole or a row of holes in a strip[3–5] and in a semi-infinite plate,[1] circular arrays of holes,[2] and semicircular notches in a strip.[3]

A method devised by Hengst has been applied to the case of a hole in a square plate[6] under equal tension in both directions, and under shear[7] when the hole is plain or reinforced.

Solutions have been obtained for the infinite plate with a circular

[1] *Trans. Roy. Soc.* (*London*), ser. A, vol. 229, p. 49, 1930. The numerous references to solutions and evaluations for circular, and other, holes, may be traced through *Applied Mechanics Reviews*. The following books should be consulted: R. E. Peterson, "Stress Concentration Factors in Design," New York, 1953; J. N. Goodier and P. G. Hodge, "Elasticity and Plasticity," New York, 1958; G. N. Savin, "Stress Concentration around Holes," New York, 1961 (a translation of the Russian original of 1951). The principal books on photoelasticity (see Chap. 5) contain many useful experimental evaluations.

[2] Z. Tuzi, *Phil. Mag.*, February, 1930, p. 210; also *Sci. Papers Inst. Phys. Chem. Res.* (*Tokyo*), vol. 9, p. 65, 1928. The corresponding problem for an *elliptical* hole was solved earlier by K. Wolf, *Z. Tech. Physik*, 1922, p. 160. The circular hole in a strip is discussed by R. C. J. Howland and A. C. Stevenson, *Trans. Roy. Soc.* (*London*), ser. A, vol. 232, p. 155, 1933. A proof of convergence of the series solutions is given by R. C. Knight, *Quart. J. Math., Oxford Series*, vol. 5, p. 255, 1934.

[3] K. J. Schulz, *Proc. Nederl. Akad. van Wetenschappen*, vol. 45, pp. 233, 341, 457, and 524, 1942, vol. 48, pp. 282 and 292, 1945.

[4] Chih-Bing Ling, "Collected Papers in Elasticity and Mathematics," Institute of Mathematics, Academia Sinica, Tapei, Taiwan, China, 1963.

[5] M. Isida, *Bull. Japan. Soc. Mech. Engr.*, vol. 3, pp. 259–266, 1960. M. Isida and S. Tagami, *Proc. 9th Japan Nat. Congr. Appl. Mech.*, pp. 51–54, 1959. Many related papers by M. Isida are found in these publications.

[6] H. Hengst, *Z. Angew. Math. Mech.*, vol. 18, p. 44, 1938.

[7] C. K. Wang, *J. Appl. Mech.*, vol. 13, p. A-77, 1946.

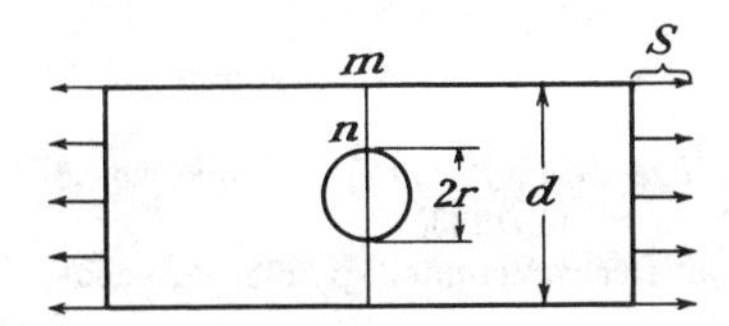

Fig. 52

hole when forces are applied to the boundary of the hole,[1] for the corresponding problem of the strip,[2] and for a row of holes parallel and near to the straight edge of a semi-infinite plate[3] (row of rivet holes).

If an elliptical hole is made in an infinite plate under tension S, with one of the principal axes parallel to the tension, the stresses at the ends of the axis of the hole perpendicular to the direction of the tension are

$$\sigma = S\left(1 + 2\frac{a}{b}\right) \tag{64}$$

where $2a$ is the axis of the ellipse perpendicular to the tension, and $2b$ is the other axis. This and other problems concerning ellipses, hyperbolas, and two circles are discussed in Chap. 6, where references will be found.

A very slender hole (a/b large) perpendicular to the direction of the tension causes a very high stress concentration.[4] This explains why cracks transverse to applied forces tend to spread. The spreading can be stopped by drilling holes at the ends of the crack to eliminate the sharp curvature responsible for the high stress concentration.

When a hole is filled with material that is rigid or has elastic constants different from those of the plate (plane stress) or body (plane strain) itself, we have the problem of the rigid or elastic *inclusion*. This has been solved for circular[5] and elliptic inclusions.[6] The results for the rigid circular inclusion have been confirmed by the photoelastic method[7] (see Chap. 5).

The stresses given by Eqs. (61) for the problem indicated by Fig. 49 are the same for plane strain as for plane stress. In plane strain, however, the axial stress

$$\sigma_z = \nu(\sigma_r + \sigma_\theta)$$

must act on the plane ends, which are parallel to the xy plane, in order to make ϵ_z zero. Removal of these stresses from the ends, to arrive at the condition of free ends, will produce further stress that will not be of a two-dimensional (plane stress or plane strain) character. If the

[1] W. G. Bickley, *Trans. Roy. Soc. (London)*, ser. A, vol. 227, p. 383, 1928.

[2] R. C. Knight, *Phil. Mag.*, ser. 7, vol. 19, p. 517, 1935.

[3] C. B. Ling and M. C. Hsu, *Bur. Aeron. Res. Tech. Rept.* 16, Chengtu, China, February, 1945. See also n. 4, p. 95.

[4] The problem of a narrow slot was discussed by M. Sadowsky, *Z. Angew. Math. Mech.*, vol. 10, p. 77, 1930.

[5] K. Sezawa and G. Nishimura, *Rept. Aeron. Res. Inst., Tokyo Imp. Univ.*, vol. 6, no. 25, 1931; J. N. Goodier, *Trans. ASME*, vol. 55, p. 39, 1933.

[6] L. H. Donnell, "Theodore von Kármán Anniversary Volume," p. 293, Pasadena, 1941.

[7] W. E. Thibodeau and L. A. Wood, *J. Res. Nat. Bur. Std.*, vol. 20, p. 393, 1938.

hole is small in diameter compared with the thickness between the ends, the disturbance will be confined to the neighborhood of the ends. But if the diameter and the thickness are of the same order of magnitude, the problem must be treated as essentially three-dimensional throughout. Investigations of this kind[1] have shown that σ_θ remains the largest stress component and its value is very close to that given by two-dimensional theory.

36 | Concentrated Force at a Point of a Straight Boundary

Let us now consider a concentrated vertical force P acting on a horizontal straight boundary AB of an infinitely large plate (Fig. 53*a*). The distribution of the load along the thickness of the plate is uniform, as indicated in Fig. 53*b*. The thickness of the plate is taken as unity, so that P is the load per unit thickness.

The distribution of stress depends on whatever forces act on the complete closed boundary, for example, $ABnm$, and not only on the conditions on AB. This is true even when $ABmn$ are removed to infinity.

There is a basic solution[2] called the simple radial distribution. Any element C at a distance r from the point of application of the load is subjected to a simple compression in the radial direction. The stress com-

[1] A. E. Green, *Trans. Roy. Soc. (London)*, ser. A, vol. 193, p. 229, 1948; E. Sternberg and M. Sadowsky, *J. Appl. Mech.*, vol. 16, p. 27, 1949.

[2] The solution of this problem was obtained by way of the three-dimensional solution of J. Boussinesq (p. 399) by Flamant, *Compt. Rend.*, vol. 114, p. 1465, 1892, Paris. The extension of the solution to the case of an inclined force is due to Boussinesq, *Compt. Rend.*, vol. 114, p. 1510, 1892. See also the paper by J. H. Michell, *Proc. London Math. Soc.*, vol. 32, p. 35, 1900. The experimental investigation that suggested the above theoretical work was done by Carus Wilson, *Phil. Mag.*, vol. 32, p. 481, 1891.

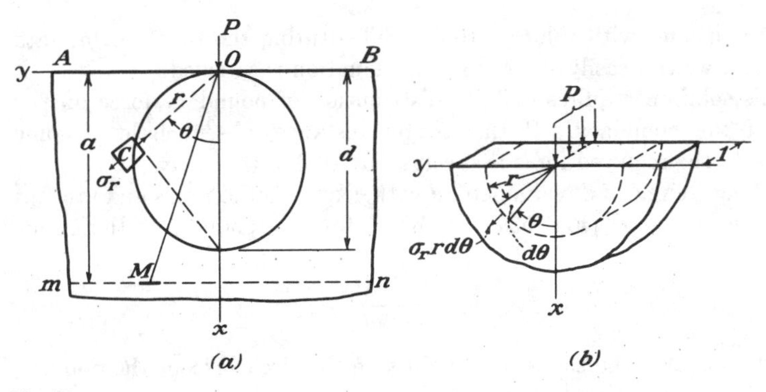

Fig. 53

ponents are

$$\sigma_r = -\frac{2P}{\pi}\frac{\cos\theta}{r} \qquad \sigma_\theta = \tau_{r\theta} = 0 \tag{65}$$

The tangential stress σ_θ and the shearing stress $\tau_{r\theta}$ are zero. It is easy to see that these values of the stress components satisfy the equations of equilibrium (37).

The boundary conditions on AB are also satisfied because σ_θ and $\tau_{r\theta}$ are zero along the straight edge of the plate, which is free from external forces except at the point of application of the load ($r = 0$). Here σ_r becomes infinite. The resultant of the forces acting on a cylindrical surface of radius r (Fig. 53b) must balance P. It is obtained by summing the vertical components $\sigma_r r\,d\theta \cos\theta$ acting on each element $r\,d\theta$ of the surface. In this manner we find

$$2\int_0^{\frac{\pi}{2}} \sigma_r\, r\cos\theta\, d\theta = -\frac{4P}{\pi}\int_0^{\frac{\pi}{2}} \cos^2\theta\, d\theta = -P$$

To prove that (65) is the exact solution of the problem we must consider also the equation of compatibility (39). The above solution is derived from the stress function

$$\phi = -\frac{P}{\pi} r\theta \sin\theta \tag{a}$$

We can verify this by using Eqs. (38) as follows:

$$\begin{aligned}
\sigma_r &= \frac{1}{r}\frac{\partial\phi}{\partial r} + \frac{1}{r^2}\frac{\partial^2\phi}{\partial\theta^2} = -\frac{2P}{\pi}\frac{\cos\theta}{r} \\
\sigma_\theta &= \frac{\partial^2\phi}{\partial r^2} = 0 \\
\tau_{r\theta} &= -\frac{\partial}{\partial r}\left(\frac{1}{r}\frac{\partial\phi}{\partial\theta}\right) = 0
\end{aligned} \tag{65'}$$

which coincides with solution (65). Substituting the function (a) into Eq. (39), we can easily show that this equation is satisfied.

This solution requires definite distribution of boundary force on the rest of the boundary. If this is, for instance, a semicircle of some radius R, the required force is specified by (65) with $r = R$.

Taking a circle of any diameter d with center on the x axis and tangent to the y axis at O (Fig. 53a), we have, for any point C of the circle, $d\cos\theta = r$. Hence, from Eq. (65),

$$\sigma_r = -\frac{2P}{\pi d}$$

i.e., the stress is the same at all points on the circle, except the point O, the point of application of the load.

Taking a horizontal plane mn at a distance a from the straight edge of the plate, the normal and shearing components of the stress on this plane at any point M (Fig. 53a) are calculated from the simple compression in the radial direction,

$$
\begin{aligned}
\sigma_x &= \sigma_r \cos^2 \theta = -\frac{2P}{\pi}\frac{\cos^3 \theta}{r} = -\frac{2P}{\pi a}\cos^4 \theta \\
\sigma_y &= \sigma_r \sin^2 \theta = -\frac{2P}{\pi a}\sin^2 \theta \cos^2 \theta \\
\tau_{xy} &= \sigma_r \sin \theta \cos \theta = -\frac{2P}{\pi}\frac{\sin \theta \cos^2 \theta}{r} = -\frac{2P}{\pi a}\sin \theta \cos^3 \theta
\end{aligned}
\tag{66}
$$

In Fig. 54 the distribution of stresses σ_x and τ_{xy} along the horizontal plane mn is represented graphically.

At the point of application of the load the stress is theoretically infinitely large because a finite force is acting on an infinitely small area. Actually, the load will be distributed over an area of finite, small width. Plastic flow may occur locally. Even so, the plastic zone may be imagined cut out by a circular cylindrical surface of small radius as shown in Fig. 53b. The equations of elasticity can then be applied to the remaining portion of the plate.

An analogous solution can be obtained for a horizontal force P applied to the straight boundary of the semi-infinite plate (Fig. 55). The stress components for this case are obtained from the same Eqs. (65′); it is only necessary to measure the angle θ from the direction of the force, as shown in the figure. By calculating the resultant of the forces acting

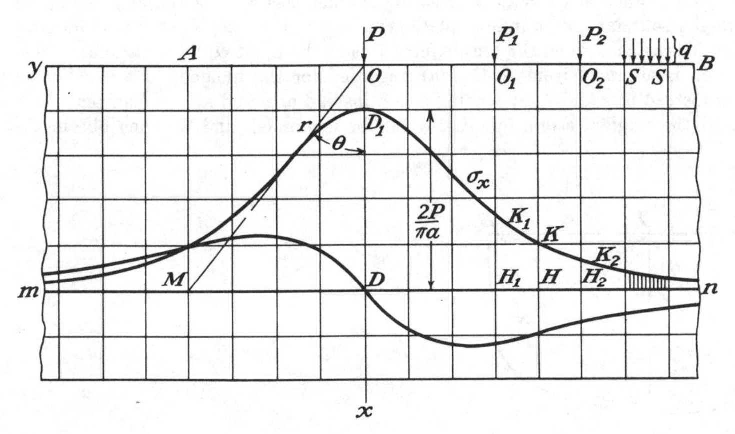

Fig. 54

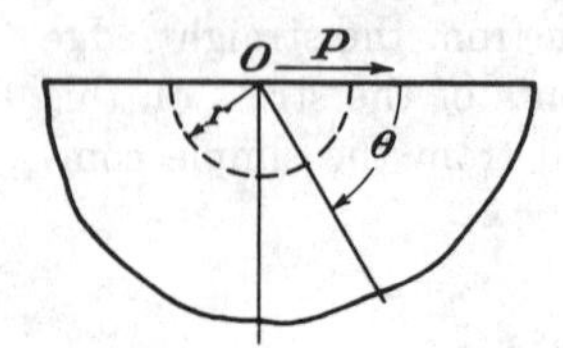

Fig. 55

on a cylindrical surface, shown in Fig. 55 by the broken line, we find

$$-\frac{2P}{\pi}\int_0^\pi \cos^2\theta\, d\theta = -P$$

This resultant balances the external force P, and, as the stress components $\tau_{r\theta}$ and σ_θ at the straight edge are zero, solution (65′) satisfies the boundary conditions.

Having the solutions for vertical and horizontal concentrated forces, solutions for inclined forces are obtained by superposition. Resolving the inclined force P into two components, $P\cos\alpha$ vertically and $P\sin\alpha$ horizontally (Fig. 56), the radial stress at any point C is, from Eqs. (65′),

$$\sigma_r = -\frac{2}{\pi r}\left[P\cos\alpha\cos\theta + P\sin\alpha\cos\left(\frac{\pi}{2}+\theta\right)\right]$$

$$= -\frac{2P}{\pi r}\cos(\alpha+\theta) \qquad (67)$$

Hence Eqs. (65′) can be used for any direction of the force, provided in each case we measure the angle θ from the direction of the force.

The stress function (a) may be used also in the case when a couple is acting on the straight boundary of an infinite plate (Fig. 57a). It is easy to see that the stress function for the case when the tensile force P is at the point O_1, at a distance a from the origin, is obtained from ϕ, Eq. (a), regarded for the moment as a function of x and y instead of r and θ, by writing $y + a$ instead of y and also $-P$ instead of P. This and the original stress function ϕ can be combined, and we then obtain the

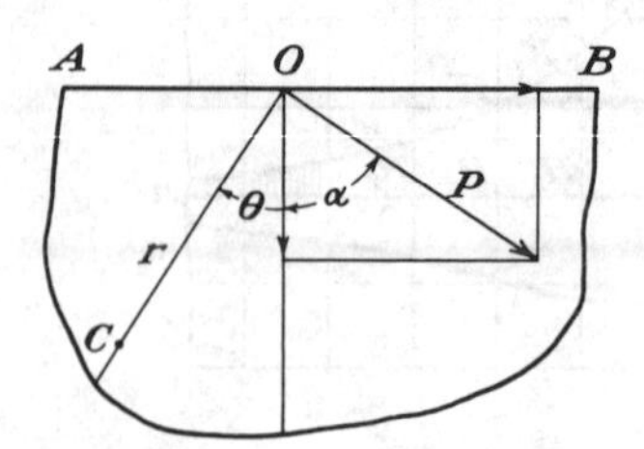

Fig. 56

stress function for the two equal and opposite forces applied at O and O_1, in the form

$$-\phi(x,y+a)+\phi(x,y)$$

When a is very small, this approaches the value

$$\phi_1 = -a\frac{\partial \phi}{\partial y} \tag{b}$$

Substituting (a) in Eq. (b), and noting (see page 67) that

$$\frac{\partial \phi}{\partial y} = \frac{\partial \phi}{\partial r}\sin\theta + \frac{\partial \phi}{\partial \theta}\frac{\cos\theta}{r}$$

we find

$$\phi_1 = \frac{Pa}{\pi}(\theta + \sin\theta\cos\theta) = \frac{M}{\pi}(\theta+\sin\theta\cos\theta) \tag{68}$$

in which M is the moment of the applied couple.

Reasoning in the same manner, we find that by differentiation of ϕ_1, we obtain the stress function ϕ_2 for the case when two equal and opposite couples M are acting at two points O and O_1 a very small distance apart (Fig. 57b). We thus find that

$$\phi_2 = \phi_1 - \left(\phi_1 + \frac{\partial \phi_1}{\partial y}a\right) = -a\frac{\partial \phi_1}{\partial y} = -\frac{2Ma}{\pi r}\cos^3\theta \tag{69}$$

If the directions of the couples are changed, it is only necessary to change the sign of the function (69).

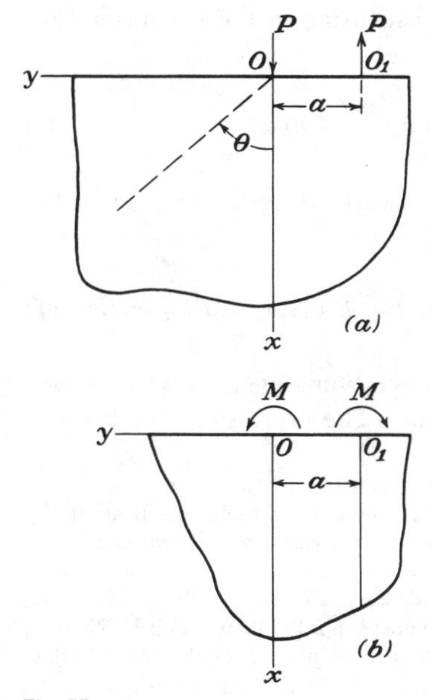

Fig. 57

A series of stress functions obtained by successive differentiation has been employed to solve the problem of stress concentration due to a semicircular notch in a semi-infinite plate in tension parallel to the edge.[1] The maximum tensile stress is slightly greater than three times the undisturbed tensile stress away from the notch. The strip with a semicircular notch in each edge has also been investigated.[2] The stress-concentration factor (ratio of maximum to mean stress at minimum section) falls below three, approaching unity as the notches are made larger.

Having the distribution of stresses, the corresponding displacements can be obtained in the usual way by applying Eqs. (48) to (50). For a force normal to the straight boundary (Fig. 53) we have

$$\begin{aligned}\epsilon_r &= \frac{\partial u}{\partial r} = -\frac{2P}{\pi E}\frac{\cos\theta}{r} \\ \epsilon_\theta &= \frac{u}{r} + \frac{\partial v}{r\,\partial\theta} = \nu\frac{2P}{\pi E}\frac{\cos\theta}{r} \\ \gamma_{r\theta} &= r\frac{\partial u}{\partial\theta} + \frac{\partial v}{\partial r} - \frac{v}{r} = 0\end{aligned} \tag{c}$$

Integrating the first of these equations, we find

$$u = -\frac{2P}{\pi E}\cos\theta\log r + f(\theta) \tag{d}$$

where $f(\theta)$ is a function of θ only. Substituting in the second of Eqs. (*c*) and integrating it, we obtain

$$v = \frac{2\nu P}{\pi E}\sin\theta + \frac{2P}{\pi E}\log r\sin\theta - \int f(\theta)\,d\theta + F(r) \tag{e}$$

in which $F(r)$ is a function of r only. Substituting (*d*) and (*e*) in the third of Eqs. (*c*), we conclude that

$$f(\theta) = -\frac{(1-\nu)P}{\pi E}\theta\sin\theta + A\sin\theta + B\cos\theta \qquad F(r) = Cr \tag{f}$$

where A, B, and C are constants of integration which are to be determined from the conditions of constraint. The expressions for the dis-

[1] F. G. Maunsell, *Phil. Mag.*, vol. 21, p. 765, 1936.

[2] Theoretical results and photoelastic measurements for tension and bending, by several authors, are compared by M. Isida, *Sci. Papers Fac. Eng., Tokushima Univ., Japan*, (in English) vol. 4, no. 1, pp. 67–69, January, 1953. These include M. M. Frocht, R. Guernsey, Jr., and D. Landsberg, *J. Appl. Mech.*, vol. 19, p. 124, 1952; C. B. Ling, *J. Appl. Mech.*, pp. 141–146, and vol. 14, pp. 275–280, 1947; H. Neuber, "Kerbspannungslehre," pp. 35–37, 1937 (1st ed.) or pp. 42–44, Springer-Verlag, OHG, Berlin, 1958 (2d ed.). See also H. Poritsky, H. D. Snively, and C. R. Wylie, *J. Appl. Mech.*, vol. 6, p. 63, 1939.

placements, from Eqs. (*d*) and (*e*), are

$$u = -\frac{2P}{\pi E}\cos\theta\log r - \frac{(1-\nu)P}{\pi E}\theta\sin\theta + A\sin\theta + B\cos\theta$$

$$v = \frac{2\nu P}{\pi E}\sin\theta + \frac{2P}{\pi E}\log r\sin\theta - \frac{(1-\nu)P}{\pi E}\theta\cos\theta + \frac{(1-\nu)P}{\pi E}\sin\theta + A\cos\theta - B\sin\theta + Cr \qquad (g)$$

Assume that the constraint of the semi-infinite plate (Fig. 53) is such that the points on the x axis have no lateral displacement. Then $v = 0$, for $\theta = 0$, and we find from the second of Eqs. (*g*) that $A = 0$, $C = 0$. With these values of the constants of integration the vertical displacements of points on the x axis are

$$(u)_{\theta=0} = -\frac{2P}{\pi E}\log r + B \qquad (h)$$

To find the constant B let us assume that a point of the x axis at a distance d from the origin does not move vertically. Then from Eq. (*h*) we find

$$B = \frac{2P}{\pi E}\log d$$

Having the values of all the constants of integration, the displacements of any point of the semi-infinite plate can be calculated from Eqs. (*g*).

Let us consider, for instance, the displacements of points on the straight boundary of the plate. The horizontal displacements are obtained by putting $\theta = \pm\pi/2$ in the first of Eqs. (*g*). We find

$$(u)_{\theta=\pi/2} = -\frac{(1-\nu)P}{2E} \qquad (u)_{\theta=-\pi/2} = -\frac{(1-\nu)P}{2E} \qquad (70)$$

The straight boundary on each side of the origin thus has a constant displacement (70), at all points, directed toward the origin. We may regard such a displacement as a physical possibility, if we remember that around the point of application of the load P we removed the portion of material bounded by a cylindrical surface of a small radius (Fig. 53*b*) within which portion the equations of elasticity do not hold. Actually, of course, this material would be plastically deformed and thus could permit displacement (70) along the straight boundary. The vertical displacements on the straight boundary are obtained from the second of Eqs. (*g*). Remembering that v is positive if the displacement is in the direction of increasing θ, and that the deformation is symmetrical with respect to the x axis, we find for the vertical displacements in the downward direction at a distance r from the origin

$$(v)_{\theta=-(\pi/2)} = -(v)_{\theta=\pi/2} = \frac{2P}{\pi E}\log\frac{d}{r} - \frac{(1+\nu)P}{\pi E} \tag{71}$$

At the origin this equation gives an infinitely large displacement. To remove this difficulty, we must assume as before that a portion of material around the point of application of the load is cut out by a cylindrical surface of small radius. For other points of the boundary, Eq. (71) gives finite displacements.

37 | Any Vertical Loading of a Straight Boundary

The curves for σ_x and τ_{xy} of the preceding article (Fig. 54) can be used as *influence lines*. We assume that these curves represent the stresses for P equal to a unit force, say 1 lb. Then for any other value of the force P the stress σ_x at any point H of the plane mn is obtained by multiplying the ordinate $\overline{HK}$ by P.

If several vertical forces $P, P_1, P_2, \ldots$, act on the horizontal straight boundary AB of the semi-infinite plate, the stresses on the horizontal plane mn are obtained by superposing the stresses produced by each of these forces. For each of them, the σ_x and τ_{xy} curves are obtained by shifting the σ_x and τ_{xy} curves, constructed for P, to the new origins $O_1, O_2, \ldots$. From this it follows that the stress σ_x produced, for instance, by the force P_1 on the plane mn at the point D is obtained by multiplying the ordinate $\overline{H_1K_1}$ by P_1. In the same manner the σ_x stress at D produced by P_2 is $\overline{H_2K_2} \cdot P_2$, and so on. The total normal stress at D on the plane mn produced by $P, P_1, P_2, \ldots$, is

$$\sigma_x = \overline{DD_1} \cdot P + \overline{H_1K_1} \cdot P_1 + \overline{H_2K_2} \cdot P_2 + \cdots$$

Hence the σ_x curve shown in Fig. 54 is the *influence line* for the normal stress σ_x at the point D. In the same manner, we conclude that the τ_{xy} curve is the influence line for the shearing stress on the plane mn at the point D.

Having these curves, the stress components at D for any kind of vertical loading of the edge AB of the plate can easily be obtained.

If instead of concentrated forces we have a uniform load of intensity q, distributed over a portion $\overline{ss}$ of the straight boundary (Fig. 54), the normal stress σ_x produced by this load at the point D is obtained by multiplying by q the corresponding *influence area* shaded in the figure.

The problem of the uniformly distributed load can be solved in another manner by means of a stress function in the form

$$\phi = Ar^2\theta \tag{a}$$

in which A is a constant. The corresponding stress components are

$$\begin{aligned}\sigma_r &= \frac{1}{r}\frac{\partial \phi}{\partial r} + \frac{1}{r^2}\frac{\partial^2 \phi}{\partial \theta^2} = 2A\theta \\ \sigma_\theta &= \frac{\partial^2 \phi}{\partial r^2} = 2A\theta \\ \tau_{r\theta} &= -\frac{\partial}{\partial r}\left(\frac{1}{r}\frac{\partial \phi}{\partial \theta}\right) = -A\end{aligned} \tag{b}$$

Applying this to the semi-infinite plate we arrive at the load distribution shown in Fig. 58a. On the straight edge of the plate there acts a uniformly distributed shearing force of intensity $-A$ and a uniformly distributed normal load of the intensity $A\pi$, abruptly changing sign at the origin O. The directions of the forces follow from the positive directions of the stress components acting on an element C.

By shifting the origin to O_1 and changing the sign of stress function

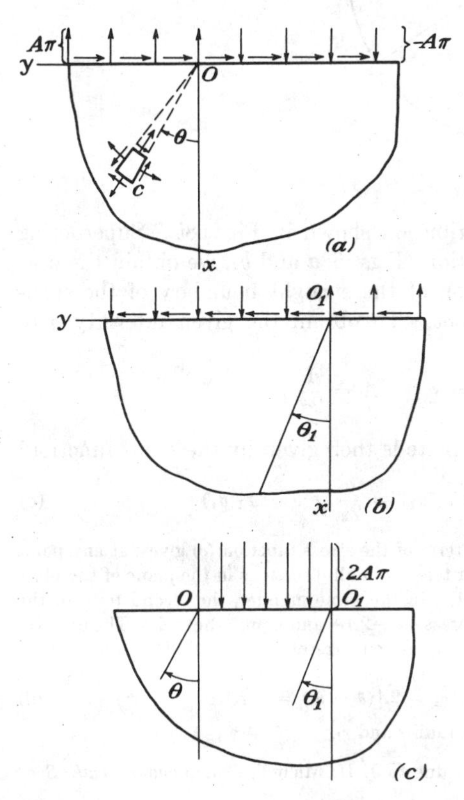

Fig. 58

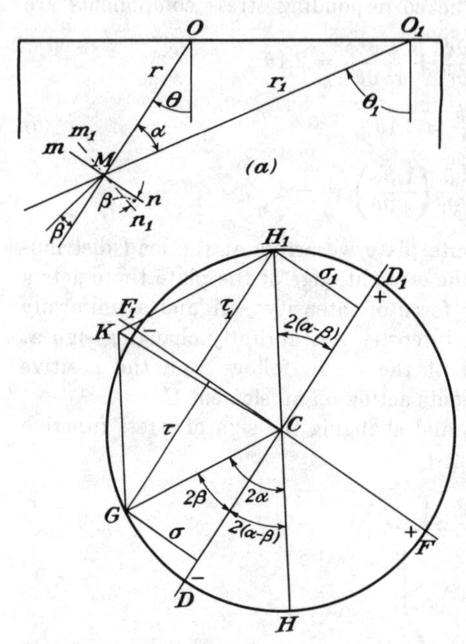

Fig. 59

ϕ, we arrive at the load distribution shown in Fig. 58*b*. Superposing the two cases of load distribution (Figs. 58*a* and *b*), we obtain the case of uniform loading of a portion of the straight boundary of the semi-infinite plate shown in Fig. 58*c*. To obtain the given intensity q of uniform load, we take

$$2A\pi = q \qquad A = \frac{1}{2\pi} q$$

The stress at any point of the plate is then given by the stress function[1]

$$\phi = A(r^2\theta - r_1{}^2\theta_1) = \frac{q}{2\pi}(r^2\theta - r_1{}^2\theta_1) \tag{c}$$

From Eqs. (*b*) we see that the first term of the stress function (*c*) gives, at any point M of the plate (Fig. 59*a*), a uniform tension in all directions in the plane of the plate equal to $2A\theta$ and a pure shear $-A$. In the same manner, the second term of the stress function gives a uniform compression $-2A\theta_1$ and a pure shear A. The uniform tension and compression can be simply added together and we find a uniform compressive stress

$$p = 2A\theta - 2A\theta_1 = 2A(\theta - \theta_1) = -2A\alpha \tag{d}$$

in which α is the angle between the radii r and r_1.

[1] This solution of the problem is due to J. H. Michell, *Proc. London Math. Soc.*, vol. 34, p. 134, 1902.

In superposing the two kinds of pure shear, one corresponding to the direction r and the other to the direction r_1, we shall use Mohr's circle (Fig. 59b), which in this case has a radius equal to the numerical value of the pure shears A. By taking the two diameters, DD_1 parallel to r and FF_1 perpendicular to r, as the τ and σ axes, we have a representation of the pure shear corresponding to the r direction. The radii CF and CF_1 represent the principal stresses A and $-A$ making angles $\pi/4$ with r at the point M, corresponding to this pure shear, and the radius CD represents the shearing stress $-A$ on the plane mn perpendicular to r. For any plane m_1n_1 inclined at an angle β to mn (Fig. 59a), the stress components are given by the coordinates σ and τ of the point G of the circle, with the angle GCD equal to 2β.

The same circle can be used also to get the stress components due to pure shear in the direction r_1 (see page 21). Considering again the plane m_1n_1, and noting that the normal to this plane makes an angle $\alpha - \beta$ with the direction r_1 (Fig. 59a), it appears that the stress components are given by the coordinates of the point H of the circle. To take care of the sign of the pure shear corresponding to the r_1 direction, we must change the signs of the stress components, and we obtain in this manner the point H_1 on the circle. The total stress acting on the plane m_1n_1 is given by the vector CK, the components of which give the normal stress $-(\sigma + \sigma_1)$ and the shearing stress $\tau_1 - \tau$. The vector CK has the same magnitude for all values of β, since the lengths of its components CH_1 and CG, and the angle between them, $\pi - 2\alpha$, are independent of β. Hence, by combining two pure shears we obtain again a pure shear (see page 22).

When $\tau_1 - \tau = 0$, the angle β determines the direction of one of the principal stresses at M. From the figure we see that τ and τ_1 are numerically equal if

$$2\beta = 2(\alpha - \beta),$$

from which $\beta = \alpha/2$. The direction of the principal stress therefore bisects the angle between the radii r and r_1. The magnitudes of the principal stresses are therefore

$$\pm 2\sigma = \pm 2A \sin 2\beta = \pm 2A \sin \alpha \tag{e}$$

Combining this with the uniform compression (d) we find, for the total values of the principal stresses at any point M,

$$-2A(\alpha + \sin \alpha) \qquad -2A(\alpha - \sin \alpha) \tag{f}$$

Along any circle through O and O_1 the angle α remains constant, and so the principal stresses (f) are also constant. At the boundary, between the points O and O_1 (Fig. 59a), the angle α is equal to π, and we find, from (f), that both principal stresses are equal to $-2\pi A = -q$. For the remaining portions of the boundary $\alpha = 0$, and both principal stresses are zero.

Hence if an arbitrary load distribution (Fig. 60) is regarded as composed of a

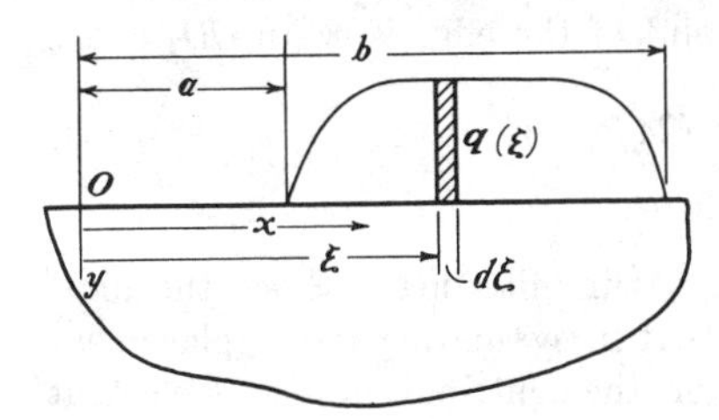

Fig. 60

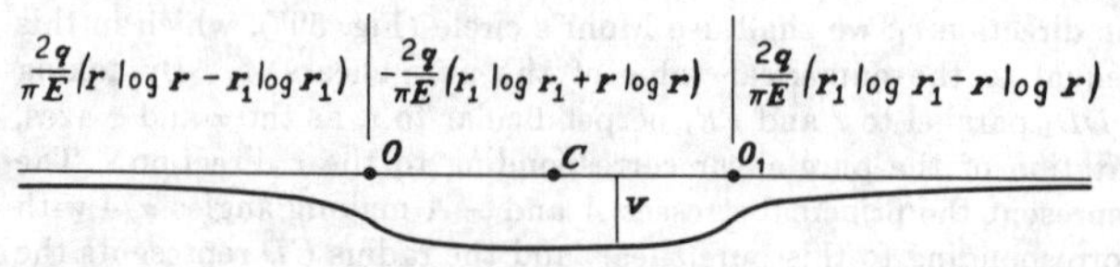

Fig. 61

large number of loads of varying intensities on short elements of the boundary, the horizontal stress σ_x under one such load element (as indicated in Fig. 60) is entirely due to that element, and

$$\sigma_x = \sigma_y = -q \tag{g}$$

all along the straight boundary.

Several other cases of distributed load on a straight boundary of the semi-infinite plate were discussed by S. D. Carothers.[1] Another manner of solving this problem will be discussed later (see page 141).

The *displacements* corresponding to the stress components given by Eqs. (*b*) are easily found by direct integration for u and v in the same manner as in Art. 31. Omitting rigid-body terms, the results are

$$u = \frac{2A}{E}(1 - \nu)r\theta \qquad v = -\frac{4A}{E}r\log r \tag{h}$$

Applying these in the superposition represented by Eq. (*c*), we can find expressions for the downward vertical displacement of each point of the originally straight horizontal edge of the plate. By definition, v is, in relation to its own $r\theta$ system, the displacement in the direction θ-increasing. To find the *downward* edge displacement for Fig. 58*c* we take v for any point to the right of O, and $-v$ for any point to the left. The contribution from the system based on O_1, corresponding to the term $-r_1^2\theta_1$ in Eq. (*c*), changes sign similarly at O_1. The plane-stress result for the downward displacement for Fig. 58*c* is shown in Fig. 61. An arbitrary rigid-body displacement can of course be added. The expressions in Fig. 61 as they stand make the slope of the edge zero at the midpoint and also at infinity. At O and O_1 the slope is unbounded, and in this sense these are singular points (compare Prob. 18, page 147).

The displacement at C, the midpoint of the edge is, with $OO_1 = 2a$,

$$v_c = -\frac{2q}{\pi E}(2a\log a) \tag{i}$$

If we now consider this load as a load element in a nonuniform distribution (Fig. 60), the width $2a$ becomes infinitesimal. Since the limit of $a \log a$ is zero for $a \to 0$, we find that in evaluating the displacement under any load element in this manner, the contribution of this element

[1] *Proc. Roy. Soc.* (*London*), ser. A, vol. 97, p. 110, 1920.

itself may be disregarded. The displacement due to the load elements elsewhere (see Fig. 60) is obtained, for any point x on the edge $y = 0$, as

$$v(x) = -\frac{2}{\pi E}\int_{\xi=a}^{\xi=b} q(\xi)\log|x-\xi|\,d\xi \tag{j}$$

the symbol $|x - \xi|$ representing the (positive) distance between the load element at ξ and the point of observation at x. Again rigid-body displacement terms can be added.

The integrand is singular at $x = \xi$, that is, for the element of load, if any, over the point x. We have seen, however, that this element makes no contribution. The integral is therefore to be taken as the Cauchy principal value.

Equation (j) can be used also for finding the intensity q of load distribution, which produces a given deflection at the straight boundary. Assuming, for instance, that the deflection is constant along the loaded portion of the straight boundary (Fig. 62), it can be shown that the distribution of pressure along this portion is given by the equation[1]

$$q = \frac{P}{\pi\sqrt{a^2 - x^2}}$$

38 | Force Acting on the End of a Wedge

The simple radial stress distribution discussed in Art. 36 can also represent the stresses in a wedge due to a concentrated force at its apex. Let us consider a symmetrical case, as shown in Fig. 63. The thickness of

[1] M. Sadowsky, Z. *Angew. Math. Mech.*, vol. 8, p. 107, 1928. For nonuniform deflection imposed at a contact see (1) N. I. Muskhelishvili, "Some Basic Problems of the Mathematical Theory of Elasticity," (translated by J. R. M. Radok), Erven P. Noordhoff, NV, Groningen, Netherlands, 1963. (2) L. A. Galin "Contact Problems in the Theory of Elasticity," [translated by H. Moss, I. N. Sneddon (ed.)], Departments of Mathematical and Engineering Research, North Carolina State College, Raleigh, N.C., 1961. For the effects of local rounding at sharp corners, see J. N. Goodier and C. B. Loutzenheiser, *J. Appl. Mech.*, vol. 32, pp. 462–463, 1965.

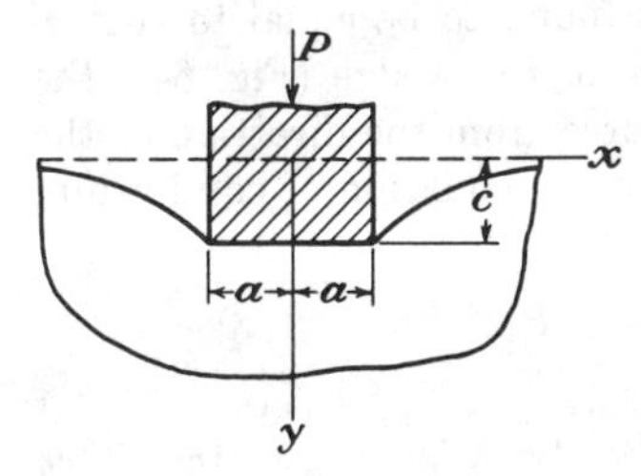

Fig. 62

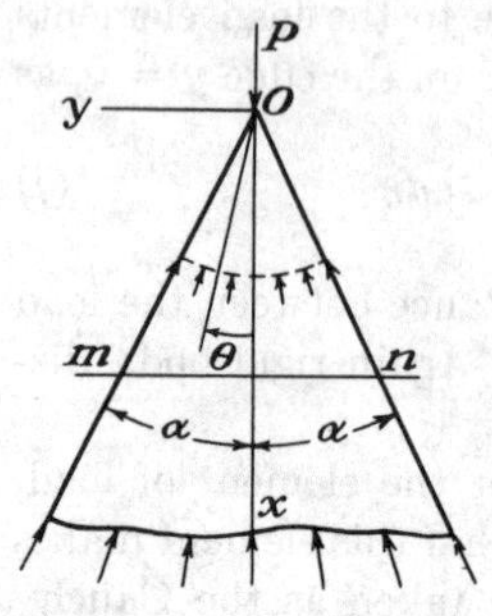

Fig. 63

the wedge in the direction perpendicular to the xy plane is taken as unity. The conditions along the faces, $\theta = \pm\alpha$, of the wedge are satisfied by taking for the stress components the values

$$\sigma_r = -\frac{kP\cos\theta}{r} \qquad \sigma_\theta = 0 \qquad \tau_{r\theta} = 0 \tag{a}$$

The constant k will now be adjusted so as to satisfy the condition of equilibrium at the point O. Making the resultant of the pressures on the cylindrical surface (shown by the dotted line) equal to $-P$ we find

$$-2\int_0^\alpha \frac{kP\cos^2\theta}{r}\, r\, d\theta = -kP\left(\alpha + \frac{1}{2}\sin 2\alpha\right) = -P$$

from which

$$k = \frac{1}{\alpha + \frac{1}{2}\sin 2\alpha}$$

Then, from Eqs. (a),[1]

$$\sigma_r = -\frac{P\cos\theta}{r(\alpha + \frac{1}{2}\sin 2\alpha)} \tag{72}$$

By making $\alpha = \pi/2$ we arrive at solution (65) for a semi-infinite plate, which has already been discussed. It may be seen that the distribution of normal stresses over any cross section mn is not uniform, and the ratio of the normal stress at the points m or n to the maximum stress at the center of the cross section is found to be equal to $\cos^4\alpha$.

If the force is perpendicular to the axis of the wedge (Fig. 64), the same solution (a) can be used if θ is measured from the direction of the force. The constant factor k is found from the equation of equilibrium

$$\int_{\frac{\pi}{2}-\alpha}^{\frac{\pi}{2}+\alpha} \sigma_r\, r\cos\theta\, d\theta = -P$$

[1] This solution is due to Michell, *loc. cit.* See also A. Mesnager, *Ann. Ponts Chaussées*, 1901.

from which

$$k = \frac{1}{\alpha - \frac{1}{2} \sin 2\alpha}$$

and the radial stress is

$$\sigma_r = -\frac{P \cos \theta}{r(\alpha - \frac{1}{2} \sin 2\alpha)} \tag{73}$$

The normal and shearing stresses over any cross section mn are

$$\begin{aligned} \sigma_y &= -\frac{Pyx \sin^4 \theta}{y^3(\alpha - \frac{1}{2} \sin 2\alpha)} \\ \tau_{xy} &= -\frac{Px^2 \sin^4 \theta}{y^3(\alpha - \frac{1}{2} \sin 2\alpha)} \end{aligned} \tag{b}$$

In the case of a small angle α, we can put

$$2\alpha - \sin 2\alpha = \frac{(2\alpha)^3}{6}$$

Then writing I for the moment of inertia of the cross section mn, we find from (b) that

$$\begin{aligned} \sigma_y &= -\frac{Pyx}{I} \cdot \left[\left(\frac{\tan \alpha}{\alpha}\right)^3 \sin^4 \theta\right] \\ \tau_{xy} &= -\frac{Px^2}{I} \cdot \left[\left(\frac{\tan \alpha}{\alpha}\right)^3 \sin^4 \theta\right] \end{aligned} \tag{c}$$

For small values of α, the factor $(\tan \alpha/\alpha)^3 \sin^4 \theta$ can be taken as nearly unity. Then the expression for σ_y coincides with that given by the elementary beam formula. The maximum shearing stress occurs at the points m and n and is twice as great as that given by the elementary theory for the centroid of a rectangular cross section of a beam.

Since we have solutions for the two cases represented in Figs. 63 and 64, we can deal with any direction of the force P in the xy plane by resolving the force into two components and using the method of super-

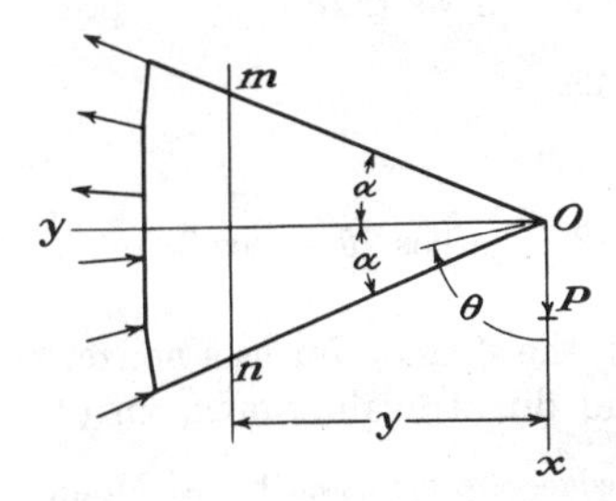

Fig. 64

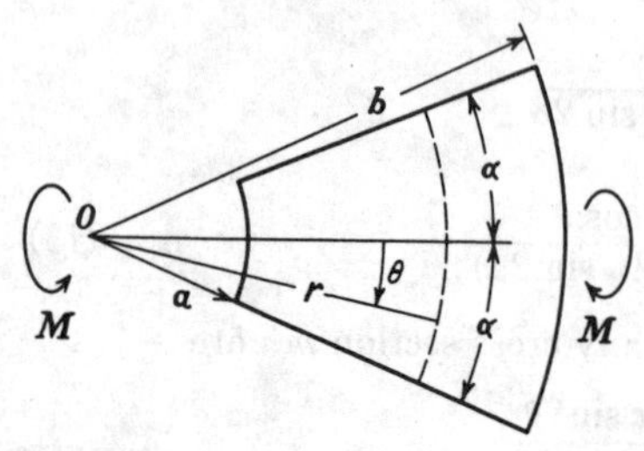

Fig. 65

position.[1] It should be noted that solutions (72) and (73) represent an exact solution only in the case when, at the supported end, the wedge is held by radially directed forces distributed in the manner given by the solutions. Otherwise the solutions are accurate only at points at large distances from the supported end.

39 | Bending Couple Acting on the End of a Wedge

The stress function

$$\phi_1 = C_1 \sin 2\theta \tag{a}$$

gives

$$\sigma_r = -4C_1 \frac{1}{r^2} \sin 2\theta \qquad \sigma_\theta = 0 \qquad \tau_{r\theta} = 2C_1 \frac{1}{r^2} \cos 2\theta \tag{b}$$

The function

$$\phi_2 = C_2\theta \tag{c}$$

gives

$$\sigma_r = 0 \qquad \sigma_\theta = 0 \qquad \tau_{r\theta} = \frac{C_2}{r^2} \tag{d}$$

Combining the two, we have

$$\sigma_r = -4C_1 \frac{1}{r^2} \sin 2\theta \qquad \sigma_\theta = 0 \qquad \tau_{r\theta} = \frac{1}{r^2}(2C_1 \cos 2\theta + C_2) \tag{e}$$

Evidently the faces $\theta = \pm\alpha$ will be free of load if we take

$$C_2 = -2C_1 \cos 2\alpha$$

Then the stress is

$$\sigma_r = -4C_1 \frac{1}{r^2} \sin 2\theta \qquad \sigma_\theta = 0 \qquad \tau_{r\theta} = 2C_1 \frac{1}{r^2}(\cos 2\theta - \cos 2\alpha) \tag{f}$$

On a cylindrical surface of radius r (Fig. 65), the stress σ_r forms a nonzero horizontal force. But so also does $\tau_{r\theta}$, and direct evaluation from (*f*)

[1] Several examples of stress distribution in wedges are discussed by A. Miura, "Spannungskurven in Rechteckigen und Keilförmigen Trägern," Berlin, 1928.

shows that the total force is zero. The resultant is a couple M given, for unit thickness, by

$$M = \int_{-\alpha}^{\alpha} \tau_{r\theta} r^2 \, d\theta = 2C_1(\sin 2\alpha - 2\alpha \cos 2\alpha) \tag{g}$$

Evidently the stress (f), in the material between the ends $r = a$, $r = b$, corresponds to bending by couples M, Fig. 65, C_1 being given in terms of M and α by (g). The inner radius a can be as small as we please.[1]

The expression in parentheses in (g)

$$\sin 2\alpha - 2\alpha \cos 2\alpha$$

cannot vanish for any value of α appropriate to a wedge. For $0 < 2\alpha < 2\pi$ it vanishes only when $2\alpha = 257.4°$, but then the region is almost three-quarters of a ring (Fig. 66). The couple M in (g) becomes zero, the contributions from ϕ_1 and ϕ_2 being equal and opposite. The loading on the arc $r = a$ is self-equilibrating, and so of course is the loading on $r = b$.

When $2\alpha > 257.4°$ the couple M is again nonzero, and (g) determines C_1. But there are now other ways of loading the arc $r = a$, preserving a couple M, which will give stress diminishing with r *less* rapidly[2] than r^{-2} [as in (f)]; and, in fact, this becomes true as soon as 2α exceeds 180°. The usefulness of (f), with (g), is limited to wedge regions of rather small angle, in which the effects of a change of load distribution on $r = a$ or $r = b$ can be localized.

40 | Concentrated Force Acting on a Beam

The problem of stress distribution in a beam subjected to the action of a concentrated force is of great practical interest. It was shown before (Art. 23) that in continuously loaded beams of narrow rectangular cross

[1] The solution was given by S. D. Carothers, *Proc. Roy. Soc. Edinburgh*, sect. A, vol. 23, pp. 292–306, 1912, and independently by C. E. Inglis, *Trans. Inst. Nav. Arch. London*, vol. 64, p. 253, 1922.

[2] E. Sternberg and W. T. Koiter, *J. Appl. Mech.*, vol. 25, pp. 575–581, 1958.

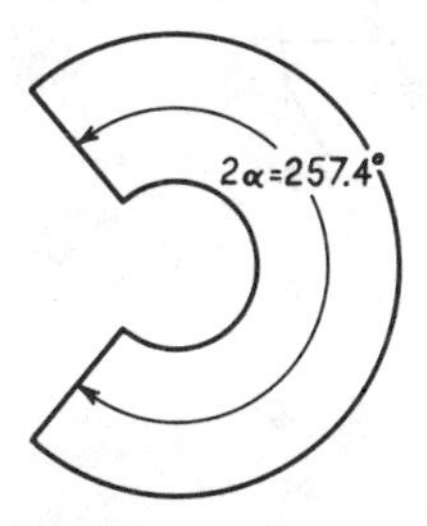

Fig. 66

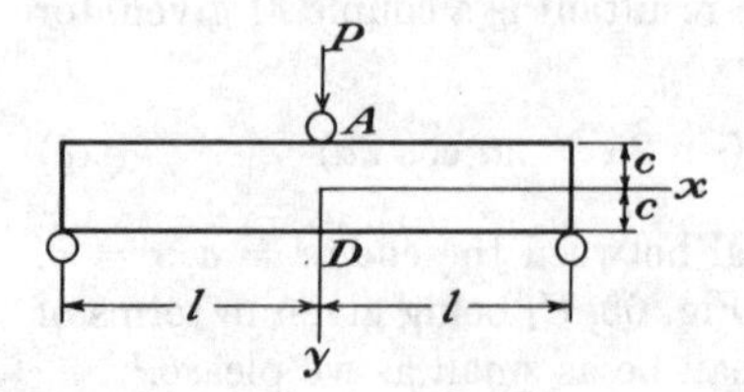

Fig. 67

section the stress distribution is obtained with satisfactory accuracy by the usual elementary theory of bending. Near the point of application of a concentrated force, however, a serious local perturbation in stress distribution should be expected, and a further investigation of the problem is necessary. The first study of these local stresses was made experimentally by Carus Wlison.[1] Experimenting with a rectangular beam of glass on two supports (Fig. 67) loaded at the middle, and using polarized light (see page 151), he showed that at the point A, where the load is applied, the stress distribution approaches that produced in a semi-infinite plate by a normal concentrated force. Along the cross section AD the normal stress σ_x does not follow a linear law, and at the point D, opposite to A, the tensile stress is smaller than would be expected from the ele-

[1] *Loc. cit.*

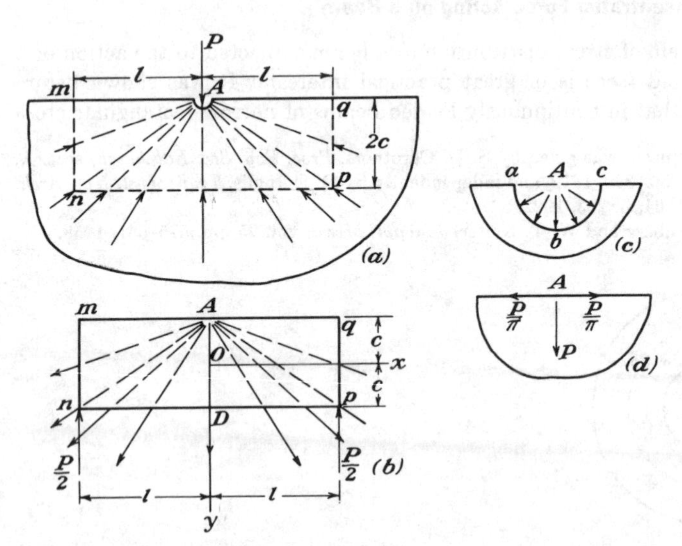

Fig. 68

mentary beam theory. These results were explained on the basis of certain empirical assumptions by G. G. Stokes.[1] The system represented in Fig. 67 can be obtained by superposing two systems shown in Fig. 68. The radial compressive stresses acting on the sections *mn*, *np*, and *pq* of a semi-infinite plate (Fig. 68*a*) are removed by equal radial tensile stresses acting on the sides of the rectangular beam supported at n and p (Fig. 68*b*). The stresses in this beam should be superposed on the stresses in the semi-infinite plate in order to get the case discussed by Stokes.

In calculating the stresses in the beam, the elementary beam formula will be applied. The bending moment at the middle cross section AD of the beam is obtained by taking the moment of the reaction $P/2$ and subtracting the moment of all the radially directed tensile forces applied to one-half of the beam. This latter moment is easily calculated if we observe that the radially distributed tensile forces are statically equivalent to the pressure distribution over the quadrant *ab* of the cylindrical surface *abc* at the point A (Fig. 68*c*) or, using Eq. (65), are equivalent to a horizontal force P/π and a vertical force $P/2$, applied at A (Fig. 68*d*). Then the bending moment, that is, the moment about the point O, is

$$\frac{P}{2}l - \frac{P}{\pi}c$$

and the corresponding bending stresses are[2]

$$\sigma_x' = \frac{P}{I}\left(\frac{l}{2} - \frac{c}{\pi}\right)y = \frac{3P}{2c^3}\left(\frac{l}{2} - \frac{c}{\pi}\right)y$$

To these bending stresses the uniformly distributed tensile stress $P/2\pi c$ produced by the tensile force P/π should be added. The normal stresses over the cross section AD, as obtained in this elementary way, are therefore

$$\sigma_x = \frac{3P}{2c^3}\left(\frac{l}{2} - \frac{c}{\pi}\right)y + \frac{P}{2\pi c}$$

This coincides with the formula given by Stokes. Its validity within appropriate limitations has been confirmed experimentally[3] by modern photoelastic techniques.

A better approximation is obtained if we observe that a continuously

[1] Wilson, *loc. cit.;* also G. G. Stokes, "Mathematical and Physical Papers," vol. 5, p. 238.

[2] As before we take P as the force per unit thickness of the plate.

[3] M. M. Frocht, "Photoelasticity," vol. 2, pp. 104–107, John Wiley & Sons, Inc., New York, 1948; C. Saad and A. W. Hendry, *Proc. Soc. Exptl. Stress Anal.*, vol. 18, pp. 192–198, 1961. For an application to stress in short beams due to impact see A. A. Betser and M. M. Frocht, *J. Appl. Mech.*, vol. 24, pp. 509–514, 1957.

distributed load is applied to the bottom of the beam (Fig. 68b) and use Eqs. (36′). The intensity of this load at the point D, from Eq. (65), is $P/\pi c$. Substituting this in (36′) and combining with the value of σ_x above, we obtain, as a second approximation,

$$\sigma_x = \frac{3P}{2c^3}\left(\frac{l}{2} - \frac{c}{\pi}\right)y + \frac{P}{2\pi c} + \frac{P}{\pi c}\left(\frac{y^3}{2c^3} - \frac{3}{10}\frac{y}{c}\right)$$
$$\sigma_y = \frac{P}{2\pi c} + \frac{P}{\pi c}\left(\frac{3y}{4c} - \frac{y^3}{4c^3}\right) \tag{a}$$

These stresses should be superposed on the stresses

$$\sigma_x = 0 \qquad \sigma_y = -\frac{2P}{\pi(c + y)} \tag{b}$$

as for a semi-infinite plate, in order to obtain the total stresses along the section AD.

A comparison with a more accurate solution, given below (see table, page 119), shows that Eqs. (a) and (b) give the stresses with very good accuracy at all points except the point D at the bottom of the beam, at which the correction to the simple beam formula is given as

$$-\frac{3P}{2\pi c} + \frac{P}{2\pi c} + \frac{1}{5}\frac{P}{\pi c} = -0.254\frac{P}{c}$$

while the more accurate solution gives only $-0.133(P/c)$.

A solution of the problem by means of trigonometric series was obtained by L. N. G. Filon.[1] He applied this solution to the case of concentrated loads and made calculations for several particular cases (see Art. 24), which are in good agreement with more recent investigations.

Further progress was made by H. Lamb,[2] who considered an infinite beam loaded at equal intervals by equal concentrated forces acting in the upward and downward directions alternately and obtained for several cases expressions for the deflection curves. These show that the elementary Bernoulli-Euler theory of bending is very accurate if the depth of the beam is small in comparison with its length. It was shown also that the correction for shearing force as given by Rankine's and Grashof's elementary theory (see page 50) is somewhat exaggerated and should be diminished to about 0.75 of its value.[3]

A more detailed study of the stress distribution and of the curvature near the point of application of a concentrated load was made by T. v. Kármán and F. Seewald.[4]

[1] L. N. G. Filon, *Trans. Roy. Soc. (London)*, ser. A, vol. 201, p. 63, 1903.
[2] *Atti IV Congr. Intern. Mat.*, vol. 3, p. 12, Rome, 1909.
[3] Filon came to the same conclusion in his paper (*loc. cit*).
[4] *Abhandl. Aerodynam. Inst., Tech. Hochschule, Aachen*, vol. 7, 1927.

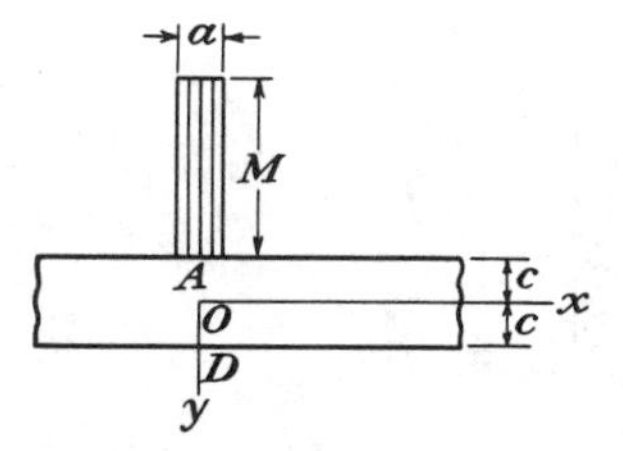

Fig. 69

Kármán considers an infinitely long beam and makes use of the solution for a semi-infinite plate with two equal and opposite couples acting on two neighboring points of its straight boundary (Fig. 57*b*). The stresses along the bottom of the beam which are introduced by this procedure can be removed by using a solution in the form of a trigonometric series (Art. 24) which, for an infinitely long beam, will be represented by a Fourier integral. In this manner Kármán arrives at the stress function

$$\phi = \frac{Ma}{\pi} \int_0^\infty \frac{(\alpha c \cosh \alpha c + \sinh \alpha c) \cdot \cosh \alpha y - \sinh \alpha c \sinh \alpha y \cdot \alpha y}{\sinh 2\alpha c + 2\alpha c} \cos \alpha x \, d\alpha$$
$$- \frac{Ma}{\pi} \int_0^\infty \frac{(\alpha c \sinh \alpha c + \cosh \alpha c) \cdot \sinh \alpha y - \cosh \alpha c \cosh \alpha y \cdot \alpha y}{\sinh 2\alpha c - 2\alpha c} \cos \alpha x \, d\alpha \qquad (c)$$

This function gives the stress distribution in the beam when the bending-moment diagram consists of a very narrow rectangle, as shown in Fig. 69. For the most general loading of the beam by vertical forces applied at the top of the beam,[1] the corresponding bending-moment diagram can be divided into elementary rectangles such as the one shown in Fig. 69, and the corresponding stress function will be obtained by integrating expression (*c*) along the length of the beam.

This method of solution was applied by Seewald to the case of a beam loaded by a concentrated force P (Fig. 67). He shows that the stress σ_x can be split into two parts: one that can be calculated by the usual elementary beam formula; and another that represents the local effect near the point of application of the load. This latter part, called σ_x', can be represented in the form $\beta(P/c)$, in which β is a numerical factor depending on the position of the point for which the local stresses are calculated. The values of this factor are given in Fig. 70. The two other stress components σ_y and τ_{xy} can also be represented in the form $\beta(P/c)$. The corresponding values of β are given in Figs. 71 and 72. It can be seen from the figures that the local stresses decrease very rapidly with increase of distance from the point of application of the load and at a distance equal to the depth of the beam are usually negligible. Using the values of the factor β for $x = 0$, the local stresses at five points of the cross section AD under the load (Fig. 67) are tabulated below. For comparison, the local stresses,[2]

[1] The case of a concentrated load applied halfway between the top and bottom of the beam was discussed by R. C. J. Howland, *Proc. Roy. Soc.* (*London*), vol. 124, p. 89, 1929 (see p. 131 below), and pairs of forces within the beam by K. Girkmann, *Ingenieur-Arch.*, vol. 13, p. 273, 1943. Concentrated longitudinal forces in the web of an I beam are considered by Girkmann in *Oesterr. Ingenieur-Arch.*, vol. 1, p. 420, 1946.

[2] That is, stresses that must be superposed on those obtained from the ordinary beam formula.

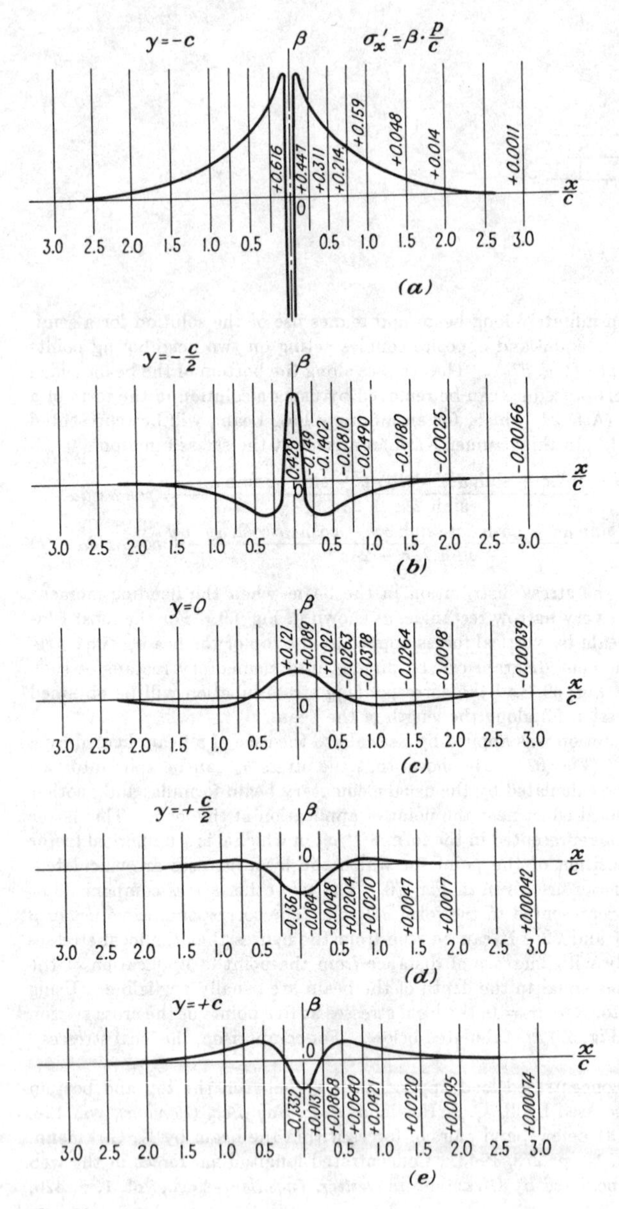

y=-c
β
σx' = β · p/c
+0.616
+0.447
+0.341
+0.311
+0.214
+0.159
+0.048
+0.014
+0.0011
0
x/c
3.0 2.5 2.0 1.5 1.0 0.5
0.5 1.0 1.5 2.0 2.5 3.0
(a)
y=-c/2
β
0.428
-0.149
-0.1402
-0.0810
-0.0490
-0.0180
-0.0025
-0.00166
0
x/c
3.0 2.5 2.0 1.5 1.0 0.5
0.5 1.0 1.5 2.0 2.5 3.0
(b)
y=0
β
+0.121
+0.089
+0.021
-0.0263
-0.0378
-0.0264
-0.0098
-0.00039
0
x/c
3.0 2.5 2.0 1.5 1.0 0.5
0.5 1.0 1.5 2.0 2.5 3.0
(c)
y=+c/2
β
0
-0.136
-0.084
-0.0048
+0.0204
+0.0210
+0.0047
+0.0017
+0.00042
x/c
3.0 2.5 2.0 1.5 1.0 0.5
0.5 1.0 1.5 2.0 2.5 3.0
(d)
y=+c
β
0
-0.1332
+0.0131
+0.0868
+0.0640
+0.0421
+0.0220
+0.0095
+0.00074
x/c
3.0 2.5 2.0 1.5 1.0 0.5
0.5 1.0 1.5 2.0 2.5 3.0
(e)

Fig. 70

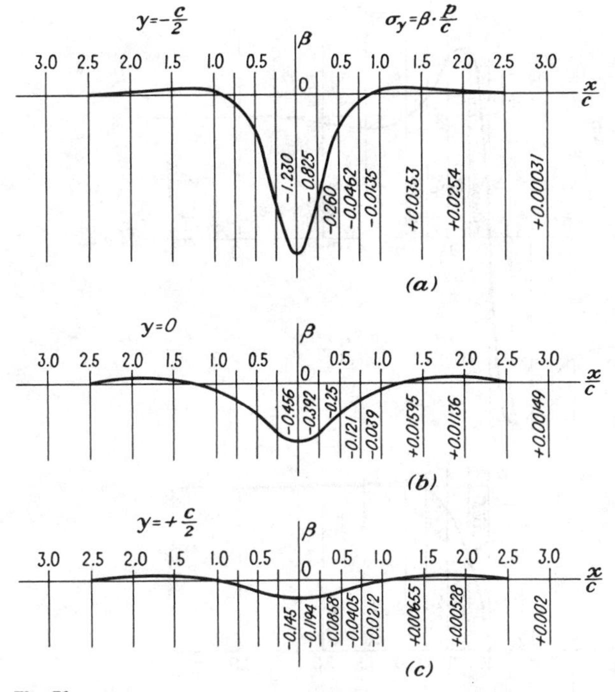

Fig. 71

as obtained from Eqs. (*a*) and (*b*) (page 116), are also given. It is seen that these equations give the local stresses with sufficient accuracy.

Knowing the stresses, the curvature and the deflection of the beam can be calculated without any difficulty. These calculations show that the curvature of the

Table of Factors β for the Midsection AD (Fig. 67)

$y =$	$-c$	$-\frac{c}{2}$	0	$\frac{c}{2}$	c
		Exact solution			
$\sigma_x' =$	.....	0.428	0.121	−0.136	−0.133
$\sigma_y =$	∞	−1.23	−0.456	−0.145	0
		Approximate solution			
$\sigma_x' =$	0.573	0.426	0.159	−0.108	−0.254
$\sigma_y =$	∞	−1.22	−0.477	−0.155	0

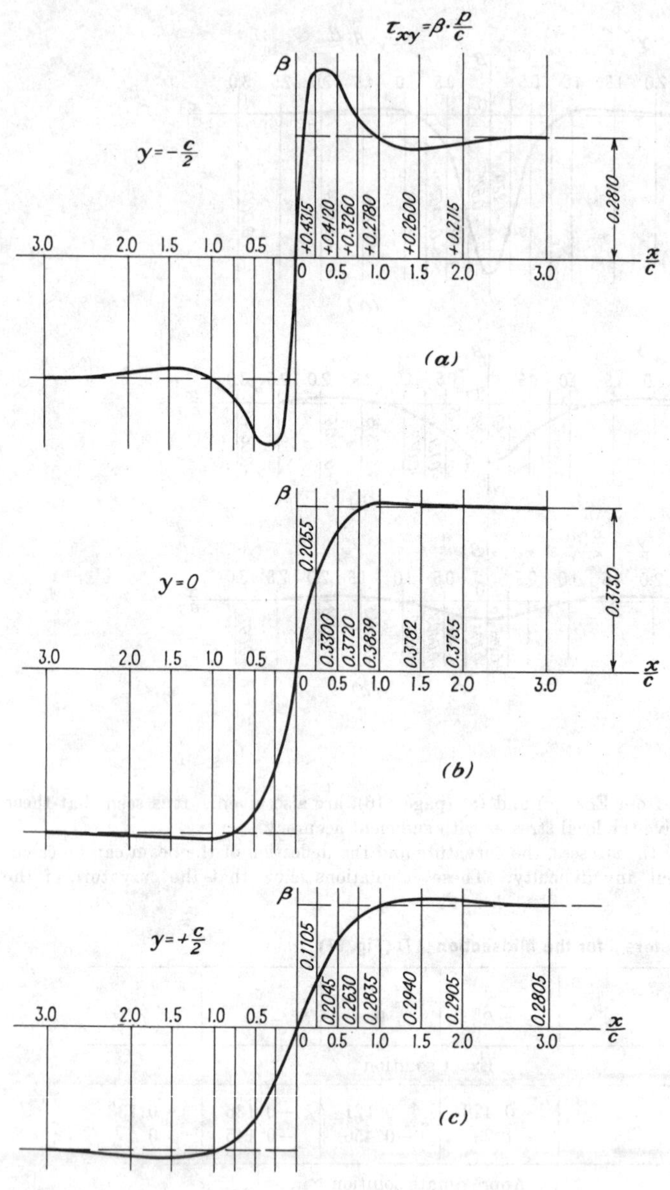

$\tau_{xy} = \beta \cdot \frac{p}{c}$
β
$y = -\frac{c}{2}$
+0.4315
+0.4120
+0.3260
+0.2780
+0.2600
+0.2715
0.2810
3.0
2.0
1.5
1.0
0.5
0
$\frac{x}{c}$
(a)
$y = 0$
0.2055
0.3300
0.3720
0.3839
0.3782
0.3755
0.3750
(b)
$y = +\frac{c}{2}$
0.1105
0.2045
0.2630
0.2835
0.2940
0.2905
0.2805
(c)

Fig. 72

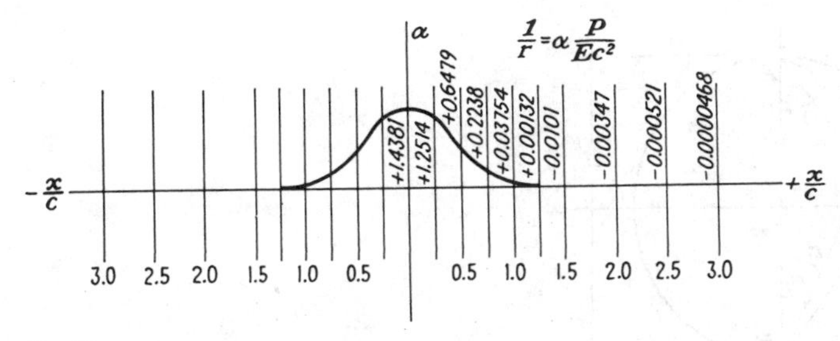

Fig. 73

deflection curve can also be split into two parts—one as given by the elementary beam theory and the other representing the local effect of the concentrated load P. This additional curvature of the centerline can be represented by the formula

$$\frac{1}{r} = \alpha \frac{P}{Ec^2} \tag{d}$$

in which α is a numerical factor varying along the length of the beam. Several values of this factor are given in Fig. 73. It is seen that at cross sections at a distance greater than half of the depth of the beam the additional curvature is negligible.

On account of this localized effect on the curvature, the two branches of the deflection curve AB and AC (Fig. 74) may be considered to meet at an angle equal to

$$\gamma = \frac{P}{c}\left(\frac{3}{4G} - \frac{3}{10E} - \frac{3\nu}{4E}\right) \tag{e}$$

The corresponding deflection at the middle is

$$\delta_1 = \frac{\gamma l}{4} = \frac{Pl}{4c}\left(\frac{3}{4G} - \frac{3}{10E} - \frac{3\nu}{4E}\right) \tag{f}$$

From this deflection a small further correction δ_2, removing the sharp change of slope at A, should be subtracted. This quantity was also calculated by Seewald and is equal to

$$\delta_2 = 0.21 \frac{P}{E}$$

Denoting now by δ_0 the deflection as calculated by using the elementary theory, the total deflection under the load is

$$\delta = \delta_0 + \delta_1 - \delta_2 = \frac{Pl^3}{48EI} + \frac{Pl}{4c}\left(\frac{3}{4G} - \frac{3}{10E} - \frac{3\nu}{4E}\right) - 0.21\frac{P}{E} \tag{74}$$

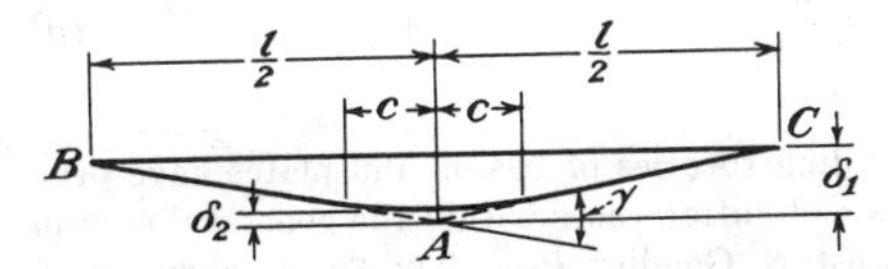

Fig. 74

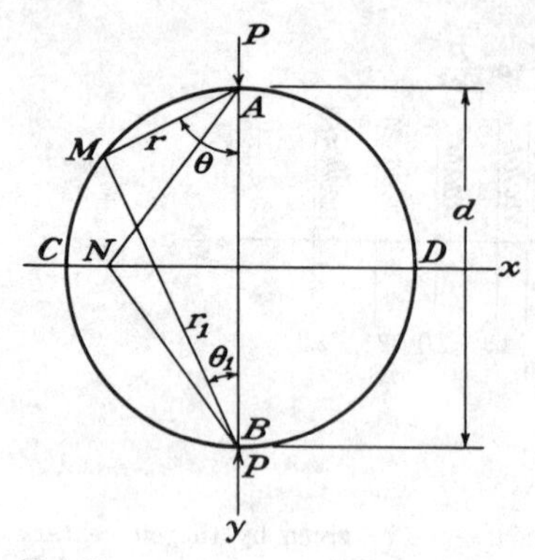

Fig. 75

Taking $\nu = 0.3$, this gives

$$\delta = \frac{Pl^3}{48EI}\left[1 + 2.85\left(\frac{2c}{l}\right)^2 - 0.84\left(\frac{2c}{l}\right)^3\right] \tag{74'}$$

The elementary Rankine-Grashof theory (see page 50) gives for this case

$$\delta = \frac{Pl^3}{48EI}\left[1 + 3.90\left(\frac{2c}{l}\right)^2\right] \tag{g}$$

It appears that Eq. (g) gives an exaggerated value for the correction due to shear.[1] In these formulas the deflection due to local deformation at the supports is not taken into account.

41 | Stresses in a Circular Disk

Let us begin with the simple case of two equal and opposite forces P acting along a diameter AB (Fig. 75). Assuming that each of the forces produces a simple radial stress distribution [Eqs. (65)], we can find what forces should be applied at the circumference of the disk in order to maintain such a stress distribution. At any point M of the circumference we have compressions in the directions of r and r_1 equal to $(2P/\pi)(\cos\theta/r)$ and $(2P/\pi)(\cos\theta_1/r_1)$, respectively. Since r and r_1 are perpendicular to each other and

$$\frac{\cos\theta}{r} = \frac{\cos\theta_1}{r_1} = \frac{1}{d} \tag{a}$$

[1] Corrections to the elementary bending theories of beams and plates have been investigated by deriving these theories as limiting cases within the general theory of three-dimensional linear elasticity. See J. N. Goodier, *Proc. Roy. Soc. Canada*, ser. 3, sec. 3, vol. 32, pp. 1–25, 1938.

where d is the diameter of the disk, we conclude that the two principal stresses at M are two equal compressive stresses of magnitude $2P/\pi d$. Hence the same compressive stress is acting on any plane through M perpendicular to the plane of the disk, and normal compressive forces of the constant intensity $2P/\pi d$ should be applied to the circumference of the disk in order to maintain the assumed pair of simple radial stress distributions.

If the boundary of the disk is free from external forces, the stress at any point is therefore obtained by superposing a uniform tension in the plane of the disk of the magnitude $2P/\pi d$ on the above two simple radial stress distributions. Let us consider the stress on the horizontal diametral section of the disk at N. From symmetry it can be concluded that there will be no shearing stress on this plane. The normal stress produced by the two equal radial compressions is

$$-2\left(\frac{2P}{\pi}\frac{\cos\theta}{r}\right)\cos^2\theta$$

in which r is the distance AN, and θ the angle between AN and the vertical diameter. Superposing on this the uniform tension $2P/\pi d$, the total normal stress on the horizontal plane at N is

$$\sigma_y = -\frac{4P}{\pi}\frac{\cos^3\theta}{r} + \frac{2P}{\pi d}$$

or, using the fact that

$$\cos\theta = \frac{d}{\sqrt{d^2 + 4x^2}}$$

we find

$$\sigma_y = \frac{2P}{\pi d}\left[1 - \frac{4d^4}{(d^2 + 4x^2)^2}\right] \tag{b}$$

The maximum compressive stress along the diameter CD is at the center of the disk, where

$$\sigma_y = -\frac{6P}{\pi d}$$

At the ends of the diameter the compressive stress σ_y vanishes.

Consider now the case of two equal and opposite forces acting along a chord AB (Fig. 76). Assuming again two simple radial distributions radiating from A and B, the stress on a plane tangential to the circumference at M is obtained by superposing the two radial compressions $(2P/\pi)(\cos\theta/r)$ and $(2P/\pi)(\cos\theta_1/r_1)$ acting in the directions r and r_1, respectively. The normal MN to the tangent at M is the diameter of the disk; hence MAN and MBN are right-angled triangles and the angles which the normal MO makes with r and r_1 are $\pi/2 - \theta_1$ and $\pi/2 - \theta$, respectively. The normal and shearing stresses on an element of the

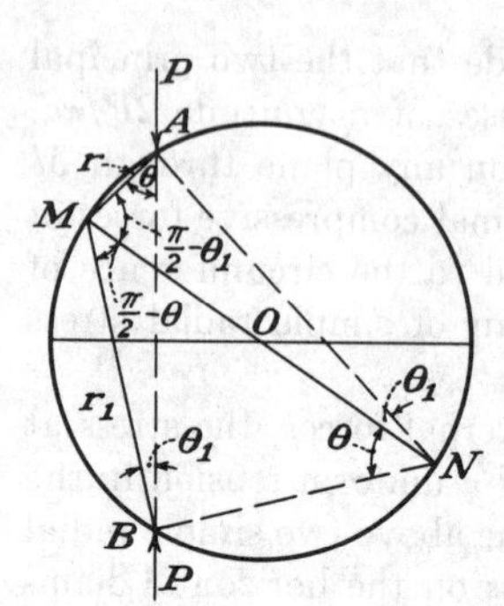

Fig. 76

boundary at M are then

$$\begin{aligned}\sigma &= -\frac{2P}{\pi}\frac{\cos\theta}{r}\cos^2\left(\frac{\pi}{2}-\theta_1\right) - \frac{2P}{\pi}\frac{\cos\theta_1}{r_1}\cos^2\left(\frac{\pi}{2}-\theta\right)\\ &= -\frac{2P}{\pi}\left(\frac{\cos\theta\sin^2\theta_1}{r} + \frac{\cos\theta_1\sin^2\theta}{r_1}\right)\\ \tau &= -\frac{2P}{\pi}\left(\frac{\cos\theta}{r}\sin\theta_1\cos\theta_1 - \frac{\cos\theta_1}{r_1}\sin\theta\cos\theta\right)\end{aligned} \tag{c}$$

These equations can be simplified if we observe that, from the triangles MAN and MBN,

$$r = d\sin\theta_1 \qquad r_1 = d\sin\theta$$

Substituting in Eqs. (c), we find

$$\sigma = -\frac{2P}{\pi d}\sin(\theta+\theta_1) \qquad \tau = 0 \tag{d}$$

From Fig. 76 it may be seen that $\sin(\theta+\theta_1)$ remains constant around the boundary. Hence uniformly distributed compressive forces of the intensity $2P/\pi d \sin(\theta+\theta_1)$ should be applied to the boundary in order to maintain the assumed radial stress distributions. To obtain the solution for a disk with its boundary free from uniform compression it is only necessary to superpose on the above two simple radial distributions a uniform tension of the intensity $2P/\pi d \sin(\theta+\theta_1)$.

The problem of the stress distribution in a disk can be solved for the more general case when any system of forces in equilibrium is acting on the boundary of the disk.[1]

[1] The problems discussed in this article were solved by H. Hertz, *Z. Math. Physik*, vol. 28, 1883, or "Gesammelte Werke," vol. 1, p. 283; and J. H. Michell, *Proc. London Math. Soc.*, vol. 32, p. 44, 1900, and vol. 34, p. 134, 1901. The problem corresponding to Fig. 75 when the disk is replaced by a rectangle is considered by J. N. Goodier, *Trans. ASME*, vol. 54, p. 173, 1932, including the effects of distribution of the load over small segments of the boundary.

Let us take one of these forces, acting at A in the direction of the chord AB (Fig. 77). Assuming again a simple radial stress distribution, we have at point M a simple radial compression of the magnitude $(2P/\pi) \cos \theta_1/r_1$ acting in the direction of AM. Let us take as origin of polar coordinates the center O of the disk, and measure θ as shown in the figure. The normal and the shearing components of the stress acting on an element tangential to the boundary at M can then be easily calculated if we observe that the angle between the normal MO to the element and the direction r_1 of the compression is equal to $\pi/2 - \theta_2$. Then

$$\sigma_r = -\frac{2P}{\pi}\frac{\cos \theta_1}{r_1}\sin^2 \theta_2$$

$$\tau_{r\theta} = -\frac{2P}{\pi}\frac{\cos \theta_1}{r_1}\sin \theta_2 \cos \theta_2 \tag{e}$$

Since, from the triangle AMN, $r_1 = d \sin \theta_2$, Eqs. (e) can be written in the form

$$\sigma_r = -\frac{P}{\pi d}\sin(\theta_1 + \theta_2) - \frac{P}{\pi d}\sin(\theta_2 - \theta_1)$$

$$\tau_{r\theta} = -\frac{P}{\pi d}\cos(\theta_1 + \theta_2) - \frac{P}{\pi d}\cos(\theta_2 - \theta_1) \tag{f}$$

This stress acting on the element tangential to the boundary at point M can be obtained by superposing the following three stresses on the element.

1. A normal stress uniformly distributed along the boundary:

$$-\frac{P}{\pi d}\sin(\theta_1 + \theta_2) \tag{g}$$

2. A shearing stress uniformly distributed along the boundary:

$$-\frac{P}{\pi d}\cos(\theta_1 + \theta_2) \tag{h}$$

3. A stress of which the normal and shearing components are

$$-\frac{P}{\pi d}\sin(\theta_2 - \theta_1) \quad \text{and} \quad -\frac{P}{\pi d}\cos(\theta_2 - \theta_1) \tag{k}$$

Observing that the angle between the force P and the tangent at M is $\theta_1 - \theta_2$, it can be concluded that the stress (k) is of magnitude $P/\pi d$ and acts in the direction opposite to the direction of the force P.

Assume now that there are several forces acting on the disk and each of them

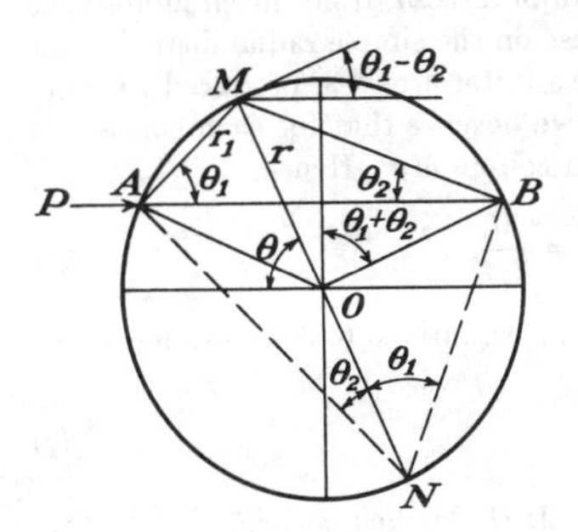

Fig. 77

produces a simple radial stress distribution. Then the forces to be applied at the boundary in order to maintain such a stress distribution are:

1. A normal force uniformly distributed along the boundary, of intensity

$$-\sum \frac{P}{\pi d} \sin(\theta_1 + \theta_2) \tag{l}$$

2. Shearing forces of intensity

$$-\sum \frac{P}{\pi d} \cos(\theta_1 + \theta_2) \tag{m}$$

3. A force, the intensity and direction of which are obtained by vectorial summation of expressions (k). The summation must extend over all forces acting on the boundary.

The moment of all the external forces with respect to O, from Fig. 77, is

$$\sum \frac{P \cos(\theta_1 + \theta_2) d}{2}$$

and, as this moment must be zero for a system in equilibrium, we conclude that the shearing forces (m) are zero. The force obtained by summation of the stresses (k), proportional to the vectorial sum of the external forces, is also zero for a system in equilibrium. Hence it is only necessary to apply at the boundary of the disk a uniform compression (l) in order to maintain the simple radial distributions. If the boundary is free from uniform compression, the stress at any point of the disk is obtained by superposing a uniform tension of magnitude

$$\sum \frac{P}{\pi d} \sin(\theta_1 + \theta_2)$$

on the simple radial distributions.

By using this general method, various other cases of stress distribution in disks can easily be solved.[1] We may select, for instance, the case of a couple acting on the disk (Fig. 78), balanced by a couple applied at the center of the disk. Assuming two equal radial stress distributions at A and B, we see that, in this case, (l) and the summation of (k) are zero and only shearing forces (m) need be applied at the boundary in order to maintain the simple radial stress distributions. The intensity of these forces, from (m), is

$$-\frac{2P}{\pi d} \cos(\theta_1 + \theta_2) = -\frac{2M_t}{\pi d^2} \tag{n}$$

where M_t is the moment of the couple. To free the boundary of the disk from shearing forces and transfer the couple balancing the pair of forces P from the circumference of the disk to its center, it is necessary to superpose on the simple radial distributions the stresses of the case shown in Fig. 78b. These latter stresses, produced by pure circumferential shear, can easily be calculated if we observe that for each concentric circle of radius r the shearing stresses must give a couple M_t. Hence,

$$\tau_{r\theta} 2\pi r^2 = M_t \qquad \tau_{r\theta} = \frac{M_t}{2\pi r^2} \tag{p}$$

These stresses may also be derived from the general equations (38) by taking as the stress function

$$\phi = \frac{M_t \theta}{2\pi} \tag{q}$$

[1] Several interesting examples are discussed by J. H. Michell, *loc. cit.*

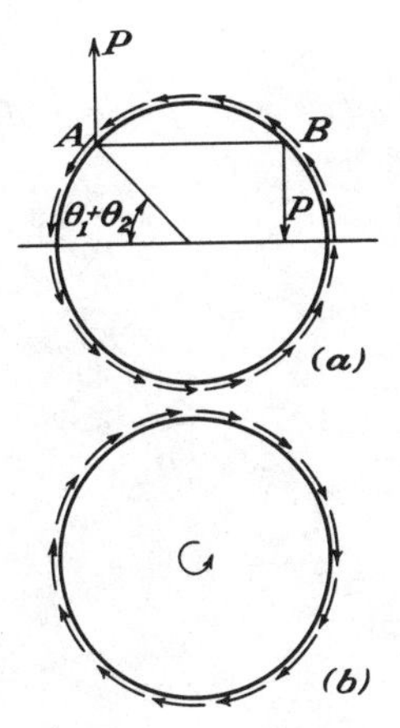

Fig. 78

from which

$$\sigma_r = \sigma_\theta = 0 \qquad \tau_{r\theta} = \frac{M_t}{2\pi r^2}$$

42 | Force at a Point of an Infinite Plate

If a force P acts in the middle plane of an infinite plate (Fig. 79*a*), the stress distribution can easily be obtained by superposition of systems that we have already discussed. We cannot, however, construct a solution by simple superposition of two solutions for a semi-infinite plate as shown in Figs. 79*b* and *c*. Although the vertical displacements are the same in both these cases, the horizontal displacements along the straight boundaries are different. While in the case 79*b* this displacement is away from the point O, in the case 79*c* it is toward the point O. The magnitudes of these displacements in both cases, from Eqs. (70), is

$$\frac{1-\nu}{4E}P \tag{a}$$

This difference in the horizontal displacements may be eliminated by combining the cases 79*b* and *c* with the cases 79*d* and *e* in which shearing forces act along the straight boundaries. The displacements for these latter cases can be obtained from the problem of bending of a curved bar, shown in Fig. 46. Making the inner radius of this bar approach zero and the outer radius increase indefinitely, we arrive at the case of a semi-infinite plate. The displacement along the straight boundary of this plate in the direction of the shearing force acting on the boundary is, from Eq. (60),

$$\frac{D\pi}{E} \tag{b}$$

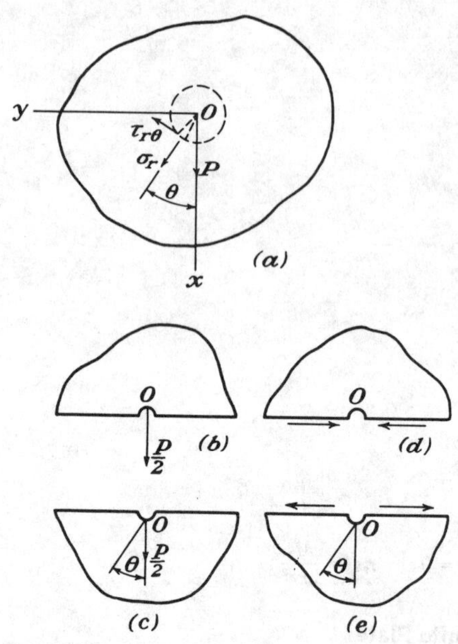

Fig. 79

The constant of integration D must now be adjusted so as to make the displacement resulting from (*a*) and (*b*) vanish. Then

$$\frac{D\pi}{E} = \frac{1-\nu}{4E}P \qquad D = \frac{1-\nu}{4\pi}P \tag{c}$$

With this adjustment the result of superposing cases 79*b*, *c*, *d*, and *e* is an infinite plate loaded at a point, Fig. 79*a*.

The stress distribution in the plate is now easily obtained by superposing the stresses in a semi-infinite plate produced by a normal load $P/2$ at the boundary (see Art. 36) on the stresses in the curved bar containing the constant of integration D. Observing the difference in measuring the angle ϑ in Figs. 46 and 79 and using Eqs. (59), the stresses in the curved bar are, for θ as in Fig. 79,

$$\sigma_r = \frac{D\cos\theta}{r} = \frac{1-\nu}{4\pi}\frac{P\cos\theta}{r}$$

$$\sigma_\theta = \frac{D\cos\theta}{r} = \frac{1-\nu}{4\pi}\frac{P\cos\theta}{r}$$

$$\tau_{r\theta} = \frac{D\sin\theta}{r} = \frac{1-\nu}{4\pi}\frac{P\sin\theta}{r}$$

Combining this with stresses (65) calculated for the load $P/2$, we obtain the following stress distribution in the infinite plate:

$$\sigma_r = \frac{1-\nu}{4\pi}\frac{P\cos\theta}{r} - \frac{P\cos\theta}{\pi r} = -\frac{(3+\nu)}{4\pi}\frac{P\cos\theta}{r}$$
$$\sigma_\theta = \frac{1-\nu}{4\pi}\frac{P\cos\theta}{r} \tag{75}$$
$$\tau_{r\theta} = \frac{1-\nu}{4\pi}\frac{P\sin\theta}{r}$$

By cutting out from the plate at the point O (Fig. 79a) a small element bounded by a cylindrical surface of radius r, and projecting the forces acting on the cylindrical boundary of the element on the x and y axes, we find

$$X = 2\int_0^\pi (\sigma_r\cos\theta - \tau_r\theta\sin\theta)r\,d\theta = P$$
$$Y = 2\int_0^\pi (\sigma_r\sin\theta + \tau_r\theta\cos\theta)r\,d\theta = 0$$

that is, the forces acting on the boundary of the cylindrical element represent the load P applied at the point O. By using Eqs. (13) the stress components, in cartesian coordinates, are found from Eqs. (75):

$$\sigma_x = \frac{P}{4\pi}\frac{\cos\theta}{r}[-(3+\nu) + 2(1+\nu)\sin^2\theta]$$
$$\sigma_y = \frac{P}{4\pi}\frac{\cos\theta}{r}[1-\nu-2(1+\nu)\sin^2\theta] \tag{76}$$
$$\tau_{xy} = -\frac{P}{4\pi}\frac{\sin\theta}{r}[1-\nu+2(1+\nu)\cos^2\theta]$$

From solution (76), for one concentrated force, solutions for other kinds of loading can be obtained by superposition. Take, for instance, the case shown in Fig. 80, in which two equal and opposite forces acting on an infinite plate are applied at two points O and O_1 a very small distance

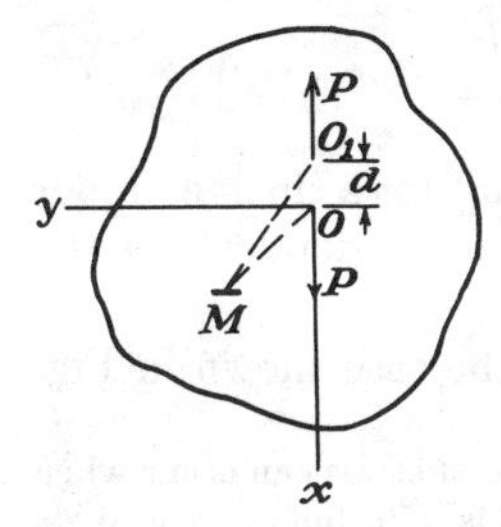

Fig. 80

d apart. The stress at any point M is obtained by superposing on the stress produced by the force at O the stress produced by the other force at O_1. Considering, for instance, an element at M perpendicular to the x axis and denoting by σ_x the normal stress produced on the element by the force at O, the normal stress σ_x' produced by the two forces shown in the figure is

$$\sigma_x' = \sigma_x - \left(\sigma_x + \frac{\partial \sigma_x}{\partial x} d\right) = -d \frac{\partial \sigma_x}{\partial x} = -d \left(\frac{\partial \sigma_x}{\partial r} \cos \theta - \frac{\partial \sigma_x}{\partial \theta} \frac{\sin \theta}{r}\right)$$

Thus the stress components for the case of Fig. 80 are obtained from Eqs. (76) by differentiation. In this manner we find

$$\begin{aligned} \sigma_x &= \frac{dP}{4\pi r^2} [-(3 + \nu) \cos^2 \theta + (1 - \nu) \sin^2 \theta + 8(1 + \nu) \sin^2 \theta \cos^2 \theta] \\ \sigma_y &= \frac{dP}{4\pi r^2} [(1 - \nu) \cos^2 \theta + (1 + 3\nu) \sin^2 \theta - 8(1 + \nu) \sin^2 \theta \cos^2 \theta] \\ \tau_{xy} &= \frac{dP}{4\pi r^2} [-(6 + 2\nu) + 8(1 + \nu) \sin^2 \theta] \sin \theta \cos \theta \end{aligned} \tag{77}$$

It can be seen that the stress components decrease rapidly, as r increases, and are negligible when r is large in comparison with d. Such a result is to be expected in accordance with Saint-Venant's principle if we have two forces in equilibrium applied very near to each other.[1]

By superposing two stress distributions such as given by Eqs. (77), we can obtain the solution of the problem shown in Fig. 81. The stress components for this case are

$$\begin{aligned} \sigma_x &= -2(1 - \nu) \frac{dP}{4\pi r^2} (1 - 2 \sin^2 \theta) \\ \sigma_y &= 2(1 - \nu) \frac{dP}{4\pi r^2} (1 - 2 \sin^2 \theta) \\ \tau_{xy} &= -2(1 - \nu) \frac{dP}{4\pi r^2} \sin 2\theta \end{aligned}$$

The same stress distribution expressed in polar coordinates is

$$\sigma_r = -2(1 - \nu) \frac{dP}{4\pi r^2} \qquad \sigma_\theta = 2(1 - \nu) \frac{dP}{4\pi r^2} \qquad \tau_{r\theta} = 0 \tag{78}$$

This solution can be made to agree with solution (45) for a thick cylinder submitted to the action of internal pressure if the outer diameter of the cylinder is taken as infinitely great.

In the same manner we can get a solution for the case shown in Fig.

[1] It should be observed, however, that similar diminution of stress can occur when neighboring forces are not in equilibrium. Fig. 82*a* and Eqs. (79) following provide an example. See E. Sternberg, *Quart. Appl. Math.*, vol. 11, pp. 393–404, 1954, and the paper by R. von Mises there cited.

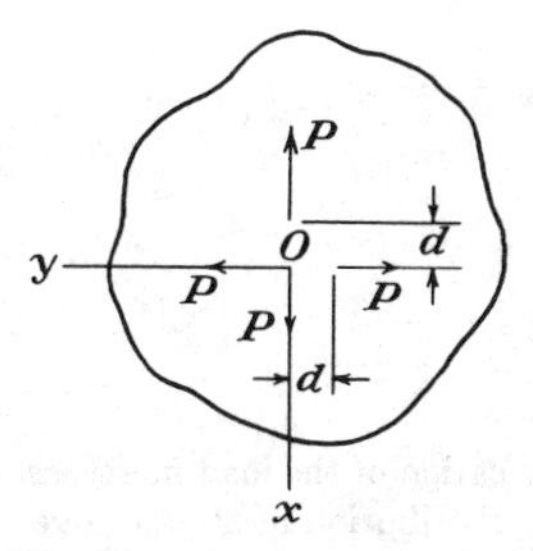

Fig. 81

82*a*. The stress components are[1]

$$\sigma_r = \sigma_\theta = 0 \qquad \tau_{r\theta} = -\frac{M}{2\pi r^2} \tag{79}$$

They represent the stresses produced by a couple M applied at the origin (Fig. 82*b*).

If instead of an infinite plate we have to deal with an infinitely long strip subjected to the action of a longitudinal force P (Fig. 83), we may begin with solution (76) as if the plate were infinite in all directions. The stresses along the edges of the strip resulting from this procedure can be annulled by superposing an equal and opposite system. The stresses produced by this corrective system can be determined by using the general method described in Art. 24. Calculations made by R. C. J. Howland[2] show that the local stresses produced by the concentrated force

[1] A. E. H. Love, "Theory of Elasticity," *op. cit.*, p. 214.

[2] *Loc. cit.* See also a paper by E. Melan, *Z. Angew. Math. Mech.*, vol. 5, p. 314, 1925.

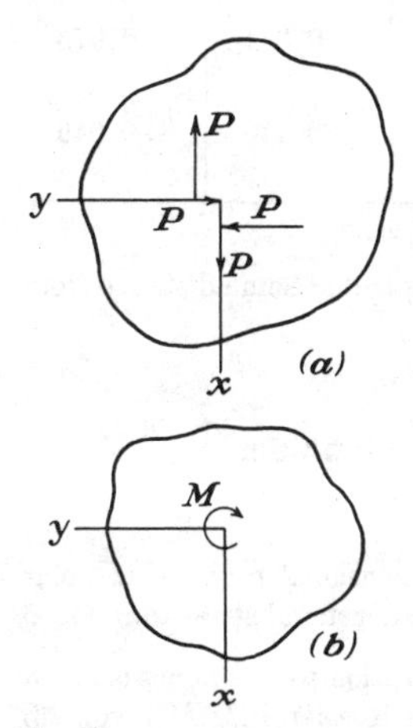

Fig. 82

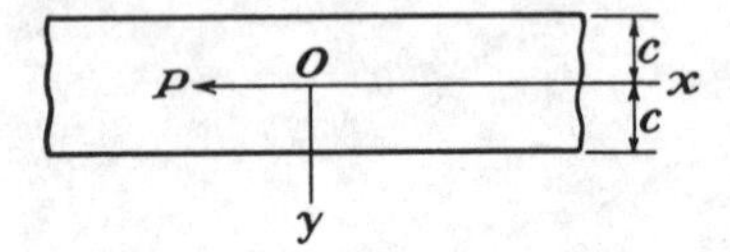

Fig. 83

P diminish rapidly as the distance from the point of application of the load increases, and at distances greater than the width of the strip the distribution of stresses over the cross section is practically uniform. In the table below several values of the stresses σ_x and σ_y are given, calculated on the assumption that the strip is fixed at the end $x = +\infty$ and Poisson's ratio is $\frac{1}{4}$.

	$\frac{x}{c} =$	$-\frac{\pi}{3}$	$-\frac{\pi}{9}$	$-\frac{\pi}{18}$	$-\frac{\pi}{30}$	0
$y = 0$	$\frac{\sigma_x 2c}{P} =$	-0.118	-0.992	....	....	∞
$y = c$	$\frac{\sigma_x 2c}{P} =$	$+0.159$	$+0.511$	0.532	0.521	0.500
$y = 0$	$\frac{\sigma_y 2c}{P} =$	0.110	0.364	....	....	....
	$\frac{x}{c} =$	$\frac{\pi}{30}$	$\frac{\pi}{18}$	$\frac{\pi}{9}$	$\frac{\pi}{3}$	$\frac{\pi}{2}$
$y = 0$	$\frac{\sigma_x 2c}{P} =$	....	....	1.992	1.118	1.002
$y = c$	$\frac{\sigma_x 2c}{P} =$	0.479	0.468	0.489	0.841	0.973
$y = 0$	$\frac{\sigma_y 2c}{P} =$	....	....	-0.364	-0.110	-0.049

Stresses produced in a semi-infinite plate by a force applied at some distance from the edge have been discussed by E. Melan.[1]

43 | Generalized Solution of the Two-dimensional Problem in Polar Coordinates

Having discussed various particular cases of the two-dimensional problem in polar coordinates we are now in a position to write down a more general stress function ϕ

[1] *Z. Angew. Math. Mech.*, vol. 12, p. 343, 1932. A correction to this paper is given by L. M. Kurshin, *Appl. Math. Mech.* (translation of the Russian *P.M.M.*), vol. 23, p. 1403, 1959.

in series form as[1]

$$\begin{aligned}\phi = a_0 \log r + b_0 r^2 + c_0 r^2 \log r + d_0 r^2\theta + a_0'\theta \\ + \frac{a_1}{2} r\theta \sin \theta + (b_1 r^3 + a_1' r^{-1} + b_1' r \log r) \cos \theta \\ - \frac{c_1}{2} r\theta \cos \theta + (d_1 r^3 + c_1' r^{-1} + d_1' r \log r) \sin \theta \\ + \sum_{n=2}^{\infty} (a_n r^n + b_n r^{n+2} + a_n' r^{-n} + b_n' r^{-n+2}) \cos n\theta \\ + \sum_{n=2}^{\infty} (c_n r^n + d_n r^{n+2} + c_n' r^{-n} + d_n' r^{-n+2}) \sin n\theta \qquad (80)\end{aligned}$$

The first three terms in the first line of this expression represent the solution for the stress distribution symmetrical with respect to the origin of coordinates (see Art. 28). The fourth term gives the stress distribution for the case shown in Fig. 58. The fifth term gives the solution for pure shear (Fig. 78b). The first term in the second line is the simple radial distribution for a load in the direction $\theta = 0$. The remaining terms of the second line represent the solution for a portion of a circular ring bent by a radial force (Fig. 46). By a combination of all the terms of the second line the solution for a force acting on an infinite plate was obtained (Art. 42). Analogous solutions are obtained also from the third line of expression (80), the only difference being that the direction of the force is changed by $\pi/2$. The further terms of (80) represent solutions for shearing and normal forces, proportional to $\sin n\theta$ and $\cos n\theta$, acting on the inner and outer boundaries of a circular ring. We had an example of this kind in discussing the stress distribution around a small circular hole (Art. 35).

The boundary conditions alone are not always sufficient for the determination of all the coefficients in the series (80). Certain additional investigations of the displacements are sometimes necessary. We shall consider the complete ring with the intensities of the normal and tangential forces given by the following Fourier series:

$$\begin{aligned}(\sigma_r)_{r=a} &= A_0 + \sum_{n=1}^{\infty} A_n \cos n\theta + \sum_{n=1}^{\infty} B_n \sin n\theta \\ (\sigma_r)_{r=b} &= A_0' + \sum_{n=1}^{\infty} A_n' \cos n\theta + \sum_{n=1}^{\infty} B_n' \sin n\theta \\ (\tau_{r\theta})_{r=a} &= C_0 + \sum_{n=1}^{\infty} C_n \cos n\theta + \sum_{n=1}^{\infty} D_n \sin n\theta \\ (\tau_{r\theta})_{r=b} &= C_0' + \sum_{n=1}^{\infty} C_n' \cos n\theta + \sum_{n=1}^{\infty} D_n' \sin n\theta\end{aligned} \qquad (a)$$

in which the constants $A_0, A_n, B_n, \ldots$, are to be calculated in the usual manner from the given distribution of forces at the boundaries (see page 56). Calculating the stress components from expression (80) by using Eqs. (38), and comparing the values of these components for $r = a$ and $r = b$ with those given by Eqs. (a), we obtain a sufficient number of equations to determine the coefficients in all cases with $n \geq 2$. For $n = 0$, that is, for the terms in the first line of expression (80), and for $n = 1$, that is, for the terms in the second and third lines, further investigations are necessary.

[1] This solution was given by J. H. Michell, *Proc. London Math. Soc.*, vol. 31, p. 100, 1899. See also A. Timpe, *Z. Math. Physik, loc. cit.* An analogous solution for the case of an elliptical ring was given by A. Timpe, *Math. Z.*, vol. 17, p. 189, 1923.

Taking the first line of expression (80) as a stress function, the constant a_0' is determined by the magnitude of the shearing forces uniformly distributed along the boundaries (see page 126). The stress distribution given by the term with d_0 is *many valued* (see page 105) and, for the complete ring, we take[1] $d_0 = 0$. For the determination of the remaining three constants a_0, b_0, and c_0 we have only two equations:

$$(\sigma_r)_{r=a} = A_0 \qquad \text{and} \qquad (\sigma_r)_{r=b} = A_0'$$

The additional equation for determining these constants is obtained from the consideration of displacements. The displacements in a complete ring should be *single-valued* functions of θ. Our previous investigation shows (see Art. 28) that this condition is fulfilled if we put $c_0 = 0$. Then the remaining two constants a_0 and b_0 are determined from the two boundary conditions stated above.

Let us consider now, in more detail, the terms for which $n = 1$. For determining the eight constants $a_1, b_1, \ldots, d_1'$ entering into the second and the third lines of expression (80), we calculate the stress components σ_r and $\tau_{r\theta}$ using this portion of ϕ. Then using conditions (a) and equating corresponding coefficients of sin $n\theta$ and cos $n\theta$ at the inner and outer boundaries, we obtain the following eight equations:

$$\begin{aligned}
(a_1 + b_1')a^{-1} + 2b_1a - 2a_1'a^{-3} &= A_1 \\
(a_1 + b_1')b^{-1} + 2b_1b - 2a_1'b^{-3} &= A_1' \\
(c_1 + d_1')a^{-1} + 2d_1a - 2c_1'a^{-3} &= B_1 \\
(c_1 + d_1')b^{-1} + 2d_1b - 2c_1'b^{-3} &= B_1'
\end{aligned} \tag{b}$$

$$\begin{aligned}
2d_1a - 2c_1'a^{-3} + d_1'a^{-1} &= -C_1 \\
2d_1b - 2c_1'b^{-3} + d_1'b^{-1} &= -C_1' \\
2b_1a - 2a_1'a^{-3} + b_1'a^{-1} &= D_1 \\
2b_1b - 2a_1'b^{-3} + b_1'b^{-1} &= D_1'
\end{aligned} \tag{c}$$

Comparing Eqs. (b) with (c) it can be seen that they are compatible only if

$$\begin{aligned}
a_1a^{-1} &= A_1 - D_1 \\
a_1b^{-1} &= A_1' - D_1' \\
c_1a^{-1} &= B_1 + C_1 \\
c_1b^{-1} &= B_1' + C_1'
\end{aligned} \tag{d}$$

from which it follows that

$$a(A_1 - D_1) = b(A_1' - D_1') \qquad a(B_1 + C_1) = b(B_1' + C_1') \tag{e}$$

It can be shown that Eqs. (e) are always fulfilled if the forces acting on the ring are in equilibrium. Taking, for instance, the sum of the components of all the forces in the direction of the x axis as zero, we find

$$\int_0^{2\pi} \{[b(\sigma_r)_{r=b} - a(\sigma_r)_{r=a}] \cos\theta - [b(\tau_{r\theta})_{r=b} - a(\tau_{r\theta})_{r=a}] \sin\theta\}\, d\theta = 0$$

Substituting for σ_r and $\tau_{r\theta}$ from (a), we arrive at the first of Eqs. (e). In the same manner, by resolving all the forces along the y axis, we obtain the second of Eqs. (e).

[1] Stress functions giving stress discontinuities in a ring can be interpreted as solutions for a ring with a cut. See J. N. Goodier and J. C. Wilhoit, Jr., *Proc. 4th Ann. Conf. Solid Mech., Univ. Texas*, Austin, Texas, pp. 152–170, 1959.

When a_1 and c_1 are determined from Eqs. (d) the two systems of Eqs. (b) and (c) become identical, and we have only four equations for determining the remaining six constants. The necessary two additional equations are obtained by considering the displacements. The terms in the second line in expression (80) represent the stress function for a combination of a simple radial distribution and the bending stresses in a curved bar (Fig. 46). By superposing[1] the general expressions for the displacements in these two cases, namely Eqs. (g) (page 103) and Eqs. (q) (page 87), and substituting $a_1/2$ for $-P/\pi$ in Eqs. (g) and b_1' for D in Eqs. (q), we find the following many-valued terms in the expressions for the displacements u and v, respectively:

$$\frac{a_1}{2}\frac{1-\nu}{E}\theta \sin\theta + \frac{2b_1'}{E}\theta \sin\theta$$

$$\frac{a_1}{2}\frac{1-\nu}{E}\theta \cos\theta + \frac{2b_1'}{E}\theta \cos\theta$$

These terms must vanish for a complete ring, when the stress is due to boundary loading only. Hence,

$$\frac{a_1}{2}\frac{1-\nu}{E} + \frac{2b_1'}{E} = 0$$

or

$$b_1' = -\frac{a_1(1-\nu)}{4} \tag{f}$$

Considering the third line of expression (80) in the same manner, we find

$$d_1' = -\frac{c_1(1-\nu)}{4} \tag{g}$$

Equations (f) and (g), together with Eqs. (b) and (c), are now sufficient for determining all the constants in the stress function represented by the second and the third lines of expression (80).

We conclude that in the case of a complete ring the boundary conditions (a) are not sufficient for the determination of the stress distribution, and it is necessary to consider the displacements. The displacements in a complete ring must be single valued and to satisfy this condition we must have

$$c_0 = 0 \qquad b_1' = -\frac{a_1(1-\nu)}{4} \qquad d_1' = -\frac{c_1(1-\nu)}{4} \tag{81}$$

We see that the constants b_1' and d_1' depend on Poisson's ratio. Accordingly, the stress distribution in a complete ring will usually depend on the elastic properties of the material. It becomes independent of the elastic constants only when a_1 and c_1 vanish so that, from Eq. (81), $b_1' = d_1' = 0$. This particular case occurs if [see Eqs. (d)]

$$A_1 = D_1 \qquad \text{and} \qquad B_1 = -C_1$$

We have such a condition when the resultant of the forces applied to each boundary of the ring vanishes. Take, for instance, the resultant component in the x direction of forces applied to the boundary $r = a$. This component, from (a), is

$$\int_0^{2\pi} (\sigma_r \cos\theta - \tau_{r\theta}\sin\theta)a\,d\theta = a\pi(A_1 - D_1)$$

[1] It should be noted that $\theta + (\pi/2)$ must be substituted for θ if the angle is measured from the vertical axis, as in Fig. 53, instead of from the horizontal axis, as in Fig. 46.

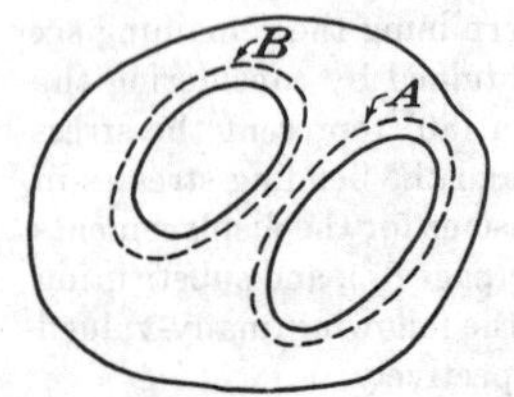

Fig. 84

If it vanishes we find $A_1 = D_1$. In the same manner, by resolving the forces in the y direction, we obtain $B_1 = -C_1$ when the y component is zero. From this we may conclude that the stress distribution in a complete ring is independent of the elastic constants of the material if the resultant of the forces applied to each boundary is zero. The moment of these forces need not be zero.

These conclusions for the case of a circular ring hold also in the most general case of the two-dimensional problem for a *multiply-connected* body. From general investigations made by J. H. Michell,[1] it follows that, for multiply-connected bodies (Fig. 84), equations analogous to Eqs. (81) and expressing the condition that the displacements are single valued should be derived for each *independent circuit* such as the circuits A and B in the figure. The stress distributions in such bodies generally depend on the elastic constants of the material. They are independent of these constants only if the resultant force on each boundary vanishes.[2] Quantitatively the effect of the moduli on the maximum stress is usually very small, and in practice it can be neglected.[3] This conclusion is of practical importance. We shall see later that in the case of transparent materials, such as glass or Bakelite, it is possible to determine the stresses by an optical method, using polarized light (see page 150), and this conclusion means that the experimental results obtained with a transparent material can be applied immediately to any other material such as steel if the external forces are the same.

44 | Applications of the Generalized Solution in Polar Coordinates

As a first application of the series solution of the two-dimensional problem in polar coordinates let us consider a circular ring compressed by two equal and opposite forces acting along a diameter[4] (Fig. 85*a*). We begin with the solution for a solid disk (Art. 41). By cutting out a concentric hole of radius a in this disk, we are left with normal and shearing forces distributed around the edge of the hole. These forces can be annulled by superposing an equal and opposite system of forces. This latter system can be represented with sufficient accuracy by using the first few terms of a Fourier series. Then the corresponding stresses in the ring are obtained by using the general solution of the previous article. These stresses together with the stresses calculated as for a solid disk constitute the total stresses in the ring. The ratios $\sigma_\theta : 2P/\pi b$,

[1] *Loc. cit.*

[2] It must be remembered that the body forces were taken as zero.

[3] An investigation of this subject is given by L. N. G. Filon, *Brit. Assoc. Advan. Sci. Rept.*, 1921. See E. G. Coker and L. N. G. Filon, "Photo-elasticity," Arts. 6.07 and 6.16, Cambridge University Press, New York, 1931.

[4] See S. Timoshenko, *Bull. Polytech. Inst. Kiew*, 1910, and *Phil. Mag.*, vol. 44, p. 1014, 1922. See also K. Wieghardt, *Sitzber. Akad. Wiss. Wien*, vol. 124, Abt. II, p. 1119, 1915.

calculated in this manner for various points of the cross sections mn and m_1n_1 for the case $b = 2a$, are given in the table below.[1,2]

$r =$	b	$0.9b$	$0.8b$	$0.7b$	$0.6b$	$0.5b$
			Exact theory			
mn	2.940	1.477	−0.113	−2.012	−4.610	−8.942
m_1n_1	−3.788	−2.185	−0.594	1.240	4.002	10.147
			Hyperbolic stress distribution			
mn	2.885	1.602	0.001	−2.060	−4.806	−8.653
m_1n_1	−7.036	−5.010	−2.482	0.772	5.108	11.18
			Linear stress distribution			
mn	3.90	1.71	−0.48	−2.67	−4.86	−7.04
m_1n_1	−8.67	−5.20	−1.73	1.73	5.20	8.67

For comparison we give the values of the same stresses calculated from the two elementary theories based on the following assumptions: (1) that cross sections remain plane, in which case the normal stresses over the cross section follow a hyperbolic law; (2) that the stresses are distributed according to a linear law. The table shows that

[1] The thickness of the plate is taken as unity.

[2] Stress evaluations on m_1n_1 for $b = 2a$, $2.5a$, $3.33a$, $5a$, $10a$, are given by E. A. Ripperger and N. Davids, *Trans. ASCE*, vol. 112, pp. 619–628, 1947.

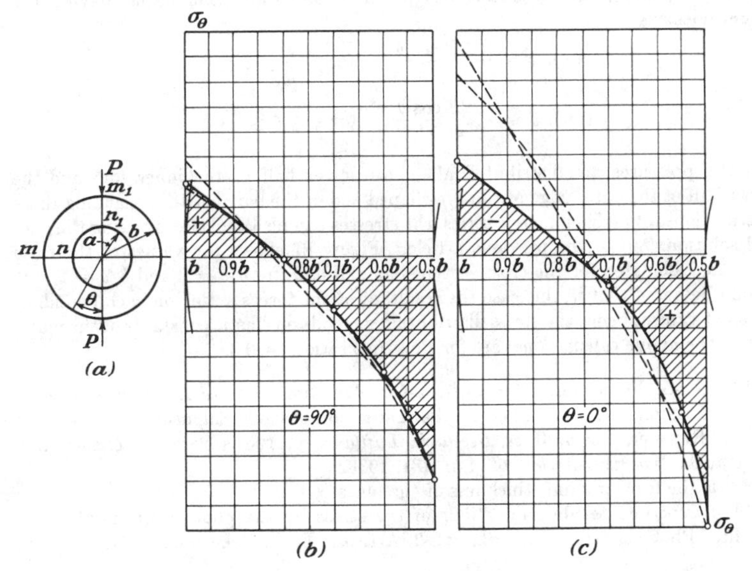

Fig. 85

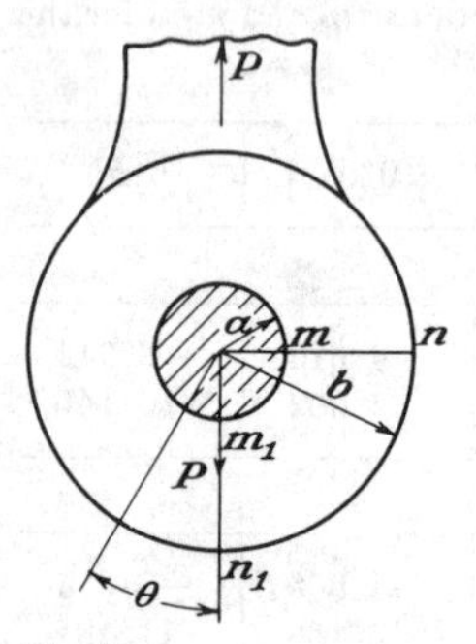

Fig. 86

for the cross section mn, which is at a comparatively large distance from the points of application of the loads P, the hyperbolic stress distribution gives results which are very nearly exact. The error in the maximum stress is only about 3 percent. For the cross section m_1n_1 the errors of the approximate solution are much larger. It is interesting to note that the resultant of the normal stresses over the cross section m_1n_1 is P/π. This is to be expected if we remember the wedge action of the concentrated force illustrated by Fig. 68*d*. The distribution of normal stresses over the cross section mn and m_1n_1 calculated by the three above methods is shown in Figs. 85*b* and *c*. The method applied above to the case of two equal and opposite forces can be used for the general case of loading of a circular ring by concentrated forces.[1]

As a second example we consider the end of an eyebar[2] (Fig. 86). The distribution of pressures along the edge of the hole depends on the amount of clearance between the bolt and the hole. The following results are obtained on the assumption that there are only normal pressures acting on the inner and outer boundaries having the magnitudes:[3]

$$(\sigma_r)_{r=a} = -\frac{2P}{\pi}\frac{\cos\theta}{a} \qquad \text{for } -\frac{\pi}{2} \leq \theta \leq \frac{\pi}{2}$$

$$(\sigma_r)_{r=b} = -\frac{2P}{\pi}\frac{\cos\theta}{b} \qquad \text{for } \frac{\pi}{2} \leq \theta \leq \frac{3\pi}{2}$$

i.e., the pressures are distributed along the lower half of the inner edge and the upper half of the outer edge of the eye-shaped end of the bar. After expanding these distributions into trigonometric series, the stresses can be calculated by using the general solution (80) of the previous article. Figure 87 shows the values of the ratio $\sigma_\theta : P/2a$, calculated for the cross sections mn and m_1n_1 for $b/a = 4$ and $b/a = 2$. It should be noted that in this case the resultant of the forces acting on each boundary does not vanish; hence the stress distribution depends on elastic constants of the material. The above calculations[4] are for Poisson's ratio $\nu = 0.3$.

[1] L. N. G. Filon, The Stresses in a Circular Ring, *Selected Engineering Papers*, No. 12, London, 1924, published by the Institution of Civil Engineers.

[2] H. Reissner, *Jahrb. Wiss. Gesellsch. Luftfahrt*, p. 126, 1928; H. Reissner, and F. Strauch, *Ingenieur-Arch.*, vol. 4, p. 481, 1933.

[3] P is the force per unit thickness of the plate.

[4] For experimental determinations of the stress by the photoelastic method see Frocht, "Photoelasticity," *op. cit.*, vol. 2, Art. 6.4; E. G. Coker and L. N. G. Filon,

45 | A Wedge Loaded along the Faces

The general solution (80) can be used also for polynomial distributions of load on the faces of a wedge.[1] By calculating the stress components from Eq. (80) in the usual way, and taking only the terms containing r^n with $n \geq 0$, we find the following expressions for the stress components in ascending powers of r:

$$
\begin{aligned}
\sigma_\theta = {} & 2b_0 + 2d_0\theta + 2a_2 \cos 2\theta + 2c_2 \sin 2\theta \\
& + 6r(b_1 \cos \theta + d_1 \sin \theta + a_3 \cos 3\theta + c_3 \sin 3\theta) \\
& \qquad + 12r^2(b_2 \cos 2\theta + d_2 \sin 2\theta + a_4 \cos 4\theta + c_4 \sin 4\theta) \\
& \cdots\cdots\cdots\cdots\cdots\cdots\cdots\cdots\cdots\cdots \\
& (n+2)(n+1)r^n[b_n \cos n\theta + d_n \sin n\theta + a_{n+2} \cos (n+2)\theta \\
& \qquad + c_{n+2} \sin (n+2)\theta] \\
& \cdots\cdots\cdots\cdots\cdots\cdots\cdots\cdots\cdots\cdots \\
\tau_{r\theta} = {} & -d_0 + 2a_2 \sin 2\theta - 2c_2 \cos 2\theta \\
& + r(2b_1 \sin \theta - 2d_1 \cos \theta + 6a_3 \sin 3\theta - 6c_3 \cos 3\theta) \\
& \qquad + r^2(6b_2 \sin 2\theta - 6d_2 \cos 2\theta + 12a_4 \sin 4\theta - 12c_4 \cos 4\theta) \\
& \cdots\cdots\cdots\cdots\cdots\cdots\cdots\cdots\cdots\cdots \\
& + r^n[n(n+1)b_n \sin n\theta - n(n+1)d_n \cos n\theta + (n+1)(n+2) \\
& \qquad a_{n+2} \sin (n+2)\theta - (n+1)(n+2)c_{n+2} \cos (n+2)\theta] \\
& \cdots\cdots\cdots\cdots\cdots\cdots\cdots\cdots\cdots\cdots
\end{aligned}
\tag{82}
$$

"Photoelasticity," *op. cit.*, Art. 6.18; K. Takemura and Y. Hosokawa, *Tokyo Imp. Univ. Aeron. Res. Inst. Rept.* 12, 1926. The stress in steel eyebars was investigated by J. Mathar, *Forschungsarbeiten*, no. 306, 1928. Further theory is given by P. S. Theocaris, *J. Appl. Mech.*, vol. 23, pp. 85–90, 1956.

[1] See S. Timoshenko, "Theory of Elasticity," Russian edition, p. 119, St. Petersburg, 1914.

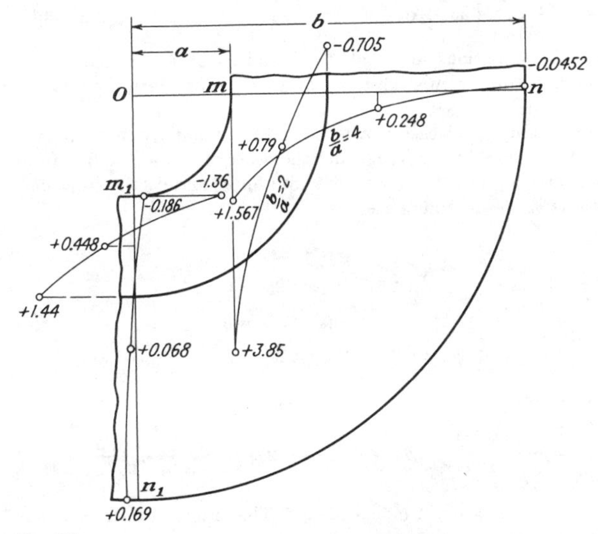

Fig. 87

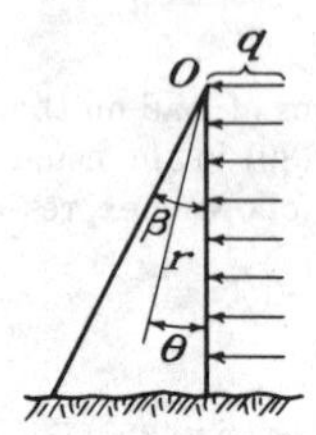

Fig. 88

Thus each power of r is associated with four arbitrary parameters so that, if the applied stresses on the boundaries, $\theta = \alpha$ and $\theta = \beta$, are given as polynomials in r, the stresses in the wedge included between these boundaries are determined.

If, for instance, the boundary conditions are[1]

$$\begin{aligned} (\sigma_\theta)_{\theta=\alpha} &= N_0 + N_1 r + N_2 r^2 + \cdots \\ (\sigma_\theta)_{\theta=\beta} &= N_0' + N_1' r + N_2' r^2 + \cdots \\ (\tau_{r\theta})_{\theta=\alpha} &= S_0 + S_1 r + S_2 r^2 + \cdots \\ (\tau_{r\theta})_{\theta=\beta} &= S_0' + S_1' r + S_2' r^2 + \cdots \end{aligned} \tag{a}$$

we have, by equating coefficients of powers of r,

$$\begin{aligned} 2(b_0 + d_0\alpha + a_2 \cos 2\alpha + c_2 \sin 2\alpha) &= N_0 \\ 6(b_1 \cos \alpha + d_1 \sin \alpha + a_3 \cos 3\alpha + c_3 \sin 3\alpha) &= N_1 \end{aligned} \tag{b}$$

and generally

$$(n+2)(n+1)[b_n \cos n\alpha + d_n \sin n\alpha + a_{n+2} \cos (n+2)\alpha + c_{n+2} \sin (n+2)\alpha] = N_n$$

with three other groups of equations for σ_θ at $\theta = \beta$ and $\tau_{r\theta}$ at $\theta = \alpha$ and $\theta = \beta$. These equations are sufficient for calculating the constants entering into the solution (82).

Let us consider, as an example, the case shown in Fig. 88. A uniform normal pressure q is acting on the face $\theta = 0$ of the wedge and the other face $\theta = \beta$ is free from forces. Using only the first lines in the expressions (82) for σ_θ and $\tau_{r\theta}$ the equations for determining constants b_0, d_0, a_2, and c_2 are

$$\begin{aligned} 2b_0 + 2a_2 &= -q \\ 2b_0 + 2d_0\beta + 2a_2 \cos 2\beta + 2c_2 \sin 2\beta &= 0 \\ -d_0 - 2c_2 &= 0 \\ -d_0 + 2a_2 \sin 2\beta - 2c_2 \cos 2\beta &= 0 \end{aligned}$$

from which (writing $k = \tan \beta - \beta$) we find

$$c_2 = \frac{q}{4k} \qquad a_2 = -\frac{q \tan \beta}{4k} \qquad d_0 = -\frac{q}{2k} \qquad 2b_0 = -q + \frac{q \tan \beta}{2k}$$

[1] The terms N_0, N_0', S_0, S_0' are not independent. They represent stress at the corner $r = 0$ and only three can be assigned.

Substituting in Eqs. (82), we obtain[1]

$$
\begin{aligned}
\sigma_\theta &= \frac{q}{k}\left(-k + \frac{1}{2}\tan\beta - \theta - \frac{1}{2}\tan\beta\cos 2\theta + \frac{1}{2}\sin 2\theta\right) \\
\tau_{r\theta} &= \frac{q}{k}\left(\frac{1}{2} - \frac{1}{2}\tan\beta\sin 2\theta - \frac{1}{2}\cos 2\theta\right) \\
\sigma_r &= \frac{q}{k}\left(-k + \frac{1}{2}\tan\beta - \theta - \frac{1}{2}\sin 2\theta + \frac{1}{2}\tan\beta\cos 2\theta\right)
\end{aligned}
\qquad (c)
$$

The stress components for any other term in the polynomial load distribution (a) may be obtained in a similar manner.

The method developed above for calculating stresses in a wedge is applicable to a semi-infinite plate by making the angle β of the wedge equal to π. The stresses for the case shown in Fig. 89, for instance, are obtained from Eqs. (c) by substituting $\beta = \pi$ in these equations. Then

$$
\begin{aligned}
\sigma_\theta &= -\frac{q}{\pi}\left(\pi - \theta + \frac{1}{2}\sin 2\theta\right) \\
\tau_{r\theta} &= -\frac{q}{2\pi}(1 - \cos 2\theta) \\
\sigma_r &= -\frac{q}{\pi}\left(\pi - \theta - \frac{1}{2}\sin 2\theta\right)
\end{aligned}
\qquad (d)
$$

These satisfy the conditions on the straight edge, and also specific conditions on a closing boundary such as a semicircle $r = b$.

46 | Eigensolutions for Wedges and Notches

In Art. 45 the stress components (82) were taken in positive integer powers of r, corresponding to a stress function of similar form. But if we return to the series in $\cos n\theta$, $\sin n\theta$, in the stress function (80), it is readily verified that each term is a stress function whether n is integer or not. In fact the differential equation (39) is satisfied regardless of the value of n. The value may be a complex number, but in that case we can use either the real or the imaginary part as a real stress function. Thus,

[1] This solution was obtained by another method by M. Levy, *Compt. Rend.*, vol. 126, p. 1235, 1898. See also P. Fillunger, *Z. Math. Physik*, vol. 60, 1912. An application of stress functions of this type to tapered box beams is given by E. Reissner, *J. Aeron. Sci.*, vol. 7, p. 353, 1940. Other loads on wedges are considered by C. J. Tranter, *Quart. J. Mech. Appl. Math.*, vol. 1, p. 125, 1948.

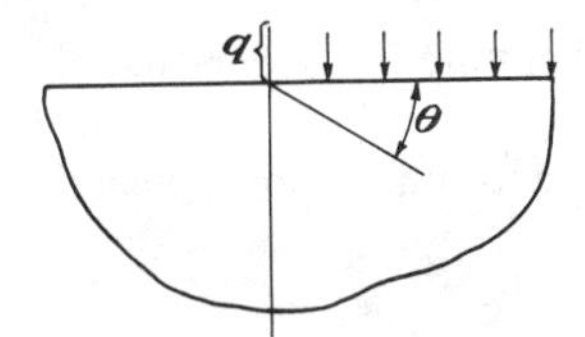

Fig. 89

writing $\lambda + 1$ in place of n, we can take

$$\phi = r^{\lambda+1}f(\theta) \tag{a}$$

where

$$f(\theta) = C_1 \sin (\lambda + 1)\theta + C_2 \cos (\lambda + 1)\theta + C_3 \sin (\lambda - 1)\theta + C_4 \cos (\lambda - 1)\theta \tag{b}$$

with C_1, C_2, C_3, C_4 arbitrary constants.

The stress and displacement components (ignoring rigid-body terms) are given by

$$\sigma_r = r^{\lambda-1}[f''(\theta) + (\lambda + 1)f(\theta)] \tag{c}$$

$$\sigma_\theta = r^{\lambda-1}[\lambda(\lambda + 1)f(\theta)] \tag{d}$$

$$\tau_{r\theta} = -r^{\lambda-1}\lambda f'(\theta) \tag{e}$$

$$2Gu = r^{\lambda}[-(\lambda + 1)f(\theta) + (1 + \nu)^{-1}g'(\theta)] \tag{f}$$

$$2Gv = r^{\lambda}[-f'(\theta) + (1 + \nu)^{-1}(\lambda - 1)g(\theta)] \tag{g}$$

where

$$g(\theta) = 4(\lambda - 1)^{-1}[C_3 \cos (\lambda - 1)\theta + C_4 \sin (\lambda - 1)\theta] \tag{h}$$

and the displacements are for plane stress.

We now consider an application of this for a wedge-shaped region bounded by radii $\theta = \pm\alpha$, which are to be free of load, so that

$$\sigma_\theta = 0 \qquad \tau_{r\theta} = 0 \qquad \text{at } \theta = \pm\alpha \tag{i}$$

By (*d*) and (*e*) this means that

$$f(\alpha) = 0 \qquad f(-\alpha) = 0 \qquad f'(\alpha) = 0 \qquad f'(-\alpha) = 0 \tag{j}$$

and, from (*b*), these become four equations involving the constants C_1, C_2, C_3, C_4. By simple additions and subtractions these are easily seen to be equivalent to

$$\begin{aligned} C_1 \sin (\lambda + 1)\alpha + C_3 \sin (\lambda - 1)\alpha &= 0 \\ (\lambda + 1)C_1 \cos (\lambda + 1)\alpha + (\lambda - 1)C_3 \cos (\lambda - 1)\alpha &= 0 \end{aligned} \tag{k}$$

$$\begin{aligned} C_2 \cos (\lambda + 1)\alpha + C_4 \cos (\lambda - 1)\alpha &= 0 \\ (\lambda + 1)C_2 \sin (\lambda + 1)\alpha + (\lambda - 1)C_4 \sin (\lambda - 1)\alpha &= 0 \end{aligned} \tag{l}$$

Each pair of equations is homogeneous, and therefore if we were merely to choose a number λ arbitrarily, the four constants would be zero. But the values of C_1, C_3 in (*k*) can be nonzero if the determinant of the coefficients vanishes, i.e., if

$$(\lambda - 1) \sin (\lambda + 1)\alpha \cos (\lambda - 1)\alpha - (\lambda + 1) \sin (\lambda - 1)\alpha \cos (\lambda + 1)\alpha = 0$$

This reduces to

$$\lambda \sin 2\alpha - \sin 2\lambda\alpha = 0 \tag{m}$$

If we take a value of λ satisfying this equation, C_1 and C_3 can be nonzero. The ratio C_3/C_1 is determined by either of the two equations (k). But C_1 itself can be retained as an arbitrary constant.

Considering the other two equations (l) similarly, we find that C_2 and C_4 can be nonzero provided λ satisfies

$$\lambda \sin 2\alpha + \sin 2\lambda\alpha = 0 \tag{n}$$

Examining (m) and (n) together, it is evident that the only value of λ which satisfies both is $\lambda = 0$, and this is of no interest. Thus if one equation is satisfied, the other will not be. Consequently if C_1, C_3 are not zero, C_2 and C_4 must be zero, and conversely.

Taking the converse (symmetrical) case, (n) is satisfied, and from the first of (l)

$$\frac{C_4}{C_2} = \frac{-\cos(\lambda + 1)\alpha}{\cos(\lambda - 1)\alpha} \tag{o}$$

The stress function (a) becomes

$$\phi = r^{\lambda+1}C_2\left[\cos(\lambda + 1)\theta - \frac{\cos(\lambda + 1)\alpha}{\cos(\lambda - 1)\alpha}\cos(\lambda - 1)\theta\right] \tag{p}$$

there being one such (complex) function for each root of (n), yielding two real solutions.

Examination of the roots[1] of (n) shows that for wedge-shaped regions, that is, $2\alpha < \pi$, there is an infinite set having positive real part, these real parts all exceeding unity. The corresponding stress functions then yield stress and displacement, by equations (c) through (h), which approach zero with r. But if λ is a root of (n), so is $-\lambda$. Thus there is another set of roots having negative real part. These make both stress and displacement increase without bound as r approaches zero. The tip of the wedge can therefore not be regarded as unloaded, even though the resultant force and couple are zero. For the antisymmetrical case governed by the equation (m) the conclusion is the same. For $2\alpha > \pi$, that is, notched plates, the roots of (n) change character.[2] A change[2] in the roots of (m) occurs at the angle $2\alpha = 257.4°$.

Results of this kind have also been worked out[1] for different conditions on the edges $\theta = \pm\alpha$. For both edges clamped ($u = v = 0$), the conclusions for the wedge-shaped region ($2\alpha < \pi$) are qualitatively similar. For one edge clamped, the other free ($\sigma_\theta = \tau_{r\theta} = 0$), there are stress functions that have displacement approaching zero with r, but stress increas-

[1] M. L. Williams, *J. Appl. Mech.*, vol. 19, p. 526, 1952.

[2] E. Sternberg and W. T. Koiter, *loc. cit.*

ing without bound, when $2\alpha > 63°$ approximately (for $\nu = 0.3$). The quadrant, $2\alpha = \pi/2$, is a special case of interest in showing the character of the singularities in the problem of the tension strip with one end clamped.[1]

In specific problems of wedges and notches, not excluding loaded radial edges, the transform methods[2] lead directly to the appropriate combinations[3] of special solutions, such as those of Arts. 38 and 39, with the eigensolutions considered in the present article.

PROBLEMS

1. Derive the three relations (b) on p. 67. Choose for each a suitable "triangular" element from which the relation can be written down at once as an equation of equilibrium.
2. Verify Eq. (d) of Art. 27 in the case

$$\phi = x^4 - y^4 = (x^2 + y^2)(x^2 - y^2) = r^4 \cos 2\theta$$

3. Examine the significance of the stress function $C\theta$ where C is a constant. Apply it to a ring $a < r < b$, and to an infinite plate.

   A ring is fixed at $r = a$ and subjected to a uniform circumferential shear at $r = b$ forming a couple M. Using Eqs. (48), (49), (50), find an expression for the circumferential displacement v at $r = b$.
4. Show that in the problem of Fig. 45, if the inner radius a is small compared with the outer radius b, the value of σ_θ at the inside is given by

$$\frac{\alpha E}{4\pi}\left(1 - 2 \log \frac{b}{a}\right)$$

   and so is large, and negative when α is positive (the gap is being closed).

   What is the largest gap (value of α) that can be closed without exceeding the elastic limit, if $b/a = 10$, $E = 3 \times 10^7$ psi, elastic limit $= 4 \times 10^4$ psi?
5. Find by superposition from Eqs. (61) the stresses in the infinite plate with a hole when the undisturbed stress at infinity is uniform tension S in both the x and y directions. The results should correspond with Eqs. (44) for the special case $b/a \to \infty$, $p_i = 0$, $p_o = -S$. Use this as a check.
6. Find expressions for the displacements corresponding to the stresses (61), and verify that they are single valued.
7. Convert the stress function (a) of Art. 36 to cartesian coordinates and hence derive the values of σ_x, σ_y, τ_{xy} which are equivalent to the stress distribution of Eqs. (65′). Show that these values approach zero as the distance from the force increases in any direction.

[1] See footnote 1 on p. 62.

[2] See the papers by Sternberg and Koiter, footnote on p. 113, and Benthem, footnote 1 on p. 62, and the references they give to earlier papers.

[3] Other combinations are considered in relation to experimental results (from the photoelastic method described in Chap. 5) by G. Sonntag in a group of papers. See *Forsch. Ing.-Wes.*, vol. 29, pp. 197–203, 1963, and references there. Also H. Neuber, *Z. Angew. Math. Mech.*, vol. 43, pp. 211–228, 1963.

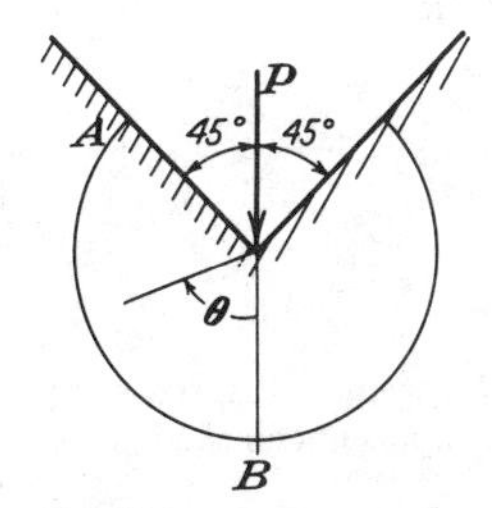

Fig. 90

8. Verify that in the special case of $\alpha = \pi/2$ the stress components (f) of Art. agree with Eq. (68), and investigate whether this stress distribution tends to agree with elementary bending theory for small α.
9. Show by evaluating the force resultants that the stress distribution (f), page 112, does in fact correspond to loading by a pure couple M at the tip of the wedge.
10. A force P per unit thickness is applied by a knife-edge to the bottom of a 90° notch in a large plate as indicated in Fig. 90. Evaluate the stresses, and the horizontal force transmitted across an arc AB, corresponding to the type of distribution represented by (a) in Art. 38.
11. Find an expression for the stress σ_x on the section mn indicated in Fig. 91. The wedge theory of the present chapter and the cantilever theory of Chap. 3 give different stress distributions for the junction rs. Comment on this.
12. Determine the value of the constant C in the stress function

$$\phi = C[r^2(\alpha - \theta) + r^2 \sin\theta \cos\theta - r^2 \cos^2\theta \tan\alpha]$$

required to satisfy the conditions on the upper and lower edges of the triangular plate shown in Fig. 92. Evaluate the stress components σ_x, τ_{xy} for a vertical section mn. Draw curves for the case $\alpha = 20°$ and draw also for comparison the curves given by elementary beam theory.

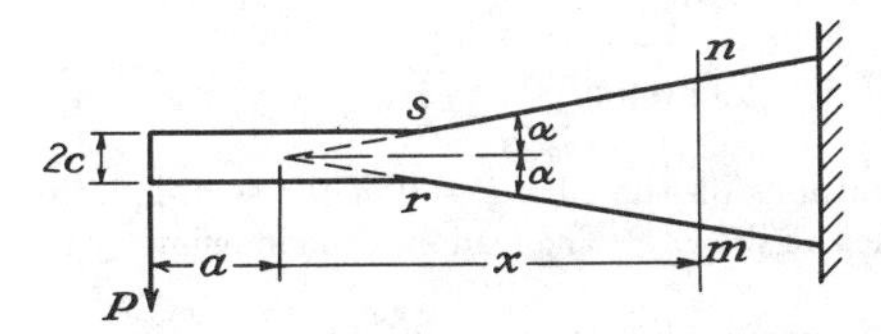

Fig. 91

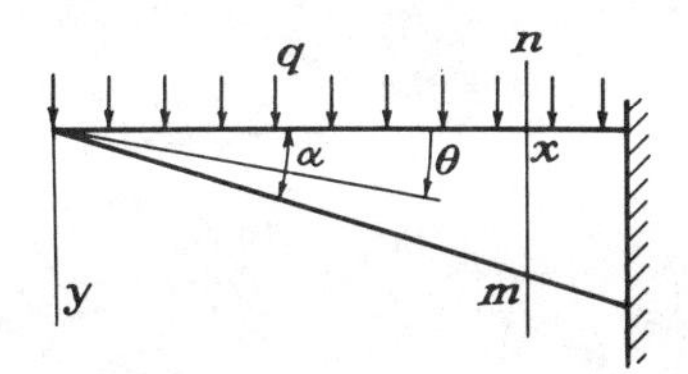

Fig. 92

13. Determine the value of the constant C in the stress function

$$\phi = Cr^2(\cos 2\theta - \cos 2\alpha)$$

required to satisfy the conditions

$$\sigma_\theta = 0 \qquad \tau_{r\theta} = s \qquad \text{on } \theta = \alpha$$
$$\sigma_\theta = 0 \qquad \tau_{r\theta} = -s \qquad \text{on } \theta = -\alpha$$

corresponding to uniform shear loading on each edge of a wedge, directed away from the vertex. Verify that no concentrated force or couple acts on the vertex.

14. Find the stress function of the type

$$a_3r^3 \cos 3\theta + b_1r^3 \cos \theta$$

that satisfies the conditions

$$\sigma_\theta = 0 \qquad \tau_{r\theta} = sr \qquad \text{on } \theta = \alpha$$
$$\sigma_\theta = 0 \qquad \tau_{r\theta} = -sr \qquad \text{on } \theta = -\alpha$$

s being a constant. Sketch the loading for positive s.

15. Find the stress function of the type

$$a_4r^4 \cos 4\theta + b_2r^4 \cos 2\theta$$

that satisfies the conditions

$$\sigma_\theta = 0 \qquad \tau_{r\theta} = sr^2 \qquad \text{on } \theta = \alpha$$
$$\sigma_\theta = 0 \qquad \tau_{r\theta} = -sr^2 \qquad \text{on } \theta = -\alpha$$

s being a constant. Sketch the loading.

16. Derive the stress distribution

$$\sigma_x = -\frac{p}{\pi}\left(\arctan\frac{y}{x} + \frac{xy}{x^2 + y^2}\right) \qquad \tau_{xy} = -\frac{p}{\pi}\frac{y^2}{x^2 + y^2}$$

$$\sigma_y = -\frac{p}{\pi}\left(\arctan\frac{y}{x} - \frac{xy}{x^2 + y^2}\right)$$

from the stress function [see Eq. (a), Art. 37]

$$\phi = -\frac{p}{2\pi}\left((x^2 + y^2)\arctan\frac{y}{x} - xy\right)$$

and show that it satisfies the conditions on the edge $y = 0$ of the semi-infinite plate indicated in Fig. 93, with axes as shown. The load extends indefinitely to the left.

Examine the value of τ_{xy}, (a) approaching O along the boundary Ox, (b) approaching O along the y axis (the discrepancy is due to the *discontinuity* of loading at O).

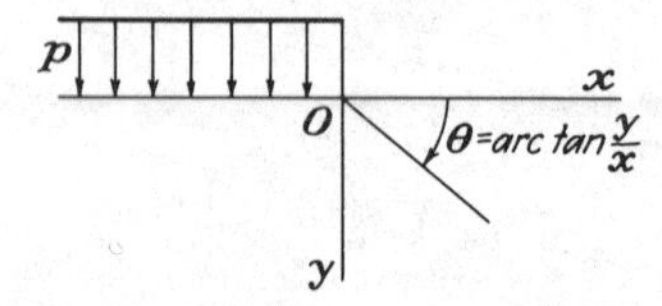

Fig. 93

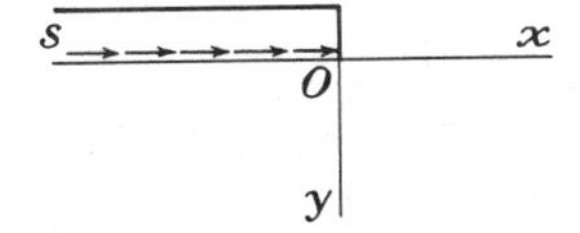

Fig. 94

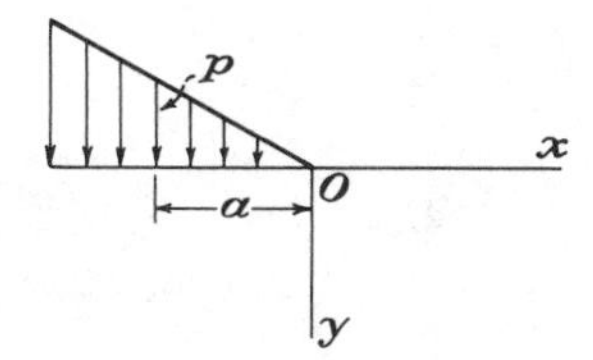

Fig. 95

17. Show that the stress function

$$\phi = \frac{s}{\pi}\left[\frac{1}{2}y^2 \log (x^2 + y^2) + xy \arctan \frac{y}{x} - y^2\right]$$

satisfies the conditions on the edge $y = 0$ of the semi-infinite plate indicated in Fig. 94, the uniform shear loading s extending from O indefinitely to the left. Show that σ_x grows without limit as O is approached from any direction. (This is due to the discontinuity of load at 0. A finite value is obtained when this is smoothed out, depending on the loading curve in the neighborhood of O.)

18. By superposition, using the results of Prob. 16, obtain σ_x, σ_y, τ_{xy}, for pressure p on a segment $-a < x < a$ of the straight edge of the semi-infinite plate. Show that the shear stress is

$$\tau_{xy} = -\frac{p}{\pi}\frac{4axy^2}{[(x-a)^2 + y^2][(x+a)^2 + y^2]}$$

and examine the behavior of this stress as the point $x = a$, $y = 0$ is approached (*a*) along the boundary, (*b*) along the line $x = a$.

19. Using the results of Prob. 17, sketch the variation of σ_x along the edge $y = 0$, for a uniform shear load s applied to the segment $-a < x < a$ of the edge.

20. Show that the stress function

$$\phi = \frac{p}{2\pi a}\left[\left(\frac{1}{3}x^3 + xy^2\right) \arctan \frac{y}{x} + \frac{1}{3}y^3 \log (x^2 + y^2) - \frac{1}{3}x^2 y\right]$$

satisfies the conditions on the edge $y = 0$ of the semi-infinite plate indicated in Fig. 95, the linearly increasing pressure load extending indefinitely to the left.

21. Show that if the pressure loading of Prob. 20 is replaced by shear loading, s replacing p, the appropriate stress function is

$$\phi = \frac{s}{2\pi a}\left[xy^2 \log (x^2 + y^2) + (x^2 y - y^3) \arctan \frac{y}{x} - 3xy^2\right]$$

22. Show how the distributions of load indicated in Fig. 96 may be obtained by superposition from loading of the type indicated in Fig. 95.

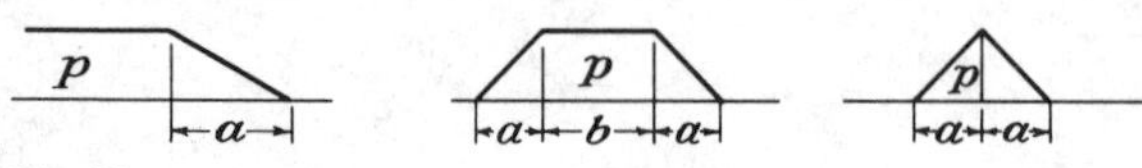

Fig. 96

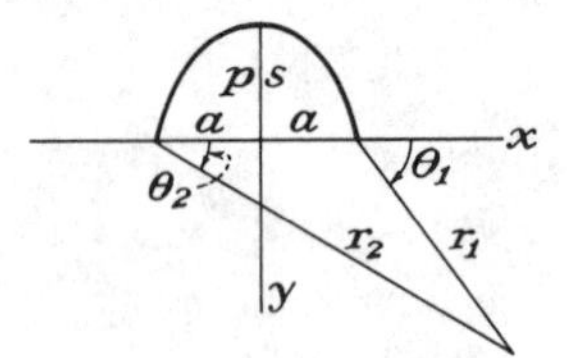

Fig. 97

23. Show that the parabolic loading indicated in Fig. 97 is given by the stress function

$$-\frac{p}{\pi}\left\{-\frac{xy^3}{3a^2}\log\frac{r_2^2}{r_1^2}-\left[\frac{a^2}{4}+\frac{1}{2}(x^2+y^2)\left(1-\frac{x^2}{6a^2}+\frac{y^2}{2a^2}\right)\right]\alpha\right.$$
$$\left.+\frac{2}{3}ax\beta+\frac{1}{2}ay\left(1-\frac{x^2}{3a^2}+\frac{y^2}{a^2}\right)\right\}$$

for pressure, and

$$\frac{s}{\pi}\left\{\frac{y^2}{6a^2}(3a^2-3x^2+y^2)\log\frac{r_2^2}{r_1^2}+\frac{2}{3}ay\beta+\frac{xy}{3a^2}(x^2-3y^2-3a^2)\alpha+\frac{4xy^2}{3a}\right\}$$

for shear, where

$$r_1^2=(x-a)^2+y^2 \qquad r_2^2=(x+a)^2+y^2$$

$$\alpha=\theta_1-\theta_2=\arctan\frac{2ay}{x^2+y^2-a^2} \qquad \beta=\theta_1+\theta_2=\arctan\frac{2xy}{x^2-y^2-a^2}$$

24. Show that in the problem of Fig. 75 there is a tensile stress $\sigma_x = 2P/\pi d$ along the vertical diameter, except at A and B. Account for the equilibrium of the semicircular part ADB by considering small semicircles about A and B in the manner of Figs. 68c and d.

25. Verify that the stress function

$$\phi=-\frac{P}{\pi}\left\{\psi r\sin\theta-\frac{1}{4}(1-\nu)r\log r\cos\theta-\frac{1}{2}r\theta\sin\theta\right.$$
$$\left.+\frac{d}{4}\log r-\frac{d^2}{32}(3-\nu)\frac{1}{r}\cos\theta\right\}$$

satisfies the boundary conditions for a force P acting in a hole in an infinite plate with zero stress at infinity, and that the circumferential stress round the hole is

$$\frac{P}{\pi d}[2+(3-\nu)\cos\theta]$$

except at A (Fig. 98).

Show that it also corresponds to single-valued displacements.

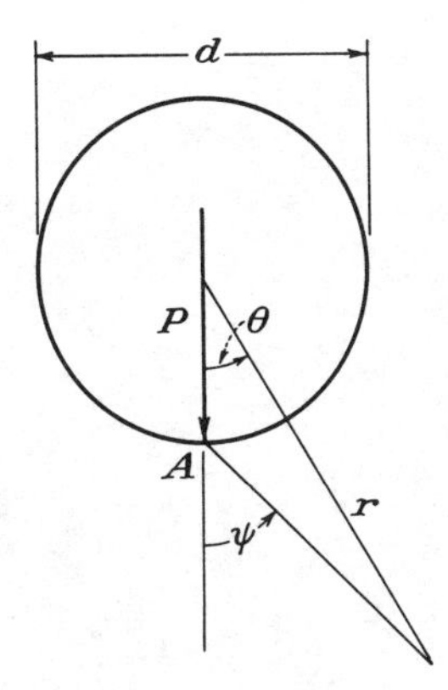

Fig. 98

26. Deduce from Prob. 25 by integration the circumferential stress round the hole due to uniform pressure p in the hole, and check the result by means of Eqs. (45).

27. Find the general form of $f(r)$ in the stress function $\theta f(r)$, and find the expressions for the stress components σ_r, σ_θ, $\tau_{r\theta}$. Could such a stress function apply to a closed ring?

chapter 5

Photoelastic and Moiré Experimental Methods

47 | Experimental Methods and Verifications

The theoretical development is resumed in Chap. 6. The present chapter is intended as an introduction to two of the principal experimental methods available for the confirmation of several features of the solutions for stress and deformation already obtained and discussed in the preceding chapters. However, the boundaries of the plates so far considered have been of simple geometrical form. For more complex shapes the difficulties of obtaining analytical solutions become formidable, but these difficulties can in most cases be avoided by resorting to numerical methods (discussed in the Appendix) or to experimental methods, such as the measurement of surface strains by strain gauges (see Art. 12), the *photoelastic* method, or the *moiré* method.

48 | Photoelastic Stress Measurement

This method is based on the discovery of David Brewster[1] that when a piece of glass is stressed and viewed by *polarized* light transmitted through it, a brilliant color pattern due to the stress is seen. He suggested that these color patterns might serve for the measurement of stresses in engineering structures such as masonry bridges, a glass model being examined in polarized light under various loading conditions. This suggestion went unheeded by engineers at the time. Comparisons of photoelastic color patterns with analytical solutions were made by the physicist Maxwell.[2] The suggestion was adopted much later by C. Wilson in a

[1] *Trans. Roy. Soc.* (*London*), 1816, p. 156.

[2] J. Clerk Maxwell, *Sci. Papers*, vol. 1, p. 30.

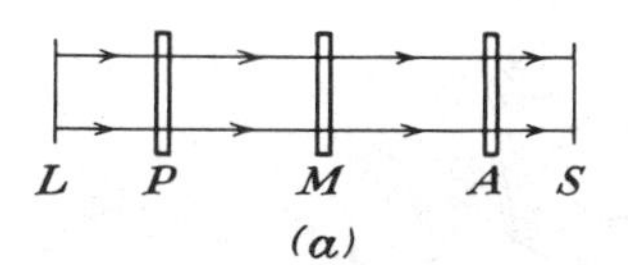

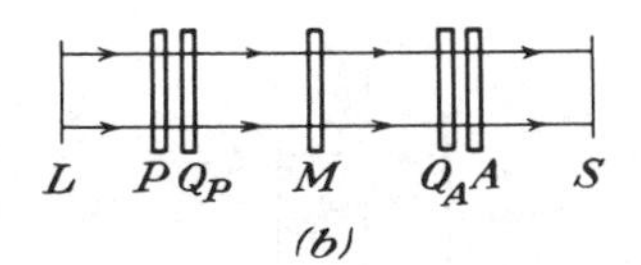

Fig. 99

study of the stresses in a beam with a concentrated load,[1] and by A. Mesnager in an investigation of arch bridges.[2] The method was devel-loped and extensively applied by E. G. Coker[3] who introduced celluloid as the model material. Later investigators have used Bakelite, Fosterite,[4] and epoxy resins.[5] For purposes of measurement, black-and-white fringe patterns obtained with monochromatic light have replaced the color patterns from white light.

In the following we consider only the simplest form of photoelastic apparatus.[6] Ordinary light is regarded as consisting of vibrations in all directions transverse to the direction of the ray. By reflection from a piece of plate glass covered on one side with black paint, or by transmission through a *polarizer*—a *Nicol prism*, or *Polaroid plate*—we obtain a more or less *polarized* beam of light in which transverse vibrations in a definite direction prevail. The plane containing this direction and a ray is the plane of *polarization*. This is the kind of light used in the photoelastic investigation of stress. We shall consider only monochromatic light.

Figure 99*a* represents diagrammatically a *plane polariscope*. A beam of light originating at L passes through a polarizer P, then through the transparent model M which modifies the light according to the stress, then through an *analyzer*—another polarizer A—to a screen S, on which a pattern of interference fringes (Figs. 101 to 105) is formed.

In Fig. 100*a*, *abcd* represents a small element of the left-hand face of

[1] *Phil. Mag.*, vol. 32, p. 481, 1891.

[2] *Ann. Ponts Chaussées*, 4ᵉ Trimestre, p. 129, 1901, and 9ᵉ Series, vol. 16, p. 135, 1913.

[3] The numerous publications of Coker are compiled in his papers: *Gen. Elec. Rev.*, vol. 23, p. 870, 1920, and *J. Franklin Inst.*, vol. 199, p. 289, 1925. See also the book by E. G. Coker and L. N. G. Filon, "Photo-elasticity," Cambridge University Press, New York, 1931.

[4] M. M. Leven, *Proc. Soc. Exptl. Stress Anal.*, vol. 6, no. 1, p. 19, 1948.

[5] See M. Hetényi, "Photoelasticity and Photoplasticity," in J. N. Goodier and N. J. Hoff (eds.), "Structural Mechanics" (*Proc. 1st Symp. Naval Structural Mech.*), pp. 483–505, Pergamon Press, New York, 1960.

[6] More complete treatments may be found in the following books: M. Hetenyi (ed.), "Handbook of Experimental Stress Analysis," John Wiley & Sons, Inc., New York, 1950; M. M. Frocht, "Photoelasticity," 2 vols., John Wiley & Sons, Inc., New York, 1941 and 1948; and the book cited in n. 3.

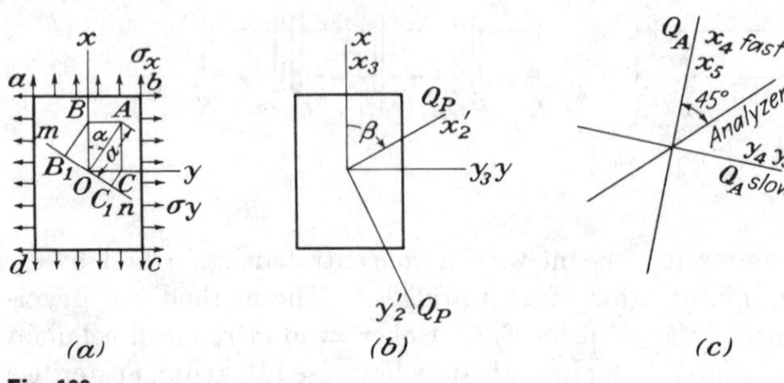

Fig. 100

the model M, the directions of the principal stresses σ_x, σ_y being drawn vertical and horizontal for convenience. A ray of light polarized in the plane OA (Fig. 100) arrives from P, the direction of the ray in Fig. 100 being through the paper. The vibration is simple-harmonic and may be represented by the transverse "displacement"

$$s = a \cos pt \tag{a}$$

in the direction OA, where p is 2π times the frequency, depending on the color of the light, and t is the time.

The displacement (a) in the plane OA is resolved into components with amplitudes $OB = a \cos \alpha$ and $OC = a \sin \alpha$ in the planes Ox, Oy respectively. The corresponding displacement components are

$$x = a \cos \alpha \cos pt \qquad y = a \sin \alpha \cos pt \tag{b}$$

The effect of the principal stresses σ_x and σ_y, acting at the point O of the plate, is to change the velocities with which these components are propagated through the plate. Let v_x and v_y denote the velocities in the planes Ox and Oy. If h is the thickness of the plate, the times required for the two components to traverse the thickness are

$$t_1 = \frac{h}{v_x} \qquad t_2 = \frac{h}{v_y} \tag{c}$$

Since the light waves are transmitted without change of form, the x displacement, x_1, of the light leaving the plate at time t corresponds to the x displacement of the light entering the plate at a time t_1 earlier. Thus,

$$x_1 = a \cos \alpha \cos p(t - t_1) \qquad y_1 = a \sin \alpha \cos p(t - t_2) \tag{d}$$

On leaving the plate, therefore, these components have a *phase difference* $\Delta = p(t_2 - t_1)$. It was established experimentally that for a given material at a given temperature, and for light of a given wavelength, this phase difference is proportional to the difference in the principal stresses.

It is also proportional to the thickness of the plate. The relationship is usually expressed in the form

$$\Delta = \frac{2\pi h}{\lambda} C(\sigma_x - \sigma_y) \tag{e}$$

where λ is the wavelength (*in vacuo*), and C the experimentally determined *stress-optical coefficient*. C depends on the wavelength and temperature as well as the material.

The analyzer A transmits only vibrations or components in its own plane of polarization. If this is at right angles to the plane of polarization of the polarizer,[1] *and if the model is removed*, no light is transmitted by A and the screen is dark. We now consider what occurs when the model is present. The components (d) on arrival at the analyzer may be represented as

$$x_2 = a \cos \alpha \cos \psi \qquad y_2 = a \sin \alpha \cos (\psi - \Delta) \tag{f}$$

since they retain the phase difference Δ in traveling from M to A. Here ψ denotes pt + constant.

The plane of polarization of A is represented by mn in Fig. 100a, for convenience. It is set perpendicular to OA. The components of the vibrations (f) which are transmitted by A are the components along Om, which are, using Eqs. (f),

$$x_2 \sin \alpha = \tfrac{1}{2}a \sin 2\alpha \cos \psi \qquad -y_2 \cos \alpha = -\tfrac{1}{2}a \sin 2\alpha \cos (\psi - \Delta)$$

The resultant vibration along mn is therefore

$$\frac{1}{2} a \sin 2\alpha \,[\cos \psi - \cos (\psi - \Delta)] = -a \sin 2\alpha \sin \frac{\Delta}{2} \sin \left(\psi - \frac{\Delta}{2}\right)$$

The factor $\sin [\psi - (\Delta/2)]$ represents the simple harmonic variation with time. The amplitude is

$$a \sin 2\alpha \sin \frac{\Delta}{2} \tag{g}$$

It follows that some light will reach the screen unless either $\sin 2\alpha = 0$ or $\sin \Delta/2 = 0$. If $\sin 2\alpha = 0$ the directions of the principal stresses are parallel to the (perpendicular) directions of polarization of P and A. Thus rays that pass through such points of M will be extinguished and the corresponding points on the screen S will be dark. These points usually lie on one or more curves, indicated by a dark band on S. Such a curve is called an "isoclinic." Very short lines parallel to the axes of P and A may be drawn at numerous points on it to record the (parallel) directions of the principal stresses at these points. By setting P and A

[1] The polarizer and analyzer are then said to be "crossed."

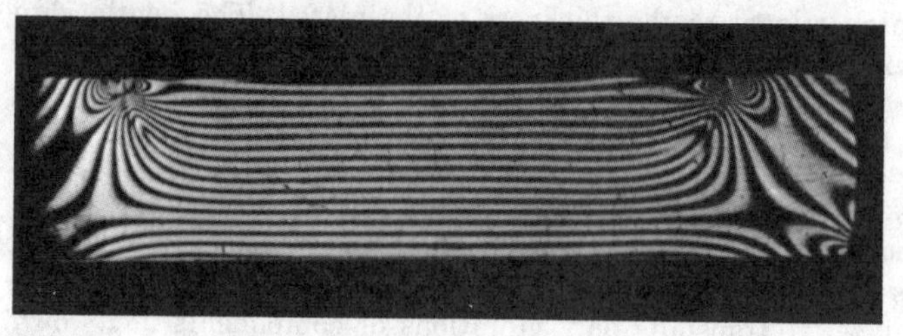

Fig. 101

in different (perpendicular) orientations, different isoclinics are obtained. The short lines then cover the field like a pattern of iron filings over a magnet, and it is possible to draw curves that are tangential at each point to the principal axes of stress. The latter lines are trajectories of the principal stresses.

If $\sin \Delta/2 = 0$, then $\Delta = 2n\pi$ where $n = 0, 1, 2, \ldots$. When $\Delta = 0$, the principal stresses are equal. Points where this occurs are called isotropic points, and will of course be dark. Points at which $n = 1$ form a dark band, or fringe, of the first order, points for which $n = 2$ form a fringe of the second order, and so on. These fringes are called *isochromatics* (because when white light is used, they correspond to extinction of a certain wavelength and therefore to a color band). It follows from Eq. (*e*) that $\sigma_x - \sigma_y$ on a fringe $n = 2$ has twice the value of $\sigma_x - \sigma_y$ on a fringe $n = 1$, and so on. To evaluate the principal stress differences, it is therefore necessary to know the order of the fringes, and the stress difference represented by the fringe of the first order, or *fringe value*.

The fringe value may be determined by loading a strip in simple tension. Since the stress is uniform there are no fringes, the whole piece appearing uniformly bright or dark on the screen. At zero load it will be dark. As the stress is increased, it will brighten, then darken as the stress difference (here simply the tensile stress) approaches the fringe value. On further increase of load it brightens once more, then darkens again when the stress is twice the fringe value, and so on.

Similar cycles of brightness and darkness will clearly occur at any point of a nonuniform stress field as the load is increased—provided the stress difference at the point reaches a multiple of the fringe value. These cycles at individual points correspond, in the view of the whole field, to gradual movement of the fringes, including the entrance of new fringes, as the load is increased. The orders of the fringes may therefore be determined by observing this movement and counting the fringes.

For instance, a strip in pure bending gives a fringe pattern as shown in Fig. 101. The parallel fringes accord with the fact that in the portion of the strip away from the points of application of the loads, the stress distribution is the same in all vertical cross sections. By watching the screen as the load is gradually increased we should observe that new fringes appear at the top and bottom of the strip and move toward the middle, the fringes as a whole becoming more and more closely packed. There will be one fringe at the neutral axis that remains dark throughout. This will clearly be the fringe of zero order ($n = 0$).

49 | Circular Polariscope

We have seen that the plane polariscope just discussed provides, for a chosen value of α, the corresponding isoclinic as well as the isochromatics or fringes. Figure 101 should therefore show darkness wherever the orientations of the principal stresses coincide with the orientations of the polarizer and analyzer. Figure 101 was actually obtained in a *circular polariscope*, which is a modification of the *plane polariscope* designed to eliminate the isoclinics.[1] It is indicated diagrammatically in Fig. 99*b*, which corresponds to Fig. 99*a* with the addition of two *quarter-wave plates* Q_P, Q_A. A quarter-wave plate is a crystal plate having two polarizing axes, which affect the light like a uniformly stressed model, introducing a phase difference Δ as in Eq. (*f*)—but the thickness of the quarter-wave plate is chosen so that $\Delta = \pi/2$. Using Eq. (*f*) with this value of Δ for the light leaving Q_P, we observe that a simple result is obtained by choosing α, now denoting the angle between the plane of polarization of P and one of the axes of Q_P, as 45°. Then we may write

$$x_2' = \frac{a}{\sqrt{2}} \cos \psi \qquad y_2' = \frac{a}{\sqrt{2}} \cos \left(\psi - \frac{\pi}{2}\right) = \frac{a}{\sqrt{2}} \sin \psi \tag{h}$$

Here x_2' corresponds to the "fast" axis of the quarter-wave plate. A point moving with these displacement components (ψ always having the form pt + constant for a given point along the ray) moves in a circle. Such light is therefore described as *circular polarized*.

The components (*h*) are along the axes of polarization of Q_P. Using β for the angle between x_2' and the direction of σ_x in the model (Fig. 100*b*), and Δ once more for the phase difference caused by the stressed element, we have, for the light leaving the model, due to x_2' only

$$x_3 = \frac{a}{\sqrt{2}} \cos \beta \cos \psi \qquad y_3 = \frac{a}{\sqrt{2}} \sin \beta \cos (\psi - \Delta) \tag{i}$$

[1] If the polarizer and analyzer rotate, their axes remaining perpendicular, the fringes remain stationary, and the isoclinics move. If the rotation is rapid the isoclinics are no longer visible. The circular polariscope achieves the same effect by purely optical means.

and for the light due to y_2' only

$$x_3 = -\frac{a}{\sqrt{2}} \sin \beta \sin \psi \qquad y_3 = \frac{a}{\sqrt{2}} \cos \beta \sin (\psi - \Delta) \tag{j}$$

Adding the components in Eqs. (*i*) and (*j*), we find for the light leaving the model

$$x_3 = \frac{a}{\sqrt{2}} \cos \psi' \qquad y_3 = \frac{a}{\sqrt{2}} \sin (\psi' - \Delta) \tag{k}$$

where $\psi' = \psi + \beta$.

Before examining the effects of Q_A and A upon the light it is convenient to represent the motion (*k*) as a superposition of two circular motions. This may be done as follows. Denoting $\psi' - (\Delta/2)$ by ψ'', and $a/\sqrt{2}$ by b, Eqs. (*k*) give

$$x_3 = b \cos \left(\psi'' + \frac{\Delta}{2}\right) = b \left(\cos \frac{\Delta}{2} \cos \psi'' - \sin \frac{\Delta}{2} \sin \psi''\right) \tag{l}$$

$$y_3 = b \sin \left(\psi'' - \frac{\Delta}{2}\right) = b \left(\cos \frac{\Delta}{2} \sin \psi'' - \sin \frac{\Delta}{2} \cos \psi''\right) \tag{m}$$

which represent the superposition of a circular motion of radius $b \cos (\Delta/2)$, clockwise in Fig. 100*b* (where the ray passes downward through the paper), and a circular motion of radius $b \sin (\Delta/2)$, counterclockwise.

We may now show that if the polarizing axis of A is set at 45° to the polarizing axes of Q_A, one of these circular motions is transmitted to the screen S, the other extinguished, and the desired result—isochromatics without isoclinics—is obtained.

The components x_3, y_3 in Eqs. (*l*) and (*m*) are along the directions of principal stress in a model. A change of axes for a circular motion will merely result in a change of the phase angle ψ'' by a constant. Thus, the clockwise circular motion can be represented by components of the form

$$x_4 = c \cos \psi \qquad y_4 = c \sin \psi \tag{n}$$

along the axes of Q_A, where ψ is again of the form pt + constant. Identifying x_4 with the *fast* axis of Q_A, we shall have on emergence from Q_A

$$x_5 = c \cos \psi \qquad y_5 = c \sin \left(\psi - \frac{\pi}{2}\right) = -c \cos \psi \tag{o}$$

ψ having again changed by a constant.

If we now set the analyzer (A) axis at 45° to Ox_4 and Oy_4 (Fig. 100*c*), the components of the displacements (*o*) along it give

$$c \cos 45° \cos \psi - c \cos 45° \cos \psi$$

or *zero*. Thus the clockwise circular motion is extinguished.

Considering in the same way the counterclockwise part of the motion of Eqs. (l) and (m), that is,

$$x_4' = -c \sin \psi \qquad y_4' = -c \cos \psi \tag{n'}$$

we find that the transmitted displacement along the analyzer axis is

$$-c \cos 45° \sin \psi - c \cos 45° \sin \psi$$

and the amplitude is thus

$$\sqrt{2}\, c \qquad \text{or} \qquad \sqrt{2}\, b \sin \frac{\Delta}{2} \qquad \text{or} \qquad a \sin \frac{\Delta}{2} \tag{p}$$

remembering that b denotes $a/\sqrt{2}$ and that a is the amplitude leaving the polarizer. No account has been taken, of course, of loss of light in the apparatus. Comparing this result with the result (g) for the plane polariscope, we observe that the factor $\sin 2\alpha$ is now absent, and therefore the isochromatics appear on the screen, but no isoclinics.

If Δ is zero, the amplitude (p) is also zero. Thus if there is no model, or if the model is unloaded, the screen is dark. We have a *dark field* setting. If the analyzer axis is turned through 90° with respect to Q_A we should have a *light field* and light fringes taking the place of the former dark fringes. The same effect is brought about in the plane polariscope by having the polarizer and analyzer axes parallel instead of at right angles.

50 | Examples of Photoelastic Stress Determination

The photoelastic method has yielded especially important results in the study of stress concentration at the boundaries of holes and reentrant corners. In such cases, the maximum stress is at the boundary, and it can be obtained directly by the optical method because one of the principal stresses vanishes at the free boundary.

Figure 102 shows the fringe pattern of a curved bar[1] bent by couples M. The outer radius is three times the inner. The order numbers of the fringes marked on the right-hand end show a maximum of 9 at both top and bottom. The regular spacing corresponds to the linear distribution of bending stress in the straight shank. The fringe orders marked along the top edge show the stress distribution in the curved part (the complete model continues above this top edge, which is its axis of symmetry), indicating a compressive stress at the inside represented by 13.5, and a tensile stress at the outside represented by 6.7. These values are

[1] E. E. Weibel, *Trans. ASME*, vol. 56, p. 637, 1934.

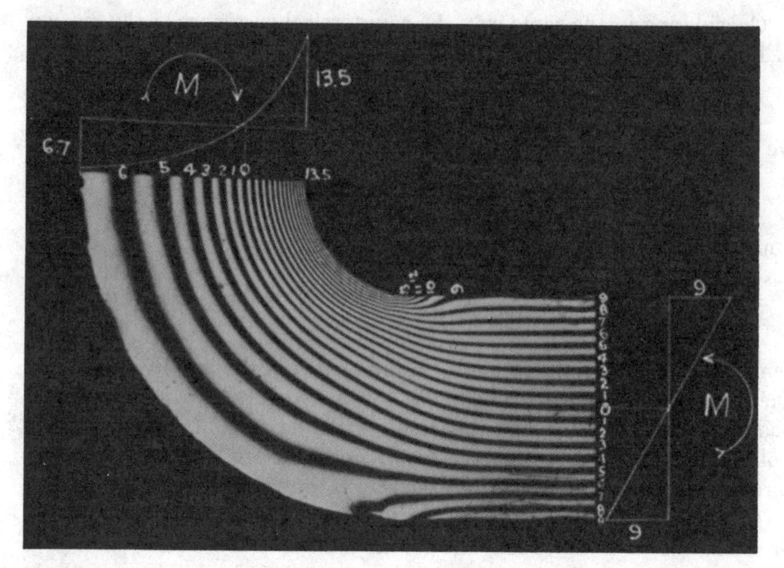

Fig. 102

in very close agreement, proportionally, with the theoretical "exact-solution" values in the last line of the table on page 74.

Figures 103 and 104 represent[1] the case of bending of a beam by a force applied at the middle. The density of distribution of dark fringes indicates high stresses near the point of application of the load. The number of fringes crossing a cross section diminishes as the distance of the cross section from the middle of the beam increases. This is due to decrease in bending moment.

Figure 105 represents the stress distribution in a plate of two different widths submitted to centrally applied tension. It is seen that the max-

[1] M. M. Frocht, *Trans. ASME*, vol. 53, 1931.

Fig. 103

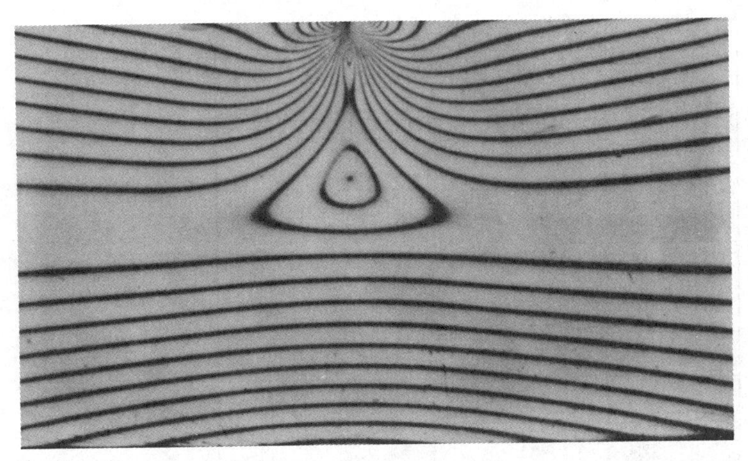

Fig. 104

imum stress occurs at the ends of the fillets. The ratio of this maximum stress to the average stress in the narrower portion of the plate is called the *stress-concentration factor*. It depends on the ratio of the radius R of the fillet to the width d of the plate. Several values of the stress-concentration factor obtained experimentally[1] are given in Fig. 106. It is seen that the maximum stress is rapidly increasing as the ratio R/d is decreasing, and when $R/d = 0.1$ the maximum stress is more than twice the average tensile stress. Figure 107 represents the same plate submitted to

[1] See paper by Weibel, *loc. cit.*

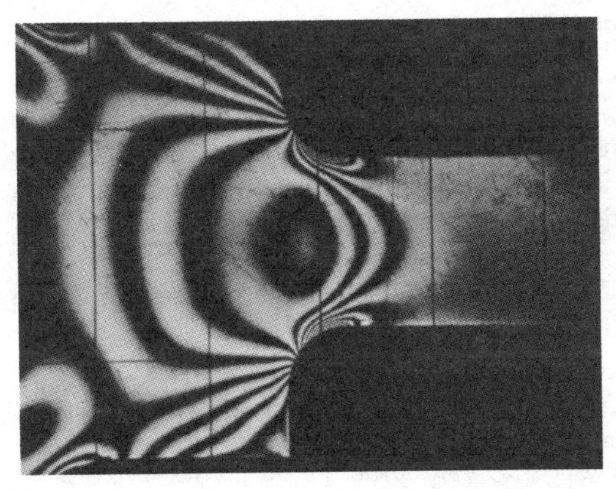

Fig. 105

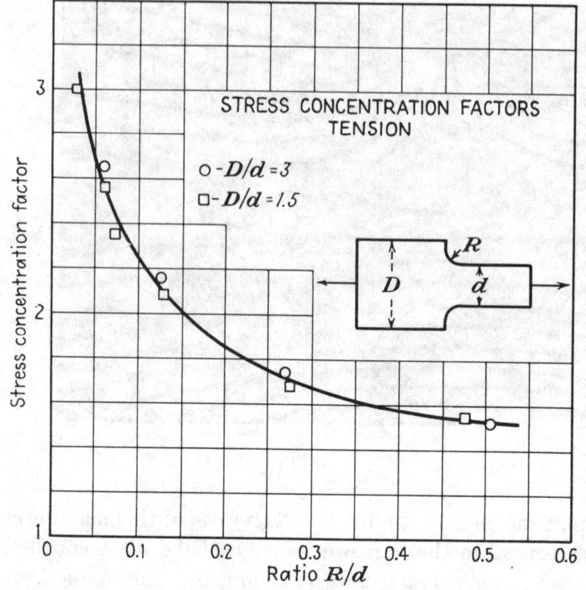

Fig. 106

pure bending by a couple applied at the end and acting in the middle plane of the plate. Figure 108 gives the stress-concentration factors,[1] defined as the ratio of greatest stress found at the fillet to the ordinary extreme fiber stress in the narrower part (d). When the design limits the space available for the fillet, an elliptic form[2] may be preferable to the circular.

51 | Determination of the Principal Stresses

The ordinary polariscope, as we have seen, determines only the *difference* of the principal stresses, and their directions. When it is required to determine the principal stresses throughout the model, or at a boundary where there is unknown loading, further measurement, or calculation, is required. Many methods have been used, or proposed. Only a brief description of some of these will be given here.[3]

The sum of the principal stresses can be found by measuring the

[1] Many similar curves for a variety of cases, obtained by theoretical and experimental methods, are given in the books R. E. Peterson, "Stress Concentration Design Factors," John Wiley & Sons, Inc., New York, 1953, and R. B. Heywood, "Designing by Photoelasticity," Chapman & Hall, Ltd., London, 1952.

[2] M. M. Frocht and D. Landsberg, *J. Appl. Mech.*, vol. 26, pp. 448–450, 1959.

[3] For further information see the references cited in footnote 6 on p. 151.

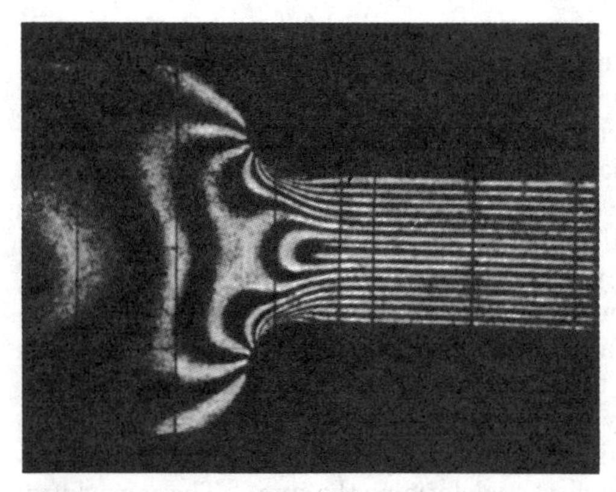

Fig. 107

changes in the thickness of the plate.[1] The decrease in thickness due to the stress is

$$\Delta h = \frac{h\nu}{E}(\sigma_x + \sigma_y) \tag{a}$$

whence $\sigma_x + \sigma_y$ may be calculated if Δh is measured at each point where the stresses are to be evaluated. Several special forms of extensometer have been designed for this purpose.[2] The pattern of interference fringes

[1] This method was suggested by Mesnager, *loc. cit.*

[2] See M. M. Frocht, "Photoelasticity," *op. cit.*, vol. 2.

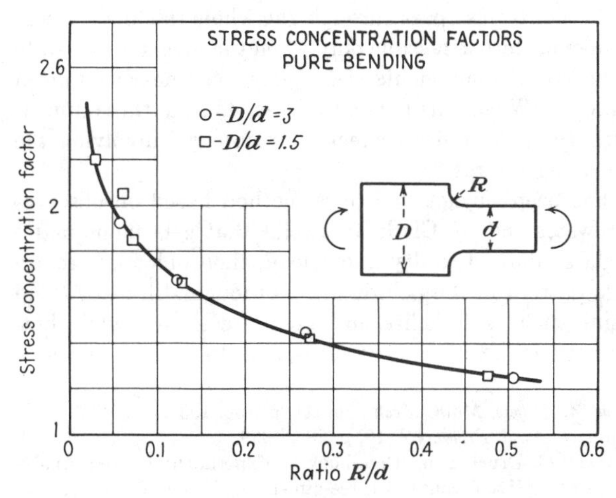

Fig. 108

formed when a model is placed against an optical flat, so as to form an air film with thickness variations determined by the thickness variations in the plate, yields the required information in a single photograph.

The differential equation satisfied by the sum of the principal stresses, Eq. (b) on page 31, is also satisfied by the deflection of a membrane of constant tension, such as a soap film, and if the boundary values are made to correspond, the deflection represents $\sigma_x + \sigma_y$ to a certain scale.[1] In many cases the boundary values of $\sigma_x + \sigma_y$ required for the construction of the membrane can be found from the photoelastic fringe pattern. The latter gives $\sigma_x - \sigma_y$. At a free boundary one principal stress, say σ_y, is zero, and $\sigma_x + \sigma_y$ becomes the same as $\sigma_x - \sigma_y$. Also at a boundary point where the loading is purely normal to the boundary and of known magnitude, it constitutes one principal stress itself, and the photoelastic measurement of the difference suffices to determine the sum. The same differential equation is satisfied by the electric potential in flow of current through a plate, and this can be made the basis of an electrical method.[2] Effective numerical methods have been developed as alternatives to these experimental procedures. These are discussed in the Appendix. The principal stresses can also be determined by purely photoelastic observations, more elaborate than those considered in Arts. 48 and 49.[2]

52 | Three-dimensional Photoelasticity

The models used in the ordinary photoelastic test are loaded at room temperature, are elastic, and the fringe pattern disappears when the load is removed. Since the light must pass through the whole thickness, interpretation of the fringe pattern is feasible only when the model is in a state of plane stress—the stress components then being very nearly uniform through the thickness. When this is not the case, as in a three-dimensional stress distribution, the optical effect is an integral involving the stress at all pcints along the ray.[3]

This difficulty has been surmounted by a method based on observations made by Brewster and by Clerk Maxwell,[4] that gelatinous materials, such as isinglass, allowed to dry under load, then unloaded, retain a permanent fringe pattern in the polariscope as though still loaded and still elastic. Resins such as Bakelite and Fosterite loaded while hot, then cooled, have been found by later investigators to possess the same

[1] J. P. Den Hartog, *Z. Angew. Math. Mech.*, vol. 11, p. 156, 1931.

[2] See R. D. Mindlin, *J. Appl. Phys.*, vol. 10, p. 282, 1939.

[3] See the article by D. C. Drucker in "Handbook of Experimental Stress Analysis," which gives a comprehensive account of three-dimensional photoelasticity.

[4] *Loc. cit.*

property. The explanation[1] is that these materials have the structure of a strong elastic skeleton, or molecular network, which is unaffected by heat, the spaces being filled by a mass of loosely bonded molecules that softens on heating. When the hot specimen is loaded, the elastic skeleton bears the load and is elastically deformed without hindrance. On cooling, the softened mass in which this skeleton is embedded becomes "frozen" and holds the skeleton almost to the same deformation even when the load is removed. The optical effect is likewise substantially retained and is not disturbed by cutting the specimen into pieces. A three-dimensional specimen may therefore be cut into thin slices, and each slice may be examined in the polariscope. The state of stress that produced the optical effect in the slice is not plane stress, but the other components τ_{xz}, τ_{yz}, σ_z are known to have no effect on a ray along the z direction, that is, normal to the slice. The fringe pattern shown in Fig. 109 was obtained from such a slice cut centrally from a round shaft (of Fosterite) with a hyperbolic groove[2] loaded in tension. The maximum stress obtained from this pattern is within two or three percent of the theoretical value.[3] Figure 110 shows another fringe pattern of the same type, obtained from a

[1] M. Hetényi, *J. Appl. Phys.*, vol. 10, p. 295, 1939.

[2] Leven, *loc. cit.*

[3] H. Neuber, "Kerbspannungslehre," p. 39, Springer-Verlag OHG, Berlin, 1958.

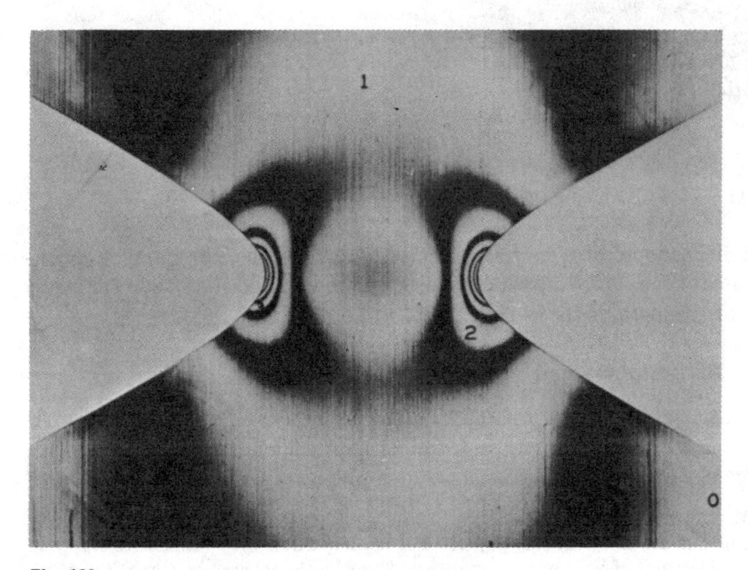

Fig. 109

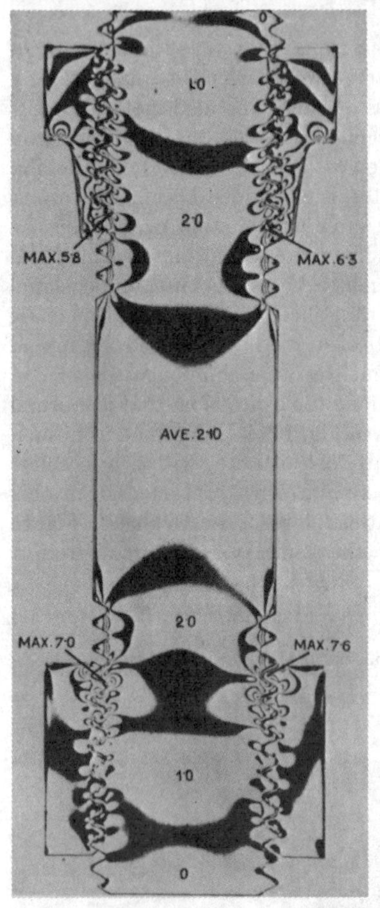

Fig. 110

(Bakelite) model of a bolt and nut fastening.[1] The lower nut is a conventional type. The upper one has a tapered lip and shows a lower stress concentration than the conventional nut.

53 | The Moiré Method

Fringe patterns directly related to *displacements* can be formed in a very simple way. As an illustration, we consider a plate that is to undergo a simple shearing deformation in its own plane. Figure 111*a* represents a

[1] M. Hetényi, *J. Appl. Mech.*, vol. 10, p. A-93, 1943. Results for several other forms of nut are given in this paper. For a comparison with fatigue tests and an extensive account of both experimental and theoretical analysis, see Chap. 7 in R. B. Heywood, *op. cit.*

grid of parallel uniformly spaced lines drawn on the plate. The shearing deformation carries these lines into tilted positions as in Fig. 111*b*. The rotation of each line is about its midpoint, which accordingly does not move. The *vertical* displacement field is evident from these lines. The original grid (Fig. 111*a*) is preserved as a copy on a transparent film, and this is superimposed on the deformed grid of Fig. 111*b*, as shown in Fig. 111*c*. We have a pattern of "points" of intersection of the two grids. Seen from a distance, or with half-closed eyes, these form rather broad vertical dark bands. Between them, as in the middle part of Fig. 111*c*, we have a stack of white "lozenge" areas, forming a vertical light band. Traversing a vertical line through the middle of the figure, we traverse a light "fringe" along which the vertical displacement is zero. We cross seven black lines, which constitute "seven-line shading." But traversing a vertical line through the adjacent dark band on the right (or left) we cross 13 black lines, and have "13-line shading." The midline of this dark fringe clearly connects points that have vertical displacement equal to one spacing (δ) of the original grid. The next vertical light fringe to the right corresponds to vertical displacement 2δ, and so on. Clearly the fringes are "contour" lines of (equal) vertical displacement.

It is also evident that if the deformation had carried the original grid into *curved* lines, the dark and light fringes would have been also curved, but they would still be contour lines of vertical displacement. If, further, the lines are stretched or contracted, the fringes are contours of vertical displacement *components*.

Similarly, a vertical grid, in place of Fig. 111*a*, will lead to fringes that are contours of horizontal displacement components.

This is, in one form, the *moiré method*.[1] The moiré fringes are some-

[1] For a general account of the optical aspects see M. Stecher, *Am. J. Phys.*, vol. 32, pp. 247–257, 1964. A comprehensive survey and bibliography is given by P. S. Theocaris in H. N. Abramson, H. Liebowitz, J. M. Crowley, and S. Juhasz (eds.), "Applied Mechanics Surveys," pp. 613–626, Spartan Books, Washington D.C., 1966.

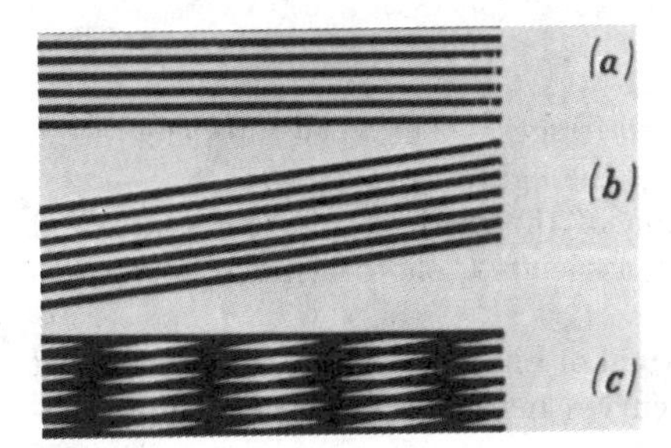

Fig. 111

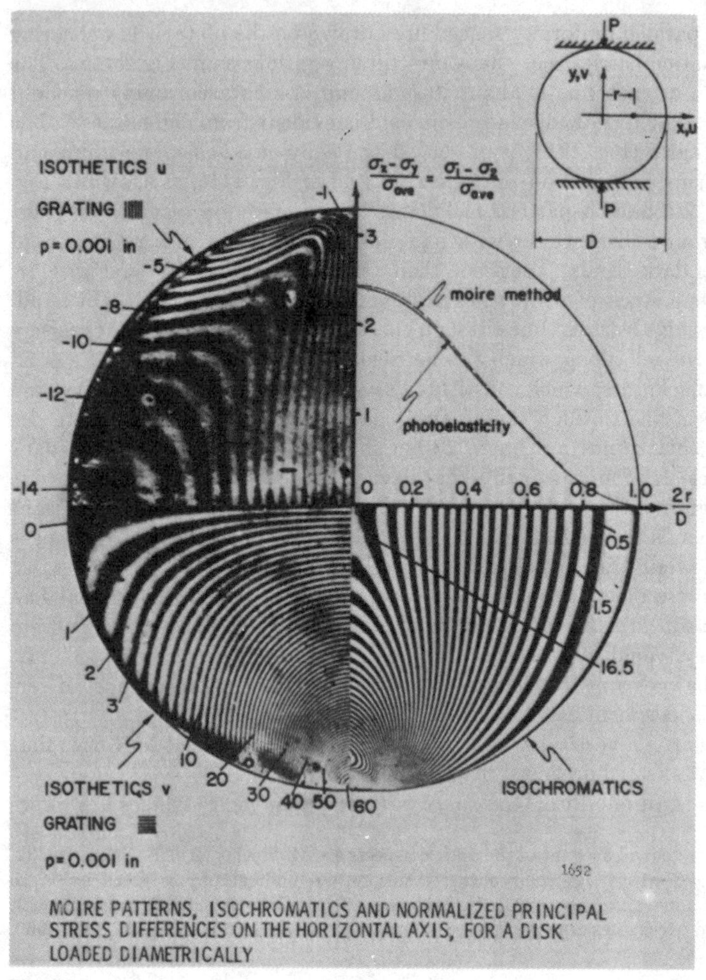

Fig. 112

times called mechanical interference fringes. The ruled lines create the dark fringes simply by blocking out the light.

Figure 112 shows an application of the method. It represents final results of a series of moiré and associated photoelastic observations conducted by A. J. Durelli.[1]

The inset diagram at the upper right of Fig. 112 shows a disk compressed by two forces P. Below it, the curves marked *moiré method* and *photo-*

[1] The authors are indebted to Prof. Durelli for this composite figure.

elasticity show the values of the principal stress difference $\sigma_1 - \sigma_2$ along the horizontal diameter, as a ratio to σ_{av}, which is the average compressive stress on this diametral section. The close agreement of the two curves indicates the accuracy attainable by the moiré method. The transition from displacement values to stress involves differentiation. The upper-left quadrant in Fig. 112 shows the moiré contours of horizontal displacement. The lower-left quadrant shows the contours of vertical displacement.

The analytical solution of the disk problem, for concentrated forces, was given in Art. 41. For the center of the disk of diameter d and unit thickness,

$$\sigma_1 = \frac{2P}{\pi d} \qquad \sigma_2 = \frac{-6P}{\pi d} \qquad \sigma_{av} = \frac{P}{d}$$

The theoretical value of $(\sigma_1 - \sigma_2)/\sigma_{av}$ is $8/\pi$, that is, 2.55. This is higher than the value shown by the experimental curves in Fig. 112, which is about 2.4. The stress at the center can be affected by the actual distribution of the forces P on the small contact areas of the model.[1]

[1] J. N. Goodier, *Trans. ASME*, vol. 54, pp. 173–183, 1932.

chapter | 6

Two-dimensional Problems in Curvilinear Coordinates

54 | Functions of a Complex Variable

For the problems solved so far, rectangular and polar coordinates have proved adequate. For other boundaries—ellipses, hyperbolas, nonconcentric circles, and less simple curves—it is usually preferable to employ different coordinates. In the consideration of these, and also in the construction of suitable stress functions, it is advantageous to use complex variables.

Two real numbers x, y form the complex number $x + iy$, with i representing $\sqrt{-1}$. Since i does not belong to the real-number system, the meaning of equality, addition, subtraction, multiplication, and division must be defined.[1] Thus, by definition, $x + iy = x' + iy'$ means $x = x'$, $y = y'$, and i^2 means -1. Otherwise the operations are defined just as for real numbers. For instance

$$(x + iy)^2 = x^2 + 2xiy + (iy)^2 = x^2 - y^2 + i2xy \qquad \text{since } i^2 = -1$$

Converting to polar coordinates, as in Fig. 113,

$$z = x + iy = r(\cos\theta + i\sin\theta) \tag{a}$$

Since

$$\cos\theta + i\sin\theta = 1 - \frac{1}{2!}\theta^2 + \frac{1}{4!}\theta^4 - \cdots + i\left(\theta - \frac{1}{3!}\theta^3 + \cdots\right)$$

and

$$i^2 = -1 \qquad i^3 = -i \qquad i^4 = 1 \qquad \text{etc.}$$

we have

$$\cos\theta + i\sin\theta = 1 + i\theta + \frac{1}{2!}(i\theta)^2 + \frac{1}{3!}(i\theta)^3 + \cdots = e^{i\theta}$$

[1] The definitions represent operations on *pairs* of *real* numbers, the use of i being merely a convenience. See for instance E. T. Whittaker and G. N. Watson, "Modern Analysis," 3d ed., pp. 6–8, Cambridge University Press, New York, 1920.

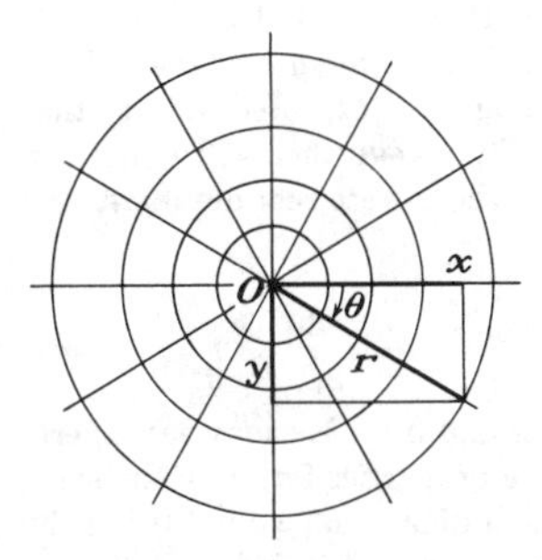

Fig. 113

the last line being the definition of the symbol $e^{i\theta}$ when θ is real. From Eq. (*a*) therefore

$$z = x + iy = re^{i\theta} \tag{b}$$

Algebraic, trigonometric, exponential, logarithmic, and other functions can be formed from z as well as from a real variable, provided an analytical rather than a geometrical definition is adopted. Thus $\sin z$, $\cos z$, and e^z may be defined by their power series. Any such function can be separated into "real" and "imaginary" parts, that is, put in the form $\alpha(x,y) + i\beta(x,y)$ where $\alpha(x,y)$, the real part, and $\beta(x,y)$, the imaginary part,[1] are ordinary real functions of x and y (they do not contain i). For instance, if the function of z, $f(z)$ is $1/z$, we have

$$f(z) = \frac{1}{x + iy} = \frac{x - iy}{(x + iy)(x - iy)} = \frac{x}{x^2 + y^2} + i\frac{(-y)}{x^2 + y^2}$$

Wherever possible, it is usually most expeditious to introduce the exponential functions in separating the real and imaginary parts. For instance

$$\begin{aligned}\sinh z &= \tfrac{1}{2}[e^{x+iy} - e^{-(x+iy)}] \\ &= \tfrac{1}{2}[(e^x - e^{-x})\cos y + (e^x + e^{-x})i \sin y] \\ &= \sinh x \cos y + i \cosh x \sin y\end{aligned}$$

and, similarly,

$$\cosh z = \cosh x \cos y + i \sinh x \sin y$$

The *conjugate* of a complex function is, by definition, obtained by substitution of $-i$ for i throughout. The product of the function and its conjugate is evidently real. In (*b*) the conjugate of $x + iy$, that is, $x - iy$, was used to obtain a real denominator. Following the same general rule we can effect the separation of $\coth z$ thus

$$\coth z = \frac{\cosh z}{\sinh z} = \frac{(e^{x+iy} + e^{-x-iy})}{(e^{x+iy} - e^{-x-iy})} \frac{(e^{x-iy} - e^{-x+iy})}{(e^{x-iy} - e^{-x+iy})}$$

Multiplying out in numerator and denominator, this reduces to

$$\coth z = \frac{\sinh 2x - i \sin 2y}{\cosh 2x - \cos 2y} \tag{c}$$

The derivative of $f(z)$ with respect to z is by definition

$$\frac{df(z)}{dz} = \lim_{\Delta z \to 0} \frac{f(z + \Delta z) - f(z)}{\Delta z} \tag{d}$$

[1] It should be observed that this is real in spite of its name.

where $\Delta z = \Delta x + i\,\Delta y$ and $\Delta z \to 0$ means of course, both $\Delta x \to 0$ and $\Delta y \to 0$. We can always think of x, y as the cartesian coordinates of a point in a plane. Then Δx, Δy represent a shift to a neighboring point. It might be expected at first that (d) could be different for different directions of the shift. Nevertheless, the limit in (d) is calculable directly in terms of z and Δz just as if these were real numbers, and the corresponding results, such as

$$\frac{d}{dz}(z^2) = 2z \qquad \frac{d}{dz}\sin z = \cos z$$

must appear, independent of the choice of Δz, and of Δx and Δy. We may say, therefore, that all the functions we may form from z to have properties formally the same as those of the real functions will have derivatives which depend on z only, being the same for all directions (of Δz) at the point z. Such functions are called *analytic.*

The quantity $x - iy$ may be regarded as a function of z, in the sense that if z is given, x and y are given, and so $x - iy$ is determined. Its derivative with respect to z is the limit of $(\Delta x - i\,\Delta y)/(\Delta x + i\,\Delta y)$ as Δx, $\Delta y \to 0$. This is not independent of the direction of the shift Δx, Δy. If we take this shift in the x direction, so that $\Delta y = 0$, we obtain 1 as the value of the limit. If we take the shift in the y direction, $\Delta x = 0$ and the limit is -1. Thus $x - iy$ is not an *analytic* function of $x + iy$. Analytic functions together with $x - iy$ will be used later in the construction of stress functions. Any function involving i will be referred to as a "complex function."

An analytic function $f(z)$ will have an indefinite integral, defined as the function having $f(z)$ as its derivative with respect to z, and written $\int f(z)\,dz$. For instance, if $f(z) = 1/z$ we have

$$\int \frac{1}{z}\,dz = \log z + C$$

the additive constant C being now a complex number $A + iB$, containing two real arbitrary constants A and B.

55 | Analytic Functions and Laplace's Equation

An analytic function $f(z)$ can be regarded as a function of x and y, having partial derivatives. Thus

$$\frac{\partial}{\partial x} f(z) = \frac{d}{dz} f(z)\frac{\partial z}{\partial x} = f'(z)\frac{\partial z}{\partial x} = f'(z) \tag{a}$$

since $\partial z/\partial x = 1$. Similarly,

$$\frac{\partial}{\partial y} f(z) = f'(z)\frac{\partial z}{\partial y} = if'(z) \tag{b}$$

since $\partial z/\partial y = i$.

But if $f(z)$ is put in the form $\alpha(x,y) + i\beta(x,y)$, or for brevity $\alpha + i\beta$, we have

$$\frac{\partial}{\partial x} f(z) = \frac{\partial \alpha}{\partial x} + i\frac{\partial \beta}{\partial x} \qquad \text{and} \qquad \frac{\partial}{\partial y} f(z) = \frac{\partial \alpha}{\partial y} + i\frac{\partial \beta}{\partial y} \tag{c}$$

Comparing Eqs. (c) with Eqs. (a) and (b) yields

$$i\left(\frac{\partial \alpha}{\partial x} + i\frac{\partial \beta}{\partial x}\right) = \frac{\partial \alpha}{\partial y} + i\frac{\partial \beta}{\partial y} \tag{d}$$

Remembering that α, β are real, $i^2 = -1$, and that the equality implies that real and imaginary parts are separately equal, we find

$$\frac{\partial \alpha}{\partial x} = \frac{\partial \beta}{\partial y} \qquad \frac{\partial \alpha}{\partial y} = -\frac{\partial \beta}{\partial x} \tag{e}$$

These are called the *Cauchy-Riemann equations*. Eliminating β by differentiating the first with respect to x, the second with respect to y, and adding, we obtain

$$\frac{\partial^2 \alpha}{\partial x^2} + \frac{\partial^2 \alpha}{\partial y^2} = 0 \tag{f}$$

An equation of this form is called *Laplace's equation* and any solution is called a *harmonic function*. In the same way, elimination of α from Eqs. (e) yields

$$\frac{\partial^2 \beta}{\partial x^2} + \frac{\partial^2 \beta}{\partial y^2} = 0 \tag{g}$$

Thus if two functions α and β of x and y are derived as the real and imaginary parts of an analytic function $f(z)$, each will be a solution of Laplace's equation. Laplace's equation is encountered in many physical problems, including those of elasticity [see for instance Eq. (b), Art. 17].

The functions α and β are called *conjugate* harmonic functions. It is evident that if we are given any harmonic function α, Eqs. (e) will, but for a constant, determine another function β, which will be the conjugate to α.

As examples of the derivation of harmonic functions from analytic functions of z, consider e^{inz}, z^n, $\log z$, n being a real constant. We have

$$e^{inz} = e^{inx}e^{-ny} = e^{-ny}\cos nx + ie^{-ny}\sin nx$$

showing that $e^{-ny}\cos nx$, $e^{-ny}\sin nx$ are harmonic functions. Changing n to $-n$ we find that $e^{ny}\cos nx$, $e^{ny}\sin nx$ are also harmonic, and it follows that

$$\sinh ny \sin nx \qquad \cosh ny \sin nx \qquad \sinh ny \cos nx \qquad \cosh ny \cos nx \tag{h}$$

are harmonic since they can be formed by addition and subtraction of the foregoing functions with factors $\frac{1}{2}$. From

$$z^n = (re^{i\theta})^n = r^n e^{in\theta} = r^n \cos n\theta + ir^n \sin n\theta$$

we find the harmonic functions

$$r^n \cos n\theta \qquad r^n \sin n\theta \qquad r^{-n} \cos n\theta \qquad r^{-n} \sin n\theta \tag{i}$$

From

$$\log z = \log re^{i\theta} = \log r + i\theta$$

we find the harmonic functions

$$\log r \qquad \theta \tag{j}$$

It is easily verified that the functions (i) and (j) satisfy Laplace's equation in polar coordinates [see Eq. (h), page 68], that is,

$$\frac{\partial^2 \psi}{\partial r^2} + \frac{1}{r}\frac{\partial \psi}{\partial r} + \frac{1}{r^2}\frac{\partial^2 \psi}{\partial \theta^2} = 0 \tag{k}$$

PROBLEMS

1. Determine the *real* functions of x and y which are the *real* and *imaginary* parts of the complex functions z^2, z^3, $\tanh z$.

   [$x^2 - y^2$, $2xy$; $x^3 - 3xy^2$, $3x^2y - y^3$;
   $\sinh 2x(\cosh 2x + \cos 2y)^{-1}$, $\sin 2y(\cosh 2x + \cos 2y)^{-1}$]

2. Determine the *real* functions of r and θ that are the *real* and *imaginary* parts of the complex functions z^{-2}, $z \log z$.

$$[r^{-2}\cos 2\theta,\ -r^{-2}\sin 2\theta;\ r\log r\cos\theta - r\theta\sin\theta,\ r\log r\sin\theta + r\theta\cos\theta]$$

3. If ζ is a complex variable, and $z = c\cosh\zeta$, find in terms of ζ

$$\frac{d}{dz}\sinh n\zeta$$

Writing $\zeta = \xi + i\eta$ find the real and imaginary parts of this derivative when c and n are real.

4. If $z = x + iy$, $\zeta = \xi + i\eta$, and $z = ia\coth\frac{1}{2}\zeta$ where a is real, show that

$$x = \frac{a\sin\eta}{\cosh\xi - \cos\eta} \qquad y = \frac{a\sinh\xi}{\cosh\xi - \cos\eta}$$

56 | Stress Functions in Terms of Harmonic and Complex Functions

If ψ is any function of x and y, we have by differentiation

$$\left(\frac{\partial^2}{\partial x^2} + \frac{\partial^2}{\partial y^2}\right)(x\psi) = x\left(\frac{\partial^2\psi}{\partial x^2} + \frac{\partial^2\psi}{\partial y^2}\right) + 2\frac{\partial\psi}{\partial x} \qquad (a)$$

If ψ is harmonic, the parenthesis on the right is zero. Also $\partial\psi/\partial x$ is a harmonic function, since

$$\left(\frac{\partial^2}{\partial x^2} + \frac{\partial^2}{\partial y^2}\right)\left(\frac{\partial\psi}{\partial x}\right) = \frac{\partial}{\partial x}\left(\frac{\partial^2\psi}{\partial x^2} + \frac{\partial^2\psi}{\partial y^2}\right) = 0$$

Thus another application of the laplacian operation to (a) yields

$$\left(\frac{\partial^2}{\partial x^2} + \frac{\partial^2}{\partial x^2}\right)\left(\frac{\partial^2}{\partial x^2} + \frac{\partial^2}{\partial y^2}\right)(x\psi) = 0 \qquad (b)$$

which is the same as

$$\left(\frac{\partial^4}{\partial x^4} + 2\frac{\partial^4}{\partial x^2\partial y^2} + \frac{\partial^4}{\partial y^4}\right)(x\psi) = 0$$

Comparison with Eq. (a), page 35, shows that $x\psi$ may be used as a stress function, ψ being harmonic. The same is true of $y\psi$, and also, of course, of the function ψ itself.

It can easily be shown by differentiation that $(x^2 + y^2)\psi$, that is, $r^2\psi$, also satisfies the same differential equation and may therefore be taken as a stress function, ψ being harmonic.

For instance, taking the two harmonic functions

$$\sinh ny \sin nx \qquad \cosh ny \sin nx$$

from the functions (h), page 171, and multiplying them by y, we arrive by superposition at the stress function (d), page 53. Taking the harmonic functions (i) and (j), page 171, as they stand or multiplied by x, y, or r^2, we can reconstruct all the terms of the stress function in polar coordinates given by Eq. (80), page 133.

The question of whether any stress function at all can be arrived at in this fashion remains open, and will be answered immediately, in the process of expressing the general stress function in terms of two arbitrary functions.

Denoting the laplacian operator

$$\frac{\partial^2}{\partial x^2} + \frac{\partial^2}{\partial y^2}$$

by ∇^2, Eq. (*a*) on page 35 can be written $\nabla^2(\nabla^2\phi) = 0$ or $\nabla^4\phi = 0$. Writing P for $\nabla^2\phi$, which represents $\sigma_x + \sigma_y$, we observe that P is a harmonic function, and so will have a conjugate harmonic function Q. Consequently $P + iQ$ is an analytic function of z, and we may write

$$f(z) = P + iQ \tag{c}$$

The integral of this function with respect to z is another analytic function, $4\psi(z)$ say. Then, writing p and q for the real and imaginary parts of $\psi(z)$, we have

$$\psi(z) = p + iq = \tfrac{1}{4}\int f(z)\, dz \tag{d}$$

so that $\psi'(z) = \frac{1}{4}f(z)$. We have also

$$\frac{\partial p}{\partial x} + i\frac{\partial q}{\partial x} = \frac{\partial}{\partial x}\psi(z) = \psi'(z)\frac{\partial z}{\partial x} = \frac{1}{4}f(z) = \frac{1}{4}(P + iQ)$$

Equating real parts of the first and last members we find

$$\frac{\partial p}{\partial x} = \frac{1}{4}P \tag{e}$$

Since p and q are conjugate functions, they satisfy Eqs. (*e*) of Art. 55, and so

$$\frac{\partial q}{\partial y} = \frac{1}{4}P \tag{f}$$

Recalling that $P = \nabla^2\phi$, Eqs. (*e*) and (*f*) enable us to show that $\phi - xp - yq$ is a harmonic function. For

$$\nabla^2(\phi - xp - yq) = \nabla^2\phi - 2\frac{\partial p}{\partial x} - 2\frac{\partial q}{\partial y} = 0 \tag{g}$$

Thus for any stress function ϕ we have

$$\phi - xp - yq = p_1$$

where p_1 is some harmonic function. Consequently,

$$\phi = xp + yq + p_1 \tag{83}$$

which shows that *any* stress function can be formed from suitably chosen conjugate harmonic functions p, q and a harmonic function p_1.

Equation (83) will prove useful later, but it may be observed that the use of both the functions p and q is not necessary. Instead of Eq. (g) we can write

$$\nabla^2(\phi - 2xp) = \nabla^2\phi - 4\frac{\partial p}{\partial x} = 0$$

showing that $\phi - 2xp$ is harmonic, say equal to p_2, so that any stress function must be expressible in the form

$$\phi = 2xp + p_2 \tag{h}$$

where p and p_2 are suitably chosen harmonic functions. Similarly, considering $\phi - 2yq$, we may show that any stress function must also be expressible in the form

$$\phi = 2yq + p_3$$

where q and p_3 are suitably chosen harmonic functions.

Returning to the form (83), let us introduce the function q_1, which is the conjugate harmonic to p_1, and write

$$\chi(z) = p_1 + iq_1$$

Then it is easily verified that the real part of

$$(x - iy)(p + iq) + p_1 + iq_1$$

is identical with the right-hand side of Eq. (83). Thus any stress function is expressible in the form[1]

$$\phi = \text{Re}\,[\bar{z}\psi(z) + \chi(z)] \tag{84}$$

where Re means "real part of," $\bar{z}$ denotes $x - iy$, and $\psi(z)$ and $\chi(z)$ are suitably chosen analytic functions. Conversely (84) yields a stress function, that is a solution of Eq. (a), page 35, for any choice of $\psi(z)$ and $\chi(z)$. It is applied later to the solution of several problems of practical interest.

Writing the "complex stress function" in brackets in (84) as

$$\bar{z}z\,\frac{\psi(z)}{z} + \chi(z)$$

and observing that $\bar{z}z = r^2$, and $\psi(z)/z$ is still a function of z, we find that any stress function can also be expressed as

$$r^2p_4 + p_5$$

where p_4, p_5 are harmonic.

[1] E. Goursat, *Bull. Soc. Math. France*, vol. 26 p. 206, 1898. N. I. Muskhelishvili, *Math. Ann.*, vol. 107, pp. 282–312, 1932.

57 | Displacement Corresponding to a Given Stress Function

It was shown in Art. 43 that the determination of the stress in a multiply connected region requires the evaluation of displacement to ensure that it is not discontinuous, that is, to ensure that the stress is not partly due to dislocations. For this reason, as well as for cases where the displacements are of interest in themselves, we require a method of finding the displacement functions u and v when a stress function is given.

The stress-strain relations for plane stress, Eqs. (22) and (23), may be written

$$E\frac{\partial u}{\partial x} = \sigma_x - \nu\sigma_y \qquad E\frac{\partial v}{\partial y} = \sigma_y - \nu\sigma_x \tag{a}$$

$$G\left(\frac{\partial v}{\partial x} + \frac{\partial u}{\partial y}\right) = \tau_{xy} \tag{b}$$

Inserting the stress function into the first, and recalling that $P = \nabla^2\phi$, we have

$$\begin{aligned} E\frac{\partial u}{\partial x} &= \frac{\partial^2\phi}{\partial y^2} - \nu\frac{\partial^2\phi}{\partial x^2} = \left(P - \frac{\partial^2\phi}{\partial x^2}\right) - \nu\frac{\partial^2\phi}{\partial x^2} \\ &= -(1+\nu)\frac{\partial^2\phi}{\partial x^2} + P \end{aligned} \tag{c}$$

and, similarly,

$$E\frac{\partial v}{\partial y} = -(1+\nu)\frac{\partial^2\phi}{\partial y^2} + P \tag{d}$$

But from Eqs. (*f*) and (*g*) of Art. 56, we can replace P in Eq. (*c*) above by $4\,\partial p/\partial x$, and in Eq. (*d*) by $4\,\partial q/\partial y$. Then, after division by $1+\nu$,

$$2G\frac{\partial u}{\partial x} = -\frac{\partial^2\phi}{\partial x^2} + \frac{4}{1+\nu}\frac{\partial p}{\partial x} \qquad 2G\frac{\partial v}{\partial y} = -\frac{\partial^2\phi}{\partial y^2} + \frac{4}{1+\nu}\frac{\partial q}{\partial y} \tag{e}$$

and these imply, by integration,

$$2Gu = -\frac{\partial\phi}{\partial x} + \frac{4}{1+\nu}p + f(y) \qquad 2Gv = -\frac{\partial\phi}{\partial y} + \frac{4}{1+\nu}q + f_1(x) \tag{f}$$

where $f(y)$ and $f_1(x)$ are arbitrary functions. If these are substituted in the left of Eq. (*b*), we obtain

$$-\frac{\partial^2\phi}{\partial x\,\partial y} + \frac{2}{1+\nu}\left(\frac{\partial p}{\partial y} + \frac{\partial q}{\partial x}\right) + \frac{1}{2}\frac{df}{dy} + \frac{1}{2}\frac{df_1}{dx} = \tau_{xy} \tag{g}$$

But the first term on the left is equal to τ_{xy}, and the parenthesis vanishes because p and q are conjugate harmonic functions satisfying the Cauchy-Riemann equations (Art. 56). Hence,

$$\frac{df}{dy} + \frac{df_1}{dx} = 0$$

which implies

$$\frac{df}{dy} = A \qquad \frac{df_1}{dx} = -A$$

where A is a constant. It follows that the terms $f(y)$ and $f_1(x)$ in Eq. (f) represent a rigid-body displacement. Discarding these terms we may write Eqs. (f) as[1]

$$2Gu = -\frac{\partial \phi}{\partial x} + \frac{4}{1+\nu}p \qquad 2Gv = -\frac{\partial \phi}{\partial y} + \frac{4}{1+\nu}q \tag{h}$$

on the understanding that a rigid-body displacement can be added. These equations enable us to find u and v when ϕ is known. We have first to find P as $\nabla^2\phi$, determine the conjugate function Q by means of the Cauchy-Riemann equations

$$\frac{\partial P}{\partial x} = \frac{\partial Q}{\partial y} \qquad \frac{\partial P}{\partial y} = -\frac{\partial Q}{\partial x},$$

form the function $f(z) = P + iQ$, and obtain p and q by integration of $f(z)$ as in Eq. (d), Art. 56. The terms of Eqs. (h) can then be evaluated.

The usefulness of Eqs. (h) will appear in later applications, for which the method of determining displacements used in Chaps. 3 and 4 is not suitable.

58 | Stress and Displacement in Terms of Complex Potentials

So far the stress and displacement components have been expressed in terms of the stress function ϕ. But since Eq. (84) expresses ϕ in terms of two functions $\psi(z)$, $\chi(z)$, it is possible to express the stress and displacement in terms of these two "complex potentials."

Any complex function $f(z)$ can be put into the form $\alpha + i\beta$ where α and β are real. To this there corresponds the *conjugate*,[2] $\alpha - i\beta$, the value taken by $f(z)$ when i is replaced, wherever it occurs in $f(z)$, by $-i$. This change is indicated by the notation

$$\bar{f}(\bar{z}) = \alpha - i\beta \tag{a}$$

Thus if $f(z) = e^{inz}$, we have

$$\bar{f}(\bar{z}) = e^{-in\bar{z}} = e^{-in(x-iy)} = e^{-inx}e^{-ny} \tag{b}$$

This may be contrasted with

$$f(\bar{z}) = e^{in\bar{z}}$$

to illustrate the significance of the bar over the f in Eq. (a).

[1] A. E. H. Love, "Mathematical Theory of Elasticity," 4th ed., Arts. 144 and 146, Cambridge University Press, New York, 1927.

[2] The word is used here with a significance quite distinct from that in the term "conjugate harmonic functions."

Evidently

$$f(z) + \bar{f}(\bar{z}) = 2\alpha = 2 \operatorname{Re} f(z)$$

In the same way, if we add its conjugate to the function in brackets in Eq. (84), the sum will be twice the real part of this function. Thus Eq. (84) may be replaced by

$$2\phi = \bar{z}\psi(z) + \chi(z) + z\bar{\psi}(\bar{z}) + \bar{\chi}(\bar{z}) \tag{85}$$

and by differentiation

$$2\frac{\partial \phi}{\partial x} = \bar{z}\psi'(z) + \psi(z) + \chi'(z) + z\bar{\psi}'(\bar{z}) + \bar{\psi}(\bar{z}) + \bar{\chi}'(\bar{z})$$

$$2\frac{\partial \phi}{\partial y} = i[\bar{z}\psi'(z) - \psi(z) + \chi'(z) - z\bar{\psi}'(\bar{z}) + \bar{\psi}(\bar{z}) - \bar{\chi}'(\bar{z})]$$

These two equations may be combined into one by multiplying the second by i and adding. Then

$$\frac{\partial \phi}{\partial x} + i\frac{\partial \phi}{\partial y} = \psi(z) + z\bar{\psi}'(\bar{z}) + \bar{\chi}'(\bar{z}) \tag{c}$$

Combining Eqs. (h) of Art. 57 in the same way, we find

$$2G(u + iv) = -\left(\frac{\partial \phi}{\partial x} + i\frac{\partial \phi}{\partial y}\right) + \frac{4}{1+\nu}(p + iq)$$

or, using Eq. (d) of Art. 56 and Eq. (c) above,

$$2G(u + iv) = \frac{3-\nu}{1+\nu}\psi(z) - z\bar{\psi}'(\bar{z}) - \bar{\chi}'(\bar{z}) \tag{86}$$

This equation determines u and v for plane stress when the complex potentials $\psi(z)$, $\chi(z)$ are given. For plane strain $\nu/(1 - \nu)$ is substituted for ν on the right of Eq. (86) in accordance with Art. 20.

The stress components σ_x, σ_y, τ_{xy} can be obtained directly from the second derivatives of Eq. (85). But, in view of later application to curvilinear coordinates, it is preferable to proceed otherwise. Differentiating Eq. (c) with respect to x, we have

$$\frac{\partial^2 \phi}{\partial x^2} + i\frac{\partial^2 \phi}{\partial x\, \partial y} = \psi'(z) + z\bar{\psi}''(\bar{z}) + \bar{\psi}'(\bar{z}) + \bar{\chi}''(\bar{z}) \tag{d}$$

Differentiating it with respect to y and multiplying by i, we have

$$i\frac{\partial^2 \phi}{\partial x\, \partial y} - \frac{\partial^2 \phi}{\partial y^2} = -\psi'(z) + z\bar{\psi}''(\bar{z}) - \bar{\psi}'(\bar{z}) + \bar{\chi}''(\bar{z}) \tag{e}$$

Simpler forms are obtained by subtracting and adding Eqs. (*d*) and (*e*). Then[1]

$$\sigma_x + \sigma_y = 2\psi'(z) + 2\bar{\psi}'(\bar{z}) = 4 \operatorname{Re} \psi'(z) \tag{87}$$

$$\sigma_y - \sigma_x - 2i\tau_{xy} = 2[z\bar{\psi}''(\bar{z}) + \bar{\chi}''(\bar{z})] \tag{88}$$

Changing i to $-i$ on both sides of Eq. (88) yields the alternative form

$$\sigma_y - \sigma_x + 2i\tau_{xy} = 2[\bar{z}\psi''(z) + \chi''(z)] \tag{89}$$

On separation of real and imaginary parts the right side of Eq. (89), or (88), gives $\sigma_x - \sigma_y$ and $2\tau_{xy}$. The two equations (87) and (89) determine the stress components in terms of the complex potentials $\psi(z)$ and $\chi(z)$. Thus by choosing definite functions for $\psi(z)$ and $\chi(z)$, we find a possible state of stress from Eqs. (87) and (89), and the displacements corresponding to this state of stress are easily obtained from Eq. (86).

As a simple illustration of this method, consider the polynomial stress system discussed on page 38. A stress function in the form of a polynomial of the fifth degree must evidently be obtained from Eq. (85) by taking

$$\psi(z) = (a_5 + ib_5)z^4 \qquad \chi(z) = (c_5 + id_5)z^5$$

where a_5, b_5, c_5, d_5 are real arbitrary coefficients. Then

$$\psi'(z) = 4(a_5 + ib_5)z^3 \qquad \chi'(z) = 5(c_5 + id_5)z^4$$

$$\psi''(z) = 12(a_5 + ib_5)z^2 \qquad \chi''(z) = 20(c_5 + id_5)z^3$$

and Eqs. (87) and (89) yield

$$\begin{aligned}
\sigma_x + \sigma_y &= 4 \operatorname{Re} 4(a_5 + ib_5)z^3 \\
&= 16 \operatorname{Re} (a_5 + ib_5)[x^3 - 3xy^2 + i(3x^2y - y^3)] \\
&= 16a_5(x^3 - 3xy^2) - 16b_5(3x^2y - y^3) \\
\sigma_y - \sigma_x + 2i\tau_{xy} &= 2[12(a_5 + ib_5)\bar{z}z^2 + 20(c_5 + id_5)z^3] \\
&= 24(a_5 + ib_5)(x - iy)(x + iy)^2 + 40(c_5 + id_5)(x + iy)^3 \\
&= [24a_5x(x^2 + y^2) - 24b_5y(x^2 + y^2) + 40c_5(x^3 - 3xy^2) \\
&\quad - 40d_5(3x^2y - y^3)] + i[24a_5y(x^2 + y^2) + 24b_5x(x^2 + y^2) \\
&\quad + 40c_5(3x^2y - y^3) + 40d_5(x^3 - 3xy^2)]
\end{aligned}$$

The expressions in brackets give $\sigma_y - \sigma_x$ and $2\tau_{xy}$, respectively. The displacement components corresponding to this stress distribution are easily obtained from Eq. (86), which yields

$$2G(u + iv) = \frac{3 - \nu}{1 + \nu}(a_5 + ib_5)z^4 - 4(a_5 - ib_5)z\bar{z}^3 - 5(c_5 - id_5)\bar{z}^4$$

[1] These results and Eq. (86) were obtained by G. Kolosoff in his doctoral dissertation, Dorpat, 1909. See his paper in *Z. Math. Physik.*, vol. 62, 1914.

It is also evident that there can be only four independent real constants in a polynomial stress function having all terms of any given degree n (>2).

59 | Resultant of Stress on a Curve. Boundary Conditions

In Fig. 114 AB is an arc of a curve drawn on the plate. The force acting on the arc ds, exerted by the material to the left on the material to the right, proceeding from A to B, may be represented by components $\bar{X}\,ds$ and $\bar{Y}\,ds$. Then, from Eqs. (12) of Art. 10,

$$\begin{aligned} \bar{X} &= \sigma_x \cos \alpha + \tau_{xy} \sin \alpha \\ \bar{Y} &= \sigma_y \sin \alpha + \tau_{xy} \cos \alpha \end{aligned} \tag{a}$$

where α is the angle between the left-hand normal N and the x axis. To ds correspond a dx and a dy as indicated in Fig. 125b. In traversing ds in the direction AB, x decreases and dx will be a negative number. The *length* of the horizontal side of the elementary triangle in Fig. 114b is therefore $-dx$. Thus

$$\cos \alpha = \frac{dy}{ds} \qquad \sin \alpha = -\frac{dx}{ds} \tag{b}$$

Inserting these, together with

$$\sigma_x = \frac{\partial^2 \phi}{\partial y^2} \qquad \sigma_y = \frac{\partial^2 \phi}{\partial x^2} \qquad \tau_{xy} = -\frac{\partial^2 \phi}{\partial x\, \partial y}$$

in Eqs. (a), we find

$$\begin{aligned} \bar{X} &= \frac{\partial^2 \phi}{\partial y^2}\frac{dy}{ds} + \frac{\partial^2 \phi}{\partial x\, \partial y}\frac{dx}{ds} = \frac{\partial}{\partial y}\left(\frac{\partial \phi}{dy}\right)\frac{dy}{ds} + \frac{\partial}{\partial x}\left(\frac{\partial \phi}{\partial y}\right)\frac{dx}{ds} = \frac{d}{ds}\left(\frac{\partial \phi}{\partial y}\right) \\ \bar{Y} &= -\frac{\partial^2 \phi}{\partial x^2}\frac{dx}{ds} - \frac{\partial^2 \phi}{\partial x\, \partial y}\frac{dy}{ds} = -\frac{d}{ds}\left(\frac{\partial \phi}{\partial x}\right) \end{aligned} \tag{c}$$

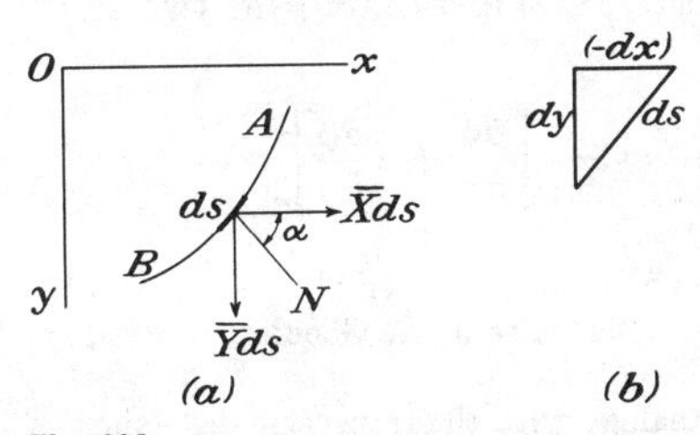

Fig. 114

The components of the resultant force on the arc AB are therefore

$$
\begin{aligned}
F_x &= \int_A^B \bar{X}\,ds = \int_A^B \frac{d}{ds}\left(\frac{\partial \phi}{\partial y}\right) ds = \left[\frac{\partial \phi}{\partial y}\right]_A^B \\
F_y &= \int_A^B \bar{Y}\,ds = -\int_A^B \frac{d}{ds}\left(\frac{\partial \phi}{\partial x}\right) ds = -\left[\frac{\partial \phi}{\partial x}\right]_A^B
\end{aligned} \tag{d}
$$

the square bracket representing the difference of the values of the enclosed quantity at B and at A.

The moment about O, clockwise, of the force on AB is, using Eqs. (*c*),

$$M = \int_A^B x\bar{Y}\,ds - \int_A^B y\bar{X}\,ds = -\int_A^B \left[xd\left(\frac{\partial \phi}{\partial x}\right) + yd\left(\frac{\partial \phi}{\partial y}\right)\right]$$

Integrating by parts yields[1]

$$M = \left[\phi\right]_A^B - \left[x\frac{\partial \phi}{\partial x} + y\frac{\partial \phi}{\partial y}\right]_A^B \tag{e}$$

It will be evident from Eqs. (*c*) that if the curve AB represents an unloaded boundary, so that $\bar{X}$ and $\bar{Y}$ are zero, $\partial\phi/\partial x$ and $\partial\phi/\partial y$ must be constant along AB. If there are prescribed loads on AB, Eqs. (*c*) show that they can be specified by giving the values of $\partial\phi/\partial x$, $\partial\phi/\partial y$ along the boundary. This is equivalent to giving the derivatives $\partial\phi/\partial s$ along, and $\partial\phi/\partial n$ normal to, AB. These are known if ϕ and $\partial\phi/\partial n$ are given along AB.[2]

Now let the arc be continued to form a closed curve, so that B coincides with A, but is still regarded as the point reached after traversing the arc, now a closed circuit, AB. Then Eqs. (*d*) and (*e*) give the resultant force and moment of the stresses acting on the piece of the plate enclosed by the circuit. If these are not zero, $\partial\phi/\partial x$ and $\partial\phi/\partial y$ do not return to their starting values (A) after completing the circuit (B). They are therefore discontinuous functions, such as the angle θ of polar coordinates. This will be the case only when loads (equal and opposite to F_x, F_y, M) are applied to the piece of the plate within the closed circuit.

In terms of the complex potentials $\psi(z)$, $\chi(z)$ of Eq. (85) the two equations (*d*) may be written as

$$F_x + iF_y = \left[\frac{\partial \phi}{\partial y} - i\frac{\partial \phi}{\partial x}\right]_A^B = -i\left[\frac{\partial \phi}{\partial x} + i\frac{\partial \phi}{\partial y}\right]_A^B$$

[1] Equations (*d*) and (*e*) serve to establish an analogy between plane stress and the slow motion of a viscous fluid in two dimensions. See J. N. Goodier, *Phil. Mag.*, ser. 7, vol. 17, pp. 554 and 800, 1934.

[2] These boundary conditions lead to an analogy with the transverse deflections of elastic plates. An account of this analogy, with references, is given by R. D. Mindlin, *Quart. Appl. Math.*, vol. 4, p. 279, 1946.

or, using Eq. (*c*) of Art. 58,

$$F_x + iF_y = -i[\psi(z) + z\bar{\psi}'(\bar{z}) + \bar{\chi}'(\bar{z})]_A^B \tag{90}$$

Equation (*e*) becomes

$$M = \text{Re}\,[-z\bar{z}\bar{\psi}'(\bar{z}) + \chi(z) - \bar{z}\bar{\chi}'(\bar{z})]_A^B \tag{91}$$

Equations (90) and (91), applied to a complete circuit round the origin, show that if $\psi(z)$ and $\chi(z)$ are taken in the form z^n where n is a positive or negative integer, F_x, F_y, and M are zero, since the functions in brackets return to their initial values when the circuit is completed. These functions by themselves could not represent stress due to loads applied at the origin. The function $\log z = \log r + i\theta$ does not return to its initial value on completing a circuit round the origin, since θ increases by 2π. Thus if $\psi(z) = C \log z$, or $\chi(z) = Dz \log z$, where C and D are (complex) constants, Eq. (90) will yield a nonzero value for $F_x + iF_y$. Similarly $\chi(z) = D \log z$ will yield a nonzero value of M if D is imaginary, but a zero value if D is real.

60 | Curvilinear Coordinates

Polar coordinates r, θ (Fig. 113) may be regarded as specifying the position of a point as the intersection of a circle (of radius r) and a radial line (at the angle θ from the initial line). A change *from* cartesian *to* polar coordinates is effected by means of the equations

$$\sqrt{x^2 + y^2} = r \qquad \arctan \frac{y}{x} = \theta \tag{a}$$

The first, when r is given various constant values, represents the family of circles. The second, when θ is given various constant values, represents the family of radial lines.

Equations (*a*) are a special case of equations of the form

$$F_1(x,y) = \xi \qquad F_2(x,y) = \eta \tag{b}$$

Giving definite constant values to ξ and η, these equations will represent two curves which will intersect, when $F_1(x,y)$, $F_2(x,y)$ are suitable functions. Different values of ξ and η will yield different curves and a different point of intersection. Thus each point in the xy plane will be characterized by definite values of ξ and η—the values which make the two curves given by Eqs. (*b*) pass through it—and ξ, η may be regarded as "coordinates" of a point. Since given values of ξ, η define the point by means of two intersecting curves, they are called *curvilinear coordinates*.[1]

[1] The general theory of curvilinear coordinates was developed by G. Lamé in the book "Leçons sur les Coordonnées Curvilignes," Gauthier-Villars, Paris, 1859.

Polar coordinates, with the associated stress components, proved very useful in Chap. 4 for problems of concentric circular boundaries. The stress and displacement on such a boundary become functions of θ only, since r is constant. If the boundaries consist of other curves, for instance ellipses, it is advantageous to use curvilinear coordinates one of which is constant on each boundary curve.

If Eqs. (b) are solved for x and y, we shall have two equations of the form

$$x = f_1(\xi,\eta) \qquad y = f_2(\xi,\eta) \tag{c}$$

and it is usually most convenient to begin with these. Consider, for example, the two equations

$$x = c \cosh \xi \cos \eta \qquad y = c \sinh \xi \sin \eta \tag{d}$$

where c is a constant. Elimination of η yields

$$\frac{x^2}{c^2 \cosh^2 \xi} + \frac{y^2}{c^2 \sinh^2 \xi} = 1$$

If ξ is constant, this is the equation of an ellipse with semiaxes $c \cosh \xi$, $c \sinh \xi$, and with foci at $x = \pm c$. For different values of ξ we obtain different ellipses with the same foci—that is, a family of *confocal ellipses* (Fig. 115). On any one of these ellipses ξ is constant and η varies (through a range 2π), as on a circle in polar coordinates r is constant and θ varies. In fact, in the present case, η is the eccentric angle[1] of a point on the ellipse.

[1] If a, θ are the polar coordinates of a point on the circle circumscribing an ellipse of semiaxes a, b, the perpendicular from this point to the x axis intersects the ellipse at the point $x = a \cos \theta$, $y = b \sin \theta$; θ is called the eccentric angle of this point on the ellipse.

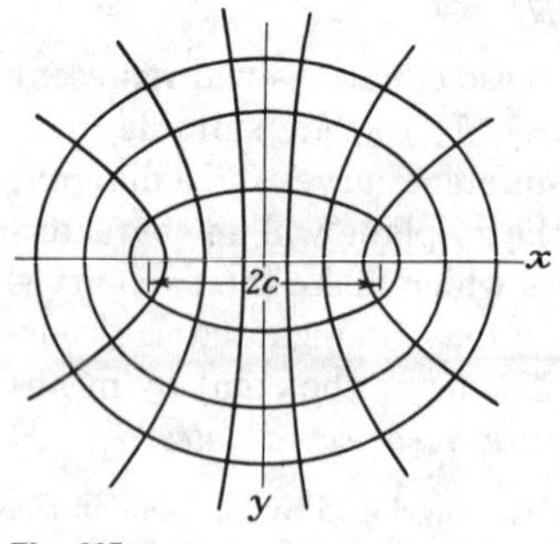

Fig. 115

If, on the other hand, we eliminate ξ from Eqs. (d), by means of the equation $\cosh^2 \xi - \sinh^2 \xi = 1$, we have

$$\frac{x^2}{c^2 \cos^2 \eta} - \frac{y^2}{c^2 \sin^2 \eta} = 1 \tag{e}$$

For a constant value of η this represents a hyperbola having the same foci as the ellipses. Thus, Eq. (e) represents a family of confocal hyperbolas, on any one of which η is constant and ξ varies. These coordinates are called *elliptic*.

The two equations (d) are equivalent to $x + iy = c \cosh (\xi + i\eta)$ or

$$z = c \cosh \zeta \tag{f}$$

where $\zeta = \xi + i\eta$. This is evidently a special case of the relation

$$z = f(\zeta) \tag{g}$$

This, besides defining z as a function of ζ, may be solved to give ζ as a function of z. Then ξ and η are the real and imaginary parts of a function of z, and therefore satisfy the Cauchy-Riemann equations (e) of Art. 55, also therefore the Laplace equations (f) and (g) of Art. 55.

The curvilinear coordinates to be used in this chapter will all be derived from equations of the form (g), and as a consequence will possess further special properties. The point x, y having the curvilinear coordinates ξ, η, a neighboring point $x + dx$, $y + dy$ will have curvilinear coordinates $\xi + d\xi$, $\eta + d\eta$, and since there will be two equations of the type (c) we may write

$$dx = \frac{\partial x}{\partial \xi} d\xi + \frac{\partial x}{\partial \eta} d\eta \qquad dy = \frac{\partial y}{\partial \xi} d\xi + \frac{\partial y}{\partial \eta} d\eta \tag{h}$$

If only ξ is varied, the increments dx, dy correspond to an element of arc ds_ξ along a curve η = constant, and

$$dx = \frac{\partial x}{\partial \xi} d\xi \qquad dy = \frac{\partial y}{\partial \xi} d\xi \tag{i}$$

Thus

$$(ds_\xi)^2 = (dx)^2 + (dy)^2 = \left[\left(\frac{\partial x}{\partial \xi}\right)^2 + \left(\frac{\partial y}{\partial \xi}\right)^2\right] (d\xi)^2 \tag{j}$$

Since $z = f(\zeta)$, we have

$$\frac{\partial z}{\partial \xi} = \frac{\partial x}{\partial \xi} + i \frac{\partial y}{\partial \xi} = \frac{d}{d\zeta} f(\zeta) \frac{\partial \zeta}{\partial \xi} = f'(\zeta) \tag{k}$$

where

$$f'(\zeta) = \frac{df(\zeta)}{d\zeta}$$

Now any complex quantity can be written in the form $J \cos \alpha + iJ \sin \alpha$, or $Je^{i\alpha}$, where J and α are real. With

$$f'(\zeta) = Je^{i\alpha} \tag{l}$$

Eq. (k) yields

$$\frac{\partial x}{\partial \xi} = J \cos \alpha \qquad \frac{\partial y}{\partial \xi} = J \sin \alpha \tag{m}$$

and then Eq. (j) gives

$$ds_\xi = J\, d\xi$$

The slope of ds_ξ is, using Eqs. (i) and (m),

$$\frac{dy}{dx} = \frac{\partial y/\partial \xi}{\partial x/\partial \xi} = \tan \alpha \tag{n}$$

Thus α, given by Eq. (l), is the angle between the tangent to the curve η = constant, in the direction ξ-increasing, and the x axis (Fig. 116). In the same way, if only η is varied, the increments dx and dy of Eqs. (h) correspond to an element of arc ds_η along a curve ξ = constant, and instead of Eqs. (i) we have

$$dx = \frac{\partial x}{\partial \eta}\, d\eta \qquad dy = \frac{\partial y}{\partial \eta}\, d\eta$$

Proceeding as above, we shall find that

$$\frac{\partial x}{\partial \eta} = -J \sin \alpha \qquad \frac{\partial y}{\partial \eta} = J \cos \alpha$$

and that $ds_\eta = J d\eta$, and

$$dy/dx = -\cot \alpha$$

Comparing this last result with Eq. (n), we see that the curves ξ = constant, η = constant, intersect at right angles, the direction η-increasing making an angle $(\pi/2) + \alpha$ with the x axis (Fig. 116).

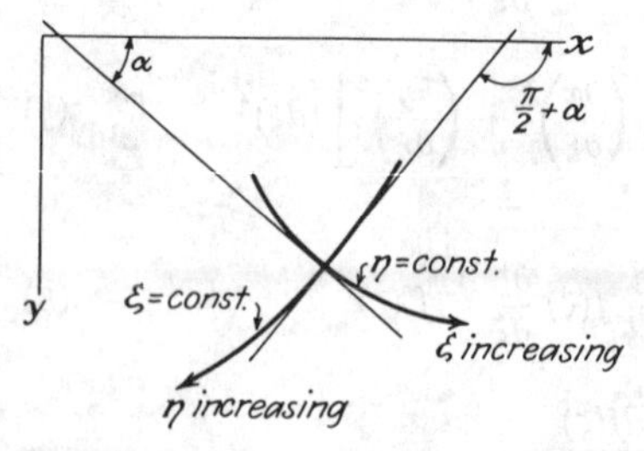

Fig. 116

Consider for instance the elliptic coordinates defined by Eq. (*f*). We have

$$f'(\zeta) = c \sinh \zeta = c \sinh \xi \cos \eta + ic \cosh \xi \sin \eta = Je^{i\alpha}$$

Comparing the real and imaginary parts of the last equality, we find

$$J \cos \alpha = c \sinh \xi \cos \eta \qquad J \sin \alpha = c \cosh \xi \sin \eta$$

and therefore

$$J^2 = c^2(\sinh^2 \xi \cos^2 \eta + \cosh^2 \xi \sin^2 \eta) = \tfrac{1}{2}c^2(\cosh 2\xi - \cos 2\eta) \tag{o}$$

$$\tan \alpha = \coth \xi \tan \eta \tag{p}$$

61 | Stress Components in Curvilinear Coordinates

Equations (86), (87), and (89) give the cartesian components of displacement and stress in terms of the complex potentials $\psi(z)$, $\chi(z)$. When curvilinear coordinates are used, the complex potentials can be taken as functions of ζ, and z itself is given in terms of ζ by the equation of the type of Eq. (*g*) of Art. 60 defining the curvilinear coordinates. There is thus no difficulty in expressing σ_x, σ_y, τ_{xy} in terms of ξ and η. It is usually convenient, however, to specify the stress as

σ_ξ, the normal component on a curve ξ = constant
σ_η, the normal component on a curve η = constant
$\tau_{\xi\eta}$, the shear component on both curves.

These components are indicated in Fig. 117. Comparing this and Fig. 116 with Fig. 12, we see that σ_ξ and $\tau_{\xi\eta}$ correspond to σ and τ in Fig. 12. We may therefore use Eqs. (13), and thus obtain

$$\sigma_\xi = \tfrac{1}{2}(\sigma_x + \sigma_y) + \tfrac{1}{2}(\sigma_x - \sigma_y) \cos 2\alpha + \tau_{xy} \sin 2\alpha$$

$$\tau_{\xi\eta} = -\tfrac{1}{2}(\sigma_x - \sigma_y) \sin 2\alpha + \tau_{xy} \cos 2\alpha$$

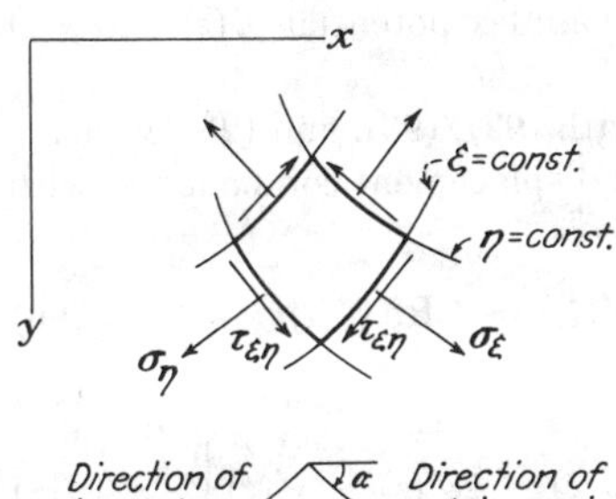

Fig. 117

Replacing α by $(\pi/2) + \alpha$, we find similarly

$$\sigma_\eta = \tfrac{1}{2}(\sigma_x + \sigma_y) - \tfrac{1}{2}(\sigma_x - \sigma_y)\cos 2\alpha - \tau_{xy}\sin 2\alpha$$

and from these we easily obtain the following equations:[1]

$$\sigma_\xi + \sigma_\eta = \sigma_x + \sigma_y \tag{92}$$

$$\sigma_\eta - \sigma_\xi + 2i\tau_{\xi\eta} = e^{2i\alpha}(\sigma_y - \sigma_x + 2i\tau_{xy}) \tag{93}$$

The factor $e^{2i\alpha}$ for curvilinear coordinates defined by $z = f(\zeta)$ can be found from Eq. (l) of Art. 60. This, with its conjugate, obtained by changing i to $-i$ throughout, gives

$$f'(\zeta) = Je^{i\alpha} \qquad \bar{f}'(\bar{\zeta}) = Je^{-i\alpha}$$

so that

$$e^{2i\alpha} = \frac{f'(\zeta)}{\bar{f}'(\bar{\zeta})} \tag{94}$$

For example, our elliptic coordinates give $f'(\zeta) = c \sinh \zeta$, and

$$e^{2i\alpha} = \frac{\sinh \zeta}{\sinh \bar{\zeta}} \tag{q}$$

With the value of $e^{2i\alpha}$ so determined, Eqs. (92) and (93) express σ_ξ, σ_η, $\tau_{\xi\eta}$ in terms of σ_x, σ_y, τ_{xy}.

The *displacement* in curvilinear coordinates is specified by means of a component u_ξ in the direction ξ-increasing (Fig. 116) and a component u_η in the direction η-increasing. If u and v are the cartesian components of the displacement, we have

$$u_\xi = u\cos\alpha + v\sin\alpha \qquad u_\eta = v\cos\alpha - u\sin\alpha$$

and therefore

$$u_\xi + iu_\eta = e^{-i\alpha}(u + iv) \tag{95}$$

Using Eq. (86) with $z = f(\zeta)$, and Eq. (94), this enables us to express u_ξ and u_η in terms of ξ and η when the complex potentials $\psi(z)$ and $\chi(z)$ have been chosen.

Combining Eqs. (86), (87), and (89) with (92), (93), and (95), we have the following equations for the stress and displacement components (with i replaced by $-i$ in the last):

$$\sigma_\xi + \sigma_\eta = 2[\psi'(z) + \bar{\psi}'(\bar{z})] = 4 \operatorname{Re} \psi'(z) \tag{96}$$

$$\sigma_\eta - \sigma_\xi + 2i\tau_{\xi\eta} = 2e^{2i\alpha}[\bar{z}\psi''(z) + \chi''(z)] \tag{97}$$

$$2G(u_\xi - iu_\eta) = e^{i\alpha}\left[\frac{3-\nu}{1+\nu}\bar{\psi}(\bar{z}) - \bar{z}\psi'(z) - \chi'(z)\right] \tag{98}$$

[1] Equations (92), (93), and (95) were obtained by Kolosoff, *loc. cit.*

We shall use these equations in the solution of several problems involving curved boundaries.

PROBLEMS

1. Show that for polar coordinates, given by $z = e^{\zeta}$, Eq. (94) becomes $e^{2i\alpha} = e^{2i\eta}$ and $\alpha = \eta = \theta$.
2. Obtain the solutions of the following problems in polar coordinates by means of the complex potentials indicated. Evaluate the stress and displacement components. Capitals denote constants, not necessarily real.
   (a) A ring ($a < r < b$) with equal and opposite couples M applied by means of shear stress to the two boundaries (Fig. 138). $\psi(z) = 0$, $\chi(z) = A \log z$.
   (b) The ring under internal pressure p_i, external pressure p_o (see page 70). $\psi(z) = Az$, $\chi(z) = B \log z$.
   (c) The pure bending of a curved bar, and the "rotational dislocation" of the ring, as in Arts. 29 and 31. $\psi(z) = Az \log z + Bz$, $\chi(z) = C \log z$.
   (d) The problem solved in Art. 33. $\psi(z) = Az^2 + B \log z$, $\chi(z) = Cz \log z + D/z$.
   (e) The plate under tension with a circular hole (Art. 35). $\psi(z) = Az + B/z$, $\chi(z) = C \log z + Dz^2 + F/z^2$.
   (f) The radial stress distribution of Art. 36. $\psi(z) = A \log z$, $\chi(z) = Bz \log z$.
   (g) The force at a point of an infinite plate (Art. 42). $\psi(z) = A \log z$, $\chi(z) = Bz \log z$.

62 | Solutions in Elliptic Coordinates. Elliptic Hole in Uniformly Stressed Plate

The elliptic coordinates ξ, η, already considered in Art. 60 and shown in Fig. 115, were defined by

$$z = c \cosh \zeta \qquad \zeta = \xi + i\eta \tag{a}$$

which give

$$x = c \cosh \xi \cos \eta \qquad y = c \sinh \xi \sin \eta$$

and

$$\frac{dz}{d\zeta} = c \sinh \zeta \qquad e^{2i\alpha} = \frac{\sinh \zeta}{\sinh \bar{\zeta}} \tag{b}$$

The coordinate ξ is constant and equal to ξ_o on an ellipse of semiaxes $c \cosh \xi_o$, $c \sinh \xi_o$. If the semiaxes are given as a and b, c and ξ_o can be found from

$$c \cosh \xi_o = a \qquad c \sinh \xi_o = b \tag{c}$$

and therefore if one member of the family of ellipses is given, the whole family of ellipses and also the family of hyperbolas (see page 182) are definite. If ξ is very small the corresponding ellipse is very slender, and in the limit $\xi = 0$ it becomes a line of length $2c$ joining the foci. Taking larger and larger positive values of ξ the ellipse becomes larger and larger, approaching an infinite circle in the limit $\xi = \infty$. A point on any one ellipse goes once around the ellipse as η goes from zero (on the positive x axis, Fig. 115) to 2π. In this respect η resembles the angle θ of polar coordinates. Continuity of displacement and stress components requires that they be periodic in η with period 2π, so that they will have the same values for $\eta = 2\pi$ as they have for $\eta = 0$.

Consider now an infinite plate in a state of uniform all-round tension S disturbed by an elliptical hole of semiaxes a and b, which is free from stress.[1] These conditions mean that

$$\sigma_x = \sigma_y = S \qquad \text{at infinity } (\xi \to \infty) \tag{d}$$

$$\sigma_\xi = \tau_{\xi\eta} = 0 \qquad \text{on the elliptical boundary of the hole, where } \xi \text{ has the value } \xi_o \tag{e}$$

From Eqs. (87) and (89) we find that the condition (d) is satisfied if

$$2 \operatorname{Re} \psi'(z) = S \qquad \bar{z}\psi''(z) + \chi''(z) = 0 \qquad \text{at infinity} \tag{f}$$

Since the stress and displacement components are, for continuity, to be periodic in η with period 2π, we are led to consider forms for $\psi(z)$ and $\chi(z)$ which will give a stress function with the same periodicity, and such forms are

$$\sinh n\zeta \qquad \text{that is, } \sinh n\xi \cos n\eta + i \cosh n\xi \sin n\eta$$
$$\cosh n\zeta \qquad \text{that is, } \cosh n\xi \cos n\eta + i \sinh n\xi \sin n\eta$$

where n is an integer. The function $\chi(z) = Bc^2\zeta$, B being a constant, is also suitable to the problem.

It is evident from (a) that for $\xi \to \infty$, ζ behaves like $\log z$, and this form of χ is required in the related circular hole problem (2b, page 187).

Taking $\psi(z) = Ac \sinh \zeta$, with A a constant, and using the first of Eqs. (b) for $d\zeta/dz$, which is the reciprocal of $dz/d\zeta$, we find

$$\psi'(z) = Ac \cosh \zeta \frac{d\zeta}{dz} = A \frac{\cosh \zeta}{\sinh \zeta} = A \coth \zeta \tag{g}$$

[1] Solutions for the plate with an elliptical hole were first given by Kolosoff, *loc. cit.*; and C. E. Inglis, *Trans. Inst. Naval Arch.*, London, 1913; *Eng.*, vol. 95, p. 415, 1913. See also T. Pöschl, *Math. Z.*, vol. 11, p. 95, 1921. The method employed here is that of Kolosoff. The same method was applied to several two-dimensional problems of elasticity by A. C. Stevenson, *Proc. Roy. Soc. (London)*, ser. A, vol. 184, pp. 129 and 218, 1945. Other references are given later in the chapter.

At an infinite distance from the origin ξ is infinite, and $\coth \zeta$ has the value unity. The first of conditions (f) is therefore satisfied if $2A = S$. From (g) we find further

$$\psi''(z) = -\frac{A}{c}\frac{1}{\sinh^3 \zeta} \tag{h}$$

and

$$\bar{z}\psi''(z) = -A\frac{\cosh \bar{\zeta}}{\sinh^3 \zeta} \tag{i}$$

Taking $\chi(z) = Bc^2\zeta$, where B is a constant, we have

$$\chi'(z) = \frac{Bc}{\sinh \zeta} \qquad \chi''(z) = -B\frac{\cosh \zeta}{\sinh^3 \zeta} \tag{j}$$

Equations (i) and (j) show that $\bar{z}\psi''(z)$ and $\chi''(z)$ each vanish at infinity. The second of conditions (f) is therefore now satisfied.

The condition (e) can be satisfied by suitable choice of the constant B. Subtracting Eq. (97) from Eq. (96) we have

$$\sigma_\xi - i\tau_{\xi\eta} = \psi'(z) + \bar{\psi}'(\bar{z}) - e^{2i\alpha}[\bar{z}\psi''(z) + \chi''(z)] \tag{k}$$

and $e^{2i\alpha}$ is given by the second of Eqs. (b). Thus

$$\begin{aligned}\sigma_\xi - i\tau_{\xi\eta} &= A\left(\frac{\cosh \zeta}{\sinh \zeta} + \frac{\cosh \bar{\zeta}}{\sinh \bar{\zeta}}\right) + \frac{\sinh \zeta}{\sinh \bar{\zeta}}\left(A\frac{\cosh \bar{\zeta}}{\sinh^3 \zeta} + B\frac{\cosh \zeta}{\sinh^3 \zeta}\right) \\ &= \frac{1}{\sinh^2 \zeta \sinh \bar{\zeta}}\{A[\sinh \zeta \sinh (\zeta + \bar{\zeta}) + \cosh \bar{\zeta}] \\ &\qquad + B\cosh \zeta\}\end{aligned} \tag{l}$$

At the boundary of the elliptical hole $\xi = \xi_0$ and $\zeta + \bar{\zeta} = 2\xi_0$, $\bar{\zeta} = 2\xi_0 - \zeta$. Then (l) reduces to

$$\frac{1}{\sinh^2 \zeta \sinh \bar{\zeta}}(A\cosh 2\xi_0 + B)\cosh \zeta$$

Condition (e) is therefore satisfied if

$$B = -A\cosh 2\xi_0 = -\tfrac{1}{2}S\cosh 2\xi_0 \tag{m}$$

We now have

$$\psi(z) = \tfrac{1}{2}Sc\sinh \zeta \qquad \chi(z) = -\tfrac{1}{2}Sc^2\cosh 2\xi_0 \cdot \zeta \tag{n}$$

All the boundary conditions have now been satisfied. But we cannot be sure that the complex potentials (n) represent the solution of our problem until we know that they imply no discontinuity in the displacement.

The cartesian components of displacement can be found from Eq. (86), which in the present case gives

$$2G(u + iv) = \frac{3 - \nu}{1 + \nu} Ac \sinh \zeta - Ac \cosh \zeta \coth \bar{\zeta} - \frac{Bc}{\sinh \bar{\zeta}} \qquad (o)$$

with $A = S/2$, and B as given by Eq. (m). The hyperbolic functions have real and imaginary parts which are periodic in η. Thus a circuit round any ellipse $\xi =$ constant, within the plate, will bring u and v back to the initial values. The complex potentials (n) therefore provide the solution of the problem.

The stress component σ_η at the hole is easily found from Eq. (96), since σ_ξ at the hole is zero. Inserting the value of $\psi'(z)$ from Eq. (g), with $A = S/2$ we have

$$\sigma_\xi + \sigma_\eta = 4 \operatorname{Re} \psi'(z) = 2S \operatorname{Re} \coth \zeta$$

But, by Eq. (c), page 169,

$$\coth \zeta = \frac{\sinh 2\xi - i \sin 2\eta}{\cosh 2\xi - \cos 2\eta}$$

Hence

$$\sigma_\xi + \sigma_\eta = \frac{2S \sinh 2\xi}{\cosh 2\xi - \cos 2\eta}$$

and at the boundary of the hole

$$(\sigma_\eta)_{\xi=\xi_o} = \frac{2S \sinh 2\xi_o}{\cosh 2\xi_o - \cos 2\eta}$$

The greatest value, occurring at the ends of the major axes, where $\eta = 0$ and π, and $\cos 2\eta = 1$, is

$$(\sigma_\eta)_{\max} = \frac{2S \sinh 2\xi_o}{\cosh 2\xi_o - 1}$$

It is easily shown from Eqs. (c) that

$$c^2 = a^2 - b^2 \qquad \sinh 2\xi_o = \frac{2ab}{c^2} \qquad \cosh 2\xi_o = \frac{a^2 + b^2}{c^2}$$

and with these we find that

$$(\sigma_\eta)_{\max} = 2S \frac{a}{b}$$

which becomes larger and larger as the ellipse is made more and more slender.

The least value of $(\sigma_\eta)_{\xi=\xi_o}$ occurs at the ends of the minor axes where $\cos 2\eta = -1$. Thus

$$(\sigma_\eta)_{\min} = \frac{2S \sinh 2\xi_o}{\cosh 2\xi_o + 1} = 2S \frac{b}{a}$$

When $a = b$, so that the ellipse becomes a circle, both $(\sigma_\eta)_{\max}$ and $(\sigma_\eta)_{\min}$ reduce to $2S$, in agreement with the value for the circular hole under uniform all-round tension found on page 92.

The problem of uniform pressure S within an elliptical hole, and zero stress at infinity, is easily obtained by combining the above solution with the state of uniform stress $\sigma_\xi = \sigma_\eta = -S$, derivable from the complex potential $\psi(z) = -Sz/2$.

63 | Elliptic Hole in a Plate under Simple Tension

As a second problem, consider the infinite plate in a state of simple tensile stress S in a direction at an angle β below the positive x axis (Fig. 118), disturbed by an elliptic hole, with its major axis along the x axis, as in the preceding problem. The elliptic hole with major axis perpendicular or parallel to the tension[1] is a special case. The more general problem is, however, no more difficult by the present method. From its solution we can find the effect of the elliptic hole on any state of uniform plane stress, specified by principal stresses at infinity in any orientation with respect to the hole.

Let Ox', Oy', be cartesian axes obtained by rotating Ox through the angle β so as to bring it parallel to the tension S. Then by Eqs. (92) and (93),

$$\sigma_{x'} + \sigma_{y'} = \sigma_x + \sigma_y \qquad \sigma_{y'} - \sigma_{x'} + 2i\tau_{x'y'} = e^{2i\beta}(\sigma_y - \sigma_x + 2i\tau_{xy})$$

Since at infinity $\sigma_{x'} = S$, $\sigma_{y'} = \tau_{x'y'} = 0$, we have

$$\sigma_x + \sigma_y = S \qquad \sigma_y - \sigma_x + 2i\tau_{xy} = -Se^{-2i\beta} \qquad \text{at infinity}$$

and so, from Eqs. (87) and (89),

$$4 \operatorname{Re} \psi'(z) = S \qquad 2[\bar{z}\psi''(z) + \chi''(z)] = -Se^{-2i\beta} \qquad \text{at infinity} \qquad (a)$$

At the boundary of the hole $\xi = \xi_o$ we must have $\sigma_\xi = \tau_{\xi\eta} = 0$.

[1] See the papers cited in n. 1, p. 188.

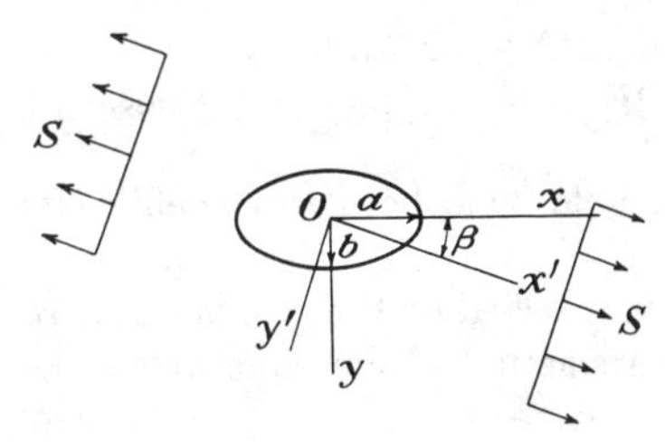

Fig. 118

All these boundary conditions can be satisfied by taking $\psi(z)$, $\chi(z)$ in the forms[1]

$$4\psi(z) = Ac \cosh \zeta + Bc \sinh \zeta$$
$$4\chi(z) = Cc^2\zeta + Dc^2 \cosh 2\zeta + Ec^2 \sinh 2\zeta$$

where A, B, C, D, E are constants to be found.

Since $z = c \cosh \zeta$, the term $Ac \cosh \zeta$ in the expression for $4\psi(z)$ is simply Az. It will contribute to the stress function [Eq. (84)] a term $\operatorname{Re} A\bar{z}z$ or $\operatorname{Re} Ar^2$. This is zero if A is imaginary, and therefore A may at once be taken as real. The constant C must also be real. For if we insert the above expressions for $\psi(z)$, $\chi(z)$ in Eq. (91), taking for the curve AB a complete circuit round the hole, we find that all terms except the term in C yield zero because the hyperbolic functions are periodic in η with period 2π. The term in C is $\operatorname{Re} [Cc^2(\xi + i\eta)]_A^B$. This vanishes for a complete circuit only if C is real.

The constants B, D, E are complex, and we may write

$$B = B_1 + iB_2 \qquad D = D_1 + iD_2 \qquad E = E_1 + iE_2 \tag{b}$$

Substitution of the above forms for $\psi(z)$, $\chi(z)$ in the conditions (a) yields

$$A + B_1 = S \qquad 2(D + E) = -Se^{-2i\beta} \tag{c}$$

Subtracting Eq. (97) from Eq. (96) to obtain $\sigma_\xi - i\tau_{\xi\eta}$, we find

$$\begin{aligned} 4(\sigma_\xi - i\tau_{\xi\eta}) = \operatorname{cosech} \bar{\zeta}[(2A + B \coth \zeta) \sinh \bar{\zeta} \\ + (\bar{B} + B \operatorname{cosech}^2 \zeta) \cosh \bar{\zeta} + (C + 2E) \operatorname{cosech} \zeta \coth \zeta \\ - 4D \sinh \zeta - 4E \cosh \zeta] \end{aligned}$$

At the boundary of the hole $\xi = \xi_o$ and $\bar{\zeta} = 2\xi_o - \zeta$. If this value of $\bar{\zeta}$ is inserted in $\sinh \bar{\zeta}$ and $\cosh \bar{\zeta}$ in the above expression, and the functions $\sinh (2\xi_o - \zeta)$, $\cosh (2\xi_o - \zeta)$ expanded, the expression in square brackets reduces to

$$\begin{aligned} (2A \sinh 2\xi_o - 2iB_2 \cosh 2\xi_o - 4E) \cosh \zeta \\ - (2A \cosh 2\xi_o - 2iB_2 \sinh 2\xi_o + 4D) \sinh \zeta \\ + (C + 2E + B \cosh 2\xi_o) \coth \zeta \operatorname{cosech} \zeta \end{aligned}$$

This, and consequently $\sigma_\xi - i\tau_{\xi\eta}$ at the hole, vanishes if the coefficients of $\cosh \zeta$, $\sinh \zeta$, $\coth \zeta \operatorname{cosech} \zeta$ vanish. We have thus three equations, together with the two equations (c), to be satisfied by the constants A, B, C, D, E. Since A and C are real, there are actually nine equations to be

[1] Stevenson, *loc. cit.*

satisfied by eight constants—A, C, and B_1, B_2, D_1, D_2, E_1, E_2, which are the real and imaginary parts of B, D, E. They are consistent, and the solution is

$$A = Se^{2\xi_o}\cos 2\beta \qquad D = -\tfrac{1}{2}Se^{2\xi_o}\cosh 2(\xi_o + i\beta)$$
$$B = S(1 - e^{2\xi_o + 2i\beta}) \qquad E = \tfrac{1}{2}Se^{2\xi_o}\sinh 2(\xi_o + i\beta)$$
$$C = -S(\cosh 2\xi_o - \cos 2\beta)$$

The complex potentials of this problem are consequently given by

$$4\psi(z) = Sc[e^{2\xi_o}\cos 2\beta \cosh \zeta + (1 - e^{2\xi_o + 2i\beta})\sinh \zeta]$$
$$4\chi(z) = -Sc^2[(\cosh 2\xi_o - \cos 2\beta)\zeta + \tfrac{1}{2}e^{2\xi_o}\cosh 2(\zeta - \xi_o - i\beta)]$$

The displacements can now be determined from Eq. (98). It may be seen at once that they are single-valued.

The stress σ_η at the hole can be obtained from Eq. (96) since at the hole σ_ξ is zero. Then

$$(\sigma_\eta)_{\xi=\xi_o} = S\,\frac{\sinh 2\xi_o + \cos 2\beta - e^{2\xi_o}\cos 2(\beta - \eta)}{\cosh 2\xi_o - \cos 2\eta}$$

When the tension S is at right angles to the major axis ($\beta = \pi/2$),

$$(\sigma_\eta)_{\xi=\xi_o} = Se^{2\xi_o}\left[\frac{\sinh 2\xi_o(1 + e^{-2\xi_o})}{\cosh 2\xi_o - \cos 2\eta} - 1\right]$$

and the greatest value, occurring at the ends of the major axis ($\cos 2\eta = 1$), reduces to

$$S\left(1 + 2\,\frac{a}{b}\right)$$

This increases without limit as the hole becomes more and more slender. When $a = b$ it agrees with the value $3S$ found for the circular hole on page 92. The least value of the stress round the elliptical hole is $-S$, at the ends of the minor axis. This is the same as for the circular hole.

When the tension S is parallel to the major axis ($\beta = 0$), the greatest value of σ_η round the hole is found at the ends of the minor axis, and is $S(1 + 2b/a)$. This approaches S when the ellipse is very slender. At the ends of the major axis the stress is $-S$ for any value of a/b.

The effect of the elliptical hole on a state of pure shear S parallel to the x and y axes is easily found by superposition of the two cases of tension S at $\beta = \pi/4$ and $-S$ at $\beta = 3\pi/4$. Then

$$(\sigma_\eta)_{\xi=\xi_o} = -2S\,\frac{e^{2\xi_o}\sin 2\eta}{\cosh 2\xi_o - \cos 2\eta}$$

This vanishes at the ends of both the major and the minor axes and has the greatest values

$$\pm S \frac{(a + b)^2}{ab}$$

at the points determined by $\tan \eta = \tanh \xi_o = b/a$. When the ellipse is very slender these values are very large, and the points at which they occur are close to the ends of the major axis.

Solutions have been found for the elliptic hole in a plate subject to pure flexure in its plane[1,2] and to a parabolic distribution of shear as in a thin rectangular beam,[2] for an elliptic hole with equal and opposite concentrated forces at the ends of the minor diameter,[3] and for rigid and elastic "inclusions" filling the hole in a plate under tension.[4] More general series forms of the real stress function ϕ in elliptic coordinates have been considered.[5] Their equivalent complex potentials can be constructed from the functions used or mentioned here, together with the analog of the simple functions quoted in the Problems on page 187, when dislocations and concentrated forces and couples are to be included. A solution for general loading on an elliptic hole is given later in Arts. 67 through 72.

Many further solutions for elliptic and other noncircular holes, or inclusions, with various loadings, have been worked out in detail.[6]

64 | Hvperbolic Boundaries. Notches

It was shown in Art. 60 that the curves η = constant in elliptic coordinates are hyperbolas, and in Art. 62 that the range of η may be taken as 0 to 2π, that of ξ being 0 to ∞.

[1] K. Wolf, *Z. Tech. Physik*, 1922, p. 160.

[2] H. Neuber, *Ingenieur-Arch.*, vol. 5, p. 242, 1934. This solution and several others relating to ellipses and hyperbolas are given in Neuber's book "Kerbspannungslehre," 2d ed., Springer-Verlag OHG, Berlin, 1958.

[3] P. S. Symonds, *J. Appl. Mech.*, vol. 13, p. A-183, 1946. A solution in finite form is given by A. E. Green, *J. Appl. Mech.*, vol. 14, p. A-246, 1947.

[4] N. I. Muskhelishvili, *Z. Angew. Math. Mech.*, vol. 13, p. 264, 1933; L. H. Donnell "Theodore von Kármán Anniversary Volume," p. 293, Pasadena, 1941.

[5] E. G. Coker and L. N. G. Filon, "Photo-elasticity," pp. 123, 535, Cambridge University Press, New York, 1931; A. Timpe, *Math. Z.*, vol. 17, p. 189, 1923.

[6] N. I. Muskhelishvili, "Some Basic Problems of the Theory of Elasticity," 4th corrected and augmented ed., 1954, translation by J. R. M. Radok, Erven P. Noordhoff, NV, Groningen, Netherlands, 1963; G. N. Savin, "Stress Concentration around Holes," 1st ed., 1951, translation edited by W. Johnson, Pergamon Press, New York, 1961; P. P. Teodorescu, One Hundred Years of Investigation in the Plane Problem of the Theory of Elasticity, in H. N. Abramson, H. Liebowitz, J. M. Crowley, and S. Juhasz (eds.), "Applied Mechanics Surveys," pp. 245–262, Spartan Books, Washington, D.C., 1966.

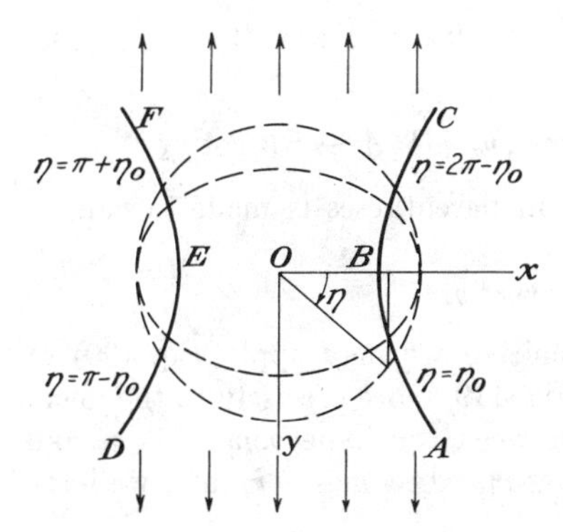

Fig. 119

Let η_o be the constant value of η along the hyperbolic arc BA of Fig. 119. It will be between 0 and $\pi/2$, since both x and y are positive along BA. Along the other half of this branch of the hyperbola, BC, the value of η is $2\pi - \eta_o$. Along the half ED of the other branch, η is $\pi - \eta_o$, and along EF it is $\pi + \eta_o$.

Consider the plate $ABCFED$ within these hyperbolic boundaries, in a state of tension in the direction Oy.[1] The tensile stress at infinity must fall to zero to preserve a finite tensile force across the waist EOB. Complex potentials that permit this, and satisfy the other necessary conditions of symmetry about Ox and Oy and freedom of the hyperbolic boundaries, are

$$\psi(z) = -\tfrac{1}{2}Ai\,\zeta \qquad \chi(z) = -\tfrac{1}{2}Ai\,\zeta z - Bci \sinh \zeta \tag{a}$$

where A and B are real constants, and $z = c \cosh \zeta$. These give

$$\psi'(z) = -\frac{iA}{2c \sinh \zeta} \qquad \chi'(z) = -\frac{1}{2}Ai\,\zeta - \left(\frac{1}{2}A + B\right) i \coth \zeta \tag{b}$$

Equation (90) of Art. 59 shows that the hyperbolic boundary $\eta = \eta_o$ will be free from force provided the function

$$\psi(z) + z\bar{\psi}'(\bar{z}) + \bar{\chi}'(\bar{z}) \tag{c}$$

is constant along it, or equivalently if the conjugate of this function is constant. The conjugate is, from Eqs. (*a*) and (*b*),

$$A\eta - \frac{1}{2}Ai\frac{\cosh \bar{\zeta}}{\sinh \zeta} - \left(\frac{1}{2}A + B\right) i \coth \zeta \tag{d}$$

[1] This problem (also the case of shear loading) was solved by A. A. Griffith, *Tech. Rept. Aeron. Res. Comm.* (Great Britain), 1927–1928, vol. 2, p. 668; and H. Neuber, *Z. Angew. Math. Mech.*, vol. 13, p. 439, 1933; or "Kerbspannungslehre," *op. cit.*, p. 35.

On the hyperbola $\eta = \eta_o$ we have $\bar{\zeta} = \zeta - 2i\eta_o$, and with this the expression becomes

$$A\eta_o - \tfrac{1}{2}A \sin 2\eta_o - (\tfrac{1}{2}A \cos 2\eta_o + \tfrac{1}{2}A + B)i \coth \zeta$$

which is a constant if the quantity in parentheses is made to vanish. Thus

$$B = -A \cos^2 \eta_o \tag{e}$$

To find the resultant force transmitted we may apply Eq. (90) of Art. 59 to the narrow section EOB, Fig. 119, more precisely to the lower part of the limiting ellipse $\xi = 0$ between the hyperbolas $\eta = \eta_0$ and $\eta = \pi - \eta_o$. On this ellipse ζ becomes $i\eta$, $\bar{\zeta}$ becomes $-i\eta$, and we have from Eqs. (90), (*c*), and (*d*)

$$\begin{aligned} F_x - iF_y &= i[A\eta - (A + B) \cot \eta]_{\eta=\eta_o}^{\eta=\pi-\eta_o} \\ &= i[A(\pi - 2\eta_o + 2 \cot \eta_o) + 2B \cot \eta_o] \end{aligned}$$

Since A and B were taken as real, F_x is zero and, using Eq. (*e*),

$$F_y = -A(\pi - 2\eta_o + \sin 2\eta_o)$$

which determines A when the total tension F_y is assigned. The stress and displacement components are easily found from Eqs. (96), (97), and (98). The first gives

$$\sigma_\xi + \sigma_\eta = -\frac{4A}{c} \frac{\cosh \xi \sin \eta}{\cosh 2\xi - \cos 2\eta}$$

The value of σ_ξ along the hyperbolic boundary is found by setting $\eta = \eta_o$ in this expression. It has a maximum, $-2A/c \sin \eta_o$, at the waist where $\xi = 0$. Neuber[1] has expressed this as a function of the radius of curvature of the hyperbola at the waist. He has solved, by another method, the problems of bending and shear of the plate as well as tension.

65 | Bipolar Coordinates

Problems involving two nonconcentric circular boundaries, including the special case of a circular hole in a semi-infinite plate, usually require the use of the *bipolar coordinates* ξ, η, defined by

$$z = ia \coth \tfrac{1}{2}\zeta \qquad \zeta = \xi + i\eta \tag{a}$$

where a is a real constant.

[1] *Loc. cit.* For a comparison of Neuber's results with photoelastic and fatigue tests of notched plates and grooved shafts see R. E. Peterson and A. M. Wahl, *J. Appl. Mech.*, vol. 3, p. 15, 1936, or S. Timoshenko, "Strength of Materials," 3d ed., vol. 2, p. 328. See also M. M. Frocht, "Photoelasticity," vol. 2, John Wiley & Sons, Inc., New York, 1948.

Replacing $\coth \tfrac{1}{2}\zeta$ by $(e^{\frac{1}{2}\zeta} + e^{-\frac{1}{2}\zeta})/(e^{\frac{1}{2}\zeta} - e^{-\frac{1}{2}\zeta})$ and solving the first equation for e^{ζ}, it is easily shown that this is equivalent to

$$\zeta = \log \frac{z + ia}{z - ia} \tag{b}$$

The quantity $z + ia$ is represented by the line joining the point $-ia$ to the point z in the xy plane, in the sense that its projections on the axes give the real and imaginary parts. The same quantity may be represented by $r_1 e^{i\theta_1}$ where r_1 is the length of the line, and θ_1 the angle it makes with the x axis (Fig. 120). Similarly $z - ia$ is the line joining the point ia to the point z, and may be represented by $r_2 e^{i\theta_2}$ (Fig. 120). Then Eq. (b) becomes

$$\xi + i\eta = \log \left(\frac{r_1}{r_2} e^{i\theta_1} e^{-i\theta_2} \right) = \log \frac{r_1}{r_2} + i(\theta_1 - \theta_2)$$

so that

$$\xi = \log \frac{r_1}{r_2} \qquad \eta = \theta_1 - \theta_2 \tag{c}$$

It may be seen from Fig. 120 that $\theta_1 - \theta_2$ is the angle between the two lines joining the "poles" $-ia$, ia to the typical point z, when this point lies to the right of the y axis, and is minus this angle when the point lies to the left. It follows that a curve η = constant is an arc of a circle passing through the poles. Several such circles are drawn in Fig. 120. From Eqs. (c) it is clear that a curve ξ = constant will be a curve for which r_1/r_2 = constant. Such a curve is also a circle. It surrounds the

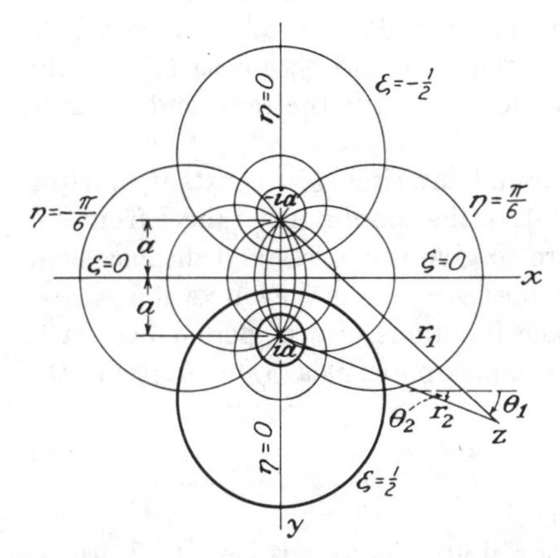

Fig. 120

pole ia if r_1/r_2 exceeds unity, that is, if ξ is positive. It surrounds the other pole $-ia$ if ξ is negative. Several such circles are drawn in Fig. 120. They form a family of coaxial circles with the two poles as limiting points.

The coordinate η changes from π to $-\pi$ on crossing the segment of the y axis joining the poles, its range for the whole plane being $-\pi$ to π. Stresses and displacements will be continuous across this segment if they are represented by periodic functions of η with period 2π.

Separation of real and imaginary parts in Eq. (a) leads to[1]

$$x = \frac{a \sin \eta}{\cosh \xi - \cos \eta} \qquad y = \frac{a \sinh \xi}{\cosh \xi - \cos \eta} \tag{d}$$

Differentiation of Eq. (a) yields

$$Je^{i\alpha} = \frac{dz}{d\zeta} = -\frac{1}{2} ia \operatorname{cosech}^2 \frac{1}{2} \zeta \tag{e}$$

and

$$e^{2i\alpha} = \frac{dz/d\zeta}{d\bar{z}/d\bar{\zeta}} = -\sinh^2 \frac{1}{2} \bar{\zeta} \operatorname{cosech}^2 \frac{1}{2} \zeta \tag{f}$$

66 | Solutions in Bipolar Coordinates

We now consider the problem of a circular disk with an eccentric hole, subject to pressure p_o round the outside and pressure p_1 round the hole.[2] The stress components obtained will also be valid for a circular thick-walled tube with eccentric bore.

Let the external boundary be that circle of the family ξ = constant for which $\xi = \xi_o$, and let the hole be the circle $\xi = \xi_1$. Two such circles are drawn in heavy lines in Fig. 120. It follows from the expression for y in Eqs. (d) of Art. 65 that these circles have radii $a \operatorname{cosech} \xi_o$, $a \operatorname{cosech} \xi_1$, and that their centers are at the distances $a \coth \xi_o$, $a \coth \xi_1$ from the origin. Thus a, ξ_o, and ξ_1 can be determined if the radii and distance between centers are given.

In going counterclockwise once round any circle ξ = constant, starting just to the left of the y axis in Fig. 120, the coordinate η ranges from $-\pi$ to π. Thus the functions that are to give the stress and displacement components must have the same values at $\eta = \pi$ as they have at $\eta = -\pi$. This is ensured if they are periodic functions of η of period 2π. It is therefore appropriate to take the complex potentials $\psi(z)$ and $\chi(z)$ in the forms

$$\cosh n\zeta \qquad \sinh n\zeta \tag{a}$$

[1] See the derivation of Eq. (c) in Art. 54.

[2] The original solution, in terms of the real stress function, is due to G. B. Jeffery, *Trans. Roy. Soc. (London)*, ser. A, vol. 221, p. 265, 1921.

with n an integer, since these are in fact periodic functions of η of period 2π. So also are their derivatives with respect to z, since $d\zeta/dz$ has the same property [Eq. (e), Art. 65].

If such functions are introduced into Eqs. (90) and (91), applied to any circle ξ = constant in the material, the corresponding force and couple will be zero, by virtue of the periodicity. This must hold for the complete solution, for equilibrium of the plate within the circle.

We shall require also the function $\chi(z) = aD\zeta$, D being a constant. Considering this in Eqs. (90) and (91) as above, we find that the moment of Eq. (91) will be zero only if D is real. We therefore take it to be so. Considering the displacement equation (86) we find that this function, as well as the functions (a) used as either $\psi(z)$ or $\chi(z)$, will give displacements free from discontinuity.

The state of uniform all-round tension or compression, which will be part of the solution, is obtained from the complex potential $\psi(z) = Az$ with A real. The corresponding real stress function is, from Eq. (84),

$$\phi = \operatorname{Re}(\bar{z}Az) = A\bar{z}z = A(x^2 + y^2)$$

This may be expressed in bipolar coordinates by means of Eqs. (d) of Art. 65, the result being

$$Aa^2 \frac{\cosh \xi + \cos \eta}{\cosh \xi - \cos \eta} \tag{b}$$

Considering functions of the form (a), with $n = 1$, we observe that since the stress distribution in the present problem is symmetrical about the y axis, we must choose them so that the corresponding stress functions have the same symmetry. Thus we may take

$$\psi(z) = iB \cosh \zeta \qquad \chi(z) = B' \sinh \zeta \tag{c}$$

with B, B' real, and

$$\psi(z) = iC \sinh \zeta \qquad \chi(z) = C' \cosh \zeta \tag{d}$$

with C, C' real.

The real stress function corresponding to (c) is, from Eq. (84),

$$aB \frac{\sinh \xi \cosh \xi \cos \eta - \sinh \xi \sin^2 \eta}{\cosh \xi - \cos \eta} + B' \frac{\sinh \xi \cosh \xi \cos \eta - \sinh \xi \cos^2 \eta}{\cosh \xi - \cos \eta}$$

If we choose $B' = aB$, the terms in $\sin^2 \eta$, $\cos^2 \eta$ in the numerators become independent of η, and the complete numerator depends on η only in the term in $\cos \eta$, just as does the function (b). The same thing is true of the complex potentials (d), if we choose $C' = aC$. We thus obtain simpler, more restricted functions which turn out to be adequate for the present problem.

Taking therefore

$$\psi(z) = iB \cosh \zeta \qquad \chi(z) = aB \sinh \zeta \tag{e}$$

we find by means of Eqs. (96), (97), and (a) and (f) of Art. 65 that the corresponding stress components are given by

$$a(\sigma_\xi + \sigma_\eta) = 2B(2 \sinh \xi \cos \eta - \sinh 2\xi \cos 2\eta) \tag{f}$$

$$\begin{aligned} a(\sigma_\eta - \sigma_\xi + 2i\tau_{\xi\eta}) = & -2B[\sinh 2\xi - 2 \sinh 2\xi \cosh \xi \cos \eta \\ & + \sinh 2\xi \cos 2\eta - i(2 \cosh 2\xi \cosh \xi \sin \eta - \cosh 2\xi \sin 2\eta)] \end{aligned} \tag{g}$$

Similarly the functions

$$\psi(z) = iC \sinh \zeta \qquad \chi(z) = aC \cosh \zeta \tag{h}$$

yield

$$a(\sigma_\xi + \sigma_\eta) = -2C(1 - 2 \cosh \xi \cos \eta + \cosh 2\xi \cos 2\eta) \tag{i}$$

$$\begin{aligned} a(\sigma_\eta - \sigma_\xi + 2i\tau_{\xi\eta}) = & \ 2C[- \cosh 2\xi + 2 \cosh 2\xi \cosh \xi \cos \eta \\ & - \cosh 2\xi \cos 2\eta + i(2 \sinh 2\xi \cosh \xi \sin \eta - \sinh 2\xi \sin 2\eta)] \end{aligned} \tag{j}$$

The stress components arising from

$$\chi(z) = aD\zeta \tag{k}$$

are given by

$$\begin{aligned} \sigma_\xi + \sigma_\eta &= 0 \\ a(\sigma_\eta - \sigma_\xi + 2i\tau_{\xi\eta}) &= D[\sinh 2\xi - 2 \sinh \xi \cos \eta \\ & \qquad - i(2 \cosh \xi \sin \eta - \sin 2\eta)] \end{aligned} \tag{l}$$

The state of uniform all-round tension given by

$$\psi(z) = Az \tag{m}$$

yields $\qquad \sigma_\xi + \sigma_\eta = 4A \qquad \sigma_\eta - \sigma_\xi + 2i\tau_{\xi\eta} = 0$

or

$$\sigma_\xi = \sigma_\eta = 2A \qquad \tau_{\xi\eta} = 0 \tag{n}$$

The solution of our problem can be obtained by superposition of the states of stress represented by the complex potentials (e), (h), (k), and (m). Collecting the terms representing $\tau_{\xi\eta}$ in Eqs. (g), (j), and (l), we find that the vanishing of $\tau_{\xi\eta}$ on the boundaries $\xi = \xi_o$, $\xi = \xi_1$ requires

$$\begin{aligned} D - 2B \cosh 2\xi_o - 2C \sinh 2\xi_o &= 0 \\ D - 2B \cosh 2\xi_1 - 2C \sinh 2\xi_1 &= 0 \end{aligned} \tag{o}$$

Solving these for B and C in terms of D, we have

$$2B = D\frac{\cosh(\xi_1 + \xi_o)}{\cosh(\xi_1 - \xi_o)} \qquad 2C = -D\frac{\sinh(\xi_1 + \xi_o)}{\cosh(\xi_1 - \xi_o)} \tag{p}$$

The normal stress σ_ξ can be found by subtracting the real part of Eq. (g) from Eq. (f) and similarly for the other pairs. On the boundary $\xi = \xi_o$ it is to take the value $-p_o$, and on the boundary $\xi = \xi_1$ the value $-p_1$. Using the values of B and C given by Eqs. (p), these conditions lead to the two equations

$$2A + \frac{D}{a}\sinh^2\xi_o \tanh(\xi_1 - \xi_o) = -p_o$$

$$2A - \frac{D}{a}\sinh^2\xi_1 \tanh(\xi_1 - \xi_o) = -p_1$$

and therefore

$$A = -\frac{1}{2}\frac{p_o\sinh^2\xi_1 + p_1\sinh^2\xi_o}{\sinh^2\xi_1 + \sinh^2\xi_o}$$

$$D = -a\frac{(p_o - p_1)\coth(\xi_1 - \xi_o)}{\sinh^2\xi_1 + \sinh^2\xi_o}$$

These with Eqs. (p) complete the determination of the complex potentials. When there is internal pressure p_1 only ($p_o = 0$), the peripheral stress at the hole is found to be

$$(\sigma_\eta)_{\xi=\xi_1} = -p_1 + 2p_1(\sinh^2\xi_1 + \sinh^2\xi_o)^{-1}(\cosh\xi_1 - \cos\eta)[\sinh\xi_1\coth(\xi_1 - \xi_o) + \cos\eta]$$

An expression for the maximum value[1] of this has already been given on page 71.

A general series form of stress function in bipolar coordinates was given by G. B. Jeffery.[2] Its equivalent complex potentials are easily found and involve the functions considered here together with the bipolar analogs of the simple functions quoted in the Problems on page 187, when dislocations and concentrated forces are included. It has been applied to the problems of a semi-infinite plate with a concentrated force at any point,[3] a semi-infinite region with a circular hole, under tension parallel to the straight edge or plane boundary,[4] and under its own weight,[5] and

[1] An exhaustive discussion of the maximum value is given by Coker and Filon, *loc. cit.*

[2] *Loc. cit.*

[3] E. Melan; see n., p. 132. Complex potentials for concentrated force and couple are given on p. 265 of A. E. Green and W. Zerna, "Theoretical Elasticity," Oxford University Press, Fair Lawn, N.J., 1954.

[4] See p. 93; also W. T. Koiter, *Quart. Appl. Math.*, vol. 15, p. 303, 1957.

[5] R. D. Mindlin, *Proc. ASCE*, p. 619, 1939.

to the infinite plate with two holes,[1] or a hole formed by two intersecting circles.[2]

Solutions have been given for the circular disk subject to concentrated forces at any point,[3] to its own weight when suspended at a point,[4] or in rotation about an eccentric axis,[5] with and without[6] the use of bipolar coordinates, and for the effect of a circular hole in a semi-infinite plate with a concentrated force on the straight edge.[7]

Other Curvilinear Coordinates The equation

$$z = e^{\zeta} + abe^{-\zeta} + ac^3e^{-3\zeta}$$

yielding

$$x = (e^{\xi} + abe^{-\xi}) \cos \eta + ac^3e^{-3\xi} \cos 3\eta$$
$$y = (e^{\xi} - abe^{-\xi}) \sin \eta - ac^3e^{-3\xi} \sin 3\eta$$

where a, b, c are constants, gives a family of curves $\xi =$constant which can be made to include various oval shapes, including a square with rounded corners. The effect of a hole of such shape in a plate under tension has been evaluated (by means of the real stress function) by M. Greenspan.[8] By means of a generalization of these coordinates A. E. Green[9] has obtained solutions for a triangular hole with rounded corners, and, by means of another coordinate transformation, for an exactly rectangular hole. In the latter case the perfectly sharp corners introduce infinite stress concentration.

The curvilinear coordinates given by

$$z = \zeta + ia_1e^{i\zeta} + ia_2e^{i2\zeta} + \cdots + ia_ne^{in\zeta}$$

$a_1, a_2, \ldots, a_n$ being real constants, have been applied by C. Weber to the semi-infinite plate with a serrated boundary,[10] as in the example of evenly spaced semicircular notches which is worked out. When the dis-

[1] T. Pöschl, *Z. Angew. Math. Mech.*, vol. 1, p. 174, 1921, and vol. 2, p. 187, 1922. Also C. Weber, *Z. Angew. Math. Mech.*, vol. 2, p. 267, 1922; E. Weinel, *ibid.*, vol. 17, p. 276, 1937; Chih Bing Ling, *J. Appl. Phys.*, vol. 19, p. 77, 1948.

[2] Ling, *ibid.*, p. 405, 1948.

[3] R. D. Mindlin, *J. Appl. Mech.*, vol. 4, p. A-115, 1937.

[4] R. D. Mindlin, *J. Appl. Phys.*, vol. 9, p. 714, 1938.

[5] R. D. Mindlin, *Phil. Mag.*, ser. 7, vol. 26, p. 713, 1938.

[6] B. Sen, *Bull. Calcutta Math. Soc.*, vol. 36, pp. 58 and 83, 1944.

[7] A. Barjansky, *Quart. Appl. Math.*, vol. 2, p. 16, 1944. Also R. M. Evan-Iwanowski, *ibid.*, vol. 19, p. 359, 1962.

[8] *Quart. Appl. Math.*, vol. 2, p. 60, 1944. See also V. Morkovin, *ibid.*, p. 350, 1945.

[9] *Proc. Roy. Soc. (London)*, ser. A, vol. 184, p. 231, 1945.

[10] *Z. Angew. Math. Mech.*, vol. 22, p. 29, 1942.

tance between notch centers is twice the notch diameter, the stress concentration, for tension, is found to be 2.13. The value for a single notch is 3.07 (see page 102).

Assignable Shapes Kikukawa has devised, and applied, methods for holes and fillets of assignable shapes.[1] Successive adjustments of an initial conformal mapping are made until an adequate approximation to the given shape is attained. The detailed results include evaluations of stress concentration in a stretched plate with (1) a hole of rhombic shape with circular arc fillets at the corners, (2) a double notch in a strip, each notch having straight parallel sides joined by a semicircle to form a U shape, and (3) circular quadrant fillets at a change from finite to infinite width. The results of case (2) agree very well with those of Neuber for the double hyperbolic notch.[2]

67 | Determination of the Complex Potentials from Given Boundary Conditions. Methods of Muskhelishvili

In the preceding articles, several specific problems were solved by judicious selection of the complex potentials in relatively simple forms having appropriate properties. But more powerful and general methods have been developed for deducing the potentials directly from given boundary conditions[3] by means of further applications of the theory of functions of the complex variable.

In Art. 59 we found that the force components F_x, F_y transmitted across an arc AB in the material were given [Eq. (90)] by

$$F_x + iF_y = -i[\psi(z) + z\bar{\psi}'(\bar{z}) + \bar{\chi}'(\bar{z})]_A^B \tag{90'}$$

The arc AB may be part of a closed boundary curve such as the hole L in Fig. 121. Then, proceeding from A to B, the material is to the left, and the forces exerted will be $-F_x$, $-F_y$. We now take A as a fixed point on the hole, and B any typical point on L. Supposing the loading on the hole given, the forces F_x, F_y are known as functions of s (Fig. 121), and we can write

$$i(F_x + iF_y) = f_1(s) + if_2(s) \tag{a}$$

[1] An outline of the method, with references, is given in J. N. Goodier and P. G. Hodge, "Elasticity and Plasticity," pp. 8–10, John Wiley & Sons, Inc., New York, 1958.

[2] See Art. 64.

[3] N. I. Muskhelishvili, "Some Basic Problems of the Theory of Elasticity." See n. 6, p. 194.

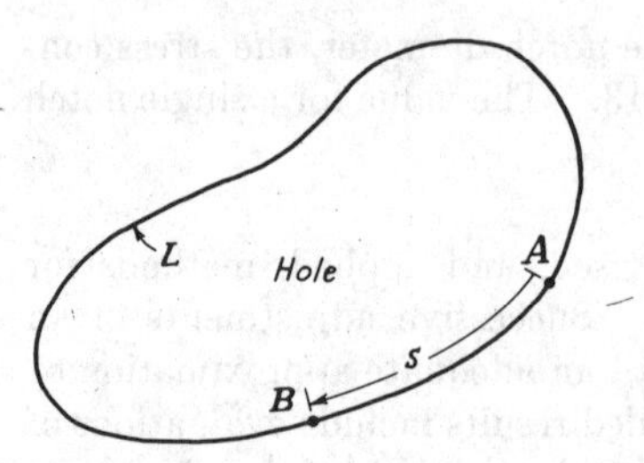

Fig. 121

where $f_1(s)$, $f_2(s)$ are real. In Eq. (90′) above, the value of the bracket at the fixed point A is some constant C. Using z for the movable point B, the boundary condition on the hole can be expressed as

$$\psi(z) + z\bar{\psi}'(\bar{z}) + \bar{\chi}'(\bar{z}) = f_1(s) + if_2(s) + C \qquad \text{on } L \tag{99}$$

In solving this equation for the two complex potentials, it will be advantageous to replace the general complex variable z for any point in the physical region by a new complex variable ζ, by a relation

$$z = \omega(\zeta) \tag{100}$$

where $\omega(\zeta)$ is a suitably chosen function of ζ. Such a relation was used previously [Eq. (g), page 183] to define a type of curvilinear coordinate. Here it is appropriate to adopt the different, though closely related, geometrical interpretation of *conformal mapping*.

A point P' given by the complex coordinate $\zeta = \xi + i\eta$ in the ζ plane (Fig. 122b) has a corresponding, or "mapped," point P in the z plane (Fig. 122a), with z evaluated by $z = \omega(\zeta)$. In general, a smooth curve $P'Q'$

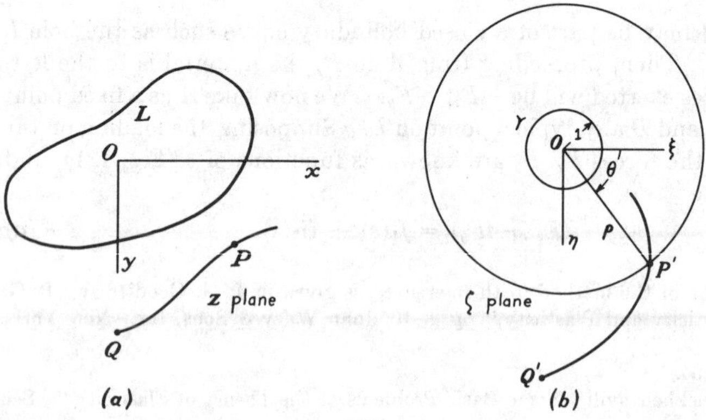

Fig. 122

maps into a different smooth curve PQ. For elasticity problems involving a single noncircular hole L in an infinite region, the conformal mapping function $\omega(\zeta)$ will be selected so that the curve L maps from the unit circle $\rho = 1$ in the ζ plane. It is then convenient to make use of the polar coordinates ρ, θ instead of the rectangular coordinates ξ, η. The function $\omega(\zeta)$ will also be selected so that a point P' (outside or on the circle) maps into only *one* point P. The function is to be analytic at every point P' which maps into a material point P. It is to admit a Laurent expansion

$$\omega(\zeta) = R\zeta + \frac{e_1}{\zeta} + \frac{e_2}{\zeta^2} + \cdots \tag{b}$$

where R, e_1, e_2, etc., are constants.

A function of z, for instance $\psi(z)$ or $\chi'(z)$, is then also a function of ζ, obtained by replacing z by $\omega(\zeta)$.

Thus,

$$\psi(z) = \psi[\omega(\zeta)] \qquad \chi'(z) = \chi'[\omega(\zeta)] \tag{c}$$

In changing over to functions of ζ, we shall adopt a change of notation, using the functional symbols ϕ and ψ for new purposes, thus:

The function $\psi[\omega(\zeta)]$ in (c) is to be written

$$\phi(\zeta) \tag{d}$$

The function $\chi'[\omega(\zeta)]$ in (c) is to be written

$$\psi(\zeta) \tag{e}$$

In converting the boundary condition (99) to the new notation, the first term on the left becomes simply $\phi(\zeta)$. The third term becomes $\bar{\psi}(\bar{\zeta})$, obtained by changing every i involved in $\psi(\zeta)$ to $-i$. For the second term on the left of (99) we replace z by $\omega(\zeta)$. For the replacement of $\bar{\psi}'(\bar{z})$ we first observe that

$$\psi'(z) = \frac{d}{dz}\psi(z) = \frac{d}{d\zeta}\psi[\omega(\zeta)]\frac{d\zeta}{dz} = \frac{d}{d\zeta}\phi(\zeta)\frac{d\zeta}{dz} = \phi'(\zeta)\frac{d\zeta}{dz} \tag{f}$$

and

$$\frac{d\zeta}{dz} = \frac{1}{dz/\partial\zeta} = \frac{1}{\omega'(\zeta)} \tag{g}$$

The second term is accordingly replaced by

$$\omega(\zeta)[\bar{\phi}'(\bar{\zeta})]\frac{1}{\bar{\omega}'(\bar{\zeta})} \tag{h}$$

On the right-hand side of (99) we have a complex function of position on L. The corresponding position on the unit circle $\rho = 1$ can be indicated by the coordinate θ, or by $e^{i\theta}$. Writing

$$\sigma = e^{i\theta} \qquad \bar{\sigma} = e^{-i\theta} \tag{i}$$

we observe that σ is in fact the value of ζ for the typical point on the unit circle. Thus it will be possible to express the right-hand side of (99) as a function of σ, and we shall write

$$f_1(s) + if_2(s) = f(\sigma) \tag{j}$$

The constant C in (99) can be eliminated by merely adding a suitable constant to ψ[or to $\chi'(z)$], a change which has no effect on the stress. The function $f(\sigma)$ is an expression of the loading applied between the points A and B, in accordance with Eq. (a), in the form $-F_y + iF_x$.

The boundary condition (99) thus becomes

$$\phi(\sigma) + \frac{\omega(\sigma)}{\bar{\omega}'(\bar{\sigma})}\bar{\phi}'(\bar{\sigma}) + \bar{\psi}(\bar{\sigma}) = f(\sigma) \tag{101}$$

This condition is the foundation of the methods of Muskhelishvili. The changed notation is in fact that of the book referred to in the footnote on page 194, to which the present introductory treatment of the methods is pointing.

68 | Formulas for the Complex Potentials[1]

The objective is now the determination of the potentials $\phi(\zeta)$, $\psi(\zeta)$, for any point ζ *outside* the unit circle, so as to satisfy the boundary condition (101).

Once chosen, ζ is fixed for the present development. Then (101) can be multiplied by $1/(\sigma - \zeta)$. Each term remains a function of σ and can be integrated all round the unit circle, indicated henceforth by γ. Then

$$\int_\gamma \frac{\phi(\sigma)\,d\sigma}{\sigma - \zeta} + \int_\gamma \frac{\omega(\sigma)}{\bar{\omega}'(\bar{\sigma})}\bar{\phi}'(\bar{\sigma})\frac{d\sigma}{\sigma - \zeta} + \int_\gamma \frac{\bar{\psi}(\bar{\sigma})\,d\sigma}{\sigma - \zeta} = \int_\gamma \frac{f(\sigma)\,d\sigma}{\sigma - \zeta} \tag{102}$$

The significance of this step is that it makes a connection with integrals of the kind well known in the Cauchy-Goursat integral theorem and the Cauchy integral formula.[2] By the theorems given later (Art. 70), the first integral in (102) is evaluated as

$$\int_\gamma \frac{\phi(\sigma)\,d\sigma}{\sigma - \zeta} = -2\pi i\phi(\zeta) \tag{a}$$

provided that $\phi(\zeta)$ is analytic at every point ζ outside γ, including infinity. The values $\phi(\sigma)$ on γ are to be continuous with the values $\phi(\zeta)$ outside.

[1] Arts. 68 and 69 assume some knowledge of complex integrals, not covered by the brief survey of complex functions given in Arts. 54 and 55. See the following footnote.

[2] See, for instance, R. V. Churchill, "Complex Variables and Applications," 2d ed., chap. 5, McGraw-Hill Book Company, New York, 1960.

The third integral in (102) is shown to vanish, provided $\psi(\zeta)$ *is analytic at every point* ζ *outside* γ, including infinity. Also $\psi(\sigma)$ is to be continuous with $\psi(\zeta)$. The second integral in (102) can be evaluated when $\omega(\zeta)$ is a rational function (the ratio of two polynomials). For the illustrative particular case in Art. 71, it is zero, and Eq. (102) gives $\phi(\zeta)$ in the form

$$\phi(\zeta) = -\frac{1}{2\pi i}\int_{\gamma}\frac{f(\sigma)\,d\sigma}{\sigma - \gamma} \tag{b}$$

Eq. (102) also leads to a similar formula for $\psi(\zeta)$, as will be seen later (Art. 72).

The requirement that $\phi(\zeta)$ and $\psi(\zeta)$ be analytic outside γ implies certain restrictions on the kind of problem that can be solved in this manner. These will now be examined.

69 | Properties of Stress and Deformation Corresponding to Complex Potentials Analytic in the Material Region around a Hole

It is understood throughout that the mapping function $\omega(\zeta)$ is analytic in the material region. Consequently, when the potentials are analytic in ζ, they are also analytic when expressed as functions of z, for any point in the material region. It follows that all their derivatives are analytic as well. "Analytic" implies that they are continuous. In particular, they return to their starting values after completion of any circuit enclosing the hole, and lying entirely in the material. It also follows that their conjugates, and their real or imaginary parts separately, are similarly continuous.[1]

Knowing this, we can make use of Eqs. (86) through (91) to establish the following characteristic properties of states representable by analytic potentials:

1. From Eqs. (87) and (88), the stress components are continuous.[2]
2. From Eq. (86), the displacement components are continuous. (Thus such a solution cannot represent a *dislocation*.)
3. From Eq. (90), the total force on any circuit is zero; therefore, the resultant force applied to the hole is zero.
4. From Eq. (91), the resultant couple of the loading applied to the hole is zero.

[1] For demonstrations of these properties see, for instance, Churchill, *op. cit.*, chap. 2.

[2] Samples of potentials representing discontinuous stress are given with applications by J. N. Goodier and J. C. Wilhoit, *Proc. 4th Ann. Midwest Conf. Solid Mech., Univ. Texas*, pp. 152–170, 1959.

Further, a function $F(z)$ analytic in the material region (including infinity)—the origin being inside the hole—has a Laurent expansion

$$F(z) = c_0 + \frac{c_1}{z} + \frac{c_2}{z^2} + \cdots$$

where c_0, c_1, etc., are constants. The potentials $\psi(z)$, $\chi(z)$ under consideration here thus have such expansions. It then follows from Eqs. (87) and (88) that

5. The stress components vanish at infinity. There is thus no loading at infinity at all, since by (3) and (4) the resultant force and couple for an infinite circuit will be zero.[1]

From these properties, it is evident that the stress and deformation represented by the analytic potentials must be attributed to self-equilibrating loading applied to the boundary of the hole.

This is not a serious restriction. The effect of an unloaded hole in an infinite region, with loading on the boundary at infinity (for instance, the problem illustrated by Fig. 118), can be found by first finding the stress when there is no hole. This implies definite loading across the curve corresponding to the hole, but by equilibrium of the material filling the hole, it is a self-equilibrating loading. We have then to determine the stress outside the hole due to equal and opposite loading on the hole and vanishing at infinity. This problem conforms to the requirements 1 to 5 above for the analytic potentials.

If it is required to have loading on the hole that has nonzero resultant force and couple, we can begin with the concentrated force solution indicated in part g of Prob. 2 on page 187, giving the force the required resultant value. To this we may add the couple solution indicated in part a of the same problem, taking b infinite and a very small. These imply loading across the curve corresponding to the hole, which has the prescribed resultant force and couple but the wrong distribution. The prescribed distribution is attained by introducing a determinable loading on the hole, and the problem presented by this conforms to the requirements for the analytic potentials.

If a dislocational solution is required, we can begin with the solutions indicated in parts e and d of the same problem, assigning the prescribed amounts of dislocational translation or rotation, thus similarly reducing the problem to one of analytic potentials.

The potentials of each part of Prob. 2 on page 187 are of course *not* both analytic everywhere in the material region, because $\log z$ is not continuous

[1] For an infinite boundary curve the vanishing of the stress components does not necessarily mean zero resultants.

for a complete circuit round the origin, and also because z, z^2 are not analytic everywhere including infinity.

70 | Theorems on Boundary Integrals

In establishing the theorems quoted in Art. 68, we begin with Eq. (a) (page 206). In the region outside γ, $\phi(\zeta)$, being analytic everywhere including infinity, will have a Laurent expansion

$$\phi(\zeta) = \frac{a_1}{\zeta} + \frac{a_2}{\zeta^2} + \cdots \qquad (a)$$

No constant term a_0 is included because it has no effect on stress.[1]

In Fig. 123 a larger circle Γ is drawn concentric with γ. Since $\phi(\zeta)$ is analytic in the region enclosed by the contour indicated by arrowheads, Cauchy's integral formula may be applied to give

$$2\pi i\phi(\zeta) = \int_\Gamma \frac{\phi(t)\,dt}{t-\zeta} - \int_\gamma \frac{\phi(\sigma)\,d\sigma}{\sigma-\zeta} \qquad (b)$$

where ζ is a point in this region, t is used for points on Γ, and σ as before for points on γ. But the first integral, round Γ, is zero. To show this, we first observe that the expansion (a) holds on Γ, with ζ replaced by t. Thus,

$$t\phi(t) = a_1 + \frac{a_2}{t} + \cdots \qquad (c)$$

[1] When such a potential is used later for the evaluation of displacements, we are free to add rigid-body terms.

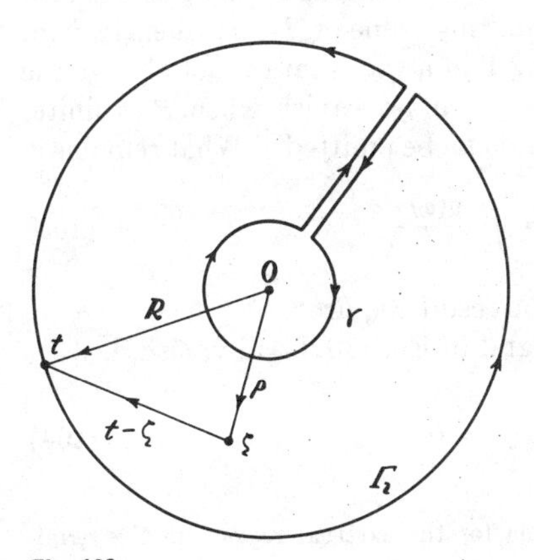

Fig. 123

and, this series being convergent, $t\phi(t)$ is bounded. We may introduce a positive constant C such that

$$|t|\,|\phi(t)| < C \tag{d}$$

where $|t|$ denotes the modulus (absolute value) of t, that is, the radius R in Fig. 123. C can be chosen so that (*d*) holds for all values of R greater than some value R_0. For instance R_0 could correspond to the greatest value of $|t|\,|\phi(t)|$.

Writing

$$I_1 = \int_\Gamma \frac{\phi(t)\,dt}{t - \zeta} \qquad |I_1| \leq \int_\Gamma \frac{|\phi(t)|\,|dt|}{|t - \zeta|} \tag{e}$$

we shall replace each of the quantities under the integral sign in (*e*) in such a way that the magnitude of the integral is increased. The first replacement is $C/|t|$ for $|\phi(t)|$, which by (*d*) means an increase of value. In replacing $|dt|$ we write

$$t = Re^{i\theta} \qquad dt = iRe^{i\theta}\,d\theta \qquad |dt| = R\,d\theta \tag{f}$$

with no change of value. In replacing $|t - \zeta|$ in the denominator, we do not decrease the value of the integral by taking the quantity $R - \rho$. From the triangle drawn in Fig. 123, it is evident that

$$\rho + |t - \zeta| \geq R \qquad R - \rho \leq |t - \zeta| \tag{g}$$

Thus, returning to (*e*),

$$|I_1| < \int_0^{2\pi} \frac{C}{R}\frac{1}{R - \rho}\,R\,d\theta = \frac{2\pi C}{R - \rho} \tag{h}$$

We can make R larger, indefinitely, without changing C, and ρ is of course fixed once ζ has been chosen. The limiting value of $|I_1|$ is evidently zero. But in enlarging R we are deforming Γ in a way that cannot change the value of the integrals in (*e*). Thus $|I_1|$ must vanish when R is finite. Now in Eq. (*b*) above, the first integral can be omitted. What remains is

$$-2\pi i\phi(\zeta) = \int_\gamma \frac{\phi(\sigma)\,d\sigma}{\sigma - \zeta} \tag{103}$$

—the same as[1] Eq. (*a*) in Art. 68, the result required.

Next we show that the third integral in Eq. (102) will vanish, i.e.,

$$\int_\gamma \frac{\psi(\bar{\sigma})\,d\sigma}{\sigma - \zeta} = 0 \tag{104}$$

[1] Essentially "Cauchy's integral formula for the external region" in the terminology of Muskhelishvili.

Since $\psi(\zeta)$ is to be analytic everywhere in the region outside γ, including infinity, it will have a Laurent expansion

$$\psi(\zeta) = \frac{b_1}{\zeta} + \frac{b_2}{\zeta^2} + \cdots \tag{i}$$

again omitting a constant term that would have no effect on the stress. For the present argument, we shall need to consider ζ not only for points outside γ but also for points inside. For clarity, we write

ζ_0 for points outside γ

ζ_1 for points inside γ

Thus in (104) and (i) we write ζ_0 for ζ. In (i) we can take the conjugates on right and left. Thus

$$\bar{\psi}(\bar{\zeta}_0) = \frac{\bar{b}_1}{\bar{\zeta}_0} + \frac{\bar{b}_2}{\bar{\zeta}_0^2} + \cdots \tag{j}$$

and this, of course, is a convergent series for any ζ_0. But

$$\zeta_0 = \rho_0 e^{i\theta} \qquad \bar{\zeta}_0 = \rho_0 e^{-i\theta} \qquad \frac{1}{\bar{\zeta}_0} = \frac{1}{\rho_0} e^{i\theta} \tag{k}$$

where clearly $\rho_0 > 1$. Thus $1/\bar{\zeta}_0$ plotted as a point in the ζ plane is *inside* γ. We can have any inside point ζ_1 in this way. From Eq. (j), therefore, we have a function $F(\zeta_1)$ equal to $\bar{\psi}(\bar{\zeta}_0)$ and represented by the convergent power series

$$F(\zeta_1) = \bar{b}_1\zeta_1 + \bar{b}_2\zeta_1^2 + \cdots \tag{l}$$

Evidently $F(\zeta_1)$ is analytic inside γ. Now using ζ for any selected point outside γ, *not* tied to ζ_1, the function

$$\frac{F(\zeta_1)}{(\zeta_1 - \zeta)}$$

is also analytic inside γ; therefore, by Cauchy's integral theorem, its integral round any contour inside γ is zero. We can bring the contour out to γ. Then

$$\int_\gamma \frac{F(\sigma)\, d\sigma}{\sigma - \zeta} = 0 \tag{m}$$

But, since $\sigma = e^{i\theta}$, we have $\sigma = 1/\bar{\sigma}$. Then from (l),

$$\begin{aligned} F(\sigma) &= \bar{b}_1\sigma + \bar{b}_2\sigma^2 + \cdots \\ &= \frac{\bar{b}_1}{\bar{\sigma}} + \frac{\bar{b}_2}{\bar{\sigma}^2} + \cdots \end{aligned} \tag{n}$$

We shall have $\zeta_0 \to \sigma$ when $\zeta_1 \to \sigma$, and then the series in (j) becomes the same as the series (n). Thus

$$F(\sigma) = \bar{\psi}(\bar{\sigma})$$

Hence, Eq. (m) leads to Eq. (104), the result required.

We have now dealt with the first and third integrals in Eq. (102) of Art. 68. The second will be considered in Art. 71 for a specific mapping function $\omega(\zeta)$.

71 | A Mapping Function $\omega(\zeta)$ for the Elliptic Hole. The Second Boundary Integral

If we take

$$z = \omega(\zeta) \qquad \text{with} \qquad \omega(\zeta) = R\left(\zeta + \frac{m}{\zeta}\right) \tag{105}$$

R being a positive constant and m a positive constant less than unity, we have

$$x = R\left(\rho + \frac{m}{\rho}\right)\cos\theta \qquad y = R\left(\rho - \frac{m}{\rho}\right)\sin\theta \tag{a}$$

The unit circle γ in the ζ plane maps into an ellipse in the z plane, with semiaxes

$$a = R(1 + m) \qquad b = R(1 - m) \tag{b}$$

and an exterior concentric circle maps into an exterior confocal ellipse.

For the second integral in Eq. (102) we observe that

$$\omega'(\zeta) = R\left(1 - \frac{m}{\zeta^2}\right) \qquad \bar{\omega}'(\bar{\sigma}) = R\left(1 - \frac{m}{\bar{\sigma}^2}\right) \tag{c}$$

Since $\sigma = 1/\bar{\sigma}$ we find

$$\frac{\omega(\sigma)}{\bar{\omega}'(\bar{\sigma})} = \frac{1}{\sigma}\frac{\sigma^2 + m}{1 - m\sigma^2} \tag{d}$$

The second integral required is thus

$$I_2 = \int_\gamma \frac{1}{\sigma}\frac{\sigma^2 + m}{1 - m\sigma^2}\bar{\phi}'(\bar{\sigma})\frac{d\sigma}{\sigma - \zeta_0} \tag{e}$$

where again we write ζ_0 for ζ to emphasize that ζ here represents some (arbitrarily) selected point outside γ. This integral is now shown to vanish by Cauchy's integral theorem. Consider the possibility that the whole integrand is the value $f(\sigma)$ on γ of a function $f(\zeta_1)$ that is analytic, ζ_1 meaning any point within γ. The continuity represented by

$$f(\sigma) = \lim_{\zeta_1 \to \sigma} f(\zeta_1)$$

is implied.

Then, in view of (e), we might have

$$f(\zeta_1) = \frac{1}{\zeta_1} \frac{\zeta_1^2 + m}{1 - m\zeta_1^2} \bar{\phi}' \left(\frac{1}{\zeta_1}\right) \frac{1}{\zeta_1 - \zeta_0} \tag{f}$$

There is no problem in the denominator terms $1 - m\zeta_1^2$, $\zeta_1 - \zeta_0$; they are nonzero because $m < 1$ and ζ_0 is outside γ. Recalling the expansion (a) of Art. 70, we have

$$\frac{1}{\zeta_1} \bar{\phi}' \left(\frac{1}{\zeta_1}\right) = -\bar{a}_1\zeta_1 - 2\bar{a}_2\zeta_1^2 - \cdots$$

and this is analytic because the differentiated series for $\phi'(\zeta)$ obtained from (a) of Art. 70 is analytic for ζ outside γ, that is, $1/\zeta$ inside. Evidently $f(\zeta_1)$ *is* analytic in γ. Hence its integral around γ, I_2 in (e), is zero by the Cauchy theorem.

This result, together with the results in Art. 70, establishes the formula (b) of Art. 68 for the elliptic-hole problem.

72 | The Elliptic Hole. Formula for $\psi(\zeta)$

The original boundary condition, Eq. (101), can be changed to the conjugate form

$$\bar{\phi}(\bar{\sigma}) + \frac{\bar{\omega}(\bar{\sigma})}{\omega'(\sigma)} \phi'(\sigma) + \psi(\sigma) = \bar{f}(\bar{\sigma}) \tag{a}$$

and then to

$$\int_\gamma \frac{\bar{\phi}(\bar{\sigma})\, d\sigma}{\sigma - \zeta} + \int_\gamma \sigma \frac{1 + m\sigma^2}{\sigma^2 - m} \phi'(\sigma) \frac{d\sigma}{\sigma - \zeta} + \int_\gamma \frac{\psi(\sigma)\, d\sigma}{\sigma - \zeta} = \int_\gamma \frac{\bar{f}(\bar{\sigma})\, d\sigma}{\sigma - \zeta} \tag{106}$$

ζ being an outside point.

Without restriction to the elliptic hole, the first integral vanishes by Eq. (j) of Art. 70, replacing ψ by ϕ, which has all the properties required of ψ in Art. 70. For the second integral we have

$$\frac{1}{2\pi i} \int_\gamma \left[\sigma \frac{1 + m\sigma^2}{\sigma^2 - m} \phi'(\sigma)\right] \frac{d\sigma}{\sigma - \zeta} = -\zeta \frac{1 + m\zeta^2}{\zeta^2 - m} \phi'(\zeta) \tag{b}$$

as an application of the theorem (103), Cauchy's integral formula for the external region. In that formula, we replace $-\phi(\zeta)$ by the function on the right of (b), observing that this function is analytic everywhere outside γ, including infinity. The third integral in (106) above becomes, by the same theorem,

$$\int_\gamma \frac{\psi(\sigma)\, d\sigma}{\sigma - \zeta} = -2\pi\psi(\zeta) \tag{c}$$

Thus (106) is reduced to

$$\psi(\zeta) = -\frac{1}{2\pi i}\int_\gamma \frac{\bar{f}(\bar{\sigma})\,d\sigma}{\sigma - \zeta} - \zeta\,\frac{1 + m\zeta^2}{\zeta^2 - m}\,\phi'(\zeta) \qquad (d)$$

Repeating (*b*) of Art. 68 as

$$\phi(\zeta) = -\frac{1}{2\pi i}\int_\gamma \frac{f(\sigma)\,d\sigma}{\sigma - \zeta} \qquad (e)$$

we have in (*e*) and (*d*) formulas for $\phi(\zeta)$ and $\psi(\zeta)$ in terms of given (self-equilibrating) loading on the hole. Of course (*d*) is restricted[1] to the elliptic hole, but (*e*) is not.

73 | The Elliptic Hole. Particular Problems

Figure 124 indicates the elliptic hole, free of loading, the stress being due to a uniform tensile stress S at infinity, at an angle β to the x axis. This problem was solved in Art. 63 by direct choice of simple complex potentials having appropriate properties. We now obtain the potentials deductiyely by Muskhelishvili's method.

In accordance with Art. 69, we shall actually find analytic potentials for superposition on the field of simple tension S prevailing throughout the region when there is no hole. The force transmitted across the arc AB in Fig. 124, with the conventions of Art. 59, is

$$F_x{}^0 + iF_y{}^0 = S(y_B' - y_A')e^{i\beta} \qquad (a)$$

With $z' = x' + iy'$ we have generally

$$z' = e^{-i\beta}z \qquad (b)$$

and on the ellipse

$$z' = e^{-i\beta}R\left(\sigma + \frac{m}{\sigma}\right) \qquad (c)$$

[1] For the generalization to any rational mapping function see p. 358 of the book by Muskhelishvili (n. 6, p. 194).

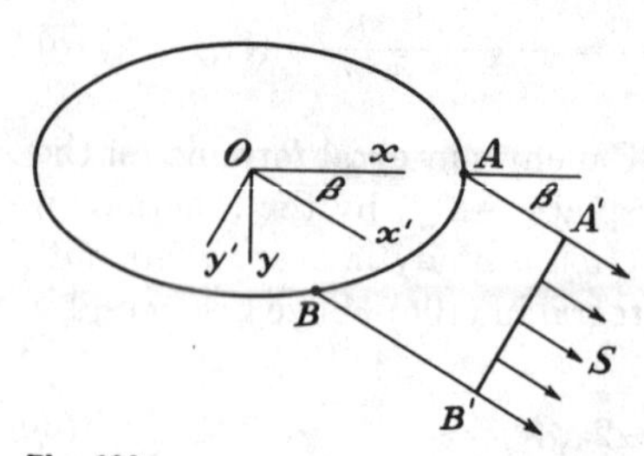

Fig. 124

Identifying σ with B, (a) can now be written as

$$F_x{}^0 + iF_y{}^0 = SR\,\frac{1}{2i}\left[\sigma + \frac{m}{\sigma} - 1 - m - e^{2i\beta}\left(\frac{1}{\sigma} + m\sigma - 1 - m\right)\right] \qquad (d)$$

The analytic potentials must provide force on the ellipse to annul this. They correspond therefore to

$$\begin{aligned} f(\sigma) &= i(F_x + iF_y) = -i(F_x{}^0 + iF_y{}^0) \\ &= -\frac{1}{2}SR[(1 - me^{2i\beta})\sigma + (m - e^{2i\beta})\frac{1}{\sigma} - (1+m)(1 - e^{2i\beta})] \end{aligned} \qquad (e)$$

This is now used in (e) of Art. 72 to determine $\phi(\zeta)$, and leads to three integrals that are easily evaluated: we have

$$\int_\gamma \frac{\sigma\,d\sigma}{\sigma - \zeta} = 0 \qquad \int_\gamma \frac{d\sigma}{\sigma - \zeta} = 0 \qquad \int_\gamma \frac{1}{\sigma}\frac{d\sigma}{\sigma - \zeta} = -2\pi i\frac{1}{\zeta} \qquad (107)$$

The first two follow at once from the Cauchy integral theorem for the unit circle and its interior, ζ being a point outside. The third follows from the Cauchy theorem for the external region, or from the residue theorem considering the internal region. Thus,

$$\phi(\zeta) = -\frac{1}{2\pi i}\int_\gamma \frac{f(\sigma)\,d\sigma}{\sigma - \zeta} = -\frac{1}{2}SR(m - e^{2i\beta})\frac{1}{\zeta} \qquad (f)$$

To determine $\psi(\zeta)$, we use (d) of Art. 72, writing from (e),

$$\bar{f}(\bar{\sigma}) = -\tfrac{1}{2}SR\left[(1 - me^{-2i\beta})\frac{1}{\sigma} + (m - e^{-2i\beta})\sigma - (1+m)(1 - e^{-2i\beta})\right] \qquad (g)$$

We can therefore make use of (107) again to evaluate the integral. Then

$$\psi(\zeta) = -\frac{1}{2}SR\left[(m - e^{2i\beta})\frac{1 + m\zeta^2}{\zeta(\zeta^2 - m)} + (1 - me^{-2i\beta})\frac{1}{\zeta}\right] \qquad (h)$$

The stress components in the xy system can now be found from the derivatives of $\phi(\zeta)$ and $\psi(\zeta)$ with respect to z. The curvilinear components corresponding to the ellipses in the z plane mapped from the circles $\rho > 1$, and to their orthogonal hyperbolas mapped from the rays θ = constant, can be found by formulas of the type (92) and (93) or (96) and (97). The displacements follow from Eqs. (86) or (98).

As a second illustrative problem we take the elliptic hole (Fig. 125) with uniform normal pressure p on the segments GAC and DEF, the segments CD and FG being unloaded. The points C, G, D, F, are z_1, $\bar{z}_1$, $-\bar{z}_1$, $-z_1$, respectively, and the corresponding points on γ in the ζ plane are σ_1, $\bar{\sigma}_1$, $-\bar{\sigma}_1$, $-\sigma_1$.

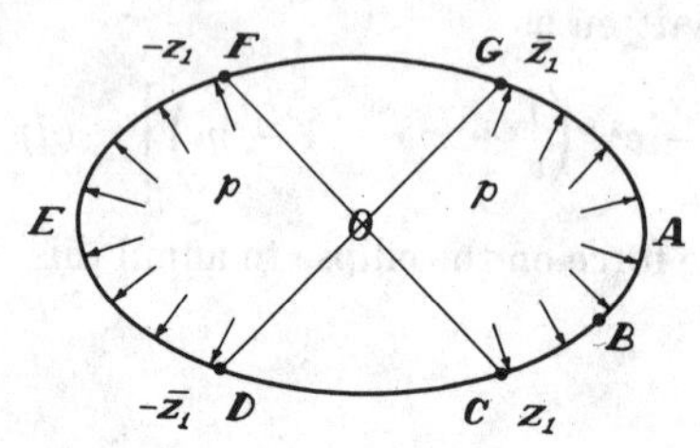

Fig. 125

Writing z for a point B on the ellipse, within GC, we have

$$F_x + iF_y = pi(z - a) \qquad \text{for } GAC$$

From C to D this force remains constant. Thus

$$F_x + iF_y = pi(z_1 - a) \qquad \text{for } CD$$

Then we have

$$F_x + iF_y = pi(z_1 - a + z + \bar{z}_1) \qquad \text{for } DEF$$

and

$$F_x + iF_y = pi(\bar{z}_1 - a) \qquad \text{for } FG$$

Accordingly, the function $f(\sigma) = i(F_x + iF_y)$ is given by

$$f(\sigma) = \begin{cases} -p\left[R\left(\sigma + \dfrac{m}{\sigma}\right) - a\right] & \text{for } GAC \\ -p\left[R\left(\sigma_1 + \dfrac{m}{\sigma_1}\right) - a\right] & \text{for } CD \\ -p\left[R\left(\sigma_1 + \dfrac{m}{\sigma_1}\right) - a + R\left(\sigma + \dfrac{m}{\sigma}\right) + R\left(\dfrac{1}{\sigma_1} + m\sigma_1\right)\right] & \text{for } DEF \\ -p\left[R\left(\dfrac{1}{\sigma_1} + m\sigma_1\right) - a\right] & \text{for } FG \end{cases} \tag{i}$$

The term $-a$ in each bracket may be ignored. Load is represented by changes in $f(\sigma)$. From (e) of Art. 72 we have

$$\begin{aligned} 2\pi i\phi(\zeta) = pR \Bigg\{ & \int_{\bar{\sigma}_1}^{\sigma_1}\left(\sigma + \frac{m}{\sigma}\right)\frac{d\sigma}{\sigma - \zeta} + \left(\sigma_1 + \frac{m}{\sigma_1}\right)\int_{\sigma_1}^{-\bar{\sigma}_1}\frac{d\sigma}{\sigma - \zeta} \\ & + \int_{-\bar{\sigma}_1}^{-\sigma_1}\left[\sigma + \frac{m}{\sigma} + (1 + m)\left(\sigma_1 + \frac{1}{\sigma_1}\right)\right]\frac{d\sigma}{\sigma - \zeta} \\ & + \left(\frac{1}{\sigma_1} + m\sigma_1\right)\int_{-\sigma_1}^{\bar{\sigma}_1}\frac{d\sigma}{\sigma - \zeta}\Bigg\} \end{aligned} \tag{j}$$

The integrals can be evaluated by writing down the indefinite integrals and inserting the limits. Then

$$\frac{2\pi i}{pR}\,\phi(\zeta) = \zeta \log \frac{\zeta^2 - \sigma_1^2}{\zeta^2 - \bar{\sigma}_1^2} - \frac{m}{\zeta}\left(4 \log \sigma_1 - \log \frac{\zeta^2 - \sigma_1^2}{\zeta^2 - \bar{\sigma}_1^2}\right)$$
$$+ \left(\sigma_1 + \frac{m}{\sigma_1}\right) \log \frac{\zeta + \bar{\sigma}_1}{\zeta - \sigma_1} + (1 + m)\left(\sigma_1 + \frac{1}{\sigma_1}\right) \log \frac{\zeta + \sigma_1}{\zeta + \bar{\sigma}_1}$$
$$+ \left(\frac{1}{\sigma_1} + m\sigma_1\right) \log \frac{\zeta - \bar{\sigma}_1}{\zeta + \sigma_1} \qquad (k)$$

This can be simplified to

$$\frac{2\pi i}{pR}\,\phi(\zeta) = -\frac{4m}{\zeta} \log \sigma_1 + \left(\zeta + \frac{m}{\zeta}\right) \log \frac{\zeta^2 - \sigma_1^2}{\zeta^2 - \bar{\sigma}_1^2}$$
$$+ \left(\sigma_1 + \frac{m}{\sigma_1}\right) \log \frac{\zeta + \sigma_1}{\zeta - \sigma_1} + \left(\frac{1}{\sigma_1} + m\sigma_1\right) \log \frac{\zeta - \bar{\sigma}_1}{\zeta + \bar{\sigma}_1} \qquad (l)$$

The function $\psi(\zeta)$ can now be found by means of (e) in Art. 72, forming $\bar{f}(\sigma)$ from (i). The integration proceeds as for (j), with the result

$$\int_\gamma \frac{\bar{f}(\bar{\sigma})\,d\sigma}{\sigma - \zeta} = -pR\left[m\zeta \log \frac{\zeta^2 - \sigma_1^2}{\zeta^2 - \bar{\sigma}_1^2} - \frac{1}{\zeta}\left(4 \log \sigma_1 - \log \frac{\zeta^2 - \sigma_1^2}{\zeta^2 - \bar{\sigma}_1^2}\right)\right.$$
$$+ \left(\frac{1}{\sigma_1} + m\sigma_1\right) \log \frac{\zeta + \bar{\sigma}_1}{\zeta - \sigma_1} + (1 + m)\left(\frac{1}{\sigma_1} + \sigma_1\right) \log \frac{\zeta + \sigma_1}{\zeta + \bar{\sigma}_1}$$
$$\left. + \left(\sigma_1 + \frac{m}{\sigma_1}\right) \log \frac{\zeta - \bar{\sigma}_1}{\zeta + \sigma_1}\right] \qquad (m)$$

Then, after some simplification,

$$\frac{2\pi i}{pR}\,\psi(\zeta) = -4 \log \sigma_1 (1 + m^2) \frac{\zeta}{\zeta^2 - m} + \left(\sigma_1 + \frac{m}{\sigma_1}\right) \log \frac{\zeta - \bar{\sigma}_1}{\zeta + \bar{\sigma}_1}$$
$$+ \left(\frac{1}{\sigma_1} + m\sigma_1\right) \log \frac{\zeta + \sigma_1}{\zeta - \sigma_1} \qquad (n)$$

With the final forms of the two complex potentials given by Eqs. (l) and (n), expressions for the displacement and stress can be obtained from the general formulas (96), (97), and (98).

PROBLEMS

1. Verify that the solution represented by (g) and (i) of Art. 73 leads to the same results as the solution given in Art. 63 for the stress at the ends of the major and minor axes of the elliptic hole.

2. Obtain, from the solution indicated in Art. 73 for the problem of Fig. 125, the potentials $\phi(\zeta)$ and $\psi(\zeta)$ for a pressure p acting at all points on the elliptic hole. Verify that the stress evaluated at the ends of the major and minor axes is consistent with the results given in Art. 62.
3. The infinite plate with an elliptic hole (as in Fig. 124) has, at infinity, uniform stress

$$\sigma_x = S_1 \qquad \sigma_y = S_2 \qquad \tau_{xy} = 0$$

(instead of S at angle β as in Fig. 124).
(*a*) Find an expression for the stress at the hole.
(*b*) Devise and apply several checks on this result, using known results for elliptic and circular holes.
(*c*) Show that if $S_2/S_1 = b/a$, the stress at the hole is the same all round the hole.[1]
(*d*) Show that when the stress at infinity is a pure shear at 45° to the axes of the ellipse, the greatest stress on the hole occurs at the ends of the major axis, and corresponds to a stress-concentration factor $2[1 + (a/b)]$.

[1] A. J. Durelli and W. M. Murray, *Proc. Soc. Exptl. Stress Anal.*, vol. 1, no. 1, 1943.

chapter 7

Analysis of Stress and Strain in Three Dimensions

74 | Introduction

The preceding chapters have been concerned with two-dimensional problems, except for the preliminary basic considerations of Chap. 1. The present chapter, and the next, are devoted to further general questions basic to the solution of further problems. In this chapter, the analysis of stress is entirely separate from the analysis of strain. No stress-strain relations are introduced. The results are applicable to stress occurring in any kind of (continuous) medium—for instance, a viscous fluid, or a plastic solid, and similarly with respect to strain.

We turn now to the general case of stress distribution in three dimensions. It was shown (see Art. 4) that the stresses acting on the six sides of a cubic element can be described by six components of stress, namely, the three normal stresses σ_x, σ_y, σ_z and the three shearing stresses $\tau_{xy} = \tau_{yx}$, $\tau_{xz} = \tau_{zx}$, $\tau_{yz} = \tau_{zy}$. If these components of stress at any point are known, the stress acting on any inclined plane through this point can be calculated from the equations of statics. Let O be a point of the stressed body, and suppose the stresses are known for the coordinate planes xy, xz, yz (Fig. 126). To get the stress for any inclined plane through O, we take a plane BCD parallel to it at a small distance from O, so that this latter plane together with the coordinate planes cuts out from the body a very small tetrahedron $BCDO$. Since the stresses vary continuously over the volume of the body, the stress acting on the plane BCD will approach the stress on the parallel plane through O as the element is made infinitesimal.

In considering the conditions of equilibrium of the elemental tetrahedron, the body forces can be neglected (see page 5). Also, because the element is very small, we can neglect the variation of the stresses over the

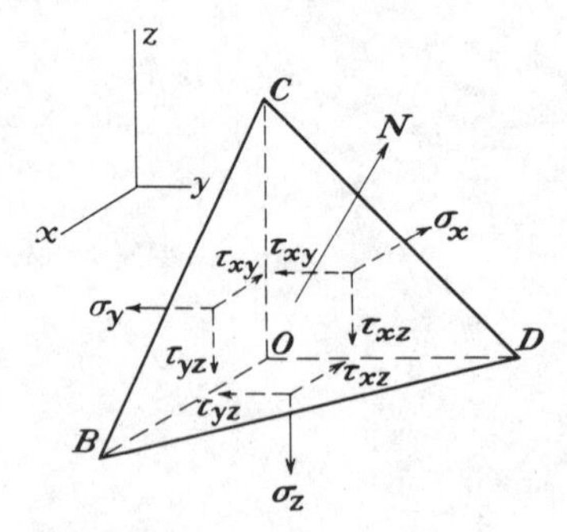

Fig. 126

sides and assume that the stresses are uniformly distributed. The forces acting on the tetrahedron can therefore be determined by multiplying the stress components by the areas of the faces. If A denotes the area of the face BCD of the tetrahedron, then the areas of the three other faces are obtained by projecting A on the three coordinate planes. If N is the normal to the plane BCD, and we write

$$\cos(Nx) = l \qquad \cos(Ny) = m \qquad \cos(Nz) = n \tag{a}$$

the areas of the three other faces of the tetrahedron are

$$Al \qquad Am \qquad An$$

If we denote by X, Y, Z the three components of stress, parallel to the coordinate axes, acting on the inclined face BCD, then the component of force acting on the face BCD in the direction of the x axis is AX. Also the components of forces in the x direction acting on the three other faces of the tetrahedron are $-Al\sigma_x$, $-Am\tau_{xy}$, $-An\tau_{xz}$. The corresponding equation of the tetrahedron is

$$AX - Al\sigma_x - Am\tau_{xy} - An\tau_{xz} = 0$$

In the same manner two other equations of equilibrium are obtained by projecting the forces on the y and z axes. After canceling the factor A, these equations of equilibrium of the tetrahedron can be written

$$\begin{aligned} X &= \sigma_x l + \tau_{xy} m + \tau_{xz} n \\ Y &= \tau_{xy} l + \sigma_y m + \tau_{zy} n \\ Z &= \tau_{xz} l + \tau_{yz} m + \sigma_z n \end{aligned} \tag{108}$$

Thus the components of stress on any plane, defined by the direction cosines l, m, n, can easily be calculated from Eqs. (108), provided the six components of stress σ_x, σ_y, σ_z, τ_{xy}, τ_{yz}, τ_{xz} at the point O are known.

75 | Principal Stresses

Let us now consider the normal component of stress σ_n acting on the plane BCD (Fig. 126). Using the notations (a) for the direction cosines, we find

$$\sigma_n = Xl + Ym + Zn$$

or, substituting the values of X, Y, Z from Eqs. (108),

$$\sigma_n = \sigma_x l^2 + \sigma_y m^2 + \sigma_z n^2 + 2\tau_{yz}mn + 2\tau_{xz}ln + 2\tau_{xy}lm \tag{109}$$

The variation of σ_n with the direction of the normal N can be represented geometrically as follows. Let us put in the direction of N a vector whose length, r, is inversely proportional to the square root of the absolute value of the stress σ_n, that is,

$$r = \frac{k}{\sqrt{|\sigma_n|}} \tag{b}$$

in which k is a constant factor. The coordinates of the end of this vector will be

$$x = lr \qquad y = mr \qquad z = nr \tag{c}$$

Substituting

$$\sigma_n = \pm \frac{k^2}{r^2} \tag{d}$$

from (b), and the values of l, m, n from (c) in Eq. (109), we find[1]

$$\pm k^2 = \sigma_x x^2 + \sigma_y y^2 + \sigma_z z^2 + 2\tau_{yz}yz + 2\tau_{zx}zx + 2\tau_{xy}xy \tag{110}$$

As the plane BCD rotates about the point O, the end of the vector r always lies on the surface of the second degree given by Eq. (110).

It is well known that in the case of a surface of the second degree, such as given by Eq. (110), it is always possible to find for the axes x, y, z such directions that the terms in this equation containing the products of coordinates vanish. This means that we can always find three perpendicular planes for which τ_{yz}, τ_{zx}, τ_{xy} vanish, that is, the resultant stresses are perpendicular to the planes on which they act. We call these stresses the *principal stresses* at the point, their directions the *principal axes*, and the planes on which they act *principal planes*. It can be seen that the stress at a point is completely defined if the directions of the principal axes and

[1] The plus-or-minus sign in Eq. (d) applies according as σ_n is tensile or compressive, and correspondingly in Eq. (110). When all three principal stresses have the same sign, only one of the alternative signs is needed, and the surface is an ellipsoid. When the principal stresses are not all of the same sign, both signs are needed and the surface, now represented by *both* Eqs. (110), consists of a hyperboloid of two sheets, together with a hyperboloid of one sheet, with a common asymptotic cone.

the magnitudes of the three principal stresses are given. The surface represented by Eq. (110) must then be the same regardless of our choice of x, y, z axes.

76 | Stress Ellipsoid and Stress-director Surface

If the coordinate axes x, y, z are taken in the directions of the principal axes, calculation of the stress on any inclined plane becomes very simple. The shearing stresses τ_{yz}, τ_{zx}, τ_{xy} are zero in this case, and Eqs. (108) become

$$X = \sigma_x l \qquad Y = \sigma_y m \qquad Z = \sigma_z n \tag{111}$$

Putting the values of l, m, n from these equations into the well-known relation $l^2 + m^2 + n^2 = 1$, we find

$$\frac{X^2}{\sigma_x^2} + \frac{Y^2}{\sigma_y^2} + \frac{Z^2}{\sigma_z^2} = 1 \tag{112}$$

This means that, if for each inclined plane through a point O the stress is represented by a vector from O with the components X, Y, Z, the ends of all such vectors lie on the surface of the ellipsoid given by Eq. (112). This ellipsoid is called the *stress ellipsoid*. Its semiaxes give the principal stresses at the point. From this it can be concluded that the maximum stress at any point is the largest of the three principal stresses at this point.

If two of the three principal stresses are numerically equal, the stress ellipsoid becomes an ellipsoid of revolution. If these numerically equal principal stresses are of the same sign, the resultant stresses on all planes through the axis of symmetry of the ellipsoid will be equal and perpendicular to the planes on which they act. In this case, the stresses on any two perpendicular planes through this axis can be considered as principal stresses. If all three principal stresses are equal and of the same sign, the stress ellipsoid becomes a sphere and any three perpendicular directions can be taken as principal axes. When one of the principal stresses is zero, the stress ellipsoid reduces to the area of an ellipse and the vectors representing the stresses on all the planes through the point lie in the same plane. This condition of stress is called *plane stress* and has already been discussed in previous sections. When two principal stresses are zero, we have the cases of simple tension or compression.

Each radius vector of the stress ellipsoid represents, to a certain scale, the stress on one of the planes through the center of the ellipsoid. To find this plane we use, together with the stress ellipsoid (112), the *stress-director surface* defined by the equation

$$\frac{x^2}{\sigma_x} + \frac{y^2}{\sigma_y} + \frac{z^2}{\sigma_z} = 1 \tag{113}$$

The stress represented by a radius vector of the stress ellipsoid acts on the plane parallel to the tangent plane to the stress-director surface at the point of its intersection with the radius vector. This can be shown as follows. The equation of the tangent plane to the stress-director surface (113) at any point x_0, y_0, z_0 is

$$\frac{xx_0}{\sigma_x} + \frac{yy_0}{\sigma_y} + \frac{zz_0}{\sigma_z} = 1 \tag{a}$$

Denoting by h the length of the perpendicular from the origin of coordinates to the above tangent plane, and by l, m, n the direction cosines of this perpendicular, the equation of this tangent plane can be written in the form

$$lx + my + nz = h \tag{b}$$

Comparing (a) and (b) we find

$$\sigma_x = \frac{x_0 h}{l} \qquad \sigma_y = \frac{y_0 h}{m} \qquad \sigma_z = \frac{z_0 h}{n} \tag{c}$$

Substituting these values in Eqs. (111) we find

$$X = x_0 h \qquad Y = y_0 h \qquad Z = z_0 h$$

i.e., the components of stress on the plane with direction cosines l, m, n are proportional to the coordinates x_0, y_0, z_0. Hence, the vector representing the stress goes through the point x_0, y_0, z_0, as was stated above.[1]

77 | Determination of the Principal Stresses

If the stress components for three coordinate planes are known, we can determine the directions and magnitudes of the principal stresses by using the property that the principal stresses are perpendicular to the planes on which they act. Let l, m, n be the direction cosines of a principal plane and S the magnitude of the principal stress acting on this plane. Then the components of this stress are

$$X = Sl \qquad Y = Sm \qquad Z = Sn$$

Substituting in Eqs. (108), we find

$$\begin{aligned} (S - \sigma_x)l - \tau_{xy}m - \tau_{xz}n &= 0 \\ -\tau_{xy}l + (S - \sigma_y)m - \tau_{yz}n &= 0 \\ -\tau_{xz}l - \tau_{yz}m + (S - \sigma_z)n &= 0 \end{aligned} \tag{a}$$

These are three homogeneous linear equations in l, m, n. They will give solutions different from zero only if the determinant of these equations is

[1] Another method of representing the stress at a point, by using circles, has been developed by O. Mohr, "Technische Mechanik," 2d ed., p. 192, 1914. See also A. Föppl and L. Föppl, "Drang und Zwang," vol. 1, p. 9, and H. M. Westergaard, *Z. Angew. Math. Mech.*, vol. 4, p. 520, 1924. Applications of Mohr's circles were made in discussing two-dimensional problems (see Art. 9).

zero. Calculating this determinant and putting it equal to zero give us the following cubic equation in S:

$$S^3 - (\sigma_x + \sigma_y + \sigma_z)S^2 + (\sigma_x\sigma_y + \sigma_y\sigma_z + \sigma_x\sigma_z - \tau_{yz}^2 - \tau_{xz}^2 - \tau_{xy}^2)S \\ - (\sigma_x\sigma_y\sigma_z + 2\tau_{yz}\tau_{xz}\tau_{xy} - \sigma_x\tau_{yz}^2 - \sigma_y\tau_{xz}^2 - \sigma_z\tau_{xy}^2) = 0 \quad (114)$$

The three roots of this equation give the values of the three principal stresses S_1, S_2, S_3. By substituting each of these stresses in Eqs. (*a*) and using the relation $l^2 + m^2 + n^2 = 1$, we can find three sets of direction cosines for the three principal planes.

78 | Stress Invariants

Regarding the state of stress, i.e., the principal stresses and principal axes, as given, we can of course represent it by components in any set of x, y, z axes. No matter what orientation is chosen for these axes, Eq. (114) must give the same three roots for S. Consequently, the coefficients must always be the same. We could choose the principal axes themselves for x, y, z axes. Then σ_x, σ_y, σ_z would be S_1, S_2, S_3 (in this or another order) and τ_{xy}, τ_{yz}, τ_{zx} would be zero. Thus the invariant values of the coefficients in Eq. (114) are given by

$$\sigma_x + \sigma_y + \sigma_z = S_1 + S_2 + S_3 \quad (a)$$

$$\sigma_x\sigma_y + \sigma_y\sigma_z + \sigma_z\sigma_x - \tau_{xy}^2 - \tau_{yz}^2 - \tau_{zx}^2 = S_1S_2 + S_2S_3 + S_3S_1 \quad (b)$$

$$\sigma_x\sigma_y\sigma_z + 2\tau_{xy}\tau_{yz}\tau_{zx} - \sigma_x\tau_{yz}^2 - \sigma_y\tau_{zx}^2 - \sigma_z\tau_{xy}^2 = S_1S_2S_3 \quad (c)$$

The expressions on the left are "stress invariants." Evidently other invariant expressions can be formed from them. Writing I_1, I_2, I_3 for the expressions on the right of (*a*), (*b*), and (*c*), respectively, it is easily verified that

$$(\sigma_x - \sigma_y)^2 + (\sigma_y - \sigma_z)^2 + (\sigma_z - \sigma_x)^2 + 6(\tau_{xy}^2 + \tau_{yz}^2 + \tau_{zx}^2) \\ = 2I_1^2 - 6I_2 \quad (d)$$

and therefore the expression on the left here is also an invariant. It occurs later in the discussion of strain energy.

79 | Determination of the Maximum Shearing Stress

Let x, y, z be the principal axes so that σ_x, σ_y, σ_z are principal stresses, and let l, m, n be the direction cosines for a given plane. Then, from Eqs. (111), the square of the total stress on this plane is

$$S^2 = X^2 + Y^2 + Z^2 = \sigma_x^2l^2 + \sigma_y^2m^2 + \sigma_z^2n^2$$

The square of the normal component of the stress on the same plane is, from Eq. (109),

$$\sigma_n{}^2 = (\sigma_x l^2 + \sigma_y m^2 + \sigma_z n^2)^2 \tag{a}$$

Then the square of the shearing stress on the same plane must be

$$\tau^2 = S^2 - \sigma_n{}^2 = \sigma_x{}^2 l^2 + \sigma_y{}^2 m^2 + \sigma_z{}^2 n^2 - (\sigma_x l^2 + \sigma_y m^2 + \sigma_z n^2)^2 \tag{b}$$

We shall now eliminate one of the direction cosines, say n, from this equation by using the relation

$$l^2 + m^2 + n^2 = 1$$

and then determine l and m so as to make τ a maximum. After substituting $n^2 = 1 - l^2 - m^2$ in expression (*b*), calculating its derivatives with respect to l and m, and equating these derivatives to zero, we obtain the following equations for determining the direction cosines of the planes for which τ is a maximum or minimum:

$$\begin{aligned} l[(\sigma_x - \sigma_z)l^2 + (\sigma_y - \sigma_z)m^2 - \tfrac{1}{2}(\sigma_x - \sigma_z)] &= 0 \\ m[(\sigma_x - \sigma_z)l^2 + (\sigma_y - \sigma_z)m^2 - \tfrac{1}{2}(\sigma_y - \sigma_z)] &= 0 \end{aligned} \tag{c}$$

One solution of these equations is obtained by putting $l = m = 0$. We can also obtain solutions different from zero. Taking, for instance, $l = 0$, we find from the second of Eqs. (*c*) that $m = \pm\sqrt{\tfrac{1}{2}}$; and taking $m = 0$, we find from the first of Eqs. (*c*) that $l = \pm\sqrt{\tfrac{1}{2}}$. There are in general no solutions of Eqs. (*c*) in which l and m are both different from zero, for in this case the expressions in brackets cannot both vanish.

Repeating the above calculations by eliminating from expression (*b*) m and then l, we finally arrive at the following table of direction cosines making τ a maximum or minimum:

Direction Cosines for Planes of τ_{max} and τ_{min}

$l =$	0	0	± 1	0	$\pm\sqrt{\tfrac{1}{2}}$	$\pm\sqrt{\tfrac{1}{2}}$
$m =$	0	± 1	0	$\pm\sqrt{\tfrac{1}{2}}$	0	$\pm\sqrt{\tfrac{1}{2}}$
$n =$	± 1	0	0	$\pm\sqrt{\tfrac{1}{2}}$	$\pm\sqrt{\tfrac{1}{2}}$	0

The first three columns give the directions of the planes of coordinates, coinciding, as was assumed originally, with the principal planes. For these planes the shearing stress is zero, i.e., expression (*b*) is a minimum. The three remaining columns give planes through each of the principal axes bisecting the angles between the two other principal axes. Substituting the direction cosines of these three planes into expression (b), we find the following values of the shearing stresses on these three planes:

$$\tau = \pm \tfrac{1}{2}(\sigma_y - \sigma_z) \qquad \tau = \pm \tfrac{1}{2}(\sigma_x - \sigma_z) \qquad \tau = \pm \tfrac{1}{2}(\sigma_x - \sigma_y) \quad (115)$$

This shows that the maximum shearing stress acts on the plane bisecting the angle between the largest and the smallest principal stresses and is equal to half the difference between these two principal stresses.

If the x, y, z axes in Fig. 126 represent the directions of principal stress, and if $OB = OC = OD$, so that the normal N to the slant face of the tetrahedron has direction cosines $l = m = n = 1/\sqrt{3}$, the normal stress on this face is given by Eq. (109) as

$$\sigma_n = \tfrac{1}{3}(\sigma_x + \sigma_y + \sigma_z) \qquad (d)$$

This is called the "mean stress." The shear stress on the face is given by Eq. (b) as

$$\tau^2 = \tfrac{1}{3}(\sigma_x^2 + \sigma_y^2 + \sigma_z^2) - \tfrac{1}{9}(\sigma_x + \sigma_y + \sigma_z)^2$$

This can also be written

$$\tau^2 = \tfrac{1}{9}[(\sigma_x - \sigma_y)^2 + (\sigma_y - \sigma_z)^2 + (\sigma_z - \sigma_x)^2]$$

and also, by using (d), as

$$\tau^2 = \tfrac{1}{3}[(\sigma_x - \sigma_n)^2 + (\sigma_y - \sigma_n)^2 + (\sigma_z - \sigma_n)^2]$$

This shear stress is called the "octahedral shear stress," because the face on which it acts is one face of a regular octahedron with vertices on the axes. It occurs frequently in the theory of plasticity.

80 | Homogeneous Deformation

We consider only small deformations, such as occur in engineering structures. The small displacements of the particles of a deformed body will usually be resolved into components u, v, w parallel to the coordinate axes x, y, z, respectively. It will be assumed that these components are very small quantities varying continuously over the volume of the body.

Consider, as an example, simple tension of a prismatical bar fixed at the upper end (Fig. 127). Let ϵ be the unit elongation of the bar in the x direction and $\nu\epsilon$ the unit lateral contraction. Then the components of displacement of a point with coordinates x, y, z are

$$u = \epsilon x \qquad v = -\nu\epsilon y \qquad w = -\nu\epsilon z$$

Denoting by x', y', z' the coordinates of the point after deformation,

$$x' = x + u = x(1 + \epsilon) \qquad y' = y + v = y(1 - \nu\epsilon) \qquad z' = z + w = z(1 - \nu\epsilon) \quad (a)$$

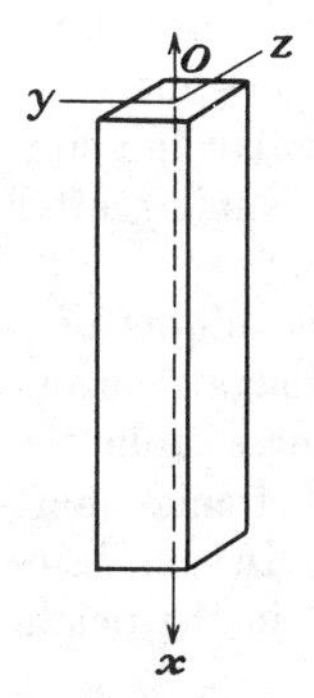

Fig. 127

If we consider a plane in the bar before deformation such as that given by the equation

$$ax + by + cz + d = 0 \tag{b}$$

the points of this plane will still be in a plane after deformation. The equation of this new plane is obtained by substituting in Eq. (*b*) the values of x, y, z from Eq. (*a*). It can easily be proved in this manner that parallel planes remain parallel after deformation and parallel lines remain parallel.

If we consider a spherical surface in the bar before deformation such as given by the equation

$$x^2 + y^2 + z^2 = r^2 \tag{c}$$

this sphere becomes an ellipsoid after deformation, the equation of which can be found by substituting in Eq. (*c*) the expressions for x, y, z obtained from Eqs. (*a*). This gives

$$\frac{x'^2}{r^2(1+\epsilon)^2} + \frac{y'^2}{r^2(1-\nu\epsilon)^2} + \frac{z'^2}{r^2(1-\nu\epsilon)^2} = 1 \tag{d}$$

Thus a sphere of radius r deforms into an ellipsoid with semiaxes $r(1+\epsilon)$, $r(1-\nu\epsilon)$, $r(1-\nu\epsilon)$.

The simple extension, and lateral contraction, considered above, represent only a particular case of a more general type of deformation in which the components of displacement, u, v, w, are linear functions of the coordinates. Proceeding as before, it can be shown that this type of deformation has all the properties found above for the case of simple tension. Planes and straight lines remain plane and straight after deformation. Parallel planes and parallel straight lines remain parallel after deformation. A sphere becomes, after deformation, an ellipsoid. This kind of deformation is called *homogeneous deformation.* It will be shown later that in this case the deformation in any given direction is the same at all points of the deformed body. Thus two geometrically similar and similarly oriented elements of a body remain geometrically similar after distortion.

In more general cases the deformation varies over the volume of a deformed body. For instance, when a beam is bent, the elongations and contractions of longitudinal fibers depend on their distances from the neutral surface; the shearing strain in elements of a twisted circular shaft is proportional to their distances from the axis of the shaft. In such cases of nonhomogeneous deformation an analysis of the strain in the neighborhood of a point is necessary.

81 | Strain at a Point

In discussing strain in the neighborhood of a point O of a deformed body (Fig. 128), let us consider a small linear element OO_1 of length r, with the direction cosines l, m, n. The small projections of this element on the coordinate axes are

$$\delta x = rl \qquad \delta y = rm \qquad \delta z = rn \tag{a}$$

They represent the coordinates of the point O_1 with respect to the x, y, z axes through O as an origin. If u, v, w are the components of the displacement of the point O during deformation of the body, the corresponding displacements of the neighboring point O_1 can be represented as follows:

$$\begin{aligned} u_1 &= u + \frac{\partial u}{\partial x}\,\delta x + \frac{\partial u}{\partial y}\,\delta y + \frac{\partial u}{\partial z}\,\delta z \\ v_1 &= v + \frac{\partial v}{\partial x}\,\delta x + \frac{\partial v}{\partial y}\,\delta y + \frac{\partial v}{\partial z}\,\delta z \\ w_1 &= w + \frac{\partial w}{\partial x}\,\delta x + \frac{\partial w}{\partial y}\,\delta y + \frac{\partial w}{\partial z}\,\delta z \end{aligned} \tag{b}$$

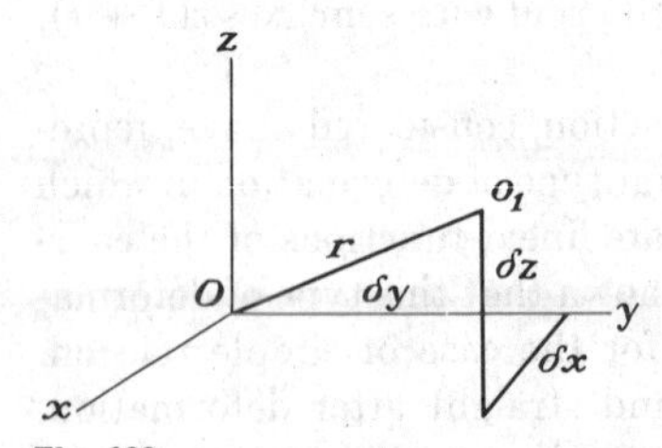

Fig. 128

It is assumed here that the quantities δx, δy, δz are small, and hence the terms with higher powers and products of these quantities can be neglected in (*b*) as small quantities of higher order. The coordinates of the point O_1 become, after deformation,

$$
\begin{aligned}
\delta x + u_1 - u &= \delta x + \frac{\partial u}{\partial x}\,\delta x + \frac{\partial u}{\partial y}\,\delta y + \frac{\partial u}{\partial z}\,\delta z \\
\delta y + v_1 - v &= \delta y + \frac{\partial v}{\partial x}\,\delta x + \frac{\partial v}{\partial y}\,\delta y + \frac{\partial v}{\partial z}\,\delta z \qquad (c) \\
\delta z + w_1 - w &= \delta z + \frac{\partial w}{\partial x}\,\delta x + \frac{\partial w}{\partial y}\,\delta y + \frac{\partial w}{\partial z}\,\delta z
\end{aligned}
$$

It will be noticed that these coordinates are linear functions of the initial coordinates δx, δy, δz; hence the deformation in a very small element of a body at a point O can be considered as *homogeneous* (Art. 80).

Let us consider the elongation of the element r, due to this deformation. The square of the length of this element after deformation is equal to the sum of the squares of the coordinates (*c*). Hence, if ϵ is the unit elongation of the element, we find

$$
\begin{aligned}
(r + \epsilon r)^2 = \left(\delta x + \frac{\partial u}{\partial x}\,\delta x + \frac{\partial u}{\partial y}\,\delta y + \frac{\partial u}{\partial z}\,\delta z\right)^2 \\
+ \left(\delta y + \frac{\partial v}{\partial x}\,\delta x + \frac{\partial v}{\partial y}\,\delta y + \frac{\partial v}{\partial z}\,\delta z\right)^2 \\
+ \left(\delta z + \frac{\partial w}{\partial x}\,\delta x + \frac{\partial w}{\partial y}\,\delta y + \frac{\partial w}{\partial z}\,\delta z\right)^2
\end{aligned}
$$

or, dividing by r^2 and using Eqs. (*a*),

$$
\begin{aligned}
(1 + \epsilon)^2 = \left[l\left(1 + \frac{\partial u}{\partial x}\right) + m\,\frac{\partial u}{\partial y} + n\,\frac{\partial u}{\partial z}\right]^2 \\
+ \left[l\,\frac{\partial v}{\partial x} + m\left(1 + \frac{\partial v}{\partial y}\right) + n\,\frac{\partial v}{\partial z}\right]^2 \\
+ \left[l\,\frac{\partial w}{\partial x} + m\,\frac{\partial w}{\partial y} + n\left(1 + \frac{\partial w}{\partial z}\right)\right]^2 \qquad (d)
\end{aligned}
$$

Remembering that ϵ and the derivatives $\partial u/\partial x, \ldots, \partial w/\partial z$ are small quantities whose squares and products can be neglected, and using $l^2 + m^2 + n^2 = 1$, Eq. (*d*) becomes

$$
\begin{aligned}
\epsilon = l^2\,\frac{\partial u}{\partial x} + m^2\,\frac{\partial v}{\partial y} + n^2\,\frac{\partial w}{\partial z} + lm\left(\frac{\partial u}{\partial y} + \frac{\partial v}{\partial x}\right) + ln\left(\frac{\partial u}{\partial z} + \frac{\partial w}{\partial x}\right) \\
+ mn\left(\frac{\partial v}{\partial z} + \frac{\partial w}{\partial y}\right) \qquad (116)
\end{aligned}
$$

Hence the elongation of an element r can be calculated provided the expressions $(\partial u/\partial x)$, . . . , $(\partial u/\partial y + \partial v/\partial x)$, . . . are known. Using the notations

$$\begin{aligned} \frac{\partial u}{\partial x} &= \epsilon_x & \frac{\partial v}{\partial y} &= \epsilon_y & \frac{\partial w}{\partial z} &= \epsilon_z \\ \frac{\partial u}{\partial y} + \frac{\partial v}{\partial x} &= \gamma_{xy} & \frac{\partial u}{\partial z} + \frac{\partial w}{\partial x} &= \gamma_{xz} & \frac{\partial v}{\partial z} + \frac{\partial w}{\partial y} &= \gamma_{yz} \end{aligned} \tag{e}$$

Eq. (116) can be presented in the form[1]

$$\epsilon = \epsilon_x l^2 + \epsilon_y m^2 + \epsilon_z n^2 + \gamma_{xy} lm + \gamma_{xz} ln + \gamma_{yz} mn \tag{117}$$

The physical meaning of such quantities as ϵ_x . . . , γ_{yz} . . . has already been discussed (see Art. 5), and it was shown that ϵ_x, ϵ_y, ϵ_z are unit elongations in the x, y, z directions and γ_{xy}, γ_{xz}, γ_{yz} the three unit shear strains related to the same directions. We now see that the elongation of any linear element through a point O can be calculated from Eq. (117), provided we know the six strain components.

In the particular case of homogeneous deformation the components u, v, w of displacement are linear functions of the coordinates, and from Eqs. (e) the components of strain are constant over the volume of the body, that is, in this case each element of the body undergoes the same strain.

In investigating strain around a point O, it is necessary sometimes to know the change in the angle between two linear elements through the point. Using Eqs. (c) and (a) and considering ϵ as a small quantity, the direction cosines of the element r (Fig. 128), after deformation, are

$$\begin{aligned} l_1 &= \frac{\delta x + u_1 - u}{r(1+\epsilon)} = l\left(1 - \epsilon + \frac{\partial u}{\partial x}\right) + m\frac{\partial u}{\partial y} + n\frac{\partial u}{\partial z} \\ m_1 &= \frac{\delta y + v_1 - v}{r(1+\epsilon)} = l\frac{\partial v}{\partial x} + m\left(1 - \epsilon + \frac{\partial v}{\partial y}\right) + n\frac{\partial v}{\partial z} \\ n_1 &= \frac{\delta z + w_1 - w}{r(1+\epsilon)} = l\frac{\partial w}{\partial x} + m\frac{\partial w}{\partial y} + n\left(1 - \epsilon + \frac{\partial w}{\partial z}\right) \end{aligned} \tag{f}$$

Taking another element r' through the same point with direction cosines l', m', n', the magnitudes of these cosines, after deformation, are given by equations analogous to (f). The cosine of the angle between the two elements after deformation is

$$\cos(rr') = l_1 l_1' + m_1 m_1' + n_1 n_1'$$

[1] This may be compared with the expression for σ_n given in Eq. (109), noting the factors 2 in the last three terms there. When the index notation is used, in particular Eqs. (f) in Art. 7, the right-hand side of Eq. (117) expressed in terms of ϵ_{ij} has corresponding factors 2. This is an advantageous form when changes of coordinates are under consideration, both stress and strain then being represented by tensors of the second rank.

Considering the elongations ϵ and ϵ' in these two directions as small quantities and using Eqs. (*f*), we find

$$\cos(rr') = (ll' + mm' + nn')(1 - \epsilon - \epsilon') + 2(\epsilon_x ll' + \epsilon_y mm' + \epsilon_z nn') + \gamma_{yz}(mn' + m'n) + \gamma_{xz}(nl' + n'l) + \gamma_{xy}(lm' + l'm) \quad (118)$$

If the directions of r and r' are perpendicular to each other, then

$$ll' + mm' + nn' = 0$$

and Eq. (118) gives the shearing strain between these directions.

82 | Principal Axes of Strain

From Eq. (117) a geometrical interpretation of the variation of strain at a point can be obtained. For this purpose let us put in the direction of each linear element such as r (Fig. 128) a radius vector of the length

$$R = \frac{k}{\sqrt{|\epsilon|}} \quad (a)$$

Then, proceeding as explained in Art. 75, it can be shown that the ends of all these radii are on the surface given by the equation

$$\pm k^2 = \epsilon_x x^2 + \epsilon_y y^2 + \epsilon_z z^2 + \gamma_{yz} yz + \gamma_{xz} xz + \gamma_{xy} xy \quad (119)$$

The shape and orientation of this surface is completely determined by the state of strain at the point and is independent of the directions of coordinates. It is always possible to take such directions of orthogonal coordinates that the terms with products of coordinates in Eq. (119) disappear, i.e., the shearing strains for such directions become zero. These directions are called *principal axes of strain*, corresponding planes the *principal planes of strain*, and the corresponding strains the *principal strains*. From the above discussion, it is evident that the principal axes of strain remain perpendicular to each other after deformation, and a rectangular parallelepiped with the sides parallel to the principal planes remains a rectangular parallelepiped after deformation. In general it will have undergone a small *rotation*.

If the x, y, and z axes are principal axes of strain, then Eq. (119) becomes

$$\pm k^2 = \epsilon_x x^2 + \epsilon_y y^2 + \epsilon_z z^2$$

In this case the elongation of any linear element with the direction cosines l, m, n becomes, from Eq. (117),

$$\epsilon = \epsilon_x l^2 + \epsilon_y m^2 + \epsilon_z n^2 \quad (120)$$

and the shearing strain corresponding to two perpendicular directions r and r' becomes, from Eq. (118),

$$\gamma_{rr'} = 2(\epsilon_x ll' + \epsilon_y mm' + \epsilon_z nn') \tag{121}$$

It can thus be seen that the strain at a point is completely determined if we know the directions of the principal axes of strain and the magnitudes of the principal extensions. The determination of the principal axes of strain and the principal extensions can be done in the same manner as explained in Art. 77. It can also be shown that the sum $\epsilon_x + \epsilon_y + \epsilon_z$ remains constant when the system of coordinates is rotated. This sum has, as we know, a simple physical meaning; it represents the unit volume expansion due to the strain at a point.

83 | Rotation

In general during the deformation of a body, any element is changed in shape, translated, and rotated. On account of the shear strain the edges do not rotate by equal amounts, and it is necessary to consider how the rotation of the whole element can be specified. Any rectangular xyz-element could have been brought into its final form, position, and orientation in the following three steps, beginning with the element in the undeformed body:

1. The strains ϵ_x, ϵ_y, ϵ_z, γ_{xy}, γ_{yz}, γ_{xz} are applied to the element, and then the element is so oriented that the directions of principal strain have not rotated.
2. The element is translated until its center occupies its final position.
3. The element is rotated into its final orientation.

The rotation in step 3 is evidently the rotation of the directions of principal strain and is therefore independent of our choice of x, y, z axes. It must be possible to evaluate it when the displacements u, v, w are given. On the other hand, it is clearly independent of the strain components.

Since the translation of the element is of no interest to us here, we may consider the displacement of a point O_1, as in Art. 81 and Fig. 128, relative to the point O, the center of the element. This relative displacement is given by Eqs. (b) of Art. 81 as

$$\begin{aligned} u_1 - u &= \frac{\partial u}{\partial x}\,\delta x + \frac{\partial u}{\partial y}\,\delta y + \frac{\partial u}{\partial z}\,\delta z \\ v_1 - v &= \frac{\partial v}{\partial x}\,\delta x + \frac{\partial v}{\partial y}\,\delta y + \frac{\partial v}{\partial z}\,\delta z \\ w_1 - w &= \frac{\partial w}{\partial x}\,\delta x + \frac{\partial w}{\partial y}\,\delta y + \frac{\partial w}{\partial z}\,\delta z \end{aligned} \tag{a}$$

Introducing the notation (*e*) of Art. 81 for the strain components, and also the notation[1]

$$\frac{1}{2}\left(\frac{\partial w}{\partial y} - \frac{\partial v}{\partial z}\right) = \omega_x \qquad \frac{1}{2}\left(\frac{\partial u}{\partial z} - \frac{\partial w}{\partial x}\right) = \omega_y \qquad \frac{1}{2}\left(\frac{\partial v}{\partial x} - \frac{\partial u}{\partial y}\right) = \omega_z \tag{122}$$

we can write Eqs. (*a*) in the form

$$\begin{aligned} u_1 - u &= \epsilon_x\,\delta x + \tfrac{1}{2}\gamma_{xy}\,\delta y + \tfrac{1}{2}\gamma_{xz}\,\delta z - \omega_z\,\delta y + \omega_y\,\delta z \\ v_1 - v &= \tfrac{1}{2}\gamma_{xy}\,\delta x + \epsilon_y\,\delta y + \tfrac{1}{2}\gamma_{yz}\,\delta z - \omega_x\,\delta z + \omega_z\,\delta x \\ w_1 - w &= \tfrac{1}{2}\gamma_{xz}\,\delta x + \tfrac{1}{2}\gamma_{yz}\,\delta y + \epsilon_z\,\delta z - \omega_y\,\delta x + \omega_x\,\delta y \end{aligned} \tag{b}$$

which express the relative displacement in two parts, one depending only on the strain components, the other depending only on the quantities ω_x, ω_y, ω_z.

We can now show that ω_x, ω_y, ω_z are in fact the components of the rotation 3. Consider the surface given by Eq. (119). The square of the radius in any direction is inversely proportional to the unit elongation of a linear element in that direction. Equation (119) is of the form

$$F(x,y,z) = \text{const} \tag{c}$$

If we consider a neighboring point $x + dx$, $y + dy$, $z + dz$ *on the surface*, we have the relation

$$\frac{\partial F}{\partial x}\,dx + \frac{\partial F}{\partial y}\,dy + \frac{\partial F}{\partial z}\,dz = 0 \tag{d}$$

The shift dx, dy, dz is in a direction whose direction cosines are proportional to dx, dy, dz. The three quantities $\partial F/\partial x$, $\partial F/\partial y$, $\partial F/\partial z$ also specify a direction, since we can take direction cosines proportional to them. The left-hand side of Eq. (*d*) is then proportional to the cosine of the angle between these two directions. Since it vanishes, the two directions are at right angles, and since dx, dy, dz represent a direction in the tangent plane to the surface at the point x, y, z, the direction represented by $\partial F/\partial x$, $\partial F/\partial y$, $\partial F/\partial z$ is normal to the surface given by Eq. (*c*).

Now $F(x,y,z)$ is in our case the function on the right-hand side of Eq. (119). Thus

$$\begin{aligned} \frac{\partial F}{\partial x} &= 2\epsilon_x x + \gamma_{xy} y + \gamma_{xz} z \\ \frac{\partial F}{\partial y} &= \gamma_{xy} x + 2\epsilon_y y + \gamma_{yz} z \\ \frac{\partial F}{\partial z} &= \gamma_{xz} x + \gamma_{yz} y + 2\epsilon_z z \end{aligned} \tag{e}$$

[1] A glance at Fig. 6 will show that $\partial v/\partial x$ and $-\partial u/\partial y$, occurring in the expression for ω_z, are the clockwise rotations of the line elements $O'A'$, $O'B'$ from their original positions OA, OB. Thus ω_z is the average of these rotations, and ω_x, ω_y have a similar significance in the yz and xz planes.

The surface given by Eq. (119) being drawn with the point O (Fig. 128) as center, we may identify δx, δy, δz in Eqs. (*b*) with x, y, z in Eqs. (*e*).

We consider now the special case when ω_x, ω_y, ω_z are zero. Then the right-hand sides of Eqs. (*e*) are the same as the right-hand sides of Eqs. (*b*) but for a factor 2. Consequently the displacement given by Eqs. (*b*) is normal to the surface given by Eq. (119). Considering the point O_1 (Fig. 128) as a point on the surface, this means that the displacement of O_1 is normal to the surface at O_1. Hence if OO_1 is one of the principal axes of strain, that is, one of the principal axes of the surface, the displacement of O_1 is in the direction of OO_1, and therefore OO_1 does not rotate. The displacement in question will correspond to step 1.

In order to complete the displacement, we must restore to Eqs. (*b*) the terms in ω_x, ω_y, ω_z. But these terms correspond to a small rigid-body rotation having components ω_x, ω_y, ω_z about the x, y, z axes. Consequently these quantities, given by (122), express the rotation of step 3—that is, the rotation of the principal axes of strain at the point O. They are called simply the *components of rotation*.

PROBLEMS

1. What is the equation, of the type $f(x,y,z) = 0$, of the surface with center at O which becomes a sphere $x'^2 + y'^2 + z'^2 = r^2$ *after* the homogeneous deformation of Art. 80? What kind of surface is it?
2. Show that if the rotation is zero throughout the body (irrotational deformation), the displacement vector is the gradient of a scalar potential function.

   Indicate one or more examples of such irrotational deformation from the problems treated in the text.

chapter | 8

General Theorems

84 | Differential Equations of Equilibrium

In the discussion of Art. 74 we considered the stress at a point of an elastic body. Let us consider now the variation of the stress as we change the position of the point. For this purpose the conditions of equilibrium of a small rectangular parallelepiped with the sides δx, δy, δz (Fig. 129) must be studied. The components of stresses acting on the sides of this small element and their positive directions are indicated in the figure. Here we take into account the small changes of the components of stress due to the small increases δx, δy, δz of the coordinates. Thus designating the midpoints of the sides of the element by 1, 2, 3, 4, 5, 6 as in Fig. 129, we distinguish between the value of σ_x at point 1, and its value at point 2, writing these $(\sigma_x)_1$ and $(\sigma_x)_2$, respectively. The symbol σ_x itself denotes, of course, the value of this stress component at the point x, y, z. In calculating the *forces* acting on the element we consider the sides as very small, and the force is obtained by multiplying the stress at the centroid of a side by the area of this side.

It should be noted that the body force acting on the element, which was neglected as a small quantity of higher order in discussing the equilibrium of a tetrahedron (Fig. 126), must now be taken into account, because it is of the same order of magnitude as the terms due to variations of the stress components, which we are now considering. If we let X, Y, Z denote the components of this force per unit volume of the element, then the equation of equilibrium obtained by summing all the forces acting on the element in the x direction is

$$[(\sigma_x)_1 - (\sigma_x)_2]\,\delta y\,\delta z + [(\tau_{xy})_3 - (\tau_{xy})_4]\,\delta x\,\delta z\,[(\tau_{xz})_5 - (\tau_{xz})_6]\,\delta x\,\delta y + X\,\delta x\,\delta y\,\delta z = 0$$

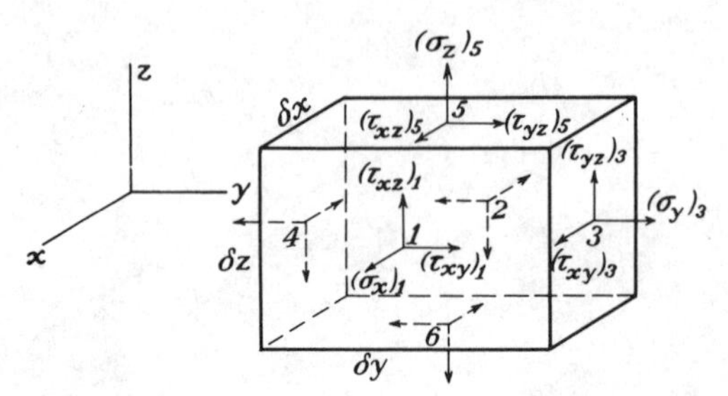

Fig. 129

The two other equations of equilibrium are obtained in the same manner. After dividing by $\delta x\ \delta y\ \delta z$ and proceeding to the limit by shrinking the element down to the point x, y, z, we find

$$
\begin{aligned}
\frac{\partial \sigma_x}{\partial x} + \frac{\partial \tau_{xy}}{\partial y} + \frac{\partial \tau_{xz}}{\partial z} + X &= 0 \\
\frac{\partial \sigma_y}{\partial y} + \frac{\partial \tau_{xy}}{\partial x} + \frac{\partial \tau_{yz}}{\partial z} + Y &= 0 \\
\frac{\partial \sigma_z}{\partial z} + \frac{\partial \tau_{xz}}{\partial x} + \frac{\partial \tau_{yz}}{\partial y} + Z &= 0
\end{aligned}
\tag{123}
$$

Equations (123) must be satisfied at all points throughout the volume of the body. The stresses vary over the volume of the body, and when we arrive at the surface they must be such as to be in equilibrium with the external forces on the surface of the body. These conditions of equilibrium at the surface can be obtained from Eqs. (108). Taking a tetrahedron $OBCD$ (Fig. 126) so that the side BCD coincides with the surface of the body, and denoting by $\bar{X}$, $\bar{Y}$, $\bar{Z}$ the components of the surface forces per unit area at this point, Eqs. (108) become

$$
\begin{aligned}
\bar{X} &= \sigma_x l + \tau_{xy} m + \tau_{xz} n \\
\bar{Y} &= \sigma_y m + \tau_{yz} n + \tau_{xy} l \\
\bar{Z} &= \sigma_z n + \tau_{xz} l + \tau_{yz} m
\end{aligned}
\tag{124}
$$

in which l, m, n are the direction cosines of the external normal to the surface of the body at the point under consideration.

If the problem is to determine the state of stress in a body submitted to the action of given forces, it is necessary to solve Eqs. (123), and the solution must be such as to satisfy the boundary conditions (124). These equations, containing six components of stress, σ_x, . . . , τ_{yz}, are not sufficient for the determination of these components. The problem is a statically indeterminate one, and in order to obtain the solution we must proceed as in the case of two-dimensional problems, i.e., the elastic deformations of the body must also be considered.

85 | Conditions of Compatibility

It should be noted that the six components of strain at each point are completely determined by the three functions u, v, w, representing the components of displacement. Hence, the components of strain cannot be taken arbitrarily as functions of x, y, z but are subject to relations that follow from Eqs. (2).

Thus, from Eqs. (2),

$$\frac{\partial^2 \epsilon_x}{\partial y^2} = \frac{\partial^3 u}{\partial x\, \partial y^2} \qquad \frac{\partial^2 \epsilon_y}{\partial x^2} = \frac{\partial^3 v}{\partial x^2\, \partial y} \qquad \frac{\partial^2 \gamma_{xy}}{\partial x\, \partial y} = \frac{\partial^3 u}{\partial x\, \partial y^2} + \frac{\partial^3 v}{\partial x^2\, \partial y}$$

from which

$$\frac{\partial^2 \epsilon_x}{\partial y^2} + \frac{\partial^2 \epsilon_y}{\partial x^2} = \frac{\partial^2 \gamma_{xy}}{\partial x\, \partial y} \tag{a}$$

Two more relations of the same kind can be obtained by cyclical interchange of the letters x, y, z.

From the derivatives

$$\frac{\partial^2 \epsilon_x}{\partial y\, \partial z} = \frac{\partial^3 u}{\partial x\, \partial y\, \partial z} \qquad \frac{\partial \gamma_{yz}}{\partial x} = \frac{\partial^2 v}{\partial x\, \partial z} + \frac{\partial^2 w}{\partial x\, \partial y}$$

$$\frac{\partial \gamma_{xz}}{\partial y} = \frac{\partial^2 u}{\partial y\, \partial z} + \frac{\partial^2 w}{\partial x\, \partial y} \qquad \frac{\partial \gamma_{xy}}{\partial z} = \frac{\partial^2 u}{\partial y\, \partial z} + \frac{\partial^2 v}{\partial x\, \partial z}$$

we find that

$$2\frac{\partial^2 \epsilon_x}{\partial y\, \partial z} = \frac{\partial}{\partial x}\left(-\frac{\partial \gamma_{yz}}{\partial x} + \frac{\partial \gamma_{xz}}{\partial y} + \frac{\partial \gamma_{xy}}{\partial z}\right) \tag{b}$$

Two more relations of the kind (*b*) can be obtained by interchange of the letters x, y, z. We thus arrive at the following six differential relations between the components of strain, which must be satisfied by virtue of Eqs. (2):

$$\begin{aligned}
&\frac{\partial^2 \epsilon_x}{\partial y^2} + \frac{\partial^2 \epsilon_y}{\partial x^2} = \frac{\partial^2 \gamma_{xy}}{\partial x\, \partial y} \qquad && 2\frac{\partial^2 \epsilon_x}{\partial y\, \partial z} = \frac{\partial}{\partial x}\left(-\frac{\partial \gamma_{yz}}{\partial x} + \frac{\partial \gamma_{xz}}{\partial y} + \frac{\partial \gamma_{xy}}{\partial z}\right)\\
&\frac{\partial^2 \epsilon_y}{\partial z^2} + \frac{\partial^2 \epsilon_z}{\partial y^2} = \frac{\partial^2 \gamma_{yz}}{\partial y\, \partial z} \qquad && 2\frac{\partial^2 \epsilon_y}{\partial x\, \partial z} = \frac{\partial}{\partial y}\left(\frac{\partial \gamma_{yz}}{\partial x} - \frac{\partial \gamma_{xz}}{\partial y} + \frac{\partial \gamma_{xy}}{\partial z}\right)\\
&\frac{\partial^2 \epsilon_z}{\partial x^2} + \frac{\partial^2 \epsilon_x}{\partial z^2} = \frac{\partial^2 \gamma_{xz}}{\partial x\, \partial z} \qquad && 2\frac{\partial^2 \epsilon_z}{\partial x\, \partial y} = \frac{\partial}{\partial z}\left(\frac{\partial \gamma_{yz}}{\partial x} + \frac{\partial \gamma_{xz}}{\partial y} - \frac{\partial \gamma_{xy}}{\partial z}\right)
\end{aligned} \tag{125}$$

These differential relations[1] are called the *conditions of compatibility*.

[1] Proofs that these six equations are sufficient to ensure the existence of a displacement corresponding to a given set of functions $\epsilon_x, \ldots, \gamma_{xy} \ldots$, may be found in

By using Hooke's law [Eqs. (3)] conditions (125) can be transformed into relations between the components of stress. Take, for instance, the condition

$$\frac{\partial^2 \epsilon_y}{\partial z^2} + \frac{\partial^2 \epsilon_z}{\partial y^2} = \frac{\partial^2 \gamma_{yz}}{\partial y \, \partial z} \tag{c}$$

From Eqs. (3) and (4), using the notation (7), we find

$$\epsilon_y = \frac{1}{E}[(1 + \nu)\sigma_y - \nu\Theta]$$

$$\epsilon_z = \frac{1}{E}[(1 + \nu)\sigma_z - \nu\Theta]$$

$$\gamma_{yz} = \frac{2(1 + \nu)\tau_{yz}}{E}$$

Substituting these expressions in (c), we obtain

$$(1 + \nu)\left(\frac{\partial^2 \sigma_y}{\partial z^2} + \frac{\partial^2 \sigma_z}{\partial y^2}\right) - \nu\left(\frac{\partial^2 \Theta}{\partial z^2} + \frac{\partial^2 \Theta}{\partial y^2}\right) = 2(1 + \nu)\frac{\partial^2 \tau_{yz}}{\partial y \, \partial z} \tag{d}$$

The right side of this equation can be transformed by using the equations of equilibrium (123). From these equations we find

$$\frac{\partial \tau_{yz}}{\partial y} = -\frac{\partial \sigma_z}{\partial z} - \frac{\partial \tau_{xz}}{\partial x} - Z$$

$$\frac{\partial \tau_{yz}}{\partial z} = -\frac{\partial \sigma_y}{\partial y} - \frac{\partial \tau_{xy}}{\partial x} - Y$$

Differentiating the first of these equations with respect to z and the second with respect to y, and adding them together, we find

$$2\frac{\partial^2 \tau_{yz}}{\partial y \, \partial z} = -\frac{\partial^2 \sigma_z}{\partial z^2} - \frac{\partial^2 \sigma_y}{\partial y^2} - \frac{\partial}{\partial x}\left(\frac{\partial \tau_{xz}}{\partial z} + \frac{\partial \tau_{xy}}{\partial y}\right) - \frac{\partial Z}{\partial z} - \frac{\partial Y}{\partial y}$$

or, by using the first of Eqs. (123),

$$2\frac{\partial^2 \tau_{yz}}{\partial y \, \partial z} = \frac{\partial^2 \sigma_x}{\partial x^2} - \frac{\partial^2 \sigma_y}{\partial y^2} - \frac{\partial^2 \sigma_z}{\partial z^2} + \frac{\partial X}{\partial x} - \frac{\partial Y}{\partial y} - \frac{\partial Z}{\partial z}$$

Substituting this in Eq. (d) and using, to simplify the writing, the symbol

$$\nabla^2 = \frac{\partial^2}{\partial x^2} + \frac{\partial^2}{\partial y^2} + \frac{\partial^2}{\partial z^2}$$

A. E. H. Love, "Mathematical Theory of Elasticity," 4th ed., p. 49, Cambridge University Press, New York, 1927, and I. S. Sokolnikoff, "Mathematical Theory of Elasticity," p. 25, 1956. The equations themselves were given by B. de Saint-Venant in his edition of the book by C. L. M. H. Navier, "Résumé des Leçons sur l'Application de la Mécanique," app. 3, Carilian-Goeury, Paris, 1864.

we find

$$(1+\nu)\left(\nabla^2\Theta - \nabla^2\sigma_x - \frac{\partial^2\Theta}{\partial x^2}\right) - \nu\left(\nabla^2\Theta - \frac{\partial^2\Theta}{\partial x^2}\right) = (1+\nu)\left(\frac{\partial X}{\partial x} - \frac{\partial Y}{\partial y} - \frac{\partial Z}{\partial z}\right) \qquad (e)$$

Two analogous equations can be obtained from the two other conditions of compatibility of the type (*c*).

Adding together all three equations of the type (*e*) we find

$$(1-\nu)\nabla^2\Theta = -(1+\nu)\left(\frac{\partial X}{\partial x} + \frac{\partial Y}{\partial y} + \frac{\partial Z}{\partial z}\right) \qquad (f)$$

Substituting this expression for $\nabla^2\Theta$ in Eq. (*e*),

$$\nabla^2\sigma_x + \frac{1}{1+\nu}\frac{\partial^2\Theta}{\partial x^2} = -\frac{\nu}{1-\nu}\left(\frac{\partial X}{\partial x} + \frac{\partial Y}{\partial y} + \frac{\partial Z}{\partial z}\right) - 2\frac{\partial X}{\partial x} \qquad (g)$$

We can obtain three equations of this kind, corresponding to the first three of Eqs. (125). In the same manner the remaining three conditions (125) can be transformed into equations of the following kind:

$$\nabla^2\tau_{yz} + \frac{1}{1+\nu}\frac{\partial^2\Theta}{\partial y\,\partial z} = -\left(\frac{\partial Z}{\partial y} + \frac{\partial Y}{\partial z}\right) \qquad (h)$$

If there are no body forces or if the body forces are constant, Eqs. (*g*) and (*h*) become

$$\begin{aligned}
(1+\nu)\nabla^2\sigma_x + \frac{\partial^2\Theta}{\partial x^2} &= 0 \qquad & (1+\nu)\nabla^2\tau_{yz} + \frac{\partial^2\Theta}{\partial y\,\partial z} &= 0 \\
(1+\nu)\nabla^2\sigma_y + \frac{\partial^2\Theta}{\partial y^2} &= 0 \qquad & (1+\nu)\nabla^2\tau_{xz} + \frac{\partial^2\Theta}{\partial x\,\partial z} &= 0 \\
(1+\nu)\nabla^2\sigma_z + \frac{\partial^2\Theta}{\partial z^2} &= 0 \qquad & (1+\nu)\nabla^2\tau_{xy} + \frac{\partial^2\Theta}{\partial x\,\partial y} &= 0
\end{aligned} \qquad (126)$$

We see that in addition to the equations of equilibrium (123) and the boundary conditions (124) the stress components in an isotropic body must satisfy the six conditions of compatibility (*g*) and (*h*) or the six conditions (126). This system of equations is generally sufficient for determining the stress components without ambiguity (see Art. 96).

The conditions of compatibility contain only second derivatives of the stress components. Hence, if the external forces are such that the equations of equilibrium (123) together with the boundary conditions (124) can be satisfied by taking the stress components either as constants or as linear functions of the coordinates, the equations of compatibility are satisfied identically and this stress system is the correct solution of the problem. Several examples of such problems will be considered in Chap. 9.

86 | Determination of Displacements

When the components of stress are found from the previous equations, the components of strain can be calculated by using Hooke's law [Eqs. (3) and (6)]. Then Eqs. (2) are used for the determination of the displacements u, v, w. Differentiating Eqs. (2) with respect to x, y, z we can obtain 18 equations containing 18 second derivatives of u, v, w, from which all these derivatives can be determined. For u, for instance, we obtain

$$\frac{\partial^2 u}{\partial x^2} = \frac{\partial \epsilon_x}{\partial x} \qquad \frac{\partial^2 u}{\partial y^2} = \frac{\partial \gamma_{xy}}{\partial y} - \frac{\partial \epsilon_y}{\partial x} \qquad \frac{\partial^2 u}{\partial z^2} = \frac{\partial \gamma_{xz}}{\partial z} - \frac{\partial \epsilon_z}{\partial x}$$

$$\frac{\partial^2 u}{\partial x\,\partial y} = \frac{\partial \epsilon_x}{\partial y} \qquad \frac{\partial^2 u}{\partial x\,\partial z} = \frac{\partial \epsilon_x}{\partial z} \qquad \frac{\partial^2 u}{\partial y\,\partial z} = \frac{1}{2}\left(\frac{\partial \gamma_{xz}}{\partial y} + \frac{\partial \gamma_{xy}}{\partial z} - \frac{\partial \gamma_{yz}}{\partial x}\right) \tag{a}$$

The second derivatives for the two other components of displacement v and w can be obtained by cyclical interchange in Eqs. (a) of the letters x, y, z.

Now u, v, w can be obtained by double integration of these second derivatives. The introduction of arbitrary constants of integration will result in adding to the values of u, v, w linear functions in x, y, z, as it is evident that such functions can be added to u, v, w without affecting such equations as (a). To have the strain components (2) unchanged by such an addition, the additional linear functions must have the form

$$\begin{aligned} u' &= a + by - cz \\ v' &= d - bx + ez \\ w' &= f + cx - ey \end{aligned} \tag{b}$$

This means that the displacements are not entirely determined by the stresses and strains. On the displacements found from the differential Eqs. (123), (124), and (126) a displacement like that of a rigid body can be superposed. The constants a, d, f in Eqs. (b) represent a translatory motion of the body, and the constants b, c, e are the three rotations of the rigid body around the coordinate axes. When there are sufficient constraints to prevent motion as a rigid body, the six constants in Eqs. (b) can easily be calculated so as to satisfy the conditions of constraint. Several examples of such calculations will be shown later.

87 | Equations of Equilibrium in Terms of Displacements

One method of solution of the problems of elasticity is to eliminate the stress components from Eqs. (123) and (124) by using Hooke's law and to express the strain components in terms of displacements by using Eqs. (2). In this manner we arrive at three equations of equilibrium containing only the three unknown functions u, v, w. Substituting in the first of Eqs. (123) from (11),

$$\sigma_x = \lambda e + 2G \frac{\partial u}{\partial x} \tag{a}$$

and from (6),

$$\begin{aligned} \tau_{xy} &= G\gamma_{xy} = G\left(\frac{\partial u}{\partial y} + \frac{\partial v}{\partial x}\right) \\ \tau_{xz} &= G\gamma_{xz} = G\left(\frac{\partial w}{\partial x} + \frac{\partial u}{\partial z}\right) \end{aligned} \tag{b}$$

we find

$$(\lambda + G)\frac{\partial e}{\partial x} + G\left(\frac{\partial^2 u}{\partial x^2} + \frac{\partial^2 u}{\partial y^2} + \frac{\partial^2 u}{\partial z^2}\right) + X = 0$$

The two other equations can be transformed in the same manner. Then, using the symbol ∇^2 (see page 238), the equations of equilibrium (123) become

$$\begin{aligned} (\lambda + G)\frac{\partial e}{\partial x} + G\,\nabla^2 u + X &= 0 \\ (\lambda + G)\frac{\partial e}{\partial y} + G\,\nabla^2 v + Y &= 0 \\ (\lambda + G)\frac{\partial e}{\partial z} + G\,\nabla^2 w + Z &= 0 \end{aligned} \tag{127}$$

and, when there are no body forces,

$$\begin{aligned} (\lambda + G)\frac{\partial e}{\partial x} + G\,\nabla^2 u &= 0 \\ (\lambda + G)\frac{\partial e}{\partial y} + G\,\nabla^2 v &= 0 \\ (\lambda + G)\frac{\partial e}{\partial z} + G\,\nabla^2 w &= 0 \end{aligned} \tag{128}$$

Differentiating these equations, the first with respect to x, the second with respect to y, and the third with respect to z, and adding them together, we find

$$(\lambda + 2G)\,\nabla^2 e = 0$$

i.e., the volume expansion e satisfies the differential equation

$$\frac{\partial^2 e}{\partial x^2} + \frac{\partial^2 e}{\partial y^2} + \frac{\partial^2 e}{\partial z^2} = 0 \tag{129}$$

The same conclusion holds also when body forces are constant throughout the volume of the body.

Substituting from such equations as (a) and (b) into the boundary conditions (124) we find

$$\bar{X} = \lambda e l + G\left(\frac{\partial u}{\partial x} l + \frac{\partial u}{\partial y} m + \frac{\partial u}{\partial z} n\right) + G\left(\frac{\partial u}{\partial x} l + \frac{\partial v}{\partial x} m + \frac{\partial w}{\partial x} n\right) \qquad (130)$$

. . . .

Equations (127) together with the boundary conditions (130) define completely the three functions u, v, w. From these the components of strain are obtained from Eqs. (2) and the components of stress from Eqs. (9) and (6). Applications of these equations will be shown in Chap. 14.

88 | General Solution for the Displacements

It is easily verified by substitution that the differential equations (128) of equilibrium in terms of displacement are satisfied by[1]

$$u = \phi_1 - \alpha \frac{\partial}{\partial x}(\phi_0 + x\phi_1 + y\phi_2 + z\phi_3)$$

$$v = \phi_2 - \alpha \frac{\partial}{\partial y}(\phi_0 + x\phi_1 + y\phi_2 + z\phi_3)$$

$$w = \phi_3 - \alpha \frac{\partial}{\partial z}(\phi_0 + x\phi_1 + y\phi_2 + z\phi_3)$$

where $4\alpha = 1/(1 - \nu)$ and the four functions ϕ_0, ϕ_1, ϕ_2, ϕ_3 are harmonic, i.e.,

$$\nabla^2\phi_0 = 0 \qquad \nabla^2\phi_1 = 0 \qquad \nabla^2\phi_2 = 0 \qquad \nabla^2\phi_3 = 0$$

It can be shown that this solution is general, even when ϕ_0 is omitted.[2]

This form of solution has been adapted to curvilinear coordinates by Neuber and applied by him in the solution of problems of solids of revolution[3] generated by hyperbolas (the hyperbolic groove on a cylinder) and ellipses (cavity in the form of an ellipsoid of revolution) transmitting tension, bending, torsion, or shear force transverse to the axis with accompanying bending.

[1] This solution was given independently by P. F. Papkovitch, *Compt. Rend.*, vol. 195, pp. 513 and 754, 1932, and by H. Neuber, *Z. Angew. Math. Mech.*, vol. 14, p. 203, 1934. Other general solutions were given by B. Galerkin, *Compt. Rend.*, vol. 190, p. 1047, 1930, and by Boussinesq and Kelvin—see Todhunter and Pearson, "History of Elasticity," vol. 2, pt. 2, p. 268. See also R. D. Mindlin, *Bull. Am. Math. Soc.*, 1936, p. 373.

[2] For discussion of the number of functions needed for completeness, see P. M. Naghdi and C. S. Hsu, *J. Math. Mech.*, vol. 10, pp. 233–246, 1961, and references given there.

[3] H. Neuber, "Kerbspannungslehre," 2d ed., Springer-Verlag OHG, Berlin, 1958. This book also contains solutions of two-dimensional problems. See Chap. 6 above.

89 | The Principle of Superposition

The solution of a problem of a given elastic solid with given surface and body forces requires us to determine stress components, or displacements, that satisfy the differential equations and the boundary conditions. If we choose to work with stress components, we have to satisfy: (1) the equations of equilibrium (123); (2) the compatibility conditions (125); (3) the boundary conditions (124). Let σ_x . . . , τ_{xy} . . . , be the stress components so determined, and due to surface forces $\bar{X}$, $\bar{Y}$, $\bar{Z}$ and body forces X, Y, Z.

Let σ_x', . . . , τ_{xy}' . . . be the stress components in the same elastic solid due to surface forces $\bar{X}'$, $\bar{Y}'$, $\bar{Z}'$ and body forces X', Y', Z'. Then the stress components $\sigma_x + \sigma_x'$, . . . , $\tau_{xy} + \tau_{xy}'$, . . . will represent the stress due to the surface forces $\bar{X} + \bar{X}'$, . . . and the body forces $X + X'$, ... This holds because all the differential equations and boundary conditions are *linear*. Thus, adding the first of Eqs. (123) to the corresponding equation

$$\frac{\partial \sigma_x'}{\partial x} + \frac{\partial \tau_{xy}'}{\partial y} + \frac{\partial \tau_{xz}'}{\partial z} + X' = 0$$

we find

$$\frac{\partial}{\partial x}(\sigma_x + \sigma_x') + \frac{\partial}{\partial y}(\tau_{xy} + \tau_{xy}') + \frac{\partial}{\partial z}(\tau_{xz} + \tau_{xz}') + X + X' = 0$$

and similarly from the first of (124) and its counterpart we have by addition

$$\bar{X} + \bar{X}' = (\sigma_x + \sigma_x')l + (\tau_{xy} + \tau_{xy}')m + (\tau_{xz} + \tau_{xz}')n$$

The compatibility conditions can be combined in the same manner. The complete set of equations shows that $\sigma_x + \sigma_x'$, . . . , $\tau_{xy} + \tau_{xy}'$ . . . satisfy all the equations and conditions determining the stress due to forces $\bar{X} + \bar{X}'$, . . . , $X + X'$, ... This is an instance of the principle of superposition. It is readily extended to other types of boundary conditions such as given displacements.

In deriving our equations of equilibrium (123) and boundary conditions (124), we made no distinction between the position and form of the element before loading, and its position and form after loading. As a consequence, our equations and the conclusions drawn from them are valid only so long as the small displacements in the deformation do not affect substantially the action of the external forces. There are cases, however, in which the deformation must be taken into account. Then the justification of the principle of superposition given above fails. The beam under simultaneous thrust and lateral load affords an example of this kind, and many others arise in considering the elastic stability of thin-walled structures.

90 | Strain Energy

When a uniform bar is loaded in simple tension, the forces on the ends do a certain amount of work as the bar stretches. Thus, if the element shown in Fig. 130 is subject to normal stresses σ_x only, we have a force $\sigma_x\, dy\, dz$ that does work on an extension $\epsilon_x\, dx$. The relation between these two quantities during loading is represented by a straight line such as OA in Fig. 130*b*, and the work done during deformation is given by the area $\frac{1}{2}(\sigma_x\, dy\, dz)(\epsilon_x\, dx)$ of the triangle OAB. Writing dV for this work we have

$$dV = \tfrac{1}{2}\sigma_x\epsilon_x\, dx\, dy\, dz \tag{a}$$

It is evident that the same amount of work is done on all such elements, if their volumes are the same. We now inquire what becomes of this work—what kind or kinds of energy is it converted into?

In the case of a gas, adiabatic compression causes a rise of temperature. When an ordinary steel bar is adiabatically compressed, there is an analogous, but quite small, rise of temperature. The original temperature can then be restored by abstracting heat. Such a change of temperature alters the strain, but only by a very small fraction of the adiabatic strain. If this were not so, there would be a significant difference between the adiabatic and isothermal moduli of elasticity. The actual differences for the common metals are very slight.[1] For instance, the adiabatic Young's modulus for iron exceeds the isothermal modulus by only 0.26 percent. Here we shall disregard such differences.[2] The work done on an element, and stored within it, will be called strain energy. It is assumed that the element remains elastic and that no kinetic energy is developed.

[1] They were evaluated by Kelvin in *Quart. J. Math.*, 1855, republished in *Phil. Mag.*, ser. 5, no. 5, pp. 4–27, 1878. For earlier references see Love, *op cit.*, p. 99.

[2] They are considered further in, for instance, C. E. Pearson, "Theoretical Elasticity," p. 164, Harvard University Press, Cambridge, Mass., 1959.

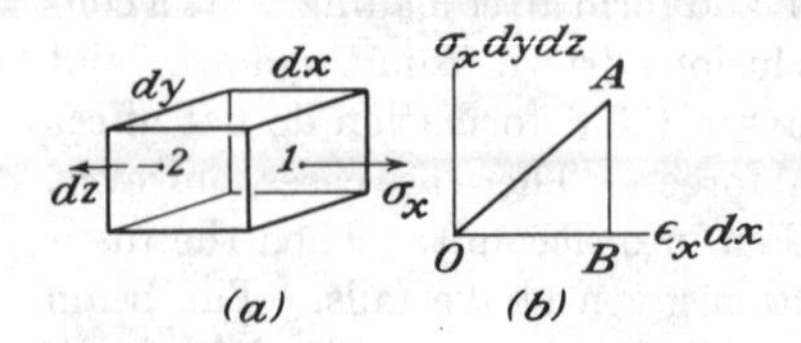

Fig. 130

The same considerations apply when the element has all six components of stress, $\sigma_x, \sigma_y, \sigma_z, \tau_{xy}, \tau_{yz}, \tau_{xz}$ acting on it (Fig. 3). Conservation of energy requires that the work cannot depend on the order in which the forces are applied but only on the final magnitudes. Otherwise we could load in one order and unload in another order corresponding to a larger amount of work. Thus a net amount of work would have been gained from the element in a complete cycle.

The calculation of the work done is simplest if the forces, or stresses, all increase simultaneously in the same ratio. Then the relation between each force and the corresponding displacement is still linear, as in Fig. 130*b*, and the work done by all the forces is

$$dV = V_0\, dx\, dy\, dz \tag{b}$$

where

$$V_0 = \tfrac{1}{2}(\sigma_x\epsilon_x + \sigma_y\epsilon_y + \sigma_z\epsilon_z + \tau_{xy}\gamma_{xy} + \tau_{yz}\gamma_{yz} + \tau_{xz}\gamma_{xz}) \tag{c}$$

Thus V_0 is the amount of work per unit volume, or strain energy per unit volume.

In the preceding discussion the stresses were regarded as the same on opposite faces of the element, and there was no body force. Let us now reconsider the work done on the element when the stress varies through the body and body force is included. Considering first the force $\sigma_x\, dy\, dz$ on the face 1 of the element in Fig. 130*a*, it does work on the displacement u of this face, of amount $\frac{1}{2}(\sigma_x u)_1\, dy\, dz$, where the subscript 1 indicates that the functions σ_x, u must be evaluated at the point 1. The force $\sigma_x\, dy\, dz$ on the face 2 does work $-\frac{1}{2}(\sigma_x u)_2\, dy\, dz$. The total for the two faces

$$\tfrac{1}{2}[(\sigma_x u)_1 - (\sigma_x u)_2]\, dy\, dz$$

is the same, in the limit, as

$$\frac{1}{2}\frac{\partial}{\partial x}(\sigma_x u)\, dx\, dy\, dz \tag{d}$$

Computing the work done by the shear stresses τ_{xy}, τ_{xz} on the faces 1 and 2, and adding to (*d*), we have the work done on the two faces by all three components of stress as

$$\frac{1}{2}\frac{\partial}{\partial x}(\sigma_x u + \tau_{xy} v + \tau_{xz} w)\, dx\, dy\, dz$$

where v and w are the components of displacement in the y and z directions. The work done on the other two pairs of faces can be similarly expressed. We find, for the total work done by the stresses on the faces,

$$\frac{1}{2}\left[\frac{\partial}{\partial x}(\sigma_x u + \tau_{xy} v + \tau_{xz} w) + \frac{\partial}{\partial y}(\sigma_y v + \tau_{yz} w + \tau_{xy} u) + \frac{\partial}{\partial z}(\sigma_z w + \tau_{xz} u + \tau_{yz} v)\right] dx\, dy\, dz \tag{e}$$

As the body is loaded the body forces $X\, dx\, dy\, dz$ etc. do work

$$\tfrac{1}{2}(Xu + Yv + Zw)\, dx\, dy\, dz \tag{f}$$

The total work done on the element is the sum of (*e*) and (*f*). On carrying out the differentiations in (*e*), we find that the total work becomes

$$\frac{1}{2}\left[\sigma_x \frac{\partial u}{\partial x} + \sigma_y \frac{\partial v}{\partial y} + \sigma_z \frac{\partial w}{\partial z} + \tau_{xy}\left(\frac{\partial v}{\partial x} + \frac{\partial u}{\partial y}\right) + \tau_{yz}\left(\frac{\partial w}{\partial y} + \frac{\partial v}{\partial z}\right) + \tau_{xz}\left(\frac{\partial u}{\partial z} + \frac{\partial w}{\partial x}\right)\right.$$
$$+ u\left(\frac{\partial \sigma_x}{\partial x} + \frac{\partial \tau_{xy}}{\partial y} + \frac{\partial \tau_{xz}}{\partial z} + X\right) + v\left(\frac{\partial \sigma_y}{\partial y} + \frac{\partial \tau_{yz}}{\partial z} + \frac{\partial \tau_{xy}}{\partial x} + Y\right)$$
$$\left. + w\left(\frac{\partial \sigma_z}{\partial z} + \frac{\partial \tau_{xz}}{\partial x} + \frac{\partial \tau_{yz}}{\partial y} + Z\right)\right] dx\,dy\,dz$$

But on account of the equations of equilibrium (123) derived in Art. 84 the brackets multiplying u, v, w are zero. The quantities multiplying the stress components are, from Eqs. (2), $\epsilon_x, \ldots, \gamma_{xy}, \ldots$. Thus the total work done on the element reduces to the value given by (*b*) and (*c*). These formulas therefore continue to give the work done on the element, or strain energy stored in it, when the stress is not uniform and body forces are included.

By means of Hooke's law, Eqs. (3) and (6), we can express V_0, given by Eq. (*c*), as a function of the stress components only. Then

$$V_0 = \frac{1}{2E}(\sigma_x^2 + \sigma_y^2 + \sigma_z^2) - \frac{\nu}{E}(\sigma_x\sigma_y + \sigma_y\sigma_z + \sigma_z\sigma_x) + \frac{1}{2G}(\tau_{xy}^2 + \tau_{yz}^2 + \tau_{xz}^2) \tag{131}$$

It is easily verified that

$$V_0 = \frac{1}{2E}[I_1^2 - 2(1+\nu)I_2]$$

where I_1, I_2 are the stress invariants of Art. 79.

Alternatively we may use Eqs. (11) and express V_0 as a function of the strain components only. Then

$$V_0 = \tfrac{1}{2}\lambda e^2 + G(\epsilon_x^2 + \epsilon_y^2 + \epsilon_z^2) + \tfrac{1}{2}G(\gamma_{xy}^2 + \gamma_{yz}^2 + \gamma_{xz}^2) \tag{132}$$

in which

$$e = \epsilon_x + \epsilon_y + \epsilon_z \qquad \lambda = \frac{E\nu}{(1+\nu)(1-2\nu)}$$

This form shows at once that V_0 is always positive.

It is easy to show that the derivative of V_0, as given by (132), with respect to any strain component gives the corresponding stress component. Thus taking the derivative with respect to ϵ_x and using Eq. (11), we find

$$\frac{\partial V_0}{\partial \epsilon_x} = \lambda e + 2G\epsilon_x = \sigma_x \tag{g}$$

For the case of plane stress, in which $\sigma_z = \tau_{xz} = \tau_{yz} = 0$, we have from (131)

$$V_0 = \frac{1}{2E}(\sigma_x^2 + \sigma_y^2) - \frac{\nu}{E}\sigma_x\sigma_y + \frac{1}{2G}\tau_{xy}^2 \tag{133}$$

or, in terms of strain,

$$V_0 = \frac{E}{2(1-\nu^2)}(\epsilon_x^2 + \epsilon_y^2 + 2\nu\epsilon_x\epsilon_y) + \frac{G}{2}\gamma_{xy}^2 \tag{134}$$

The total strain energy V of a deformed elastic body is obtained from the strain energy per unit volume, V_0, by integration. Writing $d\tau$ for the element of volume, we have

$$V = \int V_0\, d\tau \tag{135}$$

It represents the total work done against internal forces during loading. If we think of the body as consisting of a very large number of particles interconnected by springs, it would represent the work done in stretching or contracting the springs. For the work done *on* the particles by the internal forces we must reverse the sign.

The quantity of strain energy stored per unit volume of the material is sometimes used as a basis for determining the limiting stress at which failure occurs.[1] In order to bring this theory into agreement with the fact that isotropic materials can sustain very large hydrostatic pressures without yielding, it has been proposed to split the strain energy into two parts, one due to the change in volume and the other due to the distortion, and consider only the second part in determining the strength.[2]

We know that the volume change is proportional to the sum of the three normal stress components [Eq. (8)], so if this sum is zero the deformation consists of distortion only. We may resolve each stress component into two parts,

$$\sigma_x = \sigma_x' + p \qquad \sigma_y = \sigma_y' + p \qquad \sigma_z = \sigma_z' + p$$

where

$$p = \tfrac{1}{3}(\sigma_x + \sigma_y + \sigma_z) = \tfrac{1}{3}\Theta \tag{h}$$

Since, from this,

$$\sigma_x' + \sigma_y' + \sigma_z' = 0$$

the stress condition σ_x', σ_y', σ_z' produces only distortion, and the change in volume depends entirely[3] on the magnitude of the uniform tension p. The part of the total energy due to this change in volume is, from Eq. (8),

$$\frac{ep}{2} = \frac{3(1-2\nu)}{2E}p^2 = \frac{1-2\nu}{6E}(\sigma_x + \sigma_y + \sigma_z)^2 \tag{i}$$

Subtracting this from (131), and using the identity

$$\sigma_x\sigma_y + \sigma_y\sigma_z + \sigma_z\sigma_x = -\tfrac{1}{2}[(\sigma_x - \sigma_y)^2 + (\sigma_y - \sigma_z)^2 + (\sigma_z - \sigma_x)^2] + (\sigma_x^2 + \sigma_y^2 + \sigma_z^2)$$

[1] The various strength theories are discussed in S. Timoshenko, "Strength of Materials," vol. 2, New York, 1956.

[2] M. T. Huber, *Czasopismo Technizne*, Lwów, 1904. See also R. von Mises, *Göttingen Nachrichten, Math.-Phys. Klasse*, 1913, p. 582, and F. Schleicher, *Z. Angew. Math. Mech.*, vol. 5, p. 199, 1925. For experimental comparisons see R. Hill, "Plasticity," Oxford University Press, Fair Lawn, N.J., 1950.

[3] The shearing components τ_{xy}, τ_{yz}, τ_{xz} produce shearing strains that do not involve any change of volume, to the first order in the small strains.

we can present the part of the total energy due to distortion in the form

$$V_0 - \frac{1-2\nu}{6E}(\sigma_x + \sigma_y + \sigma_z)^2 = \frac{1+\nu}{6E}[(\sigma_x - \sigma_y)^2 + (\sigma_y - \sigma_z)^2 + (\sigma_z - \sigma_x)^2] + \frac{1}{2G}(\tau_{xy}^2 + \tau_{xz}^2 + \tau_{yz}^2) \qquad (136)$$

In the case of simple tension in the x direction, σ_x alone is different from zero, and the strain energy of distortion (136) is $(1 + \nu)\sigma_x^2/3E$. In the case of pure shear, say between the xz and yz planes, τ_{xy} alone is different from zero and the energy of distortion is $(1/2G)\tau_{xy}^2$. If it is true that whatever the stress system, failure occurs when the strain energy of distortion reaches a certain limit (characteristic of the material), the ratio between the critical value for tensile stress alone and for shearing stress alone is found from the equation

$$\frac{1}{2G}\tau_{xy}^2 = \frac{1+\nu}{3E}\sigma_x^2$$

from which

$$\tau_{xy} = \frac{1}{\sqrt{3}}\sigma_x = 0.557\sigma_x \qquad (j)$$

Experiments with steel show[1] that the ratio between the yield point in tension and the yield point in shear is in very good agreement with that given by (j). Saint-Venant's principle (see page 40) can, by consideration of strain energy, be connected with conservation of energy.[2] The principle is equivalent to the statement that a self-equilibrating distribution of force on a small part of an elastic solid produces only local stress.

Such a distribution of force does work during its application only because there is deformation of the loaded region. Let one surface element of this region be fixed in position and orientation. If p denotes the order of magnitude (e.g., average) of the force per unit area and a a representative linear dimension (e.g., diameter) of the loaded part, the strain components are of order p/E and the relative displacements within the loaded part are of order pa/E. The work done is of order $pa^2(pa/E)$ or p^2a^3/E.

On the other hand, stress components of order p imply strain energy of order p^2/E per unit volume. The work done is therefore sufficient only for a volume of order a^3, in accordance with the statement of the principle.

It has been supposed here that the body obeys Hooke's law and is of solid form. The former restriction may be dispensed with, E in the above argument then denoting merely the order of magnitude of the slopes of the stress-strain curves of the material. If the body is not a solid form, as for instance a beam with a very thin web, or a thin cylindrical shell, a self-equilibrating distribution of force on one end may make itself felt at distances many times the depth or diameter.[3]

The above argument can be repeated without change for a load with nonzero resultant so long as there is a fixed surface element within or near the loaded part.

[1] See the papers by W. Lode, *Z. Physik*, vol. 36, p. 913, 1926, and *Forschungsarbeiten*, no. 303, Berlin, 1928.

[2] J. N. Goodier, *Phil. Mag.*, ser. 7, vol. 24, p. 325, 1937; *J. Appl. Phys.*, vol. 13, p. 167, 1942.

[3] V. Z. Vlasov, "Thin Walled Elastic Bars," Moscow, 1940; J. N. Goodier and M. V. Barton, *J. Appl. Mech.*, vol. 11, p. A-35, 1944; N. J. Hoff, *J. Aeron. Sci.*, vol. 12, p. 455, 1945. L. H. Donnell, *J. Appl. Mech.*, vol. 29, pp. 792–793, 1962.

Thus, if a deformable material is bonded to a rigid one, pressure applied to a small part of the former near the attachment will produce only local stress.[1]

91 | Strain Energy of an Edge Dislocation

In Art. 36 it was shown that the edge dislocation with displacement δ indicated in Fig. 48b would require a pair of forces P. The relation, between P and δ, given by (b) of Art. 36 with (g) of Art. 35, is

$$P = \frac{N}{a^2 + b^2} \frac{E}{4\pi} \delta \tag{a}$$

where
$$N = a^2 - b^2 + (a^2 + b^2) \log \frac{b}{a} \tag{b}$$

The total strain energy of the ring is equal to the work done by the pair of forces P during the loading process. This is, using (a),

$$V = \frac{1}{2} P\delta = \frac{E}{8\pi} \frac{N}{a^2 + b^2} \delta^2 \tag{c}$$

per unit plate thickness.

This is for plane stress. In plane strain, for the same stress function ϕ, the strains ϵ_x, ϵ_y, γ_{xy}, and therefore the displacements u,v, differ from those of plane stress by the change of elastic constants explained in Art. 20. Thus, to convert to plane strain (with $\epsilon_z = 0$), we have to replace E in (e) by $E/(1 - \nu^2)$. Then in place of (c) we have, using (b),

$$V = \frac{\delta^2}{8\pi} \frac{E}{1 - \nu^2} \left(\log \frac{b}{a} - \frac{b^2 - a^2}{b^2 + a^2} \right) \tag{d}$$

for the strain energy per unit axial length (along Oz). This formula is in common use in materials science for the dislocation energy of a crystal.[2] Both a and b must be given finite values; otherwise the energy is infinite. The outer radius b is related to the overall dimensions of the crystal, the inner radius to the atomic spacing in the crystal lattice.

In the boundary-value problems of elasticity, the boundaries are normally *definite*. But the dislocation center in a crystal can move through the crystal—as though the inner-boundary circle $r = a$ could be translated while the outer-boundary circle $r = b$ does not move. If there are two dislocations present simultaneously, one positive (e.g., δ positive) and one negative (e.g., δ negative), there is a net total strain energy so long as their centers are separate. If they coincide, the two dislocations annul one another. There is then no stress or strain and no energy. Evidently

[1] J. N. Goodier, *J. Appl. Phys.*, *loc. cit.*

[2] See for instance, A. H. Cottrell, "Dislocations and Plastic Flow in Crystals," p. 38, Oxford University Press, Fair Lawn, N.J., 1953.

the approach of the two must reduce the total strain energy. Since under the present circumstances this energy represents the whole potential energy of the system, the two centers will appear to attract[1] one another, and at coalescence the energy has been converted to another form—for instance, wave motion throughout the crystal.

92 | Principle of Virtual Work

In the solution of problems of elasticity it is sometimes advantageous to use the *principle of virtual work*. In the case of a particle, this principle states that if a particle is in equilibrium, the total work of all forces acting on the particle in any *virtual displacement* vanishes.

If δu, δv, δw are components of a virtual displacement in the x, y, and z directions and ΣX, ΣY, ΣZ are the sums of projections on the same directions of forces, acting on the particle, the principle of virtual work gives

$$\delta u\,\Sigma X = 0 \qquad \delta v\,\Sigma Y = 0 \qquad \delta w\,\Sigma Z = 0 \tag{a}$$

These equations are satisfied for any virtual displacement if

$$\Sigma X = 0 \qquad \Sigma Y = 0 \qquad \Sigma Z = 0 \tag{b}$$

Conversely, given Eqs. (*b*), we can multiply them by arbitrary δu, δv, δw and obtain (*a*). In fact, *virtual displacement* is only a name for such arbitrary multipliers. The forces remain exactly as they were.

An elastic body at rest, with its surface and body forces, constitutes a system of particles on each of which acts a set of forces in equilibrium. In any virtual displacement the total work done by the forces on any particle vanishes, and therefore the total work done by all the forces of the system vanishes.

A virtual displacement in the case of an elastic body may be taken as any small[2] displacement compatible with the condition of continuity of the material and with the conditions for the displacements at the surface of the body, if such conditions are prescribed. If it is given, for instance, that a certain portion of the surface of the body, say a built-in end of a beam, is immovable or has a given displacement, the virtual displacement for this portion may be taken as zero.

Denote by u, v, w the components of the actual displacement due to the loads and by δu, δv, δw the components of a virtual displacement. These latter are arbitrary continuous functions of x, y, z, small in absolute value.

[1] G. I. Taylor, *Proc. Roy. Soc.* (*London*), *Ser. A*, vol. 134, pp. 362–387, 1934.

[2] Smallness (in comparison with actual displacements) is a matter of convenience.

The virtual displacements δu, δv, δw correspond to increments of the six strain components, indicated by

$$\delta\epsilon_x = \frac{\partial}{\partial x}\,\delta u, \; . \; . \; . \; , \; \delta\gamma_{xy} = \frac{\partial}{\partial x}\,\delta v + \frac{\partial}{\partial y}\,\delta u, \; . \; . \; . \qquad (c)$$

and the associated virtual work for a volume element is

$$(\sigma_x\,\delta\epsilon_x + \sigma_y\,\delta\epsilon_y + \sigma_z\,\delta\epsilon_z + \tau_{xy}\,\delta\gamma_{xy} + \tau_{yz}\,\delta\gamma_{yz} + \tau_{zx}\,\delta\gamma_{zx})\,dz\,dy\,dz \qquad (d)$$

This, by (*g*) on page 246, is the same as

$$\delta V_0\,dx\,dy\,dz \qquad (e)$$

V_0 being taken as a function of the strain components as in (132).

As already stated, this change in strain energy measures the work done *against* the mutual forces between the particles (as in stretching springs). To get the work done *by* the mutual forces *on* the particles, the sign must be reversed.

The external forces consist of (1) boundary surface forces $\bar{X}\,dS$, $\bar{Y}\,dS$, $\bar{Z}\,dS$, on each surface element dS, and (2) body forces $X\,d\tau$, $Y\,d\tau$, $Z\,d\tau$ on each volume element $d\tau$, or $dx\,dy\,dz$.

The statement that the total virtual work for the whole body is zero now takes the form

$$\int(\bar{X}\,\delta u + \bar{Y}\,\delta v + \bar{Z}\,\delta w)\,dS + \int(X\,\delta u + Y\,\delta v + Z\,\delta w)\,d\tau - \int\delta V_0\,d\tau = 0 \qquad (137)$$

Since the given external forces and the actual stress components are held unchanged in forming (137), the variational symbol δ may be put before the integral signs. Then, with a change of sign throughout,

$$\delta\left[\int V_0\,d\tau - \int(Xu + Yv + Zw)\,d\tau - \int(\bar{X}u + \bar{Y}v + \bar{Z}w)\,dS\right] = 0 \qquad (137')$$

on the understanding that δ does not affect the forces written explicitly. The first integral in the bracket is the strain energy, and, as energy available on unloading, can be called potential energy of deformation. The second integral is the potential energy of the body forces (fixed at their actual values regardless of u, v, w), this energy being reckoned as zero when $u = v = w = 0$. Similarly, the third integral is the potential energy of the surface forces. The complete expression in brackets is (by definition) the total potential energy of the system. Then (137′) is the statement that the actual displacements u, v, w under the given external forces (and given mode of support) are such that the first-order variation of the total potential energy is zero for any virtual displacement—or, briefly, the total potential energy is *stationary*.

The terms *virtual displacement* and *virtual work*, although entrenched in historical usage, imply nothing more than the use of the arbitrary multipliers represented here by δu, δv, δw, with equations of equilibrium. It is convenient, as in the preceding paragraphs, to regard them as *variations* of the actual displacements u, v, w.

To consider *stability* of equilibrium, we may contemplate impulsive disturbances followed by actual variations of the equilibrium displacements. There being no dissipation of energy, the sum of the potential and kinetic energies remains constant. If on departing from the equilibrium configuration the potential energy necessarily increases, the kinetic energy must decrease. But if the potential energy necessarily decreases, the kinetic energy must increase. These two cases are described respectively as stable or unstable with respect to small disturbances. Stability evidently implies that the potential energy is a minimum in the equilibrium position, and instability that it is a maximum. In this use of potential energy it is assumed that in the motion following the disturbance (1) the body and surface forces go with the material elements on which they act in the equilibrium configuration and (2) they remain unchanged in magnitude and direction.

Consider again the strain energy per unit volume, for plane stress, in the form (134). Following the impulsive disturbance, the equilibrium strain components are supposed increased, after a short time, by $\delta\epsilon_x$, $\delta\epsilon_y$, $\delta\gamma_{xy}$. The complete new value of V_0 is then, from (134),

$$\frac{E}{2(1-\nu^2)}\left[(\epsilon_x+\delta\epsilon_x)^2+2\nu(\epsilon_x+\delta\epsilon_x)(\epsilon_y+\delta\epsilon_y)+(\epsilon_y+\delta\epsilon_y)^2\right] + \frac{G}{2}(\gamma_{xy}+\delta\gamma_{xy})^2$$

Subtracting the equilibrium value given directly by (134), we find for the complete increase

$$\frac{E}{2(1-\nu^2)}\left[2\epsilon_x\,\delta\epsilon_x+2\nu(\epsilon_x\,\delta\epsilon_y+\epsilon_y\,\delta\epsilon_x)+2\epsilon_y\,\delta\epsilon_y\right]+\frac{G}{2}2\gamma_{xy}\,\delta\gamma_{xy}$$
$$+\frac{E}{2(1-\nu^2)}\left[(\delta\epsilon_x)^2+2\nu\,\delta\epsilon_x\,\delta\epsilon_y+(\delta\epsilon_y)^2\right]+\frac{G}{2}(\delta\gamma_{xy})^2$$

The first line here represents the *first-order increment* and corresponds exactly to (e), except for the volume factor $dx\,dy$. The second line is the *second-order increment*, and it is positive since V_0 in (134) is positive for any values of ϵ_x, ϵ_y, γ_{xy}.

In the total potential energy represented by the terms in the brackets in (137′), under postulates 1 and 2 above, there is no second-order increment from the body and boundary forces. Its first-order increment vanishes, since the actual displacements δu, δv, δw in the disturbance can be taken as virtual displacements. Since the second-order increment is necessarily positive, we have *stability* in the sense defined. It will be seen that this conclusion depends on the use of Hooke's law[1] as well as on postulates 1 and 2. For nonlinear stress-strain relations the higher-order increments would not stop at the second.

General considerations of the total energy of a system were applied by A. A. Griffith in developing his theory of rupture of brittle materials.[2] It is known that materials always show a strength much smaller than might be expected from the molecular forces. For a certain glass Griffith found a theoretical strength in tension of the order of 1.6×10^6 psi, while tensile tests with glass rods gave only 26×10^3 psi. He showed that this discrepancy between theory and experiments can be explained if we assume that in such materials as glass there exist microscopic cracks or flaws producing high stress concentrations and consequent spreading of the cracks. For purposes of calculation Griffith takes a crack in the form of a very narrow elliptical hole, the major axis of which is perpendicular to the direction of the tensile force. Consider a plate fixed along the sides ab and cd, and stretched by uniformly distributed tensile stress S, acting along the same sides (Fig. 131). If a microscopic elliptical hole AB of length l is made in the plate, ab and cd remaining fixed, the initial strain

[1] In buckling theory the material may obey Hooke's law and yet a column or plate under a compressive load in excess of the Euler critical value will not have *stability* in the present sense. But buckling problems are excluded from the *linearized* theory of elasticity, by its assumptions of smallness. For instance, the boundary conditions for the problem of Fig. 37, for the vertical edges, are taken as $\sigma_x = \tau_{xy} = 0$ on $x = \pm l$. The exact boundary conditions would be that the *deformed* edges are free of normal and tangential loading.

[2] *Trans. Roy. Soc. (London), Ser. A*, vol. 221, pp. 163–198, 1921; and *Proc. Intern. Congr. Appl. Mech., Delft*, pp. 55–63, 1924.

For references to books and survey articles on the mechanics of fracture, see A. H. Cottrell, "The Mechanical Properties of Matter," chap. 11, John Wiley & Sons, Inc., New York, 1964; D. C. Drucker and J. J. Gilman (eds.), "Fracture in Solids," Academic Press Inc., New York, 1962; and the *Intern. J. Fracture Mech.* (started March, 1965).

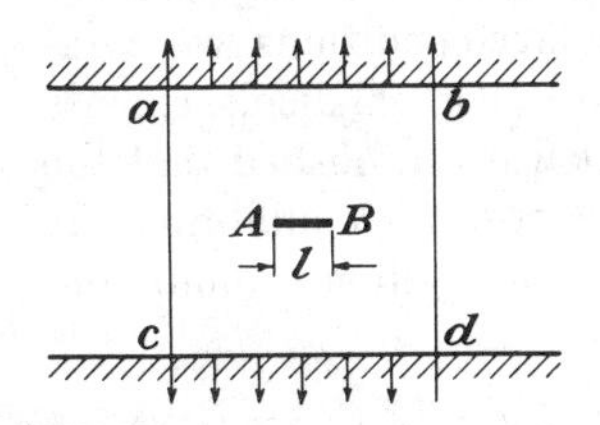

Fig. 131

energy due to the tensile stresses S will be reduced. This reduction can be calculated by using the solution for an elliptical hole,[1] and for a plate of unit thickness it is equal to

$$V = \frac{\pi l^2 S^2}{4E} \tag{f}$$

If the crack lengthens, there is a further reduction of strain energy stored in the plate. However, the lengthening of the crack means an increase of *surface energy*, since the surfaces of solids possess a surface tension just as liquids do. Griffith found, for instance, that for the kind of glass used in his experiments the surface energy T per unit surface area was of the order 3.12×10^{-3} in. lb per sq in. Now if the lengthening of the crack requires an increase of surface energy that can be supplied by the reduction of the strain energy, the lengthening can occur without increase of the total energy. The condition that the crack extends spontaneously is that these two quantities of energy are equal, or using (f),

$$\frac{dV}{dl}\,dl = \frac{\pi l S_{cr}^2}{2E}\,dl = 2dl\,T$$

from which

$$S_{cr} = \sqrt{\frac{4ET}{\pi l}} \tag{g}$$

Experiments in which cracks of known length were formed with a glass-cutter's diamond showed a very satisfactory agreement with Eq. (g). It was also shown experimentally that, if precautions are taken to eliminate microscopic cracks, a much higher strength than usual can be obtained. Some glass rods tested by Griffith showed an ultimate strength of the order of 900,000 psi, which is more than half the theoretical strength mentioned above.

The questionable aspect of the infinite stress at the ends of the crack in Griffith's theory has been removed by Barenblatt, who introduces instead large but finite stress to represent atomic cohesion.[2]

93 | Castigliano's Theorem

In the previous article the equilibrium configuration of an elastic body submitted to given body forces and given boundary conditions was compared with neighboring configurations arrived at by virtual displacements δu, δv, δw from the position of equilibrium. It was established that the true displacements corresponding to the position of stable equilibrium are those which make the total potential energy of the system a minimum.

[1] See p. 191.

[2] G. I. Barenblatt, "Advances in Applied Mechanics," vol. 7, pp. 55–129, Academic Press Inc., New York, 1962.

Let us consider now, instead of displacements, the stresses corresponding to the position of equilibrium. We know that the differential equations of equilibrium (123), together with the boundary conditions (124), are not sufficient for determining the stress components. We may find many different stress distributions satisfying the equations of equilibrium and the boundary conditions, and the question arises: What distinguishes the true stress distribution from all the other statically possible stress distributions?

Let σ_x, etc., be the true stress components corresponding to the position of equilibrium, and $\delta\sigma_x$, etc., small variations of these components such that the new stress components $\sigma_x + \delta\sigma_x$, etc., satisfy the same equations of equilibrium (123). Then, by subtracting the equations for one set from those of the other, we find that the changes in the stress components satisfy three equations of the type

$$\frac{\partial\,\delta\sigma_x}{\partial x} + \frac{\partial\,\delta\tau_{xy}}{\partial y} + \frac{\partial\,\delta\tau_{xz}}{\partial z} = 0 \qquad (a)$$

Corresponding to this variation of stress components there will be some variation in the surface forces. Let $\delta\bar{X}$, $\delta\bar{Y}$, and $\delta\bar{Z}$ be these small changes in boundary forces; then, from the boundary conditions (124), we find three of the type

$$\delta\sigma_x\, l + \delta\tau_{xy}\, m + \delta\tau_{xz}\, n = \delta\bar{X} \qquad (b)$$

Consider now the change in strain energy of the body due to the above changes in stress components. Taking the strain energy per unit volume as a function of the stress components (131), the change of this energy is

$$\delta V_0 = \frac{\partial V_0}{\partial\sigma_x}\delta\sigma_x + \cdots + \frac{\partial V_0}{\partial\tau_{xy}}\delta\tau_{xy} + \cdots \qquad (c)$$

the right-hand side indicating six terms, in which

$$\frac{\partial V_0}{\partial\sigma_x} = \frac{1}{E}[\sigma_x - \nu(\sigma_y + \sigma_z)] = \epsilon_x$$

etc., with

$$\frac{\partial V_0}{\partial\tau_{xy}} = \frac{1}{G}\tau_{xy} = \gamma_{xy}$$

etc., giving

$$\delta V_0 = \epsilon_x\,\delta\sigma_x + \cdots + \gamma_{xy}\,\delta\tau_{xy} + \cdots$$

and the total change in the strain energy due to changes of stress components is

$$\delta V = \int \delta V_0\, d\tau = \int (\epsilon_x\,\delta\sigma_x + \cdots + \gamma_{xy}\,\delta\tau_{xy} + \cdots)\, d\tau \qquad (d)$$

Let us examine this change in energy. To take into consideration the boundary conditions (b), we shall require the theorem usually known as the divergence theorem,[1] or Gauss' theorem, or Green's lemma. In a region bounded by a surface S, with l, m, n for direction cosines of the outward normal, we have three functions U, V, W of position. The theorem is

$$\int \left(\frac{\partial U}{\partial x} + \frac{\partial V}{\partial y} + \frac{\partial W}{\partial z} \right) d\tau = \int (lU + mV + nW)\, dS \tag{138}$$

the volume integral on the left being taken over the whole volume bounded by S, and the surface integral on the right over the whole bounding surface. S may also represent an outer surface together with one or more inner (cavity) surfaces. For the present purpose it is advantageous to choose first

$$U = u\, \delta\sigma_x \qquad V = u\, \delta\tau_{xy} \qquad W = u\, \delta\tau_{xz} \tag{e}$$

The theorem (138) then gives

$$\int \left[\frac{\partial}{\partial x} (u\, \delta\sigma_x) + \frac{\partial}{\partial y} (u\, \delta\tau_{xy}) + \frac{\partial}{\partial z} (u\, \delta\tau_{xz}) \right] d\tau = \int u(l\, \delta\sigma_x + m\, \delta\tau_{xy} + n\, \delta\tau_{xz})\, dS \tag{f}$$

Within the brackets on the left we may carry out the differentiations to obtain

$$u \left(\frac{\partial\, \delta\sigma_x}{\partial x} + \frac{\partial\, \delta\tau_{xy}}{\partial y} + \frac{\partial\, \delta\tau_{xz}}{\partial z} \right) + \frac{\partial u}{\partial x}\, \delta\sigma_x + \frac{\partial u}{\partial y}\, \delta\tau_{xy} + \frac{\partial u}{\partial z}\, \delta\tau_{xz} \tag{g}$$

The expression in parentheses vanishes by (a). Now (f) becomes

$$\int \left(\frac{\partial u}{\partial x}\, \delta\sigma_x + \frac{\partial u}{\partial y}\, \delta\tau_{xy} + \frac{\partial u}{\partial z}\, \delta\tau_{xz} \right) d\tau = \int u\, \delta\bar{X}\, dS \tag{h}$$

In the same way, by starting with the choice, in place of (e), of

$$V = v\, \delta\sigma_y \qquad W = v\, \delta\tau_{yz} \qquad U = v\, \delta\tau_{yx}$$

written in this order by cyclic interchange in (e), we obtain the result, written down by cyclic interchange in (h),

$$\int \left(\frac{\partial v}{\partial y}\, \delta\sigma_y + \frac{\partial v}{\partial z}\, \delta\tau_{yz} + \frac{\partial v}{\partial x}\, \delta\tau_{yx} \right) d\tau = \int v\, \delta\bar{Y}\, dS \tag{i}$$

Further, by one more cyclic interchange, we have

$$\int \left(\frac{\partial w}{\partial z}\, \delta\sigma_z + \frac{\partial w}{\partial x}\, \delta\tau_{zx} + \frac{\partial w}{\partial y}\, \delta\tau_{zy} \right) d\tau = \int w\, \delta\bar{Z}\, dS \tag{j}$$

[1] Proofs and conditions of validity are given in almost all books on advanced calculus, or vector field analysis. See for instance I. S. Sokolnikoff and R. M. Redheffer, "Mathematics of Physics and Modern Engineering," p. 389, McGraw-Hill Book Company, New York, 1958.

By addition of (*h*), (*i*), and (*j*) and use of the strain displacement relations (2), we find

$$\int(\epsilon_x\,\delta\sigma_x + \epsilon_y\,\delta\sigma_y + \epsilon_z\,\delta\sigma_z + \gamma_{xy}\,\delta\tau_{xy} + \gamma_{yz}\,\delta\tau_{yz} + \gamma_{zx}\,\delta\tau_{zx})\,d\tau = \int(u\,\delta\bar{X} + v\,\delta\bar{Y} + w\,\delta\bar{Z})\,dS \qquad (k)$$

The left-hand side of this is δV, as in (*d*). Thus the variation of strain energy in the form (131), corresponding to variations of the stress components which preserve equilibrium, is given by

$$\delta V = \int(u\,\delta\bar{X} + v\,\delta\bar{Y} + w\,\delta\bar{Z})\,dS \qquad (139)$$

The true stresses are those which satisfy this equation. Such variations are mathematical not physical. Physical stress variations induced by variations of boundary loading are subject to more restrictions than those of equilibrium expressed by (*a*). But mathematically the integral in (136), with V_0 as a function of six variables, the six stress components in (131), has a variation whenever these six variables change, no matter how.

In theory of structures, the strain energy of a linearly elastic structure under a set of concentrated forces $P_1, P_2, \ldots,$ can be expressed as a quadratic function V of these forces. Then

$$\delta V = \frac{\partial V}{\partial P_1}\,\delta P_1 + \frac{\partial V}{\partial P_2}\,\delta P_2 + \cdots$$

and we have Castigliano's theorem[1] for the corresponding displacement components $d_1, d_2 \ldots$

$$d_1 = \frac{\partial V}{\partial P_1} \qquad d_2 = \frac{\partial V}{\partial P_2} \quad \cdots$$

from a demonstration that

$$\delta V = d_1\,\delta P_1 + d_2\,\delta P_2 + \cdots \qquad (140)$$

The analogy between (140) and (139) will be evident. The theorem (140) is also referred to as Castigliano's theorem.

Returning to (139), we observe that the stress variations may be such that they leave the boundary forces $\bar{X}, \bar{Y}, \bar{Z}$ unchanged. Then $\delta\bar{X}, \delta\bar{Y}, \delta\bar{Z}$ in the three conditions of the type (*b*) are zero, and (139) becomes simply

$$\delta V = 0 \qquad (141)$$

Thus for such variations V is *stationary*. We have considered [starting with (*c*)] only the first-order increments or variations. From considera-

[1] See, for instance, S. Timoshenko and D. H. Young, "Theory of Structures," p. 234, McGraw-Hill Book Company, New York, 1965.

tion of the second-order increments, it can be shown that V is in fact a minimum. The theorem (141) is sometimes referred to as the *principle of least work*, like its counterpart for concentrated forces in theory of structures.

For plane strain or plane stress we have $w = 0$ or $\delta\bar{Z} = 0$ and (139) reduces at once to

$$\delta V = \int (u\,\delta\bar{X} + v\,\delta\bar{Y})\,ds \tag{142}$$

where V is taken in the appropriate form, e.g., (133) for plane stress, and the integral is, for unit thickness, a line integral round the bounding curve with arc element ds.

More general variational principles have been formulated, both displacement and stress being subject to variation.[1]

94 | Applications of the Principle of Least Work—Rectangular Plates

As an example let us consider a rectangular plate. Previously (page 53) it has been shown that by using trigonometric series the conditions on two sides of a rectangular plate can be satisfied. Solutions obtained in this way may be of practical interest when applied to a plate whose width is small in comparison with its length. If both dimensions of a plate are of the same order, the conditions on all four sides must be considered. In the solution of problems of this kind, the principle of minimum energy can sometimes be successfully applied.

Let us consider the case of a rectangular plate in tension, when the tensile forces at the ends are distributed according to a parabolic law[2] (Fig. 132). The boundary conditions in this case are:

For $x = \pm a$,

$$\tau_{xy} = 0 \qquad \sigma_x = S\left(1 - \frac{y^2}{b^2}\right)$$

For $y = \pm b$,

$$\tau_{xy} = 0 \qquad \sigma_y = 0 \tag{a}$$

[1] E. Reissner, "On Some Variational Theorems in Elasticity," pp. 370–381 in "Some Problems of Continuum Mechanics," N. I. Muskhelishvili 70th Anniversary Volume, Society for Industrial and Applied Mathematics, Philadelphia, 1961.

[2] See S. Timoshenko, *Phil. Mag.*, vol. 47, p. 1095, 1924.

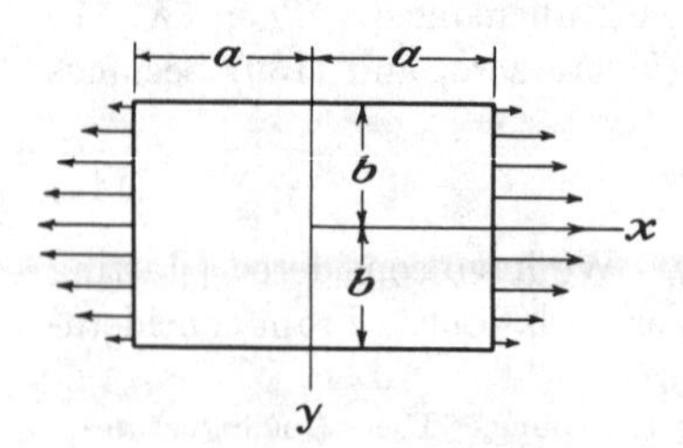

Fig. 132

The strain energy for a plate of unit thickness is, from Eq. (133),

$$V = \frac{1}{2E} \iint [\sigma_x^2 + \sigma_y^2 - 2\nu\sigma_x\sigma_y + 2(1+\nu)\tau_{xy}^2]\, dx\, dy \tag{b}$$

It should be noted that for a simply connected boundary, such as we have in the present case, the stress distribution does not depend on the elastic constants of the material (see page 136) and further calculations can therefore be simplified by taking Poisson's ratio ν as zero. Then, introducing the stress function ϕ, and substituting in (*b*)

$$\sigma_x = \frac{\partial^2 \phi}{\partial y^2} \qquad \sigma_y = \frac{\partial^2 \phi}{\partial x^2} \qquad \tau_{xy} = -\frac{\partial^2 \phi}{\partial x\, \partial y} \qquad \nu = 0$$

we find

$$V = \frac{1}{2E} \iint \left[\left(\frac{\partial^2 \phi}{\partial y^2} \right)^2 + \left(\frac{\partial^2 \phi}{\partial x^2} \right)^2 + 2 \left(\frac{\partial^2 \phi}{\partial x\, \partial y} \right)^2 \right] dx\, dy \tag{c}$$

The correct expression for the stress function is that satisfying conditions (*a*) and making the strain energy (*c*) a minimum.

If we apply variational calculus to determine the minimum of (*c*), we shall arrive at Eq. (30) for the stress function ϕ. Instead of this we shall use the following procedure[1] for an approximate solution of the problem. We take the stress function in the form of a series,

$$\phi = \phi_0 + \alpha_1\phi_1 + \alpha_2\phi_2 + \alpha_3\phi_3 + \cdots \tag{d}$$

such that the boundary conditions (*a*) are satisfied, $\alpha_1, \alpha_2, \alpha_3, \ldots$ being constants to be determined later. Substituting this series in expression (*c*), we find V as a function of the second degree in $\alpha_1, \alpha_2, \alpha_3, \ldots$. The magnitude of the constants can then be calculated from the conditions

$$\frac{\partial V}{\partial \alpha_1} = 0 \qquad \frac{\partial V}{\partial \alpha_2} = 0 \qquad \frac{\partial V}{\partial \alpha_3} = 0 \qquad \cdots \tag{e}$$

which will be linear equations in $\alpha_1, \alpha_2, \alpha_3 \ldots$.

By a suitable choice of the functions $\phi_1, \phi_2, \ldots$ we can usually get a satisfactory approximate solution by using only a few terms in the series (*d*). In our case the boundary conditions (*a*) are satisfied by taking

$$\phi_0 = \frac{1}{2} S y^2 \left(1 - \frac{1}{6} \frac{y^2}{b^2} \right)$$

since this gives

$$\sigma_y = \frac{\partial^2 \phi_0}{\partial x^2} = 0 \qquad \tau_{xy} = -\frac{\partial^2 \phi_0}{\partial x\, \partial y} = 0 \qquad \sigma_x = \frac{\partial^2 \phi_0}{\partial y^2} = S\left(1 - \frac{y^2}{b^2} \right)$$

The remaining functions $\phi_1, \phi_2, \ldots$ must be chosen so that the stresses corresponding to them vanish at the boundary. To ensure this we take the expression $(x^2 - a^2)^2(y^2 - b^2)^2$ as a factor in all these functions; the second derivative of this expression with respect to x vanishes at the sides $y = \pm b$, and the second derivative with respect to y vanishes at the sides $x = \pm a$; the second derivative $\partial^2/\partial x\, \partial y$ vanishes on all four

[1] The Ritz, or Rayleigh-Ritz, method. See W. Ritz, *J. Reine Angew. Math.*, vol. 135, pp. 1–61, 1908; or W. Ritz, "Gesammelte Werke," Gauthier-Villars, Paris, pp. 192–250, 1911.

sides of the plate. The stress function can then be taken as

$$\phi = \frac{1}{2} Sy^2 \left(1 - \frac{1}{6}\frac{y^2}{b^2}\right) + (x^2 - a^2)^2(y^2 - b^2)^2(\alpha_1 + \alpha_2 x^2 + \alpha_3 y^2 + \cdots) \qquad (f)$$

Only even powers of x and y are taken in the series because the stress distribution is symmetrical with respect to the x and y axes. Limiting ourselves to the first term α_1 in parentheses, we have

$$\phi = \frac{1}{2} Sy^2 \left(1 - \frac{1}{6}\frac{y^2}{b^2}\right) + \alpha_1 (x^2 - a^2)^2(y^2 - b^2)^2$$

The first of Eqs. (*e*) then becomes

$$\alpha_1 \left(\frac{64}{7} + \frac{256}{49}\frac{b^2}{a^2} + \frac{64}{7}\frac{b^4}{a^4}\right) = \frac{S}{a^4 b^2}$$

For a square plate ($a = b$) we find

$$\alpha_1 = 0.04253 \frac{S}{a^6}$$

and the stress components are

$$\begin{aligned}
\sigma_x &= S\left(1 - \frac{y^2}{a^2}\right) - 0.1702S\left(1 - \frac{3y^2}{a^2}\right)\left(1 - \frac{x^2}{a^2}\right)^2 \\
\sigma_y &= -0.1702S\left(1 - \frac{3x^2}{a^2}\right)\left(1 - \frac{y^2}{a^2}\right)^2 \\
\tau_{xy} &= -0.6805S\frac{xy}{a^2}\left(1 - \frac{x^2}{a^2}\right)\left(1 - \frac{y^2}{a^2}\right)
\end{aligned}$$

The distribution of σ_x on the cross section $x = 0$ is represented by curve II[1] (Fig. 133)

To obtain a closer approximation, we now take three terms in the series (*f*). Then

[1] Curve I represents the parabolic stress distribution at the ends of the plate.

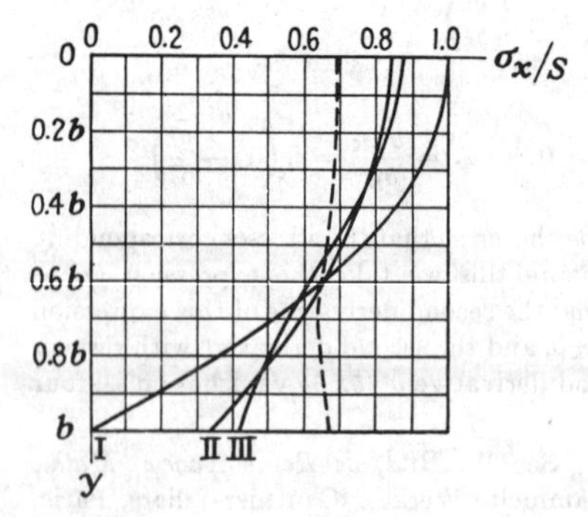

Fig. 133

Eqs. (e), for calculating the constants α_1, α_2, α_3, are

$$\alpha_1\left(\frac{64}{7}+\frac{256}{49}\frac{b^2}{a^2}+\frac{64}{7}\frac{b^4}{a^4}\right)+\alpha_2a^2\left(\frac{64}{77}+\frac{64}{49}\frac{b^4}{a^4}\right)+\alpha_3a^2\left(\frac{64}{49}\frac{b^2}{a^2}+\frac{64}{77}\frac{b^6}{a^6}\right)=\frac{S}{a^4b^2}$$

$$\alpha_1\left(\frac{64}{11}+\frac{64}{7}\frac{b^4}{a^4}\right)+\alpha_2a^2\left(\frac{192}{143}+\frac{256}{77}\frac{b^2}{a^2}+\frac{192}{7}\frac{b^4}{a^4}\right)+\alpha_3a^2\left(\frac{64}{77}\frac{b^2}{a^2}+\frac{64}{77}\frac{b^6}{a^6}\right)=\frac{S}{a^4b^2} \qquad (g)$$

$$\alpha_1\left(\frac{64}{7}+\frac{64}{11}\frac{b^4}{a^4}\right)+\alpha_2a^2\left(\frac{64}{77}+\frac{64}{77}\frac{b^4}{a^4}\right)+\alpha_3a^2\left(\frac{192}{7}\frac{b^2}{a^2}+\frac{256}{77}\frac{b^4}{a^4}+\frac{192}{143}\frac{b^6}{a^6}\right)=\frac{S}{a^4b^2}$$

For a square plate these give

$$\alpha_1=0.04040\frac{S}{a^6}\qquad \alpha_2=\alpha_3=0.01174\frac{S}{a^8}$$

The distribution of σ_x on the cross section $x = 0$ is given by

$$(\sigma_x)_{x=0}=S\left(1-\frac{y^2}{a^2}\right)-0.1616S\left(1-3\frac{y^2}{a^2}\right)+0.0235\left(1-12\frac{y^2}{a^2}+15\frac{y^4}{a^4}\right)$$

In Fig. 133 this stress distribution is shown by the curve III.[1]

As the length of the plate increases, the stress distribution over the cross section $x = 0$ becomes more and more uniform. If we take for instance $a = 2b$, we find, from Eqs. (g),

$$\alpha_1=0.07983\frac{S}{a^4b^2}\qquad \alpha_2=0.1250\frac{S}{a^6b^2}\qquad \alpha_3=0.01826\frac{S}{a^6b^2}$$

The corresponding values of σ_x over the cross section $x = 0$ are given below:

$\frac{y}{b}$	$= 0$	0.2	0.4	0.6	0.8	1.0
σ_x	$= 0.690S$	$0.684S$	$0.669S$	$0.653S$	$0.649S$	$0.675S$

This distribution is represented in Fig. 133 by the dotted line. We see that in this case the deviation from the average stress, $\frac{2}{3}S$, is very small.

To deal with other symmetrical distributions of forces over the edges $x = \pm a$ we have only to change the form of the function ϕ_0 in expression (f). Only the right-hand expressions in Eqs. (g) have to be changed.

As an example of stress distribution nonsymmetrical with respect to the x axis, let us consider the case of bending shown in Fig. 134[2] in which the forces applied at the ends are $(\sigma_x)_{x=\pm a} = Ay^3$ (curve b in Fig. 134b). Clearly, the stress system will be odd with respect to the x axis and even with respect to the y axis. These conditions are satisfied by taking a stress function in the form

$$\phi=\tfrac{1}{20}Ay^5+(x^2-a^2)^2(y^2-b^2)^2(\alpha_1y+\alpha_2yx^2+\alpha_3y^3+\alpha_4x^2y^3+\cdots) \qquad (h)$$

The first term, as before, satisfies the boundary conditions for ϕ. Using Eq. (h) with

[1] Similar results were obtained by C. E. Inglis, *Proc. Roy. Soc. (London), Ser. A*, vol. 103, 1923; and by G. Pickett, *J. Appl. Mech.*, vol. 11, p. 176, 1944.

[2] These calculations are taken from J. N. Goodier's doctoral thesis, Michigan University, 1931. See also *Trans. ASME*, vol. 54, p. 173, 1932.

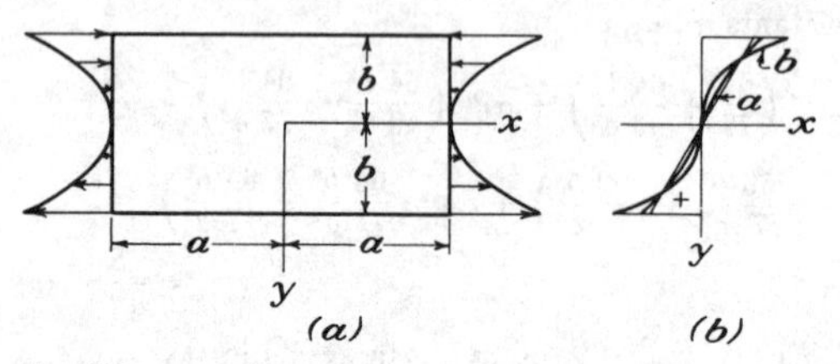

Fig. 134

four coefficients $\alpha_1, \ldots, \alpha_4$ in Eqs. (*e*), we find for a square plate ($a = b$)

$$
\begin{aligned}
\sigma_x = \frac{\partial^2 \phi}{\partial y^2} = 2Aa^3 \bigg\{ \frac{1}{2}\eta^3 - (1 - \xi^2)^2[0.08392(5\eta^3 - 3\eta) \\
+ 0.004108(21\eta^5 - 20\eta^3 + 3\eta)] - \xi^2(1 - \xi^2)^2[0.07308(5\eta^3 - 3\eta) \\
+ 0.04179(21\eta^5 - 20\eta^3 + 3\eta)] \bigg\}
\end{aligned}
\qquad (k)
$$

where $\xi = x/a$ and $\eta = y/b$. The distribution at the middle cross section $x = 0$ is not far from being linear. It is shown in Fig. 134*b* by curve *a*.

95 | Effective Width of Wide Beam Flanges

As another example of the application of the minimum-energy principle to two-dimensional problems of rectangles, let us consider a beam with very wide flanges (Fig. 135). Such beams are encountered very often in reinforced concrete structures and in the structures of hulls of ships. The elementary theory of bending assumes that the bending stresses are proportional to the distance from the neutral axis, i.e., that the stresses do not change along the width of the flange. But if this width is very large it is known that parts of the flanges at a distance from the web do not take their full share in resisting bending moment, and the beam is weaker than the elementary theory of bending indicates. It is the usual practice in calculating stresses in such beams to replace the actual width of the flanges by a certain reduced width, such that the elementary theory of bending applied to such a transformed beam cross section gives the correct value of maximum bending stress. This reduced width of flange is called the *effective width*. In the following discussion a theoretical basis for determining the effective width is given.[1]

To make the problem as simple as possible it is assumed that we have an infinitely long continuous beam on equidistant supports. All spans are equally loaded by loads symmetrical with respect to the middle of the spans. One of the supports of the span shown in Fig. 135 is taken as the origin of coordinates, with the x axis in the direction of the axis of the beam. Due to symmetry, only one span and one half of the flange, say that corresponding to positive y, need be considered. The

[1] The subject was investigated by T. v. Kármán; see "Festschrift August Föppls," p. 114, 1923. Also G. Schnadel, *Werft und Reederei*, vol. 9, p. 92, 1928; E. Reissner, *Der Stahlbau*, 1934, p. 206; E. Chwalla, *Der Stahlbau*, 1936; L. Beschkine, *Publs. Intern. Assoc. Bridge Structural Eng.*, vol. 5, p. 65, 1938. Further considerations and references may be found in Thein Wah (ed.), "A Guide for the Analysis of Ship Structures," pp. 370–391, Office of Technical Services, U.S. Department of Commerce, Washington D.C., 1960.

width of the flange is assumed infinitely large and its thickness h very small in comparison with the depth of the beam. Bending of the flange as a thin plate can then be neglected, and it can be assumed that during bending of the beam the forces are transmitted to the flange in its middle plane so that the stress distribution in the flange presents a two-dimensional problem. The corresponding stress function ϕ, satisfying the differential equation

$$\frac{\partial^4 \phi}{\partial x^4} + 2 \frac{\partial^4 \phi}{\partial x^2 \, \partial y^2} + \frac{\partial^4 \phi}{\partial y^4} = 0 \tag{a}$$

can be taken for our symmetrical case in the form of the series

$$\phi = \sum_{n=1}^{n=\infty} f_n(y) \cos \frac{n\pi x}{l} \tag{b}$$

in which $f_n(y)$ are functions of y only. Substituting in Eq. (a), we find the following expression for $f_n(y)$:

$$f_n(y) = A_n e^{-n\pi y/l} + B_n \left(1 + \frac{n\pi y}{l}\right) e^{-n\pi y/l} + C_n e^{n\pi y/l} + D_n \left(1 + \frac{n\pi y}{l}\right) e^{n\pi y/l} \quad \ldots \tag{c}$$

To satisfy the condition that stresses must vanish for an infinite value of y, we take $C_n = D_n = 0$. The expression for the stress function is then

$$\phi = \sum_{n=1}^{\infty} \left[A_n e^{-n\pi y/l} + B_n \left(1 + \frac{n\pi y}{l}\right) e^{-n\pi y/l} \right] \cos \frac{n\pi x}{l} \tag{d}$$

The coefficients A_n and B_n will now be determined from the condition that the true stress distribution is that making the strain energy of the flange together with that

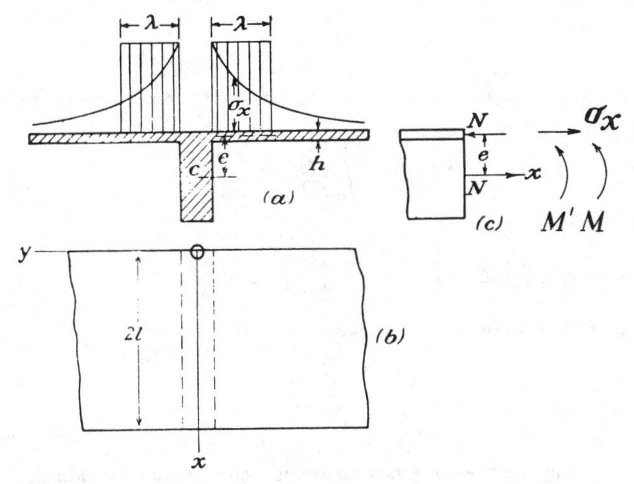

Fig. 135

of the web a minimum. Substituting

$$\sigma_x = \frac{\partial^2 \phi}{\partial y^2} \qquad \sigma_y = \frac{\partial^2 \phi}{\partial x^2} \qquad \tau_{xy} = -\frac{\partial^2 \phi}{\partial x\, \partial y}$$

in the expression for strain energy

$$V_1 = 2\frac{h}{2E}\int_0^\infty \int_0^{2l} [\sigma_x{}^2 + \sigma_y{}^2 - 2\nu\sigma_x\sigma_y + 2(1+\nu)\tau_{xy}{}^2]\, dx\, dy$$

and using Eq. (*d*) for the stress function, the strain energy of the flange is[1]

$$V_1 = 2h \sum_{n=1}^{\infty} \frac{n^3\pi^3}{l^2}\left(\frac{B_n{}^2}{E} + \frac{A_nB_n}{2G} + \frac{A_n{}^2}{2G}\right) \tag{e}$$

In considering the strain energy of the web alone, let A be its cross-sectional area, I its moment of inertia about the horizontal axis through the centroid C, and e the distance from the centroid of the web to the middle plane of the flange (Fig. 135). The total bending moment transmitted at any cross section by the web together with the flange can be represented for our symmetrical case by the series

$$M = M_0 + M_1 \cos\frac{\pi x}{l} + M_2 \cos\frac{2\pi x}{l} + \cdots \tag{f}$$

In this series M_0 is a statically indeterminate quantity depending on the magnitude of the bending moment at the supports, and the other coefficients $M_1, M_2, \ldots$ are to be calculated from the conditions of loading. Letting N denote the compressive force in the flange (Fig. 135*c*), the bending moment M can be divided into two parts: a part M' taken by the web and a part M'', equal to Ne, due to the longitudinal forces N in the web and flange. From statics the normal stresses over any cross section of the complete beam give a couple M; hence

$$\begin{aligned} N + 2h\int_0^\infty \sigma_x\, dy &= 0 \\ M' - 2he\int_0^\infty \sigma_x\, dy &= M \end{aligned} \tag{g}$$

where $-2he\int_0^\infty \sigma_x\, dy = M''$ is the part of the bending moment taken by the flange. The strain energy of the web is

$$V_2 = \int_0^{2l} \frac{N^2\, dx}{2AE} + \int_0^{2l} \frac{M'^2\, dx}{2EI} \tag{h}$$

From the first of Eqs. (*g*) we find

$$N = -2h\int_0^\infty \sigma_x\, dy = -2h\int_0^\infty \frac{\partial^2\phi}{\partial y^2}\, dy = 2h\left|\frac{\partial \phi}{\partial y}\right|_\infty^0$$

From expression (*d*) for the stress function it may be seen that

$$\left(\frac{\partial\phi}{\partial y}\right)_{y=\infty} = 0 \qquad \left(\frac{\partial\phi}{\partial y}\right)_{y=0} = \sum_{n=1}^{\infty} \frac{n\pi}{l} A_n \cos\frac{n\pi x}{l}$$

[1] The integrals entering into the expression for strain energy are calculated in the paper by Kármán, *loc. cit.*

Hence

$$N = 2h \sum_{n=1}^{\infty} \frac{n\pi}{l} A_n \cos \frac{n\pi x}{l}$$

$$M' = M + 2he \int_0^\infty \sigma_x \, dy = M - Ne = M - 2he \sum_{n=1}^{\infty} \frac{n\pi}{l} A_n \cos \frac{n\pi x}{l}$$

or, using the notation

$$2h \frac{n\pi}{l} A_n = X_n$$

we may write

$$\begin{aligned} N &= \sum_{n=1}^{\infty} X_n \cos \frac{n\pi x}{l} \\ M' &= M - e \sum_{n=1}^{\infty} X_n \cos \frac{n\pi x}{l} = M_0 + \sum_{n=1}^{\infty} (M_n - eX_n) \cos \frac{n\pi x}{l} \end{aligned} \tag{k}$$

Substituting in (h) and noting that

$$\int_0^{2l} \cos^2 \frac{n\pi x}{l} \, dx = l \qquad \int_0^{2l} \cos \frac{n\pi x}{l} \cos \frac{m\pi x}{l} \, dx = 0 \qquad \text{when } m \neq n$$

we obtain

$$V_2 = \frac{l}{2AE} \sum_{n=1}^{\infty} X_n^2 + \frac{M_0^2 l}{EI} + \frac{l}{2EI} \sum_{n=1}^{\infty} (M_n - eX_n)^2$$

Adding this to the strain energy (e) of the flange, and introducing in this latter the notations

$$2h \frac{n\pi}{l} A_n = X_n \qquad 2h \frac{n\pi}{l} B_n = Y_n$$

we find the following expression for the total strain energy:

$$V = \frac{\pi}{2hE} \sum_{n=1}^{\infty} n[Y_n^2 + (1 + \nu)X_nY_n + (1 + \nu)X_n^2] + \frac{l}{2AE} \sum_{n=1}^{\infty} X_n^2 + \frac{M_0^2 l}{EI} + \frac{l}{2EI} \sum_{n=1}^{\infty} (M_n - eX_n)^2 \tag{l}$$

The quantities M_0, X_n, Y_n are to be determined from the minimum condition of the strain energy (l). It can be seen that M_0 appears only in the term $M_0^2 l/EI$, and from the minimum requirement for (l) it follows that $M_0 = 0$.

From the condition

$$\frac{\partial V}{\partial Y_n} = 0$$

it follows that

$$2Y_n + (1 + \nu)X_n = 0$$

$$Y_n = -\frac{1+\nu}{2} X_n$$

Substituting this and $M_0 = 0$ in Eq. (l) we get the following expression for the strain energy:

$$V = \frac{\pi}{2hE} \frac{3 + 2\nu - \nu^2}{4} \sum_{n=1}^{\infty} nX_n^2 + \frac{l}{2AE} \sum_{n=1}^{\infty} X_n^2 + \frac{l}{2EI} \sum_{n=1}^{\infty} (M_n - eX_n)^2 \qquad (m)$$

From the condition that X_n should make V a minimum it follows that

$$\frac{\partial V}{\partial X_n} = 0$$

from which we find

$$X_n = \frac{M_n}{e} \frac{1}{1 + (I/Ae^2) + (n\pi I/hle^2)[(3 + 2\nu - \nu^2)/4]} \qquad (n)$$

Let us consider a particular case when the bending-moment diagram is a simple cosine line, say $M = M_1 \cos (\pi x/l)$. Then, from Eq. (n),

$$X_1 = \frac{M_1}{e} \frac{1}{1 + (I/Ae^2) + (\pi I/he^2l)[(3 + 2\nu - \nu^2)/4]}$$

and, from Eq. (k), the moment due to the force N of the flange is

$$M'' = eN = eX_1 \cos \frac{\pi x}{l} = \frac{M}{1 + (I/Ae^2) + (\pi I/he^2l)[(3 + 2\nu - \nu^2)/4]} \qquad (p)$$

The distribution of the stress σ_x along the width of the flange can now be calculated from (d) by taking all coefficients A_n and B_n, except A_1 and B_1, equal to zero, and by putting (following our notations)

$$A_1 = \frac{lX_1}{2\pi h} \qquad B_1 = -\frac{1+\nu}{2} A_1 = -\frac{(1+\nu)lX_1}{4\pi h}$$

This distribution of σ_x is shown by the curves in Fig. 135a. The stress σ_x diminishes as the distance from the web increases.

Let us now determine a width 2λ of the flange (Fig. 135a), of a T beam, such that a uniform stress distribution over the cross section of the flange, shown by the shaded area, gives the moment M'' calculated above, Eq. (p). This will then be the effective width of the flange.

Denoting, as before, by M' and M'' the portions of the bending moment taken by the web and by the flange, by σ_c the stress at the centroid C of the web, and by σ_e the stress at the middle plane of the flange, we find, from the elementary theory of bending

$$\sigma_e = \sigma_c - \frac{M'e}{I} \qquad (q)$$

and, from the equations of statics,

$$2\lambda h\sigma_e + \sigma_c A = 0 \qquad (r)$$
$$-2\lambda h\sigma_e e = M''$$

The expressions for the two portions of the bending moment, from Eqs. (q) and (r), are

$$M' = -\frac{I}{e}(\sigma_e - \sigma_c) = -\frac{I}{e}\left(1 + \frac{2\lambda h}{A}\right)\sigma_e$$

$$M'' = -2\lambda he\sigma_e$$

The ratio of M'' to the total bending moment is

$$\frac{M''}{M' + M''} = \frac{2\lambda he\sigma_e}{2\lambda he\sigma_e + (I/e)[1 + (2\lambda h/A)]\sigma_e} = \frac{1}{1 + (I/Ae^2) + (I/2\lambda he^2)} \tag{s}$$

To make this ratio equal to the ratio M''/M obtained from the exact solution (p), we must take

$$\frac{I}{2\lambda he^2} = \frac{\pi I}{he^2 l}\frac{3 + 2\nu - \nu^2}{4}$$

From this we obtain the following expression for the *effective width* 2λ:

$$2\lambda = \frac{4l}{\pi(3 + 2\nu - \nu^2)}$$

Taking, for instance, $\nu = 0.3$, we find

$$2\lambda = 0.181(2l)$$

i.e., for the assumed bending-moment diagram the effective width of the flange is approximately 18 percent of the span.

In the case of a continuous beam with equal concentrated forces at the middle of the spans, the bending-moment diagram will be as shown in Fig. 136. Representing this bending-moment diagram by a Fourier series and using the general method developed above, we find that the effective width at the supports is

$$2\lambda = 0.85\frac{4l}{\pi(3 + 2\nu - \nu^2)}$$

i.e., somewhat less than it is for the case of a moment diagram in the form of a cosine line.

A problem of the same general nature as that discussed in Art. 94 occurs in stiffened thin-walled structures. Consider a *box beam*, Fig. 137, formed from two channels $ABFE$ and $DCGH$ to which are attached thin sheets $ABCD$ and $EFGH$, by riveting or welding along the edges. If the whole beam is built in at the left-hand end, and loaded as a cantilever by two forces P applied to the channels at the other end, the elementary bending theory will give a tensile bending stress in the sheet $ABCD$ uniform across any section parallel to BC. Actually, however, the sheet acquires its tensile stress from shear stresses on its edges communicated to it by the channels, as

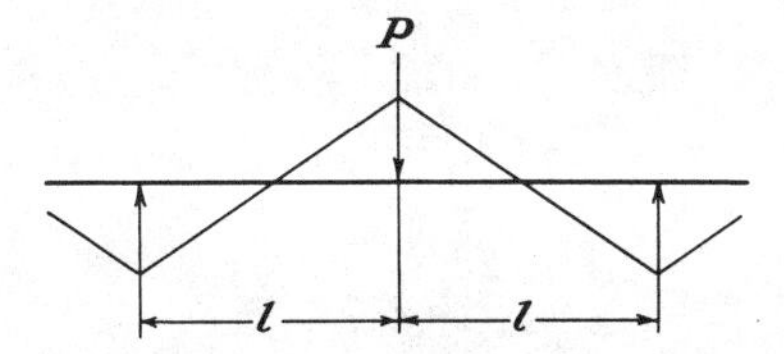

Fig. 136

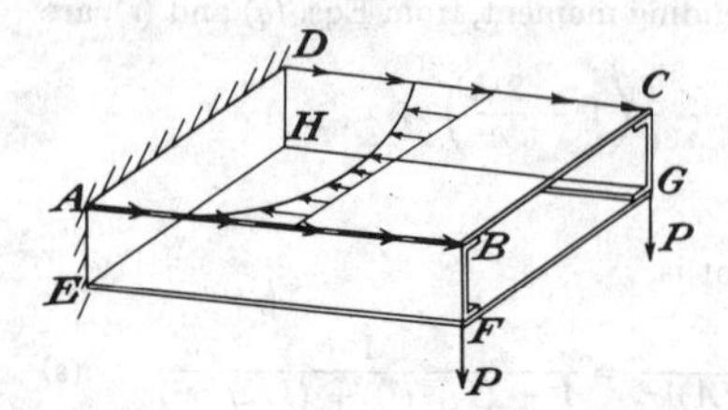

Fig. 137

indicated in Fig. 137, and the distribution of tensile stress across the width will not be uniform, but, as in Fig. 137, higher at the edges than at the middle. This departure from the uniformity assumed by the elementary theory is known as *shear lag*, since it involves a shear deformation in the sheets. The problem has been analyzed by strain-energy and other methods, with the help of simplifying assumptions.[1]

PROBLEMS

1. For what reason should we expect that the isothermal Young's modulus of any of the common metals will be *smaller* than the adiabatic modulus?
2. Find an expression in terms of σ_x, σ_y, τ_{xy} for the strain energy V per unit thickness of a cylinder or prism in plane strain ($\epsilon_z = 0$).
3. Write down the integral for the strain energy V in terms of polar coordinates and polar stress components for the case of plane stress [cf. Eq. (*b*), Art. 93].

   The stress distribution given by Eqs. (79) solves the problem indicated in Fig. 138, a couple M being applied by uniform shear to the inside of a ring, and a balancing couple to the outside. Evaluate the strain energy in the ring, and by equating this to the work done during loading deduce the rotation of the outside circle when the ring is fixed at the inside (cf. Prob. 3, page 144).

[1] E. Reissner, *Quart. Appl. Math.*, vol. 4, p. 268, 1946; J. Hadji-Argyris, *Brit. Aeron. Res. Council Repts. Mem.* 2038, 1944; J. Hadji-Argyris and H. L. Cox, *Brit. Aeron. Res. Council Repts. Mem.* 1969, 1944. References to earlier investigations are given in these papers. See also n. 1, p. 262.

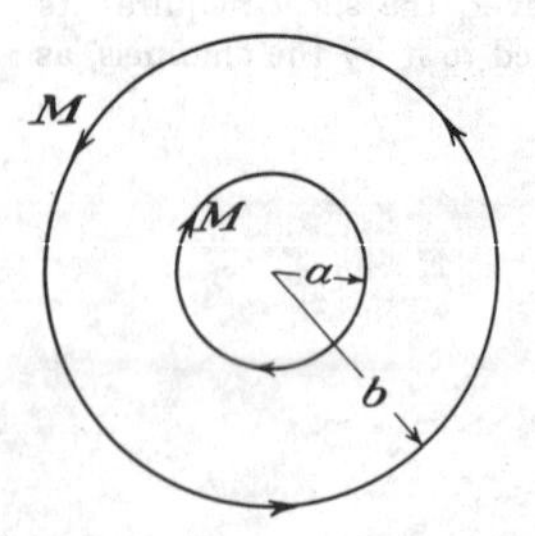

Fig. 138

4. Evaluate the strain energy per unit length of a cylinder $a < r < b$ subjected to internal pressure p_i. The ends of the cylinder are free ($\sigma_z = 0$).
5. Interpret the equation

$$\iint V_0 \, dx \, dy = \tfrac{1}{2}\iint (Xu + Yv) \, dx \, dy + \tfrac{1}{2}\int (\bar{X}u + \bar{Y}v) \, ds$$

and give the justification of the factors ½ on the right.
6. Show from Eq. (131) that if we have a case of plane stress and a corresponding case of plane strain ($\epsilon_z = 0$) in which the stresses σ_x, σ_y, τ_{xy} are the same, the strain energy is greater (per unit thickness) for the plane stress.
7. In Fig. 139, (*a*) represents a strip under compression, in which the stress therefore extends throughout. In (*b*) the deformable strip is bonded to rigid plates on its top and bottom edges. Will there be stress throughout the strip or only locally at the ends? In (*c*) the upper edge is free, as in (*a*), but the lower edge is fixed, as in (*b*). Will the stress be local or not?
8. From the principle that a system in stable equilibrium has less potential energy than that corresponding to any neighboring configuration, show without calculation that the strain energy of the plate in Fig. 131 must either decrease or remain the same when a fine cut AB is made.
9. State the Castigliano theorem expressed by Eq. (142) in a form suitable for use in polar coordinates, the boundary forces $\bar{X}$ and $\bar{Y}$ being replaced by radial and tangential components $\bar{R}$ and $\bar{T}$, and the displacement components by the polar components u and v of Chap. 4.
10. "Equation (142) is valid when δV, $\delta\bar{X}$, $\delta\bar{Y}$ result from any small changes in the stress components which satisfy the conditions of equilibrium (*a*) in Art. 93, whether these changes violate the conditions of compatibility (Art. 16) or not. In the latter case the changes in the stress are those which actually occur when the boundary forces are changed by $\delta\bar{X}$, $\delta\bar{Y}$." Is this statement correct?
11. Equation (*g*) on p. 246 refers to a material obeying Hooke's law. Suppose the material does not obey Hooke's law but has a strain-energy function V_0 which is a function of the strain components more elaborate than (132). Show that the (nonlinear) stress-strain relations are still given by relations of the types

$$\sigma_x = \partial V_0/\partial \epsilon_x \qquad \tau_{xy} = \partial V_0/\partial \gamma_{xy}$$

(Consider an increment of one strain component, the others remaining unchanged.)

96 | Uniqueness of Solution

We consider now whether our equations can have more than one solution corresponding to given surface and body forces.

Let σ_x' . . . , τ_{xy}' . . . represent a solution for loads $\bar{X}$ . . . , X . . . , and let σ_x'' . . . , τ_{xy}'' . . . represent a *second* solution for the *same* loads

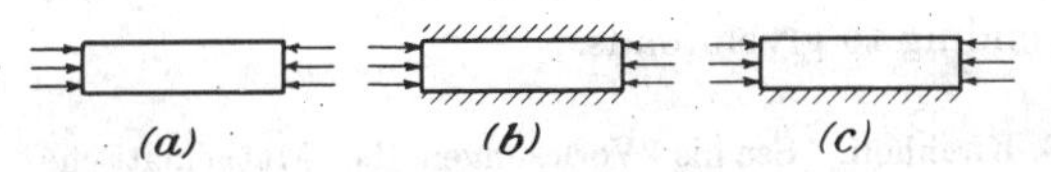

Fig. 139

$\bar{X} \ldots , X \ldots .$ Then for the first solution we have such equations as

$$\frac{\partial \sigma_x'}{\partial x} + \frac{\partial \tau_{xy}'}{\partial y} + \frac{\partial \tau_{xz}'}{\partial z} + X = 0$$

$$\cdots\cdots\cdots\cdots\cdots\cdots$$

$$\bar{X} = \sigma_x' l + \tau_{xy}' m + \tau_{xz}' n$$

$$\cdots\cdots\cdots\cdots\cdots\cdots$$

and also the conditions of compatibility.

For the second solution we have

$$\frac{\partial \sigma_x''}{\partial x} + \frac{\partial \tau_{xy}''}{\partial y} + \frac{\partial \tau_{xz}''}{\partial z} + X = 0$$

$$\cdots\cdots\cdots\cdots\cdots\cdots$$

$$\bar{X} = \sigma_x'' l + \tau_{xy}'' m + \tau_{xz}'' n$$

$$\cdots\cdots\cdots\cdots\cdots\cdots$$

and also the conditions of compatibility.

By subtraction we find that the stress distribution given by the differences $\sigma_x' - \sigma_x'', \ldots , \tau_{xy}' - \tau_{xy}''$ satisfies the equations

$$\frac{\partial (\sigma_x' - \sigma_x'')}{\partial x} + \frac{\partial (\tau_{xy}' - \tau_{xy}'')}{\partial y} + \frac{\partial (\tau_{xz}' - \tau_{xz}'')}{\partial z} = 0$$

$$\cdots\cdots\cdots\cdots\cdots\cdots\cdots\cdots$$

$$0 = (\sigma_x' - \sigma_x'')l + (\tau_{xy}' - \tau_{xy}'')m + (\tau_{xz}' - \tau_{xz}'')n$$

$$\cdots\cdots\cdots\cdots\cdots\cdots\cdots\cdots$$

in which all external forces vanish. The conditions of compatibility (125) will also be satisfied by the corresponding strain components $\epsilon_x' - \epsilon_x''$, $\ldots , \gamma_{xy}' - \gamma_{xy}'', \ldots .$

Thus this stress distribution is one that corresponds to zero surface and body forces. The work done by these forces during loading is zero, and it follows that $\iiint V_0 \, dx \, dy \, dz$ vanishes. But, as Eq. (132) shows, V_0 is positive for all states of strain, and therefore the integral can vanish only if V_0 vanishes at all points of the body. This requires that each of the strain components $\epsilon_x' - \epsilon_x'', \ldots , \gamma_{xy}' - \gamma_{xy}'', \ldots$ should be zero. The two states of strain $\epsilon_x' \ldots , \gamma_{xy}' \ldots$ and $\epsilon_x'' \ldots , \gamma_{xy}'' \ldots ,$ and consequently the two states of stress $\sigma_x' \ldots , \tau_{xy}' \ldots$ and σ_x'' $\ldots , \tau_{xy}'' \ldots$ are therefore identical. That is, the equations can yield only one solution corresponding to given loads.[1]

[1] This theorem is due to G. Kirchhoff. See his "Vorlesungen über Mathematische Physik, Mechanik."

The proof of uniqueness of solution was based on the assumption that the strain energy, and hence stresses, in a body disappear when it is freed of external forces. However there are cases when *initial stresses* may exist in a body while external forces are absent. An example of this kind was encountered in studying the circular ring (see Art. 43). If a portion of the ring between two adjacent cross sections is cut out, and the ends of the ring are joined again by welding or other means, a ring with initial stresses is obtained.[1] Several examples of this kind were discussed in considering two-dimensional problems.

We can also have initial stresses in a simply connected body due to some nonelastic deformations during the process of forming the body. We may have, for instance, considerable initial stresses in large forgings due to nonuniform cooling and also in rolled metallic bars due to the plastic flow produced by cold work. For determining these initial stresses the equations of elasticity are not sufficient, and additional information regarding the process of forming the body is necessary.

It should be noted that in all cases in which the principle of superposition can be used the deformations and stresses produced by external forces are not affected by initial stresses and can be calculated in exactly the same manner as if there were no initial stresses. Then the total stresses are obtained by superposing the stresses produced by external forces on the initial stresses. In cases when the principle of superposition is not applicable, the stresses produced by external loads cannot be determined without knowing the initial stresses. We cannot, for instance, calculate bending stresses produced by lateral loads in a thin bar, if the bar has an initial axial tension or compression, without knowing the magnitude of this initial stress.

97 | The Reciprocal Theorem

We now consider a given elastic body under one set of given surface forces $\bar{X}'$, $\bar{Y}'$, $\bar{Z}'$ and body forces X', Y', Z' and regard the displacements, strains, and stresses as known. These will be denoted by u', ϵ_x', γ_{xy}', σ_x', τ_{xy}', etc. Then, independently, we consider a second set of forces $\bar{X}''$, etc., X'', etc.,

[1] The ring represents the simplest example of multiply-connected bodies. In the case of such bodies general equations of elasticity, expressed in terms of stress components, are not sufficient for determining stresses, and to get a complete solution an additional investigation of displacements is necessary. The first investigations of this kind were made by J. H. Michell, *Proc. London Math. Soc.*, vol. 31, p. 103, 1899. See also L. N. G. Filon, *Brit. Assoc. Advanc. Sci. Rept.*, 1921, p. 305, and V. Volterra, Sur l'équilibre des corps élastiques multiplement connexés, *Ann. École Norm.*, Paris, ser. 3, vol. 24, pp. 401–517, 1907. Further references on initial stresses are given in the paper by P. Neményi, *Z. Angew. Math. Mech.*, vol. 11, p. 59, 1931.

and indicate the results for this second problem as u'', ϵ_x'', γ_{xy}'', σ_x'', τ_{xy}'', etc.

We have then two distinct solutions of two distinct problems. But the fact that they refer to the same elastic body is a relation between them. Here we establish one aspect of this relation—the *reciprocal theorem*.[1]

From the two solutions we can form, purely as a mathematical operation, the quantity $'T''$ defined by

$$'T'' = \int (\bar{X}'u'' + \bar{Y}'v'' + \bar{Z}'w'')\,dS + \int (X'u'' + Y'v'' + Z'w'')\,d\tau \tag{a}$$

Interchanging single and double primes throughout we can also form

$$''T' = \int (\bar{X}''u' + \cdot\cdot\cdot + \cdot\cdot\cdot)\,dS + \int (X''u' + \cdot\cdot\cdot + \cdot\cdot\cdot)\,d\tau \tag{b}$$

The theorem states that

$$'T'' = ''T' \tag{c}$$

For proof we require again the divergence theorem (138). Consider in (*a*) the term

$$\int \bar{X}'u''\,dS \tag{d}$$

which is the same as

$$\int (l\sigma_x' + m\tau_{xy}' + n\tau_{xz}')u''\,dS \tag{e}$$

In (138) we may put

$$U = u''\sigma_x' \qquad V = u''\tau_{xy}' \qquad W = u''\tau_{xz}' \tag{f}$$

to make the right-hand side of (138) the same as the surface integral (*e*).

We then proceed by steps analogous to (*f*) and (*g*) in Art. 93, using the three equilibrium equations of the type

$$\frac{\partial \sigma_x'}{\partial x} + \frac{\partial \tau_{xy}'}{\partial y} + \frac{\partial \tau_{xz}'}{\partial z} + X' = 0 \tag{g}$$

In place of (*k*) of Art. 93, we obtain

$$\begin{aligned} &\int (\epsilon_x''\sigma_x' + \cdot\cdot\cdot + \cdot\cdot\cdot + \gamma_{xy}''\tau_{xy}' + \cdot\cdot\cdot + \cdot\cdot\cdot)\,d\tau \\ &\qquad = \int (\bar{X}'u'' + \bar{Y}'v'' + \bar{Z}'w'')\,dS + \int (X'u'' + Y'v'' + Z'w'')\,d\tau \end{aligned} \tag{h}$$

showing that (*a*) can be converted into

$$'T'' = \int (\epsilon_x''\sigma_x' + \cdot\cdot\cdot + \cdot\cdot\cdot + \gamma_{xy}''\tau_{xy}' + \cdot\cdot\cdot + \cdot\cdot\cdot)\,d\tau \tag{i}$$

We can express (*i*) entirely in terms of stress, or entirely in terms of strain. Choosing the latter, it is convenient to use Hooke's law in the form of (11)

[1] E. Betti, *Il nuovo Cimento*, ser. 2, vols. 7 and 8, 1872. There are theorems of the same kind in other subjects—see Rayleigh, *Proc. London Math. Soc.*, vol. 4, 1873, and his "Theory of Sound," Dover Publications, New York; and H. Lamb, "Higher Mechanics," Cambridge University Press, Inc., New York, 1920.

and (6). Then for the integrand in (*i*) we find

$$\epsilon_x''\sigma_x' + \cdots + \cdots + \gamma_{xy}''\tau_{xy}' + \cdots + \cdots$$
$$= \lambda\epsilon'\epsilon'' + 2G(\epsilon_x'\epsilon_x'' + \epsilon_y'\epsilon_y'' + \epsilon_z'\epsilon_z'')$$
$$+ G(\gamma_{xy}'\gamma_{xy}'' + \gamma_{yz}'\gamma_{yz}'' + \gamma_{zx}'\gamma_{zx}'') \quad (j)$$

where

$$\epsilon' = \epsilon_x' + \epsilon_y' + \epsilon_z' \qquad \epsilon'' = \epsilon_x'' + \epsilon_y'' + \epsilon_z''$$

The result (*j*) is clearly unaffected if the single and double primes are interchanged. But this interchange, in (*i*), is all we need to express $''T'$ instead of $'T''$. Hence the theorem (*c*) is established.

The right-hand side of (*a*) is often referred to as the work of the forces of the first state (single prime) on the displacements of the second state (double prime). Then (*b*) is called the work of the forces of the second state on the displacements of the first, and these two works are equal.

The theorem can be immediately extended to the dynamical case by including the inertia forces as body forces.

The statical form has many important applications. We give here two simple illustrative examples. Further applications, to thermal stress problems, occur in Chap. 13.

Consider first a uniform bar compressed by two equal and opposite forces[1] P, Fig. 140*a*. The problem of finding the stresses produced by these forces is a complicated one; but suppose we are interested not in the stresses but in the total elongation δ of the bar. This question can be answered at once by using the theorem. For this purpose we consider in addition to the given stress condition represented in Fig. 140*a* the simple axial tension of the bar shown in Fig. 140*b*. For this second case we find the lateral contraction, equal to $\delta_1 = \nu(Qh/AE)$, where A is the cross-sectional area of the bar. Then the reciprocal theorem gives us the equation

$$P\nu\frac{Qh}{AE} = Q\delta$$

[1] We may suppose that the forces are distributed over a small area, to avoid singularities.

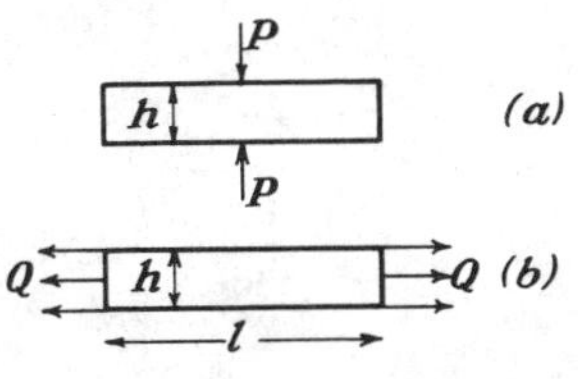

Fig. 140

and the elongation of the bar, produced by two forces P in Fig. 140a, is

$$\delta = \frac{\nu Ph}{AE}$$

and is independent of the shape of the cross section.

As a second example let us calculate the reduction Δ in volume of an elastic body produced by two equal and opposite forces P, Fig. 141a. As a second state we take the same body submitted to the action of uniformly distributed pressure p. In this latter case we will have at each point of the body a uniform compression in all directions of the magnitude $(1 - 2\nu)p/E$ [see Eq. (8)] and the distance l between the points of application A and B will be diminished by the amount $(1 - 2\nu)pl/E$. The reciprocal theorem applied to the two states[1] of Fig. 141 will then give

$$P\frac{(1 - 2\nu)pl}{E} = \Delta p$$

and the reduction in the volume of the body is therefore

$$\Delta = \frac{Pl(1 - 2\nu)}{E}$$

98 | Approximate Character of the Plane Stress Solutions

It was pointed out on page 31 that the set of equations we found sufficient for plane stress problems under the assumptions made ($\sigma_z = \tau_{xz} = \tau_{yz} = 0$, σ_x, σ_y, τ_{xy} independent of z) did not ensure satisfaction of all the conditions of compatibility. These assumptions imply that ϵ_x, ϵ_y, ϵ_z, γ_{xy} are independent of z, and that γ_{xz}, γ_{yz} are zero. The first of the conditions of compatibility (125) was included in the plane stress theory, as Eq. (21). It is easily verified that the other five are satisfied only if ϵ_z is a linear function of x and y, which is the exception rather than the rule in the plane stress solutions obtained in Chaps. 3 to 6. Evidently these solutions cannot be exact, but we shall now see that they are close approximations for thin plates.

Let us seek exact solutions of the three-dimensional equations for which[2]

$$\sigma_z = \tau_{xz} = \tau_{yz} = 0$$

[1] For other applications of this kind see Love, *op. cit.*, pp. 174–176.
[2] A. Clebsch, "Elasticität," art. 39. See also Love, *op. cit.*, p. 206.

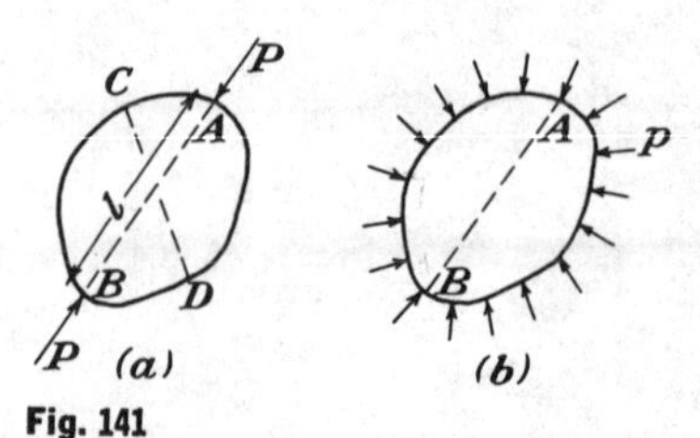

Fig. 141

taking body force as zero. Such solutions must satisfy the equations of equilibrium (123) and the compatibility conditions (126).

Since σ_z, τ_{xz}, τ_{yz} are zero, the third, fourth, and fifth of Eqs. (126) (reading by columns) give

$$\frac{\partial}{\partial z}\left(\frac{\partial \Theta}{\partial z}\right) = 0 \qquad \frac{\partial}{\partial y}\left(\frac{\partial \Theta}{\partial z}\right) = 0 \qquad \frac{\partial}{\partial x}\left(\frac{\partial \Theta}{\partial z}\right) = 0$$

which mean that $\partial\Theta/\partial z$ is a constant. Writing this k, we have, by integration with respect to z,

$$\Theta = kz + \Theta_0 \tag{a}$$

where Θ_0 is so far an arbitrary function of x and y.

The third of Eqs. (123) is identically satisfied, and the first two reduce to the two-dimensional forms

$$\frac{\partial \sigma_x}{\partial x} + \frac{\partial \tau_{xy}}{\partial y} = 0 \qquad \frac{\partial \sigma_y}{\partial y} + \frac{\partial \tau_{xy}}{\partial x} = 0$$

which are satisfied, as before, by

$$\sigma_x = \frac{\partial^2 \phi}{\partial y^2} \qquad \sigma_y = \frac{\partial^2 \phi}{\partial x^2} \qquad \tau_{xy} = -\frac{\partial^2 \phi}{\partial x\, \partial y} \tag{b}$$

but ϕ is now a function of x, y, and z.

Returning to Eqs. (126) we observe that by addition of the three equations on the left, recalling that $\Theta = \sigma_x + \sigma_y + \sigma_z$, we have

$$\nabla^2\Theta = 0 \tag{c}$$

and therefore, from (a),

$$\nabla_1^2\Theta_0 = 0 \tag{d}$$

where

$$\nabla_1^2 = \frac{\partial^2}{\partial x^2} + \frac{\partial^2}{\partial y^2}$$

Also, since σ_z is zero, and σ_x and σ_y are given by the first two of Eqs. (b), we can write $\nabla_1^2\phi = \Theta$, and therefore, using (a),

$$\nabla_1^2\phi = kz + \Theta_0 \tag{e}$$

where Θ_0 is a function of x and y satisfying Eq. (d). Using (a) and the first of (b), the first of Eqs. (126) becomes

$$(1 + \nu)\nabla^2 \frac{\partial^2 \phi}{\partial y^2} + \frac{\partial^2 \Theta_0}{\partial x^2} = 0 \tag{f}$$

But

$$\nabla^2 \frac{\partial^2 \phi}{\partial y^2} = \frac{\partial^2}{\partial y^2}\nabla^2\phi = \frac{\partial^2}{\partial y^2}\left(\nabla_1^2\phi + \frac{\partial^2 \phi}{\partial z^2}\right) = \frac{\partial^2}{\partial y^2}\left(\Theta_0 + \frac{\partial^2 \phi}{\partial z^2}\right)$$

where Eq. (e) has been used in the last step. Also, on account of (d), we can replace $\partial^2\Theta_0/\partial x^2$ in (f) by $-\partial^2\Theta_0/\partial y^2$. Then (f) becomes

$$(1 + \nu)\frac{\partial^2}{\partial y^2}\left(\Theta_0 + \frac{\partial^2 \phi}{\partial z^2}\right) - \frac{\partial^2 \Theta_0}{\partial y^2} = 0$$

or

$$\frac{\partial^2}{\partial y^2}\left(\frac{\partial^2 \phi}{\partial z^2} + \frac{\nu}{1 + \nu}\Theta_0\right) = 0 \tag{g}$$

This equation may be used in place of the first of (126). Similarly, the second and last can be replaced by

$$\frac{\partial^2}{\partial x^2}\left(\frac{\partial^2 \phi}{\partial z^2} + \frac{\nu}{1+\nu}\Theta_0\right) = 0 \qquad \frac{\partial^2}{\partial x\,\partial y}\left(\frac{\partial^2 \phi}{\partial z^2} + \frac{\nu}{1+\nu}\Theta_0\right) = 0$$

These, with (*g*), show that all three second derivatives with respect to x and y of the function (of x, y, and z) in brackets vanish. Thus this function must be linear in x and y, and we can write

$$\frac{\partial^2 \phi}{\partial z^2} + \frac{\nu}{1+\nu}\Theta_0 = a + bx + cy \tag{h}$$

where a, b, and c are arbitrary functions of z. Integrating this equation twice with respect to z, we find

$$\phi = -\frac{1}{2}\frac{\nu}{1+\nu}\Theta_0 z^2 + A + Bx + Cy + \phi_1 z + \phi_0 \tag{i}$$

where A, B, C are functions of z obtained by repeated integration of a, b, c, and ϕ_1, ϕ_0 are functions of x and y, as yet arbitrary.

If we evaluate σ_x, σ_y, τ_{xy} from (*i*) by means of the formulas (*b*), the terms

$$A + Bx + Cy$$

make no difference. We may therefore set A, B, and C equal to zero, corresponding to taking a, b, c, zero in (*h*).

If we restrict ourselves to problems in which the stress distribution is symmetrical about the middle plane of the plate, $z = 0$, the term $\phi_1 z$ must also be zero. So also must k in Eq. (*a*).

Then (*i*) reduces to

$$\phi = \phi_0 - \frac{1}{2}\frac{\nu}{1+\nu}\Theta_0 z^2 \tag{j}$$

However ϕ and Θ_0 are related by (*e*) in which we can now take $k = 0$. Thus, substituting (*j*) in (*e*) and using (*d*), we have

$$\nabla_1^2 \phi_0 = \Theta_0 \tag{k}$$

and therefore, from (*d*),

$$\nabla_1^4 \phi_0 = 0 \tag{l}$$

The remaining equations of (126) are satisfied on account of Eq. (*a*) and the vanishing of σ_z, τ_{xz}, τ_{yz}.

We can now obtain a stress distribution by choosing a function ϕ_0 of x and y which satisfies Eq. (*l*), finding Θ_0 from Eq. (*k*) and ϕ from Eq. (*j*). The stresses are then found by the formulas (*b*). Each will consist of two parts, the first derived from ϕ_0 in Eq. (*j*), the second from the term $-(1/2)(\nu/1 + \nu)\Theta_0 z^2$. In view of Eq. (*l*), the first part is exactly like the plane stress components determined in Chaps. 3 to 6. The second part, being proportional to z^2, may be made as small as we please compared with the first by restricting ourselves to plates that are sufficiently thin. Hence the conclusion that our solutions in Chaps. 3 to 6, which do not satisfy all the compatibility conditions, are nevertheless good approximations for thin plates.

The "exact" solutions, represented by stress functions of the form (*j*), will require that the stresses at the boundary, as elsewhere, have a parabolic variation over the thickness. However any change from this distribution, so long as it does not alter the intensity of force per unit length of boundary curve, will only alter the stress in the immediate neighborhood of the edge, by Saint-Venant's principle (page 40). The type of solution considered above will always represent the actual stress, and the components σ_z, τ_{xz}, τ_{yz} will in fact be negligible, except close to the edges.[1]

PROBLEMS

1. Show that

$$\epsilon_x = k(x^2 + y^2) \qquad \epsilon_y = k(y^2 + z^2) \qquad \gamma_{xy} = k'xyz \qquad \epsilon_z = \gamma_{xz} = \gamma_{yz} = 0$$

where k, k' are small constants, is *not* a possible state of strain.

2. A solid is heated nonuniformly to temperature T, a function of x, y, and z. If it is supposed that each element has unrestrained thermal expansion, the strain components will be

$$\epsilon_x = \epsilon_y = \epsilon_z = \alpha T \qquad \gamma_{xy} = \gamma_{yz} = \gamma_{xz} = 0$$

where α is the constant coefficient of thermal expansion.

Prove that this can only occur when T is a linear function of x, y, and z. (The stress and consequent further strain arising when T is not linear are discussed in Chap. 13.)

3. A disk or cylinder of the shape shown in Fig. 141*a* is compressed by forces P at C and D, along CD, causing extension of AB. It is then compressed by forces P along AB (Fig. 141*a*) causing extension of CD. Show that these extensions are equal.

4. In the general solution of Art. 88 what choice of the functions ϕ_0, ϕ_1, ϕ_2, ϕ_3 will give the general solution for plane strain ($w = 0$)?

5. Consider Eq. (*f*) in Art. 85, in relation to Eq. (25) in Art. 16. Show that the former can be reduced to the latter under the plane stress postulates $\sigma_z = \tau_{xz} = \tau_{yz} = 0$, $Z = 0$ as used in the exact theory of Art. 82.

[1] Accordingly, the device of averaging over the plate thickness, the basis of "generalized plane stress," offers little advantage. Except near the edge the simple parabolic variation prevails. Near the edge the variation with z is different but depends on the z variation of the boundary loading.

chapter 9

Elementary Problems of Elasticity in Three Dimensions

99 | Uniform Stress

In discussing the equations of equilibrium (123) and the boundary conditions (124), it was stated that the true solution of a problem must satisfy not only Eqs. (123) and (124) but also the compatibility conditions (see Art. 85). These latter conditions contain, if no body forces are acting, or if the body forces are constant, only second derivatives of the stress components. If, therefore, Eqs. (123) and conditions (124) can be satisfied by taking the stress components either as constants or as linear functions of the coordinates, the compatibility conditions are satisfied identically and these stresses are the correct solution of the problem.

As a very simple example we may take tension of a bar in the axial direction (Fig. 142). Body forces are neglected. The equations of equilibrium are satisfied by taking

$$\sigma_x = \text{const} \qquad \sigma_y = \sigma_z = \tau_{xy} = \tau_{xz} = \tau_{yz} = 0 \tag{a}$$

It is evident that boundary conditions (124) for the lateral surface of the bar, which is free of external forces, are satisfied, because all stress components, except σ_x, are zero. The boundary conditions for the ends reduce to

$$\sigma_x = \bar{X} \tag{b}$$

that is, we have a uniform distribution of tensile stresses over cross sections of a bar if the tensile stresses are uniformly distributed over the ends. In this case solution (a) satisfies Eqs. (123) and (124) and is the correct solution of the problem because the compatibility conditions (126) are identically satisfied.

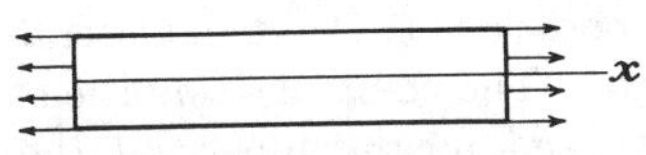

Fig. 142

If the tensile stresses are not uniformly distributed over the ends, solution (a) is no longer the correct solution because it does not satisfy the boundary conditions at the ends. The true solution becomes more complicated because the stresses on a cross section are no longer uniformly distributed. Examples of such nonuniform distribution occurred in the discussion of two-dimensional problems (see pages 58 and 258).

As a second example consider the case of a uniform hydrostatic compression with no body forces. The equations of equilibrium (123) are satisfied by taking

$$\sigma_x = \sigma_y = \sigma_z = -p \qquad \tau_{xy} = \tau_{xz} = \tau_{yz} = 0 \tag{c}$$

The ellipsoid of stress in this case is a sphere. Any three perpendicular directions can be considered as principal directions, and the stress on any plane is a normal compressive stress equal to p. The surface conditions (124) will evidently be satisfied if the pressure p is uniformly distributed over the surface of the body.

100 | Stretching of a Prismatical Bar by Its Own Weight

If ρg is the weight per unit volume of the bar (Fig. 143), the body forces are

$$X = Y = 0 \qquad Z = -\rho g \tag{a}$$

The differential equations of equilibrium (123) are satisfied by putting

$$\sigma_z = \rho g z \qquad \sigma_x = \sigma_y = \tau_{xy} = \tau_{yz} = \tau_{xz} = 0 \tag{b}$$

that is, by assuming that on each cross section we have a uniform tension produced by the weight of the lower portion of the bar.

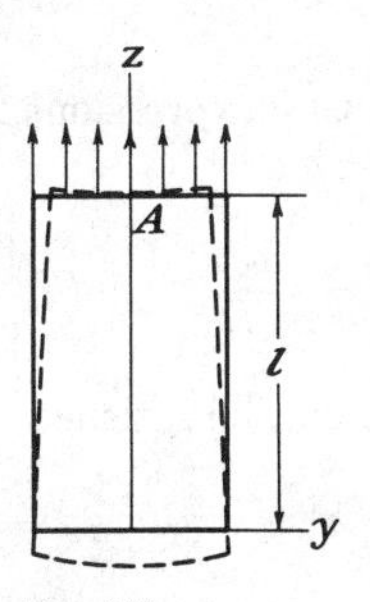

Fig. 143

It can easily be seen that the boundary conditions (124) at the lateral surface, which is free from forces, are satisfied. The boundary conditions give zero stresses for the lower end of the bar, and, for the upper end, the uniformly distributed tensile stress $\sigma_z = \rho g l$, in which l is the length of the bar.

The compatibility equations (126) are also satisfied by the solution (*b*); hence, it is the correct solution of the problem for a uniform distribution of forces at the top. It coincides with the solution that is usually given in elementary books on the strength of materials.

Let us consider now the displacements (see Art. 86). From Hooke's law, using Eqs. (3) and (6), we find

$$\epsilon_z = \frac{\partial w}{\partial z} = \frac{\sigma_z}{E} = \frac{\rho g z}{E} \tag{c}$$

$$\epsilon_x = \epsilon_y = \frac{\partial u}{\partial x} = \frac{\partial v}{\partial y} = -\nu \frac{\rho g z}{E} \tag{d}$$

$$\gamma_{xy} = \gamma_{xz} = \gamma_{yz} = \frac{\partial u}{\partial y} + \frac{\partial v}{\partial x} = \frac{\partial u}{\partial z} + \frac{\partial w}{\partial x} = \frac{\partial v}{\partial z} + \frac{\partial w}{\partial y} = 0 \tag{e}$$

The displacements u, v, w can now be found by integrating Eqs. (*c*), (*d*), and (*e*). Integration of Eq. (*c*) gives

$$w = \frac{\rho g z^2}{2E} + w_0 \tag{f}$$

where w_0 is a function of x and y, to be determined later. Substituting (*f*) in the second and third of Eqs. (*e*), we find

$$\frac{\partial w_0}{\partial x} + \frac{\partial u}{\partial z} = 0 \qquad \frac{\partial w_0}{\partial y} + \frac{\partial v}{\partial z} = 0$$

from which

$$u = -z\frac{\partial w_0}{\partial x} + u_0 \qquad v = -z\frac{\partial w_0}{\partial y} + v_0 \tag{g}$$

in which u_0 and v_0 are functions of x and y only. Substituting expressions (*g*) into Eqs. (*d*), we find

$$-z\frac{\partial^2 w_0}{\partial x^2} + \frac{\partial u_0}{\partial x} = -\nu\frac{\rho g z}{E} \qquad -z\frac{\partial^2 w_0}{\partial y^2} + \frac{\partial v_0}{\partial y} = -\nu\frac{\rho g z}{E} \tag{h}$$

Remembering that u_0 and v_0 do not depend on z, Eqs. (*h*) can be satisfied only if

$$\frac{\partial u_0}{\partial x} = \frac{\partial v_0}{\partial y} = 0 \qquad \frac{\partial^2 w_0}{\partial x^2} = \frac{\partial^2 w_0}{\partial y^2} = \frac{\nu \rho g}{E} \tag{k}$$

Substituting expressions (g) for u and v into the first of Eqs. (e), we find

$$-2z \frac{\partial^2 w_0}{\partial x\, \partial y} + \frac{\partial u_0}{\partial y} + \frac{\partial v_0}{\partial x} = 0$$

and, since u_0 and v_0 do not depend on z, we must have

$$\frac{\partial^2 w_0}{\partial x\, \partial y} = 0 \qquad \frac{\partial u_0}{\partial y} + \frac{\partial v_0}{\partial x} = 0 \tag{l}$$

From Eqs. (k) and (l) general expressions can now be written for the functions u_0, v_0, w_0. It is easy to show that all these equations are satisfied by

$$u_0 = \delta y + \delta_1$$

$$v_0 = -\delta x + \gamma_1$$

$$w_0 = \frac{\nu \rho g}{2E}(x^2 + y^2) + \alpha x + \beta y + \gamma$$

in which α, β, γ, δ, δ_1, γ_1 are arbitrary constants. Now, from Eqs. (f) and (g), the general expressions for the displacements are

$$u = -\frac{\nu \rho g x z}{E} - \alpha z + \delta y + \delta_1$$

$$v = -\frac{\nu \rho g y z}{E} - \beta z - \delta x + \gamma_1 \tag{m}$$

$$w = \frac{\rho g z^2}{2E} + \frac{\nu \rho g}{2E}(x^2 + y^2) + \alpha x + \beta y + \gamma$$

The six arbitrary constants must be determined from the conditions at the support. The support must be such as to prevent any movement of the bar as a rigid body. To prevent a translatory motion of the bar, let us fix the centroid A of the upper end of the bar so that $u = v = w = 0$ for $x = y = 0$ and $z = l$. To eliminate rotation of the bar about axes through the point A, parallel to the x and y axes, let us fix an element of the z axis at A. Then $\partial u/\partial z = \partial v/\partial z = 0$ at that point. The possibility of rotation about the z axis is eliminated by fixing an elemental area through A, parallel to the zx plane. Then $\partial v/\partial x = 0$ at the point A. Using Eqs. (m) the above six conditions at the point A become

$$-\alpha l + \delta_1 = 0 \qquad -\beta l + \gamma_1 = 0 \qquad \frac{\rho g l^2}{2E} + \gamma = 0$$

$$\alpha = 0 \qquad \beta = 0 \qquad \delta = 0$$

Hence

$$\delta_1 = 0 \qquad \gamma_1 = 0 \qquad \gamma = -\frac{\rho g l^2}{2E}$$

and the final expressions for the displacements are

$$u = -\frac{\nu\rho g x z}{E}$$

$$v = -\frac{\nu\rho g y z}{E}$$

$$w = \frac{\rho g z^2}{2E} + \frac{\nu\rho g}{2E}(x^2 + y^2) - \frac{\rho g l^2}{2E}$$

It may be seen that points on the z axis have only vertical displacements

$$w = -\frac{\rho g}{2E}(l^2 - z^2)$$

Other points of the bar, on account of lateral contraction, have not only vertical but also horizontal displacements. Lines that were parallel to the z axis before deformation become inclined to this axis after deformation, and the form of the bar after deformation is as indicated in Fig. 143 by dotted lines. Cross sections of the bar perpendicular to the z axis after deformation are curved to the surface of a paraboloid. Points on the cross section $z = c$, for instance, after deformation will be on the surface

$$z = c + w = c + \frac{\rho g c^2}{2E} + \frac{\nu\rho g}{2E}(x^2 + y^2) - \frac{\rho g l^2}{2E}$$

This surface is perpendicular to all longitudinal fibers of the bar, these being inclined to the z axis after deformation, so that there is no shearing strain γ_{xy} or γ_{xz}.

101 | Twist of Circular Shafts of Constant Cross Section

The elementary theory of twist of circular shafts states that the shearing stress τ at any point of the cross section (Fig. 144) is perpendicular to the radius r and proportional to the length r and to the angle of twist θ per unit length of the shaft:

$$\tau = G\theta r \tag{a}$$

where G is the modulus of rigidity. Resolving this stress into two components parallel to the x and y axes, we find

$$\begin{aligned} \tau_{yz} &= G\theta r \frac{x}{r} = G\theta x \\ \tau_{xz} &= -G\theta r \frac{y}{r} = -G\theta y \end{aligned} \tag{b}$$

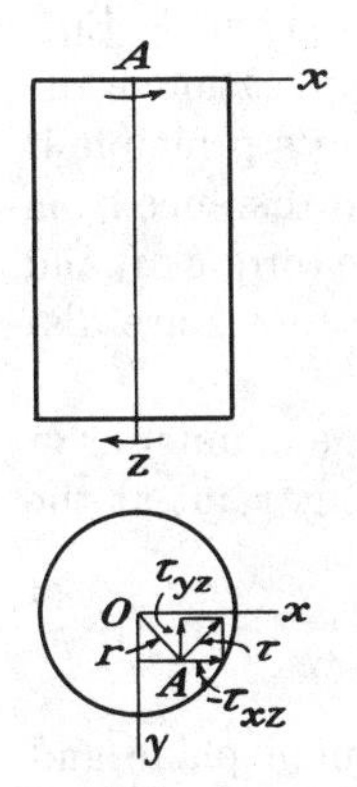

Fig. 144

The elementary theory also assumes that

$$\sigma_x = \sigma_y = \sigma_z = \tau_{xy} = 0$$

We can show that this elementary solution is the exact solution under certain conditions. Since the stress components are all either linear functions of the coordinates or zero, the equations of compatibility (126) are satisfied, and it is only necessary to consider the equations of equilibrium (123) and the boundary conditions (124). Substituting the above expressions for stress components into Eqs. (123), we find that these equations are satisfied, provided there are no body forces. The lateral surface of the shaft is free from forces, and the boundary conditions (124), remembering that for the cylindrical surface $\cos(Nz) = n = 0$, reduce to

$$0 = \tau_{xz} \cos(Nx) + \tau_{yz} \cos(Ny) \tag{c}$$

For the case of a circular cylinder we have also

$$\cos(Nx) = \frac{x}{r} \qquad \cos(Ny) = \frac{y}{r} \tag{d}$$

Substituting these and expressions (*b*) for the stress components into Eq. (*c*), it is evident that this equation is satisfied. It is also evident that for cross sections other than circular, for which Eqs. (*d*) do not hold, the stress components (*b*) do not satisfy the boundary condition (*c*), and therefore solution (*a*) cannot be applied. These more complicated problems of twist will be considered later (see Chap. 10).

Considering now the boundary conditions for the ends of the shaft, we see that the surface shearing forces must be distributed in exactly the same manner as the stresses τ_{xz} and τ_{yz} over any intermediate cross section of the shaft. Only for this case is the stress distribution given by Eqs. (*b*) an exact solution of the problem. But the practical application of the solution is not limited to such cases. From Saint-Venant's principle it can be concluded that in a long twisted bar, at a sufficient distance from the ends, the stresses depend only on the magnitude of the torque M_t and are practically independent of the manner in which the forces are distributed over the ends.

The displacements for this case can be found in the same manner as in the previous article. Assuming the same condition of constraint at the point A as in the previous problem we find

$$u = -\theta yz \qquad v = \theta xz \qquad w = 0$$

This means that the assumption that cross sections remain plane and radii remain straight, which is usually made in the elementary derivation of the theory of twist, is correct.

102 | Pure Bending of Prismatical Bars

Consider a prismatical bar bent in one of its principal planes by two equal and opposite couples M (Fig. 145). Taking the origin of the coordinates at the centroid of the cross section and the xz plane in the principal plane of bending, the stress components given by the usual elementary theory of bending are

$$\sigma_z = \frac{Ex}{R} \qquad \sigma_y = \sigma_x = \tau_{xy} = \tau_{xz} = \tau_{yz} = 0 \tag{a}$$

in which R is the radius of curvature of the bar after bending. Substituting expressions (*a*) for the stress components in the equations of equilibrium (123), it is found that these equations are satisfied if there are no body forces. The boundary conditions (124) for the lateral surface of the bar, which is free from external forces, are also satisfied. The boundary conditions (124) at the ends require that the surface forces must be distributed over the ends in the same manner as the stresses σ_z. Only under

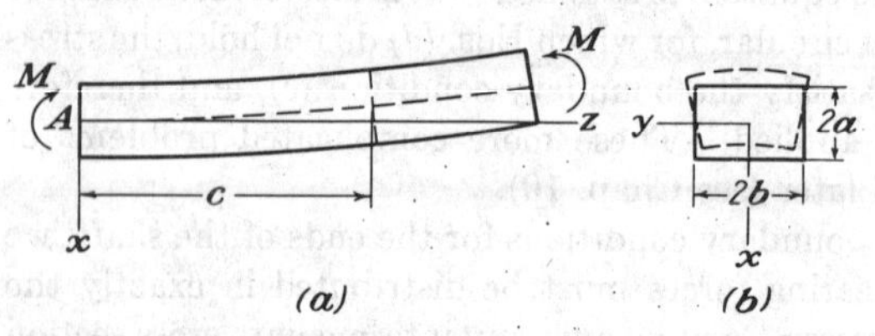

Fig. 145

this condition do the stresses (a) represent the exact solution of the problem. The bending moment M is given by the equation

$$M = \int \sigma_z x \, dA = \int \frac{Ex^2 \, dA}{R} = \frac{EI_y}{R}$$

in which I_y is the moment of inertia of the cross section of the beam with respect to the neutral axis parallel to the y axis. From this equation we find

$$\frac{1}{R} = \frac{M}{EI_y}$$

which is a well-known formula of the elementary theory of bending.

Let us consider now the displacements for the case of pure bending. Using Hooke's law and Eqs. (2) we find, from solution (a),

$$\epsilon_z = \frac{\partial w}{\partial z} = \frac{x}{R} \tag{b}$$

$$\epsilon_x = \frac{\partial u}{\partial x} = -\nu \frac{x}{R} \qquad \epsilon_y = \frac{\partial v}{\partial y} = -\nu \frac{x}{R} \tag{c}$$

$$\frac{\partial u}{\partial y} + \frac{\partial v}{\partial x} = \frac{\partial u}{\partial z} + \frac{\partial w}{\partial x} = \frac{\partial v}{\partial z} + \frac{\partial w}{\partial y} = 0 \tag{d}$$

By using these differential equations, and taking into consideration the fastening conditions of the bar, the displacements can be obtained in the same manner as in Art. 100.

From Eq. (b) we find

$$w = \frac{xz}{R} + w_0$$

in which w_0 is a function of x and y only. The second and third of Eqs. (d) give

$$\frac{\partial u}{\partial z} = -\frac{z}{R} - \frac{\partial w_0}{\partial x} \qquad \frac{\partial v}{\partial z} = -\frac{\partial w_0}{\partial y}$$

from which

$$u = -\frac{z^2}{2R} - z\frac{\partial w_0}{\partial x} + u_0 \qquad v = -z\frac{\partial w_0}{\partial y} + v_0 \tag{e}$$

Here u_0 and v_0 denote unknown functions of x and y, which will be determined later. Substituting expressions (e) in Eqs. (c),

$$-z\frac{\partial^2 w_0}{\partial x^2} + \frac{\partial u_0}{\partial x} = -\frac{\nu x}{R} \qquad -z\frac{\partial^2 w_0}{\partial y^2} + \frac{\partial v_0}{\partial y} = -\nu\frac{x}{R}$$

These equations must be satisfied for any value of z, hence

$$\frac{\partial^2 w_0}{\partial x^2} = 0 \qquad \frac{\partial^2 w_0}{\partial y^2} = 0 \tag{f}$$

and by integration

$$u_0 = -\frac{\nu x^2}{2R} + f_1(y) \qquad v_0 = -\frac{\nu xy}{R} + f_2(x) \tag{g}$$

Now substituting (*e*) and (*g*) into the first of Eqs. (*d*), we find

$$2z \frac{\partial^2 w_0}{\partial x \, \partial y} - \frac{\partial f_1(y)}{\partial y} - \frac{\partial f_2(x)}{\partial x} + \frac{\nu y}{R} = 0$$

Noting that only the first term in this equation depends on z, we conclude that it is necessary to have

$$\frac{\partial^2 w_0}{\partial x \, \partial y} = 0 \qquad \frac{\partial f_1(y)}{\partial y} + \frac{\partial f_2(x)}{\partial x} - \frac{\nu y}{R} = 0$$

These equations and Eqs. (*f*) require that

$$w_0 = mx + ny + p$$

$$f_1(y) = \frac{\nu y^2}{2R} + \alpha y + \gamma$$

$$f_2(x) = -\alpha x + \beta$$

in which $m, n, p, \alpha, \beta, \gamma$ are arbitrary constants. The expressions for the displacements now become

$$u = -\frac{z^2}{2R} - mz - \frac{\nu x^2}{2R} + \frac{\nu y^2}{2R} + \alpha y + \gamma$$

$$v = -nz - \frac{\nu xy}{R} - \alpha x + \beta$$

$$w = \frac{xz}{R} + mx + ny + p$$

The arbitrary constants are determined from the conditions of fastening. Assuming that the point A, the centroid of the left end of the bar, together with an element of the z axis and an element of the xz plane, are fixed, we have for $x = y = z = 0$

$$u = v = w = 0 \qquad \frac{\partial u}{\partial z} = \frac{\partial v}{\partial z} = \frac{\partial v}{\partial x} = 0$$

These conditions are satisfied by taking all the arbitrary constants equal to zero. Then

$$u = -\frac{1}{2R}[z^2 + \nu(x^2 - y^2)] \qquad v = -\frac{\nu xy}{R} \qquad w = \frac{xz}{R} \tag{h}$$

To get the deflection curve of the axis of the bar we substitute in the above

Eqs. (h) $x = y = 0$. Then

$$u = -\frac{z^2}{2R} = -\frac{Mz^2}{2EI_y} \qquad v = w = 0$$

This is the same deflection curve as is given by the elementary theory of bending.

Let us consider now any cross section $z = c$, a distance c from the left end of the bar. After deformation, the points of this cross section will be in the plane

$$z = c + w = c + \frac{cx}{R}$$

i.e., in pure bending the cross section remains plane as is assumed in the elementary theory. To examine the deformation of the cross section in its plane, consider the sides $y = \pm b$ (Fig. 145b). After bending we have

$$y = \pm b + v = \pm b\left(1 - \frac{\nu x}{R}\right)$$

The sides become inclined as shown in the figure by dotted lines.

The other two sides of the cross section $x = \pm a$ are represented after bending by the equations

$$x = \pm a + u = \pm a - \frac{1}{2R}[c^2 + \nu(a^2 - y^2)]$$

They are therefore bent to parabolic curves, which can be replaced with sufficient accuracy by an arc of a circle of radius R/ν, when the deformation is small. In considering the upper or lower sides of the bar it is evident that while the curvature of these sides after bending is convex down in the lengthwise direction, the curvature in the crosswise direction is convex upward. Contour lines for this anticlastic surface will be as shown in Fig. 146a. By taking x and u constant in the first of Eqs. (h) we find that the equation for the contour lines is

$$z^2 - \nu y^2 = \text{const}$$

They are therefore hyperbolas with the asymptotes

$$z^2 - \nu y^2 = 0$$

From this equation the angle α (Fig. 146a) is found from

$$\tan^2 \alpha = \frac{1}{\nu}$$

This equation has been used for determining Poisson's ratio ν.[1] If the

[1] A. Cornu, *Compt. Rend.*, vol. 69, p. 333, 1869. See also R. Straubel, *Wied. Ann.*, vol. 68, p. 369, 1899.

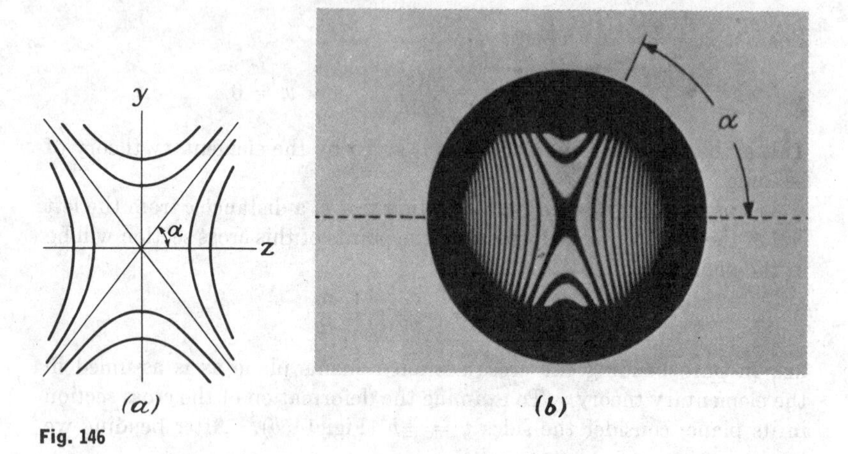

Fig. 146

upper surface of the beam is polished and a glass plate put over it, there will be, after bending, an air gap of variable thickness between the glass plate and the curved surface of the beam. This variable thickness can be measured optically. A beam of monochromatic light, say yellow sodium light, perpendicular to the glass plate, will be reflected partially by the plate and partially by the surface of the beam. The two reflected rays of light interfere with each other at points where the thickness of the air gap is such that the difference between the paths of the two rays is equal to an uneven number of half wavelengths of the light. The picture shown in Fig. 146*b*, representing the hyperbolic contour lines, was obtained by this means.

103 | Pure Bending of Plates

The result of the previous article can be applied in discussing the bending of plates of uniform thickness. If stresses $\sigma_x = Ez/R$ are distributed over the edges of the plate parallel to the y axis (Fig. 147), the surface of the plate will become[1] an anticlastic surface, the curvature of which in planes

[1] It is assumed that deflections are small in comparison with the thickness of the plate.

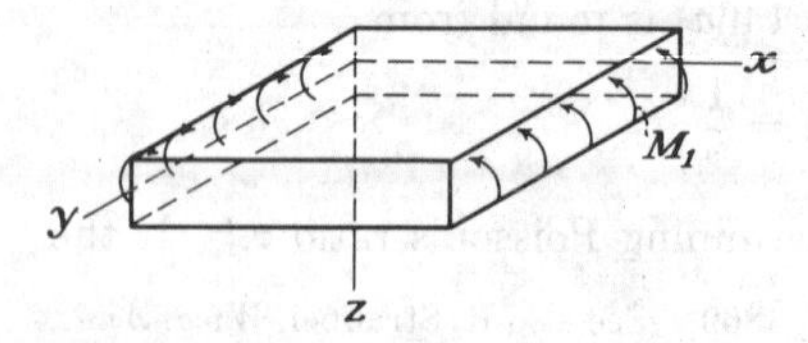

Fig. 147

parallel to the xz plane is $1/R$ and in the perpendicular direction is $-\nu/R$ If h denotes the thickness of the plate, M_1 the bending moment per unit length on the edges parallel to the y axis, and

$$I_y = \frac{h^3}{12}$$

the moment of inertia per unit length, the relation between M_1 and R, from the previous article, is

$$\frac{1}{R} = \frac{M_1}{EI_y} = \frac{12M_1}{Eh^3} \tag{a}$$

When we have bending moments in two perpendicular directions (Fig. 148), the curvatures of the deflection surface may be obtained by superposition. Let $1/R_1$ and $1/R_2$ be the curvatures of the deflection surface in planes parallel to the coordinate planes zx and zy, respectively; and let M_1 and M_2 be the bending moments per unit length on the edges parallel to the y and x axes, respectively. Then, using Eq. (a) and applying the principle of superposition, we find

$$\begin{aligned} \frac{1}{R_1} &= \frac{12}{Eh^3}(M_1 - \nu M_2) \\ \frac{1}{R_2} &= \frac{1}{Eh^3}(M_2 - \nu M_1) \end{aligned} \tag{b}$$

The moments are considered positive if they produce a deflection of the plate which is convex down. Solving Eqs. (b) for M_1 and M_2, we find

$$\begin{aligned} M_1 &= \frac{Eh^3}{12(1-\nu^2)}\left(\frac{1}{R_1} + \nu\frac{1}{R_2}\right) \\ M_2 &= \frac{Eh^3}{12(1-\nu^2)}\left(\frac{1}{R_2} + \nu\frac{1}{R_1}\right) \end{aligned} \tag{c}$$

For small deflections we can use the approximations

$$\frac{1}{R_1} = -\frac{\partial^2 w}{\partial x^2} \qquad \frac{1}{R_2} = -\frac{\partial^2 w}{\partial y^2}$$

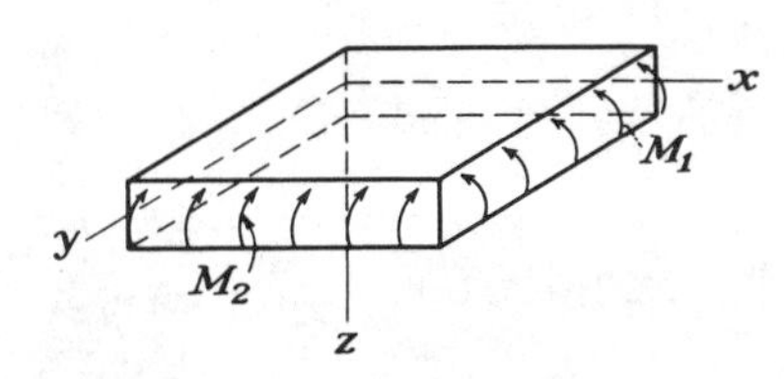

Fig. 148

Then, writing

$$\frac{Eh^3}{12(1-\nu^2)} = D \tag{143}$$

we find

$$\begin{aligned} M_1 &= -D\left(\frac{\partial^2 w}{\partial x^2} + \nu\frac{\partial^2 w}{\partial y^2}\right) \\ M_2 &= -D\left(\frac{\partial^2 w}{\partial y^2} + \nu\frac{\partial^2 w}{\partial x^2}\right) \end{aligned} \tag{144}$$

The constant D is called the *flexural rigidity* of a plate. In the particular case when the plate is bent to a cylindrical surface with generators parallel to the y axis we have $\partial^2 w/\partial y^2 = 0$, and, from Eqs. (144),

$$\begin{aligned} M_1 &= -D\frac{\partial^2 w}{\partial x^2} \\ M_2 &= -\nu D\frac{\partial^2 w}{\partial x^2} \end{aligned} \tag{145}$$

For the particular case in which $M_1 = M_2 = M$, we have

$$\frac{1}{R_1} = \frac{1}{R_2} = \frac{1}{R}$$

The plate is bent to a spherical surface and the relation between the curvature and the bending moment is, from Eq. (c),

$$M = \frac{Eh^3}{12(1-\nu)}\frac{1}{R} = \frac{D(1+\nu)}{R} \tag{146}$$

We shall have use for these results later.

The formulas (144) are used in the theory of plates when the bending moments are not uniform, and are accompanied by shear forces and surface pressures. For these circumstances they can be deduced from the general equations of Chap. 8 as approximations valid when the plate is thin. The elementary theory of bending of bars can be related to the general equations in a similar manner.[1]

[1] J. N. Goodier, *Trans. Roy. Soc. Can.*, sect. III, 3d ser., vol. 32, p. 65, 1938.

chapter 10

Torsion

104 Torsion of Straight Bars

It has already been shown (Art. 101) that the exact solution of the torsional problem for a circular shaft is obtained if we assume that the cross sections of the bar remain plane and rotate without any distortion during twist. This theory, developed by Coulomb,[1] was applied later by Navier[2] to bars of noncircular cross sections. Making the above assumption he arrived at the erroneous conclusions that, for a given torque, the angle of twist of bars is inversely proportional to the centroidal polar moment of inertia of the cross section, and that the maximum shearing stress occurs at the points most remote from the centroid of the cross section.[3] It is easy to see that the above assumption is in contradiction with the boundary conditions. Take, for instance, a bar of rectangular cross section (Fig. 149). From Navier's assumption it follows that at any point A on the boundary the shearing stress should act in the direction perpen-

[1] "Histoire de l'Académie," 1784, pp. 229–269, Paris, 1787.

[2] Navier, "Résumé des Leçons sur l'Application de la Mécanique," 3d ed., Paris, 1864, edited by Saint-Venant.

[3] These conclusions are correct for a thin elastic layer, corresponding to a slice of the bar between two cross sections, attached to rigid plates. See J. N. Goodier, *J. Appl. Phys.*, vol. 13, p. 167, 1942.

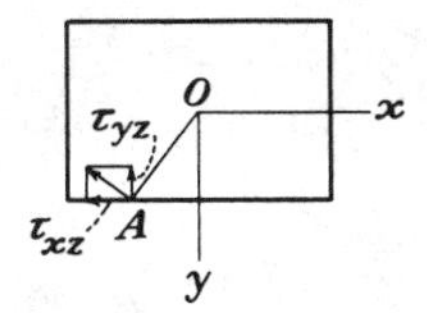

Fig. 149

dicular to the radius OA. Resolving this stress into two components τ_{xz} and τ_{yz}, it is evident that there should be a complementary shearing stress, equal to τ_{yz}, on the element of the lateral surface of the bar at the point A (see page 5), which is in contradiction with the assumption that the lateral surface of the bar is free from external forces, the twist being produced by couples applied at the ends. A simple experiment with a rectangular bar, represented in Fig. 150, shows that the cross sections of the bar do not remain plane during torsion, and that the distortions of rectangular elements on the surface of the bar are greatest at the middles of the sides, i.e., at the points which are nearest to the axis of the bar.

The correct solution of the problem of torsion of bars by couples applied at the ends was given by Saint-Venant.[1]

[1] "Mémoires Savants Etrangers," vol. 14, 1855. See also Saint-Venant's note to Navier's book, *loc. cit.*, and I. Todhunter and K. Pearson, "History of the Theory of Elasticity," vol. 2.

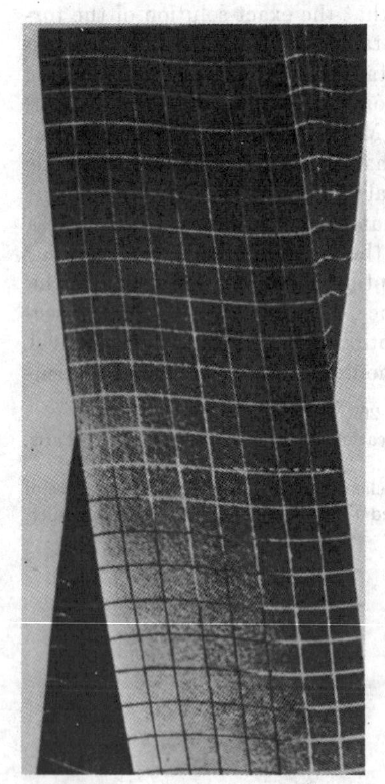

Fig. 150

He used the so-called *semi-inverse method*. That is, at the start he made certain assumptions as to the deformation of the twisted bar and showed that with these assumptions he could satisfy the equations of equilibrium (123) and the boundary conditions (124). Then from the uniqueness of solutions of the elasticity equations (Art. 96) it follows that the assumptions made at the start are correct and the solution obtained is the exact solution of the torsion problem, provided that the torques on the ends are applied as shear stress in exactly the manner required by the solution itself.

Consider a uniform bar of any cross section twisted by couples applied at the ends, Fig. 151. Guided by the solution for a circular shaft (page 282), Saint-Venant assumes that the deformation of the twisted shaft consists (1) of rotations of cross sections of the shaft as in the case of a circular shaft and (2) of *warping* of the cross sections which is the same for all cross sections. Taking the origin of coordinates in an end cross section (Fig. 151), we find that the displacements corresponding to rotation of cross sections are

$$u = -\theta zy \qquad v = \theta zx \tag{a}$$

where θz is the angle of rotation of the cross section at a distance z from the origin.

The warping of cross sections is defined by a function ψ by writing

$$w = \theta\psi(x,y) \tag{b}$$

With the assumed[1] displacements (*a*) and (*b*) we calculate the components of strain from Eqs. (2), which give

$$\begin{aligned} \epsilon_x &= \epsilon_y = \epsilon_z = \gamma_{xy} = 0 \\ \gamma_{xz} &= \frac{\partial w}{\partial x} + \frac{\partial u}{\partial z} = \theta\left(\frac{\partial \psi}{\partial x} - y\right) \\ \gamma_{yz} &= \frac{\partial w}{\partial y} + \frac{\partial v}{\partial z} = \theta\left(\frac{\partial \psi}{\partial y} + x\right) \end{aligned} \tag{c}$$

[1] It has been shown that no other form of displacement linear in the twist θ could exist if every thin slice of the bar is in the same state. See J. N. Goodier and W. S. Shaw, *J. Mech. Phys. Solids*, vol. 10, pp. 35–52, 1962.

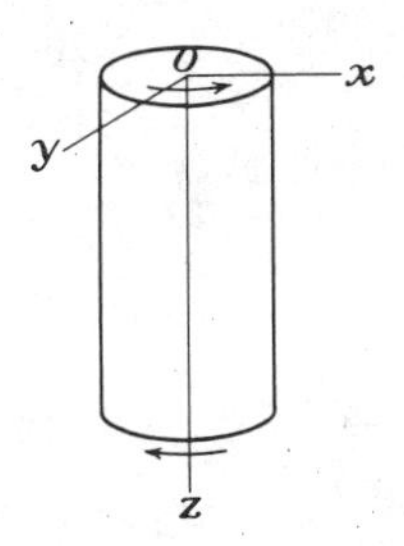

Fig. 151

The corresponding components of stress, from Eqs. (3) and (6), are

$$\sigma_x = \sigma_y = \sigma_z = \tau_{xy} = 0$$

$$\tau_{xz} = G\theta\left(\frac{\partial\psi}{\partial x} - y\right)$$

$$\tau_{yz} = G\theta\left(\frac{\partial\psi}{\partial y} + x\right) \tag{d}$$

It can be seen that with the assumptions (*a*) and (*b*) regarding the deformation, there will be no normal stresses acting between the longitudinal fibers of the shaft or in the longitudinal direction of those fibers. There also will be no distortion in the planes of cross sections, since ϵ_x, ϵ_y, γ_{xy} vanish. We have at each point pure shear, defined by the components τ_{xz} and τ_{yz}. The function $\psi(x,y)$, defining warping of cross section, must now be determined in such a way that equations of equilibrium (123) will be satisfied. Substituting expressions (*d*) in these equations and neglecting body forces, we find that the function ψ must satisfy the equation

$$\frac{\partial^2\psi}{\partial x^2} + \frac{\partial^2\psi}{\partial y^2} = 0 \tag{147}$$

Consider now the boundary conditions (124). For the lateral surface of the bar, which is free from external forces and has normals perpendicular to the z axis, we have $\bar{X} = \bar{Y} = \bar{Z} = 0$ and $\cos Nz = n = 0$. The first two of Eqs. (124) are identically satisfied and the third gives

$$\tau_{xz}l + \tau_{yz}m = 0 \tag{e}$$

which means that the resultant shearing stress at the boundary is directed along the tangent to the boundary, Fig. 152. It was shown before (see page 292) that this condition must be satisfied if the lateral surface of the bar is free from external forces.

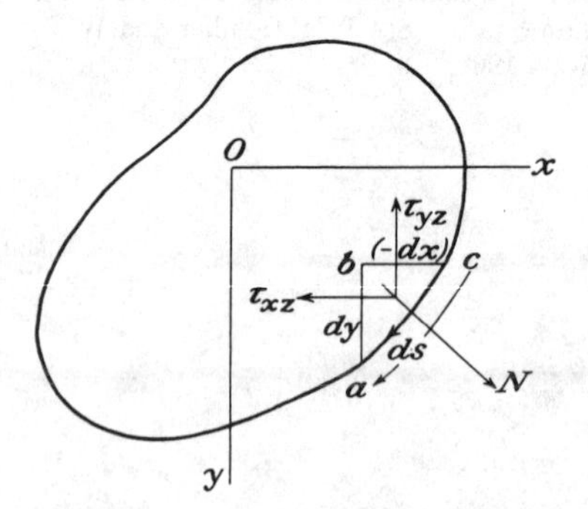

Fig. 152

Considering an infinitesimal element *abc* at the boundary and assuming that s is increasing in the direction from c to a, we have

$$l = \cos Nx = \frac{dy}{ds} \qquad m = \cos Ny = -\frac{dx}{ds}$$

and Eq. (*e*) becomes

$$\left(\frac{\partial \psi}{\partial x} - y\right)\frac{dy}{ds} - \left(\frac{\partial \psi}{\partial y} + x\right)\frac{dx}{ds} = 0 \tag{148}$$

Thus each problem of torsion is reduced to the problem of finding a function ψ satisfying Eq. (147) and the boundary condition (148).

An alternative procedure, which has the advantage of leading to a simpler boundary condition, is as follows. In view of the vanishing of σ_x, σ_y, σ_z, τ_{xy} [Eqs. (*d*)], the equations of equilibrium (123) reduce to

$$\frac{\partial \tau_{xz}}{\partial z} = 0 \qquad \frac{\partial \tau_{yz}}{\partial z} = 0 \qquad \frac{\partial \tau_{xz}}{\partial x} + \frac{\partial \tau_{yz}}{\partial y} = 0$$

The first two are already satisfied since τ_{xz} and τ_{yz}, as given by Eqs. (*d*), are independent of z. The third means that we can express τ_{xz} and τ_{yz} as

$$\tau_{xz} = \frac{\partial \phi}{\partial y} \qquad \tau_{yz} = -\frac{\partial \phi}{\partial x} \tag{149}$$

where ϕ is a function of x and y, called the *stress function.*[1]

From Eqs. (149) and (*d*) we have

$$\frac{\partial \phi}{\partial y} = G\theta\left(\frac{\partial \psi}{\partial x} - y\right) \qquad -\frac{\partial \phi}{\partial x} = G\theta\left(\frac{\partial \psi}{\partial y} + x\right) \tag{f}$$

Eliminating ψ by differentiating the first with respect to y, the second with respect to x, and subtracting from the first, we find that the stress function must satisfy the differential equation

$$\frac{\partial^2 \phi}{\partial x^2} + \frac{\partial^2 \phi}{\partial y^2} = F \tag{150}$$

where

$$F = -2G\theta \tag{151}$$

The boundary condition (*e*) becomes, introducing Eqs. (149),

$$\frac{\partial \phi}{dy}\frac{dy}{ds} + \frac{\partial \phi}{\partial x}\frac{dx}{ds} = \frac{d\phi}{ds} = 0 \tag{152}$$

This shows that the stress function ϕ must be constant along the boundary of the cross section. In the case of singly connected sections, e.g., for solid bars, this constant can be chosen arbitrarily, and in the following discussion we shall take it equal to zero. Thus the determina-

[1] It was introduced by L. Prandtl. See *Physik. Z.*, vol. 4, 1903.

tion of the stress distribution over a cross section of a twisted bar consists in finding the function ϕ that satisfies Eq. (150) and is zero at the boundary. Several applications of this general theory to particular shapes of cross sections will be shown later.

Let us consider now the conditions at the ends of the twisted bar. The normals to the end cross sections are parallel to the z axis. Hence $l = m = 0$, $n = \pm 1$ and Eqs. (124) become

$$\bar{X} = \pm \tau_{xz} \qquad \bar{Y} = \pm \tau_{yz} \tag{g}$$

in which the + sign should be taken for the end of the bar for which the external normal has the direction of the positive z axis, as for the lower end of the bar in Fig. 151. We see that over the ends the shearing forces are distributed in the same manner as the shearing stresses over the cross sections of the bar. It is easy to prove that these forces give us a torque. Substituting in Eqs. (g) from (149) and observing that ϕ at the boundary is zero, we find

$$\iint \bar{X}\,dx\,dy = \iint \tau_{xz}\,dx\,dy = \iint \frac{\partial \phi}{\partial y}\,dx\,dy = \int dx \int \frac{\partial \phi}{\partial y}\,dy = 0$$

$$\iint \bar{Y}\,dx\,dy = \iint \tau_{yz}\,dx\,dy = -\iint \frac{\partial \phi}{\partial x}\,dx\,dy = -\int dy \int \frac{\partial \phi}{\partial x}\,dx = 0$$

Thus the resultant of the forces distributed over the ends of the bar is zero, and these forces represent a couple the magnitude of which is

$$M_t = \iint (\bar{Y}x - \bar{X}y)\,dx\,dy = -\iint \frac{\partial \phi}{\partial x}\,x\,dx\,dy - \iint \frac{\partial \phi}{\partial y}\,y\,dx\,dy \tag{h}$$

Integrating this by parts, and observing that $\phi = 0$ at the boundary, we find

$$M_t = 2\iint \phi\,dx\,dy \tag{153}$$

each of the integrals in the last member of Eqs. (h) contributing one half of this torque. Thus we find that half the torque is due to the stress component τ_{xz} and the other half to τ_{yz}.

We see that by assuming the displacements (a) and (b), and determining the stress components τ_{xz}, τ_{yz} from Eqs. (149), (150), and (152), we obtain a stress distribution that satisfies the equations of equilibrium (123), leaves the lateral surface of the bar free from external forces, and sets up at the ends the torque given by Eq. (153). The compatibility conditions (126) need not be considered. The stress has been derived from the displacements (a) and (b). The question of compatibility reduces to the existence of the single displacement function ψ, which is ensured by (150) as the result of eliminating ψ from (f). Thus all the equations of elasticity are satisfied and the solution obtained in this manner is the exact solution of the torsion problem.

It was pointed out that the solution requires that the forces at the ends of the bar should be distributed in a definite manner. But the practical application of the solution is not limited to such cases. From Saint-Venant's principle it follows that in a long twisted bar, at a sufficient distance from the ends, the stresses depend only on the magnitude of the torque M_t and are practically independent of the manner in which the tractions are distributed over the ends.

105 | Elliptic Cross Section

Let the boundary of the cross section (Fig. 153) be given by the equation

$$\frac{x^2}{a^2} + \frac{y^2}{b^2} - 1 = 0 \tag{a}$$

Then Eq. (150) and the boundary condition (152) are satisfied by taking the stress function in the form

$$\phi = m\left(\frac{x^2}{a^2} + \frac{y^2}{b^2} - 1\right) \tag{b}$$

in which m is a constant. Substituting (*b*) into Eq. (150), we find

$$m = \frac{a^2b^2}{2(a^2 + b^2)} F$$

Hence

$$\phi = \frac{a^2b^2F}{2(a^2 + b^2)}\left(\frac{x^2}{a^2} + \frac{y^2}{b^2} - 1\right) \tag{c}$$

The magnitude of the constant F will now be determined from Eq. (153). Substituting in this equation from (*c*), we find

$$M_t = \frac{a^2b^2F}{a^2 + b^2}\left(\frac{1}{a^2}\iint x^2\,dx\,dy + \frac{1}{b^2}\iint y^2\,dx\,dy - \iint dx\,dy\right) \tag{d}$$

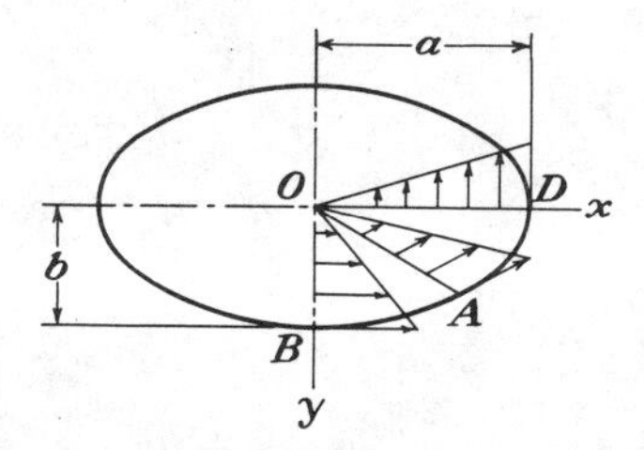

Fig. 153

Since

$$\iint x^2\,dx\,dy = I_y = \frac{\pi ba^3}{4} \qquad \iint y^2\,dx\,dy = I_x = \frac{\pi ab^3}{4} \qquad \iint dx\,dy = \pi ab$$

we find, from (*d*),

$$M_t = -\frac{\pi a^3 b^3 F}{2(a^2+b^2)}$$

from which

$$F = -\frac{2M_t(a^2+b^2)}{\pi a^3 b^3} \tag{e}$$

Then, from (*c*),

$$\phi = -\frac{M_t}{\pi ab}\left(\frac{x^2}{a^2}+\frac{y^2}{b^2}-1\right) \tag{f}$$

Substituting in Eqs. (149), the stress components are

$$\tau_{xz} = -\frac{2M_t y}{\pi ab^3} \qquad \tau_{yz} = \frac{2M_t x}{\pi a^3 b} \tag{154}$$

The ratio of the stress components is proportional to the ratio y/x and hence is constant along any radius such as OA (Fig. 153). This means that the resultant shearing stress along any radius OA has a constant direction that evidently coincides with the direction of the tangent to the boundary at the point A. Along the vertical axis OB the stress component τ_{yz} is zero, and the resultant stress is equal to τ_{xz}. Along the horizontal axis OD the resultant shearing stress is equal to τ_{yz}. It is evident that the maximum stress is at the boundary, and it can easily be proved that this maximum occurs at the ends of the minor axis of the ellipse. Substituting $y = b$ in the first of Eqs. (154), we find that the absolute value of this maximum is

$$\tau_{\max} = \frac{2M_t}{\pi ab^2} \tag{155}$$

For $a = b$ this formula coincides with the well-known formula for a circular cross section.

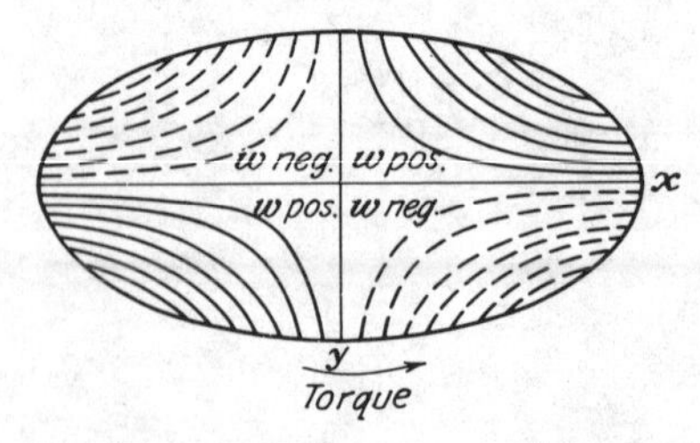

Fig. 154

Substituting (*e*) in Eq. (151) we find the expression for the angle of twist

$$\theta = M_t \frac{a^2 + b^2}{\pi a^3 b^3 G} \tag{156}$$

The factor by which we divide torque to obtain the twist per unit length is called the *torsional rigidity*. Denoting it by C, its value for the elliptic cross section, from (156), is

$$C = \frac{\pi a^3 b^3 G}{a^2 + b^2} = \frac{G}{4\pi^2} \frac{(A)^4}{I_p} \tag{157}$$

in which

$$A = \pi ab \qquad I_p = \frac{\pi ab^3}{4} + \frac{\pi ba^3}{4}$$

are the area and centroidal moment of inertia of the cross section.

Having the stress components (154) we can easily obtain the displacements. The components u and v are given by Eqs. (*a*) of Art. 104. The displacement w is found from Eqs. (*d*) and (*b*) of Art. 104. Substituting from Eqs. (154) and (156) and integrating, we find

$$w = M_t \frac{(b^2 - a^2)xy}{\pi a^3 b^3 G} \tag{158}$$

This shows that the contour lines for the warped cross section are hyperbolas having the principal axes of the ellipse as asymptotes (Fig. 154).

106 | Other Elementary Solutions

In studying the torsional problem, Saint-Venant discussed several solutions of Eq. (150) in the form of polynomials. To solve the problem, let us represent the stress function in the form

$$\phi = \phi_1 + \frac{F}{4}(x^2 + y^2) \tag{a}$$

Then, from Eq. (150),

$$\frac{\partial^2 \phi_1}{\partial x^2} + \frac{\partial^2 \phi_1}{\partial y^2} = 0 \tag{b}$$

and along the boundary, from Eq. (152),

$$\phi_1 + \frac{F}{4}(x^2 + y^2) = \text{const} \tag{c}$$

Thus the torsional problem is reduced to obtaining solutions of Eq. (*b*) satisfying the boundary condition (*c*). To get solutions in the form of polynomials, we take the function of the complex variable

$$(x + iy)^n \tag{d}$$

The real and the imaginary parts of this expression are each solutions of Eq. (*b*) (see page 171). Taking, for instance, $n = 2$ we obtain the solutions $x^2 - y^2$ and $2xy$. With $n = 3$ we obtain solutions $x^3 - 3xy^2$ and $3x^2y - y^3$. With $n = 4$, we arrive at solutions in the form of homogeneous functions of the fourth degree, and so on. Combining such solutions we can obtain various solutions in the form of polynomials.

Taking, for instance,

$$\phi = \frac{F}{4}(x^2 + y^2) + \phi_1 = \frac{F}{2}\left[\frac{1}{2}(x^2 + y^2) - \frac{1}{2a}(x^3 - 3xy^2) + b\right] \tag{e}$$

we obtain a solution of Eq. (150) in the form of a polynomial of the third degree with constants a and b which will be adjusted later. This polynomial is a solution of the torsional problem if it satisfies the boundary condition (152), i.e., if the boundary of the cross section of the bar is given by the equation

$$\frac{1}{2}(x^2 + y^2) - \frac{1}{2a}(x^3 - 3xy^2) + b = 0 \tag{f}$$

By changing the constant b in this equation, we obtain various shapes of the cross section.

Taking $b = -\frac{2}{27}a^2$ we arrive at the solution for the equilateral triangle. Equation (f) in this case can be presented in the form

$$(x - \sqrt{3}\,y - \tfrac{2}{3}a)(x + \sqrt{3}\,y - \tfrac{2}{3}a)(x + \tfrac{1}{3}a) = 0$$

which is the product of the three equations of the sides of the triangle shown in Fig. 155. Observing that $F = -2G\theta$ and substituting

$$\phi = -G\theta\left[\frac{1}{2}(x^2 + y^2) - \frac{1}{2a}(x^3 - 3xy^2) - \frac{2}{27}a^2\right] \tag{g}$$

into Eqs. (149), we obtain the stress components τ_{xz} and τ_{yz}. Along the x axis, $\tau_{xz} = 0$, from symmetry, and we find, from (g),

$$\tau_{yz} = \frac{3G\theta}{2a}\left(\frac{2ax}{3} - x^2\right) \tag{h}$$

The largest stress is found at the middle of the sides of the triangle, where, from (h),

$$\tau_{\max} = \frac{G\theta a}{2} \tag{k}$$

At the corners of the triangle the shearing stress is zero (see Fig. 155).

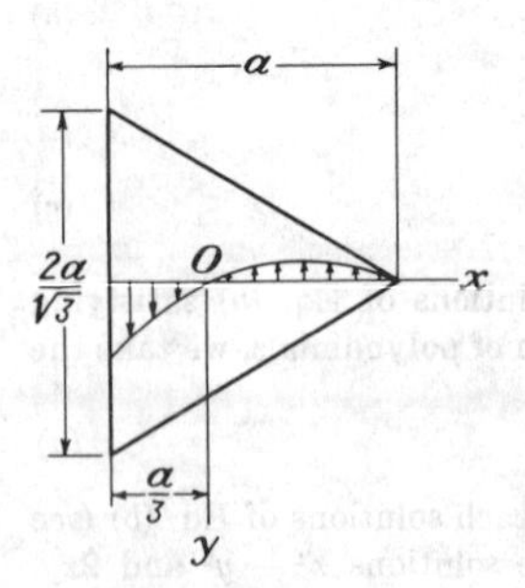

Fig. 155

Substituting (*g*) into Eq. (153), we find

$$M_t = \frac{G\theta a^4}{15\sqrt{3}} = \frac{3}{5}\theta G I_p \tag{l}$$

Taking a solution of Eq. (150) in the form of a polynomial of the fourth degree containing only even powers of x and y, we obtain the stress function

$$\phi = -G\theta\left[\frac{1}{2}(x^2 + y^2) - \frac{a}{2}(x^4 - 6x^2y^2 + y^4) + \frac{1}{2}(a - 1)\right]$$

The boundary condition (152) is satisfied if the boundary of the cross section is given by the equation

$$x^2 + y^2 - a(x^4 - 6x^2y^2 + y^4) + a - 1 = 0$$

By changing a, Saint-Venant obtained the family of cross sections shown in Fig. 156*a*, Combining solutions in the form of polynomials of the fourth and eighth degrees. he arrived at the cross section shown in Fig. 156*b*.

On the basis of his investigations, Saint-Venant drew certain general conclusions of practical interest. He showed that, in the case of singly connected sections and for a given cross-sectional area, the torsional rigidity increases if the polar moment of inertia of the cross section decreases. Thus, for a given amount of material the circular shaft gives the largest torsional rigidity. Similar conclusions can be drawn regarding the maximum shearing stress. For a given torque and cross-sectional area the maximum stress is the smallest for the cross section with the smallest polar moment of inertia.

Comparing various singly connected cross sections, Saint-Venant found that the torsional rigidity can be calculated approximately by using Eq. (157), i.e., by replacing the given shaft by the shaft

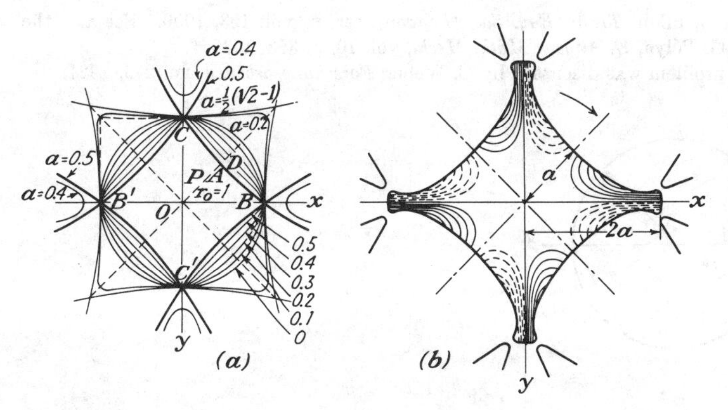

Fig. 156

of an elliptic cross section having the same cross-sectional area and the same polar moment of inertia as the given shaft has.

The maximum stress in all cases discussed by Saint-Venant was obtained at the boundary at the points which are the nearest to the centroid of the cross section. A more detailed investigation of this question by Filon[1] showed that there are cases where the points of maximum stress, although always at the boundary, are not the nearest points to the centroid of the cross section.

Taking $n = 1$ and $n = -1$ in expression (*d*), and using polar coordinates r and ψ, we obtain the following solutions of Eq. (*b*):

$$\phi_1 = r \cos \psi \qquad \phi_1 = \frac{1}{r} \cos \psi$$

Then the stress function (*a*) can be taken in the form

$$\phi = \frac{F}{4}(x^2 + y^2) - \frac{Fa}{2} r \cos \psi + \frac{Fb^2}{2}\frac{a}{r} \cos \psi - \frac{F}{4} b^2 \tag{m}$$

in which a and b are constants. It will satisfy the boundary condition (152) if at the boundary of the cross section we have $\phi = 0$, or, from (*m*),

$$r^2 - b^2 - 2a(r^2 - b^2)\frac{\cos \psi}{r} = 0 \tag{n}$$

or

$$(r^2 - b^2)\left(1 - \frac{2a \cos \psi}{r}\right) = 0 \tag{o}$$

which represents the equation of the boundary of the cross section shown[2] in Fig. 157. By taking

$$r^2 - b^2 = 0$$

we obtain a circle of radius b with the center at the origin; and by taking

$$1 - \frac{2a \cos \psi}{r} = 0$$

[1] L. N. G. Filon, *Trans. Roy. Soc.* (*London*), ser. A, vol. 193, 1900. See also the paper by G. Pólya, *Z. Angew. Math. Mech.*, vol. 10, p. 353, 1930.

[2] This problem was discussed by C. Weber, *Forschungsarbeiten*, no. 249, 1921.

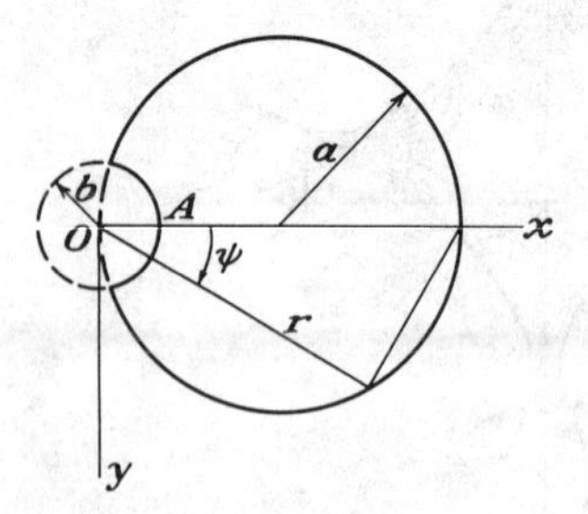

Fig. 157

we have a circle of radius a touching the y axis at the origin. The maximum shearing stress is at the point A and is

$$\tau_{\max} = G\theta(2a - b) \tag{p}$$

When b is very small in comparison with a, that is, when we have a semicircular longitudinal groove of very small radius, the stress at the bottom of the groove is twice as great as the maximum stress in the circular shaft of radius a without the groove.

107 | Membrane Analogy

In the solution of torsional problems the *membrane analogy*, introduced by L. Prandtl,[1] has proved very valuable. Imagine a homogeneous membrane (Fig. 158) supported at the edges, with the same outline as that of the cross section of the twisted bar, subjected to a uniform tension at the edges and a uniform lateral pressure. If q is the pressure per unit area of the membrane and S is the uniform tension per unit length of its boundary, the tensile forces acting on the sides ad and bc of an infinitesimal element $abcd$ (Fig. 158) give, in the case of small deflections of the membrane, a resultant in the upward direction $-S(\partial^2 z/\partial x^2)\, dx\, dy$. In the same manner, the tensile forces acting on the other two sides of the element give the resultant $-S(\partial^2 z/\partial y^2)\, dx\, dy$ and the equation of equilibrium of the ele-

[1] *Physik. Z.*, vol. 4, 1903. See also Anthes, *Dinglers Polytech. J.*, p. 342, 1906. Further development of the analogy and applications in various cases are given in the papers by A. A. Griffith and G. I. Taylor, *Tech. Rept. Adv. Comm. Aeron.*, vol. 3, pp. 910 and 938, 1917–1918.

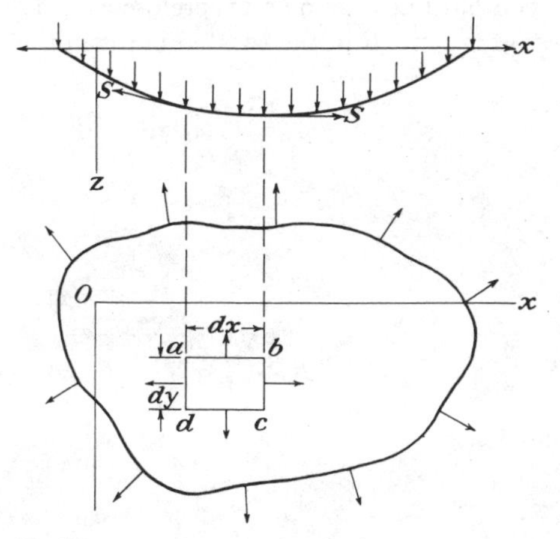

Fig. 158

ment is

$$q\,dx\,dy + S\frac{\partial^2 z}{\partial x^2}\,dx\,dy + S\frac{\partial^2 z}{\partial y^2}\,dx\,dy = 0$$

from which

$$\frac{\partial^2 z}{\partial x^2} + \frac{\partial^2 z}{\partial y^2} = -\frac{q}{S} \tag{159}$$

At the boundary the deflection of the membrane is zero. Comparing Eq. (159) and the boundary condition for the deflections z of the membrane with Eq. (150) and the boundary condition (152) (see page 295) for the stress function ϕ, we conclude that these two problems are identical. Hence, from the deflections of the membrane we can obtain values of ϕ by replacing the quantity $-(q/S)$ of Eq. (159) with the quantity $F = -2G\theta$ of Eq. (150).

Having the deflection surface of the membrane represented by contour lines (Fig. 159), several important conclusions regarding stress distribution in torsion can be obtained. Consider any point B on the membrane. The deflection of the membrane along the contour line through this point is constant, and we have

$$\frac{\partial z}{\partial s} = 0$$

The corresponding equation for the stress function ϕ is

$$\frac{\partial \phi}{\partial s} = \left(\frac{\partial \phi}{\partial y}\frac{dy}{ds} + \frac{\partial \phi}{\partial x}\frac{dx}{ds}\right) = \tau_{xz}\frac{dy}{ds} - \tau_{yz}\frac{dx}{ds} = 0$$

This expresses that the projection of the resultant shearing stress at a point B on the normal N to the contour line is zero and therefore we may conclude that the shearing stress at a point B in the twisted bar is in the

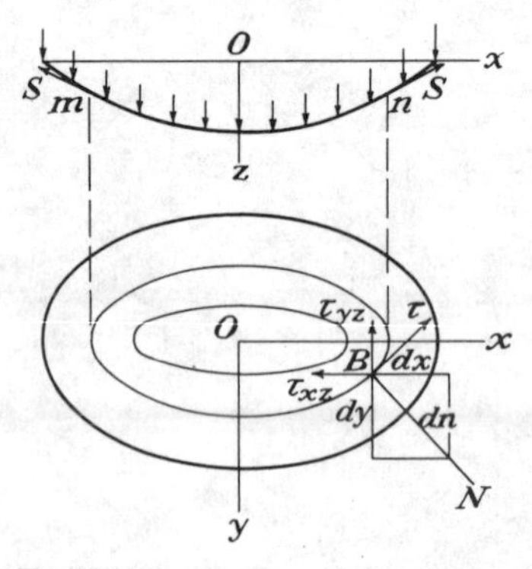

Fig. 159

direction of the tangent to the contour line through this point. The curves drawn in the cross section of a twisted bar, in such a manner that the resultant shearing stress at any point of the curve is in the direction of the tangent to the curve, are called *lines of shearing stress.* Thus, the contour lines of the membrane are the lines of shearing stress for the cross section of the twisted bar.

The magnitude of the resultant stress τ at B (Fig. 159) is obtained by projecting on the tangent the stress components τ_{xz} and τ_{yz}. Then

$$\tau = \tau_{yz} \cos (Nx) - \tau_{xz} \cos (Ny)$$

Substituting

$$\tau_{xz} = \frac{\partial \phi}{\partial y} \qquad \tau_{yz} = -\frac{\partial \phi}{\partial x} \qquad \cos (Nx) = \frac{dx}{dn} \qquad \cos (Ny) = \frac{dy}{dn}$$

we obtain

$$\tau = -\left(\frac{\partial \phi}{\partial x}\frac{dx}{dn} + \frac{\partial \phi}{\partial y}\frac{dy}{dn}\right) = -\frac{d\phi}{dn}$$

Thus, the magnitude of the shearing stress at B is given by the maximum slope of the membrane at this point. It is only necessary in the expression for the slope to replace q/S by $2G\theta$. From this it can be concluded that the maximum shear acts at the points where the contour lines are closest to each other.

From Eq. (153) it can be concluded that double the volume bounded by the deflected membrane and the xy plane (Fig. 159) represents the torque, provided q/S is replaced by $2G\theta$.

It may be observed that the form of the membrane, and therefore the stress distribution, is the same no matter what point in the cross section is taken for origin in the torsion problem. This point, of course, represents the axis of rotation of the cross sections. It is at first sight surprising that the cross sections can rotate about a different (parallel) axis when still subjected to the same torque. The difference, however, is merely a matter of rigid-body rotation. Consider, for instance, a circular cylinder twisted by rotations about the central axis. A generator on the surface becomes inclined to its original direction but can be brought back by a rigid-body rotation of the whole cylinder about a diameter. The final positions of the cross sections then correspond to torsional rotations about this generator as a fixed axis. The cross sections remain plane but become inclined to their original planes by virtue of the rigid-body rotation of the cylinder. In an arbitrary section there will be warping, and with a given choice of axis the inclination of a given element of area in the end section is definite, $\partial w/\partial x$ and $\partial w/\partial y$ being given by Eqs. (*d*) and (*b*) of Art. 104. Such an element can be brought back to its original orientation by a rigid-body rotation about an axis in the end section. This rotation will change the axis of the torsional rotations to a parallel axis. Thus, a definite axis or center of torsional rotation, or *center of torsion,* can be identified provided the final orientation of an element of area in the end section is specified—as, for instance, if the element is completely fixed.

Let us consider now the equilibrium condition of the portion mn of the membrane bounded by a contour line (Fig. 159). The slope of the membrane along this line is proportional at each point to the shearing stress τ and equal to $\tau(q/S)1/2G\theta$. Then denoting by A the horizontal projection of the portion mn of the membrane, the equation of equilibrium of this portion is

$$\int S\left(\tau\frac{q}{S}\frac{1}{2G\theta}\right)ds = qA$$

or

$$\int \tau\, ds = 2G\theta A \tag{160}$$

From this the average value of the shearing stress along a contour line can be obtained.

By taking $q = 0$, that is, considering a membrane without lateral load, we arrive at the equation

$$\frac{\partial^2 z}{\partial x^2} + \frac{\partial^2 z}{\partial y^2} = 0 \tag{161}$$

which coincides with Eq. (b) of the previous article for the function ϕ_1. Taking the ordinates of the membrane at the boundary so that

$$z + \frac{F}{4}(x^2 + y^2) = \text{const} \tag{162}$$

the boundary condition (c) of the previous article is also satisfied. Thus we can obtain the function ϕ_1 from the deflection surface of an unloaded membrane, provided the ordinates of the membrane surface have definite values at the boundary. It will be shown later that both loaded and unloaded membranes can be used for determining stress distributions in twisted bars by experiment.

The membrane analogy is useful not only when the bar is twisted within the elastic limit but also when the material yields in certain portions of the cross section.[1] Assuming that the shearing stress remains constant during yielding, the stress distribution in the elastic zone of the cross section is represented by the membrane as before, but in the plastic zone the stress will be given by a surface having a constant maximum slope corresponding to the yield stress. Imagine such a surface constructed as a roof on the cross section of the bar and the membrane stretched and loaded as explained before. On increasing the pressure we arrive at the condition when the membrane begins to touch the roof. This corresponds to the beginning of plastic flow in the twisted bar. As the pressure is increased, certain portions of the membrane come into contact with the roof. These portions of contact give us the regions of plastic

[1] This was indicated by L. Prandtl; see A. Nádai, *Z. Angew. Math. Mech.*, vol. 3, p. 442, 1923. See also E. Trefftz, *Z. Angew. Math. Mech.*, vol. 5, p. 64, 1925.

flow in the twisted bar. Interesting experiments illustrating this theory were made by A. Nádai.[1]

108 | Torsion of a Bar of Narrow Rectangular Cross Section

In the case of a narrow rectangular cross section the membrane analogy gives a very simple solution of the torsional problem. Neglecting the effect of the short sides of the rectangle and assuming that the surface of the slightly deflected membrane is cylindrical (Fig. 160*b*), we obtain the deflection of the membrane from the elementary formula for the parabolic deflection curve of a uniformly loaded string:

$$\delta = \frac{qc^2}{8S} \tag{a}$$

From the known properties of parabolic curves, the maximum slope, which occurs in the middle portions of the long sides of the rectangle, is equal to

$$\frac{4\delta}{c} = \frac{qc}{2S} \tag{b}$$

The volume bounded by the deflected membrane and the xy plane, calculated as for a parabolic cylinder, is

$$V = \frac{2}{3}c\,\delta b = \frac{qbc^3}{12S} \tag{c}$$

Now using the membrane analogy and substituting $2G\theta$ for q/S in (*b*) and (*c*), we find

$$\tau_{\max} = cG\theta \qquad M_t = \tfrac{1}{3}bc^3G\theta \tag{d}$$

[1] See *Trans. ASME,* Applied Mechanics Division, 1930. See also A. Nádai, "Theory of Flow and Fracture of Solids," chaps. 35 and 36, 1950.

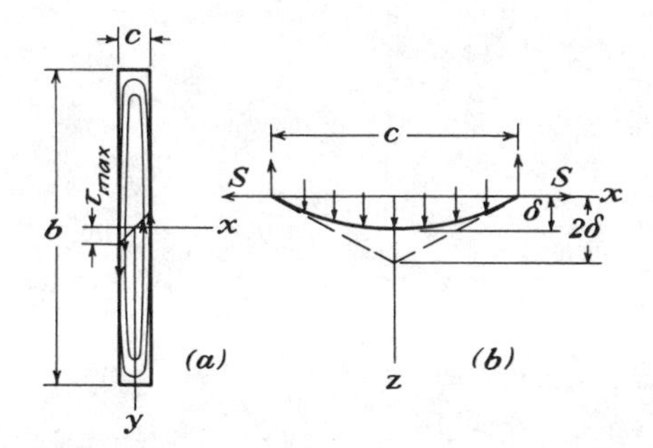

Fig. 160

from which

$$\theta = \frac{M_t}{\frac{1}{3}bc^3G} \tag{163}$$

$$\tau_{\max} = \frac{M_t}{\frac{1}{3}bc^2} \tag{164}$$

From the parabolic deflection curve (Fig. 160*b*)

$$z = \frac{4\delta}{c^2}\left(\frac{c^2}{4} - x^2\right)$$

and the slope of the membrane at any point is

$$\frac{dz}{dx} = -\frac{8\delta x}{c^2} = -\frac{q}{S}x$$

The corresponding stress in the twisted bar is

$$\tau_{yz} = 2G\theta x$$

The stress distribution follows a linear law as shown in Fig. 160*a*. Calculating the magnitude of the torque corresponding to this stress distribution, we find

$$\frac{\tau_{\max}}{4}\,c\,\frac{2}{3}\,cb = \frac{1}{6}bc^2\tau_{\max}$$

This is only one half of the total torque given by Eq. (164). The second half is given by the stress components τ_{xz}, which were entirely neglected when we assumed that the surface of the deflected membrane is cylindrical. Although these stresses have an appreciable magnitude only near the short sides of the rectangle and their maximum values are smaller than $\tau_{\max}$ as

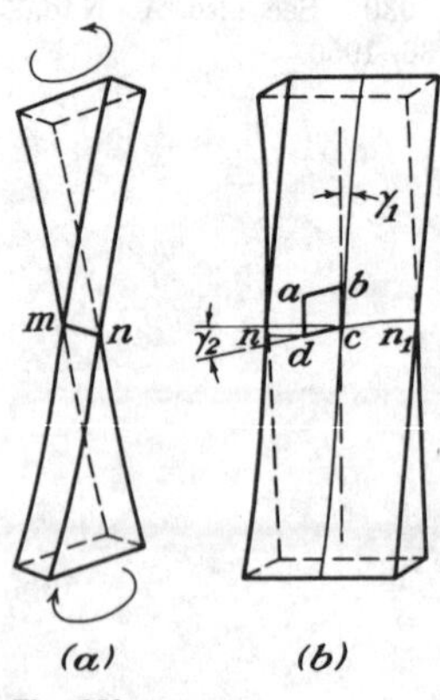

Fig. 161

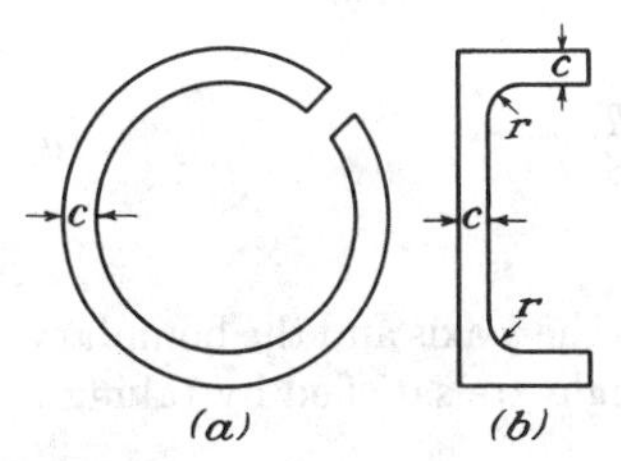

Fig. 162

calculated above, they act at a greater distance from the axis of the bar and their moment represents the second half of the torque M_t.*

It is interesting to note that the τ_{max} given by the first of Eqs. (*d*) is twice as great as in the case of a circular shaft with diameter equal to c and subjected to the same twist θ. This can be explained if we consider the warping of the cross sections. The sides of cross sections such as nn_1 (Fig. 161) remain normal to the longitudinal fibers of the bar at the corners, as is shown at the points n and n_1. The total shear of an element such as *abcd* consists of two parts: the part γ_1 due to rotation of the cross section about the axis of the bar and equal to the shear in the circular bar of diameter c; and the part γ_2 due to warping of the cross section. In the case of a narrow rectangular cross section $\gamma_2 = \gamma_1$, and the resultant shear is twice as great as in the case of a circular cross section of the diameter c.

Equations (163) and (164), obtained above for a narrow rectangle, can also be used in the cases of thin-walled bars of such cross sections as shown in Fig. 162 by setting b equal to the developed length of the cross section. This follows from the fact that if the thickness c of a slotted tube (Fig. 162*a*) is small in comparison with the diameter, the maximum slope of the membrane and the volume bounded by the membrane will be nearly the same as for a narrow rectangular cross section of the width c and of the same length as the circumference of the middle surface of the tube. An analogous conclusion can be made also for a channel (Fig. 162*b*). It should be noted that in this latter case a considerable stress concentration takes place at the reentrant corners, depending on the magnitude of the radius r of the fillets, and Eq. (164) cannot be applied at these points. A more detailed discussion of this subject will be given in Art. 112.

109 | Torsion of Rectangular Bars

Using the membrane analogy, the problem reduces to finding the deflections of a uniformly loaded rectangular membrane as shown in Fig. 163.

* This question was cleared up by Lord Kelvin; see Kelvin and Tait, "Natural Philosophy," vol. 2, p. 267.

These deflections must satisfy the Eq. (159)

$$\frac{\partial^2 z}{\partial x^2} + \frac{\partial^2 z}{\partial y^2} = -\frac{q}{S} \tag{a}$$

and be zero at the boundary.

The condition of symmetry with respect to the y axis and the boundary conditions at the sides $x = \pm a$ of the rectangle are satisfied by taking z in the form of a series,

$$z = \sum_{n=1,3,5,\ldots}^{\infty} b_n \cos \frac{n\pi x}{2a} Y_n \tag{b}$$

in which $b_1, b_3, \ldots$ are constant coefficients and $Y_1, Y_3, \ldots$ are functions of y only. Substituting (*b*) in Eq. (*a*), and observing that the constant on the right of (*a*) can be represented for $-a < x < a$ by the Fourier series

$$-\frac{q}{S} = -\sum_{n=1,3,5,\ldots}^{\infty} \frac{q}{S}\frac{4}{n\pi}(-1)^{(n-1)/2} \cos \frac{n\pi x}{2a} \tag{c}$$

we arrive at the following equation for determining Y_n:

$$Y_n'' - \frac{n^2\pi^2}{4a^2} Y_n = -\frac{q}{S}\frac{4}{n\pi b_n}(-1)^{(n-1)/2} \tag{d}$$

from which

$$Y_n = A \sinh \frac{n\pi y}{2a} + B \cosh \frac{n\pi y}{2a} + \frac{16qa^2}{Sn^3\pi^3 b_n}(-1)^{(n-1)/2} \tag{e}$$

From the condition of symmetry of the deflection surface of the membrane with respect to the x axis, it follows that the constant of integration A must be zero. The constant B is determined from the condition that the deflections of the membrane are zero for $y = \pm b$, that is, $(Y_n)_{y=\pm b} = 0$,

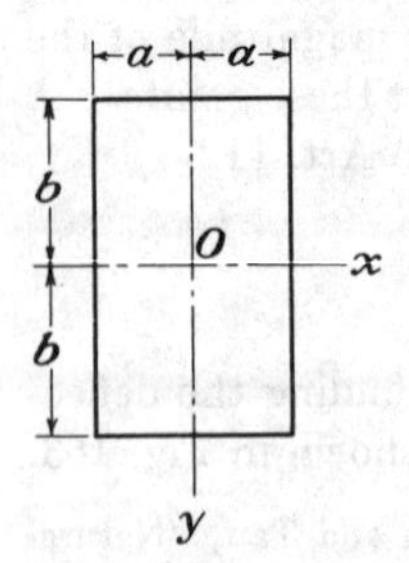

Fig. 163

which gives

$$Y_n = \frac{16qa^2}{Sn^3\pi^3 b_n}(-1)^{(n-1)/2}\left[1 - \frac{\cosh(n\pi y/2a)}{\cosh(n\pi b/2a)}\right] \tag{f}$$

and the general expression for the deflection surface of the membrane, from (*b*), becomes

$$z = \frac{16qa^2}{S\pi^3}\sum_{n=1,3,5,\ldots}^{\infty}\frac{1}{n^3}(-1)^{(n-1)/2}\left[1 - \frac{\cosh(n\pi y/2a)}{\cosh(n\pi b/2a)}\right]\cos\frac{n\pi x}{2a}$$

Replacing q/S by $2G\theta$, we obtain for the stress function

$$\phi = \frac{32G\theta a^2}{\pi^3}\sum_{n=1,3,5,\ldots}^{\infty}\frac{1}{n^3}(-1)^{(n-1)/2}\left[1 - \frac{\cosh(n\pi y/2a)}{\cosh(n\pi b/2a)}\right]\cos\frac{n\pi x}{2a} \tag{g}$$

The stress components are now obtained from Eqs. (159) by differentiation. For instance,

$$\tau_{yz} = -\frac{\partial\phi}{\partial x} = \frac{16G\theta a}{\pi^2}\sum_{n=1,3,5,\ldots}^{\infty}\frac{1}{n^2}(-1)^{(n-1)/2}\left[1 - \frac{\cosh(n\pi y/2a)}{\cosh(n\pi b/2a)}\right]\sin\frac{n\pi x}{2a} \tag{h}$$

Assuming that $b > a$, the maximum shearing stress, corresponding to the maximum slope of the membrane, is at the middle points of the long sides $x = \pm a$ of the rectangle. Substituting $x = a$, $y = 0$ in (*h*), we find

$$\tau_{\max} = \frac{16G\theta a}{\pi^2}\sum_{n=1,3,5,\ldots}^{\infty}\frac{1}{n^2}\left[1 - \frac{1}{\cosh(n\pi b/2a)}\right]$$

or, observing that[1]

$$1 + \frac{1}{3^2} + \frac{1}{5^2} + \cdots = \frac{\pi^2}{8}$$

we have

$$\tau_{\max} = 2G\theta a - \frac{16G\theta a}{\pi^2}\sum_{n=1,3,5,\ldots}^{\infty}\frac{1}{n^2\cosh(n\pi b/2a)} \tag{165}$$

The infinite series on the right side, for $b > a$, converges very rapidly and there is no difficulty in calculating $\tau_{\max}$ with sufficient accuracy for any particular value of the ratio b/a. For instance, in the case of a very narrow rectangle, b/a becomes a large number, so that the sum of the infinite series in (165) can be neglected, and we find

$$\tau_{\max} = 2G\theta a$$

This coincides with the first of the Eqs. (*d*) of the previous article.

[1] See, for instance, H. S. Carslaw, "Fourier Series and Integrals," 3d ed., p. 235, Dover Publications, Inc., New York, 1930.

In the case of a square cross section, $a = b$; and we find, from Eq. (165),

$$\tau_{\max} = 2G\theta a \left\{1 - \frac{8}{\pi^2}\left[\frac{1}{\cosh(\pi/2)} + \frac{1}{9\cosh(3\pi/2)} + \cdots\right]\right\}$$
$$= 2G\theta a \left[1 - \frac{8}{\pi^2}\left(\frac{1}{2.509} + \frac{1}{9 \times 55.67} + \cdots\right)\right] = 1.351G\theta a \quad (166)$$

In general we obtain

$$\tau_{\max} = k2G\theta a \quad (167)$$

in which k is a numerical factor depending on the ratio b/a. Several values of this factor are given in the table below.

Table of Constants for Torsion of a Rectangular Bar

$\frac{b}{a}$	k	k_1	k_2	$\frac{b}{a}$	k	k_1	k_2
1.0	0.675	0.1406	0.208	3	0.985	0.263	0.267
1.2	0.759	0.166	0.219	4	0.997	0.281	0.282
1.5	0.848	0.196	0.231	5	0.999	0.291	0.291
2.0	0.930	0.229	0.246	10	1.000	0.312	0.312
2.5	0.968	0.249	0.258	∞	1.000	0.333	0.333

Let us calculate now the torque M_t as a function of the twist θ. Using Eq. (153) for this purpose, we find

$$M_t = 2\int_{-a}^{a}\int_{-b}^{b} \phi\, dx\, dy = \frac{64G\theta a^2}{\pi^3}\int_{-a}^{a}\int_{-b}^{b}\left\{\sum_{n=1,3,5,\ldots}^{\infty} \frac{1}{n^3}(-1)^{(n-1)/2}\right.$$
$$\left.\left[1 - \frac{\cosh(n\pi y/2a)}{\cosh(n\pi b/2a)}\right]\cos\frac{n\pi x}{2a}\right\} dx\, dy = \frac{32G\theta(2a)^3(2b)}{\pi^4}\sum_{n=1,3,5,\ldots}^{\infty}\frac{1}{n^4}$$
$$- \frac{64G\theta(2a)^4}{\pi^5}\sum_{n=1,3,5,\ldots}^{\infty}\frac{1}{n^5}\tanh\frac{n\pi b}{2a}$$

or, observing that

$$\frac{1}{1} + \frac{1}{3^4} + \frac{1}{5^4} + \cdots = \frac{\pi^4}{96}$$

we have

$$M_t = \frac{1}{3}G\theta(2a)^3(2b)\left(1 - \frac{192}{\pi^5}\frac{a}{b}\sum_{n=1,3,5,\ldots}^{\infty}\frac{1}{n^5}\tanh\frac{n\pi b}{2a}\right) \quad (168)$$

The series on the right side converges very rapidly, and M_t can easily be evaluated for any value of the ratio a/b. In the case of a narrow rectangle we can take

$$\tanh \frac{n\pi b}{2a} = 1$$

Then

$$M_t = \frac{1}{3} G\theta(2a)^3(2b)\left(1 - 0.630\,\frac{a}{b}\right) \tag{169}$$

In the case of a square, $a = b$; and (168) gives

$$M_t = 0.1406G\theta(2a)^4 \tag{170}$$

In general, the torque can be represented by the equation

$$M_t = k_1 G\theta(2a)^3(2b) \tag{171}$$

in which k_1 is a numerical factor depending on the magnitude of the ratio b/a. Several values of this factor are given in the table on page 312.

Substituting the value of θ from Eq. (171) into Eq. (167), we obtain the maximum shearing stress as a function of the torque in the form

$$\tau_{\max} = \frac{M_t}{k_2(2a)^2(2b)} \tag{172}$$

where k_2 is a numerical factor the values of which can be taken from the table on page 312.

110 | Additional Results

By using infinite series as in the previous article, the torsional problem can be solved for several other shapes of cross sections.

In the case of a *sector of a circle*[1] (Fig. 164) the boundaries are given by $\psi = \pm\alpha/2$,

[1] This problem was discussed by Saint-Venant, *Compt. Rend.*, vol. 87, pp. 849 and 893, 1878. See also A. G. Greenhill, *Messenger of Math.*, vol. 10, p. 83, 1880. Another method of solution by using Bessel's function was given by A. Dinnik, *Bull. Don Polytech. Inst., Novotcherkassk*, vol. 1, p. 309. See also A. Föppl and L. Föppl, "Drang und Zwang," p. 96, 1928.

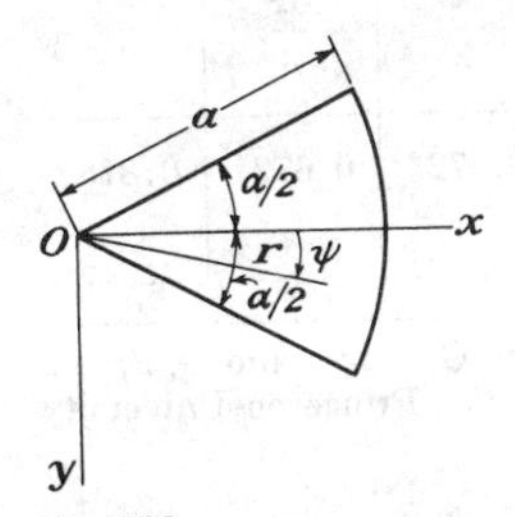

Fig. 164

$r = 0$, $r = a$. We take a stress function in the form

$$\phi = \phi_1 + \frac{F}{4}(x^2 + y^2) = \phi_1 - \frac{G\theta r^2}{2}$$

The function ϕ_1 must satisfy Laplace's equation (see Art. 106). Taking a solution of this equation in the form of the series

$$\phi_1 = \frac{G\theta}{2}\left[\frac{r^2 \cos 2\psi}{\cos\alpha} + a^2 \sum_{n=1,3,5,\ldots}^{\infty} A_n \left(\frac{r}{a}\right)^{n\pi/\alpha} \cos\frac{n\pi\psi}{\alpha}\right]$$

we arrive at the stress function

$$\phi = \frac{G\theta}{2}\left[-r^2\left(1 - \frac{\cos 2\psi}{\cos\alpha}\right) + a^2 \sum_{n=1,3,5,\ldots}^{\infty} A_n \left(\frac{r}{a}\right)^{n\pi/\alpha} \cos\frac{n\pi\psi}{\alpha}\right]$$

This expression is zero at the boundaries:

$$\psi = \pm\frac{\alpha}{2}$$

To make it vanish also along the circular boundary $r = a$, we must put

$$\sum_{n=1,3,5,\ldots}^{\infty} A_n \cos\frac{n\pi\psi}{\alpha} = 1 - \frac{\cos 2\psi}{\cos\alpha}$$

from which we obtain, in the usual way,

$$A_n = \frac{16\alpha^2}{\pi^3}(-1)^{(n+1)/2}\frac{1}{n(n + 2\alpha/\pi)(n - 2\alpha/\pi)}$$

The stress function is therefore

$$\phi = \frac{G\theta}{2}\left[-r^2\left(1 - \frac{\cos 2\psi}{\cos\alpha}\right) + \frac{16a^2\alpha^2}{\pi^3} \sum_{n=1,3,5,\ldots}^{\infty} (-1)^{(n+1)/2}\left(\frac{r}{a}\right)^{n\pi/\alpha} \frac{\cos(n\pi\psi/\alpha)}{n(n + 2\alpha/\pi)(n - 2\alpha/\pi)}\right]$$

Substituting into Eq. (153), we find $M_t = 2\iint \phi r\, d\psi\, dr = kGa^4\theta$, in which k is a factor depending on the angle α of the sector. Several values of k, calculated by Saint-Venant, are given below.

$\alpha =$	$\frac{\pi}{4}$	$\frac{\pi}{3}$	$\frac{\pi}{2}$	$\frac{2\pi}{3}$	π	$\frac{3\pi}{2}$	$\frac{5\pi}{3}$	2π
$k =$	0.0181	0.0349	0.0825	0.148	0.298*	0.572*	0.672†	0.878†
$k_1 =$	......	0.452	......	0.622	0.728‡	......	......	......
$k_2 =$	......	0.490	......	0.652	0.849	......	......	......

* These figures have been corrected by M. Aissen. See G. Pólya and G. Szegö, "Isoperimetric Inequalities in Mathematical Physics," p. 261, Princeton University Press, Princeton, N.J., 1951.

† Corrected by Dinnik, *loc. cit.*

‡ The authors are indebted to G. Szegö for this correction.

The maximum shearing stresses along the circular and along the radial boundaries are given by the formulas $k_1 G a \theta$ and $k_2 G a \theta$, respectively. Several values of k_1 and k_2 are given in the table on page 314.

The solution for a curvilinear rectangle bounded by two concentric circular arcs and two radii can be obtained in the same manner.[1]

In the case of an isosceles right-angled triangle[2] the angle of twist is given by the equation

$$\theta = 38.3 \frac{M_t}{Ga^4}$$

in which a is the length of the equal sides of the triangle. The maximum shearing stress is at the middle of the hypotenuse and is equal to

$$\tau_{max} = 18.02 \frac{M_t}{a^3}$$

By introducing curvilinear coordinates several other cross sections have been investigated. Taking elliptic coordinates (see page 188) and using conjugate functions ξ and η, determined by the equation

$$x + iy = c \cosh (\xi + i\eta)$$

we arrive at cross sections bounded by confocal ellipses and hyperbolas.[3] By using the equation[4]

$$x + iy = \tfrac{1}{2}(\xi + i\eta)^2$$

we obtain cross sections bounded by orthogonal parabolas.

Solutions have been found for many other sections,[5] solid and hollow, including polygons, angles, cardioids, lemniscates,[6] and circles with one or several eccentric holes.[7] When the section can be conformally mapped into the unit circle a solution can always be written down in terms of a complex integral.[8]

111 | Solution of Torsional Problems by Energy Method[9]

We have seen that the solution of torsional problems is reduced in each particular case to the determination of the stress function satisfying the

[1] Saint-Venant, *loc. cit.* See also A. E. H. Love, "Theory of Elasticity," 4th ed., p. 319, 1927; A. G. Greenhill, *Messenger of Math.*, vol. 9, p. 35, 1879.

[2] B. G. Galerkin, *Bull. Acad. des Sci. de Russ.*, p. 111, 1919; G. Kolosoff, *Compt. Rend.*, vol. 178, p. 2057, 1924.

[3] A. G. Greenhill, *Quart. J. Math.*, vol. 16, 1879. See also Filon, *loc. cit.*

[4] E. W. Anderson and D. L. Holl, *Iowa State Coll. J. Sci.*, vol. 3, p. 231, 1929.

[5] A compilation is given by T. J. Higgins, *Am. J. Phys.*, vol. 10, p. 248, 1942.

[6] References to papers giving exact solutions for such sections, too numerous to include here, may be found by consulting *Applied Mechanics Reviews*, *Science Abstracts A*, *Mathematical Reviews*, and *Zentralblatt für Mechanik*. Most of the references on p. 369 refer to or include the corresponding torsion problem.

[7] See C. B. Ling, *Quart. Appl. Math.*, vol. 5, p. 168, 1947.

[8] Due to N. I. Muskhelishvili (see n. 6, p. 194). See also I. S. Sokolnikoff, "Mathematical Theory of Elasticity," 2d ed., p. 151, McGraw-Hill Book Company, New York, 1956. Further examples and later references are given by W. A. Bassali, *J. Mech. Phys. Solids*, vol. 8, pp. 87–99, 1960.

[9] For a survey, with references, of this and other approximate methods see T. J. Higgins, *J. Appl. Phys.*, vol. 14, p. 469, 1943.

differential equation (150) and the boundary condition (152). In deriving an approximate solution of the problem it is useful, instead of working with the differential equation, to determine the stress function from the minimum condition of a certain integral,[1] which can be obtained from consideration of the strain energy of the twisted bar. The strain energy of the twisted bar per unit length, from (136), is

$$V = \frac{1}{2G} \iint (\tau_{xz}^2 + \tau_{yz}^2)\, dx\, dy = \frac{1}{2G} \iint \left[\left(\frac{\partial \phi}{\partial x} \right)^2 + \left(\frac{\partial \phi}{\partial y} \right)^2 \right] dx\, dy$$

If we give to the stress function ϕ any small variation $\delta\phi$, vanishing at the boundary,[2] the variation of the strain energy is

$$\frac{1}{2G}\, \delta \iint \left[\left(\frac{\partial \phi}{\partial x} \right)^2 + \left(\frac{\partial \phi}{\partial y} \right)^2 \right] dx\, dy$$

and the variation of the torque is, from Eq. (153),

$$2 \iint \delta\phi\, dx\, dy$$

Then by reasoning analogous to that used in developing equation (142), we conclude that

$$\frac{1}{2G}\, \delta \iint \left[\left(\frac{\partial \phi}{\partial x} \right)^2 + \left(\frac{\partial \phi}{\partial y} \right)^2 \right] dx\, dy = 2\theta \iint \delta\phi\, dx\, dy$$

or

$$\delta \iint \left\{ \frac{1}{2} \left[\left(\frac{\partial \phi}{\partial x} \right)^2 + \left(\frac{\partial \phi}{\partial y} \right)^2 \right] - 2G\theta\phi \right\} dx\, dy = 0$$

Thus, the true expression for the stress function ϕ is that which makes zero the variation of the integral

$$U = \iint \left\{ \frac{1}{2} \left[\left(\frac{\partial \phi}{\partial x} \right)^2 + \left(\frac{\partial \phi}{\partial y} \right)^2 \right] - 2G\theta\phi \right\} dx\, dy \tag{173}$$

We come also to the same conclusion by using the membrane analogy and the principle of virtual work (Art. 92). If S is the uniform tension in the membrane, the increase in strain energy of the membrane due to deflection is obtained by multiplying the tension S by the increase of the surface of the membrane. In this manner we obtain

$$\frac{1}{2} S \iint \left[\left(\frac{\partial z}{\partial x} \right)^2 + \left(\frac{\partial z}{\partial y} \right)^2 \right] dx\, dy$$

where z is the deflection of the membrane. If we take now a virtual displacement of the membrane from the position of equilibrium, the change in the strain energy

[1] This method was proposed by W. Ritz, who used it in the solution of problems of bending and vibration of rectangular plates. See *J. Reine Angew. Math.*, vol. 135, 1908, and *Ann. Physik*, ser. 4, vol. 28, p. 737, 1909.

[2] If $\delta\phi$ is taken equal to zero at the boundary, no forces on the lateral surface of the bar will be introduced by variation of ϕ.

of the membrane due to this displacement must be equal to the work done by the uniform load q on the virtual displacement. Thus we obtain

$$\frac{1}{2} S\delta \iint \left[\left(\frac{\partial z}{\partial x}\right)^2 + \left(\frac{\partial z}{\partial y}\right)^2 \right] dx\, dy = \iint q\, \delta z\, dx\, dy$$

and the determination of the deflection surface of the membrane is reduced to finding an expression for the function z which makes the integral

$$\iint \left\{ \frac{1}{2} \left[\left(\frac{\partial z}{\partial x}\right)^2 + \left(\frac{\partial z}{\partial y}\right)^2 \right] - \frac{q}{S} z \right\} dx\, dy$$

a minimum. If we substitute in this integral $2G\theta$ for q/S, we arrive at the integral (173) above.

In the approximate solution of torsional problems we replace the above problem of variational calculus by a simple problem of finding a minimum of a function. We take the stress function in the form of a series

$$\phi = a_0\phi_0 + a_1\phi_1 + a_2\phi_2 + \cdots \tag{a}$$

in which $\phi_0, \phi_1, \phi_2, \ldots$ are functions satisfying the boundary condition, that is, vanishing at the boundary. In choosing these functions we should be guided by the membrane analogy and take them in a form suitable for representing the function ϕ. The quantities $a_0, a_1, a_2, \ldots$ are numerical factors to be determined from the minimum condition of the integral (173). Substituting the series (a) in this integral we obtain, after integration, a function of the second degree in $a_0, a_1, a_2, \ldots$ and the minimum condition of this function is

$$\frac{\partial U}{\partial a_0} = 0 \qquad \frac{\partial U}{\partial a_1} = 0 \qquad \frac{\partial U}{\partial a_2} = 0 \cdots \tag{b}$$

Thus we obtain a system of linear equations from which the coefficients $a_0, a_1, a_2, \ldots$ can be determined. By increasing the number of terms in the series (a) we increase the accuracy of our approximate solution, and by using infinite series we may arrive at an exact solution of the torsional problem.[1]

Take as an example the case of a rectangular cross section[2] (Fig. 163). The boundary is given by the equations $x = \pm a$, $y = \pm b$, and the function $(x^2 - a^2)(y^2 - b^2)$ is zero at the boundary. The series (a) can be taken in the form

$$\phi = (x^2 - a^2)(y^2 - b^2)\Sigma\Sigma a_{mn}x^m y^n \tag{c}$$

in which, from symmetry, m and n must be even.

[1] The convergence of this method of solution was investigated by Ritz, *loc. cit.* See also E. Trefftz, "Handbuch der Physik," vol. 6, p. 130, 1928.

[2] See S. Timoshenko, *Bull. Inst. Ways of Communication*, St. Petersburg, 1913, and *Proc. London Math. Soc.*, ser. 2, vol. 20, p. 389, 1921.

Assuming that we have a square cross section and limiting ourselves to the first term of the series (*c*), we take

$$\phi = a_0(x^2 - a^2)(y^2 - a^2) \qquad (d)$$

Substituting this in (173), we find from the minimum condition that

$$a_0 = \frac{5}{8}\frac{G\theta}{a^2}$$

The magnitude of the torque, from Eq. (153), is then

$$M_t = 2\iint \phi \, dx \, dy = {}^{20}\!/_{9} G\theta a^4 = 0.1388(2a)^4 G\theta$$

Comparing this with the correct solution (170), we see that the error in the torque is about 1⅓ percent.

To get a closer approximation we take the three first terms in the series (*c*). Then, by using the condition of symmetry, we obtain

$$\phi = (x^2 - a^2)(y^2 - a^2)[a_0 + a_1(x^2 + y^2)] \qquad (e)$$

Substituting this in (173) and using Eqs. (*b*), we find

$$a_0 = \frac{5}{8}\frac{259}{277}\frac{G\theta}{a^2} \qquad a_1 = \frac{5}{8}\frac{3}{2}\frac{35}{277}\frac{G\theta}{a^4}$$

Substituting in expression (153) for the torque, we obtain

$$M_t = {}^{20}\!/_{9}[{}^{259}\!/_{277} + ({}^{2}\!/_{5})({}^{3}\!/_{2})({}^{35}\!/_{277})]G\theta a^4 = 0.1404 G\theta(2a)^4$$

This value is only 0.15 percent less than the correct value.

A much larger error is found in the magnitude of the maximum stress. Substituting (*e*) into expressions (149) for the stress components, we find that the error in the maximum stress is about 4 percent, and to get a better accuracy more terms of the series (*c*) must be taken.

It can be seen from the membrane analogy that in proceeding as explained above, we generally get smaller values for the torque than the correct value. A perfectly flexible membrane, uniformly stretched at the boundary and uniformly loaded, is a system with an infinite number of degrees of freedom. Limiting ourselves to a few terms of the series (*c*) is equivalent to introducing into the system certain constraints, which reduce it to a system with a few degrees of freedom only. Such constraints can only reduce the flexibility of the system and diminish the volume bounded by the deflected membrane. Hence the torque, obtained from this volume, will generally be smaller than its true value.

E. Trefftz suggested[1] another method of approximate determination

[1] *Proc. 2d Intern. Congr. Appl. Mech.*, Zürich, 1926, p. 131. See also N. M. Basu, *Phil. Mag.*, vol. 10, p. 886, 1930.

of the stress function ϕ. With this method the approximate magnitude of the torque is larger than its true value. Hence by using the Ritz and the Trefftz methods together the limits of error of the approximate solution can be established.

In using Ritz's method we are not limited to polynomials (*c*). We can take the functions $\phi_0, \phi_1, \phi_2, \ldots$ of the series (*a*) in other forms suitable for the representation of the stress function ϕ. Taking, for instance, trigonometric functions, and observing the conditions of symmetry (Fig. 163), we obtain

$$\phi = \sum_{n=1,3,5,\ldots}^{\infty} \sum_{m=1,3,5,\ldots}^{\infty} a_{mn} \cos \frac{m\pi x}{2a} \cos \frac{n\pi y}{2b} \tag{f}$$

Substituting in (173) and performing the integration, we find that

$$U = \frac{\pi^2 ab}{8} \sum_{m=1,3,5,\ldots}^{\infty} \sum_{n=1,3,5,\ldots}^{\infty} a_{mn}{}^2 \left(\frac{m^2}{a^2} + \frac{n^2}{b^2}\right) - 2G\theta \sum_{m=1,3,5,\ldots}^{\infty} \sum_{n=1,3,5,\ldots}^{\infty} a_{mn} \frac{16ab}{mn\pi^2} (-1)^{[(m+n)/2]-1}$$

Equations (*b*) become

$$\frac{\pi^2 ab}{4} a_{mn} \left(\frac{m^2}{a^2} + \frac{n^2}{b^2}\right) - 2G\theta \frac{16ab}{mn\pi^2} (-1)^{[(m+n)/2]-1} = 0$$

and we find

$$a_{mn} = \frac{128G\theta b^2(-1)^{[(m+n)/2]-1}}{\pi^4 mn(m^2\alpha^2 + n^2)}$$

where $\alpha = b/a$. Substituting in (*f*), we obtain the exact solution of the problem in the form of an infinite trigonometric series. The torque will then be

$$M_t = 2 \int_{-a}^{a} \int_{-b}^{b} \phi \, dx \, dy = \sum_{m=1,3,\ldots}^{\infty} \sum_{n=1,3,\ldots}^{\infty} \frac{128G\theta b^2}{\pi^4 mn(m^2\alpha^2 + n^2)} \frac{32ab}{mn\pi^2} \tag{g}$$

This expression is brought into coincidence with expression (168) given before if we observe that

$$\frac{1}{m^2} \sum_{n=1,3,5,\ldots}^{\infty} \frac{1}{n^2(m^2\alpha^2 + n^2)} = \frac{\pi^4}{96m^2} \frac{\tanh (m\alpha\pi/2) - (m\alpha\pi/2)}{-\frac{1}{3}(m\alpha\pi/2)^3}$$

As another example, in the case of a narrow rectangle, when b is very large in comparison with a (Fig. 163), we may take, as a first approximation,

$$\phi = G\theta(a^2 - x^2) \tag{h}$$

which coincides with the solution discussed before (Art. 108). To get a better approximation satisfying the boundary condition at the short sides of the rectangle, we may take

$$\phi = G\theta(a^2 - x^2)[1 - e^{-\beta(b-y)}] \tag{i}$$

and choose the quantity β in such a manner as to make the integral (173) a minimum. In this way we find

$$\beta = \frac{1}{a}\sqrt{\frac{5}{2}}$$

Due to the exponential term in the brackets of expression (i) we obtain a stress distribution which practically coincides with that of the solution (h) at all points a considerable distance from the short sides of the rectangle. Near these sides the stress function (i) satisfies the boundary condition (152). Substituting (i) into equation (153) for the torque, we find

$$M_t = 2\int_{-a}^{a}\int_{-b}^{b} \phi \, dx \, dy = \frac{1}{3} G\theta(2a)^3(2b)\left(1 - 0.632\frac{a}{b}\right)$$

which is in very good agreement with Eq. (169) obtained before by using infinite series.

A polynomial expression for the stress function, analogous to expression (c) taken above for a rectangle, can be used successfully in all cases of cross sections bounded by a convex polygon. If

$$a_1x + b_1y + c_1 = 0 \qquad a_2x + b_2y + c_2 = 0, \ . \ . \ .$$

are the equations of the sides of the polygon, the stress function can be taken in the form

$$\phi = (a_1x + b_1y + c_1)(a_2x + b_2y + c_2) \cdot \cdot \cdot (a_nx + b_ny + c_n)\Sigma\Sigma a_{mn}x^ny^m$$

and the first few terms of the series are usually sufficient to get a satisfactory accuracy.

The energy method is also useful when the boundary of the cross section (Fig. 165) is given by two curves[1]

$$y = a\psi\left(\frac{x}{b}\right) \qquad \text{and} \qquad y = -a_1\psi\left(\frac{x}{b}\right)$$

where

$$\psi\left(\frac{x}{b}\right) = \psi(t) = (t)^m[1 - (t)^p]^q$$

[1] Such problems were discussed by L. S. Leibenson. See his book "Variational Methods for Solving Problems of the Theory of Elasticity," Moscow, 1943. See also W. J. Duncan, *Phil. Mag.*, ser. 7, vol. 25, p. 634, 1938.

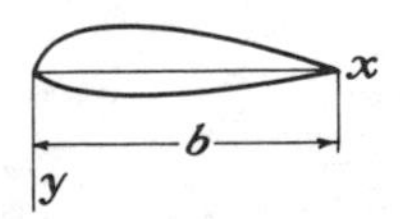

Fig. 165

The boundary conditions will be satisfied if we take for the stress function an approximate expression

$$\phi = A(y - a\psi)(y + a_1\psi)$$

Substituting into the integral (173) we find, from the equation $dI/dA = 0$,

$$A = -\frac{G\theta}{1 + \alpha(a^2 + a_1^2 + aa_1)/b^2}$$

where

$$\alpha = \frac{\int_0^1 \psi^3 (d\psi/dt)^2 \, dt}{\int_0^1 \psi^3 \, dt}$$

From Eq. (153) we find the torque

$$M_t = -A \frac{b(a + a_1)^3}{3} \int_0^1 \psi^3 \, dt$$

In the particular case when $m = \frac{1}{2}$, $p = q = 1$, $a = a_1$, we have $y = \pm a\psi(x/b) = \pm \sqrt{x/b}\,[1 - (x/b)]$, and we obtain

$$A = -\frac{G\theta}{1 + (11/13)(a^2/b^2)} \qquad M_t = 0.0736 \frac{G\theta b a^3}{1 + (11/13)(a^2/b^2)}$$

An approximate solution, and a comparison with tests, for sections bounded by a circle and a chord has been given by A. Weigand.[1] Numerical methods are discussed in the Appendix.

112 | Torsion of Rolled Profile Sections

In investigating the torsion of rolled sections such as angles, channels, and I beams, the formulas derived for narrow rectangular bars (Art. 108) can be used. If the cross section is of constant thickness, as in Fig. 166*a*, the angle of twist is obtained with sufficient accuracy from Eq. (163) by putting, instead of b, in this equation the developed length of the centerline,[2] namely, $b = 2a - c$. In the case of a channel section (Fig. 166*b*) a rough approximation for the angle of twist is obtained by taking for the flanges

[1] *Luftfahrt-Forsch.*, vol. 20, 1944, translated as *NACA Tech. Mem.* 1182, 1948.

[2] A more elaborate formula, taking account of the increased stiffness resulting from the junctions of the rectangles, was developed on the basis of soap film and torsion tests by G. W. Trayer and H. W. March, *Natl. Advisory Comm. Aeron. Rept.* 334, 1930.

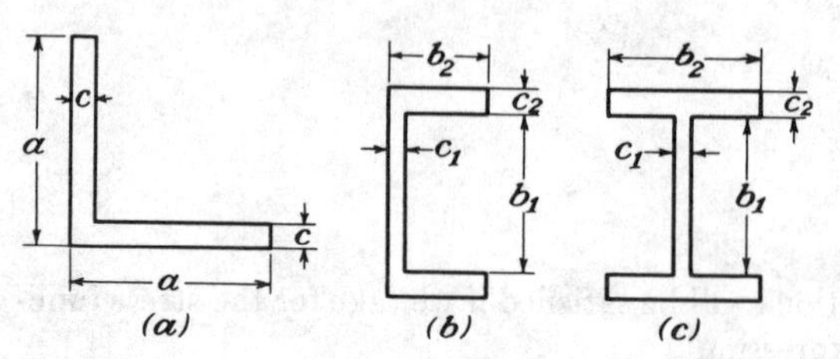

Fig. 166

an average thickness c_2, subdividing the cross section into the three rectangles, and substituting in Eq. (163), $b_1c_1^3 + 2b_2c_2^3$ instead of bc^3, that is, assuming that the torsional rigidity of the channel is equal to the sum of the torsional rigidities of the three rectangles.[1] Then

$$\theta = \frac{3M_t}{(b_1c_1^3 + 2b_2c_2^3)G} \tag{a}$$

To calculate the stress at the boundary at points a considerable distance from the corners of the cross section, we can use once more the equation for a narrow rectangle and take

$$\tau = c\theta G$$

Then, from Eq. (a), we obtain for the flanges of the channel

$$\tau = \frac{3M_tc_2}{b_1c_1^3 + 2b_2c_2^3} \tag{b}$$

The same approximate equations can be used for an I beam (Fig. 166c).

At reentrant corners there is a considerable stress concentration, the magnitude of which depends on the radius of the fillets. A rough approximation for the maximum stress at these fillets can be obtained from the membrane analogy. Let us consider a cross section in the form of an angle of constant thickness c (Fig. 167) and with radius a of the fillet of the reentrant corner. Assuming that the surface of the membrane at the bisecting line OO_1 of the fillet is approximately a surface of revolution, with axis perpendicular to the plane of the figure at O, and using polar coordinates, the Eq. (159) of the deflection surface of the membrane becomes (see page 68)

$$\frac{d^2z}{dr^2} + \frac{1}{r}\frac{dz}{dr} = -\frac{q}{S} \tag{c}$$

[1] Comparison of torsional rigidities obtained in this manner with those obtained by experiments is given for several types of rolled sections and for various dimensions in the paper by A. Föppl, *Sitzber. Bayer. Akad. Wiss., München*, p. 295, 1921. See also *Bauingenieur*, ser. 5, vol. 3, p. 42, 1922.

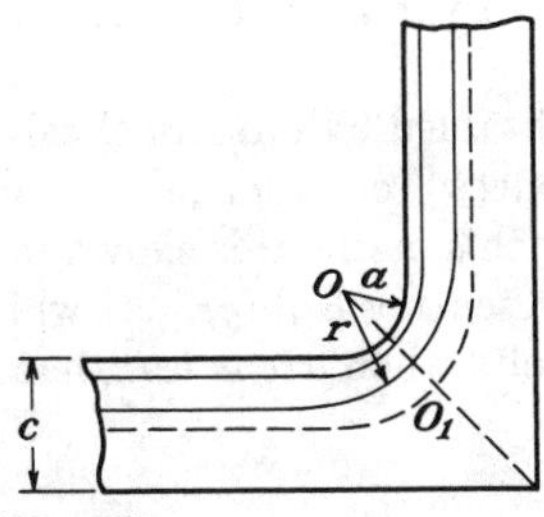

Fig. 167

Remembering that the slope of the membrane dz/dr gives the shearing stress τ when q/S is replaced by $2G\theta$, we find from (*c*) the following equation for the shearing stress:

$$\frac{d\tau}{dr} + \frac{1}{r}\tau = -2G\theta \tag{d}$$

The corresponding equation in the arms of the angle at a considerable distance from the corners, where the membrane has a nearly cylindrical surface, is

$$\frac{d\tau}{dn} = -2G\theta \tag{e}$$

in which n is the normal to the boundary. Denoting by τ_1 the stress at the boundary we find from (*e*) the previously found solution for a narrow rectangle $\tau_1 = G\theta c$. Using this, we obtain from (*d*)

$$\frac{d\tau}{dr} + \frac{1}{r}\tau = -\frac{2\tau_1}{c} \tag{d'}$$

from which, by integration,

$$\tau = \frac{A}{r} - \frac{\tau_1 r}{c} \tag{f}$$

where A is a constant of integration. For the determination of this constant, let us assume that the shearing stress becomes zero at point O_1 at a distance $c/2$ from the boundary (Fig. 167). Then, from (*f*),

$$\frac{A}{a + (c/2)} - \frac{\tau_1[a + (c/2)]}{c} = 0 \qquad \text{and} \qquad A = \frac{\tau_1}{c}\left(a + \frac{c}{2}\right)^2$$

Substituting in (*f*) and taking $r = a$, we find

$$\tau_{max} = \tau_1\left(1 + \frac{c}{4a}\right) \tag{g}$$

For $a = \frac{1}{2}c$, as in Fig. 167, we have $\tau_{max} = 1.5\tau_1$. For a very small radius of fillet the maximum stress becomes very high. Taking, for instance, $a = 0.1c$ we find $\tau_{\max} = 3.5\tau_1$.

More accurate and complete results can be obtained by numerical calculations based on the method of finite differences (see Appendix). A curve of $\tau_{\max}/\tau_1$ as a function of a/c obtained by this method[1] is shown in Fig. 168 (curve A), together with the curve representing Eq. (g). It will be seen that this simple formula gives good results when a/c is less than 0.3.

113 | Experimental Analogies

We have seen that the membrane analogy is very useful in enabling us to visualize the stress distribution over the cross section of a twisted bar. Membranes in the form of soap films have also been used for direct measurements of stresses.[2] The films were formed on holes cut to the required shapes in flat plates. To make possible the direct determination of stresses, it was found necessary to have in the same plate a circular hole to represent a circular section for comparison. Submitting both films to the

[1] By J. H. Huth, *J. Appl. Mech.*, vol. 17, p. 388, 1950. The rise of the curve toward the right is required by the limiting case as the fillet radius is increased in relation to the leg thickness. References to earlier attempts to solve this problem including soap-film measurements are given by I. Lyse and B. G. Johnston, *Proc. ASCE*, 1935, p. 469, and in the above paper.

[2] See papers by Griffith and Taylor, *loc. cit.*; also the paper by Trayer and March, *loc. cit.* A survey of this and other analogies for torsion, with references, is given by T. J. Higgins, *Exp. Stress Anal.*, vol. 2, no. 2, p. 17, 1945.

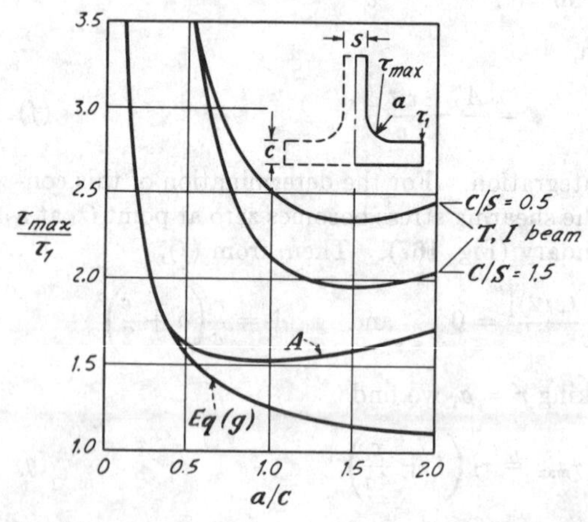

Fig. 168

same pressure, we have the same values of q/S,[1] which correspond to the same values of $G\theta$ for the two bars under twist. Hence, by measuring the slopes of the two soap films, we can compare the stresses in the bar of the given cross section with those in a circular shaft under the condition that they have the same angle of twist θ per unit length and the same G. The corresponding ratio of the torques is determined by the ratio of the volumes between the soap films and the plane of the plate.[2]

At points of stress concentration, as at fillets of small radius, the soap film is likely to yield inaccurate results.[3] More reliable values can be found from the *conducting-sheet analogy*.[4] An electrically conducting sheet is cut to the shape of the cross section of the twisted bar. If constant current density i (per unit area) is fed into the surface over its whole area, the electric potential V throughout the sheet will satisfy the equation

$$\nabla^2 V = -\rho i$$

where ρ is the uniform resistivity of the sheet. If the boundary is kept at a constant electric potential (by a bus bar fitted to it), we have a complete analogy to the torsion problem represented by Eqs. (150) and (151), with the condition (152) that ϕ is constant on the (single) boundary curve. The curves marked T, I beam on Fig. 168 were obtained[5] by this method, and the curve A for the angle section was confirmed.

114 | Hydrodynamical Analogies

There are several analogies between the torsional problem and the hydrodynamical problem of the motion of fluid in a tube. Lord Kelvin[6] pointed out that the function ϕ_1 [see Eq. (a), Art. 106] which is sometimes used in the solution of torsional problems is identical with the *stream function* of a certain irrotational motion of "ideal fluid" contained in a vessel of the same cross section as the twisted bar.

[1] It is assumed that the surface tension is the same in both films. This was proved with sufficient accuracy by the tests.

[2] For a more detailed account see M. Hetényi (ed.), "Handbook of Experimental Stress Analysis," chap. 16, John Wiley & Sons, Inc., New York, 1950.

[3] See the paper by C. B. Biezeno and J. M. Rademaker, *De Ingenieur*, no. 52, 1931. See also papers by P. A. Cushman, *Trans. ASME*, 1932, H. Quest, *Ingenieur-Arch.*, vol. 4, p. 510, 1933, and J. H. Huth, *loc. cit.*

[4] N. S. Waner and W. W. Soroka, *Proc. Soc. Exptl. Stress Anal.*, vol. II(L), pp. 19–26, 1953; also W. W. Soroka, "Analog Methods in Computation and Simulation."

[5] C. W. Beadle and H. D. Conway (1) *Exp. Mech.*, pp. 198–200, August, 1963; (2) *J. Appl. Mech.*, vol. 30, pp. 138–141, 1963. The latter paper gives further results by an analytical approximation method.

[6] Kelvin and Tait, "Natural Philosophy," pt. 2, p. 242.

Another analogy was indicated by J. Boussinesq.[1] He showed that the differential equation and the boundary condition for determining the stress function ϕ [see Eqs. (150) and (152)] are identical with those for determining velocities in a laminar motion of viscous fluid along a tube of the same cross section as the twisted bar.[2]

Greenhill showed that the stress function ϕ is mathematically identical with the stream function of a motion of ideal fluid circulating with uniform *vorticity*,[3] in a tube of the same cross section as the twisted bar.[4] Let u and v be the components of the velocity of the circulating fluid at a point A (Fig. 169). Then from the condition of incompressibility of the ideal fluid we have

$$\frac{\partial u}{\partial x} + \frac{\partial v}{\partial y} = 0 \tag{a}$$

The condition of uniform vorticity is

$$\frac{\partial v}{\partial x} - \frac{\partial u}{\partial y} = \text{const} \tag{b}$$

By taking

$$u = \frac{\partial \phi}{\partial y} \qquad v = -\frac{\partial \phi}{\partial x} \tag{c}$$

we satisfy Eq. (a), and from Eq. (b) we find

$$\frac{\partial^2 \phi}{\partial x^2} + \frac{\partial^2 \phi}{\partial y^2} = \text{const} \tag{d}$$

which coincides with Eq. (150) for the stress function in torsion.

At the boundary the velocity of the circulating fluid is in the direction of the tangent to the boundary, and the boundary condition for the hydro-

[1] *J. Math. Pure Appl.*, ser. 2, vol. 16, 1871.

[2] This analogy was used by M. Paschoud, *Compt. Rend.*, vol. 179, p. 451, 1924. See also *Bull. Tech. Suisse Rom.* (Lausanne), November, 1925.

[3] The analytical expression for vorticity is the same as for rotation ω_z discussed on p. 233, provided u and v denote the components of the velocity of the fluid.

[4] A. G. Greenhill, Hydromechanics, an article in the Encyclopaedia Britannica, 11th ed., p. 115, 1910.

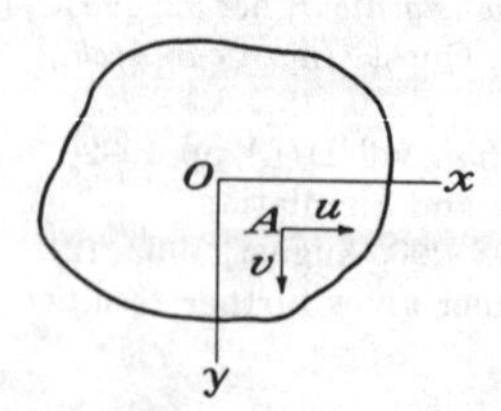

Fig. 169

dynamical problem is the same as the condition (152) for the torsional problem. Hence, the velocity distribution in the hydrodynamical problem is mathematically identical with the stress distribution in torsion, and some practically important conclusions can be drawn by using the known solutions of hydrodynamics.

As a first example we take the case of a small *circular hole* in a twisted circular shaft[1] (Fig. 170). The effect of this hole on the stress distribution is similar to that of introducing a stationary solid cylinder of the same diameter as the hole into the stream of circulating fluid of the hydrodynamical model. Such a cylinder greatly changes the velocity of the fluid in its immediate neighborhood. The velocities at the front and rear points are reduced to zero, whereas those at the side points m and n are doubled. A hole of this kind therefore doubles the shearing stress in the portion of the shaft in which it is located. A small *semicircular* groove on the surface parallel to the length of the shaft (Fig. 170) has the same effect. The shearing stress at the bottom of the groove, the point m, is about twice the shearing stress at the surface of the shaft far away from the groove.

The same hydrodynamical analogy explains the effect of a small *hole of elliptic cross section* or of a groove of *semi-elliptic cross section*. If one of the principal axes a of the small elliptical hole is in the radial direction and the other principal axis is b, the stresses at the edge of the hole at the ends of the a axis are increased in the proportion $(1 + a/b):1$. The maximum stress produced in this case thus depends upon the magnitude of the ratio a/b. The effect of the hole on the stress is greater when the major axis of the ellipse is in the radial direction than when it runs circumferentially. This explains why a radial crack has such a weakening effect on the strength of a shaft. Similar effects on the stress distribution are produced by a semi-elliptic groove on the surface, parallel to the axis of the shaft.

From the hydrodynamical analogy it can be concluded also that at the projecting corners of a cross section of a twisted bar the shearing stress becomes zero and that at reentrant corners this stress becomes theoretically infinitely large, i.e., even the smallest torque will produce yielding

[1] See J. Larmor, *Phil. Mag.*, vol. 33, p. 76, 1892.

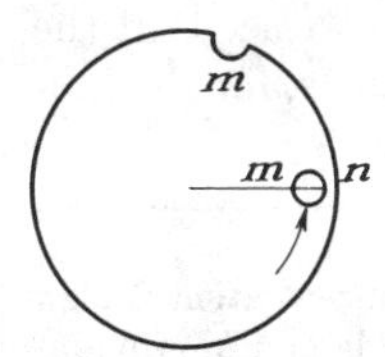

Fig. 170

of material or a crack at such a corner. In the case of a rectangular *keyway*, therefore, a high stress concentration takes place at the reentrant corners at the bottom of the keyway. These high stresses can be reduced by rounding the corners.[1]

115 | Torsion of Hollow Shafts

So far the discussion has been limited to shafts whose cross sections are bounded by single curves. Let us consider now hollow shafts whose cross sections have two or more boundaries. The simplest problem of this kind is a hollow shaft with an inner boundary coinciding with one of the *stress lines* (see page 305) of the solid shaft, having the same boundary as the outer boundary of the hollow shaft.

Take, for instance, an elliptic cross section (Fig. 153). The stress function for the solid shaft is

$$\phi = \frac{a^2b^2F}{2(a^2+b^2)}\left(\frac{x^2}{a^2}+\frac{y^2}{b^2}-1\right) \tag{a}$$

The curve

$$\frac{x^2}{(ak)^2}+\frac{y^2}{(bk)^2}=1 \tag{b}$$

is an ellipse that is geometrically similar to the outer boundary of the cross section. Along this ellipse the stress function (a) remains constant, and hence, for k less than unity, *this ellipse* is a stress line for the solid elliptic shaft. The shearing stress at any point of this line is in the direction of the tangent to the line. Imagine now a cylindrical surface generated by this stress line with its axis parallel to the axis of the shaft. Then, from the above conclusion regarding the direction of the shearing stresses, it follows that there will be no stresses acting across this cylindrical surface. We can imagine the material bounded by this surface removed without changing the stress distribution in the outer portion of the shaft. Hence the stress function (a) applies to the hollow shaft also.

For a given angle θ of twist the stresses in the hollow shaft are the same as in the corresponding solid shaft. But the torque will be smaller by the amount which in the case of the solid shaft is carried by the portion of the cross section corresponding to the hole. From Eq. (156) we see that the latter portion is in the ratio k^4: 1 to the total torque. Hence, for the hol-

[1] The stresses at the keyway were investigated by the soap-film method. See the paper by Griffith and Taylor, *loc. cit.*, p. 938.

For design formulas and charts, see R. E. Peterson, "Stress Concentration Design Factors," John Wiley & Sons, Inc., New York, 1953; also M. Nisida and M. Hondo, *Proc. Japan Nat. Congr. Appl. Mech.*, vol. 2, pp. 129–132, 1959.

low shaft, instead of Eq. (156), we will have

$$\theta = \frac{M_t}{1 - k^4} \frac{a^2 + b^2}{\pi a^3 b^3 G}$$

and the stress function (*a*) becomes

$$\phi = - \frac{M_t}{\pi ab(1 - k^4)} \left(\frac{x^2}{a^2} + \frac{y^2}{b^2} - 1 \right)$$

The formula for the maximum stress will be

$$\tau_{\max} = \frac{2M_t}{\pi ab^2} \frac{1}{1 - k^4}$$

In the membrane analogy the middle portion of the membrane, corresponding to the hole in the shaft (Fig. 171), must be replaced by the horizontal plate *CD*. We note that the uniform pressure distributed over the portion *CFD* of the membrane is statically equivalent to the pressure of the same magnitude uniformly distributed over the plate *CD* and the tensile forces *S* in the membrane acting along the edge of this plate are in equilibrium with the uniform load on the plate. Hence, in the case under consideration the same experimental soap-film method as before can be employed because the replacement of the portion *CFD* of the membrane by the plate *CD* causes no changes in the configuration and equilibrium conditions of the remaining portion of the membrane.

Let us consider now the more general case when the boundaries of the holes are no longer stress lines of the solid shaft. From the general theory of torsion we know (see Art. 104) that the stress function must be constant along each boundary, but these constants cannot be chosen arbitrarily. In discussing multiply-connected boundaries in two-dimensional problems, it was shown that recourse must be had to the expressions for the displacements, and the constants of integration should be found in such

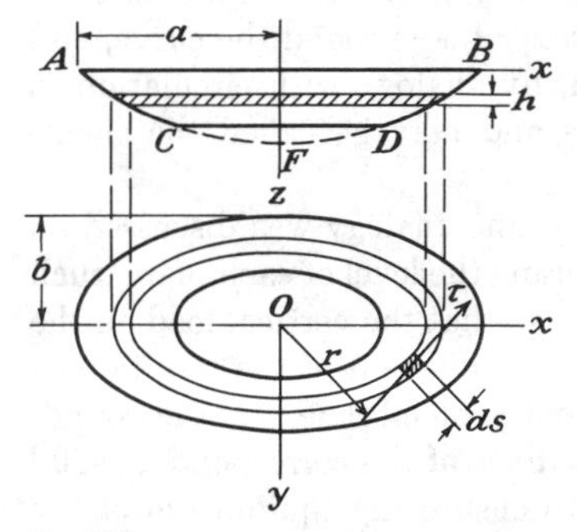

Fig. 171

a manner as to make these expressions single-valued. An analogous procedure is necessary in dealing with the torsion of hollow shafts. The constant values of the stress function along the boundaries should be determined in such a manner as to make the displacements single-valued. A sufficient number of equations for determining these constants will then be obtained.

From Eqs. (*b*) and (*d*) of Art. 104 we have

$$\tau_{xz} = G\left(\frac{\partial w}{\partial x} - \theta y\right) \qquad \tau_{yz} = G\left(\frac{\partial w}{\partial y} + \theta x\right) \tag{c}$$

Let us now calculate the integral

$$\int \tau \, ds \tag{d}$$

along each boundary. Using (*c*) and resolving the total stress into its components we find

$$\begin{aligned} \int \tau \, ds &= \int \left(\tau_{xz}\frac{dx}{ds} + \tau_{yz}\frac{dy}{ds}\right) ds \\ &= G\int\left(\frac{\partial w}{\partial x}dx + \frac{\partial w}{\partial y}dy\right) - \theta G\int (y\,dx - x\,dy) \end{aligned} \tag{174}$$

The first integral must vanish from the condition that the integration is taken round a closed curve and that w is a single-valued function. Hence,

$$\int \tau \, ds = \theta G \int (x\,dy - y\,dx)$$

The integral on the right side is equal to double the area enclosed. Then,

$$\int \tau \, ds = 2G\theta A \tag{175}$$

Thus, we must determine the constant values of the stress function along the boundaries of the holes so as to satisfy Eq. (175) for each boundary.

For any closed curve drawn in the cross section (lying wholly in the material) the first and second members of (174) represent the line integral of the tangential component of shear stress τ taken round the curve, and this may be called the shear circulation, by analogy with circulation in fluid dynamics. Then (175) still holds and may be called the shear-circulation theorem.

The significance of (175) for the membrane analogy was discussed on page 306. It indicates that in the membrane the level of each plate, such as the plate *CD* (Fig. 171), must be taken so that the vertical load on the plate is equal and opposite to the vertical component of the resultant of the tensile forces on the plate produced by the membrane. If the boundaries of the holes coincide with the stress lines of the corresponding solid shaft, the above condition is sufficient to ensure the equilibrium of the plates. In the general case this condition is not sufficient, and to keep the plates in equilibrium in a horizontal position special guiding devices become necessary. This makes the soap-film experiments for hollow shafts more complicated.

To remove this difficulty the following procedure may be adopted.[1] We make a hole in the plate corresponding to the outer boundary of the shaft. The interior boundaries, corresponding to the holes, are mounted each on a vertical sliding column so that their heights can be easily adjusted. Taking these heights arbitrarily and stretching the film over the boundaries, we obtain a surface that satisfies Eq. (150) and boundary conditions (152), but the Eq. (175) above generally will not be satisfied and the film does not represent the stress distribution in the hollow shaft. Repeating such an experiment as many times as the number of boundaries, each time with another adjustment of heights of the interior boundaries and taking measurements on the film each time, we obtain sufficient data for determining the correct values of the heights of the interior boundaries and can finally stretch the soap film in the required manner. This can be proved as follows: If i is the number of boundaries and $\phi_1, \phi_2, \ldots, \phi_i$ are the film surfaces obtained with i different adjustments of the heights of the boundaries, then a function

$$\phi = m_1\phi_1 + m_2\phi_2 + \cdots + m_i\phi_i \tag{e}$$

in which $m_1, m_2, \ldots, m_i$ are numerical factors, is also a solution of Eq. (150), provided that

$$m_1 + m_2 + \cdots + m_i = 1$$

Observing now that the shearing stress is equal to the slope of the membrane, and substituting (e) into Eqs. (175) we obtain i equations of the form

$$\int \frac{\partial \phi}{\partial n}\, ds = 2G\theta A_i$$

from which the i factors $m_1, m_2, \ldots, m_i$ can be obtained as functions of θ. Then the true stress function is obtained from (e).[2] This method was applied by Griffith and Taylor in determining stresses in a hollow circular shaft having a keyway in it. It was shown in this manner that the maximum stress can be considerably reduced and the strength of the shaft increased by throwing the bore in the shaft off center.

The torque in the shaft with one or more holes is obtained using twice the volume under the membrane and the flat plates. To see this we calculate the torque produced by the shearing stresses distributed over an elemental ring between two adjacent stress lines, as in Fig. 171 (now taken to represent an arbitrary hollow section). Denoting by δ the variable width of the ring and considering an element such as that shaded in the figure, the shearing force acting on this element is $\tau\delta\, ds$ and its moment with respect to O is $r\tau\delta\, ds$. Then the torque on the elemental ring is

$$dM_t = \int r\tau\delta\, ds \tag{f}$$

[1] Griffith and Taylor, *loc. cit.*, p. 938.

[2] Griffith and Taylor concluded from their experiments that instead of *constant-pressure* films it is more convenient to use *zero-pressure* films (see p. 306) in studying the stress distribution in hollow shafts. A detailed discussion of the calculation of factors $m_1, m_2, \ldots$ is given in their paper.

in which the integration must be extended over the length of the ring. Denoting by A the area bounded by the ring and observing that τ is the slope, so that $\tau\delta$ is the difference in level h of the two adjacent contour lines, we find, from (f),

$$dM_t = 2hA \tag{g}$$

i.e., the torque corresponding to the elemental ring is given by twice the volume shaded in the figure. The total torque is given by the sum of these volumes, i.e., twice the volume between AB, the membrane AC and DB, and the flat plate CD. The conclusion follows similarly for several holes.

116 | Torsion of Thin Tubes

An approximate solution of the torsional problem for thin tubes can easily be obtained by using the membrane analogy. Let AB and CD (Fig. 172) represent the levels of the outer and the inner boundaries, and AC and DB be the cross section of the membrane stretched between these boundaries. In the case of a thin wall, we can neglect the variation in the slope of the membrane across the thickness and assume that AC and BD are straight lines. This is equivalent to the assumption that the shearing stresses are uniformly distributed over the thickness of the wall. Then denoting by h the difference in level of the two boundaries and by δ the variable thickness of the wall, the stress at any point, given by the slope of the membrane, is

$$\tau = \frac{h}{\delta} \tag{a}$$

It is inversely proportional to the thickness of the wall and thus greatest where the thickness of the tube is least.

To establish the relation between the stress and the torque M_t, we apply again the membrane analogy and calculate the torque from the volume

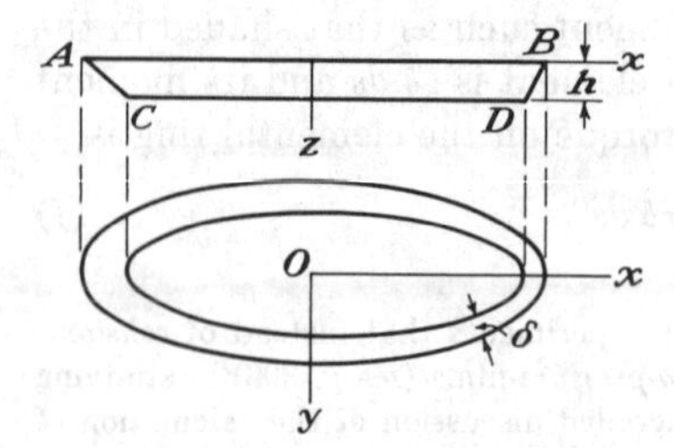

Fig. 172

ACDB. Then

$$M_t = 2Ah = 2A\delta\tau \tag{b}$$

in which A is the mean of the areas enclosed by the outer and the inner boundaries of the cross section of the tube. From (*b*) we obtain a simple formula for calculating shearing stresses:

$$\tau = \frac{M_t}{2A\delta} \tag{176}$$

For determining the angle of twist θ, we apply Eq. (160). Then

$$\tau\, ds = \frac{M_t}{2A} \int \frac{ds}{\delta} = 2G\theta A \tag{c}$$

from which[1]

$$\theta = \frac{M_t}{4A^2G} \int \frac{ds}{\delta} \tag{177}$$

In the case of a tube of uniform thickness, δ is constant and (177) gives

$$\theta = \frac{M_t s}{4A^2G\delta} \tag{178}$$

in which s is the length of the centerline of the ring section of the tube.

If the tube has reentrant corners, as in the case represented in Fig. 173, a considerable stress concentration may take place at these corners. The maximum stress is larger than the stress given by Eq. (176) and depends on the radius a of the fillet of the reentrant corner (Fig. 173*b*). In calculating this maximum stress, we shall use the membrane analogy as we

[1] Equations (176) and (177) for thin tubular sections were obtained by R. Bredt, *VDI*, vol. 40, p. 815, 1896.

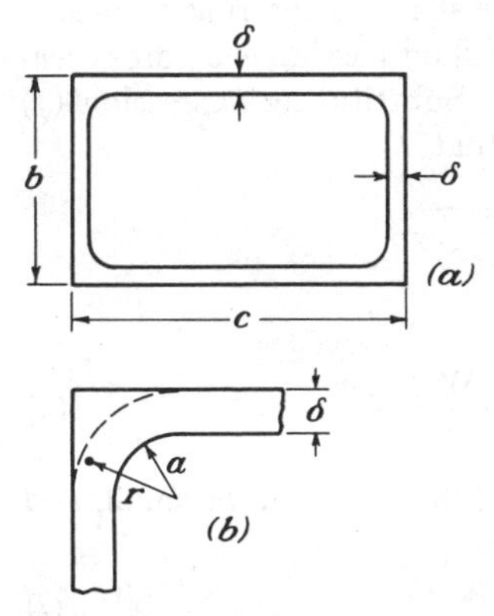

Fig. 173

did for the reentrant corners of rolled sections (Art. 112). The equation of the membrane at the reentrant corner may be taken in the form

$$\frac{d^2z}{dr^2} + \frac{1}{r}\frac{dz}{dr} = -\frac{q}{S}$$

Replacing q/S by $2G\theta$ and noting that $\tau = -dz/dr$ (see Fig. 172), we find

$$\frac{d\tau}{dr} + \frac{1}{r}\tau = 2G\theta \tag{d}$$

Assuming that we have a tube of constant thickness δ and denoting by τ_0 the stress at a considerable distance from the corner calculated from Eq. (176), we find, from (*c*),

$$2G\theta = \frac{\tau_0 s}{A}$$

Substituting in (*d*),

$$\frac{d\tau}{dr} + \frac{1}{r}\tau = \frac{\tau_0 s}{A} \tag{e}$$

The general solution of this equation is

$$\tau = \frac{C}{r} + \frac{\tau_0 s r}{2A} \tag{f}$$

Assuming that the projecting angles of the cross section have fillets with the radius a, as indicated in the figure, the constant of integration C can be determined from the equation

$$\int_a^{a+\delta} \tau\, dr = \tau_0 \delta \tag{g}$$

which follows from the hydrodynamical analogy (Art. 114), viz.: if an ideal fluid circulates in a channel having the shape of the ring cross section of the tubular member, the quantity of fluid passing each cross section of the channel must remain constant. Substituting expression (*f*) for τ into Eq. (*g*), and integrating, we find that

$$C = \tau_0 \delta \frac{1 - (s/4A)(2a + \delta)}{\log_e (1 + \delta/a)}$$

and, from Eq. (*f*), that

$$\tau = \frac{\tau_0 \delta}{r} \frac{1 - (s/4A)(2a + \delta)}{\log_e (1 + \delta/a)} + \frac{\tau_0 s r}{2A} \tag{h}$$

For a thin-walled tube the ratios $s(2a + \delta)/A$, sr/A, will be small, and (*h*) reduces to

$$\tau = \frac{\tau_0 \delta / r}{\log_e (1 + \delta/a)} \tag{i}$$

Substituting $r = a$ we obtain the stress at the reentrant corner. This is plotted in Fig. 174. The other curve[1] (A in Fig. 174) was obtained by the method of finite differences, without the assumption that the membrane at the corner has the form of a surface of revolution. It confirms the accuracy of Eq. (i) for small fillets—say up to $a/\delta = \frac{1}{4}$. For larger fillets the values given by Eq. (i) are too high.

Let us consider now the case when the cross section of a tubular member has more than two boundaries. Taking, for example, the case shown in Fig. 175 and assuming that the thickness of the wall is very small, the shearing stresses in each portion of the wall, from the membrane analogy, are

$$\tau_1 = \frac{h_1}{\delta_1} \qquad \tau_2 = \frac{h_2}{\delta_2} \qquad \tau_3 = \frac{h_1 - h_2}{\delta_3} = \frac{\tau_1\delta_1 - \tau_2\delta_2}{\delta_3} \tag{j}$$

in which h_1 and h_2 are the levels of the inner boundaries CD and EF.[2]

The magnitude of the torque, determined by the volume $ACDEFB$, is

$$M_t = 2(A_1h_1 + A_2h_2) = 2\,A_1\delta_1\tau_1 + 2A_2\delta_2\tau_2 \tag{k}$$

where A_1 and A_2 are areas indicated in the figure by dotted lines.

Further equations for the solution of the problem are obtained by applying Eq. (160) to the closed curves indicated in the figure by dotted lines. Assuming that the thicknesses δ_1, δ_2, δ_3 are constant and denoting

[1] Huth, *op. cit.*

[2] It is assumed that the plates are guided so as to remain horizontal (see p. 331).

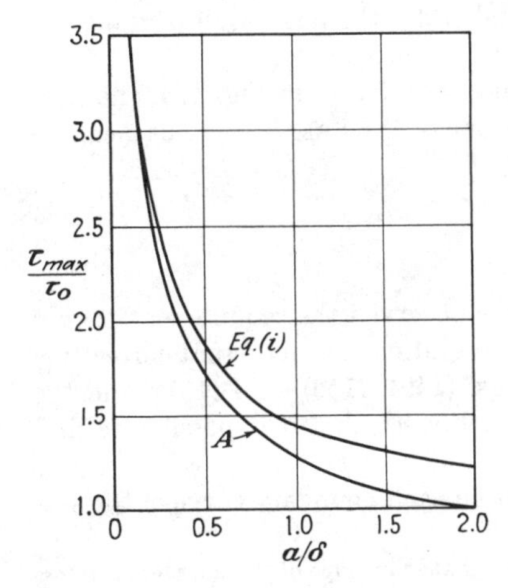

Fig. 174

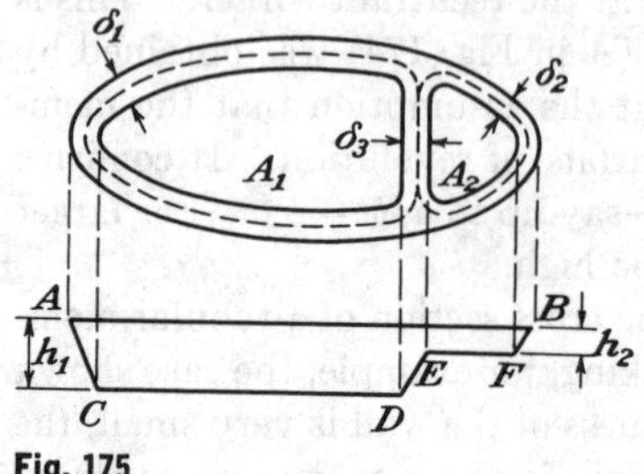

Fig. 175

by s_1, s_2, s_3 the lengths of corresponding dotted curves, we find, from Fig. 175,

$$\begin{aligned} \tau_1 s_1 + \tau_3 s_3 &= 2G\theta A_1 \\ \tau_2 s_2 - \tau_3 s_3 &= 2G\theta A_2 \end{aligned} \tag{l}$$

By using the last of the Eqs. (*j*) and Eqs. (*k*) and (*l*), we find the stresses τ_1, τ_2, τ_3 as functions of the torque:

$$\tau_1 = \frac{M_t[\delta_3 s_2 A_1 + \delta_2 s_3 (A_1 + A_2)]}{2[\delta_1 \delta_3 s_2 A_1^2 + \delta_2 \delta_3 s_1 A_2^2 + \delta_1 \delta_2 s_3 (A_1 + A_2)^2]} \tag{m}$$

$$\tau_2 = \frac{M_t[\delta_3 s_1 A_2 + \delta_1 s_3 (A_1 + A_2)]}{2[\delta_1 \delta_3 s_2 A_1^2 + \delta_2 \delta_3 s_1 A_2^2 + \delta_1 \delta_2 s_3 (A_1 + A_2)^2]} \tag{n}$$

$$\tau_3 = \frac{M_t(\delta_1 s_2 A_1 - \delta_2 s_1 A_2)}{2[\delta_1 \delta_3 s_2 A_1^2 + \delta_2 \delta_3 s_1 A_2^2 + \delta_1 \delta_2 s_3 (A_1 + A_2)^2]} \tag{o}$$

In the case of a symmetrical cross section, $s_1 = s_2$, $\delta_1 = \delta_2$, $A_1 = A_2$, and $\tau_3 = 0$. In this case the torque is taken by the outer wall of the tube, and the web remains unstressed.[1]

To get the twist for any section like that shown in Fig. 175, one substitutes the values of the stresses in one of the Eqs. (*l*). Thus θ can be obtained as a function of the torque M_t.

117 | Screw Dislocations

In the two preceding articles, we have observed the requirement that w must be a single-valued function if the solution is to represent correctly a state of torsion. On reexamining Eqs. (149), (150), and (151), and the boundary condition (152), we can quickly see that it is possible to find states of stress corresponding to $\theta = 0$. The stress function ϕ is to satisfy Laplace's equation and to be constant on each boundary curve of the sec-

[1] The small stresses corresponding to the change in slope of the membrane across the thickness of the web are neglected in this derivation.

tion. But we must use w rather than the form $\theta\psi(x,y)$ of Eq. (b) on page 293. Then Eqs. (f) of page 295 are replaced by

$$\frac{\partial \phi}{\partial y} = G\frac{\partial w}{\partial x} \qquad -\frac{\partial \phi}{\partial x} = G\frac{\partial w}{\partial y} \tag{a}$$

These are Cauchy-Riemann equations (see page 171) for the functions Gw and ϕ. Therefore, $Gw + i\phi$ is an analytic function of $x + iy$. Thus,

$$Gw + i\phi = f(x + iy) \tag{b}$$

Once the function f is chosen, we have a definite state, in which w will be the only nonzero displacement component.

Let r, ψ now represent polar coordinates in the cross section. The choice

$$f(x + iy) = -iA \log(x + iy) = A\psi - iA \log r \tag{c}$$

where A is a real constant, is of particular interest in the dislocation theory of plastic deformation (see Art. 34). From (b), we now have

$$Gw = A\psi \qquad \phi = -A \log r \tag{d}$$

The corresponding shear stress is in the circumferential direction and is given by the polar components

$$\tau_{z\psi} = -\frac{\partial \phi}{\partial r} = \frac{A}{r} \qquad \tau_{zr} = 0 \tag{e}$$

Any cylindrical boundary surface r = constant is free from loading. But the displacement w is not continuous. We can apply the solution to a hollow circular cylinder $a < r < b$ as in Fig. 176, which has an axial cut. One face is moved axially along the other by the uniform relative displacement

$$w(r,2\pi) - w(r,0) = \frac{2\pi A}{G} \tag{f}$$

obtained from the first of (d). The stress (e) can be regarded as induced

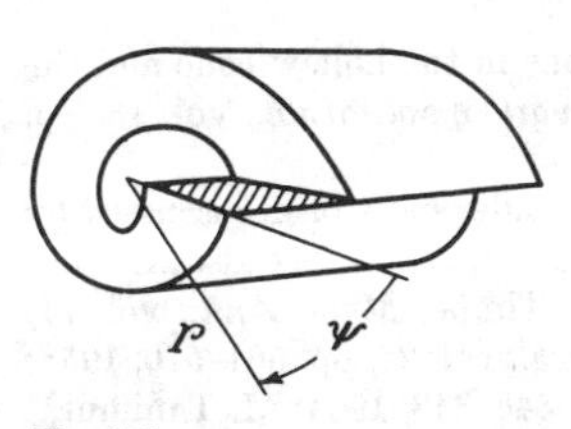

Fig. 176

by imposing this relative displacement, together with the shear loading on the ends implied by (*e*). This loading forms a torque

$$2\pi \int_a^b \tau_{z\psi} r^2 \, dr = \pi(b^2 - a^2)A$$

An equal and opposite torque can be introduced by superposing a state of simple torsion (Art. 101) with

$$\tau_{z\psi} = Br \qquad \tau_{zr} = 0 \qquad B = \frac{-2A}{a^2 + b^2}$$

and with $w = 0$. Finally, then, we have stress

$$\tau_{z\psi} = A\left(\frac{1}{r} - \frac{2r}{a^2 + b^2}\right) \tag{g}$$

which may be attributed to the relative displacement (*f*), the end torque being zero. There is still, of course, a distribution of shear stress on the ends, represented by (*g*). Since its resultant vanishes, its removal would have only a local effect, according to Saint-Venant's principle.

This final state, as applied in materials science, is called a *screw dislocation.*[1] A hollow cylinder with a cut has six distinct types of dislocation, in each of which the strain is continuous across the cut. The screw dislocation, the edge dislocation of Art. 34, the parallel gap dislocation of Art. 34 applied to the same cut, and the angular gap dislocation of Art. 31 (Fig. 45), account for four of the six.[2]

118 | Torsion of a Bar in Which One Cross Section Remains Plane

In discussing torsional problems, it has always been assumed that the torque is applied by means of shearing stresses distributed over the ends of a bar in a definite manner, obtained from the solution of Eq. (150) and satisfying the boundary condition (152). If the distribution of stresses at the ends is different from this, a local irregularity in stress distribution results and the solution of Eqs. (150) and (152) can be applied with satisfactory accuracy only in regions at some distance from the ends of the bar.[3]

[1] See, for instance, A. H. Cottrell, "Dislocations and Plastic Flow in Crystals," Oxford University Press, Fair Lawn, N.J., 1953.

[2] See references in n. 2, p. 89. For screw dislocations in the hollow cone and the hollow sphere, see J. N. Goodier and J. C. Wilhoit, *Quart. Appl. Math.*, vol. 13, pp. 263–269, 1955.

[3] The local irregularities at the ends of a circular cylinder have been discussed by F. Purser, *Proc. Roy. Irish Acad.*, Dublin, ser. A, vol. 26, p. 54, 1906. See also K. Wolf, *Sitzber. Akad. Wiss. Wien*, vol. 125, p. 1149, 1916; A. Timpe, *Math. Ann.*, vol. 71, p. 480, 1912; G. Horvay and J. A. Mirabel, *J. Appl. Mech.*, vol. 25, pp. 561–570, 1958; H. D. Conway and J. R. Moynihan, *ibid.*, vol. 31, pp. 346–348, 1964; M. Tanimura, *Tech. Repts. Osaka Univ.*, vol. 12, no. 497, pp. 93–104, 1962.

A similar irregularity occurs if a cross section of a twisted bar is prevented from warping by some constraint. We encounter problems of this kind occasionally in engineering.[1] A simple example is shown in Fig. 177. From symmetry it can be concluded that the middle cross section of the bar remains plane during torsion. Hence the stress distribution near this cross section must be different from that obtained above for rectangular bars (Art. 109). In discussing these stresses, let us consider first the case of a very narrow rectangle[2] and assume that the dimension a is large in comparison with b. If cross sections are free to warp, the stresses, from Art. 108, are

$$\tau_{xz} = -2G\theta y \qquad \tau_{yz} = 0 \tag{a}$$

and the corresponding displacements, from Eqs. (a), (b), and (d) of Art. 104, are

$$u = -\theta yz \qquad v = \theta xz \qquad w = -\theta xy \tag{b}$$

To prevent the warping of the cross sections, designated as displacement w, normal stresses σ_z must be distributed over the cross sections. We obtain an approximate solution by assuming that σ_z is proportional to w and that it diminishes with increase of distance z from the middle cross section. These assumptions are satisfied by taking

$$\sigma_z = -mE\theta e^{-mz}xy \tag{c}$$

in which m is a factor to be determined later. Due to the factor e^{-mz} the stress σ_z diminishes with increase of z and becomes negligible within a certain distance depending upon the magnitude of m.

The remaining stress components must now be chosen in such a manner as to satisfy the differential equations of equilibrium (123) and the boundary conditions.

[1] Torsion of I beams under such conditions was discussed by S. Timoshenko, *Z. Math. Physik*, vol. 58, p. 361, 1910. See also C. Weber, *Z. Angew. Math. Mech.*, vol. 6, p. 85, 1926.

[2] See S. Timoshenko, *Proc. London Math. Soc.*, ser. 2, vol. 20, p. 389, 1921.

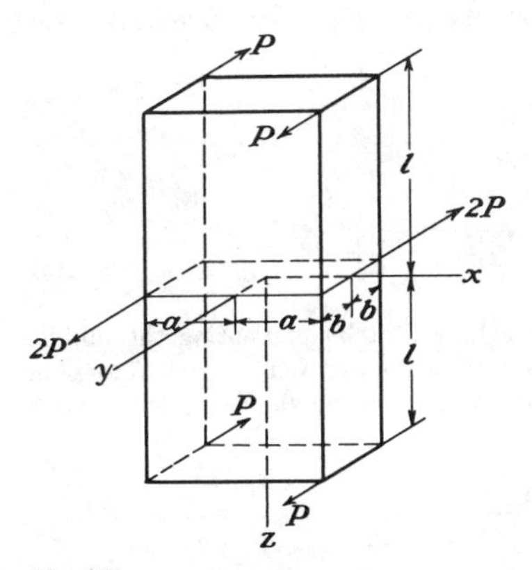

Fig. 177

It is easy to prove that these requirements are satisfied by taking

$$
\begin{aligned}
\sigma_x &= \sigma_y = 0 \\
\tau_{xy} &= -\tfrac{1}{8}Em^3\theta e^{-mz}(a^2 - x^2)(b^2 - y^2) \qquad (d) \\
\tau_{xz} &= \tfrac{1}{4}Em^2\theta e^{-mz}(a^2 - x^2)y - 2G\theta y \\
\tau_{yz} &= \tfrac{1}{4}Em^2\theta e^{-mz}(b^2 - y^2)x
\end{aligned}
$$

For large values of z this stress distribution approaches the stresses (*a*) for simple torsion. The stress component τ_{xy} becomes zero at the boundary $x = \pm a$ and $y = \pm b$; τ_{xz} and τ_{yz} are zero for $x = \pm a$ and $y = \pm b$, respectively. Hence the boundary conditions are satisfied and the lateral surface of the bar is free from forces.

For determining the factor m, we consider the strain energy of the bar and calculate m to make this energy a minimum. By using Eq. (131), we find

$$
V = \frac{1}{2G}\int_{-l}^{l}\int_{-a}^{a}\int_{-b}^{b}\left[\tau_{xy}^2 + \tau_{xz}^2 + \tau_{yz}^2 + \frac{1}{2(1+\nu)}\sigma_z^2\right]dx\,dy\,dz
$$

Substituting from (*d*), and noting that for a long bar we can with sufficient accuracy put

$$
\int_0^l e^{-mz}\,dz = \frac{1}{m}
$$

we get

$$
V = \frac{1}{9}E\theta^2a^3b^3\left\{-3m + (1+\nu)\left[\frac{2}{25}a^2b^2m^5 + \frac{1}{5}(a^2 + b^2)m^3 + \frac{12}{(1+\nu)^2}\frac{l}{a^2}\right]\right\} \qquad (e)
$$

The minimum condition gives us the following equation for determining m:

$$
(1+\nu)[\tfrac{2}{5}a^2b^2m^4 + \tfrac{3}{5}(a^2 + b^2)m^2] = 3
$$

which, for a narrow rectangle, reduces approximately to

$$
m^2 = \frac{5}{(1+\nu)a^2} \qquad (f)
$$

Substituting this value of m in (*c*) and (*d*), we find the stress distribution for the case when the middle cross section of the bar remains plane.

For calculating the angle of twist ψ, we put the potential energy (*e*) equal to the work done by the torque M_t,

$$
\frac{M_t\psi}{2} = V
$$

from which the angle of twist is

$$
\psi = \frac{3M_t}{16Gab^3}\left[l - \frac{\sqrt{5(1+\nu)}}{6}a\right] \qquad (g)
$$

Comparing this result with Eq. (163), we conclude that by preventing the middle cross section from warping we increase the rigidity of the bar with respect to torsion. The effect of the local irregularity in stress distribution on the value of ψ is the same as the influence of a diminution of the length l by

$$
a\,\frac{\sqrt{5(1+\nu)}}{6}
$$

Taking $\nu = 0.30$, this reduction in l becomes $0.425a$. We see that the effect of the constraint of the middle cross section on the angle of twist is small if the dimension a is small in comparison with l.

The twist of a bar of an elliptic cross section can be discussed in an analogous manner.[1] Of greater effect is the constraint of the middle cross section in the case of torsion of a bar of I cross section. An approximate method for calculating the angle of twist in this case is obtained by considering bending of the flanges during torsion.[2]

119 | Torsion of Circular Shafts of Variable Diameter

Let us consider a shaft in the form of a body of revolution twisted by couples applied at the ends (Fig. 178). We may take the axis of the shaft

[1] A. Föppl, *Sitzber. Bayer. Akad. Wiss., Math.-Phys. Klasse,* München, 1920, p. 261.

[2] See S. Timoshenko, *Z. Math. Physik.*, vol. 58, p. 361, 1910; or "Strength of Materials," vol. 2, p. 260, 1956. For similar approximation methods for other thin-walled sections, open and closed, see the article on Torsion in pt. 4 of W. Flügge (ed.), "Handbook of Engineering Mechanics," McGraw-Hill Book Company, New York, 1962.

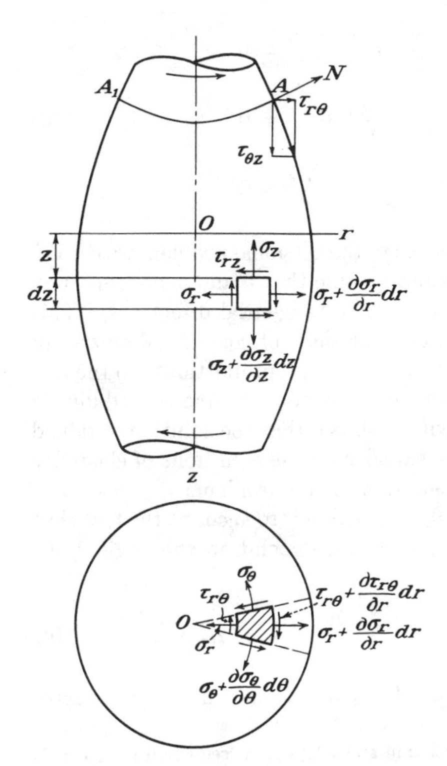

Fig. 178

as the z axis and use polar coordinates r and θ for defining the position of an element in the plane of a cross section. The notations for stress components in such a case are $\sigma_r, \sigma_\theta, \sigma_z, \tau_{rz}, \tau_{r\theta}, \tau_{\theta z}$. The components of displacements in the radial and tangential directions we may denote by u and v and the component in the z direction by w. Then, using the formulas obtained previously for two-dimensional problems (Art. 30), we find the following expressions for the strain components:

$$
\begin{aligned}
\epsilon_r &= \frac{\partial u}{\partial r} & \epsilon_\theta &= \frac{u}{r} + \frac{\partial v}{r\,\partial \theta} & \epsilon_z &= \frac{\partial w}{\partial z} \\
\gamma_{r\theta} &= \frac{\partial u}{r\,\partial \theta} + \frac{\partial v}{\partial r} - \frac{v}{r} & \gamma_{rz} &= \frac{\partial u}{\partial z} + \frac{\partial w}{\partial r} & \gamma_{z\theta} &= \frac{\partial v}{\partial z} + \frac{\partial w}{r\,\partial \theta}
\end{aligned}
\tag{179}
$$

Writing down the equations of equilibrium of an element (Fig. 178), as was done before for the case of two-dimensional problems (Art. 27), and assuming that there are no body forces, we arrive at the following differential equations of equilibrium:[1]

$$
\begin{aligned}
\frac{\partial \sigma_r}{\partial r} + \frac{1}{r}\frac{\partial \tau_{r\theta}}{\partial \theta} + \frac{\partial \tau_{rz}}{\partial z} + \frac{\sigma_r - \sigma_\theta}{r} &= 0 \\
\frac{\partial \tau_{rz}}{\partial r} + \frac{1}{r}\frac{\partial \tau_{\theta z}}{\partial \theta} + \frac{\partial \sigma_z}{\partial z} + \frac{\tau_{rz}}{r} &= 0 \\
\frac{\partial \tau_{r\theta}}{\partial r} + \frac{1}{r}\frac{\partial \sigma_\theta}{\partial \theta} + \frac{\partial \tau_{\theta z}}{\partial z} + \frac{2\tau_{r\theta}}{r} &= 0
\end{aligned}
\tag{180}
$$

In the application of these equations to the torsional problem we use the *semi-inverse method* (see page 293) and assume that u and w are zero, i.e., that during twist the particles move only in tangential directions. This assumption differs from that for a circular shaft of constant diameter in that these tangential displacements are no longer proportional to the distance from the axis, i.e., the radii of a cross section become curved during twist. In the following pages it will be shown that the solution obtained on the basis of such an assumption satisfies all the equations of elasticity and therefore represents the true solution of the problem.

Substituting in (179) $u = w = 0$, and taking into account the fact that from symmetry the displacement v does not depend on the angle θ, we find that

$$
\epsilon_r = \epsilon_\theta = \epsilon_z = \gamma_{rz} = 0 \qquad \gamma_{r\theta} = \frac{\partial v}{\partial r} - \frac{v}{r} \qquad \gamma_{\theta z} = \frac{\partial v}{\partial z} \tag{a}
$$

Hence, of all the stress components, only $\tau_{r\theta}$ and $\tau_{\theta z}$ are different from zero.

[1] These equations were obtained by Lamé and Clapeyron; see *Crelle's J.*, vol. 7, 1831.

The first two of Eqs. (180) are identically satisfied, and the third of these equations gives

$$\frac{\partial \tau_{r\theta}}{\partial r} + \frac{\partial \tau_{\theta z}}{\partial z} + \frac{2\tau_{r\theta}}{r} = 0 \tag{b}$$

This equation can be written in the form

$$\frac{\partial}{\partial r}(r^2\tau_{r\theta}) + \frac{\partial}{\partial z}(r^2\tau_{\theta z}) = 0 \tag{c}$$

It is seen that this equation is satisfied by using a stress function ϕ of r and z, such that

$$r^2\tau_{r\theta} = -\frac{\partial \phi}{\partial z} \qquad r^2\tau_{\theta z} = \frac{\partial \phi}{\partial r} \tag{d}$$

To satisfy the compatibility conditions, it is necessary to consider the fact that $\tau_{r\theta}$ and $\tau_{\theta z}$ are functions of the displacement v. From Eqs. (a) and (d) we find

$$\begin{aligned}
\tau_{r\theta} &= G\gamma_{r\theta} = G\left(\frac{\partial v}{\partial r} - \frac{v}{r}\right) = Gr\frac{\partial}{\partial r}\left(\frac{v}{r}\right) = -\frac{1}{r^2}\frac{\partial \phi}{\partial z} \\
\tau_{\theta z} &= G\gamma_{\theta z} = G\frac{\partial v}{\partial z} = Gr\frac{\partial}{\partial z}\left(\frac{v}{r}\right) = \frac{1}{r^2}\frac{\partial \phi}{\partial r}
\end{aligned} \tag{e}$$

From these equations it follows that

$$\frac{\partial}{\partial r}\left(\frac{1}{r^3}\frac{\partial \phi}{\partial r}\right) + \frac{\partial}{\partial z}\left(\frac{1}{r^3}\frac{\partial \phi}{\partial z}\right) = 0 \tag{f}$$

or

$$\frac{\partial^2 \phi}{\partial r^2} - \frac{3}{r}\frac{\partial \phi}{\partial r} + \frac{\partial^2 \phi}{\partial z^2} = 0 \tag{g}$$

Let us consider now the boundary conditions for the function ϕ. From the condition that the lateral surface of the shaft is free from external forces, we conclude that at any point A at the boundary of an axial section (Fig. 178) the total shearing stress must be in the direction of the tangent to the boundary and its projection on the normal N to the boundary must be zero. Hence

$$\tau_{r\theta}\frac{ds}{dz} - \tau_{\theta z}\frac{dr}{ds} = 0$$

where ds is an element of the boundary. Substituting from (d), we find that

$$\frac{\partial \phi}{\partial z}\frac{dz}{ds} + \frac{\partial \phi}{\partial r}\frac{dr}{ds} = 0 \tag{h}$$

from which we conclude that ϕ is constant along the boundary of the axial section of the shaft.

Equation (g) together with the boundary condition (h) completely determines the stress function ϕ, from which we may obtain the stresses satisfying the equations of equilibrium, the compatibility equations, and the condition at the lateral surface of the shaft.[1]

The magnitude of the torque is obtained by taking a cross section and calculating the moment given by the shearing stresses $\tau_{\theta z}$. Then

$$M_t = \int_0^a 2\pi r^2 \tau_{\theta z}\, dr = 2\pi \int_0^a \frac{\partial \phi}{\partial r}\, dr = 2\pi |\phi|_0^a \tag{i}$$

where a is the outer radius of the cross section. The torque is thus easily obtained if we know the difference between the values of the stress function at the outer boundary and at the center of the cross section.

In discussing displacements during twist of the shaft, let us use the notation $\psi = v/r$ for the angle of rotation of an elemental ring of radius r in a cross section of the shaft. This ring can be considered as the cross section of one of a number of thin elemental tubes into which the shaft is subdivided. Then ψ is the angle of twist of such a tube. From the fact that the radii of the cross section become curved, it follows that ψ varies with r and the angles of twist of elemental tubes are not equal for the same cross section of the shaft. Equations (e) can now be written in the form

$$Gr^3 \frac{\partial \psi}{\partial r} = -\frac{\partial \phi}{\partial z} \qquad Gr^3 \frac{\partial \psi}{\partial z} = \frac{\partial \phi}{\partial r}$$

from which

$$\frac{\partial}{\partial r}\left(r^3 \frac{\partial \psi}{\partial r}\right) + \frac{\partial}{\partial z}\left(r^3 \frac{\partial \psi}{\partial z}\right) = 0$$

or

$$\frac{\partial^2 \psi}{\partial r^2} + \frac{3}{r}\frac{\partial \psi}{\partial r} + \frac{\partial^2 \psi}{\partial z^2} = 0 \tag{j}$$

A solution of this equation[2] gives us the angle of twist as a function of r

[1] This general solution of the problem is due to J. H. Michell, *Proc. London Math. Soc.*, vol. 31, p. 141, 1899. See also A. Föppl, *Sitzber. Bayer. Akad. Wiss.*, München, vol. 35, pp. 249 and 504, 1905. The book "Kerbspannungslehre" by H. Neuber, Springer-Verlag OHG, Berlin, 1958, gives solutions for the hyperboloid of revolution, and for a cavity in the form of an ellipsoid of revolution, by a different method. Reviews of the literature on the subject have been given by T. Pöschl, *Z. Angew. Math. Mech.*, vol. 2, p. 137, 1922, and T. J. Higgins, *Exp. Stress Anal.*, vol. 3, no. 1, p. 94, 1945.

[2] Solutions in the cylindrical coordinates r, z are given by H. Reissner and G. J. Wennagel, *J. Appl. Mech.*, vol. 17, pp. 275–282, 1950. Product solutions in spherical coordinates are given by H. Poritsky, *Proc. Symp. Appl. Math., Am. Math. Soc.*, vol. 3, pp. 163–186, 1951; and J. C. Wilhoit, Jr., *Quart. Appl. Math.*, vol. 11, pp. 499–501, 1954.

and z. If we put

$$\psi = \text{const} \tag{k}$$

in this solution, we obtain a surface in which all the points have the same angle of twist. In Fig. 178, AA_1 represents the intersection of such a surface with the axial section of the shaft. From symmetry it follows that the surfaces given by Eq. (k) are surfaces of revolution and AA_1 is a meridian of the surface going through the point A. During twist these surfaces rotate about the z axis without any distortion, exactly in the same manner as the plane cross sections in the case of circular cylindrical shafts. Hence the total strain at any point of the meridian AA_1 is pure shearing strain in the plane perpendicular to the meridian, and the corresponding shearing stress in the axial section of the shaft has the direction normal to the meridian. At the boundary this stress is tangent to the boundary and the meridians are normal to the boundary of the axial section. If we go from the surface ψ = constant to an adjacent surface, the rate of change of ψ along the boundary of the axial section of the shaft is $d\psi/ds$, and in the same manner as for a cylindrical shaft of circular cross section (Art. 101) we have

$$\tau = Gr\frac{d\psi}{ds} \tag{l}$$

where

$$\tau = \tau_{r\theta}\frac{dr}{ds} + \tau_{\theta z}\frac{dz}{ds}$$

is the resultant shearing stress at the boundary. It is seen that the value of this shearing stress is easily obtained if we find by experiment the values of $d\psi/ds$.[1]

Let us consider now a particular case of a conical shaft[2] (Fig. 179). In this case the ratio

$$\frac{z}{(r^2 + z^2)^{1/2}}$$

is constant at the boundary of the axial section and equal to $\cos\alpha$. Any function of this ratio will satisfy the boundary condition (h). In order to satisfy also Eq. (g), we take

$$\phi = c\left\{\frac{z}{(r^2 + z^2)^{1/2}} - \frac{1}{3}\left[\frac{z}{(r^2 + z^2)^{1/2}}\right]^3\right\} \tag{m}$$

where c is a constant. Then by differentiation we find

$$\tau_{\theta z} = \frac{1}{r^2}\frac{\partial\phi}{\partial r} = -\frac{crz}{(r^2 + z^2)^{5/2}} \tag{n}$$

[1] Such experiments were made by R. Sonntag, *Z. Angew. Math. Mech.*, vol. 9, p. 1, 1929.

[2] See Föppl, *loc. cit.*, n. 1, p. 344.

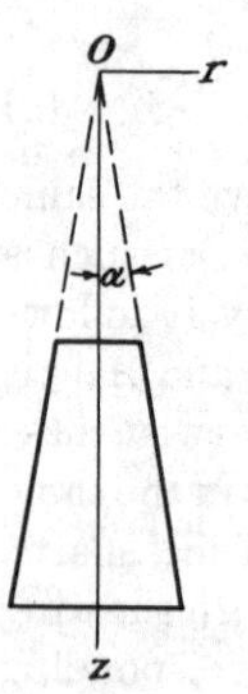

Fig. 179

The constant c is obtained from Eq. (*i*). Substituting (*m*) in this equation, we find

$$c = -\frac{M_t}{2\pi(\frac{2}{3} - \cos\alpha + \frac{1}{3}\cos^3\alpha)}$$

To calculate the angle of twist we use Eqs. (*e*), from which the expression for ψ, satisfying Eq. (*j*) and the boundary condition, is

$$\psi = \frac{c}{3G(r^2 + z^2)^{3/2}} \tag{o}$$

It will be seen that the surfaces of equal angle of twist are spherical surfaces with their center at the origin O.

The case of a shaft in the form of an ellipsoid, hyperboloid, or paraboloid of revolution can be discussed in an analogous manner.[1]

The problems encountered in practice are of a more complicated nature. The diameter of the shaft usually changes abruptly, as shown in Fig. 180*a*. The first investigation of such problems was made by A. Föppl. C. Runge suggested a numerical method for the approximate solution of these problems,[2] and it was shown that considerable stress concentration takes place at such points as m and n, and that the magnitude of the maximum stress

[1] See papers by E. Melan, *Tech. Blätter*, Prag, 1920; A. N. Dinnik, *Bull. Don Polytech. Inst., Novotcherkask*, 1912; W. Arndt, "Die Torsion von Wellen mit achsensymmetrischen Bohrungen und Hohlräumen," dissertation, Göttingen, 1916; A. Timpe, *Math. Ann.*, 1911, p. 480. Further references are given in a review by Higgins, *loc. cit.*, n. 1, p. 344. Design curves are given by R. E. Peterson, *op. cit.* The shaft with a sharp corner at m and n, Fig. 177*a*, or a rectangular groove instead of semicircular as in Fig. 177*b*, is treated by M. Tanimura, *Tech. Repts. Osaka Univ.*, vol. 12, no. 498, pp. 105–122, 1962.

[2] See F. A. Willers, *Z. Math. Physik*, vol. 55, p. 225, 1907. Another approximate method was developed by L. Föppl, *Sitzber. Bayer. Akad. Wiss., München*, vol. 51, p. 61, 1921, and by R. Sonntag, *Z. Angew. Math. Mech.*, *loc. cit.*

for a shaft of two different diameters d and D (Fig. 180a) depends on the ratio of the radius a of the fillet to the diameter d of the shaft and on the ratio d/D.

In the case of a semicircular groove of very small radius a, the maximum stress at the bottom of the groove (Fig. 180b) is twice as great as at the surface of the cylindrical shaft without the groove.[1]

In discussing stress concentration at the fillets and grooves of twisted circular shafts, an electrical analogy has proved very useful.[2] The general equation for the flow of an electric current in a thin homogeneous plate of variable thickness is

$$\frac{\partial}{\partial x}\left(h\frac{\partial\psi}{\partial x}\right) + \frac{\partial}{\partial y}\left(h\frac{\partial\psi}{\partial y}\right) = 0 \tag{p}$$

in which h is the variable thickness of the plate and ψ the potential function.

Let us assume that the plate has the same boundary as the axial section of the shaft (Fig. 181), that the x and y axes coincide with the z and r axes, and that the thickness of the plate is proportional to the cube of the radial distance r, so that $h = \alpha r^3$. Then Eq. (p) becomes

$$\frac{\partial^2\psi}{\partial z^2} + \frac{3}{r}\frac{\partial\psi}{\partial r} + \frac{\partial^2\psi}{\partial r^2} = 0$$

This coincides with equation (j), and we conclude that the equipotential lines of the plate are determined by the same equation as the lines of equal angles of twist in the case of a shaft of variable diameter.

[1] For larger grooves, see the book by R. E. Peterson, *op. cit.;* also Flügge, *op. cit.,* p. 36–26.

[2] See paper by L. S. Jacobsen, *Trans. ASME,* vol. 47, p. 619, 1925, and the survey given by T. J. Higgins, *loc. cit.,* n. 1, p. 344. Discrepancies between results obtained from this and other methods are discussed in the latter paper. For further comparisons and strain-gauge measurements extending Fig. 182 to $2a/d = 0.50$ see A. Weigand, *Luftfahrt-Forsch.,* vol. 20, p. 217, 1943, translated in *NACA Tech. Mem.* 1179, September, 1947.

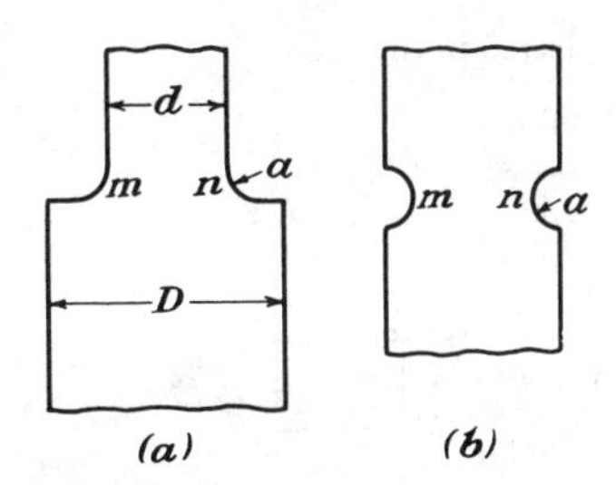

Fig. 180

Assuming that the ends of the plate, corresponding to the ends of the shaft, are maintained at a certain difference of potential so that the current flows along the z axis, the equipotential lines are normal to the lateral sides of the plate, i.e., we have the same boundary conditions as for lines of constant angle of twist. If the differential equations and the boundary conditions are the same for these two kinds of lines, the lines are identical. Hence, by investigating the distribution of potential in the plate, valuable information regarding the stress distribution in the twisted shaft can be obtained.

The maximum stress is at the surface of the shaft, and we obtain this stress by using Eq. (l). From this equation, by applying the electrical analogy, it follows that the stress is proportional to the rate of drop of potential along the edge of the plate.

Actual measurements were made on a steel model 24 in. long by 6 in. wide at the larger end and 1 in. maximum thickness (Fig. 181). The drop of potential along the edge $mnpq$ of the model was investigated by using a sensitive galvanometer, the terminals of which were connected to two sharp needles fastened in a block at a distance 2 mm apart. By touching the plate with the needles, the drop in potential over the distance between the needle points was indicated by the galvanometer. By moving the needles along the fillet it is possible to find the place of, and measure, the maximum voltage gradient. The ratio of this maximum to the voltage gradient at a remote point m (Fig. 181a) gives the magnitude of the factor of stress concentration k* in the equation

$$\tau_{\max} = k \frac{16M_t}{\pi d^3}$$

The results of such tests in one particular case are represented in Fig. 181c, in which the potential drop measured at each point is indicated by the length on the normal to the edge of the plate at this point. From this figure the factor of stress concentration is found to be 1.54. The magnitudes of this factor obtained with various proportions of shafts are given in Fig. 182, in which the abscissas represent the ratios $2a/d$ of the radius

* Small variations in radius r [Eq. (l)] can be neglected in this case.

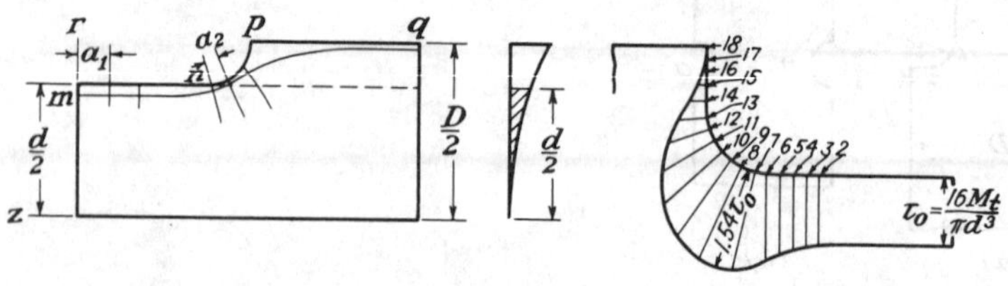

Fig. 181

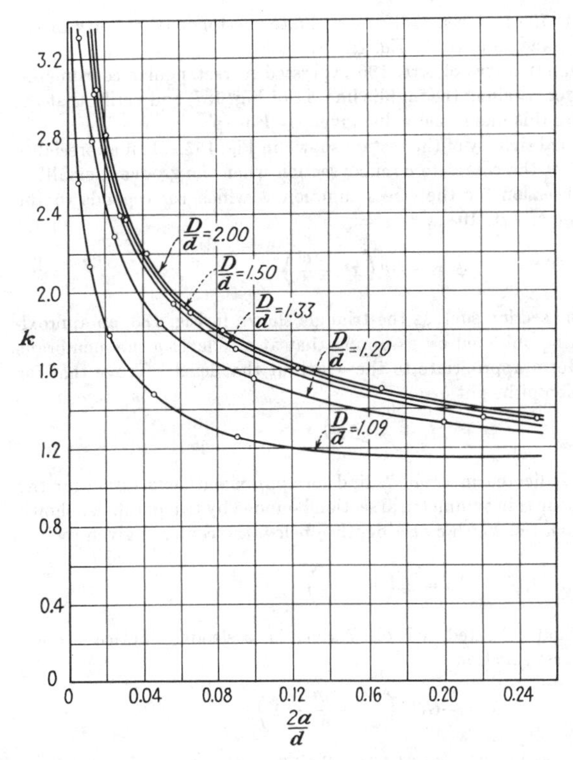

Fig. 182

of the fillet to the radius of the smaller shaft and the ordinates the factor of stress concentration k for various values of the ratio D/d (see Fig. 180).

PROBLEMS

1. Show by considering the equilibrium of the whole bar that when all stress components vanish except τ_{xz}, τ_{yz}, the loading must consist of torsional couples only [cf. Eqs. (*h*), Art. 104].
2. Show that $\phi = A(r^2 - a^2)$ solves the torsion problem for the solid or hollow circular shaft. Determine A in terms of $G\theta$. Using Eqs. (149) and (153) evaluate the maximum shearing stress and the torsional rigidity in terms of M_t for the solid shaft, and verify that the results are in agreement with those given in any text on strength of materials.
3. Show that for the same twist, the elliptic section has a greater shearing stress than the inscribed circular section (radius equal to the minor axis b of the ellipse). Which takes the greater torque for the same allowable stress?

4. Use Eq. (g) of Art. 106 and Eq. (153) to evaluate the torsional rigidity of the equilateral triangle, and thus verify Eq. (l), Art. 106.

5. Using the stress function (m) of Art. 106 expressed in rectangular coordinates, find an expression for τ_{yz} along the middle line Ax of Fig. 157, and verify that the greatest value along this line is the value given by Eq. (p).

6. Evaluate the torsional rigidity of the section shown in Fig. 157. Is it appreciably different from that of the complete circular section when the groove is small?

7. Show that the expression for the stress function ϕ which corresponds to the parabolic membrane of Art. 108 is

$$\phi = -G\theta\left(x^2 - \frac{c^2}{4}\right)$$

In a narrow tapered section such as the triangle shown in Fig. 183, an approximate[1] solution can be obtained by assuming that at any level y the membrane has the parabolic form appropriate to the width at that level. Prove that for the triangular section of height b

$$M_t = \tfrac{1}{12}G\theta bc_o^3$$

approximately.

8. Using the method indicated in Prob. 7, find an approximate expression for the torsional rigidity of the thin symmetrical section bounded by two parabolas shown in Fig. 184, for which the width c at a depth y below the center is given by

$$c = c_o\left(1 - \frac{y^2}{b^2}\right)$$

9. Show that the method indicated in Prob. 7 gives for a slender elliptical section the approximate stress function

$$\phi = -G\theta b^2\left(\frac{x^2}{a^2} + \frac{y^2}{b^2} - 1\right)$$

the ellipse being that of Fig. 153 with b/a small. Show that the exact solution of Art. 105 approaches this as b/a is made small.

Derive the approximate formulas

$$M_t = \pi ab^3G\theta \qquad \tau_{\max} = 2G\theta b = \frac{2M_t}{\pi ab^2}$$

[1] The error is examined by W. J. Carter, *J. Appl. Mech.*, vol. 25, pp. 115–121, 1958.

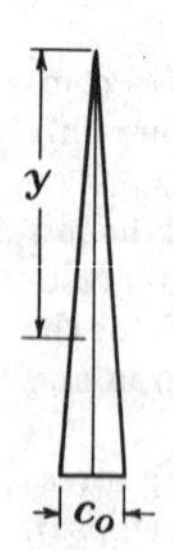

Fig. 183

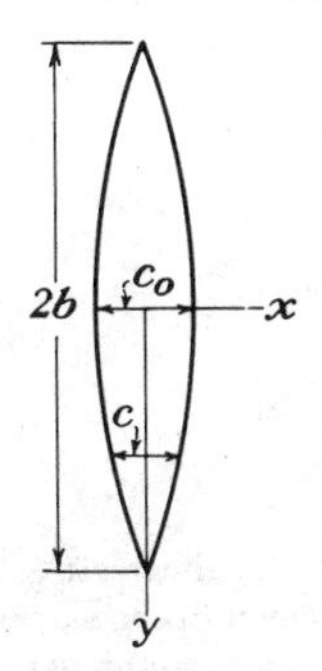

Fig. 184

for the slender elliptical section, and compare with the corresponding formulas for the thin rectangular section of length $2a$ and thickness $2b$.

10. Apply the method given at the end of Art. 111 to find an approximation to the torsional rigidity of the section described in Prob. 8.

11. A section has a single hole, and the stress function ϕ is determined so that it vanishes on the outside boundary and has a constant value ϕ_H on the boundary of the hole. By adapting the calculation indicated on page 296 for Eq. (153), prove that the total torque is given by twice the volume under the ϕ surface plus twice the volume under a flat roof at height ϕ_H covering the hole (cf. page 331).

12. A closed thin-walled tube has a perimeter l and a uniform wall thickness δ. An open tube is made by making a fine longitudinal cut in it. Show that when the maximum shear stress is the same in both closed and open tubes,

$$\frac{M_{t\,\text{open}}}{M_{t\,\text{closed}}} = \frac{l\delta}{6A} \qquad \frac{\theta_{\text{open}}}{\theta_{\text{closed}}} = \frac{2A}{l\delta}$$

and that the ratio of the torsional rigidities is $l^2\delta^2/12A^2$, A being the area of the "hole."

Evaluate these ratios for a circular tube of 1 in. radius, $\frac{1}{10}$ in. wall thickness.

13. A thin-walled tube has the cross section shown in Fig. 185, with uniform wall thickness δ. Show that there will be no stress in the central web when the tube is twisted.

Find formulas for (*a*) the shear stress in the walls, away from the corners, (*b*) the unit twist θ, in terms of the torque.

14. Find expressions for the shear stresses in a tube of the section shown in Fig. 186, the wall thickness δ being uniform.

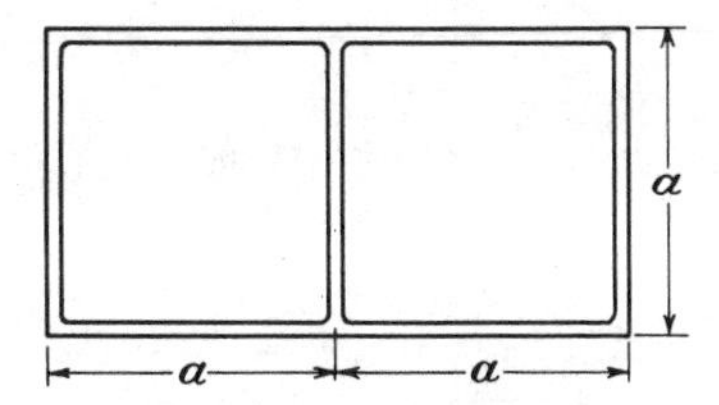

Fig. 185

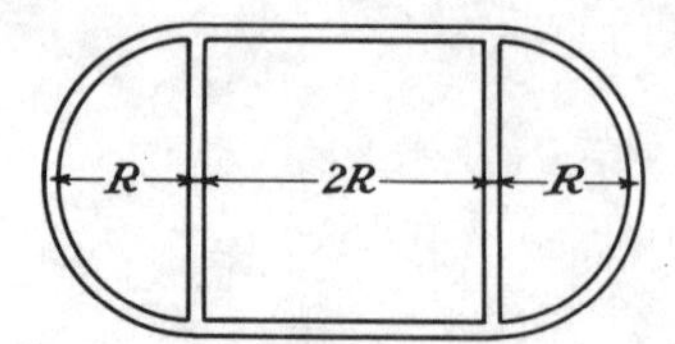

Fig. 186

15. In discussing thin-walled closed sections, it was assumed that the shear stress is constant across the wall thickness, corresponding to constant membrane slope across the thickness. Show that this cannot be strictly true for a straight part of the wall (e.g., Fig. 173*a*) and that in general the correction to this shear stress consists of the shear stress in a tube made "open" by longitudinal cuts (cf. Prob. 12).

16. The theory of Art. 119 includes the uniform circular shaft as a special case. What are the corresponding forms for the functions ϕ and ψ? Show that these functions give the correct relation between torque and unit twist.

17. Prove that

$$\phi = \frac{z}{R} + \frac{Az^3}{R^3} \qquad \text{where} \qquad R = (r^2 + z^2)^{1/2}$$

satisfies Eq. (*g*) of Art. 119 only if the constant A is $-\frac{1}{3}$ [cf. Eq. (*m*)].

18. At any point of an axial section of a shaft of variable diameter, line elements ds and dn (at right angles) in the section are chosen arbitrarily as shown in Fig. 187. The shear stress is expressed by components τ_s, τ_n along these. Show that

$$\tau_s = \frac{1}{r^2}\frac{\partial \phi}{\partial n} \qquad \tau_n = -\frac{1}{r^2}\frac{\partial \phi}{\partial s} \qquad \tau_s = Gr\frac{\partial \psi}{\partial s} \qquad \tau_n = Gr\frac{\partial \psi}{\partial n}$$

and deduce the boundary condition satisfied by ψ.

Show without calculation that the function given by Eq. (*q*) of Art. 119 satisfies this boundary condition for a conical boundary of any angle.

19. Verify that Eq. (*o*) of Art. 119 gives correctly the function ψ corresponding to the function ϕ in Eq. (*m*).

20. If the theory of Art. 119 is modified by discarding the boundary condition ϕ = constant, the stress will be due to certain "rings of shear" on the boundary,

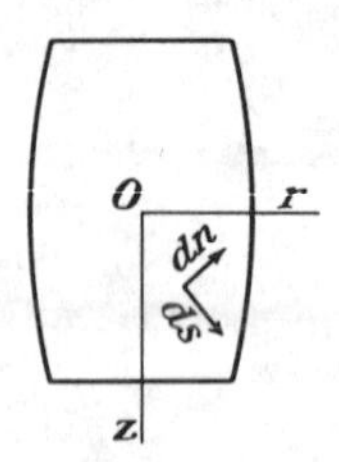

Fig. 187

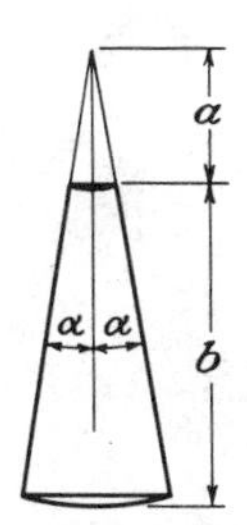

Fig. 188

as well as end torques. Considering the uniform circular shaft, describe the problem solved by $\phi = Czr^4$ where C is a constant, for $0 < z < l$.

21. Prove that the relative rotation of the ends of the (conical) tapered shaft shown in Fig. 188 due to torque M_t is

$$\frac{M}{2\pi(\frac{2}{3} - \cos\alpha + \frac{1}{3}\cos^3\alpha)}\frac{1}{3G}\left(\frac{1}{a^3} - \frac{1}{b^3}\right)$$

If a and b are both made large, with $b - a = l$, and α is made small, the above result should approach the relative rotation of the ends of a uniform shaft of length l, and radius αa, due to torque M_t. Show that it does so.

22. Use the functions given by Eqs. (m) and (o) of Art. 119 to find, in terms of M_t, the relative rotation of the ends of the hollow conical shaft shown in Fig. 189. The ends are spherical surfaces of radii a, b, center O.

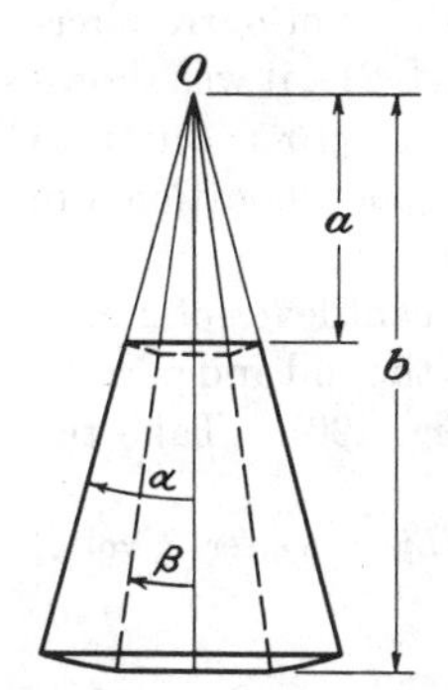

Fig. 189

chapter | 11

Bending of Bars

120 | Bending of a Cantilever

In discussing pure bending (Art. 102), it was shown that if a bar is bent in one of its principal planes by two equal and opposite couples applied at the ends, the deflection occurs in the same plane, and of the six components of stress only the normal stress parallel to the axis of the bar is different from zero. This stress is proportional to the distance from the neutral axis. Thus the exact solution coincides in this case with the elementary theory of bending. In discussing bending of a cantilever of narrow rectangular cross section by a force applied at the end (Art. 21), it was shown that in addition to normal stresses, proportional in each cross section to the bending moment, there will act also shearing stresses proportional to the shearing force.

Consider now a more general case of bending of a cantilever of a constant cross section of any shape by a force P applied at the end and parallel to one of the principal axes of the cross section[1] (Fig. 190). Take the

[1] This problem was solved by Saint-Venant, *J. Mathémat.* (*Liouville*), ser. 2, vol. 1, 1856.

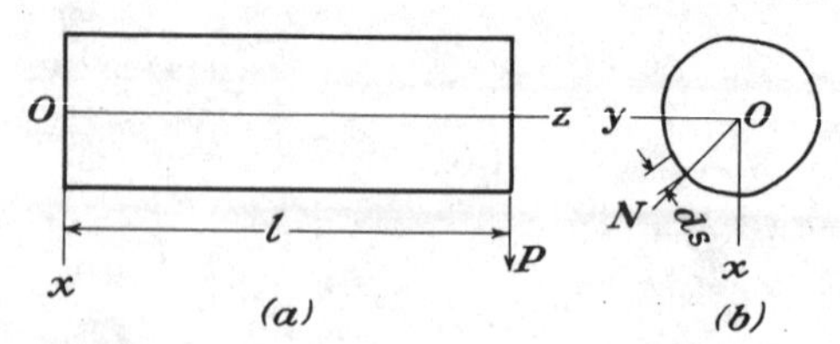

Fig. 190

origin of the coordinates at the centroid of the fixed end. The z axis coincides with the centerline of the bar, and the x and y axes coincide with the principal axes of the cross section. In the solution of the problem we apply Saint-Venant's semi-inverse method and at the very beginning make certain assumptions regarding stresses. We assume that normal stresses over a cross section at a distance z from the fixed end are distributed in the same manner as in the case of pure bending:

$$\sigma_z = -\frac{P(l-z)x}{I} \tag{a}$$

We assume also that there are shearing stresses, acting on the same cross sections, which we resolve at each point into components τ_{xz} and τ_{yz}. We assume that the remaining three stress components σ_x, σ_y, τ_{xy} are zero. It will now be shown that by using these assumptions we arrive at a solution which satisfies all of the equations of the theory of elasticity and which is therefore the exact solution of the problem, provided that the load P is distributed on the end $z = l$, and the reaction on $z = 0$, in the manner required by the solution.

With these assumptions, neglecting body forces, the differential equations of equilibrium (123) become

$$\frac{\partial \tau_{xz}}{\partial z} = 0 \qquad \frac{\partial \tau_{yz}}{\partial z} = 0 \tag{b}$$

$$\frac{\partial \tau_{xz}}{\partial x} + \frac{\partial \tau_{yz}}{\partial y} = -\frac{Px}{I} \tag{c}$$

From (*b*) we conclude that shearing stresses do not depend on z and are the same in all cross sections of the bar.

Considering now the boundary conditions (124) and applying them to the lateral surface of the bar, which is free from external forces, we find that the first two of these equations are identically satisfied and the third one gives

$$\tau_{xz} l + \tau_{yz} m = 0$$

From Fig. 190*b* we see that

$$l = \cos(Nx) = \frac{dy}{ds} \qquad m = \cos(Ny) = -\frac{dx}{ds}$$

in which ds is an element of the bounding curve of the cross section. Then the condition at the boundary is

$$\tau_{xz}\frac{dy}{ds} - \tau_{yz}\frac{dx}{ds} = 0 \tag{d}$$

Turning to the compatibility equations (126), we see that the first three of these equations, containing normal stress components, and the last equation, containing τ_{xy}, are identically satisfied. The system (126) then reduces to the two equations

$$\nabla^2 \tau_{yz} = 0 \qquad \nabla^2 \tau_{xz} = -\frac{P}{I(1+\nu)} \tag{e}$$

Thus the solution of the problem of bending of a prismatical cantilever of any cross section reduces to finding, for τ_{xz} and τ_{yz}, functions of x and y that satisfy the equation of equilibrium (*c*), the boundary condition (*d*), and the compatibility equations (*e*).

121 | Stress Function

In discussing the bending problems, we shall again make use of a stress function $\phi(x,y)$. It is easy to see that the differential equations of equilibrium (*b*) and (*c*) of the previous article are satisfied by taking

$$\tau_{xz} = \frac{\partial \phi}{\partial y} - \frac{Px^2}{2I} + f(y) \qquad \tau_{yz} = -\frac{\partial \phi}{\partial x} \tag{181}$$

in which ϕ is the stress function of x and y, and $f(y)$ is a function of y only, which will be determined later from the boundary condition.

Substituting (181) in the compatibility equations (*e*) of the previous article, we obtain

$$\frac{\partial}{\partial x}\left(\frac{\partial^2 \phi}{\partial x^2} + \frac{\partial^2 \phi}{\partial y^2}\right) = 0$$

$$\frac{\partial}{\partial y}\left(\frac{\partial^2 \phi}{\partial x^2} + \frac{\partial^2 \phi}{\partial y^2}\right) = \frac{\nu}{1+\nu}\frac{P}{I} - \frac{d^2 f}{dy^2}$$

From these equations we conclude that

$$\frac{\partial^2 \phi}{\partial x^2} + \frac{\partial^2 \phi}{\partial y^2} = \frac{\nu}{1+\nu}\frac{Py}{I} - \frac{df}{dy} + c \tag{a}$$

where c is a constant of integration. This constant has a very simple physical meaning. Consider the rotation of an element of area in the plane of a cross section of the cantilever. This rotation ω_z is expressed by the equation (see page 233)

$$2\omega_z = \frac{\partial v}{\partial x} - \frac{\partial u}{\partial y}$$

The rate of change of this rotation in the direction of the z axis can be written in the following manner:

$$\frac{\partial}{\partial z}\left(\frac{\partial v}{\partial x} - \frac{\partial u}{\partial y}\right) = \frac{\partial}{\partial x}\left(\frac{\partial v}{\partial z} + \frac{\partial w}{\partial y}\right) - \frac{\partial}{\partial y}\left(\frac{\partial u}{\partial z} + \frac{\partial w}{\partial x}\right) = \frac{\partial \gamma_{yz}}{\partial x} - \frac{\partial \gamma_{xz}}{\partial y}$$

and, by using Hooke's law and expressions (181) for the stress components, we find

$$\frac{\partial}{\partial z}(2\omega_z) = \frac{1}{G}\left(\frac{\partial \tau_{yz}}{\partial x} - \frac{\partial \tau_{xz}}{\partial y}\right) = -\frac{1}{G}\left(\frac{\partial^2 \phi}{\partial x^2} + \frac{\partial^2 \phi}{\partial y^2} + \frac{df}{dy}\right)$$

Substituting in Eq. (*a*),

$$-G\frac{\partial}{\partial z}(2\omega_z) = \frac{\nu}{1+\nu}\frac{Py}{I} + c \tag{b}$$

If the x axis is an axis of symmetry of the cross section, bending by a force P in this axis will result in a symmetrical pattern of rotation ω_z of elements of the cross section (corresponding to anticlastic curvature), with a mean value of zero for the whole cross section. The mean value of $\partial\omega_z/\partial z$ will then also be zero, and this requires that c in Eq. (*b*) be taken as zero. If the cross section is not symmetrical, we can *define*[1] bending without torsion by means of the zero mean value of $\partial\omega_z/\partial z$, again of course requiring the zero value for c. Then Eq. (*b*) shows that $\partial\omega_z/\partial z$ vanishes for the elements of cross sections at the centroids—that is, these elements along the axis have zero relative rotation, and if one is fixed the others have no rotation—about the axis. With c zero, Eq. (*a*) becomes

$$\frac{\partial^2 \phi}{\partial x^2} + \frac{\partial^2 \phi}{\partial y^2} = \frac{\nu}{1+\nu}\frac{Py}{I} - \frac{df}{dy} \tag{182}$$

Substituting (181) in the boundary condition (*d*) of the previous article we find

$$\frac{\partial \phi}{\partial y}\frac{dy}{ds} + \frac{\partial \phi}{\partial x}\frac{dx}{ds} = \frac{\partial \phi}{\partial s} = \left[\frac{Px^2}{2I} - f(y)\right]\frac{dy}{ds} \tag{183}$$

From this equation, the values of the function ϕ along the boundary of the cross section can be calculated if the function $f(y)$ is chosen. Equation (182), together with the boundary condition (183), determines the stress function ϕ.

In the problems that will be discussed later we shall take function $f(y)$ in such a manner as to make the right side of Eq. (183) equal to zero.[2] ϕ is then constant along the boundary. Taking this constant equal to zero, we reduce the bending problem to the solution of the differential equation (182) with the condition $\phi = 0$ at the boundary. This problem is analogous to that of the deflection of a membrane uniformly stretched, having the same boundary as the cross section of the bent bar and loaded

[1] J. N. Goodier, *J. Aeron. Sci.*, vol. 11, p. 273, 1944. A different definition was proposed by E. Trefftz, *Z. Angew. Math. Mech.*, vol. 15, p. 220, 1935.

[2] See S. Timoshenko, *Bull. Inst. Eng. Ways Communication*, St. Petersburg, 1913. See also *Proc. London Math. Soc.*, ser. 2, vol. 20, p. 398, 1922.

by a continuous load given by the right side of Eq. (182). Several applications of this analogy will now be shown.

122 | Circular Cross Section

Let the boundary of the cross section be given by the equation

$$x^2 + y^2 = r^2 \tag{a}$$

The right side of the boundary condition (183) becomes zero if we take

$$f(y) = \frac{P}{2I}(r^2 - y^2) \tag{b}$$

Substituting this into Eq. (182), the stress function ϕ is then determined by the equation

$$\frac{\partial^2 \phi}{\partial x^2} + \frac{\partial^2 \phi}{\partial y^2} = \frac{1 + 2\nu}{1 + \nu}\frac{Py}{I} \tag{c}$$

and the condition that $\phi = 0$ at the boundary. Thus the stress function is given by the deflections of a membrane with circular boundary of radius r, uniformly stretched and loaded by a transverse load of intensity proportional to

$$-\frac{1 + 2\nu}{1 + \nu}\frac{Py}{I}$$

It is easy to see that Eq. (c) and the boundary condition are satisfied in this case by taking

$$\phi = m(x^2 + y^2 - r^2)y \tag{d}$$

where m is a constant factor. This function is zero at the boundary (a) and satisfies Eq. (c) if we take

$$m = \frac{(1 + 2\nu)P}{8(1 + \nu)I}$$

Equation (d) then becomes

$$\phi = \frac{(1 + 2\nu)P}{8(1 + \nu)I}(x^2 + y^2 - r^2)y \tag{e}$$

The stress components are now obtained from Eqs. (181):

$$\begin{aligned}\tau_{xz} &= \frac{(3 + 2\nu)P}{8(1 + \nu)I}\left(r^2 - x^2 - \frac{1 - 2\nu}{3 + 2\nu}y^2\right)\\ \tau_{yz} &= -\frac{(1 + 2\nu)Pxy}{4(1 + \nu)I}\end{aligned} \tag{184}$$

The vertical shearing-stress component τ_{xz} is an even function of x and y

and the horizontal component τ_{yz} is an odd function of the same variables. Hence the distribution of stresses (184) gives a resultant along the vertical diameter of the circular cross section.

Along the horizontal diameter of the cross section, $x = 0$; and we find, from (184),

$$\tau_{xz} = \frac{(3 + 2\nu)P}{8(1 + \nu)I}\left(r^2 - \frac{1 - 2\nu}{3 + 2\nu}\, y^2\right) \qquad \tau_{yz} = 0 \tag{f}$$

The maximum shearing stress is obtained at the center ($y = 0$), where

$$(\tau_{xz})_{\max} = \frac{(3 + 2\nu)Pr^2}{8(1 + \nu)I} \tag{g}$$

The shearing stress at the ends of the horizontal diameter ($y = \pm r$) is

$$(\tau_{xz})_{y=\pm r} = \frac{(1 + 2\nu)Pr^2}{4(1 + \nu)I} \tag{h}$$

It will be seen that the magnitude of the shearing stresses depends on the magnitude of Poisson's ratio. Taking $\nu = 0.3$, (g) and (h) become

$$(\tau_{xz})_{\max} = 1.38\,\frac{P}{A} \qquad (\tau_{xz})_{y=\pm r} = 1.23\,\frac{P}{A} \tag{k}$$

where A is the cross-sectional area of the bar. The elementary beam theory, based on the assumption that the shearing stress τ_{xz} is uniformly distributed along the horizontal diameter of the cross section, gives

$$\tau_{xz} = \frac{4}{3}\frac{P}{A}$$

The error of the elementary solution for the maximum stress is thus in this case about 4 percent.

123 | Elliptic Cross Section

The method of the previous article can also be used in the case of an elliptic cross section. Let

$$\frac{x^2}{a^2} + \frac{y^2}{b^2} - 1 = 0 \tag{a}$$

be the boundary of the cross section. The right side of Eq. (183) will vanish if we take

$$f(y) = -\frac{P}{2I}\left(\frac{a^2}{b^2}\, y^2 - a^2\right) \tag{b}$$

Substituting into Eq. (182), we find

$$\frac{\partial^2 \phi}{\partial x^2} + \frac{\partial^2 \phi}{\partial y^2} = \frac{Py}{I}\left(\frac{a^2}{b^2} + \frac{\nu}{1+\nu}\right) \qquad (c)$$

This equation together with the condition $\phi = 0$ at the boundary determines the stress function ϕ. The boundary condition and Eq. (c) are satisfied by taking

$$\phi = \frac{(1+\nu)a^2 + \nu b^2}{2(1+\nu)(3a^2+b^2)} \frac{P}{I}\left(x^2 + \frac{a^2}{b^2} y^2 - a^2\right) y \qquad (d)$$

When $a = b$, this solution coincides with solution (c) of the previous article.

Substituting (b) and (d) in Eqs. (181), we find the stress components

$$\tau_{xz} = \frac{2(1+\nu)a^2 + b^2}{(1+\nu)(3a^2+b^2)} \frac{P}{2I}\left[a^2 - x^2 - \frac{(1-2\nu)a^2}{2(1+\nu)a^2 + b^2} y^2\right] \qquad (185)$$

$$\tau_{yz} = -\frac{(1+\nu)a^2 + \nu b^2}{(1+\nu)(3a^2+b^2)} \frac{Pxy}{I}$$

For the horizontal axis of the elliptic cross section ($x = 0$), we find

$$\tau_{xz} = \frac{2(1+\nu)a^2 + b^2}{(1+\nu)(3a^2+b^2)} \frac{P}{2I}\left[a^2 - \frac{(1-2\nu)a^2}{2(1+\nu)a^2 + b^2} y^2\right]$$

$$\tau_{yz} = 0$$

The maximum stress is at the center ($y = 0$) and is given by equation

$$(\tau_{xz})_{\max} = \frac{Pa^2}{2I}\left[1 - \frac{a^2 + \nu b^2/(1+\nu)}{3a^2 + b^2}\right]$$

If b is very small in comparison with a, we can neglect the terms containing b^2/a^2, in which case

$$(\tau_{xz})_{\max} = \frac{Pa^2}{3I} = \frac{4}{3}\frac{P}{A}$$

which coincides with the solution of the elementary beam theory. If b is very large in comparison with a, we obtain

$$(\tau_{xz})_{\max} = \frac{2}{1+\nu}\frac{P}{A}$$

The stress at the ends of the horizontal diameter ($y = \pm b$) for this case is

$$\tau_{xz} = \frac{4\nu}{1+\nu}\frac{P}{A}$$

The stress distribution along the horizontal diameter is in this case very far from uniform and depends on the magnitude of Poisson's ratio ν.

Taking $\nu = 0.30$, we find

$$(\tau_{xz})_{\max} = 1.54\frac{P}{A} \qquad (\tau_{xz})_{x=0,\ y=b} = 0.92\frac{P}{A}$$

The maximum stress is about 14 percent larger than that given by the elementary formula.

124 | Rectangular Cross Section

The equation for the boundary line in the case of the rectangle shown in Fig. 191 is

$$(x^2 - a^2)(y^2 - b^2) = 0 \tag{a}$$

If we substitute into Eq. (183) the constant $Pa^2/2I$ for $f(y)$, the expression $Px^2/2I - Pa^2/2I$ becomes zero along the sides $x = \pm a$ of the rectangle. Along the vertical sides $y = \pm b$ the derivative dy/ds is zero. Thus the right side of Eq. (183) is zero along the boundary line and we can take $\phi = 0$ at the boundary. Differential equation (182) becomes

$$\frac{\partial^2\phi}{\partial x^2} + \frac{\partial^2\phi}{\partial y^2} = \frac{\nu}{1+\nu}\frac{Py}{I} \tag{b}$$

This equation together with the boundary condition determines completely the stress function. The problem reduces to the determination of the deflections of a uniformly stretched rectangular membrane produced by a continuous load, the intensity of which is proportional to

$$-\frac{\nu}{1+\nu}\frac{Py}{I}$$

The curve mnp in Fig. 191 represents the intersection of the membrane with the yz plane.

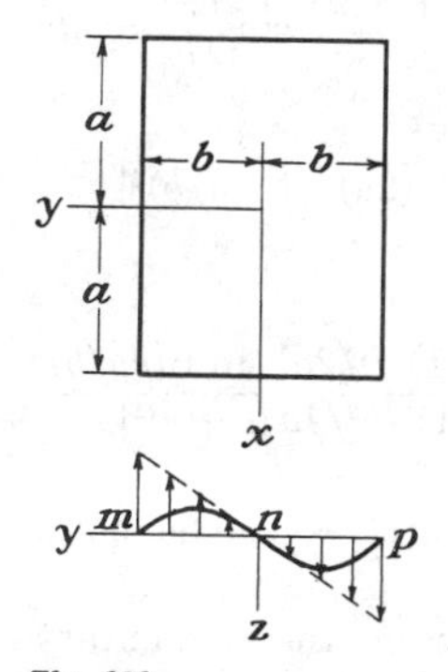

Fig. 191

From Eqs. (181) we see that shearing stresses can be resolved into the two following systems:

$$\begin{aligned} &(1) \qquad \tau_{xz}' = \frac{P}{2I}(a^2 - x^2) \qquad \tau_{yz}' = 0 \\ &(2) \qquad \tau_{xz}'' = \frac{\partial \phi}{\partial y} \qquad \tau_{yz}'' = -\frac{\partial \phi}{\partial x} \end{aligned} \tag{c}$$

The first system represents the parabolic stress distribution given by the usual elementary beam theory. The second system, depending on the function ϕ, represents the necessary corrections to the elementary solution. The magnitudes of these corrections are given by the slopes of the membrane. Along the y axis, $\partial\phi/\partial x = 0$, from symmetry, and the corrections to the elementary theory are vertical shearing stresses given by the slope $\partial\phi/\partial y$. From Fig. 191, τ_{xz}'' is positive at the points m and p and negative at n. Thus, along the horizontal axis of symmetry, the stress τ_{xz} is not uniform as in the elementary theory but has maxima at the ends, m and p, and a minimum at the center n.

From the condition of loading of the membrane it can be seen that ϕ is an even function of x and an odd function of y. This requirement and also the boundary condition are satisfied by taking the stress function ϕ in the form of the Fourier series,

$$\phi = \sum_{m=0}^{m=\infty} \sum_{n=1}^{n=\infty} A_{2m+1,n} \cos \frac{(2m+1)\pi x}{2a} \sin \frac{n\pi y}{b} \tag{d}$$

Substituting this into Eq. (b) and applying the usual method of calculating the coefficients of a Fourier series, we arrive at the equations

$$A_{2m+1,n}\pi^2 ab \left[\left(\frac{2m+1}{2a}\right)^2 + \left(\frac{n}{b}\right)^2\right] = -\frac{\nu}{1+\nu}\frac{P}{I}\int_{-a}^{a}\int_{-b}^{b} y \cos \frac{(2m+1)\pi x}{2a} \sin \frac{n\pi y}{b}\, dx\, dy$$

$$A_{2m+1,n} = -\frac{\nu}{1+\nu}\frac{P}{I}\frac{8b(-1)^{m+n-1}}{\pi^4(2m+1)n[(2m+1)^2/(2a)^2 + (n/b)^2]}$$

Substituting in (d), we find

$$\phi = -\frac{\nu}{1+\nu}\frac{P}{I}\frac{8b^3}{\pi^4}\sum_{m=0}^{m=\infty}\sum_{n=1}^{n=\infty}\frac{(-1)^{m+n-1}\cos[(2m+1)\pi x/2a]\sin(n\pi y/b)}{(2m+1)n[(2m+1)^2(b^2/4a^2) + n^2]}$$

Having this stress function, the components of shearing stress can be found from Eqs. (c).

Let us derive the corrections to the stress given by the elementary theory along the y axis. It may be seen from the deflection of the membrane (Fig. 191) that along this axis the corrections have the largest values, and therefore the maximum stress occurs at the middle points of the sides $y = \pm b$. Calculating the derivative $\partial\phi/\partial y$ and taking $x = 0$, we find that

$$(\tau_{xz}'')_{x=0} = -\frac{\nu}{1+\nu}\frac{P}{I}\frac{8b^2}{\pi^3}\sum_{m=0}^{m=\infty}\sum_{n=1}^{n=\infty}\frac{(-1)^{m+n-1}\cos(n\pi y/b)}{(2m+1)[(2m+1)^2(b^2/4a^2)+n^2]}$$

From this we find the following formulas for the center of the cross section ($y = 0$) and for the middle of the vertical sides of the rectangle:

$$(\tau_{xz}'')_{x=0,y=0} = -\frac{\nu}{1+\nu}\frac{P}{I}\frac{8b^2}{\pi^3}\sum_{m=0}^{m=\infty}\sum_{n=1}^{n=\infty}\frac{(-1)^{m+n-1}}{(2m+1)[(2m+1)^2(b^2/4a^2)+n^2]}$$

$$(\tau_{xz}'')_{x=0,y=b} = -\frac{\nu}{1+\nu}\frac{P}{I}\frac{8b^2}{\pi^3}\sum_{m=0}^{m=\infty}\sum_{n=1}^{n=\infty}\frac{(-1)^{m-1}}{(2m+1)[(2m+1)^2(b^2/4a^2)+n^2]}$$

The summation of these series is greatly simplified if we use the known formulas

$$\sum_{n=1}^{n=\infty}\frac{1}{n^2} = \frac{\pi^2}{6} \qquad \sum_{n=1}^{n=\infty}\frac{(-1)^n}{n^2} = -\frac{\pi^2}{12}$$

$$\sum_{m=0}^{m=\infty}\frac{(-1)^m}{(2m+1)[(2m+1)^2+k^2]} = \frac{\pi^3}{32}\frac{[1-\operatorname{sech}(k\pi/2)]}{\frac{1}{2}(k\pi/2)^2}*$$

* This formula can be obtained in the following way: For a simply-supported tie rod of length l under tension S, bent by a couple M at the end $x = 0$, we can find the deflection y as the Fourier series

$$y = \frac{2Ml^2}{EI\pi^3}\sum_{n=1}^{n=\infty}\frac{\sin(n\pi x/l)}{n(n^2+k^2)} \qquad \text{where} \qquad k^2 = \frac{Sl^2}{EI\pi^2}$$

The deflection at $x = \frac{1}{2}l$ is

$$\delta = \frac{2Ml^2}{EI\pi^3}\sum_{m=0}^{m=\infty}\frac{(-1)^m}{(2m+1)[(2m+1)^2+k^2]} \tag{a}$$

The same deflection obtained by integration of the differential equation of the deflection curve is

$$\delta = \frac{Ml^2}{2EI\pi^2k^2}\left(1-\operatorname{sech}\frac{k\pi}{2}\right) \tag{b}$$

The above formula follows from comparison of (a) and (b).

Then

$$
\begin{aligned}
(\tau_{xz}'')_{x=0,y=0} &= -\frac{\nu}{1+\nu}\frac{3P}{2A}\frac{b^2}{a^2}\left[\frac{1}{3}+\frac{4}{\pi^2}\sum_{n=1}^{n=\infty}\frac{(-1)^n}{n^2\cosh(n\pi a/b)}\right] \\
(\tau_{xz}'')_{x=0,y=b} &= \frac{\nu}{1+\nu}\frac{3P}{2A}\frac{b^2}{a^2}\left[\frac{2}{3}-\frac{4}{\pi^2}\sum_{n=1}^{n=\infty}\frac{1}{n^2\cosh(n\pi a/b)}\right]
\end{aligned}
\tag{186}
$$

in which $A = 4ab$ is the cross-sectional area. These series converge rapidly, and it is not difficult to calculate corrections τ_{xz}'' for any value of the ratio a/b. These corrections must be added to the value $3P/2A$ given by

Point	$\frac{a}{b} =$	2	1	$\frac{1}{2}$	$\frac{1}{4}$
$x = 0, y = 0$	Exact	0.983	0.940	0.856	0.805
	Approximate	0.981	0.936	0.856	0.826
$x = 0, y = b$	Exact	1.033	1.126	1.396	1.988
	Approximate	1.040	1.143	1.426	1.934

the elementary formula. In the first lines of the table above, numerical factors are given by which the approximate value of the shearing stress $3P/2A$ must be multiplied in order to obtain the exact values of the stress.[1] The Poisson's ratio ν is taken equal to one-fourth in this calculation. It is seen that the elementary formula gives very accurate values for these stresses when $a/b \geq 2$. For a square cross section, the error in the maximum stress obtained by the elementary formula is about 10 percent.

If both sides of the rectangle are of the same order of magnitude, we can obtain an approximate solution for the stress distribution in a polynomial form by taking the stress function in the form

$$
\phi = (x^2 - a^2)(y^2 - b^2)(my + ny^3) \tag{e}
$$

Calculating the coefficients m and n from the condition of minimum energy we find[2]

$$
m = -\frac{\nu}{1+\nu}\frac{P}{8Ib^2}\frac{1/11 + 8a^2/b^2}{(1/7 + 3a^2/5b^2)(1/11 + 8a^2/b^2) + 1/21 + 9a^2/35b^2}
$$

$$
n = -\frac{\nu}{1+\nu}\frac{P}{8Ib^4}\frac{1}{(1/7 + 3a^2/5b^2)(1/11 + 8a^2/b^2) + 1/21 + 9a^2/35b^2}
$$

[1] The figures of this table are somewhat different from those given by Saint-Venant. Checking of Saint-Venant's results showed that there is a numerical error in his calculations.

[2] S. Timoshenko; see n. 2, p. 357.

The shearing stresses, calculated from (*e*), are

$$(\tau_{xz})_{x=0,y=0} = \frac{Pa^2}{2I} + ma^2b^2$$
$$(\tau_{xz})_{x=0,y=b} = \frac{Pa^2}{2I} - 2a^2b^2(m + nb^2) \tag{f}$$

The approximate values of the shearing stresses given on the second lines of the table (see page 364) were calculated by using these formulas. It will be seen that the approximate formulas (*f*) give satisfactory accuracy in this range of values of a/b.

The membrane analogy provides further useful approximate formulas for calculating these shearing stresses. If a is large in comparison with b (Fig. 191), we can assume that at points sufficiently distant from the short sides of the rectangle, the surface of the membrane is practically cylindrical. Then Eq. (*b*) becomes

$$\frac{d^2\phi}{dy^2} = \frac{\nu}{1+\nu}\frac{Py}{I}$$

and we find

$$\phi = \frac{\nu}{1+\nu}\frac{P}{6I}(y^3 - b^2y) \tag{g}$$

Substituting in Eqs. (*c*), the stresses along the y axis are

$$\tau_{xz} = \frac{P}{2I}\left[a^2 + \frac{\nu}{1+\nu}\left(y^2 - \frac{b^2}{3}\right)\right] \tag{h}$$

It will be seen that for a narrow rectangle the correction to the elementary formula, given by the second term in the brackets, is always small.

If b is large in comparison with a, the deflections of the membrane at points distant from the short sides of the rectangle[1] can be taken as a linear function of y, and from Eq. (*b*) we find

$$\frac{\partial^2\phi}{\partial x^2} = \frac{\nu}{1+\nu}\frac{Py}{I}$$
$$\phi = \frac{\nu}{1+\nu}\frac{Py}{2I}(x^2 - a^2) \tag{i}$$

Substituting in Eqs. (*c*), the shearing stress components are

$$\tau_{xz} = \frac{1}{1+\nu}\frac{P}{2I}(a^2 - x^2) \qquad \tau_{yz} = -\frac{\nu}{1+\nu}\frac{P}{I}xy$$

[1] Approximations for other slender shapes, and comparisons with more accurate results from finite difference calculations, are given by W. J. Carter, *J. Appl. Mech.*, vol. 25, pp. 115–121, 1958.

At the centroid of the cross section ($x = y = 0$),

$$\tau_{xz} = \frac{1}{1+\nu}\frac{Pa^2}{2I} \qquad \tau_{yz} = 0$$

In comparison with the usual elementary solution the stress at this point is reduced in the ratio $1/(1 + \nu)$.

But for very wide rectangles (b much greater than a) maximum stress values much larger than the value $3P/2A$ of the elementary theory are found elsewhere in the section. Moreover, if b/a exceeds 15, the maximum stress is no longer the component τ_{xz} at $x = 0$, $y = \pm b$, the midpoints of the vertical sides. It is the *horizontal* component τ_{yz} at points $x = a$, $y = \pm\eta$ on the top and bottom edges near the corners. Values of these stresses for $\nu = \frac{1}{4}$ are given in the table[1] below. The values of η are given in the form $(b - \eta)/2a$ in the last column, $b - \eta$ being the distance of the maximum point from the corner.

$\frac{b}{a}$	$\frac{(\tau_{xz})_{x=0,\ y=b}}{3P/2A}$	$\frac{(\tau_{yz})_{x=a,\ y=\eta}}{3P/2A}$	$\frac{b-\eta}{2a}$
0	1.000	0.000	0.000
2	1.39(4)	0.31(6)	0.31(4)
4	1.988	0.968	0.522
6	2.582	1.695	0.649
8	3.176	2.452	0.739
10	3.770	3.226	0.810
15	5.255	5.202	0.939
20	6.740	7.209	1.030
25	8.225	9.233	1.102
50	15.650	19.466	1.322

125 | Additional Results

Let us consider a cross section the boundary of which consists of two vertical sides

[1] E. Reissner and G. B. Thomas, *J. Math. Phys.*, vol. 25, p. 241, 1946.

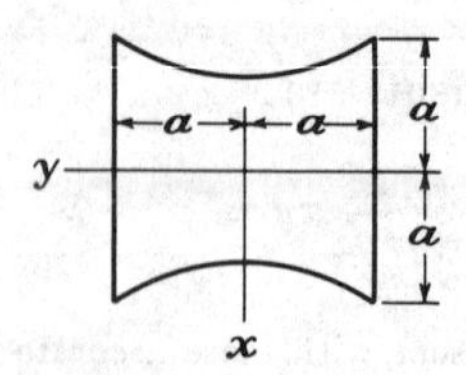

Fig. 192

$y = \pm a$ (Fig. 192) and two hyperbolas[1]

$$(1 + \nu)x^2 - \nu y^2 = a^2 \tag{a}$$

It is easy to show that this makes the right side of Eq. (183) zero at the boundary if we take

$$f(y) = \frac{P}{2I}\left(\frac{\nu}{1+\nu}\, y^2 + \frac{a^2}{1+\nu}\right)$$

Substituting into Eq. (182), we find

$$\frac{\partial^2 \phi}{\partial x^2} + \frac{\partial^2 \phi}{\partial y^2} = 0$$

This equation and the boundary condition (183) are satisfied by taking $\phi = 0$. Then the shearing-stress components, from Eq. (181), are

$$\tau_{xz} = \frac{P}{2I}\left(-x^2 + \frac{\nu}{1+\nu}\, y^2 + \frac{a^2}{1+\nu}\right)$$
$$\tau_{yz} = 0$$

At each point of the cross section the shearing stress is vertical. The maximum of this stress is at the middle of the vertical sides of the cross section and is equal to

$$\tau_{\max} = \frac{Pa^2}{2I}$$

The problem can also be easily solved if the boundary of the cross section is given by the equation

$$\left(\pm \frac{y}{b}\right)^{1/\nu} = \left(1 - \frac{x^2}{a^2}\right) \qquad a > x > -a \tag{b}$$

For $\nu = \frac{1}{4}$, this cross-section curve has the shape shown in Fig. 193.

By taking

$$f(y) = \frac{Pa^2}{2I}\left[1 - \left(\pm \frac{y}{b}\right)^{1/\nu}\right]$$

the left side of the boundary condition (183) vanishes, that is, ϕ must be constant along the boundary. Equation (182) becomes

$$\frac{\partial^2 \phi}{\partial x^2} + \frac{\partial^2 \phi}{\partial y^2} = \frac{\nu}{1+\nu}\frac{Py}{I} \pm \frac{Pa^2}{2bI\nu}\left(\pm \frac{y}{b}\right)^{(1-\nu)/\nu}$$

[1] This problem was discussed by F. Grashof, "Elastizität und Festigkeit," p. 246, 1878.

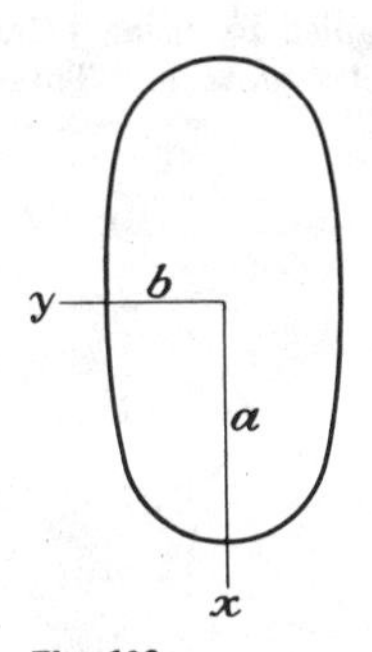

Fig. 193

This equation and the boundary condition are satisfied by taking

$$\phi = \frac{Pa^2\nu}{2(1+\nu)I}\left[y\left(\frac{x^2}{a^2}-1\right) \pm b\left(\pm\frac{y}{b}\right)^{(1+\nu)/\nu}\right]$$

Substituting in Eqs. (181), we find

$$\tau_{xz} = \frac{P}{2(1+\nu)I}(a^2 - x^2) \qquad \tau_{yz} = -\frac{P\nu}{(1+\nu)I}xy \tag{c}$$

We can arrive at the same result in a different way. In discussing stresses in a rectangular beam the width of which is large in comparison with the depth, we used as an approximate solution for the stress function [Eq. (*g*), Art. 124] the expression

$$\phi = \frac{\nu}{1+\nu}\frac{Py}{2I}(x^2 - a^2)$$

from which the expressions (*c*) for stress components may be derived. The equation of the boundary can now be found from the condition that at the boundary the direction of shearing stress coincides with the tangent to the boundary. Hence

$$\frac{dx}{\tau_{xz}} = \frac{dy}{\tau_{yz}}$$

Substituting from (*c*) and integrating, we arrive at the equation of the boundary,

$$y = b(a^2 - x^2)^{\nu}$$

By using the energy method (Art. 124) we may arrive at an approximate solution in many other cases. Let us consider, for instance, the cross section shown in Fig. 194. The vertical sides of the boundary are given by the equation $y = \pm b$, and the other two sides are arcs of the circle

$$x^2 + y^2 - r^2 = 0 \tag{d}$$

The right side of Eq. (183) vanishes if we take

$$f(y) = \frac{P}{2I}(r^2 - y^2)$$

Then an approximate expression for the stress function is

$$\phi = (y^2 - b^2)(x^2 + y^2 - r^2)(Ay + By^3 + \cdots)$$

in which the coefficients $A, B, \ldots$ are to be calculated from the condition of minimum energy.

Solutions for many shapes of cross section have been obtained by using polar and other curvilinear coordinates and functions of the complex variable. These

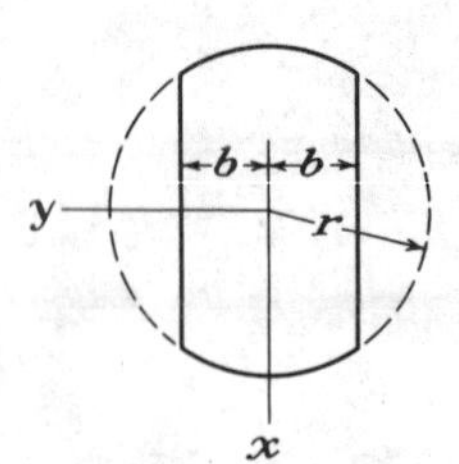

Fig. 194

include sections bounded by two circles, concentric[1] or nonconcentric,[2] a circle with radial slits,[3] a cardioid,[4] a limaçon,[5] an elliptic limaçon,[6] two confocal ellipses,[7] an ellipse and confocal hyperbolas,[8] triangles and polygons[9] including a rectangle with slits,[10] and a sector of a circular ring.[11]

126 | Nonsymmetrical Cross Sections

As a first example let us consider the case of an isosceles triangle (Fig. 195). The boundary of the cross section is given by the equation

$$(y - a)[x + (2a + y) \tan \alpha][x - (2a + y) \tan \alpha] = 0$$

The right side of Eq. (183) is zero if we take

$$f(y) = \frac{P}{2I} (2a + y)^2 \tan^2 \alpha$$

Equation (182) for determining the stress function ϕ then becomes

$$\frac{\partial^2 \phi}{\partial x^2} + \frac{\partial^2 \phi}{\partial y^2} = \frac{\nu}{1 + \nu} \frac{Py}{I} - \frac{P}{I} (2a + y) \tan^2 \alpha \tag{a}$$

An approximate solution may be obtained by using the energy method. In the

[1] A solution is given in A. E. H. Love's "Mathematical Theory of Elasticity," 4th ed., p. 335, and in I. S. Sokolnikoff's "Mathematical Theory of Elasticity," 2d ed., p. 228, 1956.

[2] B. R. Seth, *Proc. Indian Acad. Sci.*, vol. 4, sec. A, p. 531, 1936, and vol. 5, p. 23, 1937.

[3] W. M. Shepherd, *Proc. Roy. Soc. (London)*, ser. A, vol. 138, p. 607, 1932; L. A. Wigglesworth, *Proc. London Math. Soc.*, ser. 2, vol. 47, p. 20, 1940, and *Proc. Roy. Soc. (London)*, ser. A, vol. 170, p. 365, 1939.

[4] W. M. Shepherd, *Proc. Roy. Soc. (London)*, ser. A, vol. 154, p. 500, 1936.

[5] D. L. Holl and D. H. Rock, *Z. Angew. Math. Mech.*, vol. 19, p. 141, 1939.

[6] A. C. Stevenson, *Proc. London Math. Soc.*, ser. 2, vol. 45, p. 126, 1939.

[7] Love, *op. cit.*, p. 336.

[8] B. G. Galerkin, *Bull. Inst. Eng. Ways Communication*, St. Petersburg, vol. 96, 1927. See also S. Ghosh, *Bull. Calcutta Math. Soc.*, vol. 27, p. 7, 1935.

[9] B. R. Seth, *Phil. Mag.*, vol. 22, p. 582, 1936, and vol. 23, p. 745, 1937.

[10] D. F. Gunder, *Phys.*, vol. 6, p. 38, 1935.

[11] M. Seegar and K. Pearson, *Proc. Roy. Soc. (London)*, ser. A, vol. 96, p. 211, 1920.

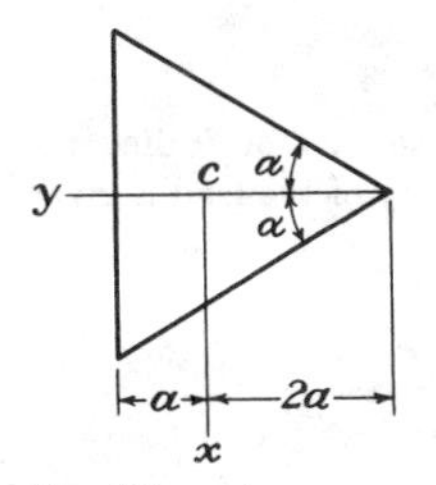

Fig. 195

particular case when

$$\tan^2 \alpha = \frac{\nu}{1+\nu} = \frac{1}{3} \tag{b}$$

an exact solution of Eq. (a) is obtained by taking for the stress function the expression

$$\phi = \frac{P}{6I}\left[x^2 - \frac{1}{3}(2a+y)^2\right](y-a)$$

The stress components are then obtained from Eqs. (181):

$$\tau_{xz} = \frac{\partial \phi}{\partial y} - \frac{Px^2}{2I} + \frac{P}{6I}(2a+y)^2 = \frac{2\sqrt{3}P}{27a^4}[-x^2 + a(2a+y)] \tag{c}$$

$$\tau_{yz} = -\frac{\partial \phi}{\partial x} = \frac{2\sqrt{3}P}{27a^4}x(a-y)$$

Along the y axis, $x = 0$, and the resultant shearing stress is vertical and is represented by the linear function

$$(\tau_{xz})_{x=0} = \frac{2\sqrt{3}P}{27a^3}(2a+y)$$

The maximum value of this stress, at the middle of the vertical side of the cross section, is

$$\tau_{\max} = \frac{2\sqrt{3}P}{9a^2} \tag{d}$$

By calculating the moment with respect to the z axis of the shearing forces given by the stresses (c), it can be shown that in this case the resultant shearing force passes through the centroid C of the cross section.

Let us consider next the more general case of a cross section with a horizontal axis of symmetry (Fig. 196), the lower and upper portions of the boundary being given by the equations

$$x = \psi(y) \qquad \text{for } x > 0$$

$$x = -\psi(y) \qquad \text{for } x < 0$$

Then the function

$$[x + \psi(y)][x - \psi(y)] = x^2 - [\psi(y)]^2$$

vanishes along the boundary and in our expressions for stress components (181) we can take

$$f(y) = \frac{P}{2I}[\psi(y)]^2$$

With this assumption, the stress function has to satisfy the differential equation

$$\frac{\partial^2 \phi}{\partial x^2} + \frac{\partial^2 \phi}{\partial y^2} = \frac{\nu}{1+\nu}\frac{Py}{I} - \frac{P}{I}\psi(y)\frac{d\psi}{dy}$$

and be constant at the boundary. The problem is reduced to that of finding the deflections of a uniformly stretched membrane when the intensity of the load is given

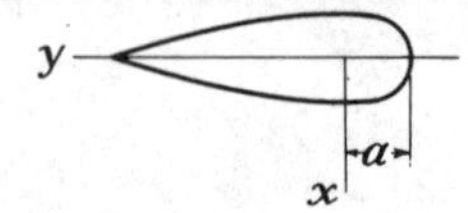

Fig. 196

by the right-hand side of the above equation. This latter problem can usually be solved with sufficient accuracy by using the energy method as was shown in the case of the rectangular cross section (page 364).

The case shown in Fig. 197 can be treated in a similar manner. Assume, for example, that the cross section is a parabolic segment and that the equation of the parabola is

$$x^2 = A(y + a)$$

Then we take

$$f(y) = \frac{P}{2I} A(y + a)$$

With this expression for $f(y)$, the first factor on the right-hand side of Eq. (183) vanishes along the parabolic portion of the boundary. The factor dy/ds vanishes along the straight-line portion of the boundary. Thus we find again that the stress function is constant along the boundary and the problem can be treated by using the energy method.

127 | Shear Center

In discussing the cantilever problem we chose for z axis the centroidal axis of the bar and for x and y axes the principal centroidal axes of the cross section. We assumed that the force P is parallel to the x axis and at such a distance from the centroid that twisting of the bar does not occur. This distance, which is of importance in practical calculations, can readily be found once the stresses represented by Eqs. (181) are known. For this purpose we evaluate the moment about the centroid produced by the shear stresses τ_{xz} and τ_{yz}. This moment evidently is

$$M_z = \iint (\tau_{xz} y - \tau_{yz} x)\, dx\, dy \tag{a}$$

Observing that the stresses distributed over the end cross section of the beam are statically equivalent to the acting force P, we conclude that the distance d of the force P from the centroid of the cross section is

$$d = \frac{|M_z|}{P} \tag{b}$$

For positive M_z the distance d must be taken in the direction of positive y. In the preceding discussion the assumption was made that the force is acting parallel to the x axis.

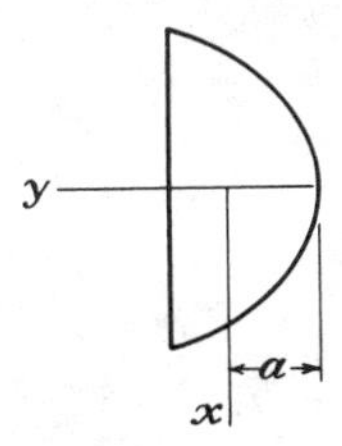

Fig. 197

When the force P is parallel to the y axis instead of the x axis we can, by a similar calculation, establish the position of the line of action of P for which no rotation of centroidal elements of cross sections occurs. The intersection point of the two lines of action of the bending forces has an important significance. If a force, perpendicular to the axis of the beam, is applied at that point, we can resolve it into two components parallel to the x and y axes; and on the basis of the above discussion we conclude that it does not produce rotation of centroidal elements of cross sections of the beam. This point is called the *shear center*—sometimes also the center of flexure, or flexural center.

If the cross section of the beam has two axes of symmetry, we can conclude at once that the shear center coincides with the centroid of the cross section. When there is only one axis of symmetry we conclude, from symmetry, that the shear center will be on that axis. Taking the symmetry axis for y axis, we calculate the position of the shear center from Eq. (*b*).

Let us consider, as an example, a semicircular cross section[1] as shown in Fig. 198. To find the shearing stresses we can utilize the solution developed for circular beams (see page 358). In that case there are no stresses acting on the vertical diametral section xz. We can imagine the beam divided by the xz plane into two halves each of which represents a semicircular beam bent by the force $P/2$. The stresses are given by Eq. (184). Substituting into Eq. (*a*), integrating, and dividing M_z by $P/2$, we find for the distance of the bending force from the origin O the value

$$e = \frac{2M_z}{P} = \frac{8}{15\pi}\,\frac{3 + 4\nu}{1 + \nu}\,r$$

This defines the position of the force for which the cross-sectional element at point O, the center of the circle, does not rotate. At the same time an element at the centroid of the semicircular cross section will rotate by the

[1] See S. Timoshenko, *Bull. Inst. Eng. Ways Communication, op. cit.* It seems that the displacement of the bending force from the centroid of the cross section was investigated in this paper for the first time.

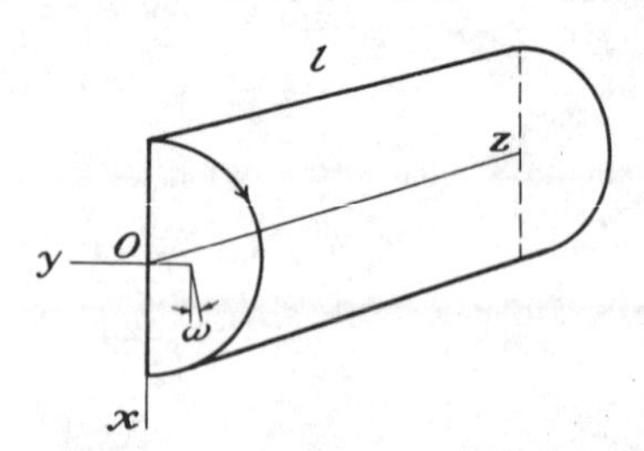

Fig. 198

amount [see Eq. (b), page 357]

$$\omega = \frac{\nu P(l - z)}{EI} 0.424r$$

where $0.424r$ is the distance from the origin O to the centroid of the semicircle. To eliminate this rotation, a torque as shown in Fig. 198 must be applied. The magnitude of this torque is found by using the table on page 314, which gives for a semicircular cross section the angle of twist per unit length

$$\theta = \frac{M_t}{0.298Gr^4}$$

Then the condition that centroidal elements of cross sections do not rotate gives

$$\frac{M_t(l - z)}{0.298Gr^4} = \frac{\nu P(l - z)}{EI} 0.424r \quad \text{and} \quad M_t = \frac{\nu P(0.298r^4)0.424r}{2(1 + \nu)I}$$

This torque will be produced by shifting the bending force $P/2$ toward the z axis by the amount

$$\delta = \frac{2M_t}{P} = \frac{8\nu(0.298)0.424r}{2(1 + \nu)\pi}$$

This quantity must be subtracted from the previously calculated distance e to obtain the distance of the shear center from the center O of the circle. Assuming $\nu = 0.3$, we obtain

$$e - \delta = 0.548r - 0.037r = 0.511r$$

In sections as in Fig. 196 the shearing-stress components are

$$\tau_{xz} = \frac{\partial \phi}{\partial y} - \frac{P}{2I}[x^2 - \psi^2(y)] \qquad \tau_{yz} = -\frac{\partial \phi}{\partial x}$$

Hence

$$M_z = \iint \left(\frac{\partial \phi}{\partial y} y + \frac{\partial \phi}{\partial x} x \right) dx\, dy - \frac{P}{2I} \iint [x^2 - \psi^2(y)] y\, dx\, dy \qquad (c)$$

Integrating by parts and observing that ϕ vanishes at the boundary $x = \pm\psi(y)$, we obtain

$$\iint \left(\frac{\partial \phi}{\partial y} y + \frac{\partial \phi}{\partial x} x \right) dx\, dy = -2 \iint \phi\, dx\, dy$$

$$\int [x^2 - \psi^2(y)]\, dx = \tfrac{2}{3}\psi^3(y) - 2\psi^3(y) = -\tfrac{4}{3}\psi^3(y)$$

$$\iint [x^2 - \psi^2(y)] y\, dx\, dy = -\tfrac{4}{3} \int y \psi^3(y)\, dy$$

$$I = \iint x^2\, dx\, dy = \tfrac{2}{3} \int \psi^3(y)\, dy$$

Substituting in (*c*) and dividing by P we find

$$d = \frac{|M_z|}{P} = \left| -\frac{2}{P} \iint \phi\, dx\, dy + \frac{\int y\psi^3(y)\, dy}{\int \psi^3(y)\, dy} \right|$$

Knowing $\psi(y)$ and using the membrane analogy for finding ϕ, we can always calculate[1] with sufficient accuracy the position of the shear center for these cross sections.

The question of the shear center is especially important in the case of thin-walled open sections. Its position can be easily determined for such sections with sufficient accuracy by assuming that the shearing stresses are uniformly distributed over the thickness of the wall and are parallel to the middle surface of the wall.[2]

The location of the shear center in the cross section is determined by the shape of the section only. On the other hand, the location of the center of twist (see page 305) is dependent on the manner in which the bar is supported. By choosing this manner of support suitably, the axis of twist can be made to coincide with the axis of shear centers. It can be shown that this occurs when the bar is so supported that the integral $\iint w^2\, dx\, dy$ over the cross section is a minimum,[3] w being the warping displacement of torsion (indeterminate by a linear function of x and y before this condition is applied). In practice the fixing will usually disturb the stress distribution near the fixed end—as, for instance, when it prevents displacements in the end section completely. In that case, if we regard the bending force as a concentrated load at the shear center, producing zero rotation, the reciprocal theorem (page 272) shows that a torque will produce zero deflection of the shear center. This indicates that the center of twist will coincide with the shear center.[4] The argument is of an approximate character because the existence of a center of twist depends on absence of deformation of cross sections in their planes, and this will not hold in the disturbed zone near the fixed end.

128 | The Solution of Bending Problems by the Soap-film Method

The exact solutions of bending problems are known for only a few special cases in which the cross sections have certain simple forms. For practical purposes it is important to have means of solving the problem for any assigned shape of the cross section. This can be accomplished by numerical calculations based on equations of finite differences as explained in the

[1] Examples of such calculations can be found in the book by L. S. Leibenson, "Variational Methods for Solving Problems of the Theory of Elasticity," Moscow, 1943.

[2] References may be found in S. Timoshenko, "Strength of Materials," 3d ed., vol. 1, p. 240, 1955.

[3] R. Kappus, *Z. Angew. Math. Mech.*, vol. 19, p. 347, 1939; A. Weinstein, *Quart. Appl. Math.*, vol. 5, p. 79, 1947.

[4] See R. V. Southwell, "Introduction to the Theory of Elasticity," p. 29; W. J. Duncan, D. L. Ellis, and C. Scruton, *Phil. Mag.*, vol. 16, p. 201, 1933.

Appendix, or experimentally by the soap-film method,[1] analogous to that used in solving torsional problems (see page 303). For deriving the theory of the soap-film method we use Eqs. (181), (182), and (183). Taking

$$f(y) = \frac{\nu}{2(1+\nu)} \frac{Py^2}{I}$$

Eq. (182) for the stress function is

$$\frac{\partial^2 \phi}{\partial x^2} + \frac{\partial^2 \phi}{\partial y^2} = 0 \tag{a}$$

This is the same equation as for an unloaded and uniformly stretched membrane (see page 306). The boundary condition (183) becomes

$$\frac{\partial \phi}{\partial s} = \left[\frac{Px^2}{2I} - \frac{\nu}{2(1+\nu)} \frac{Py^2}{I}\right] \frac{dy}{ds} \tag{b}$$

Integrating along the boundary s we find the expression

$$\phi = \frac{P}{I} \int \frac{x^2\, dy}{2} - \frac{\nu}{2(1+\nu)} \frac{Py^3}{3I} + \text{const} \tag{c}$$

from which the value of ϕ for every point of the boundary can be calculated. $\int (x^2/2)\, dy$ vanishes when taken around the boundary, since it represents the moment of the cross section with respect to the y axis, which passes through the centroid of the cross section. Hence ϕ, calculated from (c), is represented along the boundary by a closed curve.

Imagine now that the soap film is stretched over this curve. Then the surface of the film satisfies Eq. (a) and boundary condition (c). Hence, the ordinates of the film represent the stress function ϕ at all points of the cross section to the scale used for representation of the function ϕ along the boundary [Eq. (c)].

The photograph[2] (Fig. 199) illustrates a method used for construction of the boundary of the soap film. A hole is cut in a thin, soft metal plate of such a shape that after the plate is bent the projection of the edge of the hole on the horizontal plane has the same shape as the boundary of the cross section of the beam. The plate is bent to make the ordinates along the edge of the hole represent to a certain scale the value of ϕ given by Eq. (c).

The analogy between the soap-film and the bending-problem equations

[1] This method was indicated first by Vening Meinesz, "De Ingenieur," p. 108, Holland, 1911. It was developed independently by A. A. Griffith and G. I. Taylor, *Advisory Comm. Aeron. Tech. Rept.*, vol. 3, p. 950, 1917–1918. The results given here are taken from this paper.

[2] By P. A. Cushman. This and other experimental methods are described in M. Hetényi (ed.), "Handbook of Experimental Stress Analysis," chap. 16, John Wiley & Sons, Inc., New York, 1950.

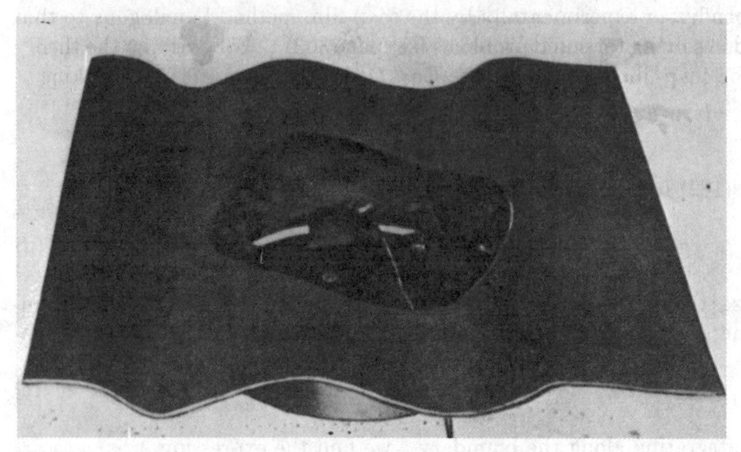

Fig. 199

holds only for small deflections of the membrane. It is desirable to have the total range of the ordinates of the film not more than one-tenth of the maximum horizontal dimension. If necessary, the range of the function along the boundary can be reduced by introducing a new function ϕ_1, instead of ϕ, by the substitution

$$\phi = \phi_1 + ax + by \tag{d}$$

where a and b are arbitrary constants. It may be seen that the function ϕ_1 also satisfies the membrane equation (*a*). The values of the function ϕ_1 along the boundary, from Eqs. (*c*) and (*d*), are given by

$$\phi_1 = \frac{P}{I} \int \frac{x^2}{2}\, dy - \frac{\nu}{2(1+\nu)} \frac{Py^3}{3I} - ax - by + \text{const}$$

The reduction of the range of the function ϕ_1 at the boundary can usually be effected by a proper adjustment of the constants a and b.

When the function ϕ_1 is obtained from the soap film, the function ϕ is calculated from Eq. (*d*). Then the shearing-stress components are obtained from Eqs. (181), which have now the form

$$\begin{aligned} \tau_{xz} &= \frac{\partial \phi}{\partial y} - \frac{Px^2}{2I} + \frac{\nu}{2(1+\nu)} \frac{Py^2}{I} \\ \tau_{yz} &= -\frac{\partial \phi}{\partial x} \end{aligned} \tag{e}$$

The stress components can now be easily calculated for every point of the cross section provided we know the values of the derivatives $\partial\phi/\partial y$ and $\partial\phi/\partial x$ at this point. These derivatives are given by the slopes of the soap film in the y and x directions. For determining slopes, we proceed as in the case of torsional problems and first map contour lines of the film surface. From the contour map the slopes may be found by drawing straight lines parallel to the coordinate axes and constructing curves representing the corresponding sections of the soap film. The slopes found in this way must now be inserted in expressions (e) for shear-stress components. The accuracy of this procedure can be checked by calculating the resultant of all the shear stresses distributed over the cross section. This resultant should be equal to the bending force P applied at the end of the cantilever.

Experiments show that a satisfactory accuracy in determining stresses can be attained by using the soap-film method. The results obtained for an I section[1] are shown in Figs. 200. From these figures it may be seen that the usual assumptions of the elementary theory, that the web of an I beam takes most of the shearing force and that the shearing stresses are constant across the thickness of the web, are fully confirmed. The maximum shearing stress at the neutral plane is in very good agreement with that calculated from the elementary theory. The component τ_{yz} is practically zero in the web and reaches a maximum at the reentrant corner. This maximum should depend on the radius of the fillet rounding the reentrant corner. For the proportions taken, it is only about one-half of the maximum stress τ_{xz} at the neutral plane. The lines of equal shearing-stress components, giving the ratio of these components to the average shearing stress P/A, are shown in the figures.

The stress concentration at the reentrant corner has been studied for the case of a T beam. The radius of the reentrant corner was increased

[1] In this case of symmetry only one-quarter of the cross section need be investigated.

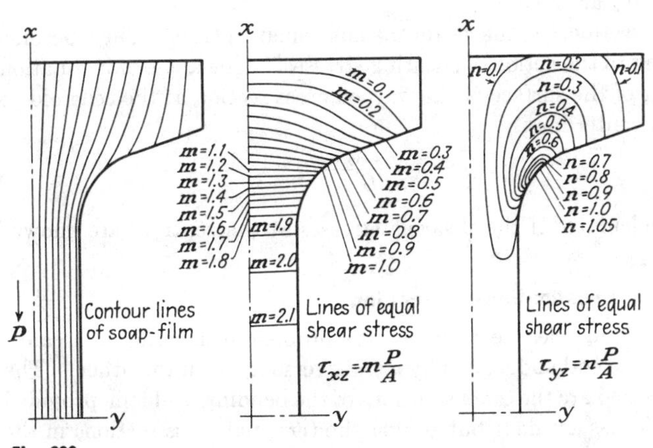

Fig. 200

in a series of steps, and contour lines were mapped for each case. It was shown in this manner that the maximum stress at the corner equals the maximum stress in the web when the radius of the fillet is about one-sixteenth of the thickness of the web.

129 | Displacements

When the stress components are found, the displacements u, v, w can be calculated in the same manner as in the case of pure bending (see page 285). Let us consider here the deflection curve of the cantilever. The curvatures of this line in the xz and yz planes are given with sufficient accuracy by the values of the derivatives $\partial^2 u/\partial z^2$ and $\partial^2 v/\partial z^2$ for $x = y = 0$. These quantities can be calculated from the equations

$$
\begin{aligned}
\frac{\partial^2 u}{\partial z^2} &= \frac{\partial \gamma_{xz}}{\partial z} - \frac{\partial \epsilon_z}{\partial x} = \frac{1}{G}\frac{\partial \tau_{xz}}{\partial z} - \frac{1}{E}\frac{\partial \sigma_z}{\partial x} = \frac{P(l-z)}{EI} \\
\frac{\partial^2 v}{\partial z^2} &= \frac{\partial \gamma_{yz}}{\partial z} - \frac{\partial \epsilon_z}{\partial y} = 0
\end{aligned}
\tag{a}
$$

We see that the centerline of the cantilever is bent in the xz plane in which the load is acting, and the curvature at any point is proportional to the bending moment at this point, as is usually assumed in the elementary theory of bending. By integration of the first of Eqs. (a), we find

$$
u = \frac{Plz^2}{2EI} - \frac{Pz^3}{6EI} + cz + d \tag{b}
$$

where c and d are constants of integration which must be determined from the conditions at the fixed end of the cantilever. If the end of the centerline is built in, u and du/dz are zero when $z = 0$, and hence constants c and d in Eq. (b) are zero.

The cross sections of the beam do not remain plane. They become warped, owing to the action of shearing stresses. The angle of inclination of an element of the surface of the warped cross section at the centroid to the deflected centerline is

$$
\frac{\pi}{2} - \frac{(\tau_{xz})_{x=0,\ y=0}}{G}
$$

and can be calculated if the shearing stresses at the centroid are known.

130 | Further Investigations of Bending

In the foregoing articles we discussed the problem of bending of a cantilever fixed at one end and loaded by a transverse force on the other. The solutions obtained are the exact solutions of the bending problem, provided the external forces are distributed over the terminal cross sections in the same manner as the stresses σ_z, τ_{xz}, τ_{yz} found in the solutions. If this condition is not fulfilled, there will be local irregularities in the stress distribution near the ends of the beam, but on the basis of Saint-Venant's principle we can assume that at a sufficient distance from the ends, say at a distance larger than cross-sectional dimensions of the beam, our solutions are sufficiently accurate. By using the same principle, we may extend the application of the above solutions to other cases of loading and supporting of beams. We may assume with sufficient accuracy that the stresses at any cross section of a beam, at sufficient distance from the loads, depend only on the magnitude of the bending moment and the shearing force at this cross section and can be calculated by superposition of the solutions obtained before for the cantilever.

If the bending forces are inclined to the principal axes of the cross section of the beam, they can always be resolved into two components acting in the direction of the principal axes and bending in each of the two principal planes can be discussed separately. The total stresses and displacements will then be obtained by using the principle of superposition.

Near the points of application of external forces there are irregularities in stress distribution that we discussed before for the particular case of a narrow rectangular cross section (see Art. 40). Analogous discussion for other shapes of cross section shows that these irregularities are of a local character.[1]

The problem of bending is solved also for certain cases of distributed load.[2] It is shown that in such cases the central line of the beam usually extends or contracts as in the case of the narrow rectangular cross section (see Art. 22) already discussed. The curvature of the centerline in these cases is no longer proportional to the bending moment, but the necessary corrections are small and can be neglected in practical problems. For instance, in the case of a circular beam bent by its own weight,[3] the curvature at the fixed end is given by the equation

$$\frac{1}{r} = \frac{M}{EI}\left[1 - \frac{7 + 12\nu + 4\nu^2}{6(1 + \nu)} \frac{a^2}{l^2}\right]$$

in which a is the radius of the cross section and l the length of the cantilever. The second term in the brackets represents the correction to the curvature arising from the distribution of the load. It is small, of the order of a^2/l^2. This conclusion holds also for beams of other shapes of cross section bent by their own weight.[4]

[1] See L. Pochhammer, "Untersuchungen über das Gleichgewicht des elastischen Stabes," Kiel, 1879. See also a paper by J. Dougall, *Trans. Roy. Soc. Edinburgh*, vol. 49, p. 895, 1914.

[2] J. H. Michell, *Quart. J. Math.*, vol. 32, 1901; also K. Pearson, *Quart. J. Math.*, vol. 24, 1889, and K. Pearson and L. N. G. Filon, *Quart. J. Math.*, vol. 31, 1900.

[3] This problem is discussed by Love, *op. cit.*, p. 362, 1927.

[4] The case of a cantilever of an elliptical cross section has been discussed by J. M. Klitchieff, *Bull. Polytech. Inst.*, St. Petersburg, p. 441, 1915.

chapter 12

Axisymmetric Stress and Deformation in a Solid of Revolution

131 | General Equations

A few problems of solids of revolution deformed by loads symmetrical about the axis have occurred in earlier chapters. The simplest examples are the circular cylinder with uniform internal pressure (Art. 28) and the rotating circular disk (Art. 32). These are examples of *torsionless axisymmetry*. By contrast, we have the torsion of a circular cylinder (cf. Prob. 2, page 349), in which the shear stress is a function of the cylindrical coordinate r only. Also, in the torsion of circular shafts of variable diameter (Art. 119), the nonzero stress components $\tau_{r\theta}$ and $\tau_{\theta z}$ are functions of r and z only, independent of θ.

The present chapter (except the two last articles, 146 and 147) is devoted to torsionless axisymmetry. In cylindrical coordinates r, θ, z, with corresponding displacement components u, v, w, the component v vanishes and u, w are independent of θ. Then the stress components are also independent of θ, two of them, $\tau_{r\theta}$ and $\tau_{\theta z}$, being zero. This may be seen from Eqs. (179), which are the general strain-displacement relations in cylindrical coordinates. These reduce to

$$\epsilon_r = \frac{\partial u}{\partial r} \qquad \epsilon_\theta = \frac{u}{r} \qquad \epsilon_z = \frac{\partial w}{\partial z} \qquad \gamma_{rz} = \frac{\partial u}{\partial z} + \frac{\partial w}{\partial r} \tag{187}$$

Equations (180) for the equilibrium of an element reduce to

$$\begin{aligned} \frac{\partial \sigma_r}{\partial r} + \frac{\partial \tau_{rz}}{\partial z} + \frac{\sigma_r - \sigma_\theta}{r} &= 0 \\ \frac{\partial \tau_{rz}}{\partial r} + \frac{\partial \sigma_z}{\partial z} + \frac{\tau_{rz}}{r} &= 0 \end{aligned} \tag{188}$$

For many problems it is again an advantage to introduce a stress function[1] ϕ.

It may be verified by substitution that Eqs. (188) are satisfied if we take

$$
\begin{aligned}
\sigma_r &= \frac{\partial}{\partial z}\left(\nu\,\nabla^2\phi - \frac{\partial^2\phi}{\partial r^2}\right) \\
\sigma_\theta &= \frac{\partial}{\partial z}\left(\nu\,\nabla^2\phi - \frac{1}{r}\frac{\partial\phi}{\partial r}\right) \\
\sigma_z &= \frac{\partial}{\partial z}\left[(2-\nu)\,\nabla^2\phi - \frac{\partial^2\phi}{\partial z^2}\right] \\
\tau_{rz} &= \frac{\partial}{\partial r}\left[(1-\nu)\,\nabla^2\phi - \frac{\partial^2\phi}{\partial z^2}\right]
\end{aligned}
\tag{189}
$$

provided that the stress function ϕ satisfies the equation

$$
\left(\frac{\partial^2}{\partial r^2} + \frac{1}{r}\frac{\partial}{\partial r} + \frac{\partial^2}{\partial z^2}\right)\left(\frac{\partial^2\phi}{\partial r^2} + \frac{1}{r}\frac{\partial\phi}{\partial r} + \frac{\partial^2\phi}{\partial z^2}\right) = \nabla^2\,\nabla^2\phi = 0 \tag{190}
$$

The symbol ∇^2 denotes the operation

$$
\frac{\partial^2}{\partial r^2} + \frac{1}{r}\frac{\partial}{\partial r} + \frac{1}{r^2}\frac{\partial^2}{\partial\theta^2} + \frac{\partial^2}{\partial z^2} \tag{a}
$$

which corresponds to Laplace's operator

$$
\frac{\partial^2}{\partial x^2} + \frac{\partial^2}{\partial y^2} + \frac{\partial^2}{\partial z^2}
$$

in rectangular coordinates [see Eq. (h), page 68]. It should be noted that the stress function ϕ does not depend on θ, so that the third term in (a) gives zero when applied to ϕ.

Displacements u, v, w corresponding to stress expressed by (189) are easily found. For u we have from (187), (189), and (a)

$$
u = r\epsilon_\theta = \frac{r}{E}\left[\sigma_\theta - \nu(\sigma_r + \sigma_z)\right] = -\frac{1+\nu}{E}\frac{\partial^2\phi}{\partial r\,\partial z} \tag{190'}
$$

From w we find $\partial w/\partial z$ from the third of (187), and $\partial w/\partial r$ from the fourth. Thus

$$
E\frac{\partial w}{\partial z} = \sigma_z - \nu(\sigma_r + \sigma_\theta) = \frac{\partial}{\partial z}\left[2(1-\nu^2)\,\nabla^2\phi - (1+\nu)\frac{\partial^2\phi}{\partial z^2}\right]
$$

[1] Love's stress function. See A. E. H. Love, "Mathematical Theory of Elasticity," 4th ed., p. 274, Cambridge University Press, New York, 1927. A survey of various functions thus used for stress and displacement is given by K. Marguerre, *Z. Angew. Math. Mech.*, vol. 35, pp. 242–263, 1955.

and therefore

$$Ew = (1 + \nu)\left[2(1 - \nu)\,\nabla^2\phi - \frac{\partial^2\phi}{\partial z^2}\right] + f(r) \tag{b}$$

where $f(r)$ is a function of r only, as yet arbitrary. Using (190′) in the fourth of (187) we have

$$G\frac{\partial w}{\partial r} = \tau_{rz} - G\frac{\partial u}{\partial z} = \frac{\partial}{\partial r}\left[(1 - \nu)\,\nabla^2\phi - \frac{1}{2}\frac{\partial^2\phi}{\partial z}\right]$$

and therefore, recalling that $2(1 + \nu)G = E$,

$$Ew = (1 + \nu)\left[2(1 - \nu)\,\nabla^2\phi - \frac{\partial^2\phi}{\partial z^2}\right] + g(z) \tag{c}$$

where $g(z)$ is a function of z only, as yet arbitrary. But since (b) and (c) must agree, $f(r)$ and $g(z)$ must be equal at all points of the region. Then

$$f(r) = g(z) = A \qquad \text{a constant}$$

This constant, in (b) or (c), corresponds to an axial rigid-body translation and will be discarded on the understanding that it can be restored whenever needed to meet a fixity condition. Accordingly, the displacement components are expressed from (190′) and (b) or (c) by

$$2Gu = -\frac{\partial^2\phi}{\partial r\,\partial z} \qquad 2Gw = 2(1 - \nu)\,\nabla^2\phi - \frac{\partial^2\phi}{\partial z^2} \tag{190″}$$

If we then begin with a displacement so expressed in terms of a function ϕ satisfying the differential equation (190), we can derive from it the strain components (187) and then the stress components (189). There is no question of the compatibility of such strain components, because they have been derived directly from the displacement components (190″). A particular problem is solved if we can find such a function ϕ that also satisfies the boundary conditions. Several such problems are treated in Arts. 133 to 144 following. Other methods are indicated in Art. 145.

In some cases it is useful to have Eq. (190) in polar coordinates R and ψ (Fig. 201) instead of cylindrical coordinates r and z. This transforma-

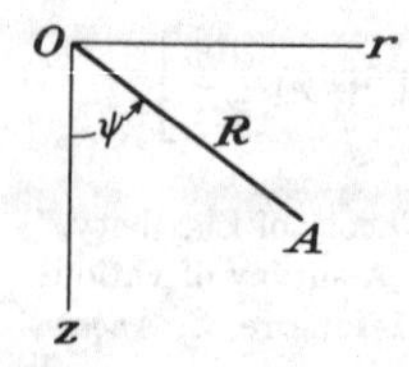

Fig. 201

tion can easily be accomplished by using the formulas of Art. 27. We find

$$\frac{\partial^2}{\partial r^2} + \frac{\partial^2}{\partial z^2} = \frac{\partial^2}{\partial R^2} + \frac{1}{R}\frac{\partial}{\partial R} + \frac{1}{R^2}\frac{\partial^2}{\partial \psi^2}$$

$$\frac{1}{r}\frac{\partial}{\partial r} = \frac{1}{R \sin \psi}\left(\frac{\partial}{\partial R}\sin \psi + \frac{\cos \psi}{R}\frac{\partial}{\partial \psi}\right) = \frac{1}{R}\frac{\partial}{\partial R} + \frac{\operatorname{ctn} \psi}{R^2}\frac{\partial}{\partial \psi}$$

Substituting in Eq. (190),

$$\left(\frac{\partial^2}{\partial R^2} + \frac{2}{R}\frac{\partial}{\partial R} + \frac{1}{R^2}\operatorname{ctn}\psi\frac{\partial}{\partial \psi} + \frac{1}{R^2}\frac{\partial^2}{\partial \psi^2}\right)\left(\frac{\partial^2 \phi}{\partial R^2} + \frac{2}{R}\frac{\partial \phi}{\partial R} + \frac{1}{R^2}\operatorname{ctn}\psi\frac{\partial \phi}{\partial \psi} + \frac{1}{R^2}\frac{\partial^2 \phi}{\partial \psi^2}\right) = 0 \tag{191}$$

We shall apply several solutions of this equation in succeeding articles to the investigation of particular problems involving axial symmetry.

132 | Solution by Polynomials

Let us consider solutions of Eq. (191), which are at the same time solutions of the Laplace equation

$$\frac{\partial^2 \phi}{\partial R^2} + \frac{2}{R}\frac{\partial \phi}{\partial R} + \frac{1}{R^2}\operatorname{ctn}\psi\frac{\partial \phi}{\partial \psi} + \frac{1}{R^2}\frac{\partial^2 \phi}{\partial \psi^2} = 0 \tag{192}$$

A particular solution of this latter equation can be taken in the form

$$\phi_n = R^n \Psi_n \tag{a}$$

in which Ψ_n is a function of the angle ψ only. Substituting (*a*) into Eq. (192), we find for Ψ_n the following ordinary differential equation:

$$\frac{1}{\sin \psi}\frac{d}{d\psi}\left(\sin \psi \frac{d\Psi_n}{d\psi}\right) + n(n+1)\Psi_n = 0 \tag{b}$$

When x is written for $\cos \psi$ and used as a new independent variable, Eq. (*b*) becomes

$$(1 - x^2)\frac{d^2\Psi_n}{dx^2} - 2x\frac{d\Psi_n}{dx} + n(n+1)\Psi_n = 0 \tag{193}$$

This is *Legendre's equation.*[1] The two basic solutions, usually symbolized by $P_n(x)$, $Q_n(x)$ are Legendre functions of the first and second kinds,

[1] See, for instance, F. B. Hildebrand, "Advanced Calculus for Applications," Prentice-Hall, Inc., Englewood Cliffs, N.J., 1962.

respectively. For $n = 0, 1, 2, 3, 4, 5, \ldots$, the $P_n(x)$ are the Legendre polynomials

$$P_0(x) = 1 \qquad P_1(x) = x \qquad P_2(x) = \tfrac{1}{2}(3x^2 - 1)$$
$$P_3(x) = \tfrac{1}{2}(5x^3 - 3x) \qquad P_4(x) = \tfrac{1}{8}(35x^4 - 30x^2 + 3)$$
$$P_5(x) = \tfrac{1}{8}(63x^5 - 70x^3 + 15x) \qquad \text{etc.}$$

Using these for ψ_n in (a), we have corresponding solutions of Eq. (192). Each may be multiplied by an arbitrary constant A_n. Restoring r and z by means of

$$x = \cos\psi \qquad Rx = z \qquad R = \sqrt{r^2 + z^2}$$

we obtain polynomial solutions of Eq. (192) in the forms

$$\begin{aligned}
\phi_0 &= A_0 \\
\phi_1 &= A_1 z \\
\phi_2 &= A_2[z^2 - \tfrac{1}{3}(r^2 + z^2)] \\
\phi_3 &= A_3[z^3 - \tfrac{3}{5}z(r^2 + z^2)] \\
\phi_4 &= A_4[z^4 - \tfrac{6}{7}z^2(r^2 + z^2) + \tfrac{3}{35}(r^2 + z^2)^2] \\
\phi_5 &= A_5[z^5 - \tfrac{10}{9}z^3(r^2 + z^2) + \tfrac{5}{21}z(r^2 + z^2)^2] \\
&\cdots\cdots\cdots\cdots
\end{aligned} \tag{194}$$

These polynomials are also solutions of Eq. (191). From them we can get further solutions of Eq. (191), which are not solutions of Eq. (192). If $R^n\Psi_n$ is a solution of Eq. (192), it can be shown that $R^{n+2}\Psi_n$ is a solution of Eq. (191). Performing the operation indicated in the parentheses of Eq. (191), we have

$$\left(\frac{\partial^2}{\partial R^2} + \frac{2}{R}\frac{\partial}{\partial R} + \frac{1}{R^2}\operatorname{ctn}\psi\frac{\partial}{\partial\psi} + \frac{1}{R^2}\frac{\partial^2}{\partial\psi^2}\right)R^{n+2}\Psi_n = 2(2n + 3)R^n\Psi_n \tag{c}$$

Repeating the same operation again, as indicated in Eq. (191), we obtain zero, since the right-hand side of (c) is a solution of Eq. (192). Hence $R^{n+2}\Psi_n$ is a solution of Eq. (191). Multiplying solutions (194) by $R^2 = r^2 + z^2$, we obtain the following new solutions:

$$\begin{aligned}
\phi_2 &= B_2(r^2 + z^2) \\
\phi_3 &= B_3 z(r^2 + z^2) \\
\phi_4 &= B_4(2z^2 - r^2)(r^2 + z^2) \\
\phi_5 &= B_5(2z^3 - 3r^2 z)(r^2 + z^2) \\
&\cdots\cdots\cdots\cdots
\end{aligned} \tag{195}$$

133 | Bending of a Circular Plate

Several problems of practical interest can be solved with the help of the foregoing solutions. Among these are various cases of the bending of symmetrically loaded circular plates (Fig. 202). Taking, for instance, the polynomials of the third degree from (194) and (195), we obtain the stress function

$$\phi = a_3(2z^3 - 3r^2z) + b_3(r^2z + z^3) \tag{a}$$

Substituting in Eqs. (189), we find

$$\begin{aligned} \sigma_r &= 6a_3 + (10\nu - 2)b_3 & \sigma_\theta &= 6a_3 + (10\nu - 2)b_3 \\ \sigma_z &= -12a_3 + (14 - 10\nu)b_3 & \tau_{rz} &= 0 \end{aligned} \tag{196}$$

The stress components are thus constant throughout the plate. By a suitable adjustment of constants a_3 and b_3 we can represent the stresses in a plate when any constant values of σ_z and σ_r at the surface of the plate are given.

Let us now take the polynomials of the fourth degree from (194) and (195), which gives us

$$\phi = a_4(8z^4 - 24r^2z^2 + 3r^4) + b_4(2z^4 + r^2z^2 - r^4) \tag{b}$$

Substituting in Eqs. (189), we find

$$\begin{aligned} \sigma_r &= 96a_4z + 4b_4(14\nu - 1)z \\ \sigma_z &= -192a_4z + 4b_4(16 - 14\nu)z \\ \tau_{rz} &= 96a_4r - 2b_4(16 - 14\nu)r \end{aligned} \tag{197}$$

Taking

$$96a_4 - 2b_4(16 - 14\nu) = 0$$

we have

$$\sigma_z = \tau_{rz} = 0 \qquad \sigma_r = 28(1 + \nu)b_4z \tag{c}$$

If z is the distance from the middle plane of the plate, the solution (c) represents pure bending of the plate by moments uniformly distributed along the boundary.

To get the solution for a circular plate uniformly loaded, we take the stress function in the form of a polynomial of the sixth power. Proceed-

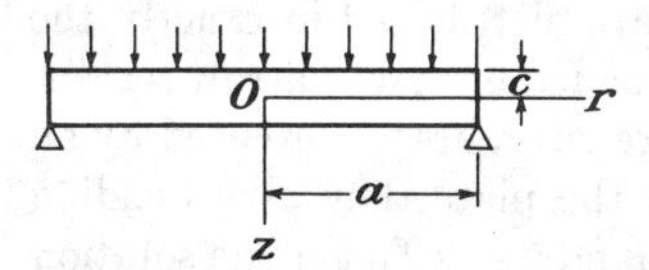

Fig. 202

ing as explained in the previous article, we find

$$\phi = \tfrac{1}{3}a_6(16z^6 - 120z^4r^2 + 90z^2r^4 - 5r^6) + b_6(8z^6 - 16z^4r^2 - 21z^2r^4 + 3r^6)$$

Substituting in (179),

$$\sigma_r = a_6(320z^3 - 720r^2z) + b_6\{64(2 + 11\nu)z^3 + [504 - 48(22\nu)]r^2z\}$$

$$\sigma_z = a_6(-640z^3 + 960r^2z) + b_6\{[-960 + 32(22)(2 - \nu)]z^3 + [384 - 48(22)(2 - \nu)]r^2z\}$$

$$\tau_{rz} = a_6(960rz^2 - 240r^3) + b_6\{[-672 + 48(22)\nu]z^2r + [432 - 12(22)\nu]r^3\}$$

To these stresses we add the stresses

$$\sigma_r = 96a_4z \qquad \sigma_z = -192a_4z \qquad \tau_{rz} = 96a_4r$$

obtained from (197) by taking $b_4 = 0$, and a uniform tension in the z direction $\sigma_z = b$, which can be obtained from (196). Thus we arrive at expressions for the stress components containing four constants a_6, b_6, a_4, b. These constants can be adjusted so as to satisfy the boundary conditions on the upper and lower surfaces of the plate (Fig. 202). The conditions are

$$\begin{aligned} \sigma_z &= 0 && \text{for } z = c \\ \sigma_z &= -q && \text{for } z = -c \\ \tau_{rz} &= 0 && \text{for } z = c \\ \tau_{rz} &= 0 && \text{for } z = -c \end{aligned} \tag{d}$$

Here q denotes the intensity of the uniform load and $2c$ is the thickness of the plate. Substituting the expressions for the stress components in these equations, we determine the four constants a_6, b_6, a_4, b. Using these values, the expressions for the stress components satisfying conditions (d) are

$$\begin{aligned} \sigma_r &= q\left[\frac{2+\nu}{8}\frac{z^3}{c^3} - \frac{3(3+\nu)}{32}\frac{r^2z}{c^3} - \frac{3}{8}\frac{z}{c}\right] \\ \sigma_z &= q\left(-\frac{z^3}{4c^3} + \frac{3}{4}\frac{z}{c} - \frac{1}{2}\right) \\ \tau_{rz} &= -\frac{3qr}{8c^3}(c^2 - z^2) \end{aligned} \tag{e}$$

It will be seen that the stresses σ_z and τ_{rz} are distributed in exactly the same manner as in the case of a uniformly loaded beam of narrow rectangular cross section (Art. 22). The radial stresses σ_r are represented by an odd function of z, and at the boundary of the plate they give bending moments uniformly distributed along the boundary. To get the solution for a simply supported plate (Fig. 202), we superpose a pure bending stress (c) and adjust the constant b_4 so as to obtain for the boundary ($r = a$),

$$\int_{-c}^{c} \sigma_r z \, dz = 0$$

Then the final expression for σ_r becomes

$$\sigma_r = q\left[\frac{2+\nu}{8}\frac{z^3}{c^3} - \frac{3(3+\nu)}{32}\frac{r^2 z}{c^3} - \frac{3}{8}\frac{2+\nu}{5}\frac{z}{c} + \frac{3(3+\nu)}{32}\frac{a^2 z}{c^3}\right] \tag{198}$$

and at the center of the plate we have

$$(\sigma_r)_{r=0} = q\left[\frac{2+\nu}{8}\frac{z^3}{c^3} - \frac{3}{8}\frac{2+\nu}{5}\frac{z}{c} + \frac{3(3+\nu)}{32}\frac{a^2 z}{c^3}\right] \tag{f}$$

The elementary theory of bending of plates, based on the assumptions that linear elements of the plate perpendicular to the *middle plane* ($z = 0$) remain straight and normal to the deflection surface of the plate[1] during bending, gives for the radial stresses at the center

$$\sigma_r = \frac{3(3+\nu)}{32}\frac{a^2 z}{c^3} q \tag{g}$$

Comparing this with (f), we see that the additional terms of the exact solution are small if the thickness of the plate, $2c$, is small in comparison with the radius a.

It should be noted that by superposing pure bending we eliminated bending moments along the boundary of the plate; but there is radial stress at the boundary, given by

$$(\sigma_r)_{r=a} = q\left(\frac{2+\nu}{8}\frac{z^3}{c^3} - \frac{3}{8}\frac{2+\nu}{5}\frac{z}{c}\right) \tag{h}$$

The resultant of these stresses per unit length of the boundary line and their moment, however, are zero. Hence, on the basis of Saint-Venant's principle, we can say that the removal of these stresses does not significantly affect the stress distribution in the plate at some distance from the edge.

By taking polynomials of higher order than the sixth for the stress function, we can investigate cases of bending of a circular plate by nonuniformly distributed loads. By including the functions $Q_n(x)$ as well as $P_n(x)$ in Art. 132, we can also find solutions for a circular plate with a central hole.[2] All these solutions are satisfactory only if the deflection

[1] This assumption is analogous to the *plane cross sections* hypothesis in the theory of bending of beams. The exact theory of bending of plates was developed by J. H. Michell, *Proc. London Math. Soc.*, vol. 31, p. 114, 1899, and Love, *op. cit.*, p. 465.

[2] A number of solutions for a circular plate symmetrically loaded have been discussed by A. Korobov, *Bull. Polytech. Inst., Kiew,* 1913. Similar solutions were obtained independently by A. Timpe, *Z. Angew. Math. Mech.*, vol. 4, 1924.

of the plate remains small in comparison with the thickness. For larger deflections the stretching of the middle plane of the plate must be considered.[1]

134 | The Rotating Disk as a Three-dimensional Problem

In our previous discussion (Art. 32) it was assumed that the stresses do not vary through the thickness of the disk. Let us now consider the same problem assuming only that the stress distribution is symmetrical with respect to the axis of rotation. The differential equations of equilibrium are obtained by including in Eqs. (188) the centrifugal force. Then

$$\begin{aligned} \frac{\partial \sigma_r}{\partial r} + \frac{\partial \tau_{rz}}{\partial z} + \frac{\sigma_r - \sigma_\theta}{r} + \rho\omega^2 r &= 0 \\ \frac{\partial \tau_{rz}}{\partial r} + \frac{\partial \sigma_z}{\partial z} + \frac{\tau_{rz}}{r} &= 0 \end{aligned} \tag{199}$$

where ρ is the mass per unit volume and ω the angular velocity of the disk.

The compatibility equations also must be changed. Instead of the system (126) we shall have three equations of the type (*g*) in Art. 85 and three equations of the type (*h*). Substituting in these equations the components of body force,

$$X = \rho\omega^2 x \qquad Y = \rho\omega^2 y \qquad Z = 0 \tag{a}$$

we find that the last three equations, containing shearing-stress components, remain the same as in the system (126), and the first three equations become

$$\begin{aligned} \nabla^2\sigma_r - \frac{2}{r^2}(\sigma_r - \sigma_\theta) + \frac{1}{1+\nu}\frac{\partial^2\Theta}{\partial r^2} &= -\frac{2\rho\omega^2}{1-\nu} \\ \nabla^2\sigma_\theta + \frac{2}{r^2}(\sigma_r - \sigma_\theta) + \frac{1}{1+\nu}\frac{1}{r}\frac{\partial\Theta}{\partial r} &= -\frac{2\rho\omega^2}{1-\nu} \\ \nabla^2\sigma_z + \frac{1}{1+\nu}\frac{\partial^2\Theta}{\partial z^2} &= -\frac{2\nu\rho\omega^2}{1-\nu} \end{aligned} \tag{b}$$

We begin with a particular solution of Eqs. (199), satisfying the compatibility equations. On this solution we superpose solutions in the form of polynomials (194) and (195) and adjust the constants of these polynomials so as to satisfy the boundary conditions of the problem. For the particular solution we take the expressions

$$\sigma_r = Br^2 + Dz^2 \qquad \sigma_z = Ar^2 \qquad \sigma_\theta = Cr^2 + Dz^2 \qquad \tau_{rz} = 0 \tag{c}$$

It can be seen that these expressions satisfy the second of the equations of equilibrium. They also satisfy the compatibility equations that contain shearing-stress components. It remains to determine the constants A, B, C, D, so as to satisfy the remaining four equations, namely, the first of Eqs. (199) and Eqs. (*b*). Substituting (*c*) in these equations, we find

$$A = \frac{\rho\omega^2(1+3\nu)}{6\nu} \qquad B = -\frac{\rho\omega^2}{3} \qquad C = 0 \qquad D = -\frac{\rho\omega^2(1+2\nu)(1+\nu)}{6\nu(1-\nu)}$$

[1] See Kelvin and Tait, "Natural Philosophy," vol. 2, p. 171, 1903.

The particular solution is then

$$
\begin{aligned}
\sigma_r &= -\frac{\rho\omega^2}{3} r^2 - \frac{\rho\omega^2(1+2\nu)(1+\nu)}{6\nu(1-\nu)} z^2 \\
\sigma_z &= \frac{\rho\omega^2(1+3\nu)}{6\nu} r^2 \\
\sigma_\theta &= -\frac{\rho\omega^2(1+2\nu)(1+\nu)}{6\nu(1-\nu)} z^2 \\
\tau_{rz} &= 0
\end{aligned}
\tag{200}
$$

This solution can be used in discussing the stresses in any body rotating about an axis of generation.

In the case of a circular disk of constant thickness, we superpose on the solution (200) the stress distribution derived from a stress function having the form of a polynomial of the fifth degree [see Eqs. (194) and (195)],

$$
\phi = a_5(8z^5 - 40r^2z^3 + 15r^4z) + b_5(2z^5 - r^2z^3 - 3r^4z) \tag{d}
$$

Then, from Eqs. (189), we find

$$
\begin{aligned}
\sigma_r &= -a_5(180r^2 - 240z^2) + b_5[(36 - 54\nu)r^2 + (1 + 18\nu)6z^2] \\
\sigma_z &= -a_5(-240r^2 + 480z^2) + b_5[(96 - 108\nu)z^2 + (-102 + 54\nu)r^2] \\
\sigma_\theta &= a_5(-60r^2 + 240z^2) + b_5[(6 + 108\nu)z^2 + (12 - 54\nu)r^2] \\
\tau_{rz} &= 480a_5rz - b_5(96 - 108\nu)rz
\end{aligned}
\tag{e}
$$

Adding this to the stresses (200) and determining the constants a_5 and b_5 so as to make the resultant stresses τ_{rz} and σ_z vanish, we find

$$
\begin{aligned}
\sigma_r &= -\rho\omega^2\left[\frac{\nu(1+\nu)}{2(1-\nu)} z^2 + \frac{3+\nu}{8} r^2\right] \\
\sigma_\theta &= -\rho\omega^2\left[\frac{(1+3\nu)}{8} r^2 + \frac{\nu(1+\nu)}{2(1-\nu)} z^2\right]
\end{aligned}
\tag{f}
$$

To eliminate the resultant radial compression along the boundary, i.e., to make

$$
\left(\int_{-c}^{c} \sigma_r \, dz\right)_{r=a} = 0
$$

we superpose on (*f*) a uniform radial tension of magnitude

$$
\frac{\rho\omega^2}{8}(3+\nu)a^2 + \rho\omega^2 \frac{\nu(1+\nu)}{2(1-\nu)} \frac{c^2}{3}
$$

Then the final stresses[1] are

$$
\begin{aligned}
\sigma_r &= \rho\omega^2\left[\frac{3+\nu}{8}(a^2 - r^2) + \frac{\nu(1+\nu)}{6(1-\nu)}(c^2 - 3z^2)\right] \\
\sigma_\theta &= \rho\omega^2\left[\frac{3+\nu}{8} a^2 - \frac{1+3\nu}{8} r^2 + \frac{\nu(1+\nu)}{6(1-\nu)}(c^2 - 3z^2)\right] \\
&\sigma_z = 0 \qquad \tau_{rz} = 0
\end{aligned}
\tag{201}
$$

Comparing this with the previous solution (54), we have here additional terms with

[1] Derived otherwise by Love, *op. cit.*, pp. 147–148. Displacement expressions, and the additional terms introduced by a central free hole, are also given.

the factor[1] $(c^2 - 3z^2)$. The corresponding stresses are small in the case of a thin disk and their resultant over the thickness of the disk is zero. If the rim of the disk is free from external forces, solution (201) represents the state of stress in parts of the disk some distance from the edge.

The stress distribution in a rotating disk having the shape of a flat ellipsoid of revolution has been discussed by C. Chree.[2]

135 | Force at a Point in an Infinite Solid

When the origin is the point of application of the force, some or all of the stress components must have singularities there. Suitable solutions can be found by returning to Eq. (*a*) on page 383 as a solution of Eq. (192). Regarding n as already chosen, we compare Eq. (*a*) with what it becomes when we take $-n - 1$ instead of n, that is,

$$\phi_{-n-1} = R^{-n-1}\Psi_{-n-1} \tag{a}$$

The coefficient $n(n + 1)$ in Eq. (*b*) on page 383 becomes $(-n - 1)(-n)$, and thus has the same value for Ψ_{-n-1} as for Ψ_n. Consequently, we may take instead of Eq. (*a*) above

$$\phi_{-n-1} = R^{-n-1}\Psi_n \tag{b}$$

Using $P_n(x)$ for Ψ_n, as on page 384, we find the following set of solutions of Eqs. (192) and (191),

$$\begin{aligned}
\phi_1 &= A_1(r^2 + z^2)^{-1/2} \\
\phi_2 &= A_2 z(r^2 + z^2)^{-3/2} \\
\phi_3 &= A_3[z^2(r^2 + z^2)^{-5/2} - \tfrac{1}{3}(r^2 + z^2)^{-3/2}] \\
&\cdots\cdots\cdots\cdots\cdots
\end{aligned} \tag{202}$$

which are also solutions of Eq. (191). Multiplying expressions (202) by $r^2 + z^2$ (see page 384), we obtain another set of solutions of Eq. (191), namely,

$$\begin{aligned}
\phi_1 &= B_1(r^2 + z^2)^{1/2} \\
\phi_2 &= B_2 z(r^2 + z^2)^{-1/2} \\
&\cdots\cdots\cdots
\end{aligned} \tag{203}$$

Each of the solutions (202) and (203), and any linear combination of them, can be taken as a stress function, and, by a suitable adjustment of the

[1] These terms are of the same nature as the terms in z^2 found in Art. 98. Equations (201) represent a state of plane stress since σ_z and τ_{rz} vanish. Body force (here centrifugal force), not included in Art. 98, does not alter the general conclusions so long as it is independent of z.

[2] *Proc. Roy. Soc.* (*London*), vol. 58, p. 39, 1895. For the general ellipsoid, see M. A. Goldberg and M. Sadowsky, *J. Appl. Mech.*, vol. 26, pp. 549–552, 1959.

constants $A_1, A_2, \ldots, B_1, B_2, \ldots$, solutions of various problems may be found.

For the case of a concentrated force we take the first of the solutions (203). Discarding the subscripts, the stress function is

$$\phi = B(r^2 + z^2)^{1/2}$$

where B is a constant to be adjusted later. Substituting in Eqs. (189), we find the corresponding stress components

$$\begin{aligned}
\sigma_r &= B[(1 - 2\nu)z(r^2 + z^2)^{-3/2} - 3r^2z(r^2 + z^2)^{-5/2}] \\
\sigma_\theta &= B(1 - 2\nu)z(r^2 + z^2)^{-3/2} \\
\sigma_z &= -B[(1 - 2\nu)z(r^2 + z^2)^{-3/2} + 3z^3(r^2 + z^2)^{-5/2}] \\
\tau_{rz} &= -B[(1 - 2\nu)r(r^2 + z^2)^{-3/2} + 3rz^2(r^2 + z^2)^{-5/2}]
\end{aligned} \tag{204}$$

All these stresses are singular at the origin of coordinates, where the concentrated force is applied. We therefore take the origin to be the center of a small spherical cavity (Fig. 203), and consider forces over the surface of the cavity as calculated from Eqs. (204). It can be shown that the resultant of these forces represents a force applied at the origin in the direction of the z axis. From the condition of equilibrium of a ring-shaped element, adjacent to the cavity (Fig. 203), the component of surface forces in the z direction is

$$\bar{Z} = -(\tau_{rz} \sin \psi + \sigma_z \cos \psi)$$

Using Eqs. (204) and the formulas

$$\sin \psi = r(r^2 + z^2)^{-1/2} \qquad \cos \psi = z(r^2 + z^2)^{-1/2}$$

we find that

$$\bar{Z} = B[(1 - 2\nu)(r^2 + z^2)^{-1} + 3z^2(r^2 + z^2)^{-2}]$$

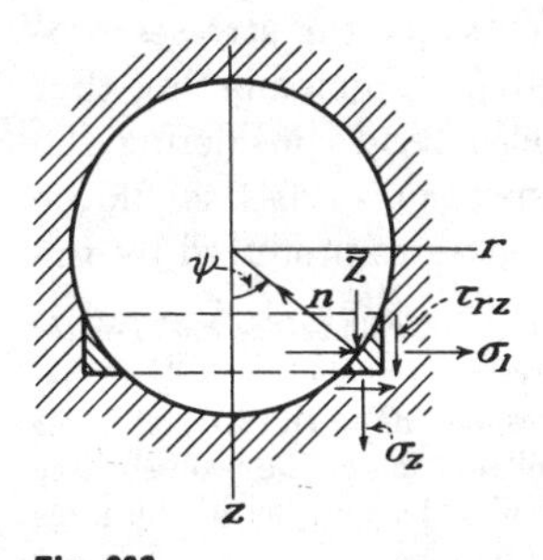

Fig. 203

The resultant of these forces over the surface of the cavity is

$$2\int_0^{\pi/2} \bar{Z}\,\sqrt{r^2+z^2}\,d\psi\,2\pi r = 8B\pi(1-\nu)$$

The resultant of the surface forces in a radial direction is zero, from symmetry. If P is the magnitude of the applied force, we have

$$P = 8B\pi(1-\nu)$$

Substituting

$$B = \frac{P}{8\pi(1-\nu)} \tag{205}$$

into Eqs. (204), we obtain the stresses produced by a force P applied at the origin in the z direction.[1] This force is balanced by the surface force on an outer spherical or other boundary, however large, as required by Eqs. (204). The solution is the three-dimensional analog of the solution of the two-dimensional problem discussed in Art. 42.

Substituting $z = 0$ in Eqs. (204), we find that there are no normal stresses acting on the coordinate plane $z = 0$. The shearing stresses over the same plane are

$$\tau_{rz} = -\frac{B(1-2\nu)}{r^2} = -\frac{P(1-2\nu)}{8\pi(1-\nu)r^2} \tag{c}$$

These stresses are inversely proportional to the square of the distance r from the point of application of the load.

136 | Spherical Container under Internal or External Uniform Pressure

By superposition we can get from the solution of the previous article some new solutions of practical interest. We begin with the case of two equal and opposite forces, a small distance d apart, applied to an indefinitely extended elastic body (Fig. 204). The stresses produced at any point by the force P applied at the origin O are determined by Eqs. (204) and (205) of the previous article. By using the same equations, the stresses produced by the force P at O_1 can also be calculated. Remembering that the second force is acting in the opposite direction and considering the distance d as infinitesimal, any term $f(r,z)$ in expressions (204) should be replaced by $-[f + (\partial f/\partial z)d]$. Superposing the stresses produced by the

[1] The solution of this problem was given by Lord Kelvin, *Cambridge and Dublin Math. J.*, 1848. See also his "Mathematical and Physical Papers," vol. 1, p. 37. From his solution it follows that the displacements corresponding to the stresses (204) are single-valued, which proves that (204) is the correct solution of the problem (see Art. 96). The solution does of course require a particular distribution of the force over the surface of the cavity, however small the latter may be.

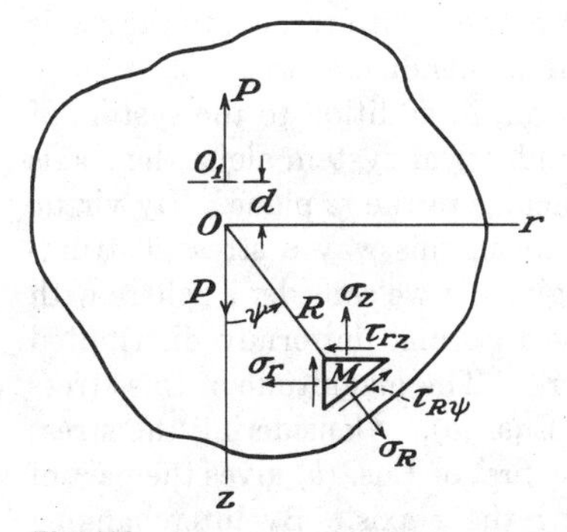

Fig. 204

two forces and using the symbol A for the product Bd, we find

$$\begin{aligned}
\sigma_r &= -A \frac{\partial}{\partial z} [(1 - 2\nu)z(r^2 + z^2)^{-3/2} - 3r^2 z(r^2 + z^2)^{-5/2}] \\
\sigma_\theta &= -A \frac{\partial}{\partial z} [(1 - 2\nu)z(r^2 + z^2)^{-3/2}] \\
\sigma_z &= A \frac{\partial}{\partial z} [(1 - 2\nu)z(r^2 + z^2)^{-3/2} + 3z^3(r^2 + z^2)^{-5/2}] \\
\tau_{rz} &= A \frac{\partial}{\partial z} [(1 - 2\nu)r(r^2 + z^2)^{-3/2} + 3rz^2(r^2 + z^2)^{-5/2}]
\end{aligned} \tag{206}$$

Let us consider (Fig. 204) the stress components σ_R and $\tau_{R\psi}$ acting at a point M on an elemental area perpendicular to the radius OM, the length of which is denoted by R. From the condition of equilibrium of a small triangular element such as indicated in the figure we find[1]

$$\begin{aligned}
\sigma_R &= \sigma_r \sin^2 \psi + \sigma_z \cos^2 \psi + 2\tau_{rz} \sin \psi \cos \psi \\
\tau_{R\psi} &= (\sigma_r - \sigma_z) \sin \psi \cos \psi - \tau_{rz}(\sin^2 \psi - \cos^2 \psi)
\end{aligned} \tag{a}$$

Using (206), and taking

$$\sin \psi = r(r^2 + z^2)^{-1/2} = \frac{r}{R} \qquad \cos \psi = z(r^2 + z^2)^{-1/2} = \frac{z}{R}$$

we obtain

$$\begin{aligned}
\sigma_R &= -\frac{2(1 + \nu)A}{R^3} \left[-\sin^2 \psi + \frac{2(2 - \nu)}{1 + \nu} \cos^2 \psi \right] \\
\tau_{R\psi} &= -\frac{2(1 + \nu)A}{R^3} \sin \psi \cos \psi
\end{aligned} \tag{b}$$

[1] The stress components σ_θ, acting on the sides of the element in the meridional sections of the body, give a small resultant of higher order and can be neglected in deriving the equations of equilibrium.

The distribution of these stresses is symmetrical with respect to the z axis and with respect to the coordinate plane perpendicular to z.

Imagine now that we have at the origin, in addition to the system of two forces P acting along the z axis, an identical system along the r axis and another one along the axis perpendicular to the rz plane. By virtue of the symmetry stated above, we obtain in this way a stress distribution symmetrical with respect to the origin. If we consider a sphere with center at the origin, there will be only a normal uniformly distributed stress acting on the surface of this sphere. The magnitude of this stress can be calculated by using the first of Eqs. (*b*). Considering the stress at points on the circle in the rz plane, the first of Eqs. (*b*) gives the part of this stress due to the *double force* along the z axis. By interchanging $\sin \psi$ and $\cos \psi$, we obtain the normal stress round the same circle produced by the double force along the r axis. The normal stress due to the double force perpendicular to the rz plane is obtained by substituting $\psi = \pi/2$ in the same equation. Combining the actions of the three perpendicular double forces, we find the following normal stress acting on the surface of the sphere:

$$\sigma_R = -\frac{4(1-2\nu)A}{R^3} \tag{c}$$

The *combination* of the three perpendicular double forces is called a *center of compression.* We see from (*c*) that the corresponding compression stress in the radial direction depends only on the distance from the center of compression and is inversely proportional to the cube of this distance. This spherically symmetrical singular solution can be used in finding the stresses in a hollow sphere[1] under given internal and external pressures, p_i and p_o (Fig. 205). Superposing on (*c*) a uniform tension or compression

[1] The problem is easily solved directly in terms of the radial displacement u_R.

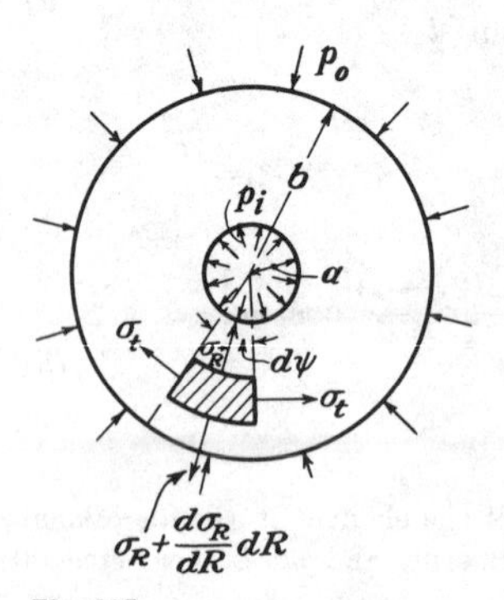

Fig. 205

in all directions, we can take a general expression for the radial normal stress in the form

$$\sigma_R = \frac{C}{R^3} + D \tag{d}$$

C and D are constants, the magnitudes of which are determined from the conditions on the inner and outer surfaces of the container, which are

$$\frac{C}{a^3} + D = -p_i \qquad \frac{C}{b^3} + D = -p_o$$

Then

$$C = \frac{(p_i - p_o)a^3b^3}{a^3 - b^3}$$

$$D = \frac{p_o b^3 - p_i a^3}{a^3 - b^3} \tag{207}$$

$$\sigma_R = \frac{p_o b^3(R^3 - a^3)}{R^3(a^3 - b^3)} + \frac{p_i a^3(b^3 - R^3)}{R^3(a^3 - b^3)}$$

The pressures p_o and p_i also produce in the sphere normal stresses σ_t in a tangential direction, the magnitude of which we find from the condition of equilibrium of an element cut out from the sphere by the two concentric spherical surfaces of radii R and $R + dR$ and by a circular cone with a small angle $d\psi$ (Fig. 205). The equation of equilibrium is

$$\sigma_t \frac{\pi R}{2}\, dR(d\psi)^2 = \frac{d\sigma_R}{dR}\frac{\pi R^2}{4}\, dR(d\psi)^2 + \sigma_R \frac{\pi R}{2}\, dR(d\psi)^2$$

from which

$$\sigma_t = \frac{d\sigma_R}{dR}\frac{R}{2} + \sigma_R \tag{e}$$

Using expression (207) for σ_R this becomes

$$\sigma_t = \frac{p_o b^3(2R^3 + a^3)}{2R^3(a^3 - b^3)} - \frac{p_i a^3(2R^3 + b^3)}{2R^3(a^3 - b^3)} \tag{208}$$

If $p_o = 0$, then

$$\sigma_t = \frac{p_i a^3}{2R^3}\frac{(2R^3 + b^3)}{b^3 - a^3}$$

It will be seen that the greatest tangential tension in this case is at the inner surface, at which

$$(\sigma_t)_{\max} = \frac{p_i}{2}\frac{2a^3 + b^3}{b^3 - a^3}$$

All these results are due to Lamé.[1]

[1] "Leçons sur la Théorie . . . de l'Élasticité," Paris, 1852.

137 | Local Stresses around a Spherical Cavity

As a second example consider the stress distribution around a small spherical cavity in a bar submitted to uniform tension of magnitude S (Fig. 206).[1] In the case of a solid bar in tension, the normal and shearing components of stress acting on a spherical surface are

$$\sigma_R = S\cos^2\psi \qquad \tau_{R\psi} = -S\sin\psi\cos\psi \tag{a}$$

To get the solution for the case of a small spherical cavity of radius a, we must superpose on the simple tension a stress system which has stress components on the spherical surface equal and opposite to those given by Eqs. (a) and which vanishes at infinity.

Taking from the previous article the stresses (b), due to a double force in the z direction, and the stresses (c), due to a center of compression, the corresponding stresses acting on the spherical surface of radius a can be presented in the following form:

$$\begin{aligned} \sigma_R' &= -\frac{2(1+\nu)A}{a^3}\left(-1+\frac{5-\nu}{1+\nu}\cos^2\psi\right) \\ \tau_{R\psi}' &= -\frac{2(1+\nu)A}{a^3}\sin\psi\cos\psi \end{aligned} \tag{b}$$

$$\sigma_R'' = \frac{B}{a^3} \qquad \tau_{R\psi}'' = 0 \tag{c}$$

where A and B are constants to be adjusted later. It is seen that, combining stresses (b) and (c), the stresses (a) produced by tension cannot be made to vanish and that an additional stress system is necessary.

Taking, from solutions (202), a stress function

$$\phi = Cz(r^2+z^2)^{-3/2}$$

[1] Solution of this problem is due to R. V. Southwell, *Phil. Mag.*, 1926. For the elastic inclusion of different material, see J. N. Goodier, *Trans. ASME*, vol. 55, p. 39, 1933. The problem of the triaxial ellipsoidal cavity is solved in E. Sternberg and M. Sadowsky, *J. Appl. Mech.*, vol. 16, p. 149, 1949. For a survey of the literature on three-dimensional stress concentration, see E. Sternberg, *Appl. Mech. Rev.*, vol. 11, pp. 1–4, 1958.

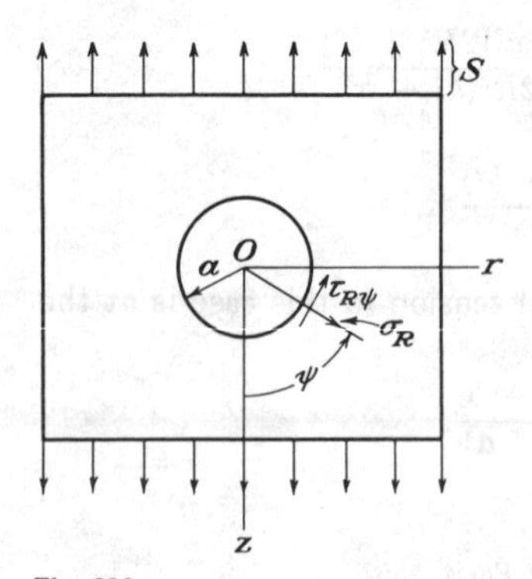

Fig. 206

the corresponding stress components, from Eqs. (189), are

$$
\begin{aligned}
\sigma_r &= \frac{3C}{R^5}(-4 + 35 \sin^2 \psi \cos^2 \psi) \\
\sigma_z &= \frac{3C}{R^5}(3 - 30 \cos^2 \psi + 35 \cos^4 \psi) \\
\sigma_\theta &= \frac{3C}{R^5}(1 - 5 \cos^2 \psi) \\
\tau_{rz} &= \frac{15C}{R^5}(-3 \sin \psi \cos \psi + 7 \sin \psi \cos^3 \psi)
\end{aligned} \tag{d}
$$

Using now Eqs. (*a*) of the previous article, the stress components acting on a spherical surface of radius a are

$$
\sigma_R''' = \frac{12C}{a^5}(-1 + 3 \cos^2 \psi) \qquad \tau_{R\psi}''' = \frac{24C}{a^5} \sin \psi \cos \psi \tag{e}
$$

Combining stress systems (*b*), (*c*), (*e*) we find

$$
\begin{aligned}
\sigma_R &= \frac{2(1+\nu)A}{a^3} - 2(5-\nu)\frac{A}{a^3}\cos^2\psi + \frac{B}{a^3} - \frac{12C}{a^5} + \frac{36C}{a^5}\cos^2\psi \\
\tau_{R\psi} &= -\frac{2(1+\nu)A}{a^3}\sin\psi\cos\psi + \frac{24C}{a^5}\sin\psi\cos\psi
\end{aligned} \tag{f}
$$

Superposing these stresses on the stresses (*a*), the spherical surface of the cavity becomes free from forces if we satisfy the conditions

$$
\begin{aligned}
\frac{2(1+\nu)A}{a^3} + \frac{B}{a^3} - \frac{12C}{a^5} &= 0 \\
-2(5-\nu)\frac{A}{a^3} + \frac{36C}{a^5} &= -S \\
-\frac{2(1+\nu)A}{a^3} + \frac{24C}{a^5} &= S
\end{aligned} \tag{g}
$$

from which

$$
\frac{A}{a^3} = \frac{5S}{2(7-5\nu)} \qquad \frac{B}{a^3} = \frac{S(1-5\nu)}{7-5\nu} \qquad \frac{C}{a^5} = \frac{S}{2(7-5\nu)} \tag{h}
$$

The complete stress at any point is now obtained by superposing on the simple tension S the stresses given by Eqs. (*d*), the stresses (206) due to the double force, and the stresses due to the center of pressure given by Eqs. (*c*) and (*e*) of the previous article.

Consider, for instance, the stresses acting on the plane $z = 0$. From the condition of symmetry there are no shearing stresses on this plane. From Eqs. (*d*), substituting $\psi = \pi/2$ and $R = r$,

$$
\sigma_z' = \frac{9C}{r^5} = \frac{9Sa^5}{2(7-5\nu)r^5} \tag{i}
$$

From Eqs. (206), for $z = 0$,

$$
\sigma_z'' = \frac{A(1-2\nu)}{r^3} = \frac{5(1-2\nu)S}{2(7-5\nu)}\frac{a^3}{r^3} \tag{j}
$$

From Eq. (*e*) of the previous article,

$$
\sigma_z''' = (\sigma_t)_{z=0} = -\frac{B}{2r^3} = -\frac{S(1-5\nu)}{2(7-5\nu)}\frac{a^3}{r^3} \tag{k}
$$

The total stress on the plane $z = 0$ is

$$\sigma_z = \sigma_z' + \sigma_z'' + \sigma_z''' + S = S\left[1 + \frac{4 - 5\nu}{2(7 - 5\nu)}\frac{a^3}{r^3} + \frac{9}{2(7 - 5\nu)}\frac{a^5}{r^5}\right] \tag{l}$$

At $r = a$, we find

$$(\sigma_z)_{\max} = \frac{27 - 15\nu}{2(7 - 5\nu)} S \tag{m}$$

Taking $\nu = 0.3$,

$$(\sigma_z)_{\max} = {}^{45}\!/\!_{22} S$$

The maximum stress is thus about twice as great as the uniform tension S applied to the bar. This increase in stress is of a highly localized character. With increase of r, the stress (l) rapidly approaches the value S. Taking, for instance, $r = 2a$, $\nu = 0.3$, we find $\sigma_z = 1.054S$.

In the same manner we find, for points in the plane $z = 0$,

$$(\sigma_\theta)_{z=0} = \frac{3C}{r^5} - \frac{A(1 - 2\nu)}{r^3} - \frac{B}{2r^3}$$

Substituting from Eqs. (h) and taking $r = a$, we find that the tensile stress along the equator $(\psi = \pi/2)$ of the cavity is

$$(\sigma_\theta)_{z=0, r=a} = \frac{15\nu - 3}{2(7 - 5\nu)} S$$

At the pole of the cavity ($\psi = 0$ or $\psi = \pi$) we have

$$\sigma_r = \sigma_\theta = \frac{2(1 - 2\nu)A}{a^3} - \frac{12C}{a^5} - \frac{B}{2a^3} = -\frac{3 + 15\nu}{2(7 - 5\nu)} S$$

Thus the longitudinal tension S produces compression at this point.

Combining a tension S in one direction with compression S in the perpendicular direction, we can obtain the solution for the stress distribution around a spherical cavity in the case of pure shear.[1] It can be shown in this way that the maximum shearing stress is

$$\tau_{\max} = \frac{15(1 - \nu)}{7 - 5\nu} S \tag{n}$$

The results of this article are of some practical interest in discussing the effect of small cavities[2] on the endurance limit of specimens submitted to the action of cyclical stresses.

138 | Force on Boundary of a Semi-infinite Body

Imagine that the plane $z = 0$ is the boundary of a semi-infinite solid and

[1] This problem was discussed by J. Larmor, *Phil. Mag.*, ser. 5, vol. 33, 1892. See also Love, *op. cit.*, p. 252. Chapter 11 of the same book contains solutions for more general problems of spherical boundaries. The axisymmetric problem of the hollow sphere is reconsidered by E. Sternberg, R. A. Eubanks, and M. A. Sadowsky, *Proc. 1st U.S. Congr. Appl. Mech.*, pp. 209–215, 1951.

[2] See, for instance, R. V. Southwell and H. J. Gough, *Phil. Mag.*, vol. 1, p. 71, 1926.

that a force P is acting on this plane along the z axis (Fig. 207).[1] It was shown in Art. 135 that the stress distribution given by Eqs. (204) and (205) can be produced in a semi-infinite body by a concentrated force at the origin and by shearing forces on the boundary plane $z = 0$, given by the equation

$$\tau_{rz} = -\frac{B(1 - 2\nu)}{r^2} \tag{a}$$

To eliminate these forces and arrive at the solution of the problem shown in Fig. 207, we use the stress distribution corresponding to the center of compression (see page 394). In polar coordinates this stress distribution is

$$\sigma_R = \frac{A}{R^3} \qquad \sigma_t = \frac{d\sigma_R}{dR}\frac{R}{2} + \sigma_R = -\frac{1}{2}\frac{A}{R^3}$$

in which A is a constant. In cylindrical coordinates (Fig. 207) we have the following expressions for the stress components:

$$\begin{aligned}
\sigma_r &= \sigma_R \sin^2\psi + \sigma_t \cos^2\psi = A(r^2 - \tfrac{1}{2}z^2)(r^2 + z^2)^{-5/2} \\
\sigma_z &= \sigma_R \cos^2\psi + \sigma_t \sin^2\psi = A(z^2 - \tfrac{1}{2}r^2)(r^2 + z^2)^{-5/2} \\
\tau_{rz} &= \tfrac{1}{2}(\sigma_R - \sigma_t)\sin 2\psi = \tfrac{3}{2}Arz(r^2 + z^2)^{-5/2} \\
\sigma_\theta &= \sigma_t = -\frac{1}{2}\frac{A}{R^3} = -\frac{1}{2}A(r^2 + z^2)^{-3/2}
\end{aligned} \tag{209}$$

[1] The solution of this problem was given by J. Boussinesq: see "Application des Potentiels a l'Etude de l'Equilibre et du Mouvement des Solides Elastiques," Gauthier-Villars, Paris, 1885. For the tangential force, and other boundary conditions on the plane, references are given in the book by Love, *op. cit.*, Art. 167. The solution for a force at an internal point of a semi-infinite body was found by R. D. Mindlin, *Phys.*, vol. 7, p. 195, 1936; also *Proc. 1st Midwestern Conf. Solid Mech.*, pp. 56–59, 1953. For the fixed plane boundary, see L. Rongved, *J. App. Mech.*, vol. 22, pp. 545–546, 1955.

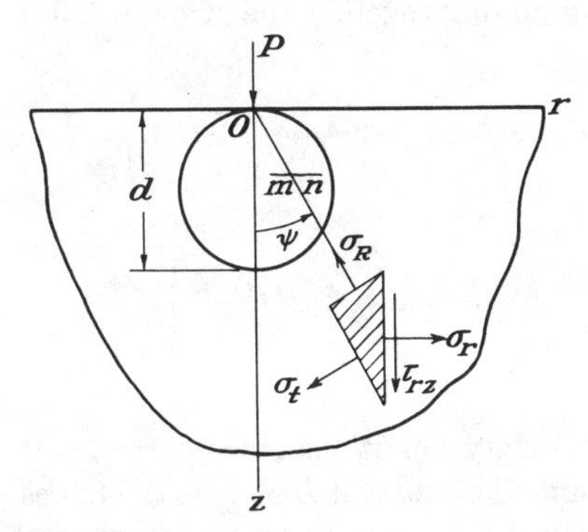

Fig. 207

Suppose now that centers of pressure are uniformly distributed along the z axis from $z = 0$ to $z = -\infty$. Then, by superposition, the stress components produced in an indefinitely extended solid are, from Eqs. (209) with a new constant A_1,

$$
\begin{aligned}
\sigma_r &= A_1 \int_z^\infty (r^2 - \tfrac{1}{2}z^2)(r^2 + z^2)^{-5/2}\, dz \\
&\qquad = \frac{A_1}{2}\left[\frac{1}{r^2} - \frac{z}{r^2}(r^2 + z^2)^{-1/2} - z(r^2 + z^2)^{-3/2}\right] \\
\sigma_z &= A_1 \int_z^\infty \left(z^2 - \frac{1}{2}r^2\right)(r^2 + z^2)^{-5/2}\, dz = \frac{A_1}{2} z(r^2 + z^2)^{-3/2} \\
\tau_{rz} &= \frac{3}{2}A_1 \int_z^\infty rz(r^2 + z^2)^{-5/2}\, dz = \frac{A_1}{2} r(r^2 + z^2)^{-3/2} \\
\sigma_\theta &= -\frac{1}{2}A_1 \int_z^\infty (r^2 + z^2)^{-3/2}\, dz = -\frac{A_1}{2}\left[\frac{1}{r^2} - \frac{z}{r^2}(r^2 + z^2)^{-1/2}\right]
\end{aligned}
\tag{210}
$$

On the plane $z = 0$ we find that the normal stress is zero, and the shearing stress is

$$
(\tau_{rz})_{z=0} = \frac{1}{2}\frac{A_1}{r^2} \tag{b}
$$

It appears now that by combining solutions (204) and (210), we can, by a suitable adjustment of the constants A and B, obtain such a stress distribution that the plane $z = 0$ will be free from stresses and a concentrated force P will act at the origin. From (a) and (b) we see that the shearing forces on the boundary plane are eliminated if

$$
-B(1 - 2\nu) + \frac{A_1}{2} = 0
$$

from which

$$
A_1 = 2B(1 - 2\nu)
$$

Substituting in expressions (210) and adding together the stresses (204) and (210), we find

$$
\begin{aligned}
\sigma_r &= B\left\{(1 - 2\nu)\left[\frac{1}{r^2} - \frac{z}{r^2}(r^2 + z^2)^{-1/2}\right] - 3r^2z(r^2 + z^2)^{-5/2}\right\} \\
\sigma_z &= -3Bz^3(r^2 + z^2)^{-5/2} \\
\sigma_\theta &= B(1 - 2\nu)\left[-\frac{1}{r^2} + \frac{z}{r^2}(r^2 + z^2)^{-1/2} + z(r^2 + z^2)^{-3/2}\right] \\
\tau_{rz} &= -3Brz^2(r^2 + z^2)^{-5/2}
\end{aligned}
\tag{c}
$$

This stress distribution satisfies the boundary conditions, since $\sigma_z = \tau_{rz} = 0$ for $z = 0$. It remains now to determine the constant B so that the forces distributed over a hemispherical surface with center at the origin are statically equivalent to the force P acting along the z axis. Considering the equilibrium of an element such as shown in Fig. 203, the component in the z direction of forces on the hemispherical surface is

$$\bar{Z} = -(\tau_{rz} \sin \psi + \sigma_z \cos \psi) = 3Bz^2(r^2 + z^2)^{-2}$$

For determining B we obtain the equation

$$P = 2\pi \int_0^{\pi/2} \bar{Z} r (r^2 + z^2)^{1/2} \, d\psi = 6\pi B \int_0^{\pi/2} \cos^2 \psi \sin \psi \, d\psi = 2\pi B$$

from which

$$B = \frac{P}{2\pi}$$

Finally, substituting in (c) we obtain the following expressions for the stress components due to a normal force P acting on the plane boundary of a semi-infinite solid:

$$\begin{aligned}
\sigma_r &= \frac{P}{2\pi} \left\{ (1 - 2\nu) \left[\frac{1}{r^2} - \frac{z}{r^2} (r^2 + z^2)^{-1/2} \right] - 3r^2 z (r^2 + z^2)^{-5/2} \right\} \\
\sigma_z &= -\frac{3P}{2\pi} z^3 (r^2 + z^2)^{-5/2} \\
\sigma_\theta &= \frac{P}{2\pi} (1 - 2\nu) \left\{ -\frac{1}{r^2} + \frac{z}{r^2} (r^2 + z^2)^{-1/2} + z (r^2 + z^2)^{-3/2} \right\} \\
\tau_{rz} &= -\frac{3P}{2\pi} r z^2 (r^2 + z^2)^{-5/2}
\end{aligned} \tag{211}$$

This solution is the three-dimensional analog of the solution for the semi-infinite plate (see Art. 36).

If we take an elemental area mn perpendicular to the z axis (Fig. 207), the ratio of the normal and shearing components of the stress on this element, from Eqs. (211), is

$$\frac{\sigma_z}{\tau_{rz}} = \frac{z}{r} \tag{d}$$

Hence the direction of the resultant stress passes through the origin O. The magnitude of this resultant stress is

$$S = \sqrt{\sigma_z^2 + \tau_{rz}^2} = \frac{3P}{2\pi} \frac{z^2}{(r^2 + z^2)^2} = \frac{3P}{2\pi} \frac{\cos^2 \psi}{(r^2 + z^2)} \tag{212}$$

The stress is thus inversely proportional to the square of the distance from the point of application of the load P. Imagine a spherical surface of diameter d, tangent to the plane $z = 0$ at the origin O. For each point of this surface,

$$r^2 + z^2 = d^2 \cos^2 \psi \tag{e}$$

Substituting in (212), we conclude that for points of the sphere the resultant stress on horizontal planes is constant and equal to $3P/2\pi d^2$.

Consider now the displacements produced in the semi-infinite solid by the load P. From Eqs. (187) for strain components,

$$u = \epsilon_\theta r = \frac{r}{E}[\sigma_\theta - \nu(\sigma_r + \sigma_z)]$$

Substituting the values for the stress components from Eqs. (211),

$$u = \frac{(1 - 2\nu)(1 + \nu)P}{2\pi E r} \left[z(r^2 + z^2)^{-1/2} - 1 + \frac{1}{1 - 2\nu} r^2 z (r^2 + z^2)^{-3/2} \right] \tag{213}$$

For determining vertical displacements w, we have, from Eqs. (187),

$$\frac{\partial w}{\partial z} = \epsilon_z = \frac{1}{E}[\sigma_z - \nu(\sigma_r + \sigma_\theta)]$$

$$\frac{\partial w}{\partial r} = \gamma_{rz} - \frac{\partial u}{\partial z} = \frac{2(1 + \nu)\tau_{rz}}{E} - \frac{\partial u}{\partial z}$$

Substituting for the stress components, and for the displacement u the values found above, we obtain

$$\frac{\partial w}{\partial z} = \frac{P}{2\pi E}\{3(1 + \nu)r^2 z(r^2 + z^2)^{-5/2} - [3 + \nu(1 - 2\nu)]z(r^2 + z^2)^{-3/2}\}$$

$$\frac{\partial w}{\partial r} = -\frac{P(1 + \nu)}{2\pi E}[2(1 - \nu)r(r^2 + z^2)^{-3/2} + 3rz^2(r^2 + z^2)^{-5/2}]$$

from which, by integration, omitting an arbitrary constant,

$$w = \frac{P}{2\pi E}[(1 + \nu)z^2(r^2 + z^2)^{-3/2} + 2(1 - \nu^2)(r^2 + z^2)^{-1/2}] \tag{214}$$

For the boundary plane ($z = 0$) the displacements are

$$(u)_{z=0} = -\frac{(1 - 2\nu)(1 + \nu)P}{2\pi E r} \qquad (w)_{z=0} = \frac{P(1 - \nu^2)}{\pi E r} \tag{215}$$

showing that the product wr is constant at the boundary. Hence the radii drawn from the origin on the boundary surface, after deformation, are hyperbolas with the asymptotes Or and Oz. At the origin the displacements and stresses become infinite. We must therefore imagine the material around the origin cut out by a hemispherical surface of small radius and the concentrated force P replaced by statically equivalent forces distributed over this surface in the manner required by the solution.

139 | Load Distributed over a Part of the Boundary of a Semi-infinite Solid

Having the solution for a concentrated force acting on the boundary of a semi-infinite solid, we can find the displacements and the stresses produced by a distributed load by superposition. Take, as a simple example, the case of a uniform load distributed over the area of a circle of radius a (Fig. 208), and consider the deflection, in the direction of the load, of a point M on the surface of the body at a distance r from the center of the circle. Taking a small element of the loaded area shown shaded in the figure, bounded by two radii including the angle $d\psi$ and two arcs of circle with the radii s and $s + ds$, all drawn from M, the load on this element is $qs\,d\psi\,ds$ and the corresponding deflection at M, from Eq. (215), is

$$\frac{(1-\nu^2)q}{\pi E}\frac{s\,d\psi\,ds}{s} = \frac{(1-\nu^2)q}{\pi E}\,d\psi\,ds$$

The total deflection is now obtained by double integration,

$$w = \frac{(1-\nu^2)q}{\pi E}\iint d\psi\,ds$$

Integrating with respect to s and taking into account the fact that the length of the chord mn is equal to $2\sqrt{a^2 - r^2\sin^2\psi}$, we find

$$w = \frac{4(1-\nu^2)q}{\pi E}\int_0^{\psi_1}\sqrt{a^2 - r^2\sin^2\psi}\,d\psi \tag{a}$$

in which ψ_1 is the maximum value of ψ, that is, the angle between r and the tangent to the circle. The calculation of the integral (a) is simplified by introducing, instead of the variable ψ, the variable angle θ. From the figure we have

$$a\sin\theta = r\sin\psi$$

from which

$$d\psi = \frac{a\cos\theta\,d\theta}{r\cos\psi} = \frac{a\cos\theta\,d\theta}{r\sqrt{1-(a^2/r^2)\sin^2\theta}}$$

Substituting in Eq. (a) and remembering that θ varies from 0 to $\pi/2$, when

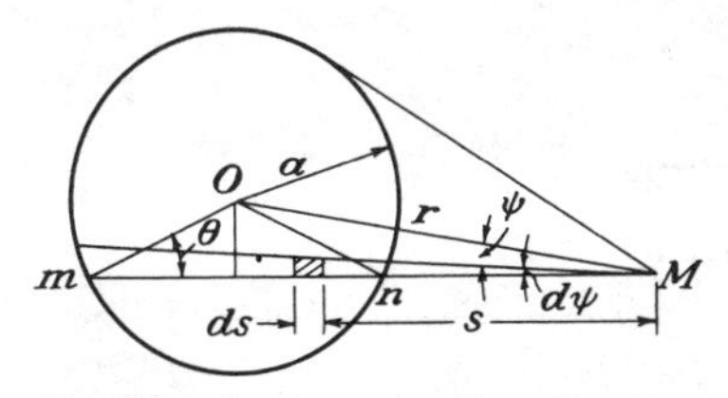

Fig. 208

ψ changes from 0 to ψ_1, we find

$$w = \frac{4(1-\nu^2)q}{\pi E} \int_0^{\pi/2} \frac{a^2 \cos^2 \theta \, d\theta}{r\sqrt{1-(a^2/r^2)\sin^2\theta}} = \frac{4(1-\nu^2)qr}{\pi E}$$

$$\left[\int_0^{\pi/2} \sqrt{1-(a^2/r^2)\sin^2\theta}\, d\theta - \left(1-\frac{a^2}{r^2}\right)\int_0^{\pi/2} \frac{d\theta}{\sqrt{1-(a^2/r^2)\sin^2\theta}}\right] \tag{216}$$

The integrals in this equation are known as *complete elliptic integrals*, and their values for any value of a/r can be taken from tables.[1]

To get the deflection at the boundary of the loaded circle, we take $r = a$ in Eq. (216) and find

$$(w)_{r=a} = \frac{4(1-\nu^2)qa}{\pi E} \tag{217}$$

If the point M is within the loaded area (Fig. 209*a*), we again consider the deflection produced by a shaded element on which the load $qs\, ds\, d\psi$ acts. Then the total deflection is

$$w = \frac{(1-\nu^2)q}{\pi E} \iint ds\, d\psi$$

The length of the chord mn is $2a \cos \theta$, and ψ varies from zero to $\pi/2$, so

$$w = \frac{4(1-\nu^2)q}{\pi E} \int_0^{\pi/2} a \cos \theta \, d\psi$$

or, since $a \sin \theta = r \sin \psi$, we have

$$w = \frac{4(1-\nu^2)qa}{\pi E} \int_0^{\pi/2} \sqrt{1-(r^2/a^2)\sin^2\psi}\, d\psi \tag{218}$$

Thus the deflection can easily be calculated for any value of the ratio r/a

[1] See, for instance, E. Jahnke, F. Emde, and F. Lösch, "Tables of Higher Functions," McGraw-Hill Book Company, New York, 1960.

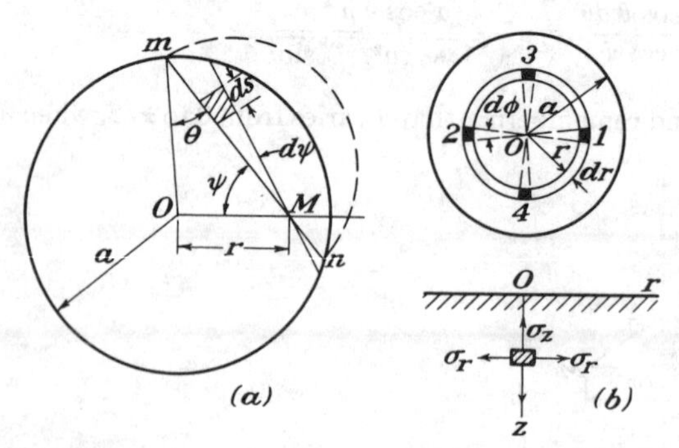

Fig. 209

by using tables of elliptic integrals. The maximum deflection occurs, of course, at the center of the circle. Substituting $r = 0$ in Eq. (218), we find

$$(w)_{\max} = \frac{2(1 - \nu^2)qa}{E} \tag{219}$$

Comparing this with the deflection at the boundary of the circle we find that the latter is $2/\pi$ times the maximum deflection.[1] It is interesting to note that for a given intensity of the load q the maximum deflection is not constant but increases in the same ratio as the radius of the loaded circle.[2]

By using superposition the stresses can also be calculated. Consider, for example, the stresses at a point on the z axis (Fig. 209b). The stress σ_z produced at such a point by a load distributed over a ring area of radius r and width dr is obtained by substituting in the second of Eqs. (211) $2\pi r\, drq$ instead of P. Then the stress σ_z produced by the uniform load distributed over the entire circular area of radius a is

$$\begin{aligned} \sigma_z = -\int_0^a 3qr\, dr z^3(r^2 + z^2)^{-5/2} &= qz^3 \left| (r^2 + z^2)^{-3/2} \right|_0^a \\ &= q\left[-1 + \frac{z^3}{(a^2 + z^2)^{3/2}}\right] \end{aligned} \tag{b}$$

This stress is equal to $-q$ at the surface of the body and gradually decreases with increase of distance z. In calculating the stresses σ_r and σ_θ at the same point, consider the two elements 1 and 2 of the loaded area (Fig. 209b) with the loads $qr\, d\phi\, dr$. The stresses produced by these two elemental loads at a point on the z axis, from the first and third of Eqs. (211), are

$$\left.\begin{aligned} d\sigma_r' &= \frac{qr\, d\phi\, dr}{\pi} \left\{ (1 - 2\nu)\left[\frac{1}{r^2} - \frac{z}{r^2}(r^2 + z^2)^{-1/2}\right] - 3r^2 z(r^2 + z^2)^{-5/2} \right\} \\ d\sigma_\theta' &= \frac{qr\, d\phi\, dr}{\pi}(1 - 2\nu)\left[-\frac{1}{r^2} + \frac{z}{r^2}(r^2 + z^2)^{-1/2} + z(r^2 + z^2)^{-3/2}\right] \end{aligned}\right\} \tag{c}$$

The normal stresses produced on the same planes by the elemental loads

[1] The solution of this problem was given by Boussinesq, *loc. cit.* See also H. Lamb, *Proc. London Math. Soc.*, vol. 34, p. 276, 1902; K. Terazawa, *J. Coll. Sci., Univ. Tokyo*, vol. 37, 1916; F. Schleicher, *Bauingenieur*, vol. 7, 1926, and *Bauingenieur*, vol. 14, p. 242, 1933. A complete investigation of this problem, also of the case in which the load is distributed over a rectangle, is given in the paper by A. E. H. Love, *Trans. Roy. Soc. (London)*, ser. A, vol. 228, 1929. Special properties of the deformation and stress in the general case are pointed out by S. Way, *J. Appl. Mech.*, vol. 7, p. A-147, 1940.

[2] This, and its generalization to a noncircular loaded area, follow from a simple dimensional analysis of the problem.

at points 3 and 4 are

$$
\begin{aligned}
d\sigma_r'' &= \frac{qr\,d\phi\,dr}{\pi}(1-2\nu)\left[-\frac{1}{r^2}+\frac{z}{r^2}(r^2+z^2)^{-1/2}+z(r^2+z^2)^{-3/2}\right] \\
d\sigma_\theta'' &= \frac{qr\,d\phi\,dr}{\pi}\left\{(1-2\nu)\left[\frac{1}{r^2}-\frac{z}{r^2}(r^2+z^2)^{-1/2}\right]-3r^2z(r^2+z^2)^{-5/2}\right\}
\end{aligned}
\tag{d}
$$

By summation of (*c*) and (*d*) we find that the four elemental loads, indicated in the figure, produce the stresses

$$
\begin{aligned}
d\sigma_r = d\sigma_\theta &= \frac{qr\,d\phi\,dr}{\pi}[(1-2\nu)z(r^2+z^2)^{-3/2}-3r^2z(r^2+z^2)^{-5/2}] \\
&= \frac{qr\,d\phi\,dr}{\pi}[-2(1+\nu)z(r^2+z^2)^{-3/2}+3z^3(r^2+z^2)^{-5/2}]
\end{aligned}
\tag{e}
$$

To get the stresses produced by the entire load uniformly distributed over the area of a circle of radius a we integrate expression (*e*) with respect to ϕ between the limits 0 and $\pi/2$, and with respect to r, from 0 to a. Then

$$
\begin{aligned}
\sigma_r = \sigma_\theta &= \frac{q}{2}\int_0^a[-2(1+\nu)z(r^2+z^2)^{-3/2}+3z^3(r^2+z^2)^{-5/2}]r\,dr \\
&= \frac{q}{2}\left[-(1+2\nu)+\frac{2(1+\nu)z}{\sqrt{a^2+z^2}}-\left(\frac{z}{\sqrt{a^2+z^2}}\right)^3\right]
\end{aligned}
\tag{f}
$$

For the point O, the center of the loaded circle, we find, from Eqs. (*b*) and (*f*),

$$
\sigma_z = -q \qquad \sigma_r = \sigma_\theta = -\frac{q(1+2\nu)}{2}
$$

Taking $\nu = 0.3$, we have $\sigma_r = \sigma_\theta = -0.8q$. The maximum shearing stress at the point O, on planes at 45° to the z axis, is equal to $0.1q$. Assuming that yielding of the material depends on the maximum shearing stress, it can be shown that the point O, considered above, is not the most unfavorable point on the z axis. The maximum shearing stress at any point on the z axis (Fig. 209*b*), from Eqs. (*b*) and (*f*), is

$$
\frac{1}{2}(\sigma_\theta-\sigma_z) = \frac{q}{2}\left[\frac{1-2\nu}{2}+(1+\nu)\frac{z}{\sqrt{a^2+z^2}}-\frac{3}{2}\left(\frac{z}{\sqrt{a^2+z^2}}\right)^3\right] \tag{g}
$$

This expression becomes a maximum when

$$
\frac{z}{\sqrt{a^2+z^2}} = \frac{1}{3}\sqrt{2(1+\nu)}
$$

from which

$$
z = a\sqrt{\frac{2(1+\nu)}{7-2\nu}} \tag{h}
$$

Substituting in expression (g),

$$\tau_{\max} = \frac{q}{2}\left[\frac{1-2\nu}{2} + \frac{2}{9}(1+\nu)\sqrt{2(1+\nu)}\right] \tag{k}$$

Assuming $\nu = 0.3$, we find, from Eqs. (h) and (k),

$$z = 0.638a \qquad \tau_{\max} = 0.33q$$

This shows that the maximum shearing stress for points on the z axis is at a certain depth, approximately equal to two-thirds of the radius of the loaded circle, and the magnitude of this maximum is about one-third of the applied uniform pressure q.

For the case of a uniform pressure distributed over the surface of a square with sides $2a$, the maximum deflection at the center is

$$w_{\max} = \frac{8}{\pi}\ln(\sqrt{2}+1)\frac{qa(1-\nu^2)}{E} = 2.24\frac{qa(1-\nu^2)}{E} \tag{220}$$

The deflection at the corners of the square is only half the deflection at the center, and the average deflection is

$$w_{\text{av}} = 1.90\frac{qa(1-\nu^2)}{E} \tag{221}$$

Analogous calculations have also been made for uniform pressure distribution over rectangles with various ratios, $\alpha = a/b$, of the sides. All the results can be put in the form[1]

$$w_{\text{av}} = m\frac{P(1-\nu^2)}{E\sqrt{A}} \tag{222}$$

in which m is a numerical factor depending on α, A is the magnitude of this area, and P is the total load. Several values of the factor m are given

Table of Factors m in Eq. (222)

	Circle	Square	Rectangles with various $\alpha = \frac{a}{b}$					
			1.5	2	3	5	10	100
$m =$	0.96	0.95	0.94	0.92	0.88	0.82	0.71	0.37

in the table. It will be seen that for a given load P and area A deflections increase when the ratio of the perimeter of the loaded area to the area

[1] See Schleicher, *loc. cit.*

decreases. Equation (222) is sometimes used in discussing deflections of foundations[1] of engineering structures. In order to have equal deflections of various portions of the structure, the average pressure on the foundation must be in a certain relation to the shape and the magnitude of the loaded area.

It was assumed in the previous discussion that the load was given, and we found the displacements produced. Consider now the case when the displacements are given and it is necessary to find the corresponding distribution of pressures on the boundary plane. Take, as an example, the case of an absolutely rigid die in the form of a circular cylinder pressed against the plane boundary of a semi-infinite elastic solid. In such a case the displacement w is constant over the circular base of the die. The distribution of pressures is not constant and its intensity is given by the equation[2]

$$q = \frac{P}{2\pi a \sqrt{a^2 - r^2}} \tag{223}$$

in which P is the total load on the die, a the radius of the die, and r the distance from the center of the circle on which the pressure acts. This distribution of pressures is obviously not uniform and its smallest value is at the center ($r = 0$), where

$$q_{\min} = \frac{P}{2\pi a^2}$$

i.e., it is equal to half the average pressure on the circular area of contact. At the boundary of the same area ($r = a$) the pressure becomes infinite. In actual cases we shall have yielding of material along the boundary. This yielding, however, is of local character and does not substantially affect the distribution of pressures (223) at points some distance from the boundary of the circle.

The displacement of the die is given by the equation

$$w = \frac{P(1 - \nu^2)}{2aE} \tag{224}$$

We see that for a given value of the average unit pressure on the boundary plane, the deflection is not constant but increases in the same ratio as the radius[3] of the die.

For comparison, we give also the average deflection for the case of a

[1] See Schleicher, *loc. cit.*

[2] This solution was given by Boussinesq, *loc. cit.*

[3] Footnote 2, p. 405, applies again here.

uniform distribution of pressures [Eq. (218)]:

$$w_{av} = \frac{\int_0^a w 2\pi r\, dr}{\pi a^2} = \frac{16}{3\pi^2} \frac{P(1-\nu^2)}{aE} = 0.54 \frac{P(1-\nu^2)}{aE} \tag{225}$$

This average deflection is not very much different from the displacement (224) for an absolutely rigid die. Many solutions for noncircular dies are available,[1] including some for the dynamical problem of the moving die.

140 | Pressure between Two Spherical Bodies in Contact

The results of the previous article can be used in discussing the pressure distribution between two bodies in contact.[2] We assume that at the point of contact these bodies have spherical surfaces with the radii R_1 and R_2 (Fig. 210). If there is no pressure between the bodies, we have contact at one point O. The distances from the plane tangent at O of points such as M and N, on a meridian section of the spheres at a very small distance[3] r from the axes z_1 and z_2, can be represented with sufficient accuracy by the formulas

$$z_1 = \frac{r^2}{2R_1} \qquad z_2 = \frac{r^2}{2R_2} \tag{a}$$

and the mutual distance between these points is

$$z_1 + z_2 = r^2\left(\frac{1}{2R_1} + \frac{1}{2R_2}\right) = \frac{r^2(R_1 + R_2)}{2R_1R_2} \tag{b}$$

In the particular case of contact between a sphere and a plane (Fig. 211*a*), $1/R_1$ is zero, and Eq. (*b*) for the distance MN gives

$$\frac{r^2}{2R_2} \tag{c}$$

[1] The section on Elasticity in J. N. Goodier and P. G. Hodge, "Elasticity and Plasticity," vol. 1 of "Surveys in Applied Mathematics," New York, 1958, gives an outline of some of these, contained in the Russian book "Contact Problems of the Theory of Elasticity," by L. A. Galin. An English translation of this book by Mrs. H. Moss, edited by I. N. Sneddon, has been issued by the Department of Mathematics and Engineering Research, North Carolina State College, Raleigh, N.C., 1961.

[2] This problem was solved by H. Hertz, *J. Math.* (*Crelle's J.*), vol. 92, 1881. See also H. Hertz, "Gesammelte Werke," vol. 1, p. 155, Leipzig, 1895. The contact is supposed frictionless. When there is no relative slip between the surfaces as contact develops, the problem is changed unless the two spheres are exactly alike. A no-slip solution for unlike spheres was given by L. E. Goodman (1962). Tangential force and twisting couple at the contact were treated by R. D. Mindlin (1949). References to these papers, and further solved problems, are given in J. L. Lubkin, Contact Problems, chap. 42 of W. Flügge (ed.), "Handbook of Engineering Mechanics," McGraw-Hill Book Company, New York, 1962.

[3] r is small in comparison with R_1 and R_2.

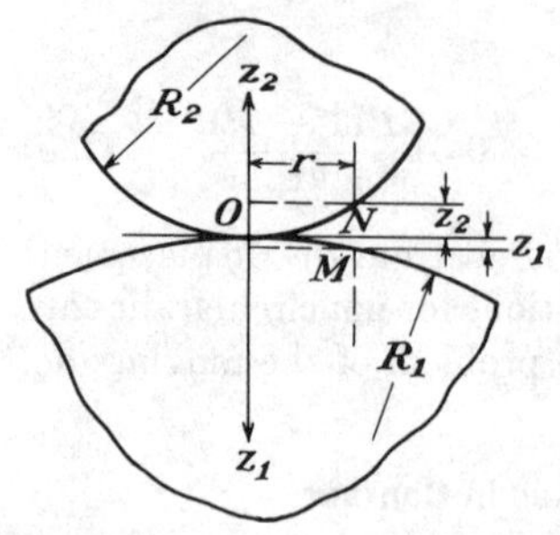

Fig. 210

In the case of contact between a ball and a spherical seat (Fig. 211*b*) R_1 is negative in Eq. (*b*), and

$$z_2 - z_1 = \frac{r^2(R_1 - R_2)}{2R_1R_2} \tag{c'}$$

If the bodies are pressed together along the normal at O by a force P, there will be a local deformation near the point of contact producing contact over a small surface with a circular boundary, called the *surface of contact.* Assuming that the radii of curvature R_1 and R_2 are very large in comparison with the radius of the boundary of the surface of contact, we can apply, in discussing local deformation, the results obtained before for semi-infinite bodies. Let w_1 denote the displacement due to the local deformation in the direction z_1 of a point such as M on the surface of the lower ball (Fig. 210) and w_2 denote the same displacement in the direction z_2 for a point such as N of the upper ball. Holding the tangent plane at O immovable during local compression, any two points of the bodies on the axes z_1 and z_2 at large distances[1] from O will approach each other by a certain amount α, and the distance between two points such as M and N (Fig. 210) will diminish by $\alpha - (w_1 + w_2)$. If finally, due to local com-

[1] The distances are such that deformations due to the compression at these points can be neglected.

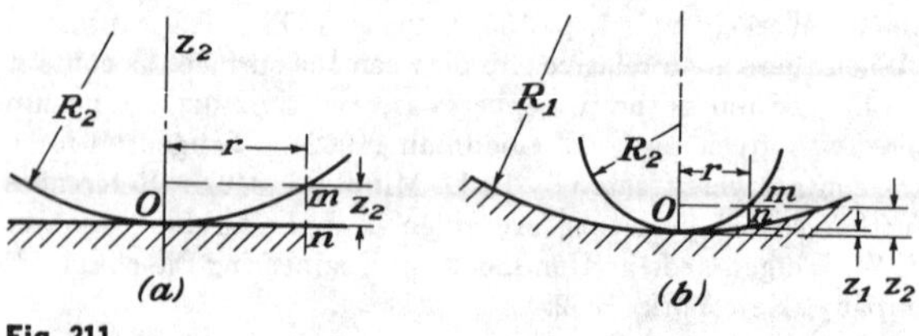

Fig. 211

pression, the points M and N come inside the surface of contact, we have

$$\alpha - (w_1 + w_2) = z_1 + z_2 = \beta r^2 \tag{d}$$

in which β is a constant depending on the radii R_1 and R_2 and given by Eq. (b), (c), or (c'). Thus from geometrical considerations we find for any point of the surface of contact,

$$w_1 + w_2 = \alpha - \beta r^2 \tag{e}$$

Let us now consider local deformations. From the condition of symmetry the intensity q of pressure between the bodies in contact and the corresponding deformation are symmetrical with respect to the center O of the surface of contact. Taking Fig. 209a to represent the surface of contact, and M as a point on the surface of contact of the lower ball, the displacement w_1 of this point, from the previous article, is

$$w_1 = \frac{(1 - \nu_1^2)}{\pi E_1} \iint q \, ds \, d\psi \tag{f}$$

in which ν_1 and E_1 are the elastic constants for the lower ball, and the integration is extended over the entire area of contact. An analogous formula is obtained also for the upper ball. Then

$$w_1 + w_2 = (k_1 + k_2) \iint q \, ds \, d\psi \tag{g}$$

in which

$$k_1 = \frac{1 - \nu_1^2}{\pi E_1} \qquad k_2 = \frac{1 - \nu_2^2}{\pi E_2} \tag{226}$$

From Eqs. (e) and (g),

$$(k_1 + k_2) \iint q \, ds \, d\psi = \alpha - \beta r^2 \tag{h}$$

Thus we must find an expression for q to satisfy Eq. (h). It will now be shown that this requirement is satisfied by assuming that the distribution of pressures q over the contact surface is represented by the ordinates of a hemisphere of radius a constructed on the surface of contact. If q_0 is the pressure at the center O of the surface of contact, then

$$q_0 = ka$$

in which $k = q_0/a$ is a constant factor representing the scale of our representation of the pressure distribution. Along a chord mn the pressure q varies, as indicated in Fig. 209 by the dotted semicircle. Performing the integration along this chord we find

$$\int q \, ds = \frac{q_0}{a} A$$

in which A is the area of the semicircle indicated by the dotted line and is equal to $\frac{1}{2}\pi\,(a^2 - r^2 \sin^2 \psi)$. Substituting in Eq. ($h$), we find that

$$\frac{\pi(k_1 + k_2)q_0}{a} \int_0^{\pi/2} (a^2 - r^2 \sin^2 \psi)\, d\psi = \alpha - \beta r^2$$

or

$$(k_1 + k_2)\,\frac{q_0 \pi^2}{4a}\,(2a^2 - r^2) = \alpha - \beta r^2$$

This equation will be fulfilled for any value of r, and hence the assumed pressure distribution is the correct one if the following relations exist for displacement α and the radius a of the surface of contact:

$$\alpha = (k_1 + k_2) q_0 \frac{\pi^2 a}{2}$$
$$a = (k_1 + k_2)\,\frac{\pi^2 q_0}{4\beta} \tag{227}$$

The value of the maximum pressure q_0 is obtained by equating the sum of the pressures over the contact area to the compressive force P. For the hemispherical pressure distribution this gives

$$\frac{q_0}{a}\,\frac{2}{3}\,\pi a^3 = P$$

from which

$$q_0 = \frac{3P}{2\pi a^2} \tag{228}$$

i.e., the maximum pressure is $1\frac{1}{2}$ times the average pressure on the surface of contact. Substituting in Eqs. (227) and taking, from Eq. (b),

$$\beta = \frac{R_1 + R_2}{2R_1R_2}$$

we find for two balls in contact (Fig. 207)

$$a = \sqrt[3]{\frac{3\pi}{4}\,\frac{P(k_1 + k_2)R_1R_2}{R_1 + R_2}}$$
$$\alpha = \sqrt[3]{\frac{9\pi^2}{16}\,\frac{P^2(k_1 + k_2)^2(R_1 + R_2)}{R_1R_2}} \tag{229}$$

Assuming that both balls have the same elastic properties and taking $\nu = 0.3$, these become

$$a = 1.109\sqrt[3]{\frac{P}{E}\,\frac{R_1R_2}{R_1 + R_2}}$$
$$\alpha = 1.23\sqrt[3]{\frac{P^2}{E^2}\,\frac{R_1 + R_2}{R_1R_2}} \tag{230}$$

The corresponding maximum pressure is

$$q_0 = \frac{3}{2}\frac{P}{\pi a^2} = 0.388\sqrt[3]{PE^2\frac{(R_1+R_2)^2}{R_1^2R_2^2}} \tag{231}$$

For a ball pressed into a plane surface, and the same elastic constants for both bodies, we find by substituting $1/R_1 = 0$ in Eqs. (230) and (231),

$$a = 1.109\sqrt[3]{\frac{PR_2}{E}} \qquad \alpha = 1.23\sqrt[3]{\frac{P^2}{E^2R_2}} \qquad q_0 = 0.388\sqrt[3]{\frac{PE^2}{R_2^2}} \tag{232}$$

By taking R_1 negative we can write down also equations for a ball in a spherical seat (Fig. 211*b*).

Having the magnitude of the surface of contact and the pressures acting on it, the stresses can be calculated by using the method developed in the previous article.[1] The results of these calculations for points along the axes Oz_1 and Oz_2 are shown in Fig. 212. The maximum pressure q_0 at the center of the surface of contact is taken as a unit of stress. In measuring the distances along the z axis, the radius a of the surface of contact is taken as the unit. The greatest stress is the compressive stress σ_z at the center of the surface of contact, but the two other principal stresses σ_r and σ_θ, at the same point, are equal to $\frac{1}{2}(1 + 2\nu)\,\sigma_z$. Hence the maximum shearing stress, on which the yielding of such material as steel depends, is comparatively small at this point. The point with maximum shearing stress is on the z axis at a depth equal to about half of the

[1] Such calculations were made by A. N. Dinnik, *Bull. Polytech. Inst., Kiew*, 1909. See also M. T. Huber, *Ann. Physik*, vol. 14, 1904, p. 153; S. Fuchs, *Physik. Z.*, vol. 14, p. 1282, 1913; M. C. Huber and S. Fuchs, *Physik. Z.*, vol. 15, p. 298, 1914; W. B. Morton and L. J. Close, *Phil. Mag.*, vol. 43, p. 320, 1922.

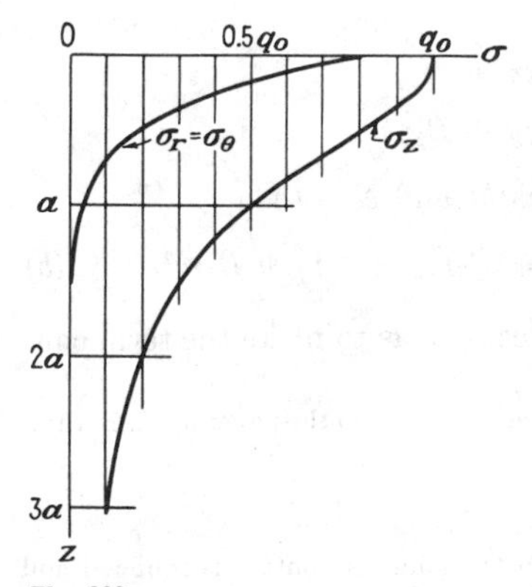

Fig. 212

radius of the surface of contact. This point must be considered as the weakest point in such material as steel. The maximum shearing stress at this point (for $\nu = 0.3$) is about $0.31q_0$.

In the case of brittle materials failure is produced by maximum tensile stress. This stress occurs at the circular boundary of the surface of contact. It acts in a radial direction and has the magnitude

$$\sigma_r = \frac{(1 - 2\nu)}{3} q_0$$

The other principal stress, acting in the circumferential direction, is numerically equal to the above radial stress but of opposite sign. Hence, along the boundary of the surface of contact, where normal pressure on the surface becomes equal to zero, we have pure shear of the amount $q_0(1 - 2\nu)/3$. Taking $\nu = 0.3$, this shear becomes equal to $0.133q_0$. This stress is much smaller than the maximum shearing stress calculated above, but it is larger than the shearing stress at the center of the surface of contact, where the normal pressure is the largest.

Many experiments have been made that verify the theoretical results of Hertz for materials that follow Hooke's law and stress within the elastic limit.[1]

141 | Pressure between Two Bodies in Contact. More General Case[2]

The general case of compression of elastic bodies in contact may be treated in the same manner as the case of spherical bodies discussed in the previous article. Consider the tangent plane at the point of contact O as the xy plane (Fig. 210). The surfaces of the bodies near the point of contact, by neglecting small quantities of higher order, can be represented by the equations[3]

$$\begin{aligned} z_1 &= A_1x^2 + A_2xy + A_3y^2 \\ z_2 &= B_1x^2 + B_2xy + B_3y^2 \end{aligned} \tag{a}$$

The distance between two points such as M and N is then

$$z_1 + z_2 = (A_1 + B_1)x^2 + (A_2 + B_2)xy + (A_3 + B_3)y^2 \tag{b}$$

We can always take for x and y such directions as to make the term con-

[1] References to the corresponding literature can be found in the paper by G. Berndt, *Z. Tech. Physik*, vol. 3, p. 14, 1922. See also "Handbuch der Physikalischen und Technischen Mechanik," vol. 3, p. 120.

[2] This theory is due to Hertz, *loc. cit.*

[3] It is assumed that the surface adjacent to the point of contact is rounded and may be considered as a surface of the second degree.

taining the product xy disappear. Then

$$z_1 + z_2 = Ax^2 + By^2 \tag{c}$$

in which A and B are constants depending on the magnitudes of the principal curvatures of the surfaces in contact and on the angle between the planes of principal curvatures of the two surfaces. If R_1 and R_1' denote the principal radii of curvature at the point of contact of one of the bodies, and R_2 and R_2' those of the other,[1] and ψ the angle between the normal planes containing the curvatures $1/R_1$ and $1/R_2$, then the constants A and B are determined from the equations

$$\begin{aligned} A + B &= \frac{1}{2}\left(\frac{1}{R_1} + \frac{1}{R_1'} + \frac{1}{R_2} + \frac{1}{R_2'}\right) \\ B - A &= \frac{1}{2}\left[\left(\frac{1}{R_1} - \frac{1}{R_1'}\right)^2 + \left(\frac{1}{R_2} - \frac{1}{R_2'}\right)^2 \right. \\ &\qquad \left. + 2\left(\frac{1}{R_1} - \frac{1}{R_1'}\right)\left(\frac{1}{R_2} - \frac{1}{R_2'}\right)\cos 2\psi\right]^{1/2} \end{aligned} \tag{d}$$

A and B in Eq. (c) are both positive since $z_1 + z_2$ must be positive. It can therefore be concluded that all points with the same mutual distance $z_1 + z_2$ lie on one ellipse. Hence, if we press the bodies together in the direction of the normal to the tangent plane at O, the surface of contact will have an elliptical boundary.

Let α, w_1, w_2 have the same meaning as in the previous article. Then, for points on the surface of contact, we have

$$\begin{aligned} w_1 + w_2 + z_1 + z_2 &= \alpha \\ \text{or} \qquad w_1 + w_2 &= \alpha - Ax^2 - By^2 \end{aligned} \tag{e}$$

This is obtained from geometrical considerations. Consider now the local deformation at the surface of contact. Assuming that this surface is very small and applying Eq. (215), obtained for semi-infinite bodies, the sum of the displacements w_1 and w_2 for points of the surface of contact is

$$w_1 + w_2 = \left(\frac{1 - \nu_1^2}{\pi E_1} + \frac{1 - \nu_2^2}{\pi E_2}\right) \iint \frac{q\, dA}{r} \tag{f}$$

where $q\, dA$ is the pressure acting on an infinitely small element of the surface of contact and r is the distance of this element from the point under consideration. The integration must be extended over the entire sur-

[1] The curvature of a body is considered positive if the radius of curvature at the point enters the material there. In Fig. 210 the curvatures of the bodies are both positive. In Fig. 211 the spherical seat has a negative curvature.

face of contact. Using notations (226), we obtain, from (*e*) and (*f*),

$$(k_1 + k_2) \iint \frac{q\,dA}{r} = \alpha - Ax^2 - By^2 \tag{g}$$

The problem now is to find a distribution of pressures q to satisfy Eq. (*g*). Hertz showed that this requirement is satisfied by assuming that the intensity of pressures q over the surface of contact is represented by the ordinates of a semi-ellipsoid constructed on the surface of contact. The maximum pressure is then clearly at the center of the surface of contact. Denoting it by q_0, and denoting by a and b the semiaxes of the elliptic boundary of the surface of contact, the magnitude of the maximum pressure is obtained from the equation

$$P = \iint q\,dA = \tfrac{2}{3}\pi ab q_0$$

from which

$$q_0 = \frac{3}{2}\frac{P}{\pi ab} \tag{233}$$

We see that the maximum pressure is $1\frac{1}{2}$ times the average pressure on the surface of contact. To calculate this pressure we must know the magnitudes of the semiaxes a and b. From an analysis analogous to that used for spherical bodies we find that

$$a = m\sqrt[3]{\frac{3\pi}{4}\frac{P(k_1 + k_2)}{(A + B)}}$$
$$b = n\sqrt[3]{\frac{3\pi}{4}\frac{P(k_1 + k_2)}{(A + B)}} \tag{234}$$

in which $A + B$ is determined from Eqs. (*d*) and the coefficients m and n are numbers depending on the ratio $(B - A):(A + B)$. Using the notation

$$\cos\theta = \frac{B - A}{A + B} \tag{h}$$

the values of m and n for various values of θ are given in the table below.[1]

θ =	30°	35°	40°	45°	50°	55°	60°	65°	70°	75°	80°	85°	90°
m =	2.731	2.397	2.136	1.926	1.754	1.611	1.486	1.378	1.284	1.202	1.128	1.061	1.000
n =	0.493	0.530	0.567	0.604	0.641	0.678	0.717	0.759	0.802	0.846	0.893	0.944	1.000

[1] The table is taken from the paper by H. L. Whittemore and S. N. Petrenko, *Natl. Bur. Std. Tech. Paper* 201, 1921. An extension for $0 < \theta < 30°$ is given by M. Kornhauser, *J. Appl. Mech.*, vol. 18, pp. 251–252, 1951.

Considering, for instance, the contact of a wheel with a cylindrical rim of radius $R_1 = 15.8$ in. and of a rail with the radius of the head $R_2 = 12$ in., we find, by substituting $1/R_1' = 1/R_2' = 0$ and $\psi = \pi/2$ into Eqs. (*d*),

$$A + B = 0.0733 \qquad B - A = 0.0099 \qquad \cos\theta = 0.135 \qquad \theta = 82°15'$$

Then, by interpolation, we find from the above table that

$$m = 1.098 \qquad n = 0.918$$

Substituting in Eqs. (234) and taking $E = 30.10^6$ psi and $\nu = 0.25$,[1] we find

$$a = 0.00946\sqrt[3]{P} \qquad b = 0.00792\sqrt[3]{P}$$

For a load $P = 1{,}000$ lb,

$$a = 0.0946 \text{ in.} \qquad b = 0.0792 \text{ in.} \qquad \text{area of contact } \pi ab = 0.0236 \text{ sq in.}$$

and the maximum pressure at the center is

$$q_0 = \frac{3}{2}\frac{P}{\pi ab} = 63{,}600 \text{ psi}$$

Knowing the distribution of pressure, the stresses at any point can be calculated.[2] It was shown in this manner that the point of maximum shearing stress is on the z axis at a certain small depth z_1, depending on the magnitude of the semiaxes a and b. For instance: $z_1 = 0.47a$, when $b/a = 1$; and $z_1 = 0.24a$, when $b/a = 0.34$. The corresponding values of maximum shearing stress (for $\nu = 0.3$) are $\tau_{max} = 0.31q_0$ and $\tau_{max} = 0.32q_0$, respectively.

Considering points on the elliptical surface of contact and taking the x and y axes in the direction of the semiaxes a and b, respectively, the principal stresses at the center of the surface of contact are

$$\begin{aligned} \sigma_x &= -2\nu q_0 - (1 - 2\nu)q_0\frac{b}{a+b} \\ \sigma_y &= -2\nu q_0 - (1 - 2\nu)q_0\frac{a}{a+b} \qquad (i) \\ \sigma_z &= -q_0 \end{aligned}$$

[1] If ν is increased from 0.25 to 0.30 the semiaxes (234) decrease about 1 percent and the maximum pressure q_0 increases about 2 percent.

[2] Such investigations have been made by N. M. Belajef; see *Bull. Inst. Eng. Ways Communication*, St. Petersburg, 1917, and "Memoirs on Theory of Structures," St. Petersburg, 1924; see also H. R. Thomas and V. A. Hoersch, *Univ. Illinois Eng. Exptl. Sta. Bull.* 212, 1930, and G. Lundberg and F. K. G. Odqvist, *Proc. Ingeniörs Vetenskaps Akad.*, no. 116, Stockholm, 1932. A compilation of formulas and curves is given in C. Lipson and R. C. Juvinall, "Handbook of Stress and Strength," chap. 7, The Macmillan Company, New York, 1963.

For the ends of the axes of the ellipse we find $\sigma_x = -\sigma_y$ and $\tau_{xy} = 0$. The tensile stress in the radial direction is equal to the compressive stress in the circumferential direction. Thus at these points there exists pure shear. The magnitude of this shear for the ends of the major axis ($x = \pm a$, $y = 0$) is

$$\tau = (1 - 2\nu) q_0 \frac{\beta}{e^2} \left(\frac{1}{e} \operatorname{arctanh} e - 1 \right) \tag{j}$$

and for the ends of the minor axis ($x = 0$, $y = \pm b$) is

$$\tau = (1 - 2\nu) q_0 \frac{\beta}{e^2} \left(1 - \frac{\beta}{e} \arctan \frac{e}{\beta} \right) \tag{k}$$

where $\beta = b/a$, $e = (1/a)\sqrt{a^2 - b^2}$. When b approaches a and the boundary of the surface of contact approaches a circular shape, the stresses given by (i), (j), and (k) approach the stresses given in the previous article for the case of compression of balls.

A more detailed investigation of stresses for all points in the surface of contact shows[1] that for $e < 0.89$ the maximum shearing stress is given by Eq. (j). For $e > 0.89$ the maximum shearing stress is at the center of the ellipse and can be calculated from Eqs. (i) above.

By increasing the ratio a/b, we obtain narrower and narrower ellipses of contact, and at the limit $a/b = \infty$ we arrive at the case of contact of two cylinders with parallel axes.[2] The surface of contact is now a narrow rectangle. The distribution of pressure q along the width of the surface of contact (Fig. 213) is represented by a semi-ellipse. If the x axis is perpendicular to the plane of the figure, b is half the width of the surface of contact, and P' the load per unit length of the surface of contact, we obtain, from the semi-elliptic pressure distribution,

$$P' = \tfrac{1}{2}\pi b q_0$$

from which

$$q_0 = \frac{2P'}{\pi b} \tag{235}$$

The investigation of local deformation gives for the quantity b the expression

$$b = \sqrt{\frac{4P'(k_1 + k_2)R_1R_2}{R_1 + R_2}} \tag{236}$$

in which R_1 and R_2 are the radii of the cylinders and k_1 and k_2 are constants defined by Eqs. (226). If both cylinders are of the same material

[1] See Belajef, *loc. cit.*

[2] A direct derivation of this case, with consideration of tangential force at the contact, is given by H. Poritsky, *J. Appl. Mech.*, vol. 17, p. 191, 1950.

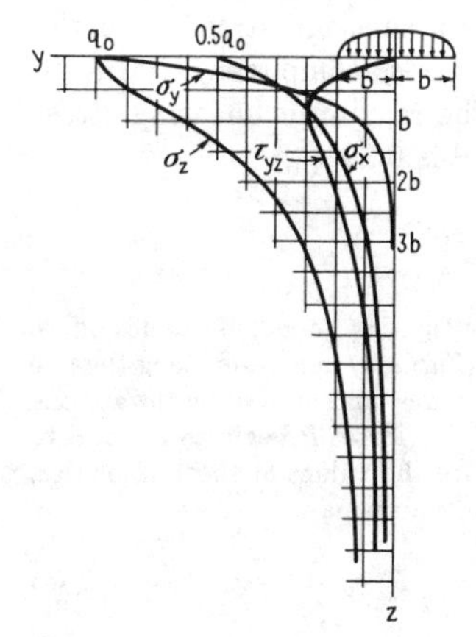

Fig. 213

and $\nu = 0.3$, then

$$b = 1.52 \sqrt{\frac{P'R_1R_2}{E(R_1 + R_2)}} \tag{237}$$

In the case of two equal radii, $R_1 = R_2 = R$,

$$b = 1.08 \sqrt{\frac{P'R}{E}} \tag{238}$$

For the case of contact of a cylinder with a plane surface,

$$b = 1.52 \sqrt{\frac{P'R}{E}} \tag{239}$$

Substituting b from Eq. (236) into Eq. (235), we find

$$q_0 = \sqrt{\frac{P'(R_1 + R_2)}{\pi^2(k_1 + k_2)R_1R_2}} \tag{240}$$

If the materials of both cylinders are the same and $\nu = 0.3$,

$$q_0 = 0.418 \sqrt{\frac{P'E(R_1 + R_2)}{R_1R_2}} \tag{241}$$

In the case of contact of a cylinder with a plane surface,

$$q_0 = 0.418 \sqrt{\frac{P'E}{R}} \tag{242}$$

Knowing q_0 and b, the stress at any point can be calculated. These calculations show[1] that the point with maximum shearing stress is on the z axis at a certain depth. The variation of stress components with the depth, for $\nu = 0.3$, is shown in Fig. 213. The maximum shearing stress is at the depth $z_1 = 0.78b$ and its magnitude[2] is $0.304q_0$.

142 | Impact of Spheres

The results of the last two articles can be used in investigating impact of elastic bodies. Consider, as an example, the impact of two spheres (Fig. 214) in motion along the line joining the centers of the spheres. As soon as the spheres, in their motion toward one another, come in contact at a point O, the compressive forces P begin to act and to change the velocities of the spheres. If v_1 and v_2 are the values of these velocities, their rates of change during impact are given by the equations

$$m_1 \frac{dv_1}{dt} = -P \qquad m_2 \frac{dv_2}{dt} = -P \tag{a}$$

in which m_1 and m_2 denote the masses of the spheres. Let α be the distance the two spheres approach one another due to local compression at O. Then the velocity of this approach is

$$\dot{\alpha} = v_1 + v_2$$

and we find, from Eqs. (a), that

$$\ddot{\alpha} = -P \frac{m_1 + m_2}{m_1 m_2} \tag{b}$$

For spheres of not very different sizes and properties, the duration of impact, i.e., the time during which the spheres remain in contact, is very long in comparison with the period of lowest mode of vibration of the spheres.[3] Vibrations can therefore be neglected, and it can be assumed that Eq. (229), which was established for statical

[1] See Belajef, *loc. cit.*

[2] The Hertz contact theory has found extensive practical application. Calculated contact pressure approaching 10^6 psi has been reported. See, for instance, J. B. Bidwell (ed.), "Rolling Contact Phenomena," Elsevier Publishing Company, Amsterdam, 1962, especially pp. 430 and 406.

[3] Lord Rayleigh, *Phil. Mag.*, ser. 6, vol. 11, p. 283, 1906. If the spheres are very different, especially if one of them is regarded as of infinite size, the period is very long, or infinite, and the statement cannot hold. Nevertheless, the measured durations agree very well with the results of this quasi-statical theory. See, for instance, J. N. Goodier, W. E. Jahsman, and E. A. Ripperger, *J. Appl. Mech.*, vol. 26, p. 3, 1959.

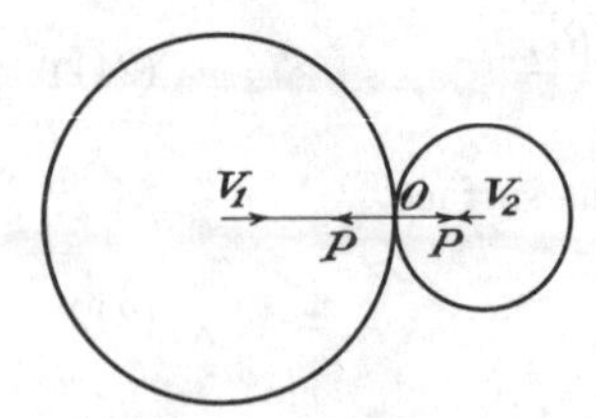

Fig. 214

conditions, holds during impact. Using the notations

$$n = \sqrt{\frac{16}{9\pi^2}\frac{R_1R_2}{(k_1 + k_2)^2(R_1 + R_2)}} \qquad n_1 = \frac{m_1 + m_2}{m_1m_2} \tag{c}$$

we find, from (229)

$$P = n\alpha^{3/2} \tag{d}$$

and Eq. (*b*) becomes

$$\ddot{\alpha} = -nn_1\alpha^{3/2} \tag{e}$$

Multiplying both sides of this equation by $\dot{\alpha}$,

$$\tfrac{1}{2}d(\dot{\alpha})^2 = -nn_1\alpha^{3/2}\,d\alpha$$

from which, by integration,

$$\tfrac{1}{2}(\dot{\alpha}^2 - v^2) = -\tfrac{2}{5}nn_1\alpha^{5/2} \tag{f}$$

where v is the velocity of approach of the two spheres at the beginning of impact. If we substitute $\dot{\alpha} = 0$ in this equation, we find the value of the approach at the instant of maximum compression, α_1, as

$$\alpha_1 = \left(\frac{5}{4}\frac{v^2}{nn_1}\right)^{2/5} \tag{g}$$

With this value we can calculate, from Eqs. (229), the value of the maximum compressive force P acting between the spheres during impact, and the corresponding radius a of the surface of contact.

For calculating the duration of impact we write Eq. (*f*) in the following form:

$$dt = \frac{d\alpha}{\sqrt{v^2 - \frac{4}{5}nn_1\alpha^{5/2}}}$$

or writing $\alpha/\alpha_1 = x$ and using Eq. (*g*), we find that

$$dt = \frac{\alpha_1}{v}\frac{dx}{\sqrt{1 - (x)^{5/2}}}$$

from which the duration of impact is

$$t = \frac{2\alpha_1}{v}\int_0^1 \frac{dx}{\sqrt{1 - (x)^{5/2}}} = 2.94\,\frac{\alpha_1}{v} \tag{243}$$

In the particular case of two equal spheres of the same material and radius R, we have, from (*g*),

$$\alpha_1 = \left(\frac{5\sqrt{2}\pi\rho}{4}\frac{1 - \nu^2}{E}v^2\right)^{2/5} R$$
$$t = 2.94\left(\frac{5\sqrt{2}\pi\rho}{4}\frac{1 - \nu^2}{E}\right)^{2/5}\frac{R}{(v)^{1/5}} \tag{244}$$

where ρ denotes the mass per unit volume of the spheres.

We see that the duration of impact is proportional to the radius of the spheres and inversely proportional to $(v)^{1/5}$. This result was verified by several experimenters.[1] In the case of long bars with spherical ends, the period of the fundamental mode of

[1] M. Hamburger, *Wied. Ann.*, vol. 28, p. 653, 1886; A. Dinnik, *J. Russ. Phys.-Chem. Soc.*, vol. 38, p. 242, 1906, and vol. 41, p. 57, 1909. Further references to the literature of the subject are given in "Handbuch der Physikalischen und Technischen Mechanik," vol. 3, p. 448, 1927.

vibration may be of the same order as the duration of impact, and in investigating local compression at the point of contact these vibrations should be considered.[1]

143 | Symmetrical Deformation of a Circular Cylinder

For a circular cylinder submitted to the action of forces applied to the lateral surface and distributed symmetrically with respect to the axis of the cylinder, we introduce a stress function ϕ in cylindrical coordinates and apply Eq. (190).[2] This equation is satisfied if we take for the stress function ϕ a solution of the equation

$$\frac{\partial^2 \phi}{\partial r^2} + \frac{1}{r}\frac{\partial \phi}{\partial r} + \frac{\partial^2 \phi}{\partial z^2} = 0 \tag{a}$$

in the form

$$\phi = f(r) \sin kz \tag{b}$$

in which f is a function of r only. Substituting (b) into Eq. (a), we arrive at the following ordinary differential equation for determining $f(r)$:

$$\frac{d^2 f}{dr^2} + \frac{1}{r}\frac{df}{dr} - k^2 f = 0 \tag{c}$$

This is the differential equation satisfied by the modified Bessel functions (of the first and second kinds) of argument kr and order zero. The solution appropriate to the solid cylinder is easily obtained directly in the series form

$$f(r) = a_0 + a_1 r^2 + a_2 r^4 + a_3 r^6 + \cdots \tag{d}$$

Substituting this series in Eq. (c), we find the following relation between the consecutive coefficients:

$$(2n)^2 a_n - k^2 a_{n-1} = 0$$

from which

$$a_1 = \frac{k^2}{2^2} a_0 \qquad a_2 = \frac{k^2}{4^2} a_1 = \frac{k^4}{2^2(4^2)} a_0 \quad \cdots$$

Substituting these in the series (d), we have

$$f(r) = a_0 \left[1 + \frac{k^2 r^2}{2^2} + \frac{k^4 r^4}{2^2(4^2)} + \frac{k^6 r^6}{2^2(4^2)6^2} + \cdots \right] \tag{e}$$

The series in the brackets of Eq. (e) is the Bessel function of zero order and of the imaginary argument ikr, usually denoted by $I_0(kr)$. In the following we shall use for this function the notation $J_0(ikr)$ and write the stress function (b) in the form

$$\phi_1 = a_0 J_0(ikr) \sin kz \tag{f}$$

[1] See p. 504. Longitudinal impact of rods with spherical surfaces at the ends has been discussed by J. E. Sears, *Proc. Cambridge Phil. Soc.*, vol. 14, p. 257, 1908, and *Trans. Cambridge Phil. Soc.*, vol. 21, p. 49, 1912. Lateral impact of rods with consideration of local compression was discussed by S. Timoshenko, *Z. Math. Physik*, vol. 62, p. 198, 1914.

[2] The problem of the deformation of a circular cylinder under the action of forces applied to the surface was discussed first by L. Pochhammer, *Crelle's J.*, vol. 81, 1876. Several problems of symmetrical deformation of cylinders were discussed by C. Chree, *Trans. Cambridge Phil. Soc.*, vol. 14, p. 250, 1889. See also the paper by L. N. G. Filon, *Trans. Roy. Soc. (London)*, ser. A, vol. 198, 1902, which contains solutions of several problems of practical interest relating to symmetrical deformation in a cylinder.

Equation (190) also has solutions different from solutions of Eq. (a). One of these solutions can be derived from the above function $J_0(ikr)$. By differentiation,

$$\frac{dJ_0(ikr)}{d(ikr)} = -\frac{ikr}{2}\left[1 + \frac{k^2r^2}{2(4)} + \frac{k^4r^4}{2(4^2)6} + \frac{k^6r^6}{2(4^2)6^2(8)} + \cdots\right] \tag{g}$$

This derivative with negative sign is called Bessel's function of the first order and is denoted by $J_1(ikr)$. Consider now the function

$$f_1(r) = r\frac{d}{dr}J_0(ikr) = -ikrJ_1(ikr) = \frac{k^2r^2}{2}\left[1 + \frac{k^2r^2}{2(4)} + \frac{k^4r^4}{2(4^2)6} + \cdots\right] \tag{h}$$

By differentiation it can be shown that

$$\left(\frac{d^2}{dr^2} + \frac{1}{r}\frac{d}{dr} - k^2\right)f_1(r) = 2k^2J_0(ikr)$$

Then, remembering that $J_0(ikr)$ is a solution of Eq. (c), it follows that $f_1(r)$ is a solution of the equation

$$\left(\frac{d^2}{dr^2} + \frac{1}{r}\frac{d}{dr} - k^2\right)\left(\frac{d^2f_1}{dr^2} + \frac{1}{r}\frac{df_1}{dr} - k^2f_1\right) = 0$$

Hence a solution of Eq. (180) can be taken in the form

$$\phi_2 = a_1 \sin kz\,(ikr)J_1(ikr) \tag{i}$$

Combining solutions (f) and (i), we can take the stress function in the form

$$\phi = \sin kz[a_0J_0(ikr) + a_1(ikr)J_1(ikr)] \tag{j}$$

Substituting this stress function in Eqs. (189), we find the following expressions for the stress components:

$$\begin{aligned}\sigma_r &= \cos kz[a_0F_1(r) + a_1F_2(r)]\\ \tau_{rz} &= \sin kz[a_0F_3(r) + a_1F_4(r)]\end{aligned} \tag{k}$$

in which $F_1(r), \ldots, F_4(r)$ are certain functions of r containing $J_0(ikr)$ and $J_1(ikr)$. By using tables of Bessel functions, the values of $F_1(r), \ldots, F_4(r)$ can easily be calculated for any value of r.

Denoting by a the external radius of the cylinder, the forces applied to the surface of the cylinder, from Eqs. (k), are given by the following values of the stress components:

$$\begin{aligned}\sigma_r &= \cos kz[a_0F_1(a) + a_1F_2(a)]\\ \tau_{rz} &= \sin kz[a_0F_3(a) + a_1F_4(a)]\end{aligned} \tag{l}$$

By a suitable adjustment of the constants k, a_0, a_1, various cases of symmetrical loading of a cylinder can be discussed. Denoting the length of the cylinder by l and taking

$$k = \frac{n\pi}{l}$$

$$\begin{aligned}a_0F_1(a) + a_1F_2(a) &= -A_n\\ a_0F_3(a) + a_1F_4(a) &= 0\end{aligned}$$

we obtain the values of the constants a_0 and a_1 for the case when normal pressures $A_n \cos(n\pi z/l)$ act on the lateral surface of the cylinder. The case when $n = 1$ is represented in Fig. 215. In an analogous manner we can get a solution for the case when tangential forces of intensity $B_n \sin(n\pi z/l)$ act on the surface of the cylinder.

By taking $n = 1, 2, 3, \ldots$ and using the superposition principle, we arrive at

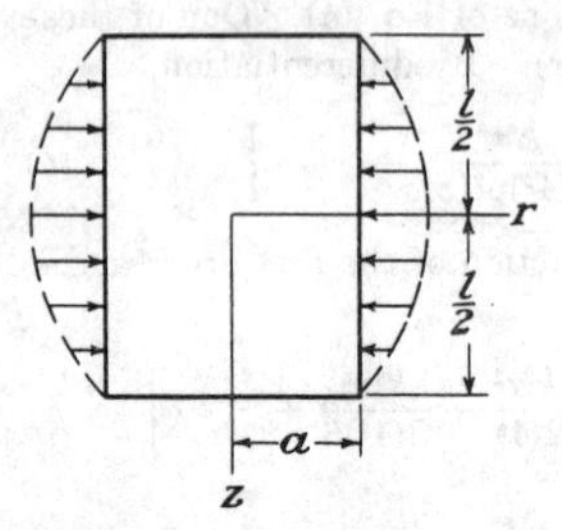

Fig. 215

solutions of problems in which the normal pressures on the surface of the cylinder can be represented by the series

$$A_1 \cos \frac{\pi z}{l} + A_2 \cos \frac{2\pi z}{l} + A_3 \cos \frac{3\pi z}{l} + \cdots \tag{m}$$

and the shearing forces by the series

$$B_1 \sin \frac{\pi z}{l} + B_2 \sin \frac{2\pi z}{l} + B_3 \sin \frac{3\pi z}{l} + \cdots \tag{n}$$

If we take for the stress function ϕ, instead of expression (b), the expression

$$\phi = f(r) \cos kz$$

and proceed as before, we find, instead of expression (j), the stress function

$$\phi = \cos kz[b_0 J_0(ikr) + b_1(ikr)J_1(ikr)] \tag{o}$$

By a suitable adjustment of the constants k, b_0, b_1, we obtain the solution for the case in which normal pressures on the cylinder are represented by a sine series and the shearing forces by cosine series. Hence, by combining solutions (j) and (o), we can get any axially symmetrical distribution of normal and shearing forces over the surface of the cylinder. At the same time there will also be certain forces distributed over the ends of the cylinder. By superposing a simple tension or compression we can always arrange that the resultant of these forces is zero, and their effect on stresses at some distance from the ends is then negligible by Saint-Venant's principle. Several

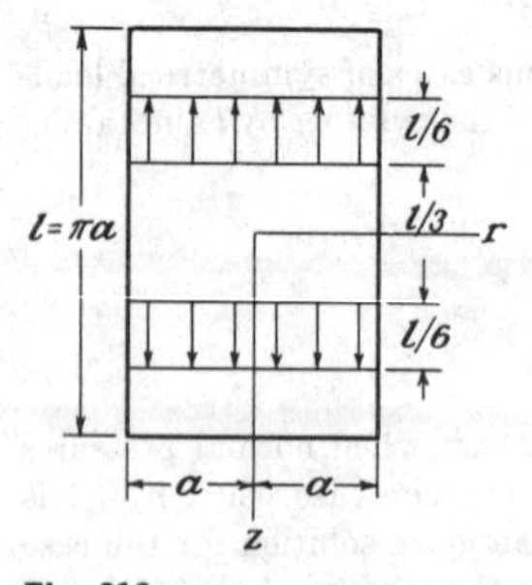

Fig. 216

examples of symmetrical loading of cylinders are discussed by L. N. G. Filon in the paper already mentioned.[1] We give here final results from his solution for the case shown in Fig. 216. A cylinder, the length of which is equal to πa, is submitted to the tensile action of shearing forces uniformly distributed over the shaded portion of the surface of the cylinder indicated in the figure. The distribution of the normal stress σ_z over cross sections of the cylinder is of practical interest, and the table below gives the ratios of these stresses to the average tensile stress, obtained by dividing the total tensile force by the cross-sectional area of the cylinder. It can be seen that local tensile stresses near the loaded portions of the surface diminish rapidly with increase of distance from these portions and approach the average value.

z	$r = 0$	$r = 0.2a$	$r = 0.4a$	$r = 0.6a$	$r = a$
0	0.689	0.719	0.810	0.962	1.117
$0.05l$	0.673	0.700	0.786	0.937	1.163
$0.10l$	0.631	0.652	0.720	0.859	1.344
$0.15l$	0.582	0.594	0.637	0.737	2.022
$0.20l$	0.539	0.545	0.565	0.617	1.368

Another application of the general solution of the problem in terms of Bessel's functions is given by A. Nádai in discussing the bending of circular plates by a force concentrated at the middle[2] (Fig. 217). The method of Hankel transforms, appropriate for thick plates, the semi-infinite solid, contact, and circular crack problems, has been extensively applied by I. N. Sneddon.[3]

144 | The Circular Cylinder with a Band of Pressure[4]

When a short collar is shrunk on a much longer shaft the simple shrink-fit formulas, valid when collar and shaft are of equal lengths, are not accurate. A much better approximation is obtained by considering the problem indicated in Fig. 218*a* of a long cylinder with a uniform[5] normal pressure p acting on the band $ABCD$ of the surface.

The required solution can evidently be obtained by superposing the effects of the two pressure distributions indicated in Fig. 218*b*. The basic problem is therefore that of pressure $p/2$ on the lower half of the cylindrical surface and $-p/2$ on the

[1] *Loc. cit.* See also G. Pickett, *J. Appl. Mech.*, vol. 11, p. 176, 1944.

[2] "Elastische Platten," p. 315, 1925.

[3] "Fourier Transforms," McGraw-Hill Book Company, New York, 1951.

[4] M. V. Barton, *J. Appl. Mech.*, vol. 8, p. A-97, 1941. A. W. Rankin, *ibid.*, vol. 11, p. A-77, 1944. C. J. Tranter and J. W. Craggs, *Phil. Mag.*, vol. 36, p. 241, 1945.

[5] An analysis of an elastic collar shrunk onto a long elastic shaft, without friction, is given by H. Okubo, *Z. Angew. Math. Mech.*, vol. 32, pp. 178–186, 1952. Evaluations for the nonuniform pressure due to a shrunk-on rigid sleeve, without and with friction, are given by H. D. Conway and K. A. Farnham, *Intern J. Eng. Sci.*, vol. 5, pp. 541–554, 1967; also W. F. Yau, *SIAM J. Appl. Math.*, vol. 15, pp. 219–227, 1967.

Rounding the sharp corners reduces the stress singularity to a local peak, for which formulas (frictionless) are given by J. N. Goodier and C. B. Loutzenheiser, *J. Appl. Mech.*, vol. 32, pp. 462–463, 1965.

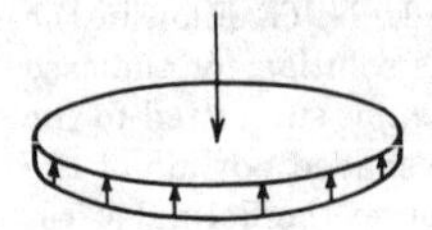

Fig. 217

upper half, the length of the cylinder being infinite, and its solution will now be given.

We begin with the stress function given by Eq. (o) of Art. 143, writing $I_0(kr)$ for $J_0(ikr)$ and $iI_1(kr)$ for $J_1(ikr)$. We also write $b_0 = \rho b_1$. Then

$$\phi = [\rho I_0(kr) - krI_1(kr)]b_1 \cos kz \tag{a}$$

This satisfies Eq. (190) no matter what value is given to k. If we consider k to take a range of values we can allow b_1 to depend on k and an increment dk by writing

$$b_1 = f(k)\, dk$$

Putting this in (a) and adding up all such stress functions, we obtain a more general stress function in the form

$$\phi = \int_0^\infty [\rho I_0(kr) - krI_1(kr)]f(k) \cos kz\, dk \tag{b}$$

We shall now see how it is possible to select the function $f(k)$ so that this stress function will give the solution to our problem.

From Eqs. (189) we find that the shear stress will be

$$\tau_{rz} = \int_0^\infty [\rho kI_0'(kr) - k^2rI_1'(kr) - kI_1(kr) - 2k(1-\nu)I_0'(kr)]k^2f(k) \cos kz\, dk \tag{c}$$

where primes denote differentiation with respect to kr. This must vanish at the surface $r = a$. Putting $r = a$ in the expression in brackets, and equating this bracket to zero, we obtain an equation for ρ which gives

$$\rho = 2(1-\nu) + ka\frac{I_0(ka)}{I_1(ka)} \tag{d}$$

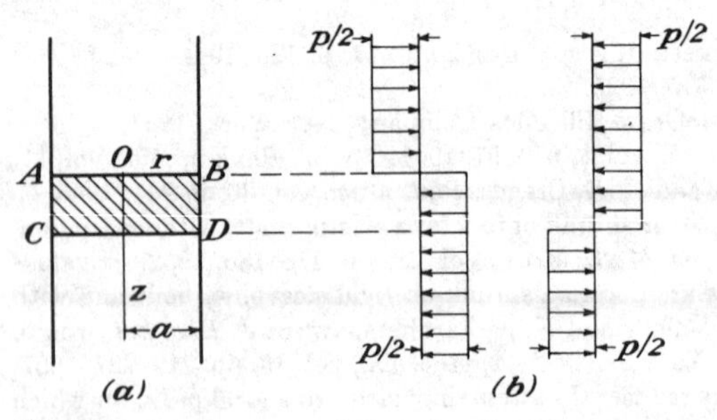

Fig. 218

The remaining boundary condition is

$$\begin{aligned} \sigma_r &= \frac{p}{2} \qquad \text{for } r = a,\ z > 0 \\ \sigma_r &= -\frac{p}{2} \qquad \text{for } r = a,\ z < 0 \end{aligned} \tag{e}$$

The value of σ_r obtained from (*b*) by Eqs. (189) is

$$\sigma_r = -\int_0^\infty \left[(1 - 2\nu - \rho) I_0(kr) + \left(kr + \frac{\rho}{kr} \right) I_1(kr) \right] k^3 f(k) \sin kz\, dk \tag{f}$$

We now make use of the fact that[1]

$$\int_0^\infty \frac{\sin kz}{k}\, dk = \begin{cases} \dfrac{\pi}{2} & \text{for } z > 0 \\ 0 & \text{for } z = 0 \\ -\dfrac{\pi}{2} & \text{for } z < 0 \end{cases} \tag{g}$$

If we multiply this by p/π, we obtain

$$\frac{p}{\pi} \int_0^\infty \frac{\sin kz}{k}\, dk = \begin{cases} \dfrac{p}{2} & \text{for } z > 0 \\ 0 & \text{for } z = 0 \\ -\dfrac{p}{2} & \text{for } z < 0 \end{cases} \tag{h}$$

in which the values on the right correspond to the boundary values for σ_r given by (*e*). The boundary conditions (*e*) are therefore satisfied if we make the right-hand side of Eq. (*f*), with $r = a$, identical with the left-hand side of Eq. (*h*). This requires

$$-\left[(1 - 2\nu - \rho) I_0 (ka) + \left(ka + \frac{\rho}{ka} \right) I_1(ka) \right] k^3 f(k) = \frac{p}{\pi} \frac{1}{k} \tag{i}$$

and this equation determines $f(k)$. The stress components are then found from the stress function (*b*) by means of the formulas (189) and will be integrals of the same general nature as that of Eq. (*f*), which gives σ_r. Values, obtained by numerical integration, are given by Rankin in the paper cited on page 425. The curves in Fig. 219 show the variation of the stresses in the axial direction for various radial distances, and also the surface displacements.

They are reproduced from the paper of Barton (see page 425) and were obtained by a different method using Fourier series. From these curves results can be obtained for the problem of Fig. 218 by superposition, as explained at the beginning of this article. Curves for the stresses and displacement for pressure bands of several widths are given in the papers cited. When the width is equal to the radius of the cylinder, the tangential stress σ_θ at the surface and at the middle of the pressure band reaches a value about 10 percent higher than the applied pressure, and is, of course, compressive. The axial stress σ_z in the surface just outside the pressure band reaches a tensile value of about 45 percent of the applied pressure. The shear stress τ_{rz} attains a greatest value, equal to 31.8 percent of the applied pressure, at the edges of the pressure band AB and CD in Fig. 218 and just below the surface.

When the pressure is applied all over the curved surface of the cylinder, of any

[1] See for instance I. S. Sokolnikoff, "Advanced Calculus," 1st ed., p. 362, McGraw-Hill Book Company, New York, 1939.

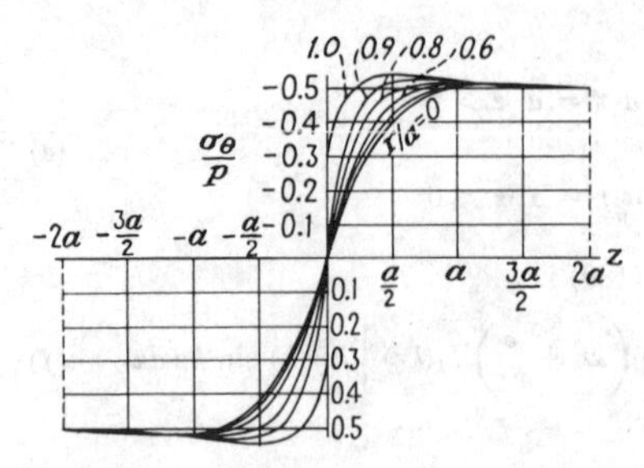

Fig. 219

length, we have simply compressive σ_r and σ_θ equal to the applied pressure, and σ_z and τ_{rz} zero.

Solutions have been obtained in a similar manner for a band of pressure in a hole in an infinite solid,[1] and for a band of pressure near one end of a solid cylinder.[2]

145 | Boussinesq's Solution in Two Harmonic Functions

The solutions obtained so far in this chapter for torsionless axisymmetry have been expressed in terms of Love's single biharmonic stress function φ. An earlier general form of solution, in terms of two harmonic functions, was given by Boussinesq.[3] It has since been employed extensively for problems somewhat more advanced than those we have already discussed, for instance, in Neuber's solutions for axial tension in solids bounded by ellipsoids and hyperboloids of revolution. These are among the solutions mentioned on page 242 (the rest being nonaxisymmetric), in connection with the Papcovitch-Neuber expressions for displacement in terms of four harmonic functions. Here we derive the Boussinesq forms from these more general expressions. For the four harmonic functions φ_0, φ_1, φ_2, φ_3 in Art. 88, we take [with $4\alpha = 1/(1-\nu)$]

$$\varphi_0 = -\frac{1}{\alpha}\Phi(r,z) \qquad \varphi_1 = \varphi_2 = 0 \qquad \varphi_3 = -\alpha\Psi(r,z) \tag{a}$$

where Φ, Ψ are independent of θ and are harmonic functions, that is,

$$\nabla^2\Phi = 0 \qquad \nabla^2\Psi = 0 \tag{b}$$

with the laplacian operator ∇^2 as in Eq. (*a*), page 381.

The displacement in Eqs. (*a*) can be separated into the contributions from φ_0, . . . , φ_3 alone. That from φ_0 is a vector $-\alpha$ grad φ_0 and is now given by the cylindrical components

$$(u_1,v_1,w_1) = \left(\frac{\partial\Phi}{\partial r}, 0, \frac{\partial\Phi}{\partial z}\right) \tag{c}$$

The displacement contributed by φ_3 is now

$$(u_2,v_2,w_2) = \left(\frac{z\partial\Psi}{\partial r}, 0, \frac{z\partial\Psi}{\partial z}\right) - (3-4\nu)(0,0,\Psi) \tag{d}$$

[1] C. J. Tranter, *Quart. Appl. Math.*, vol. 4, p. 298, 1946; O. L. Bowie, *ibid.*, vol. 5, p. 100, 1947.

[2] C. J. Tranter and J. W. Craggs, *Phil. Mag.*, vol. 38, p. 214, 1947.

[3] J. Boussinesq, "Applications des Potentiels à l'Etude de l'Equilibre et du Mouvement des Solides Elastiques," Gauthier-Villars, Paris, 1885.

The displacement

$$(u,v,w) = (u_1 + u_2, v_1 + v_2, w_1 + w_2) \tag{e}$$

is the Boussinesq form.[1]

146 | The Helical Spring under Tension (Screw Dislocation in a Ring)

In several basic problems of practical importance, the stress is axisymmetric but the displacement field is not. If the undeformed boundary is a surface of revolution, the deformed boundary is not.

As an example, we take a helical spring stretched by forces P. Any segment of the coil is in equilibrium under two equal and opposite axial forces P, as Fig. 220 indicates. Shear stress on any cross section forms an axial resultant P and can be the same for all sections. When the radius of the section, ρ_0, is not small compared with the coil radius R_0, the elementary theory well known in strength of materials becomes inadequate. It treats each thin slice between two neighboring cross sections as though it were a straight cylinder with torsional couple PR_0. The corresponding shear stress, in cylindrical coordinates, has nonzero components $\tau_{r\theta}$, $\tau_{\theta z}$, and these are independent of θ. The *pitch* of the helical centerline is ignored.

To find a solution of the general equations, taking account of the coil curvature $1/R_0$, the problem is first simplified by the Saint-Venant semi-inverse method. We consider displacements of the form

$$u = 0 \qquad v = r\Psi(r,z) \qquad w = c\theta \tag{a}$$

where c is a constant, subsequently related to P. Deriving the strain components from these by Eqs. (179) we find that ϵ_r, ϵ_θ, ϵ_z, $\gamma_{r\theta}$ are zero and that

$$\frac{\tau_{r\theta}}{G} = r\frac{\partial \Psi}{\partial r} \qquad \frac{\tau_{\theta z}}{G} = r\frac{\partial \Psi}{\partial z} + \frac{c}{r} \tag{b}$$

Thus $\tau_{r\theta}$, $\tau_{\theta z}$ are the only nonzero stress components. Being independent of θ they are the same for all cross sections. The three equilibrium equations (180) reduce to Eq. (*c*) of Art. 119, and we have again a stress function $\varphi(r,z)$ as in Eqs. (*d*) of Art. 119, which are

$$r^2\tau_{r\theta} = -\frac{\partial \varphi}{\partial z} \qquad r^2\tau_{\theta z} = \frac{\partial \varphi}{\partial r} \tag{c}$$

From (*b*) and (*c*) it is easily found that the differential equations satisfied by φ and Ψ

[1] For applications of this, and of the Papcovitch-Neuber form, see the survey article by Sternberg, *loc. cit.*

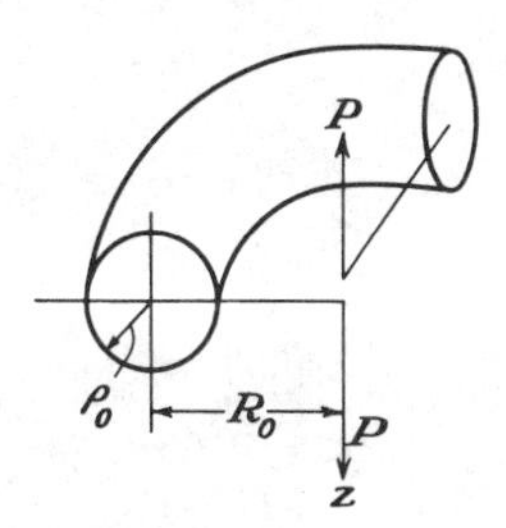

Fig. 220

separately are

$$\frac{\partial^2 \varphi}{\partial r^2} - \frac{3}{r}\frac{\partial \varphi}{\partial r} + \frac{\partial^2 \varphi}{\partial z^2} = -2Gc \tag{d}$$

and

$$\frac{\partial^2 \Psi}{\partial r^2} + \frac{3}{r}\frac{\partial \Psi}{\partial r} + \frac{\partial^2 \Psi}{\partial z^2} = 0 \tag{e}$$

In terms of the Laplace operator

$$\nabla^2 \equiv \frac{\partial^2}{\partial r^2} + \frac{1}{r}\frac{\partial}{\partial r} + \frac{\partial^2}{\partial z^2} \tag{f}$$

Eqs. (*d*) and (*e*) can be written

$$\left(\nabla^2 - \frac{4}{r^2}\right)\frac{\varphi}{r^2} = -2Gc\frac{1}{r^2} \qquad \left(\nabla^2 - \frac{1}{r^2}\right) r\Psi = 0 \tag{g}$$

The displacements (*a*) differ from those adopted in Art. 119 only by introduction of the w component[1] $c\theta$. Accordingly, the present Eqs. (*b*), (*c*), (*d*), and (*e*) reduce to the forms given in Art. 119 when we take $c = 0$.

Since Eqs. (*c*) are the same as Eqs. (*d*) of Art. 119, the condition for a free boundary surface of revolution is the same as in Eq. (*h*) of Art. 119, which is

$$\varphi = \text{const} \tag{h}$$

When pitch is suppressed, one turn of an undeformed "helical" spring of round wire becomes a tore, generated by rotating the circle in Fig. 221 about the z axis. The two ends of the turn are not joined. They carry equal and opposite shear stress distributions forming resultants P with lines of action in the z axis. They separate, as these forces are applied, by the axial relative displacement, from Eq. (*a*),

$$(w)_{\theta=2\pi} - (w)_{\theta=0} = c2\pi \tag{i}$$

This problem requires a function $\varphi(r,z)$ that satisfies Eq. (*d*) within the circle in Fig. 221 and is constant *on* the circle. The solution has been given in series form by W. Freiberger,[2] in *toroidal coordinates*. The coordinate surfaces of this system are generated by rotating the plane bipolar system of Fig. 120 about the x axis, turning the whole figure to make this axis vertical, to correspond to the z axis in Fig. 221 (the

[1] The displacements of Art. 117 also have such a component, and in fact represent the special case of Eqs. (*a*) obtained by taking $\Psi(r,z)$ to be Az.

[2] The Uniform Torsion of an Incomplete Tore, *Australian J. Sci. Res.*, ser. A, vol. 2, pp. 354–375, 1949. For further references, including solutions for wire of rectangular section, see a discussion by R. Schmidt, *J. Appl. Mech.*, vol. 31, p. 154, 1964.

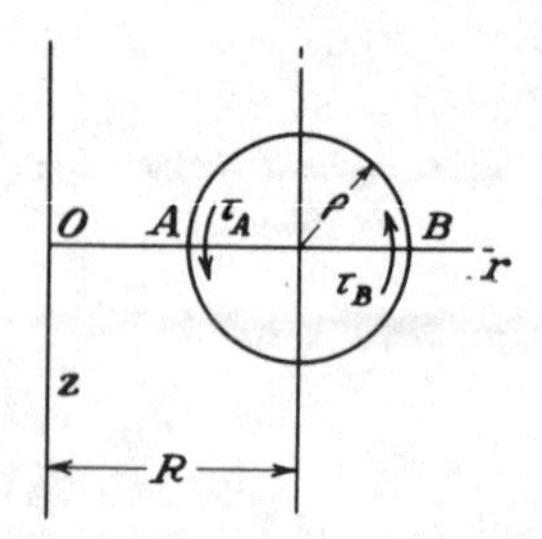

Fig. 221

third family consists of the meridional planes θ = constant). The analysis, necessarily somewhat elaborate, is not reproduced here. The principal results[1] are shown in the table below and in Fig. 222. The table gives the relation between the relative

R_0/ρ_0	3	4	5	6	8	10
δ	1.025	1.017	1.007	1.006	1.004	1.001

axial displacement $2\pi c$ of Eq. (*i*) (corresponding to the "stretch" of one turn) and the forces P causing it. P would be the axial stretching force in a one- or multi-turn "helical spring of negligible pitch." We can write

$$\delta = \frac{(P/\pi\rho_0^2)}{G(2\pi c/\rho_0)} 4\pi \left(\frac{R_0}{\rho_0}\right)^3 \tag{j}$$

which is a dimensionless ratio involving P and $2\pi c$. The values of δ for several values of R_0/ρ_0 are given in the table. For a given case, knowing R_0/ρ_0 and δ, we find from (*j*) a value for the ratio

$$\frac{P/\pi\rho_0^2}{G(2\pi c/\rho_0)}$$

and for given ρ_0 and G we then have the ratio $P/2\pi c$, the spring stiffness per turn. The elementary theory corresponds to $\delta = 1$. The value in the table for $R_0/\rho_0 = 10$ is very close to 1. Even for $R_0/\rho_0 = 3$ (thick wire) the departure from 1 is only 0.025.

The shear stress on the circular section is necessarily circumferential at the circular boundary. The greatest value is found at A, Fig. 221. The smallest, on the circle, is found at B. The actual values can each be given by means of the dimensionless coefficient K in

$$\tau = K \frac{2R_0}{\pi\rho_0^3} P$$

The variation of K with R_0/ρ_0, for points A (upper curve) and B (lower curve), is shown in Fig. 222. The elementary theory would give $K = 1$ for thin wire (R_0/ρ_0

[1] From Freiberger's paper, *ibid.*

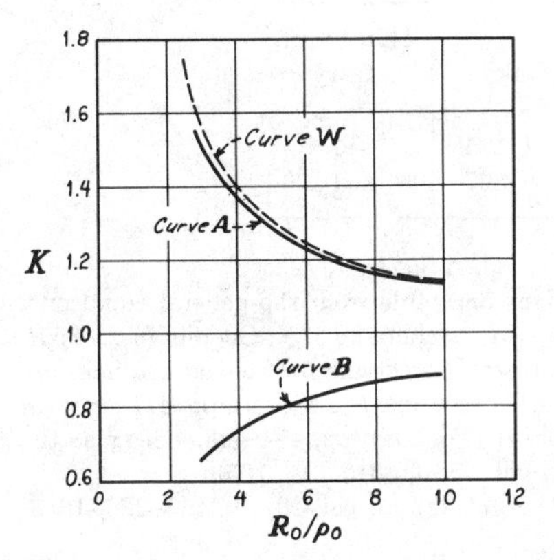

Fig. 222

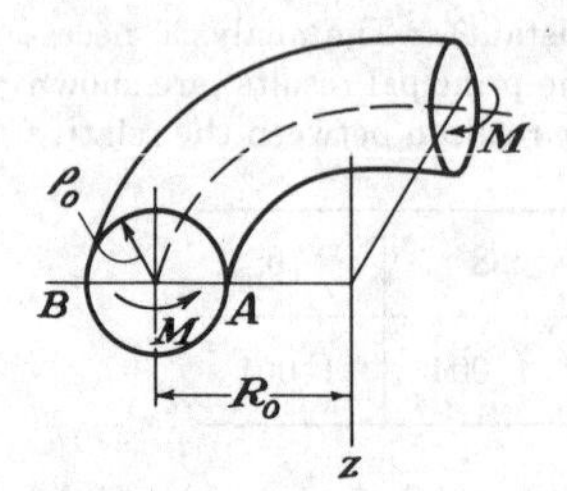

Fig. 223

large). As Fig. 222 shows, the departures from this are significant, especially for the smaller values of R_0/ρ_0 (thick wire).

The curve[1] marked W in Fig. 222 is derived from the elementary theory by making corrections for the curvature $1/R_0$ and the shear force P (the elementary theory stress represents only the "torsional" couple PR_0). For helical spring design, correction for pitch may also be significant.[2]

147 | Pure Bending of an Incomplete Ring

The problem is illustrated in Fig. 223. Again the stress is axisymmetric when the bending couples M are applied by appropriate distributions of the normal stress σ on the ends, the same distribution then occurring on any cross section made by a plane through the z axis. Approximations to such stress are given by the thin-beam theory of strength of materials and by the Winkler thick curved-bar theory. Approximations have also been obtained by Göhner from the general axisymmetric equations, by successive corrections of the thin-beam theory. The following table shows how these approximations compare with values from a series solution of the general equations obtained by Sadowsky and Sternberg, from whose paper[3] it is reproduced. This paper gives the history of the problem, and also of the problem of Art. 146, with references.

The table gives the values of $\pi\rho_0{}^3\sigma/(4M)$ at points A and B (Fig. 223) for the illustrative case $R_0/\rho_0 = 5$, $\nu = 0.3$.

	Sadowsky-Sternberg	Göhner	Winkler	Thin-beam theory
Point A	-1.273	-1.200	-1.175	-1
Point B	0.891	0.851	0.867	1

[1] See A. M. Wahl, *J. Appl. Mech.*, vol. 2, pp. A-35–A-37, 1935.

[2] A systematic investigation of corrections derivable from the general equations, for both curvature and pitch, is reported by C. J. Ancker and J. N. Goodier in *J. Appl. Mech.*, vol. 25, 1958: (1) Pitch and Curvature Corrections for Helical Springs, pp. 466–470; (2) Theory of Pitch and Curvature Corrections–I (Tension), pp. 471–483; (3) Theory of Pitch and Curvature Corrections–II (Torsion), pp. 484–495. See also the discussion by A. M. Wahl, *J. Appl. Mech.*, vol. 26, pp. 312–313, 1959.

[3] M. A. Sadowsky and E. Sternberg, *J. Appl. Mech.*, vol. 20, pp. 215–226, 1953. See also the references in n. 2 above.

chapter 13

Thermal Stress

148 | The Simplest Cases of Thermal Stress Distribution. Method of Strain Suppression

One of the causes of stress in a body is nonuniform heating. With rising temperature the elements of a body expand. Such an expansion generally cannot proceed freely in a continuous body, and stresses due to the heating are set up.

Fracture of glass when a surface is rapidly heated is attributable to such stress. Fatigue failure[1] can occur as a result of temperature fluctuations. The consequences of such thermal stress are important in many aspects of engineering design, as in turbines, jet engines, and nuclear reactors.

The simpler problems of thermal stress can easily be reduced to problems of boundary force of types already considered. As a first example let us consider a thin rectangular plate of uniform thickness in which the temperature T is an even function of y (Fig. 224) and is independent of

[1] See, for instance, L. F. Coffin, Jr., *Trans. ASME*, vol. 76, p. 931, 1954.

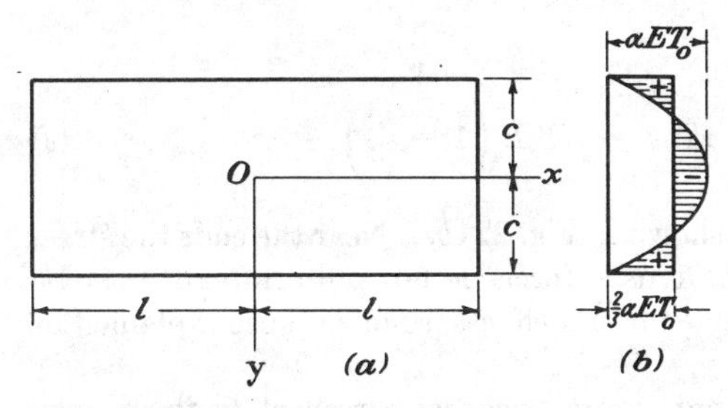

Fig. 224

x and z. The longitudinal thermal expansion αT will be entirely suppressed by applying to each element of the plate the longitudinal stress

$$\sigma_x' = -\alpha TE \tag{a}$$

which is compressive when T is positive. Since the plate is free to expand laterally, the application of the stress (a) will not produce stress in the lateral directions. To maintain the stress (a) throughout the plate, it will be necessary to distribute compressive forces of the magnitude (a) at the ends of the plate only. These compressive forces will completely suppress any expansion of the plate in the direction of the x axis due to the temperature rise T. To get the *thermal stress* in the plate, which is free from external forces, we have to superpose on the stress (a) the stresses produced in the plate by tensile forces of intensity αTE distributed at the ends. These forces have the resultant

$$\int_{-c}^{+c} \alpha TE \, dy$$

and at a sufficient distance from the ends they will produce approximately uniformly distributed tensile stress of the magnitude

$$\frac{1}{2c} \int_{-c}^{+c} \alpha TE \, dy$$

so that the thermal stress in the plate with free ends, at a sufficient distance from the ends, will be

$$\sigma_x = \frac{1}{2c} \int_{-c}^{+c} \alpha TE \, dy - \alpha TE \tag{b}$$

Assuming, for example, that the temperature is distributed parabolically and is given by the equation

$$T = T_0 \left(1 - \frac{y^2}{c^2}\right)$$

we get, from Eq. (b),

$$\sigma_x = \frac{2}{3} \alpha T_0 E - \alpha T_0 E \left(1 - \frac{y^2}{c^2}\right) \tag{c}$$

This stress distribution is shown in Fig. 224b. Near the ends the stress distribution produced by the tensile forces is not uniform and must be calculated by methods suitable for end effects, such as those explained in Arts. 28 and 93.

If the temperature T is not symmetrical with respect to the x axis, we begin again with compressive stress (a) suppressing the strain ϵ_x. In the nonsymmetrical cases this stress gives rise not only to a resultant force $-\int_{-c}^{+c} \alpha ET\,dy$ but also to a resultant couple $-\int_{-c}^{+c} \alpha ETy\,dy$; and in order to satisfy the conditions of equilibrium, we must superpose on the compressive stresses (a) a uniform tension, determined as before, and bending stresses $\sigma''_x = \sigma y/c$ determined from the condition that the moment of the forces distributed over a cross section must be zero. Then

$$\int_{-c}^{+c} \frac{\sigma y^2\,dy}{c} - \int_{-c}^{+c} \alpha ETy\,dy = 0$$

from which

$$\frac{\sigma}{c} = \frac{3}{2c^3}\int_{-c}^{+c} \alpha ETy\,dy \qquad \sigma_x'' = \frac{3y}{2c^3}\int_{-c}^{+c} \alpha ETy\,dy$$

Then the total stress is

$$\sigma_x = -\alpha ET + \frac{1}{2c}\int_{-c}^{+c} \alpha ET\,dy + \frac{3y}{2c^3}\int_{-c}^{+c} \alpha ETy\,dy \tag{d}$$

In this discussion it was assumed that the plate was thin in the z direction. Suppose now that the dimension in the z direction is large. We have then a plate with the xz plane as its middle plane and a thickness $2c$. Let the temperature T be, as before, independent of x and z, and so a function of y only.

The free thermal expansion of an element of the plate in the x and z directions will be completely suppressed by applying stresses σ_x, σ_z obtained from Eqs. (3), by putting $\epsilon_x = \epsilon_z = -\alpha T$, $\sigma_y = 0$. These equations then give

$$\sigma_x = \sigma_z = -\frac{\alpha ET}{1-\nu} \tag{e}$$

The elements can be maintained in this condition by applying the distributions of compressive force given by (e) to the edges (x = constant, z = constant). The thermal stress in the plate free from external force is obtained by superposing on the stresses (e) the stresses due to application of equal and opposite distributions of force on the edges. If T is an even function of y such that the mean value over the thickness of the plate is zero, the resultant force per unit run of edge is zero, and Saint-Venant's principle (Art. 19) indicates that it produces no stress except near the edge.

If the mean value of T is not zero, uniform tensions in the x and z directions corresponding to the resultant force on the edge must be superposed on the compressive stresses (e). If in addition to this the temperature is not symmetrical with respect to the xz plane, we must add the bending stresses. In this manner we finally arrive at the equation

$$\sigma_x = \sigma_z = -\frac{\alpha TE}{1-\nu} + \frac{1}{2c(1-\nu)}\int_{-c}^{+c} \alpha TE\,dy + \frac{3y}{2c^3(1-\nu)}\int_{-c}^{+c} \alpha TEy\,dy \qquad (f)$$

which is analogous to Eq. (*d*). From Eq. (*f*) we can easily calculate thermal stresses in a plate, if the distribution of temperature T over the thickness of the plate is known.

Consider, as an example, a plate which has initially a uniform temperature T_0 and which is being cooled down by maintaining the surfaces $y = \pm c$ at a constant temperature T_1.[1] By Fourier's theory the distribution of temperature at any instant t is

$$T = T_1 + \frac{4}{\pi}(T_0 - T_1)\left(e^{-p_1 t}\cos\frac{\pi y}{2c} - \frac{1}{3}e^{-p_3 t}\cos\frac{3\pi y}{2c} + \cdots\right) \qquad (g)$$

in which p_1, $p_3 = 3^2 p_1$, . . . , $p_n = n^2 p_1$, . . . are certain constants. Substituting in Eq. (*f*), we find

$$\sigma_x = \sigma_z = \frac{4\alpha E(T_0 - T_1)}{\pi(1-\nu)}\left[e^{-p_1 t}\left(\frac{2}{\pi} - \cos\frac{\pi y}{2c}\right) + \frac{1}{3}e^{-p_3 t}\left(\frac{2}{3\pi} + \cos\frac{3\pi y}{2c}\right) + \frac{1}{5}e^{-p_5 t}\left(\frac{2}{5\pi} - \cos\frac{5\pi y}{2c}\right) + \cdots\right] \qquad (h)$$

After a moderate time the first term acquires dominant importance, and we can assume

$$\sigma_x = \sigma_z = \frac{4\alpha E(T_0 - T_1)}{\pi(1-\nu)} e^{-p_1 t}\left(\frac{2}{\pi} - \cos\frac{\pi y}{2c}\right)$$

For $y = \pm c$ we have tensile stresses

$$\sigma_x = \sigma_z = \frac{4\alpha E(T_0 - T_1)}{\pi(1-\nu)} e^{-p_1 t}\frac{2}{\pi}$$

At the middle plane $y = 0$ we obtain compressive stresses

$$\sigma_x = \sigma_z = -\frac{4\alpha E(T_0 - T_1)}{\pi(1-\nu)} e^{-p_1 t}\left(1 - \frac{2}{\pi}\right)$$

The points with zero stresses are obtained from the equation

$$\frac{2}{\pi} - \cos\frac{\pi y}{2c} = 0$$

from which

$$y = \pm 0.560c$$

If the surfaces $y = \pm c$ of a plate are maintained at two different temperatures T_1, T_2, a steady state of heat flow is established after a certain

[1] This problem was discussed by Lord Rayleigh, *Phil. Mag.*, ser. 6, vol. 1, p. 169, 1901.

time and the temperature is then given by the linear function

$$T = \frac{1}{2}(T_1 + T_2) + \frac{1}{2}(T_1 - T_2)\frac{y}{c} \tag{i}$$

Substitution in Eq. (*f*) shows that the thermal stresses are zero,[1] provided, of course, that the plate is not restrained. If the edges are perfectly restrained against expansion and rotation, the stress induced by the heating is given by Eqs. (*e*). For instance if $T_2 = -T_1$, we have from (*i*)

$$T = T_1 \frac{y}{c} \tag{j}$$

and Eqs. (*e*) give

$$\sigma_x = \sigma_z = -\frac{\alpha E}{1 - \nu} T_1 \frac{y}{c} \tag{k}$$

The maximum stress is

$$(\sigma_x)_{\max} = (\sigma_z)_{\max} = \frac{\alpha E T_1}{1 - \nu} \tag{l}$$

The thickness of the plate does not enter in this formula, but in the case of a thicker plate a greater difference of temperature between the two surfaces usually exists. Thus a thick plate of a brittle material is more liable to break due to thermal stresses than a thin one.

If, as in many applications, one face of the plate is in contact with hot gases of periodically fluctuating temperature, the temperature T in the plate will show corresponding cyclic fluctuations superposed on a steady heat flow. The amplitude of temperature fluctuation *in* the plate material at the face is usually small compared with that of the hot gas. Moreover, the plate amplitude rapidly diminishes[2] with distance from the surface. For instance,[3] in a steel plate 3.5 cm thick, at 110 cycles per min, the plate temperature amplitude was found to be 10°C at the surface and 0.33°C at 0.5 cm below the surface, when the gas temperature amplitude was 640°C. Under such conditions, the second and third terms in (*f*) become very small compared with the first, which gives ±5360 psi for $T = \pm 10°\text{C}$, $\alpha = 1.25 \times 10^{-5}$.

[1] In general, when T is a linear function of x, y, z, the strain corresponding to free thermal expansion of each element,

$$\epsilon_x = \epsilon_y = \epsilon_z = \alpha T \qquad \gamma_{xy} = \gamma_{yz} = \gamma_{zx} = 0$$

satisfies the conditions of compatibility (125) and there will be no thermal stress (see Prob. 2, p. 277).

[2] Solutions of such temperature problems are given in books on heat conduction. See, for instance, H. S. Carslaw and J. C. Jaeger, "Heat Conduction in Solids," 2d ed., Oxford University Press, Fair Lawn, N.J., 1959.

[3] G. Eichelberg, *Forschungsarbeiten*, no. 263, 1923.

As a further elementary example, let us consider a sphere of large radius and assume that there occurs a temperature rise T in a small spherical element of radius a at the center of the large sphere. Since the element is not free to expand, a pressure p will be produced at the surface of the element. The radial and the tangential stresses due to this pressure at any point of the sphere at a radius $r > a$ can be calculated from formulas (207) and (208). Assuming the outer radius of the sphere as very large in comparison with a, we obtain from these formulas

$$\sigma_r = -\frac{pa^3}{r^3} \qquad \sigma_t = \frac{pa^3}{2r^3} \tag{m}$$

At the radius $r = a$ we obtain

$$\sigma_r = -p \qquad \sigma_t = \tfrac{1}{2}p$$

and the increase of this radius, due to pressure p, is

$$\Delta r = (a\epsilon_t)_{r=a} = \frac{a}{E}[\sigma_t - \nu(\sigma_r + \sigma_t)]_{r=a} = \frac{pa}{2E}(1 + \nu)$$

This increase must be equal to the increase of the radius of the heated spherical element produced by temperature rise and pressure p. Thus we obtain the equation

$$\alpha Ta - \frac{pa}{E}(1 - 2\nu) = \frac{pa}{2E}(1 + \nu)$$

from which

$$p = \frac{2}{3}\frac{\alpha TE}{1 - \nu} \tag{n}$$

Substituting in Eqs. (m), we obtain the formulas for the stresses outside the heated element

$$\sigma_r = -\frac{2}{3}\frac{\alpha TEa^3}{(1 - \nu)r^3} \qquad \sigma_t = \frac{1}{3}\frac{\alpha TEa^3}{(1 - \nu)r^3} \tag{o}$$

PROBLEMS

1. In deriving Eq. (d) it was assumed that E was constant and also that the thickness of the plate (in the z direction) was uniform.

   Suppose instead that E can vary with y, continuously or (as in a composite "sandwich" strip) discontinuously, and also that the thickness h may vary gradually with y. Show that Eq. (d) must now be replaced by

$$\sigma_x = E(-\alpha T + \epsilon + \beta y)$$

where

$$\epsilon = \frac{\int_{-c}^{+c} E\alpha Th\,dy}{\int_{-c}^{+c} Eh\,dy} \qquad \beta = \frac{\int_{-c}^{+c} E\alpha Thy\,dy}{\int_{-c}^{+c} Ehy^2\,dy}$$

Verify that this agrees with (d) when E and h are constants.

2. Obtain the results analogous to Eqs. (n) and (o) for a central circular heated region, at uniform temperature T, in (a) a large thin plate, and (b) a large slab in plane strain, meaning that ϵ_z, initially zero when $T = 0$, is maintained everywhere zero throughout by imposition of suitable σ_z on the faces z = constant. The heated region is

$$r^2 < a^2 \qquad \text{where } r^2 = x^2 + y^2$$

The temperature outside the heated region is zero.

149 | Longitudinal Temperature Variation in a Strip

Suppose that a strip of thin plate (Fig. 225) is nonuniformly heated so that the temperature T is a function of the longitudinal coordinate x only, being uniform across any given cross section. If the plate is cut into strips such as AB (Fig. 225), these strips expand vertically by different amounts. Due to the mutual restraint, there will be stresses set up when they are in fact attached as in the plate.

Considering the unattached strips, their vertical expansion is suppressed if they are subjected to compressive stress

$$\sigma_y = -\alpha ET \tag{a}$$

by applying such stress at the ends A and B of each strip. The strips fit together as in the unheated plate except for lateral expansion, which is supposed free to occur by rigid-body horizontal translation of each strip.

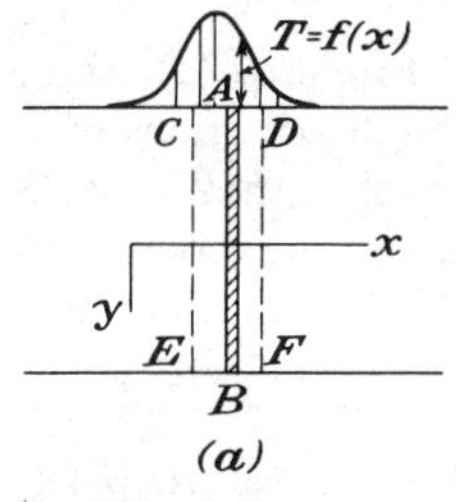

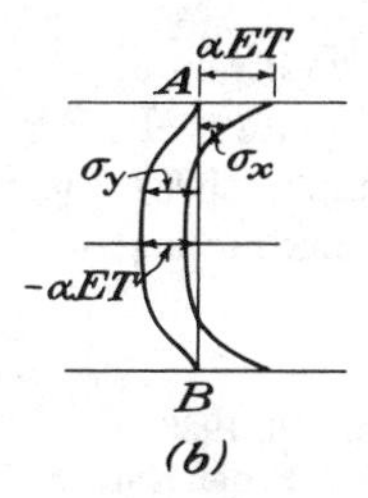

Fig. 225

To arrive at the thermal stress, we must superpose on (a) the stress due to the application of equal and opposite forces, i.e., tension of intensity αET, along the edges $y = \pm c$ of the strip.

If the heating is confined to a length of the strip short in comparison with its width $2c$, such as $CDFE$ in Fig. 225, the effect of the tensions αET will be felt only in the neighborhood of CD on the top edge and of EF on the bottom edge. Each of these neighborhoods can then be considered as presenting a problem of the type considered in Art. 37. It was pointed out on page 108 that a normal stress on a straight boundary produces a like normal stress parallel to and at the boundary. Hence, the tensions αET will produce tensile stress αET in the x direction. Both normal stresses die away as we proceed into the plate normal to the edge. On superposing these stresses on the compressive stress (a) in the y direction, we obtain curves[1] for σ_x and σ_y along a line such as AB in the hottest part of the plate of the character shown in Fig. 225b. Near the edges the prevailing stress is σ_x, with the value αET, tensile when T is positive; and near the middle the prevailing stress is σ_y, a compressive stress of magnitude αET when T is positive. The maximum stresses are of magnitude $\alpha ET_{\max}$.

If the temperature T is a periodic function of x, the application of edge tensions αET presents a problem of the type considered in Art. 24. When

$$T = T_0 \sin \alpha x \tag{b}$$

we find from Eqs. (k) of Art. 24, putting $A = B = -\alpha ET_0$ in accordance with Eq. (f),

$$\sigma_x = -2\alpha ET_0 \frac{(\alpha c \cosh \alpha c - \sinh \alpha c) \cosh \alpha y - \alpha y \sinh \alpha y \sinh \alpha c}{\sinh 2\alpha c + 2\alpha c} \times \sin \alpha x$$

$$\sigma_y = 2\alpha ET_0 \frac{(\alpha c \cosh \alpha c + \sinh \alpha c) \cosh \alpha y - \alpha y \sinh \alpha y \sinh \alpha c}{\sinh 2\alpha c + 2\alpha c} \sin \alpha x$$

$$\tau_{xy} = 2\alpha ET_0 \frac{\alpha c \cosh \alpha c \sinh \alpha y - \alpha y \cosh \alpha y \sinh \alpha c}{\sinh 2\alpha c + 2\alpha c} \cos \alpha x$$

Together with the compressive stress $\sigma_y = -\alpha ET$ from Eq. (a), these give the thermal stress in the plate.[2] In Fig. 226 the distributions of σ_x along the lines of maximum temperature for various wavelengths $2l = 2\pi/\alpha$ are shown. We see that the maximum stress increases as the wavelength diminishes, and approaches the value αET_0. Having the solution for a

[1] J. N. Goodier, *Phys.*, vol. 7, p. 156, 1936.

[2] The problem was discussed by J. P. den Hartog, *J. Franklin Inst.*, vol. 222, p. 149, 1936, in connection with the thermal stress produced in the process of welding.

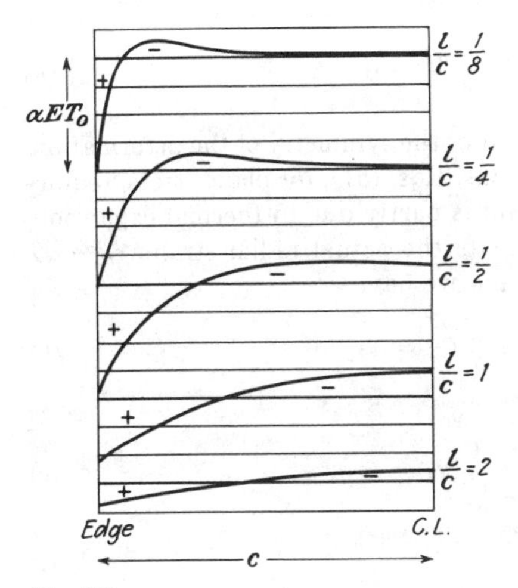

Fig. 226

sinusoidal temperature distribution, other cases in which the temperature is a periodic function of x can be treated. It can be concluded also that the maximum stress in plates of finite length can differ only slightly from the value $\alpha E T_0$ obtained for an infinite strip.

In this and the preceding articles, we have in each problem (except that of the sphere on page 438) applied forces to suppress a strain component arising from the thermal expansion. This method of strain suppression can be applied more systematically by applying forces to suppress all three strain components introduced by the expansion, and general equations for the three-dimensional problem will be derived and interpreted from this standpoint in Art. 153.

Alternatively, the problems of thermal stress can be treated independently from the beginning. The problem of imposed loading or deformation can be regarded as a special case of the more general problem admitting nonuniform temperature and imposed loading (or deformation) simultaneously. In the next three articles we consider the simplest problems of the disk, cylinder, and sphere in this way.

150 | The Thin Circular Disk: Temperature Symmetrical about Center

When the temperature T does not vary over the thickness of the disk, we may assume that the stress and displacement due to the heating also do not vary over the thickness. The stresses σ_r and σ_θ satisfy the equation of equilibrium

$$\frac{d\sigma_r}{dr} + \frac{\sigma_r - \sigma_\theta}{r} = 0 \tag{a}$$

The shear stress $\tau_{r\theta}$ is zero on account of the symmetry of the deformation.

The ordinary stress-strain relations, Eqs. (51), for plane stress, require modification because now the strain is partly due to thermal expansion, partly due to stress. If ϵ_r represents the actual radial strain, $\epsilon_r - \alpha T$ represents the part due to stress, and we have

$$\epsilon_r - \alpha T = \frac{1}{E}(\sigma_r - \nu\sigma_\theta) \tag{b}$$

and similarly

$$\epsilon_\theta - \alpha T = \frac{1}{E}(\sigma_\theta - \nu\sigma_r) \tag{c}$$

Solving (*b*) and (*c*) for σ_r, σ_θ, we find

$$\begin{aligned} \sigma_r &= \frac{E}{1-\nu^2}[\epsilon_r + \nu\epsilon_\theta - (1+\nu)\alpha T] \\ \sigma_\theta &= \frac{E}{1-\nu^2}[\epsilon_\theta + \nu\epsilon_r - (1+\nu)\alpha T] \end{aligned} \tag{d}$$

and with these Eq. (*a*) becomes

$$r\frac{d}{dr}(\epsilon_r + \nu\epsilon_\theta) + (1-\nu)(\epsilon_r - \epsilon_\theta) = (1+\nu)\alpha r\frac{dT}{dr} \tag{e}$$

If u denotes the radial displacement, we have, from Art. 30,

$$\epsilon_r = \frac{du}{dr} \qquad \epsilon_\theta = \frac{u}{r} \tag{f}$$

Substituting these in (*e*), we obtain

$$\frac{d^2u}{dr^2} + \frac{1}{r}\frac{du}{dr} - \frac{u}{r^2} = (1+\nu)\alpha\frac{dT}{dr}$$

which may be written

$$\frac{d}{dr}\left[\frac{1}{r}\frac{d(ru)}{dr}\right] = (1+\nu)\alpha\frac{dT}{dr} \tag{g}$$

Integration of this equation yields

$$u = (1+\nu)\alpha\frac{1}{r}\int_a^r Tr\,dr + C_1 r + \frac{C_2}{r} \tag{h}$$

where the lower limit a in the integral can be chosen arbitrarily. For a disk with a hole it may be the inner radius. For a solid disk we may take it as zero.

The stress components are now found by using the solution (*h*) in Eqs. (*f*) and substituting the results in Eqs. (*d*). Then

$$\sigma_r = -\alpha E \frac{1}{r^2} \int_a^r Tr\,dr + \frac{E}{1-\nu^2}\left[C_1(1+\nu) - C_2(1-\nu)\frac{1}{r^2}\right] \tag{i}$$

$$\sigma_\theta = \alpha E \frac{1}{r^2} \int_a^r Tr\,dr - \alpha ET + \frac{E}{1-\nu^2}\left[C_1(1+\nu) + C_2(1-\nu)\frac{1}{r^2}\right] \tag{j}$$

The constants C_1, C_2 are determined by the boundary conditions.

For a solid disk, we take a as zero, and observing that

$$\lim_{r\to 0} \frac{1}{r} \int_0^r Tr\,dr = 0$$

we see from Eq. (*h*) that the C_2 term must be excluded in order that u may be zero at the center. At the edge $r = b$ we must have $\sigma_r = 0$, and therefore from Eq. (*i*),

$$C_1 = (1-\nu)\frac{\alpha}{b^2} \int_0^b Tr\,dr$$

The final expressions for the stresses are consequently

$$\sigma_r = \alpha E \left(\frac{1}{b^2}\int_0^b Tr\,dr - \frac{1}{r^2}\int_0^r Tr\,dr\right) \tag{245}$$

$$\sigma_\theta = \alpha E \left(-T + \frac{1}{b^2}\int_0^b Tr\,dr + \frac{1}{r^2}\int_0^r Tr\,dr\right) \tag{246}$$

These give finite values at the center since

$$\lim_{r\to 0} \frac{1}{r^2} \int_0^r Tr\,dr = \frac{1}{2} T_0$$

where T_0 is the temperature at the center.

151 | The Long Circular Cylinder

The temperature is taken to be symmetrical about the axis and independent of the axial coordinate z.[1] We shall suppose first that w, the axial displacement, is zero throughout, and then modify the solution to the case of free ends.

Plane Strain We now have three components of stress, σ_r, σ_θ, σ_z, all three shear strains and stresses being zero on account of the symmetry about the axis and the uniformity in the axial direction. The stress-strain rela-

[1] The first solution of this problem is that of J. M. C. Duhamel, "Memoires . . . par Divers Savants," vol. 5, p. 440, Paris, 1838.

tions are

$$\begin{aligned} \epsilon_r - \alpha T &= \frac{1}{E}[\sigma_r - \nu(\sigma_\theta + \sigma_z)] \\ \epsilon_\theta - \alpha T &= \frac{1}{E}[\sigma_\theta - \nu(\sigma_r + \sigma_z)] \\ \epsilon_z - \alpha T &= \frac{1}{E}[\sigma_z - \nu(\sigma_r + \sigma_\theta)] \end{aligned} \tag{247}$$

But since $w = 0$, $\epsilon_z = 0$, and the third of Eqs. (247) gives

$$\sigma_z = \nu(\sigma_r + \sigma_\theta) - \alpha ET \tag{a}$$

On substituting this into the first two of Eqs. (247), these equations become

$$\begin{aligned} \epsilon_r - (1+\nu)\alpha T &= \frac{1-\nu^2}{E}\left(\sigma_r - \frac{\nu}{1-\nu}\sigma_\theta\right) \\ \epsilon_\theta - (1+\nu)\alpha T &= \frac{1-\nu^2}{E}\left(\sigma_\theta - \frac{\nu}{1-\nu}\sigma_r\right) \end{aligned} \tag{b}$$

It may be seen at once that these equations can be obtained from the corresponding equations of plane stress, Eqs. (*b*) and (*c*) of the preceding article, by putting, in the latter equations, $E/(1-\nu^2)$ for E, $\nu/(1-\nu)$ for ν, and $(1+\nu)\alpha$ for α.

Equations (*a*) and (*f*) of the preceding article remain valid here. The solution for u, σ_r, and σ_θ proceeds in just the same way. We may therefore write down the results by making the above substitutions in Eqs. (*h*), (*i*), and (*j*). Thus for the present problem

$$u = \frac{1+\nu}{1-\nu}\alpha\frac{1}{r}\int_a^r Tr\,dr + C_1 r + \frac{C_2}{r} \tag{c}$$

$$\sigma_r = -\frac{\alpha E}{1-\nu}\frac{1}{r^2}\int_a^r Tr\,dr + \frac{E}{1+\nu}\left(\frac{C_1}{1-2\nu} - \frac{C_2}{r^2}\right) \tag{d}$$

$$\sigma_\theta = \frac{\alpha E}{1-\nu}\frac{1}{r^2}\int_a^r Tr\,dr - \frac{\alpha ET}{1-\nu} + \frac{E}{1+\nu}\left(\frac{C_1}{1-2\nu} + \frac{C_2}{r^2}\right) \tag{e}$$

and, from Eq. (*a*),

$$\sigma_z = -\frac{\alpha ET}{1-\nu} + \frac{2\nu EC_1}{(1+\nu)(1-2\nu)} \tag{f}$$

Normal force distributed according to Eq. (*f*) must be applied to the ends of the cylinder in order to keep $w = 0$ throughout. But we can now superpose a uniform axial stress $\sigma_z = C_3$, choosing C_3 so that the resultant force on the ends is zero. The self-equilibrating distribution remaining on each end will, by Saint-Venant's principle (see page 40), give rise only to local effects at the ends.

The stresses σ_r, σ_θ will still be given by Eqs. (*d*) and (*e*). The displacement u, however, is affected by the axial stress C_3. A term $-\nu C_3 r/E$ must be added on the right of Eq. (*c*). The axial displacement is that corresponding to the uniform stress C_3.

Solid Cylinder In this case we may take a, the lower limit of the integrals in Eqs. (*c*), (*d*), and (*e*), as zero. The displacement u must vanish when $r = 0$. We must therefore exclude the terms in C_2.

The constant C_1 is found from the condition that the curved surface $r = b$ is free from force, so that $(\sigma_r)_{r=b} = 0$. Thus from Eq. (*d*), putting $C_2 = 0$, $a = 0$ we find

$$\frac{C_1}{(1+\nu)(1-2\nu)} = \frac{\alpha}{1-\nu}\frac{1}{b^2}\int_0^b Tr\,dr \tag{g}$$

The resultant of the axial stress (*f*) is

$$\int_0^b \sigma_z\, 2\pi r\,dr = -\frac{2\pi\alpha E}{1-\nu}\int_0^b Tr\,dr + \frac{2\nu E C_1}{(1+\nu)(1-2\nu)}\pi b^2$$

and the resultant of the uniform axial stress C_3 is $C_3\pi b^2$. The value of C_3 making the total axial force zero is therefore given by

$$C_3\pi b^2 = \frac{2\pi\alpha E}{1-\nu}\int_0^b Tr\,dr - \frac{2\nu E C_1}{(1+\nu)(1-2\nu)}\pi b^2 \tag{h}$$

For zero axial strain ($\epsilon_z = 0$), the final expressions for u, σ_r, σ_θ, σ_z are, from Eqs. (*c*), (*d*), (*e*), (*f*), (*g*), and (*h*),

$$u = \frac{1+\nu}{1-\nu}\alpha\left[(1-2\nu)\frac{r}{b^2}\int_0^b Tr\,dr + \frac{1}{r}\int_0^r Tr\,dr\right] \tag{248}$$

$$\sigma_r = \frac{\alpha E}{1-\nu}\left(\frac{1}{b^2}\int_0^b Tr\,dr - \frac{1}{r^2}\int_0^r Tr\,dr\right) \tag{249}$$

$$\sigma_\theta = \frac{\alpha E}{1-\nu}\left(\frac{1}{b^2}\int_0^b Tr\,dr + \frac{1}{r^2}\int_0^r Tr\,dr - T\right) \tag{250}$$

$$\sigma_z = \frac{\alpha E}{1-\nu}\left(\frac{2\nu}{b^2}\int_0^b Tr\,dr - T\right) \tag{251}$$

For zero axial force ($F_z = 0$), σ_r and σ_θ are given by (249) and (250), but for u and σ_z we have

$$u = \frac{1+\nu}{1-\nu}\alpha\left(\frac{1-3\nu}{1+\nu}\frac{r}{b^2}\int_0^b Tr\,dr + \frac{1}{r}\int_0^r Tr\,dr\right) \tag{252}$$

$$\sigma_z = \frac{\alpha E}{1-\nu}\left(\frac{2}{b^2}\int_0^b Tr\,dr - T\right) \tag{253}$$

Take, for example, a long cylinder with a constant initial temperature T_0. If, beginning from an instant $t = 0$, the lateral surface of the cylinder is maintained at a temperature zero,[1] the distribution of temperature at any instant t is given by the series[2]

$$T = T_0 \sum_{n=1}^{\infty} A_n J_0\left(\beta_n \frac{r}{b}\right) e^{-p_n t} \tag{i}$$

in which $J_0(\beta_n r/b)$ is the Bessel function of zero order and the β's are the roots of the equation $J_0(\beta) = 0$. The coefficients of the series (i) are

$$A_n = \frac{2}{\beta_n J_1(\beta_n)}$$

and the constants p_n are given by the equation

$$p_n = \frac{k}{c\rho} \frac{\beta_n^2}{b^2} \tag{j}$$

in which k is the thermal conductivity, c the specific heat of the material, and ρ the density. Substituting series (i) into Eq. (249) and taking into account the fact that[3]

$$\int_0^r J_0\left(\beta_n \frac{r}{b}\right) r\,dr = \frac{br}{\beta_n} J_1\left(\beta_n \frac{r}{b}\right)$$

we find that

$$\sigma_r = \frac{2\alpha E T_0}{1 - \nu} \sum_{n=1}^{\infty} e^{-p_n t} \left\{\frac{1}{\beta_n^2} - \frac{1}{\beta_n^2} \frac{b}{r} \frac{J_1[\beta_n(r/b)]}{J_1(\beta_n)}\right\} \tag{k}$$

In the same manner, substituting series (i) in Eq. (250), we obtain

$$\sigma_\theta = \frac{2\alpha E T_0}{1 - \nu} \sum_{n=1}^{\infty} e^{-p_n t} \left\{\frac{1}{\beta_n^2} + \frac{1}{\beta_n^2} \frac{b}{r} \frac{J_1[\beta_n(r/b)]}{J_1(\beta_n)} - \frac{J_0[\beta_n(r/b)]}{\beta_n J_1(\beta_n)}\right\} \tag{l}$$

Substituting series (i) in Eq. (253), we find

$$\sigma_z = \frac{2\alpha E T_0}{1 - \nu} \sum_{n=1}^{\infty} e^{-p_n t} \left\{\frac{2}{\beta_n^2} - \frac{J_0[\beta_n(r/b)]}{\beta_n J_1(\beta_n)}\right\} \tag{m}$$

Formulas (k), (l), and (m) represent the complete solution of the problem. Several numerical examples can be found in the papers[2] by A. Dinnik and C. H. Lees.

[1] It is assumed that the surface of the cylinder suddenly assumes the temperature zero. If the temperature of the surface is T_1 instead of zero, then $T_0 - T_1$ must be put instead of T_0 in our equations.

[2] See Byerly, "Fourier Series and Spherical Harmonics," p. 229. The calculation of thermal stresses for this case is given by A. Dinnik, "Applications of Bessel's Function to Elasticity Problems," pt. 2, p. 95, Ekaterinoslav, 1915. See also C. H. Lees, *Proc. Roy. Soc. (London)*, vol. 101, p. 411, 1922.

[3] See E. Jahnke, F. Emde, and F. Lösch, "Tables of Higher Functions," McGraw-Hill Book Company, New York, 1960.

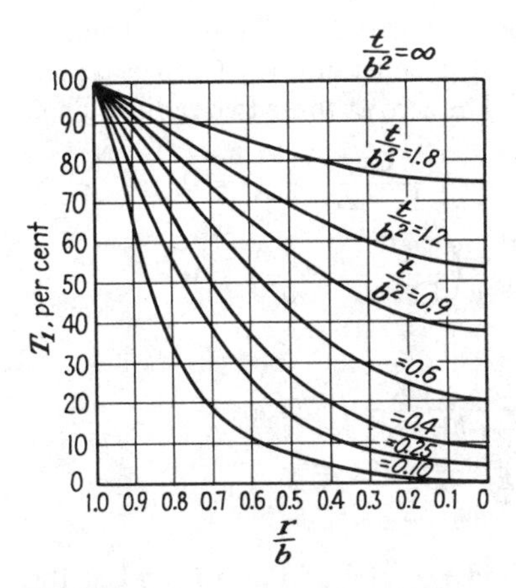

Fig. 227

Figure 227 represents[1] the distribution of temperature in a steel cylinder. It is assumed that the cylinder had a uniform initial temperature equal to zero and that beginning from an instant $t = 0$ the surface of the cylinder is maintained at a temperature T_1. The temperature distributions along the radius, for various values of t/b^2 (t is measured in seconds and b in centimeters), are represented by curves. It will be seen from Eqs. (*i*) and (*j*) that the temperature distribution for cylinders of various diameters is the same if the time of heating t is proportional to the square of the diameter. From the figure, the average temperature of the whole cylinder and also of an inner portion of the cylinder of radius r can be calculated. Having these temperatures, we find the thermal stresses from Eqs. (249), (250), and (253). If we take a very small value for t, the average temperatures, mentioned above, approach zero and we find at the surface

$$\sigma_r = 0 \qquad \sigma_\theta = \sigma_z = -\frac{\alpha E T_1}{1-\nu}$$

This is the numerical maximum of the thermal stress produced in a cylinder by heating. It is equal to the stress necessary for entire suppression of the thermal expansion *in* the surface (but not normal to the surface). During heating this stress is compression, during cooling it is tension. In order to reduce the maximum stresses it is the usual practice to begin the heating of shafts and rotors with a somewhat lower temperature than the final temperature T_1 and to increase the time of heating in proportion to the square of the diameter.

[1] The figure is taken from A. Stodola, "Dampf- und Gasturbinen," 6th ed., p. 961, 1924.

Cylinder with a Concentric Circular Hole[1] The radius of the hole being a, and the outer radius of the cylinder b, the constants C_1, C_2 in Eqs. (*c*), (*d*), and (*e*) are determined so that σ_r will be zero at these two radii. Then

$$\frac{C_1}{1-2\nu} - \frac{C_2}{a^2} = 0$$

$$-\frac{\alpha E}{1-\nu}\frac{1}{b^2}\int_a^b Tr\,dr + \frac{E}{1+\nu}\left(\frac{C_1}{1-2\nu} - \frac{C_2}{b^2}\right) = 0$$

and from these

$$\frac{EC_2}{1+\nu} = \frac{\alpha E}{1-\nu}\frac{a^2}{b^2-a^2}\int_a^b Tr\,dr$$

$$\frac{EC_1}{(1+\nu)(1-2\nu)} = \frac{\alpha E}{1-\nu}\frac{1}{b^2-a^2}\int_a^b Tr\,dr$$

Substituting these values in (*d*), (*e*), and (*f*), and adding to the last the axial stress C_3 required to make the resultant axial force zero, we find the formulas

$$\sigma_r = \frac{\alpha E}{1-\nu}\frac{1}{r^2}\left(\frac{r^2-a^2}{b^2-a^2}\int_a^b Tr\,dr - \int_a^r Tr\,dr\right) \tag{254}$$

$$\sigma_\theta = \frac{\alpha E}{1-\nu}\frac{1}{r^2}\left(\frac{r^2+a^2}{b^2-a^2}\int_a^b Tr\,dr + \int_a^r Tr\,dr - Tr^2\right) \tag{255}$$

$$\sigma_z = \frac{\alpha E}{1-\nu}\left(\frac{2}{b^2-a^2}\int_a^b Tr\,dr - T\right) \tag{256}$$

Consider, as an example, a *steady heat flow*. If T_i is the temperature on the inner surface of the cylinder and the temperature on the outer surface is zero, the temperature T at any distance r from the center is represented by the expression

$$T = \frac{T_i}{\log(b/a)}\log\frac{b}{r} \tag{n}$$

Substituting this in Eqs. (254), (255), and (256), we find the following expressions for the thermal stresses:[2]

$$\begin{aligned}
\sigma_r &= \frac{\alpha E T_i}{2(1-\nu)\log(b/a)}\left[-\log\frac{b}{r} - \frac{a^2}{(b^2-a^2)}\left(1-\frac{b^2}{r^2}\right)\log\frac{b}{a}\right]\\
\sigma_\theta &= \frac{\alpha E T_i}{2(1-\nu)\log(b/a)}\left[1-\log\frac{b}{r} - \frac{a^2}{(b^2-a^2)}\left(1+\frac{b^2}{r^2}\right)\log\frac{b}{a}\right]\\
\sigma_z &= \frac{\alpha E T_i}{2(1-\nu)\log(b/a)}\left[1-2\log\frac{b}{r} - \frac{2a^2}{(b^2-a^2)}\log\frac{b}{a}\right]
\end{aligned} \tag{257}$$

[1] R. Lorenz, *Z. Ver. Deutsch. Ing.*, vol. 51, p. 743, 1907.

[2] Charts for the rapid calculation of stresses from Eqs. (257) are given by L. Barker, *Eng.*, vol. 124, p. 443, 1927.

If T_i is positive, the radial stress is compressive at all points and becomes zero at the inner and outer surfaces of the cylinder. The stress components σ_θ and σ_z have their largest numerical values at the inner and outer surfaces of the cylinder. Taking $r = a$, we find that

$$(\sigma_\theta)_{r=a} = (\sigma_z)_{r=a} = \frac{\alpha E T_i}{2(1-\nu)\log(b/a)}\left(1 - \frac{2b^2}{b^2-a^2}\log\frac{b}{a}\right) \tag{258}$$

For $r = b$ we obtain

$$(\sigma_\theta)_{r=b} = (\sigma_z)_{r=b} = \frac{\alpha E T_i}{2(1-\nu)\log(b/a)}\left(1 - \frac{2a^2}{b^2-a^2}\log\frac{b}{a}\right) \tag{259}$$

The distribution of thermal stresses over the thickness of the wall for the particular case $a/b = 0.3$ is shown in Fig. 228. If T_i is positive, the stresses are compressive at the inner surface and tensile at the outer surface. In such materials as stone, brick, or concrete, which are weak in tension, cracks are likely to start on the outer surface of the cylinder under the above conditions.

If the thickness of the wall is small in comparison with the outer radius of the cylinder, we can simplify Eqs. (258) and (259) by putting

$$\frac{b}{a} = 1 + m \qquad \log\frac{b}{a} = m - \frac{m^2}{2} + \frac{m^3}{3} - \cdots$$

and considering m as a small quantity. Then

$$(\sigma_\theta)_{r=a} = (\sigma_z)_{r=a} = -\frac{\alpha E T_i}{2(1-\nu)}\left(1 + \frac{m}{3}\right) \tag{258'}$$

$$(\sigma_\theta)_{r=b} = (\sigma_z)_{r=b} = \frac{\alpha E T_i}{2(1-\nu)}\left(1 - \frac{m}{3}\right) \tag{259'}$$

If the temperature at the outer surface of the cylinder is different from zero, the above results can be used by substituting the difference between the inner and the outer temperatures, $T_i - T_o$, in all our equations instead of T_i.

In the case of a very thin wall we can make a further simplification and neglect the term $m/3$ in comparison with unity in Eqs. (258′) and (259′).

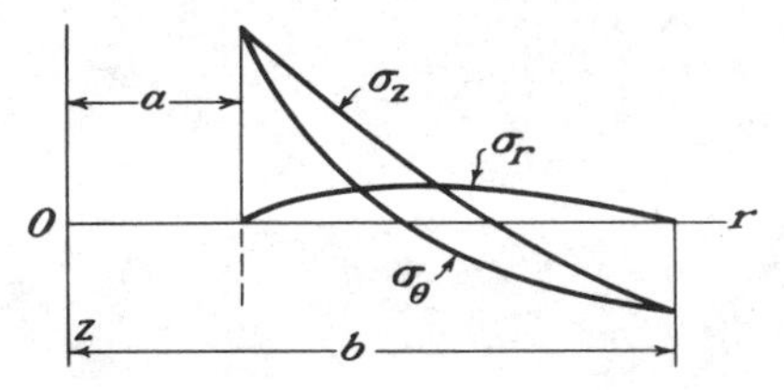

Fig. 228

Then

$$
\begin{aligned}
(\sigma_\theta)_{r=a} &= (\sigma_z)_{r=a} = -\frac{\alpha E T_i}{2(1-\nu)} \\
(\sigma_\theta)_{r=b} &= (\sigma_z)_{r=b} = \frac{\alpha E T_i}{2(1-\nu)}
\end{aligned}
\tag{260}
$$

and the distribution of thermal stresses over the thickness of the wall is the same as in the case of a flat plate of thickness $2c = b - a$, when the temperature is given by the equation (Fig. 224)

$$
T = \frac{T_i y}{(b-a)}
$$

and the edges are clamped, so that bending of the plate, due to nonuniform heating, is prevented [see Eq. (k), Art. 148].

If a high-frequency fluctuation of temperature is superposed on a steady heat flow, the thermal stresses produced by the fluctuation can be calculated in the manner explained for the case of flat plates (see Art. 148).

In the foregoing discussion it was assumed that the cylinder is very long and that we are considering stresses far away from the ends. Near the ends, the problem of thermal stress distribution is more complicated due to local irregularities. Let us consider this problem for the case of a cylinder with a thin wall. Solution (260) requires that the normal forces shown in Fig. 229a should be distributed over the ends of the cylinder. To find the stresses in a cylinder with free ends, we must superpose on the stresses (260) the stresses produced by forces equal and opposite to those shown in Fig. 229a. In the case of a thin wall of thickness h these forces can be reduced to bending moments M, as shown in Fig. 229b, uniformly distributed along the edge of the cylinder and equal to

$$
M = \frac{\alpha E T_i}{2(1-\nu)} \frac{h^2}{6} \tag{o}
$$

per unit length of the edge. To estimate the stresses produced by these moments, consider a longitudinal strip, of width equal to unity, cut out from the cylindrical

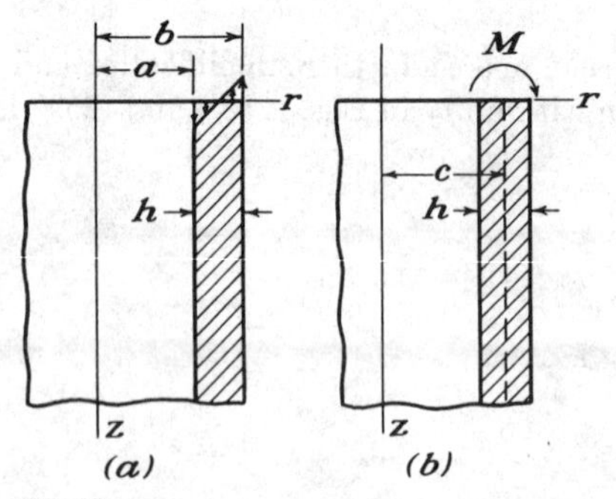

Fig. 229

shell. Such a strip can be treated as a bar on an elastic foundation. The deflection curve of this strip is given by the equation[1]

$$u = \frac{Me^{-\beta z}}{2\beta^2 D} (\cos \beta z - \sin \beta z) \tag{p}$$

in which

$$\beta = \sqrt{\frac{3(1 - \nu^2)}{c^2 h^2}} \qquad D = \frac{Eh^3}{12(1 - \nu^2)} \tag{q}$$

and c is the middle radius of the cylindrical shell. Having this deflection curve, the corresponding bending stresses σ_z and the tangential stresses σ_θ can be calculated for any value of z. The maximum deflection of the strip is evidently at the end $z = 0$, where

$$(u)_{z=0} = \frac{M}{2\beta^2 D} = \frac{\alpha c T_i \sqrt{1 - \nu^2}}{2\sqrt{3}\,(1 - \nu)}$$

The corresponding strain component in the tangential direction is

$$\epsilon_\theta = \frac{u}{c} = \frac{\alpha T_i \sqrt{1 - \nu^2}}{2\sqrt{3}\,(1 - \nu)} \tag{r}$$

The stress component in the tangential direction at the outer surface of the wall is then obtained, using Hooke's law, from the equation

$$\sigma_\theta = E\epsilon_\theta + \nu\sigma_z = \frac{\alpha E T_i \sqrt{1 - \nu^2}}{2\sqrt{3}\,(1 - \nu)} - \frac{\nu \alpha E T_i}{2(1 - \nu)}$$

Adding this stress to the corresponding stress calculated from Eqs. (260), the maximum tangential stress for a thin-walled cylinder at the free end is

$$(\sigma_\theta)_{max} = \frac{\alpha E T_i}{2(1 - \nu)} \left(\frac{\sqrt{1 - \nu^2}}{\sqrt{3}} - \nu + 1 \right) \tag{261}$$

Assuming $\nu = 0.3$, we find

$$(\sigma_\theta)_{max} = 1.25 \frac{\alpha E T_i}{2(1 - \nu)}$$

Thus the maximum tensile stress at the free end of the cylinder is 25 percent greater than that obtained from Eqs. (260) for the stress at points remote from the ends. From Eq. (p) it can be seen that the increase of stress near the free ends of the cylinder, since it depends on the deflection u, is of a local character and diminishes rapidly with increase of distance z from the end.

The approximate method of calculating thermal stresses in thin-walled cylinders, by using the deflection curve of a bar on an elastic foundation, can also be applied in the case in which the temperature varies along the axis of the cylindrical shell.[2] A suitable external pressure will annul the radial expansion of each elementary ring, axial expansion occurring freely. Removal of the pressure, the rings now being joined, presents a solvable nonthermal problem.

[1] See S. Timoshenko, "Strength of Materials," 3d ed., vol. 2, pp. 126–137, D. Van Nostrand Company, Inc., Princeton, N.J., 1956.

[2] S. Timoshenko and J. M. Lessells, "Applied Elasticity," p. 147, 1925, and C. H. Kent, *Trans. ASME*, Applied Mechanics Division, vol. 53, p. 167, 1931.

PROBLEMS

1. Evaluate σ_r and σ_θ at $r = 0$ from (249) and (250). Explain why the results *must* be equal.

   Find an expression for the displacement w in the cylinder with a concentric circular hole, under zero axial force. Explain why it is not valid near an end when the end condition is $\sigma_z = 0$.

2. *If* in (254) to (256) we put $a = 0$, the results agree with (249) and (250) and σ_r at $r = 0$ is not zero (Prob. 1). But (254) satisfies the boundary condition $\sigma_r = 0$ at $r = a$ and continues to do so however small a may be.

   Clarify this by giving an account of the state of stress in the neighborhood of the hole when it is very small. Begin with the solid cylinder and consider the removal of the σ_r traction on the hole.

152 | The Sphere

We consider here the simple case of a temperature symmetrical with respect to the center, and so a function of r, the radial distance, only.[1]

On account of the symmetry there will be three nonzero stress components, the radial component σ_r, and two tangential components σ_t, as in Art. 136, and these must satisfy the condition of equilibrium, in the radial direction, of an element [Fig. 205, Eq. (*e*), page 395],

$$\frac{d\sigma_r}{dr} + \frac{2}{r}(\sigma_r - \sigma_t) = 0 \tag{a}$$

The stress-strain relations are

$$\epsilon_r - \alpha T = \frac{1}{E}(\sigma_r - 2\nu\sigma_t) \tag{b}$$

$$\epsilon_t - \alpha T = \frac{1}{E}[\sigma_t - \nu(\sigma_r + \sigma_t)] \tag{c}$$

and, u being the radial displacement, we have

$$\epsilon_r = \frac{du}{dr} \qquad \epsilon_t = \frac{u}{r} \tag{d}$$

From (*b*) and (*c*) we find

$$\sigma_r = \frac{E}{(1+\nu)(1-2\nu)}[(1-\nu)\epsilon_r + 2\nu\epsilon_t - (1+\nu)\alpha T] \tag{e}$$

$$\sigma_t = \frac{E}{(1+\nu)(1-2\nu)}[\epsilon_t + \nu\epsilon_r - (1+\nu)\alpha T] \tag{f}$$

[1] The problem was solved by Duhamel, *loc. cit.*; F. Neuman, *Abhandl. Akad. Wiss.*, Berlin, 1841; see also his "Vorlesungen über die Theorie der Elastizität der festen Körper," Leipzig, 1885; J. Hopkinson, *Messenger Math.*, vol. 8, p. 168, 1879. Nonsymmetrical temperatures were considered by C. W. Borchardt, *Monatsber. Akad. Wiss.*, Berlin, 1873, p. 9.

Substituting these in (a), then replacing ϵ_r, ϵ_t by the values given in (d), we find the differential equation for u

$$\frac{d^2u}{dr^2} + \frac{2}{r}\frac{du}{dr} - \frac{2u}{r^2} = \frac{1+\nu}{1-\nu}\alpha\frac{dT}{dr} \tag{g}$$

which can be written

$$\frac{d}{dr}\left[\frac{1}{r^2}\frac{d}{dr}(r^2u)\right] = \frac{1+\nu}{1-\nu}\alpha\frac{dT}{dr}$$

The solution is

$$u = \frac{1+\nu}{1-\nu}\alpha\frac{1}{r^2}\int_a^r Tr^2\,dr + C_1r + \frac{C_2}{r^2} \tag{h}$$

where C_1 and C_2 are constants of integration to be determined later from boundary conditions, and a is any convenient lower limit for the integral, such as the inner radius of a hollow sphere.

This solution can be substituted in Eqs. (d), and the results used in Eqs. (e) and (f). Then

$$\sigma_r = -\frac{2\alpha E}{1-\nu}\frac{1}{r^3}\int_a^r Tr^2\,dr + \frac{EC_1}{1-2\nu} - \frac{2EC_2}{1+\nu}\frac{1}{r^3} \tag{i}$$

$$\sigma_t = \frac{\alpha E}{1-\nu}\frac{1}{r^3}\int_a^r Tr^2\,dr + \frac{EC_1}{1-2\nu} + \frac{EC_2}{1+\nu}\frac{1}{r^3} - \frac{\alpha ET}{1-\nu} \tag{j}$$

We shall now consider several particular cases.

Solid Sphere In this case the lower limit a of the integrals may be taken as zero. We must have $u = 0$ at $r = 0$, and since

$$\lim_{r\to 0}\frac{1}{r^2}\int_0^r Tr^2\,dr = 0$$

this means that in Eq. (h) we must exclude the term in C_2. The stress components given by (i) and (j) will now be finite at the center since

$$\lim_{r\to 0}\frac{1}{r^3}\int_0^r Tr^2\,dr = \frac{T_0}{3}$$

where T_0 is the temperature at the center. The constant C_1 is determined from the condition that the outer surface $r = b$ is free from force, so that $\sigma_r = 0$. Then from Eq. (i), putting $\sigma_r = 0$, $a = 0$, $C_2 = 0$, $r = b$, we find

$$\frac{EC_1}{1-2\nu} = \frac{2\alpha E}{1-\nu}\frac{1}{b^3}\int_0^b Tr^2\,dr$$

and the stress components become

$$
\begin{aligned}
\sigma_r &= \frac{2\alpha E}{1-\nu}\left(\frac{1}{b^3}\int_0^b Tr^2\,dr - \frac{1}{r^3}\int_0^r Tr^2\,dr\right) \\
\sigma_t &= \frac{\alpha E}{1-\nu}\left(\frac{2}{b^3}\int_0^b Tr^2\,dr + \frac{1}{r^3}\int_0^r Tr^2\,dr - T\right)
\end{aligned}
\tag{262}
$$

The mean temperature of the sphere within the radius r is

$$
\frac{4\pi\int_0^r Tr^2\,dr}{\frac{4}{3}\pi r^3} = \frac{3}{r^3}\int_0^r Tr^2\,dr
$$

Therefore the stress σ_r at any radius r is proportional to the difference between the mean temperature of the whole sphere and the mean temperature of a sphere of radius r. If the distribution of temperature is known, the calculation of stresses in each particular case can be carried out without difficulty.[1] An interesting example of such calculations was made by G. Grünberg[2] in connection with an investigation of the strength of isotropic materials subjected to equal tension in three perpendicular directions. If a solid sphere at a uniform initial temperature T_0 is put in a liquid at a higher temperature T_1, the outer portion of the sphere expands and produces at the center of the sphere a uniform tension in all directions. The maximum value of this tension occurs after a time

$$
t = 0.0574\,\frac{b^2 c\rho}{k} \tag{k}
$$

Here b is the radius of the sphere, k the thermal conductivity, c the specific heat of the material, and ρ the density. The magnitude of this maximum tensile stress is[3]

$$
\sigma_r = \sigma_t = 0.771\,\frac{\alpha E}{2(1-\nu)}(T_1 - T_0) \tag{l}
$$

The maximum compressive stress occurs at the surface of the sphere at the moment of application of the temperature T_1 and is equal to $\alpha E(T_1 - T_0)/(1-\nu)$. This is the same as we found before for a cylinder (see page 447). Applying Eqs. (k) and (l) to the case of steel, and taking $b = 10$ cm and $T_1 - T_0 = 100°$C, we find $\sigma_r = \sigma_t = 1{,}270$ kg per sq cm and $t = 33.4$ sec.

[1] Several examples of such calculations are given in the paper by E. Honegger, Festschrift Prof. A. Stodola, Zürich, 1929.

[2] *Z. Physik*, vol. 35, p. 548, 1925.

[3] It was assumed in the analysis that the surface of the sphere takes at once the temperature T_1 of the fluid.

Sphere with a Hole at the Center Denoting by a and b the inner and outer radii of the sphere, we determine the constants C_1 and C_2 in (i) and (j) from the conditions that σ_r is zero on the inner and outer surfaces. Then from (i) we have

$$\frac{EC_1}{1-2\nu} - \frac{2EC_2}{1+\nu}\frac{1}{a^3} = 0$$

$$-\frac{2\alpha E}{1-\nu}\frac{1}{b^3}\int_a^b Tr^2\,dr + \frac{EC_1}{1-2\nu} - \frac{2EC_2}{1+\nu}\frac{1}{b^3} = 0$$

Solving for C_1 and C_2 and inserting the results in (i) and (j) we find

$$\begin{aligned}\sigma_r &= \frac{2\alpha E}{1-\nu}\left[\frac{r^3-a^3}{(b^3-a^3)r^3}\int_a^b Tr^2\,dr - \frac{1}{r^3}\int_a^r Tr^2\,dr\right] \\ \sigma_t &= \frac{2\alpha E}{1-\nu}\left[\frac{2r^3+a^3}{2(b^3-a^3)r^3}\int_a^b Tr^2\,dr + \frac{1}{2r^3}\int_a^r Tr^2\,dr - \frac{1}{2}T\right]\end{aligned} \tag{263}$$

Thus, the stress components can be calculated if the distribution of temperature is given.

Consider, as an example, the case of steady heat flow. We denote the temperature at the inner surface by T_i and the temperature at the outer surface we take as zero. Then the temperature at any distance r from the center is

$$T = \frac{T_i a}{b-a}\left(\frac{b}{r} - 1\right) \tag{m}$$

Substituting this in expressions (263), we find

$$\sigma_r = \frac{\alpha E T_i}{1-\nu}\frac{ab}{b^3-a^3}\left[a + b - \frac{1}{r}(b^2+ab+a^2) + \frac{a^2b^2}{r^3}\right]$$

$$\sigma_t = \frac{\alpha E T_i}{1-\nu}\frac{ab}{b^3-a^3}\left[a + b - \frac{1}{2r}(b^2+ab+a^2) - \frac{a^2b^2}{2r^3}\right]$$

We see that the stress σ_r is zero for $r = a$ and $r = b$. It becomes a maximum or minimum when

$$r^2 = \frac{3a^2b^2}{a^2+ab+b^2}$$

The stress σ_t, for $T_i > 0$, increases as r increases. When $r = a$, we have

$$\sigma_t = -\frac{\alpha E T_i}{2(1-\nu)}\frac{b(b-a)(a+2b)}{b^3-a^3} \tag{n}$$

When $r = b$, we obtain

$$\sigma_t = \frac{\alpha E T_i}{2(1-\nu)}\frac{a(b-a)(2a+b)}{b^3-a^3} \tag{o}$$

In the case of a spherical shell of small thickness we put

$$b = a(1 + m)$$

where m is a small quantity. Substituting in (n) and (o) and neglecting higher powers of m, we obtain:

For $r = a$,

$$\sigma_t = -\frac{\alpha E T_i}{2(1 - \nu)} \left(1 + \frac{2}{3} m\right)$$

For $r = b$,

$$\sigma_t = \frac{\alpha E T_i}{2(1 - \nu)} \left(1 - \frac{2}{3} m\right)$$

If we neglect the quantity $\frac{2}{3}m$ we arrive at the same values for the tangential stresses as we obtained before for a thin cylindrical shell [see Eqs. (260)] and for a thin plate with clamped edges.

153 | General Equations

The differential equations (128) of equilibrium in terms of displacements can be extended to cover thermal stress and strain. The stress-strain relations for three-dimensional problems are

$$\begin{aligned} \epsilon_x - \alpha T &= \frac{1}{E} [\sigma_x - \nu(\sigma_y + \sigma_z)] \\ \epsilon_y - \alpha T &= \frac{1}{E} [\sigma_y - \nu(\sigma_x + \sigma_z)] \\ \epsilon_z - \alpha T &= \frac{1}{E} [\sigma_z - \nu(\sigma_x + \sigma_y)] \end{aligned} \tag{a}$$

$$\gamma_{xy} = \frac{\tau_{xy}}{G} \qquad \gamma_{yz} = \frac{\tau_{yz}}{G} \qquad \gamma_{xz} = \frac{\tau_{xz}}{G} \tag{b}$$

Equations (b) are not affected by the temperature because free thermal expansion does not produce angular distortion in an isotropic material.

Adding Eqs. (a) and using the notation given in Eqs. (7), we find

$$e = \frac{1}{E}(1 - 2\nu)\Theta + 3\alpha T$$

Using this, and solving for the stresses from Eqs. (a), we find

$$\sigma_x = \lambda e + 2G\epsilon_x - \frac{\alpha E T}{1 - 2\nu} \tag{c}$$

Substituting from this and Eqs. (6) into the equations of equilibrium (123), and assuming there are no body forces, we find three equations of which the first is

$$(\lambda + G)\frac{\partial e}{\partial x} + G\nabla^2 u - \frac{\alpha E}{1 - 2\nu}\frac{\partial T}{\partial x} = 0 \tag{264}$$

These equations replace Eqs. (127) in the calculation of thermal stresses. The boundary equations (124), after substituting from Eqs. (*c*) and (6) and assuming that there are no surface forces, become

$$\frac{\alpha E T}{1 - 2\nu} l = \lambda e l + G\left(\frac{\partial u}{\partial x} l + \frac{\partial u}{\partial y} m + \frac{\partial u}{\partial z} n\right) + G\left(\frac{\partial u}{\partial x} l + \frac{\partial v}{\partial x} m + \frac{\partial w}{\partial x} n\right) \tag{265}$$

. . . .

Comparing Eqs. (264) and (265) with Eqs. (127) and (130), it is seen that terms

$$-\frac{\alpha E}{1 - 2\nu}\frac{\partial T}{\partial x} \qquad -\frac{\alpha E}{1 - 2\nu}\frac{\partial T}{\partial y} \qquad -\frac{\alpha E}{1 - 2\nu}\frac{\partial T}{\partial z}$$

take the places of components X, Y, Z of the body forces, and terms

$$\frac{\alpha E T}{1 - 2\nu} l \qquad \frac{\alpha E T}{1 - 2\nu} m \qquad \frac{\alpha E T}{1 - 2\nu} n$$

replace components $\bar{X}$, $\bar{Y}$, $\bar{Z}$ of the surface forces. Thus the displacements u, v, w, produced by the temperature change T, are the same as the displacements produced by the body forces

$$X = -\frac{\alpha E}{1 - 2\nu}\frac{\partial T}{\partial x} \qquad Y = -\frac{\alpha E}{1 - 2\nu}\frac{\partial T}{\partial y} \qquad Z = -\frac{\alpha E}{1 - 2\nu}\frac{\partial T}{\partial z} \tag{d}$$

and normal tensions

$$\frac{\alpha E T}{1 - 2\nu} \tag{e}$$

distributed over the surface.

If the solution of Eqs. (264) satisfying the boundary conditions (265) is found, giving the displacements u, v, w, the shearing stresses can be calculated from Eqs. (*b*) and the normal stresses from Eqs. (*c*). From the latter equations we see that the normal stress components consist of two parts: (1) a part derived in the usual way by using the strain components, and (2) a "hydrostatic" pressure of the amount

$$\frac{\alpha E T}{1 - 2\nu} \tag{f}$$

proportional at each point to the temperature change at that point. Thus, the total stress produced by nonuniform heating is obtained by superposing hydrostatic pressure (*f*) on the stresses produced by body forces (*d*) and surface forces (*e*).

Method of Strain Suppression The same conclusion may be reached by the method of strain suppression. Imagine that the body undergoing nonuniform heating is subdivided into infinitesimal elements. Let the free thermal strains $\epsilon_x = \epsilon_y = \epsilon_z = \alpha T$ of these elements be counteracted by applying to each element a uniform pressure p, of the magnitude given by (*f*). Then the free thermal strain is annulled. The elements fit one another exactly and form a continuous body of the original shape and size. The pressure distribution (*f*) can be realized by applying certain body forces and surface pressures to the above body formed by the elements. These forces must satisfy the equations of equilibrium (123) and the boundary conditions (124). Substituting in these equations

$$\sigma_x = \sigma_y = \sigma_z = -p = -\frac{\alpha E T}{1 - 2\nu} \qquad \tau_{xy} = \tau_{xz} = \tau_{yz} = 0 \tag{g}$$

we find that, to keep the body formed by the elements in its initial shape, the necessary body forces are

$$X = \frac{\alpha E}{1 - 2\nu}\frac{\partial T}{\partial x} \qquad Y = \frac{\alpha E}{1 - 2\nu}\frac{\partial T}{\partial y} \qquad Z = \frac{\alpha E}{1 - 2\nu}\frac{\partial T}{\partial z} \tag{h}$$

and that the pressure (*f*) should be applied to the surface.

We now assume that the elements are joined together and remove the forces (*h*) and the surface pressure (*f*). Then the thermal stresses are evidently obtained by superposing on the pressures (*f*) the stresses that are produced in the elastic body by the body forces

$$X = -\frac{\alpha E}{1 - 2\nu}\frac{\partial T}{\partial x} \qquad Y = -\frac{\alpha E}{1 - 2\nu}\frac{\partial T}{\partial y} \qquad Z = -\frac{\alpha E}{1 - 2\nu}\frac{\partial T}{\partial z}$$

and a normal tension on the surface equal to

$$\frac{\alpha E T}{1 - 2\nu}$$

These latter stresses satisfy the equations of equilibrium

$$\begin{aligned}
\frac{\partial \sigma_x}{\partial x} + \frac{\partial \tau_{xy}}{\partial y} + \frac{\partial \tau_{xz}}{\partial z} - \frac{\alpha E}{1 - 2\nu}\frac{\partial T}{\partial x} &= 0 \\
\frac{\partial \sigma_y}{\partial y} + \frac{\partial \tau_{xy}}{\partial x} + \frac{\partial \tau_{yz}}{\partial z} - \frac{\alpha E}{1 - 2\nu}\frac{\partial T}{\partial y} &= 0 \\
\frac{\partial \sigma_z}{\partial z} + \frac{\partial \tau_{xz}}{\partial x} + \frac{\partial \tau_{yz}}{\partial y} - \frac{\alpha E}{1 - 2\nu}\frac{\partial T}{\partial z} &= 0
\end{aligned} \tag{266}$$

and the boundary conditions

$$\begin{aligned} \sigma_x l + \tau_{xy} m + \tau_{xz} n &= \frac{\alpha E T}{1 - 2\nu} \\ \sigma_y m + \tau_{yz} n + \tau_{xy} l &= \frac{\alpha E T}{1 - 2\nu} m \\ \sigma_z n + \tau_{xz} l + \tau_{yz} m &= \frac{\alpha E T}{1 - 2\nu} n \end{aligned} \tag{267}$$

together with the compatibility conditions discussed in Art. 85. They are accompanied by displacements u, v, w, and strains

$$\epsilon_x = \frac{\partial u}{\partial x}, \text{ etc.} \qquad \gamma_{xy} = \frac{\partial v}{\partial x} + \frac{\partial u}{\partial y}, \text{ etc.}$$

and are related to these strains by Hooke's law as in Eqs. (3) and (6). These, with (266) and (267), present an "ordinary" (isothermal) loading problem, in which the body and surface loads are specified in terms of the temperature field $T(x,y,z)$ of the original thermoelastic problem. The solution of this *ordinary problem* evidently yields the actual thermoelastic displacements.

It will now be evident that the methods and theorems already established for the ordinary problems can be taken over at once for application to the thermoelastic problems. For instance, the *uniqueness theorem* of Art. 96 assures us that for a given body with a given temperature field, there is only one solution for the stress and strain under the conditions of the linear, small-deformation theory. Buckling behavior, of course, does not come within these conditions.

The reciprocal theorem of Art. 97 is especially useful in leading to immediate solutions of thermoelastic problems by taking advantage of existing solutions of ordinary problems. We now explain this method and give a number of specific examples.

154 | Thermoelastic Reciprocal Theorem

The ordinary problem described above is now taken as the first state in the theorem of Art. 97, corresponding to the symbols carrying the single prime. The second state in that theorem is left general, exactly as in Art. 97. In the applications it is chosen so as to bring out the result desired.

Writing for brevity

$$\beta = \frac{E}{1 - 2\nu} \tag{a}$$

the theorem becomes, using u, v, w for the actual thermoelastic displacement components,

$$\int (X''u + Y''v + Z''w)\, d\tau + \int (\bar{X}''u + \bar{Y}''v + \bar{Z}''w)\, dS = -\beta \int \left(u'' \frac{\partial}{\partial x}\alpha T + v'' \frac{\partial}{\partial y}\alpha T + w'' \frac{\partial}{\partial z}\alpha T \right) d\tau + \beta \int (lu'' + mv'' + nw'')\alpha T\, dS \qquad (b)$$

The divergence theorem, Eq. (138), provides the equation

$$\int \left[\frac{\partial}{\partial x}(u''\alpha T) + \frac{\partial}{\partial y}(v''\alpha T) + \frac{\partial}{\partial z}(w''\alpha T) \right] d\tau = \int (lu'' + mv'' + nw'')\alpha T\, dS \qquad (c)$$

The integral on the right corresponds to the second on the right of (*b*). On the left of (*c*) we can write

$$\frac{\partial}{\partial x}(u''\alpha T) = u'' \frac{\partial}{\partial x}\alpha T + \frac{\partial u''}{\partial x}\alpha T, \text{ etc.}$$

and use this in the first integral on the right of (*b*). Then (*b*) becomes

$$\int (X''u + Y''v + Z''w)\, d\tau + \int (\bar{X}''u + \bar{Y}''v + \bar{Z}''w)\, dS = \int \Theta''\alpha T\, d\tau \qquad (268)$$

where [see (7) and (8)]

$$\Theta'' = \beta \left(\frac{\partial u''}{\partial x} + \frac{\partial v''}{\partial y} + \frac{\partial w''}{\partial z} \right) = \sigma_x'' + \sigma_y'' + \sigma_z'' \qquad (d)$$

Equation (268) is a *thermoelastic reciprocal theorem.*[1] The left-hand side can conveniently be called the work of the externally applied body forces (X'', . . .) and surface forces ($\bar{X}''$, . . .) of the *auxiliary* problem, or state, on the actual displacements (u,v,w) of the thermoelastic problem.

155 | Overall Thermoelastic Deformations. Arbitrary Temperature Distribution

In the following applications[2] of the preceding theorem, the auxiliary problem is one which is either elementary or has been solved earlier in the book, and in each case a simple general formula useful in design is found.

[1] The derivation from the "ordinary" reciprocal theorem of Art. 97 was given in J. N. Goodier, *Proc. 3d U.S. Nat. Congr. Appl. Mech.*, pp. 343–345, 1958. Essentially the same theorem was obtained earlier, directly from the thermoelastic equations, by W. M. Maisel, *C. R. (Doklady) Acad. Sci. USSR (N.S.)*, vol. 30, pp. 115–118, 1941.

[2] From the paper by J. N. Goodier mentioned in the preceding footnote.

Volume Change Considering a body of arbitrary shape, with or without cavities, let the auxiliary state be that induced by uniform normal (tensile) loading σ'' over the whole surface of the body, including cavities if any. Then, at any point in the material

$$\sigma_x'' = \sigma_y'' = \sigma_z'' = \sigma'' \qquad \text{and} \qquad \Theta'' = 3\sigma'' \tag{a}$$

The work of the forces of this auxiliary state on the thermoelastic displacements u, v, w corresponding to an arbitrary temperature rise[1] $T(x,y,z)$, is simply $\sigma'' \, \Delta\tau$, where $\Delta\tau$ is the thermoelastic increase of volume of solid material. The theorem (268) now yields

$$\sigma'' \, \Delta\tau = \int 3\sigma''\alpha T \, d\tau \qquad \text{that is} \qquad \Delta\tau = \int 3\alpha T \, d\tau \tag{b}$$

This means that the volume change is simply that of free thermal expansion. Although there is thermal stress and the elastic strain it induces, the corresponding increases of volume are in some parts positive and in some negative, with zero sum.[2]

Change of Cavity Volume The body contains a cavity. The volume contained by the cavity increases by $\Delta\tau_c$ when the arbitrary temperature rise $T(x,y,z)$ occurs. We can determine $\Delta\tau_c$ if the solution of the auxiliary problem of uniform internal pressure in the cavity is known. If internal pressure p_i'' alone induces stress such that

$$\Theta'' = p_i''S \tag{c}$$

the theorem (268) yields

$$p_i''\Delta\tau_c = \int p_i''S\alpha T \, d\tau \qquad \text{that is} \qquad \Delta\tau_c = \int S\alpha T \, d\tau \tag{d}$$

Consider, for instance, the hollow sphere. The solution for internal pressure (page 395) provides the sum of the three principal stresses as

$$\Theta'' = \sigma_R'' + 2\sigma_t'' = p_i'' \frac{3a^3}{b^3 - a^3} \tag{e}$$

Comparing with (*c*), we have the value of S for this case, and (*d*) then becomes

$$\Delta\tau_c = \frac{a^3}{b^3 - a^3} \int_{R=a}^{R=b} 3\alpha T \, d\tau \tag{f}$$

[1] From a state of uniform temperature taken as zero.

[2] This simple result seems to have remained undiscovered until 1954. It was given (for anisotropic bodies) by W. Nowacki, *Arch. Mech. Stos.*, vol. 6, p. 487, 1954, and independently by M. Hieke, *Z. Angew. Math. Mech.*, vol. 35, pp. 285–294, 1955. It holds more generally for any *initial stress* in a linearly elastic solid, whether due to uneven heating or not. See Art. 158.

The integral represents simply the total free thermal volumetric expansion of the material elements. When the outer radius b is infinite, the cavity volume does not change at all provided the integral remains finite.

Extension of a Bar For a bar of any (uniform) cross section, a mean extension for arbitrary temperature rise $T(x,y,z)$ can be determined, choosing for the auxiliary state simple tension with stress

$$\sigma_x'' = \sigma \qquad \sigma_y'' = \sigma_z'' = 0 \qquad \text{and} \qquad \Theta'' = \sigma$$

Defining ΔL as the extensions of lines parallel to the axis of the bar averaged over the cross-sectional area A, the theorem (268) yields

$$\Delta L = \frac{1}{A}\int \alpha T\, d\tau \tag{g}$$

In general, the bar will also have deformations other than extension.

Flexural Rotation in a Bar A mean thermoelastic flexural rotation ω of one end of the bar relative to the other can be obtained by taking pure bending (Art. 102) as the auxiliary state. Then, with bending moment M'' in the xz plane, in place of M in Fig. 145, we have

$$\sigma_z'' = \frac{M''x}{I_y} \qquad \sigma_x'' = \sigma_y'' = 0 \qquad \text{and} \qquad \Theta'' = \frac{M''x}{I_y}$$

The theorem (268) suggests the introduction of ω by writing the work erm of the left-hand side as $M''\omega$. Then, after cancellation of M'',

$$\omega = \frac{1}{I_y}\int \alpha T x\, d\tau \tag{h}$$

Deflection of a Cantilever Taking for the auxiliary problem the Saint-Venant cantilever bending of Art. 120, we have, with P'' in place of P,

$$\sigma_z'' = -P''(l - z)\frac{x}{I} \qquad \sigma_x'' = \sigma_y'' = 0 \qquad \text{and} \qquad \Theta'' = \sigma_z''$$

The left-hand side of the theorem (268) will consist of the work of the load P at the loaded end $z = l$ (Fig. 190) and the reactions at the "fixed" end $z = 0$ on the thermoelastic displacements due to T. The cantilever may be fixed by fixing one element of the end $z = 0$ in position and orientation. Displacements at this end may then be treated as small if the bar is slender, and the corresponding work may be neglected. At the loaded end $z = l$, a mean deflection δ, in the x direction, may be introduced by expressing the work on the left of the theorem (268) as $P\delta$. Then

$$\delta = -\frac{1}{I}\int \alpha T x(l - z)\, d\tau \tag{i}$$

Torsional Rotation of a Bar The auxiliary problem is taken as the Saint-Venant torsion problem of Art. 104. The stress components σ_x'', σ_y'', σ_z'' all vanish and therefore Θ'' does also. Thus the right-hand side of (268) is zero. Therefore the thermoelastic torsional rotation of one end of the bar relative to the other, in the mean sense implied here, is also zero.

In the torsion of a nonuniform bar of circular section (Michell's theory, as given in Art. 119) the normal stress components σ_r'', σ_θ'', σ_z'' vanish [see Eqs. (*a*), Art. 119], and again therefore the relative torsional rotation of the ends is zero.

156 | Thermoelastic Displacement. Maisel's Integral Solution

The thermoelastic displacement at any point can be found if we have available, for the auxiliary problem, a solution for the stress due to a concentrated force applied at the point. Figure 230 indicates the elastic body supported in some definite way (so that there can be definite displacements) and loaded by a force P_x'' in the x direction, at a point A. This means that the point is regarded as the center of a small spherical cavity, as in the problem of Art. 135. The solution of this auxiliary problem gives Θ'' as a function of position. It will be proportional to P_x'', and we may write

$$\Theta'' = P_x''\Theta_{1x}'' \qquad (a)$$

so that Θ_{1x}'' corresponds to a load P_x'' of unit value.

Turning to the theorem (268), the left-hand side will consist of the work of P_x'' on the thermoelastic displacement u at the point A, plus the work of the support reactions on the thermoelastic displacements there. But *we now require that this support work be zero.* The support points may be completely fixed, for instance. Then the theorem (268) yields at once

$$u = \int \Theta_{1x}''\alpha T \, d\tau \qquad (b)$$

Thus u is obtainable as this volume integral taken over the entire body. The singularity of Θ_{1x}'' at A causes no difficulty, as may be seen at once

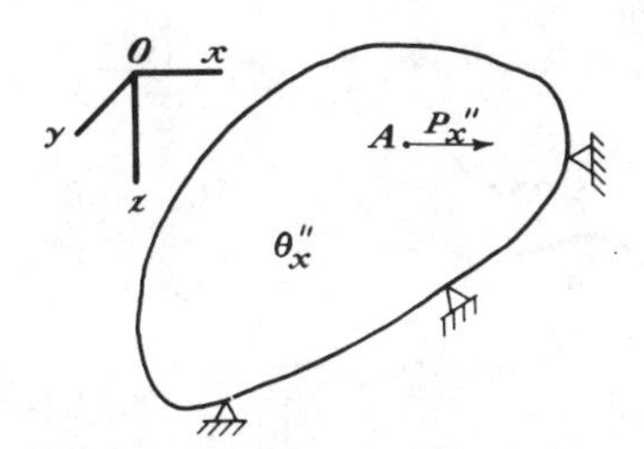

Fig. 230

by considering $d\tau$ in terms of spherical polar coordinates centered at A. Similarly, replacing the suffixes x in (a) by y, then z, to correspond to forces P_y'', P_z'' in the y and z directions, we have[1]

$$v = \int \Theta_{1y}''\alpha T\, d\tau \qquad w = \int \Theta_{1z}''\alpha T\, d\tau \tag{c}$$

It will be apparent that the point A may also be taken on the surface of the body. In the auxiliary problem the small spherical cavity is then replaced by a small hemispherical (as in the problem of Art. 138) or other open surface.

The solution represented by (b) and (c) above has very wide application,[1] since an abundance of solutions of the auxiliary concentrated-force problems is available. The approximate solutions for slender beams, curved bars, rings, thin plates, and thin shells can also be used in (b) and (c) to obtain thermoelastic results[2] of corresponding validity for the arbitrary temperature distribution. The assumption of linear variation of temperature through the thickness of a plate or shell, which has been widely used, becomes unnecessary.

As an example, we shall consider the normal component of displacement of the plane surface $z = 0$ of a semi-infinite solid $z > 0$. The temperature rise $T(x,y,z)$ is to be an even function of x, that is, symmetrical with respect to the yz plane, but is otherwise general.

In the thermoelastic problem, Fig. 231a, we can evaluate the displacement at any surface point A in the xz plane relative to the point at the origin by taking the auxiliary problem as in Fig. 231b. The force P'' at the origin can be identified with the force P in Fig. 207 and Art. 138. The stress components are given by Eqs. (211) in cylindrical coordinates, and we have

$$(\Theta_{1z}'')_0 = \sigma_r'' + \sigma_\theta'' + \sigma_z'' = -\frac{1}{\pi}(1+\nu)P''\frac{z}{R^3} \tag{d}$$

[1] Maisel, *op. cit.*

[2] Several examples are given by J. N. Goodier and G. E. Nevill, Jr., in a report to the Office of Naval Research, "Applications of a Reciprocal Theorem of Linear Thermoelasticity," 1961, and in the doctoral dissertation of G. E. Nevill, Jr., Division of Engineering Mechanics, Stanford University, 1961.

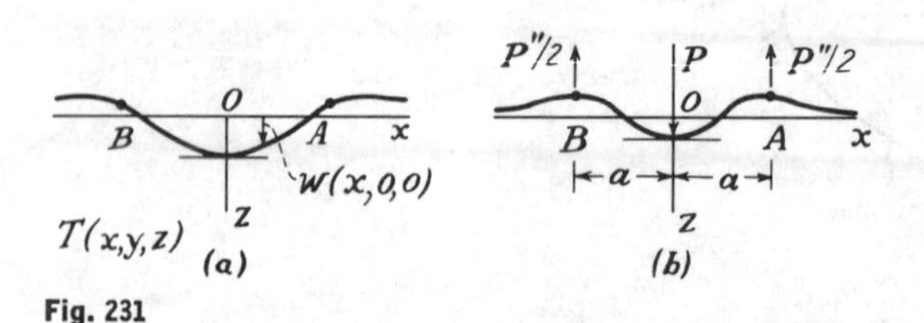

Fig. 231

where $$R^2 = x^2 + y^2 + z^2 \tag{e}$$

The two upward forces $\frac{1}{2}P''$ in Fig. 231*b* similarly induce stress with

$$(\Theta_{1z}'')_{AB} = \frac{1}{\pi}(1+\nu)\frac{1}{2}P''z\left(\frac{1}{R_A^3} + \frac{1}{R_B^3}\right) \tag{f}$$

where R_A and R_B are the positive radii from

$$R_A^2 = (x-a)^2 + y^2 + z^2 \qquad R_B^2 = (x+a)^2 + y^2 + z^2 \tag{g}$$

Thus, for the complete auxiliary problem of Fig. 231*b* we have

$$\Theta_{1z}'' = -\frac{1}{\pi}(1+\nu)\frac{1}{2}P''z\left(\frac{2}{R^3} - \frac{1}{R_A^3} - \frac{1}{R_B^3}\right) \tag{h}$$

Writing w_0 for the normal displacement at 0, and w_A for points A and B, the theorem (268) gives

$$P''w_0 - 2\frac{1}{2}P''w_A = -\frac{1}{2\pi}(1+\nu)P''\int z\left(\frac{2}{R^3} - \frac{1}{R_A^3} - \frac{1}{R_B^3}\right)\alpha T\,d\tau$$

or $$w_A - w_0 = \frac{1}{\pi}(1+\nu)\int\left(\frac{1}{R^3} - \frac{1}{R_A^3}\right)\alpha T\,d\tau$$

The requirement of zero support work imposes a condition on T. The force P'' together with the two forces $\frac{1}{2}P''$ in Fig. 231*b* induces stress components that tend to zero at infinity like R^{-3}. The work done by the corresponding loading, on an infinite hemisphere, on the thermoelastic displacements there must also tend to zero. This is assured if the thermoelastic displacements themselves tend to zero. They do so when nonzero values of T are confined to a finite volume in the neighborhood of the plane surface, as may be readily verified by the method of strain suppression explained in Art. 153, in conjunction with Saint-Venant's principle.

For interior points of the semi-infinite solid the auxiliary solution is available in the paper of R. D. Mindlin referred to in footnote 1, page 399. For interior points of the infinite solid we have the solution given in Art. 135. The thermoelastic displacement for this problem is found later (page 477) by a different method.

The two-dimensional solutions given in Chap. 4 for concentrated forces on the semi-infinite region (Art. 36), the wedge (Art. 38), and the circular region (Art. 41), and the infinite region (Art. 42) are also useful as auxiliary solutions leading immediately to thermoelastic displacement formulas.

The thermoelastic reciprocal theorem of Art. 154 can also be used to advantage in conjunction with the Fourier method for sinusoidal (instead of concentrated) loadings. Examples are given in the report and dissertation mentioned in the footnote on page 464.

PROBLEMS

1. A curved bar, as in Fig. 42, has a temperature rise $T(r,\theta)$. Assuming plane stress, obtain an integral formula for a mean thermoelastic rotation of one end relative to the other.

   What is the exact sense of "mean" here, and also in the ω of Eq. (*h*) in Art. 155?
2. Write out a sequence of steps of computation for finding the shortening of a diameter due to a temperature rise $T(x,y)$ in a disk, as in Fig. 75, in plane stress.
3. Carry out in detail the verification indicated on page 465, for a temperature distribution T which is uniform (T_0) inside the hemispherical region $R < b$ and zero outside.

157 | Initial Stress

The method of strain suppression described in Art. 153 can be applied in the more general problem of *initial stresses*. Imagine a body subdivided into small elements and suppose that each of these elements undergoes a certain permanent plastic deformation or change in shape produced by metallographical transformation. Let this deformation be defined by the strain components

$$\epsilon_x' \qquad \epsilon_y' \qquad \epsilon_z' \qquad \gamma_{xy}' \qquad \gamma_{xz}' \qquad \gamma_{yz}' \tag{a}$$

We assume that these strain components are small and are represented by continuous functions of the coordinates. If they also satisfy the compatibility conditions (125), the elements into which the body is subdivided fit each other after the permanent set (*a*) and there will be no initial stresses produced.

Let us consider now a general case when the strain components (*a*) do not satisfy the compatibility conditions so that the elements into which the body is subdivided do not fit each other after permanent set, and forces must be applied to the surface of these elements in order to make them satisfy the compatibility equations. Assuming that after the permanent set (*a*) the material remains perfectly elastic, and applying Hooke's law, we find from Eqs. (11) and (6) that the permanent set (*a*) can be eliminated by applying to each element the surface forces

$$\sigma_x' = -(\lambda e' + 2G\epsilon_x'), \ \ldots \qquad \tau_{xy}' = -G\gamma_{xy}', \ \ldots \tag{b}$$

where

$$e' = \epsilon_x' + \epsilon_y' + \epsilon_z'$$

The surface forces (*b*) can be induced by applying certain body and surface forces to the body formed by the small elements. These forces must satisfy the equations of equilibrium (123) and the boundary conditions (124). Substituting the stress components (*b*) in these equations, we find that the necessary body forces are

$$X = \frac{\partial}{\partial x}(\lambda e' + 2G\epsilon_x') + \frac{\partial}{\partial y}(G\gamma_{xy}') + \frac{\partial}{\partial z}(G\gamma_{xz}') \\ \cdots\cdots\cdots\cdots \tag{c}$$

and the surface forces are

$$\bar{X} = -(\lambda e' + 2G\epsilon_x')l - G\gamma_{xy}'m - G\gamma_{xz}'n \\ \cdots\cdots\cdots\cdots \tag{d}$$

By applying the body forces (*c*) and the surface forces (*d*), we remove the initial permanent set (*a*) so that the elements fit one another and form a continuous body. We now assume that the elements into which the body was subdivided are joined together and remove the forces (*c*) and (*d*). Then evidently the initial stresses are obtained by superposing on the stresses (*b*) the stresses that are produced in the elastic body by the body forces

$$X = -\frac{\partial}{\partial x}(\lambda e' + 2G\epsilon_x') - \frac{\partial}{\partial y}(G\gamma_{xy}') - \frac{\partial}{\partial z}(G\gamma_{xz}') \qquad (e)$$

. . . .

and the surface forces

$$\bar{X} = (\lambda e' + 2G\epsilon_x')l + G\gamma_{xy}'m + G\gamma_{xz}'n \qquad (f)$$

. . . .

Thus the problem of determining the initial stresses is reduced to the usual system of equations of the theory of elasticity in which the magnitudes of the fictitious body and surface forces are completely determined if the permanent set (*a*) is given.

In the particular case when $\epsilon_x' = \epsilon_y' = \epsilon_z' = \alpha T$ and $\gamma_{xy}' = \gamma_{xz}' = \gamma_{yz}' = 0$, the above equations coincide with those obtained before in calculating thermal stresses.

Let us consider now the reversed problem when the initial stresses are known and it is desired to determine the permanent set (*a*) that produces these stresses. In the case of transparent materials, such as glass, the initial stresses can be investigated by the photoelastic method (Chap. 5). In other cases these stresses can be determined by cutting the body into small elements and measuring the strains that occur as the result of freeing these elements from surface forces representing initial stresses in the uncut body. From the previous discussion it is clear that the initial deformation produces initial stresses only if the strain components (*a*) do not satisfy the compatibility equations; otherwise these strains may exist without producing initial stresses. From this it follows that knowledge of the initial stresses is not sufficient for determining the strain components (*a*). If a solution for these components is obtained, any permanent strain system satisfying the compatibility equations can be superposed on this solution without affecting the initial stresses.[1]

Initial stresses, producing doubly-refracting properties in glass, present great difficulties in manufacturing optical instruments. To diminish these stresses it is the usual practice to anneal the glass. The elastic limit of glass at high temperatures is very low, and the material yields under the action of the initial stresses. If a sufficient time is given, the yielding of the material at a high temperature results in a considerable release from initial stresses. Annealing has an analogous effect in the case of various metallic castings and forgings.

Cutting of large bodies into smaller pieces releases initial stresses along the surfaces of cutting and diminishes the total amount of strain energy due to initial stresses, but the magnitude of the maximum initial stress is not always diminished by such cutting. For example, suppose a circular ring (Fig. 232) has initial stresses symmetrically distributed with respect to the center and the initial stress component σ_θ' varies along a cross section *mn* according to a linear law (*ab* in the figure). Cutting the ring radially, as shown in the figure by dotted lines, releases the stresses σ_θ' along these cuts. This is equivalent to the application to the ends of each portion of the ring of two equal and opposite couples producing pure bending. The distribution of the stress σ_θ along *mn*, due to this bending, is nearly hyperbolic (see Art. 29), as shown by the curve *cde*. The residual stress along *mn* after cutting is then given by $\sigma_\theta + \sigma_\theta'$ and is shown in the

[1] The fact that permanent set (*a*) is not completely determined by the magnitudes of initial stresses is discussed in detail in the paper by H. Reissner; see *Z. Angew. Math. Mech.*, vol. 11, p. 1, 1931.

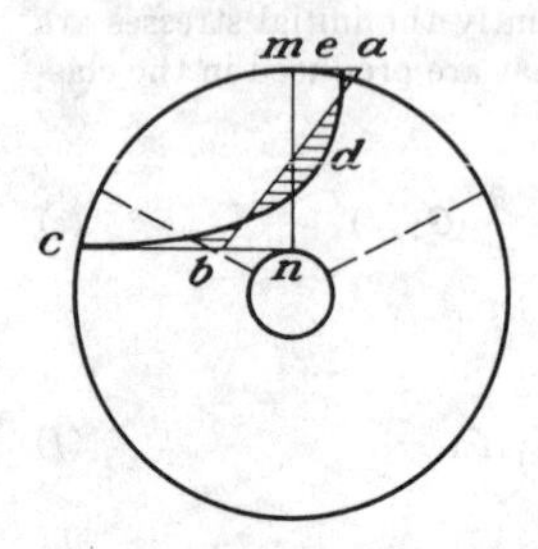

Fig. 232

figure by the shaded area. If the inner radius of the ring is small there is a high stress concentration at the inner boundary, and the maximum initial stress after cutting, represented in Fig. 232 by bc, may be larger than the maximum initial stress before cutting. This or similar reasoning explains why glass sometimes cracks after cutting.[1]

158 | Total Volume Change Associated with Initial Stress

The analysis of Art. 157 shows that the displacements u, v, w which will actually occur in the body, when the (incompatible) strain components (a) are effected in each element, are the same as those occurring in an ordinary elastic body when the body forces (e) and the surface forces (f) are applied. But certain overall features of this deformation can be deduced from equilibrium conditions, with the assumption that after the strains (a) are introduced, the elements respond to stress according to Hooke's law. Consider, for instance, a body in which there is initial stress $\sigma_x \, . \, . \, . \, , \tau_{xy} \, . \, . \, . \, ,$ the body as a whole being free from loads or constraints (Fig. 233). For the part to the right of any plane section AA parallel to the yz plane, equilibrium requires

$$\iint \sigma_x \, dy \, dz = 0 \tag{a}$$

Integrating this for a slice dx, and then over the whole volume, we have

$$\int \sigma_x \, d\tau = 0 \tag{b}$$

[1] Several examples of the calculation of initial stresses in portions cut out from a circular plate are discussed in the paper by M. V. Laue, *Z. Tech. Physik*, vol. 11, p. 385, 1930. Various methods of calculating residual stresses in cold-drawn tubes are discussed in the paper by N. Dawidenkow, *Z. Metallkunde*, vol. 14, p. 25, 1932.

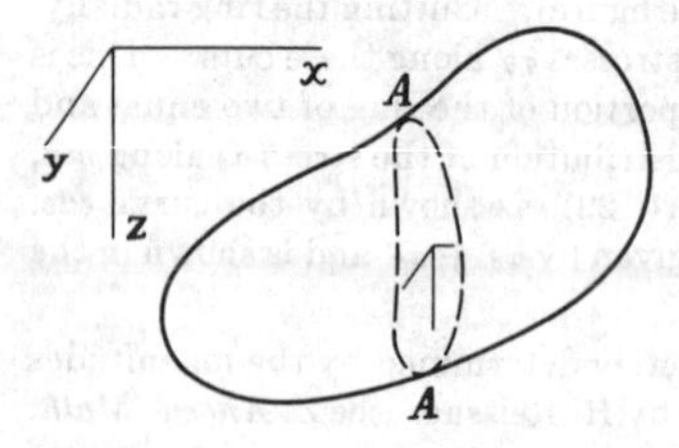

Fig. 233

Similarly, the volume integrals of each of the other stress components must vanish. Therefore

$$\int(\sigma_x + \sigma_y + \sigma_z)\,d\tau = 0 \tag{c}$$

It follows from the Hooke's law relations that the total change of volume corresponding to the strains induced by these stress components is zero. The actual total change of volume is therefore simply that due to the incompatible strain components (*a*) of Art. 157 applied to the separated elements.

Thus for simple thermal expansion, the total change of volume is

$$\Delta\tau = \int 3\alpha T\,d\tau \tag{d}$$

as already shown otherwise in Eq. (*b*) of Art. 155.

Returning to (*b*), and its companion equations, it will be evident that the volume integral, over the whole body, of any linear function of the stress components must be zero. Then any linear relation between stress and strain components insures that the volume integral of any strain component is also zero. Isotropy of the material is not required. In particular, the total volume change induced by the stress is zero.

Further relations can evidently be obtained from equilibrium of moments on such a section as AA, Fig. 233.

159 | Plane Strain and Plane Stress. Method of Strain Suppression

Plane strain will occur in a long cylindrical or prismatic body when the temperature, although varying over a cross section, does not vary along lines parallel to the axis of the cylinder or prism (the z axis). Then T is independent of z.

Beginning again with the stresses (*g*) in Art. 153 which result in zero strain, the necessary body forces are given by (*h*) where now $Z = 0$, and the pressure (*f*) must be applied to the surface, including the ends.

Then supposing the elements joined together, we remove the body forces and the surface pressure *on the curved surface only*, keeping the axial strain ϵ_z zero. The effects of this removal are obtained by solving the problem of applying body force

$$X = -\frac{\alpha E}{1 - 2\nu}\frac{\partial T}{\partial x} \qquad Y = -\frac{\alpha E}{1 - 2\nu}\frac{\partial T}{\partial y} \tag{a}$$

and a normal tensile stress of amount

$$\frac{\alpha E T}{1 - 2\nu} \tag{b}$$

on the curved surface only, as a problem in plane strain ($\epsilon_z = 0$). This problem is of the type considered at the end of Art. 17, except that we have to convert Eq. (32) from plane stress to plane strain by replacing ν by $\nu/(1 - \nu)$. Thus instead of Eqs. (31) and (32) we shall have

$$\sigma_x - \frac{\alpha E T}{1 - 2\nu} = \frac{\partial^2\phi}{\partial y^2} \qquad \sigma_y - \frac{\alpha E T}{1 - 2\nu} = \frac{\partial^2\phi}{\partial x^2} \qquad \tau_{xy} = -\frac{\partial^2\phi}{\partial x\,\partial y} \tag{c}$$

and

$$\frac{\partial^4\phi}{\partial x^4} + 2\frac{\partial^4\phi}{\partial x^2\partial y^2} + \frac{\partial^4\phi}{\partial y^4} = -\frac{\alpha E}{1 - \nu}\left(\frac{\partial^2 T}{\partial x^2} + \frac{\partial^2 T}{\partial y^2}\right) \tag{d}$$

The required stress function is that which satisfies Eq. (*d*) and gives the normal boundary tension (*b*). The stresses are then calculated from (*c*). On these we have to superpose the stresses (*g*), Art. 153.

The axial stress σ_z will consist of the term from (g) in Art. 153 together with $\nu(\sigma_x + \sigma_y)$ obtained from (c). The resultant axial force and bending couple on the ends can be removed by superposition of simple tension and bending.

Plane stress will occur in a thin plate when the temperature does not vary through the thickness. Taking the x and y plane as the middle plane of the plate, we may assume that $\sigma_z = \tau_{xz} = \tau_{yz} = 0$. We may also regard each element as free to expand in the z direction. It will be sufficient, to ensure fitting of the elements, to suppress the expansion in the x and y directions only. This requires

$$\sigma_x = \sigma_y = -\frac{\alpha ET}{1-\nu} \qquad \tau_{xy} = 0 \tag{e}$$

Substituting these in the equations of equilibrium (18), we find that the required body forces are

$$X = \frac{\alpha E}{1-\nu}\frac{\partial T}{\partial x} \qquad Y = \frac{\alpha E}{1-\nu}\frac{\partial T}{\partial y} \tag{f}$$

and the normal pressure $\alpha ET/(1-\nu)$ should be applied to the edges of the plate.

Removing these forces, we conclude that the thermal stress consists of (e) together with the plane stress due to body forces

$$X = -\frac{\alpha E}{1-\nu}\frac{\partial T}{\partial x} \qquad Y = -\frac{\alpha E}{1-\nu}\frac{\partial T}{\partial y} \tag{g}$$

and to normal tensile stress $\alpha ET/(1-\nu)$ applied round the edges. The determination of this plane stress again presents us with a problem of the type considered in Art. 17. We have only to put in Eqs. (31) and (32)

$$V = \frac{\alpha ET}{1-\nu}$$

this being the potential corresponding to the forces (g).

160 | Two-dimensional Problems with Steady Heat Flow

In steady heat flow parallel to the xy plane, as in a thin plate or in a long cylinder with no variation of temperature in the axial (z)direction, the temperature T will satisfy the equation

$$\frac{\partial^2 T}{\partial x^2} + \frac{\partial^2 T}{\partial y^2} = 0 \tag{a}$$

Consider a cylinder (not necessarily circular) in a state of plane strain, with $\epsilon_z = \gamma_{xz} = \gamma_{yz} = 0$. The stress-strain relations in cartesian coordinates are analogous to Eqs. (a) and (b) of Art. 151 in the case of plane strain. Corresponding to Eqs. (b) we shall have

$$\begin{aligned} \epsilon_x - (1+\nu)\alpha T &= \frac{1-\nu^2}{E}\left(\sigma_x - \frac{\nu}{1-\nu}\sigma_y\right) \\ \epsilon_y - (1+\nu)\alpha T &= \frac{1-\nu^2}{E}\left(\sigma_y - \frac{\nu}{1-\nu}\sigma_x\right) \end{aligned} \tag{b}$$

We now inquire whether it is possible to have σ_x, σ_y, and τ_{xy} zero. Putting $\sigma_x = \sigma_y = 0$ in Eqs. (*b*) we find

$$\epsilon_x = (1 + \nu)\alpha T \qquad \epsilon_y = (1 + \nu)\alpha T \tag{c}$$

and of course $\gamma_{xy} = 0$.

Such strain components are possible only if they satisfy the conditions of compatibility (125). Since $\epsilon_z = 0$ and the other components of strain are independent of z, all of these conditions except the first are satisfied. The first reduces to

$$\frac{\partial^2 \epsilon_x}{\partial y^2} + \frac{\partial^2 \epsilon_y}{\partial x^2} = 0$$

But on account of Eqs. (*c*) and (*a*), this equation also is satisfied. We find, therefore, that in steady heat flow the equations of equilibrium, the boundary condition that the curved surface be free from force, and the compatibility conditions, are all satisfied by taking

$$\sigma_x = \sigma_y = \tau_{xy} = 0 \qquad \sigma_z = -\alpha E T \tag{d}$$

For a *solid* cylinder the above equations and conditions are complete, and we can conclude that in a steady state of two-dimensional heat conduction there is no thermal stress except the axial stress σ_z given by Eqs. (*d*), required to maintain the plane strain condition $\epsilon_z = 0$. In the case of a long cylinder with *unrestrained* ends, we obtain an approximate solution valid except near the ends by superposing simple tension, or compression, and simple bending so as to reduce the resultant force and couple on the ends, due to σ_z, to zero.

For a hollow cylinder, however, we *cannot* conclude that the plane strain problem is solved by Eqs. (*d*). It is necessary to examine the corresponding displacements. It is quite possible that they will prove to have discontinuities, analogous to those discussed on pages 79 and 88.

For instance, suppose the cylinder is a tube and a longitudinal slit is cut, as indicated in Fig. 234*b*. If it is hotter on the inside than on the outside it will tend to uncurl, and the slit will open up. There will be a discontinuity of displacement between the two faces of the slit. Thus the displacement *should* be represented by discontinuous functions of θ. The cross section is solid, i.e., singly connected, and Eqs. (*d*) give the stress correctly for plane strain. But if the tube has no slit (Fig. 234*a*) discontinuities of displacement are physically impossible. This indicates that the assumed temperature distribution will in fact give rise to stress components σ_x, σ_y, τ_{xy}, representing the stress produced by suitably drawing together again the separated faces of the slit tube and joining them. The component σ_z will also be affected by this operation.

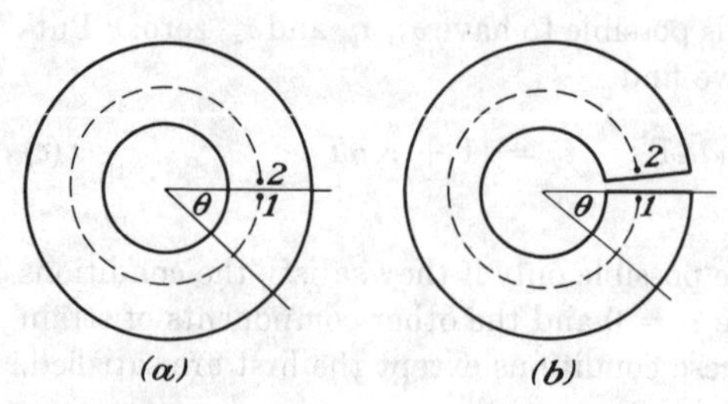

Fig. 234

To investigate this question further, we rewrite Eqs. (*c*) as

$$\frac{\partial u}{\partial x} = \epsilon' \qquad \frac{\partial v}{\partial y} = \epsilon' \tag{e}$$

where $\epsilon' = (1 + \nu)\alpha T$. Since $\gamma_{xy} = 0$, we can write

$$\frac{\partial v}{\partial x} + \frac{\partial u}{\partial y} = 0 \tag{f}$$

and

$$\frac{\partial v}{\partial x} - \frac{\partial u}{\partial y} = 2\omega_z \tag{g}$$

ω_z being a component of rotation (see page 233). Equations (*f*) and (*g*) yield

$$\frac{\partial u}{\partial y} = -\omega_z \qquad \frac{\partial v}{\partial x} = \omega_z \tag{h}$$

and these with (*e*) give

$$\frac{\partial \epsilon'}{\partial x} = \frac{\partial \omega_z}{\partial y} \qquad \frac{\partial \epsilon'}{\partial y} = -\frac{\partial \omega_z}{\partial x} \tag{i}$$

Equations (*i*) are *Cauchy-Riemann* equations, discussed in Art. (55). They show that $\epsilon' + i\omega_z$ is an analytic function of the complex variable $x + iy$. Denoting this function by Z, we have

$$Z = \epsilon' + i\omega_z \tag{j}$$

If u_1, v_1, u_2, v_2 are the values of u, v at two points 1, 2 in the cross section of the cylinder, the differences $u_2 - u_1$, $v_2 - v_1$ can be expressed as

$$u_2 - u_1 = \int_1^2 \left(\frac{\partial u}{\partial x}\, dx + \frac{\partial u}{\partial y}\, dy \right) \qquad v_2 - v_1 = \int_1^2 \left(\frac{\partial v}{\partial x}\, dx + \frac{\partial v}{\partial y}\, dy \right)$$

where the integrals are taken along any curve joining the two points and lying entirely in the material. Multiplying the second by i and adding to the first we find

$$u_2 - u_1 + i(v_2 - v_1) = \int_1^2 \left[\frac{\partial u}{\partial x}\, dx + \frac{\partial u}{\partial y}\, dy + i \left(\frac{\partial v}{\partial x}\, dx + \frac{\partial v}{\partial y}\, dy \right) \right] \tag{k}$$

and it is easily verified from Eqs. (*e*) and (*h*) that the integral on the right-hand side is the same as $\int_1^2 (\epsilon' + i\omega_z)(dx + i\, dy)$ or $\int_1^2 Z\, dz$. Thus Eq. (*k*) becomes

$$u_2 - u_1 + i(v_2 - v_1) = \int_1^2 Z\, dz \tag{l}$$

The displacements will be single-valued when this integral vanishes for a complete circuit of any closed curve, such as the broken-line circle in Fig. 234*a*, which can be drawn in the material of the cross section. We shall use this result later in solving a thermal stress problem of the hollow circular cylinder.

For the rotation ω_z (see Art. 83) we have

$$(\omega_z)_2 - (\omega_z)_1 = \int_1^2 \left(\frac{\partial \omega_z}{\partial x} dx + \frac{\partial \omega_z}{\partial y} dy \right)$$

and using Eqs. (*i*) this becomes

$$(\omega_z)_2 - (\omega_z)_1 = \int_1^2 \left(- \frac{\partial \epsilon'}{\partial y} dx + \frac{\partial \epsilon'}{\partial x} dy \right)$$

Since ϵ' is proportional to T, this integral is proportional to the amount of heat flowing per unit time, per unit axial distance, across the curve joining points 1 and 2. If this is a closed curve $(\omega_z)_2 - (\omega_z)_1$ must vanish, and therefore the total heat flow across the curve must be zero.[1] If a pipe has heat flow from inside to outside or vice versa, this condition is not fulfilled and the stress is not correctly given by Eqs. (*d*).

But if the pipe is slit, as indicated in Fig. 234*b*, the displacement or rotation at point 2 can differ from that at point 1, for instance, if the heating causes the slit to open up. The simple state of stress given by Eqs. (*d*) is then correct. To arrive at the state of stress which exists in the pipe when it is not slit, we have to superpose the stress due to closing the gap. The determination of this dislocational stress[2] involves problems of the types illustrated by Figs. 45 and 48.

Consider, for example, a hollow circular cylinder of outer radius b, with a concentric bore of radius a. If the temperature T_i at the inner surface is uniform, and the temperature at the outer surface is zero, the temperature T at any radius r is given by Eq. (*n*) of Art. 151. We may write this as

$$T = -A \log b + A \log r \qquad (m)$$

where

$$A = - \frac{T_i}{\log (b/a)} \qquad (n)$$

The constant term $-A \log b$ in Eq. (*m*) can be ignored, since a uniform change of temperature does not cause thermal stress. Then, since $\log z = \log r + i\theta$,

$$\begin{aligned} Z &= \epsilon' + i\omega_z = (1 + \nu)\alpha T + i\omega_z \\ &= (1 + \nu)\alpha A \log r + i\omega_z = (1 + \nu)\alpha A \log z \end{aligned}$$

Writing B for $(1 + \nu)\alpha A$, we have from Eq. (*l*)

$$u_2 - u_1 + i(v_2 - v_1) = B \int_1^2 \log z \, dz = B[z(\log z - 1)]_1^2 \qquad (o)$$

[1] When the heat flow is zero there can still be thermal stress in addition to (*d*). See Art. 161.

[2] The relation between thermal stress in steady heat flow and dislocational stress was established by N. Muskhelishvili, *Bull. Elec. Tech. Inst.*, St. Petersburg, vol. 13, p. 23, 1916, and independently by M. A. Biot, *Phil. Mag.*, ser. 7, vol. 19, p. 540, 1935. Thermal stresses in a hollow circular cylinder, and in a square cylinder with a circular hole, have been determined photoelastically by this method by E. E. Weibel, *Proc. 5th Intern. Cong. Appl. Mech., Cambridge, Mass.*, 1938, p. 213.

This equation applies to any curve between points 1 and 2 lying wholly in the material. It gives the relative displacement of the two points when the temperature is given by Eq. (*m*) and the stress by Eqs. (*d*).

Applying it to a circular path of radius r starting at 1 (Fig. 234), going round the hole, and ending at 2, we have, since $\theta_1 = 0$, $\theta_2 = 2\pi$,

$$[z(\log z - 1)]_1^2 = re^{i2\pi}(\log r + i2\pi) - re^{io}(\log r + i0) = i2\pi r$$

Inserting this in Eq. (*o*), we find

$$u_2 - u_1 = 0 \qquad v_2 - v_1 = B2\pi r \tag{p}$$

The relative displacement is not zero, and therefore it is necessary to consider the cylinder as slit, so that the point 2 can move away from the point 1 by the vertical displacement $2\pi rB$ (Fig. 234*b*). The movement of the upper face of the slit relative to the lower face is equivalent to a rotation $2\pi B$ in the clockwise sense about the center of the cylinder. However, B is negative if T_i is positive. Then the slit opens to a gap subtending an angle $-2\pi B$ at the center. The problem of closing such a gap was solved on page 79 for the case of plane stress. The solution can be converted to plane strain by the substitutions given on page 444. The stress components which result, combined with the axial stress $\sigma_z = -\alpha ET$ from Eqs. (*d*), become identical with those given by Eqs. (257) when the axial force is annulled.

Inside and outside temperatures that vary round the boundary circles can be represented by Fourier series

$$\begin{aligned} T_i &= A_0 + A_1 \cos\theta + A_2 \cos 2\theta + \cdots + B_1 \sin\theta + B_2 \sin 2\theta + \cdots \\ T_0 &= A_0' + A_1' \cos\theta + A_2' \cos 2\theta + \cdots + B_1' \sin\theta + B_2' \sin 2\theta + \cdots \end{aligned} \tag{q}$$

The thermal stress due to the several terms can be treated separately, that due to the uniform terms A_0, A_0' being covered by the preceding case, with $T_i = A_0 - A_0'$. Corresponding to the terms $\cos\theta$, $\sin\theta$; $\cos 2\theta$, $\sin 2\theta$; etc., the function Z will have terms proportional to

$$z,\ z^{-1} \qquad z^2,\ z^{-2} \qquad \text{etc.} \tag{r}$$

Now $\int z^n\,dz$ taken round a complete circle of radius r vanishes unless $n = -1$, for we have

$$\begin{aligned} \int z^n\,dz &= \int r^n e^{in\theta} r e^{i\theta} i\,d\theta = ir^{n+1}\int_0^{2\pi} e^{i(n+1)\theta}\,d\theta \\ &= ir^{n+1}\int_0^{2\pi} [\cos(n+1)\theta + i\sin(n+1)\theta]\,d\theta \end{aligned}$$

This is clearly zero unless $n + 1 = 0$, in which case we have

$$\int \frac{dz}{z} = 2\pi i \tag{s}$$

Thus the only term in (*r*) that will produce a nonzero integral on the right of Eq. (*l*) is the term z^{-1}. It follows that the terms in $\cos 2\theta$, $\sin 2\theta$, and higher harmonics in the temperature series (*q*) do not produce any relative displacement of the two faces of the slit in the slit tube. The net heat flow from inside to outside corresponding to such terms is zero, and the only stress they produce is that given by Eqs. (*d*).

The terms in (*q*) that give rise to a term in z^{-1} in Z are those in $\cos\theta$, $\sin\theta$. It is sufficient to consider $\cos\theta$ only, since the effects of the $\sin\theta$ terms can be deduced from those of $\cos\theta$ by changing the initial line $\theta = 0$. Accordingly, we consider only

$$T_i = A_1 \cos\theta \qquad T_0 = A_1' \cos\theta \tag{t}$$

The problem of determining the steady temperature distribution corresponding to these boundary values is solved by taking the temperature T as the real part of a function

$$\frac{C_1}{z} + C_2 z \tag{u}$$

and determining C_1 and C_2 so that the conditions (t) are satisfied. The values are

$$C_1 = \frac{a^2 b^2}{b^2 - a^2}\left(\frac{A_1}{a} - \frac{A_1'}{b}\right) \qquad C_2 = \frac{A_1' b - A_1 a}{b^2 - a^2} \tag{v}$$

The term C_1/z in (u) corresponds to the value

$$(1 + \nu)\alpha \frac{C_1}{z}$$

for the function Z. Inserting this in Eq. (l) and making use of (s), we find that the displacement discontinuity is given by

$$u_2 - u_1 + i(v_2 - v_1) = i2\pi(1 + \nu)\alpha C_1$$

and therefore

$$u_2 - u_1 = 0 \qquad v_2 - v_1 = 2\pi(1 + \nu)\alpha C_1$$

This means that the top face of the slit in Fig. 234 moves down by $2\pi(1 + \nu)\alpha C_1$ into the space occupied by the bottom face and material below it. Physically this is impossible, of course, and is prevented by forces between the faces sufficient to create a counteracting displacement. The stress set up by this counteracting displacement is determined as explained at the end of Art. 43, in the present case for plane strain, of course. We find, for plane strain ($\epsilon_z = 0$)

$$\sigma_r = \kappa \cos\theta\, r\left(1 - \frac{a^2}{r^2}\right)\left(\frac{b^2}{r^2} - 1\right)$$

$$\sigma_\theta = \kappa \cos\theta\, r\left(\frac{a^2 b^2}{r^4} + \frac{a^2 + b^2}{r^2} - 3\right)$$

$$\tau_{r\theta} = \kappa \sin\theta\, r\left(1 - \frac{a^2}{r^2}\right)\left(\frac{b^2}{r^2} - 1\right)$$

and, as in (a) of Art. 151,

$$\sigma_z = \nu(\sigma_r + \sigma_\theta) - E\alpha T$$

where

$$\kappa = -\frac{-\alpha E}{2(1 - \nu)}\left(\frac{A_1}{a} - \frac{A_1'}{b}\right)\frac{a^2 b^2}{b^4 - a^4}$$

If the ends are free the axial stress due to removal of the force and couple on each end must be considered.

161 | Thermal Plane Stress Due to Disturbance of Uniform Heat Flow by an Insulated Hole

If an otherwise uniform heat flow is disturbed by a hole, cavity, or inclusion of different material, thermal stress is induced by the diversion of the flow. The problem of an insulated circular hole in an infinite body is solvable as an application of the preceding analysis. Corresponding to an undisturbed flow in the negative y direction with temperature gradient of magnitude τ, we may write $T = \tau y$. With the hole we have

$$T = \tau\left(r + \frac{a^2}{r}\right)\sin\theta$$

This time, considering plane stress rather than plane strain and recalling the conversion rule detailed on page 444, we take

$$Z = -i\alpha\tau\left(z - \frac{a^2}{z}\right)$$

where a is the radius of the hole. The displacement discontinuity corresponding to zero stress is again given by (l), which now leads to

$$(u)_{\theta=0} - (u)_{\theta=2\pi} = 2\pi a^2\alpha\tau \qquad (v)_{\theta=0} - (v)_{\theta=2\pi} = 0$$

This is annulled by an *edge dislocation* of the type discussed in Art. 34. The final plane stress components[1] are given by

$$\sigma_r,\ \sigma_\theta,\ \tau_{r\theta} = -\frac{1}{2}E\alpha\tau a\left[\left(\frac{a}{r} - \frac{a^3}{r^3}\right)\sin\theta,\ \left(\frac{a}{r} + \frac{a^3}{r^3}\right)\sin\theta,\ -\left(\frac{a}{r} - \frac{a^3}{r^3}\right)\cos\theta\right]$$

The greatest value, $E\alpha\tau a$, is attained by σ_θ at the "poles" $\theta = \pi/2$, $\theta = 3\pi/2$, compressive at the hot pole, tensile at the cool pole. For $2\tau a = 100°$F, it is about 4,480 psi in steel.

Solutions of the corresponding problem have been given for other shapes[2] of hole, and for a circular hole in a semi-infinite region.[3] The axisymmetric problem of the spheroidal cavity has also been solved in series form.[4]

162 | Solutions of the General Equations. Thermoelastic Displacement Potential

Any particular solution we can obtain of Eqs. (264) will reduce the thermal stress problem to an ordinary problem of surface forces. The solution for u, v, w will lead by means of Eqs. (a) and (b) of Art. 153, using Eqs. (2), to values of the stress components. The surface forces required, together with the nonuniform temperature, to maintain these stresses, are then found from Eqs. (124). The removal of these forces in order to make the boundary free, so that the stress is due entirely to the nonuniform temperature, constitutes an ordinary problem of surface loading.

One way of finding particular solutions of Eqs. (264) is to take

$$u = \frac{\partial\psi}{\partial x} \qquad v = \frac{\partial\psi}{\partial y} \qquad w = \frac{\partial\psi}{\partial z} \tag{a}$$

where ψ is a function of x, y, z, and also of time t if the temperature varies with time. Such a function is called a thermoelastic displacement potential.

[1] A. L. Florence and J. N. Goodier, *J. Appl. Mech.*, vol. 26, pp. 293–294, 1959.

[2] A. L. Florence and J. N. Goodier, *J. Appl. Mech.*, vol. 27, pp. 635–639, 1960. Also H. Deresiewicz, *J. Appl. Mech.*, vol. 28, pp. 147–149, 1961.

[3] J. N. Goodier and A. L. Florence, *Quart. J. Mech. Appl. Math.*, vol. 16, pp. 273–282, 1963.

[4] A. L. Florence and J. N. Goodier, *Proc. 4th U.S. Nat. Congr.* Appl. Mech., pp. 595–602, 1962.

Using Eqs. (5) and (10), we can write Eqs. (264) in the form

$$\frac{\partial e}{\partial x} + (1 - 2\nu)\nabla^2 u = 2(1 + \nu)\alpha \frac{\partial T}{\partial x}$$

$$\cdots\cdots\cdots\cdots\cdots\cdots \tag{b}$$

Since $e = \partial u/\partial x + \partial v/\partial y + \partial w/\partial z$, Eqs. (*a*) lead to $e = \nabla^2\psi$, and Eqs. (*b*) become

$$(1 - \nu)\frac{\partial}{\partial x}\nabla^2\psi = (1 + \nu)\alpha \frac{\partial T}{\partial x}$$

$$\cdots\cdots\cdots\cdots\cdots\cdots \tag{c}$$

$\partial/\partial y$ and $\partial/\partial z$ replacing $\partial/\partial x$ in the second and third of these equations. All three of the equations are evidently satisfied if we take the function[1] ψ as a solution of the equation

$$\nabla^2\psi = \frac{1 + \nu}{1 - \nu}\alpha T \tag{d}$$

Solutions of equations of this type are considered in the theory of potential.[2] A solution can be written down as the gravitational potential of a distribution of matter of density $-(1 + \nu)\alpha T/4\pi(1 - \nu)$, which is[3]

$$\psi = -\frac{(1 + \nu)\alpha}{4\pi(1 - \nu)}\iiint T(\xi,\eta,\zeta)\frac{1}{r'}\,d\xi\,d\eta\,d\zeta \tag{e}$$

where $T(\xi,\eta,\zeta)$ is the temperature at a typical point ξ, η, ζ at which there is an element of volume $d\xi\,d\eta\,d\zeta$, and r' is the distance between this point and the point x, y, z. Equation (*e*) gives the complete solution of the thermal stress problem of an infinite solid at temperature zero except for a heated (or cooled) region.[4] The cases of such a region in the form of an ellipsoid of revolution and a semi-infinite circular cylinder, uniformly hot, have been worked out.[5] For the ellipsoid the maximum stress that can occur is $\alpha ET/1 - \nu$, and is normal to the surface of the ellipsoid at the points of sharpest curvature of the generating ellipse. This value occurs only for the two extreme cases of a very flat or very elongated ellipsoid of revolution. Intermediate cases have smaller maximum stress. For a spherical region the value is two-thirds as great.

When T is independent of z, and $w = 0$, we shall have plane strain, with ψ, u, and v independent of z. Equation (*d*) becomes

$$\frac{\partial^2\psi}{\partial x^2} + \frac{\partial^2\psi}{\partial y^2} = \frac{1 + \nu}{1 - \nu}\alpha T \tag{f}$$

[1] Functions of this kind were used by E. Almansi in the problem of the sphere. See (1) *Atti Reale Accad. Sci. Torino*, vol. 32, p. 963, 1896–1897; (2) *Mem. Reale Accad. Sci. Torino*, ser. 2, vol. 47, 1897.

[2] See, for instance, W. D. MacMillan, "Theory of the Potential," New York, 1930.

[3] This potential function was used by C. W. Borchardt in the problem of the sphere. See *Monatsber. Königl. Preuss. Akad. Wiss.*, Berlin, 1873, p. 9.

[4] J. N. Goodier, *Phil. Mag.*, vol. 23, p. 1017, 1937. The semi-infinite solid is considered by R. D. Mindlin and D. H. Cheng, *J. Appl. Phys.*, vol. 21, pp. 926, 931, 1950.

[5] N. O. Myklestad, *J. Appl. Mech.*, 1942, p. A-131. R. H. Edwards, *J. Appl. Mech.*, vol. 18, pp. 19–30, 1951, solves the problem for different elastic constants within the hot ellipsoidal region.

A particular solution is given by the *logarithmic potential*

$$\psi = \frac{1}{2\pi}\frac{1+\nu}{1-\nu}\alpha \iint T(\xi,\eta) \log r' \, d\xi \, d\eta \tag{g}$$

where $\qquad r' = [(x-\xi)^2 + (y-\eta)^2]^{1/2}$

For a thin plate, with no variation of T through the thickness, we may assume plane stress, with $\sigma_z = \tau_{xz} = \tau_{yz} = 0$, and u, v, σ_x, σ_y, τ_{xy} independent of z. We have then the stress-strain relations [cf. Eqs. (*d*) of Art. 150]

$$\begin{aligned} \sigma_x &= \frac{E}{1-\nu^2}\left[\frac{\partial u}{\partial x} + \nu\frac{\partial v}{\partial y} - (1+\nu)\alpha T\right] \\ \sigma_y &= \frac{E}{1-\nu^2}\left[\frac{\partial v}{\partial y} + \nu\frac{\partial u}{\partial x} - (1+\nu)\alpha T\right] \\ \tau_{xy} &= \frac{E}{2(1+\nu)}\left(\frac{\partial v}{\partial x} + \frac{\partial u}{\partial y}\right) \end{aligned} \tag{h}$$

Substituting these in the two equations of equilibrium (18) (with zero body force), we find the equations

$$\frac{\partial}{\partial x}\left(\frac{\partial u}{\partial x} + \frac{\partial v}{\partial y}\right) + \frac{1-\nu}{1+\nu}\left(\frac{\partial^2 u}{\partial x^2} + \frac{\partial^2 u}{\partial y^2}\right) = 2\alpha\frac{\partial T}{\partial x}$$

$$\cdots\cdots\cdots\cdots\cdots\cdots\cdots\cdots \tag{i}$$

These are satisfied by

$$u = \frac{\partial \psi}{\partial x} \qquad v = \frac{\partial \psi}{\partial y} \tag{j}$$

provided that ψ is a solution of

$$\frac{\partial^2 \psi}{\partial x^2} + \frac{\partial^2 \psi}{\partial y^2} = (1+\nu)\alpha T \tag{k}$$

Comparing with Eq. (*f*), we see that a particular solution is given by the logarithmic potential (*g*) with the factor $1-\nu$ in the denominator omitted. This gives the complete solution for local heating in an infinite plate, where the stress and deformation must tend to zero at infinity.

As a first example of this kind we consider an infinite plate at temperature zero except for a rectangular region $ABCD$ of sides $2a$, $2b$ (Fig. 235) within which the temperature is T, and uniform.[1] The required logarithmic potential is

$$\psi = \frac{1}{2\pi}(1+\nu)\alpha T \int_{-b}^{b}\int_{-a}^{a} \frac{1}{2}\log[(x-\xi)^2 + (y-\eta)^2] \, d\xi \, d\eta \tag{l}$$

The displacements are obtained by differentiation according to (*j*) and then the stress components can be found from (*h*). The results for σ_x and τ_{xy} at points such as P outside the hot rectangle can be reduced to

$$\sigma_x = E\alpha T \frac{1}{2\pi}(\psi_1 - \psi_2) \qquad \tau_{xy} = E\alpha T \frac{1}{4\pi}\log\frac{r_1 r_3}{r_2 r_4} \tag{m}$$

[1] J. N. Goodier, *Phil. Mag., loc. cit.* A solution for the three-dimensional problem of a parallelepiped as the hot region by J. Ignaczak and W. Nowacki (1958) is given in the book by W. Nowacki, "Thermoelasticity," Pergamon Press, New York, 1962.

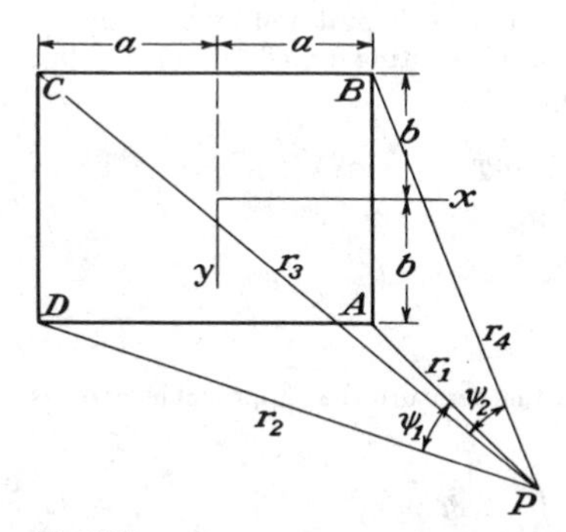

Fig. 235

the angles ψ_1, ψ_2 and the distances r_1, r_2, r_3, r_4 being those indicated in Fig. 235. The angles are those subtended at P by the two sides AD, BC of the rectangle parallel to the x axis. The expression for σ_y is obtained from the first of Eqs. (m) by using instead of ψ_1 and ψ_2 the angles subtended at P by the other two sides AB, DC of the rectangle.

The value of σ_x just below AD and just to the left of A is

$$E\alpha T \frac{1}{2\pi}\left(\pi - \arctan \frac{a}{b}\right)$$

and is greatest for a rectangle infinitely long in the y direction, when it becomes $\frac{1}{2}E\alpha T$. Both normal stress components change sharply on turning a corner of the rectangle. The shear stress τ_{xy} approaches infinity as a corner is approached. These peculiarities are, of course, a consequence of the ideally sharp corners of the heated rectangle.

If the heated area is elliptical[1] instead of rectangular, the ellipse being

$$\frac{x^2}{a^2} + \frac{y^2}{b^2} = 1$$

the value of the stress σ_y just outside the ellipse, near an end of the major axis, is

$$\frac{E\alpha T}{1 + (b/a)}$$

which approaches $E\alpha T$ for a very slender ellipse. If the heated area is circular it becomes $\frac{1}{2}E\alpha T$. The stress σ_x just beyond an end of the minor axis is

$$\frac{E\alpha T}{1 + (a/b)}$$

and approaches zero for a very slender ellipse.

The method of the present article becomes particularly simple when the temperature varies with time and satisfies the differential equation of heat conduction[2]

$$\frac{\partial T}{\partial t} = \kappa \nabla^2 T \qquad (n)$$

[1] Goodier, *loc. cit.*

[2] See, for instance, Carslaw and Jaeger, *op. cit.*

where κ is the thermal conductivity divided by the specific heat and by the density. Differentiating Eq. (*d*) with respect to t and then substituting for $\partial T/\partial t$ from Eq. (*n*), we find that the function ψ must satisfy the equation

$$\nabla^2 \frac{\partial \psi}{\partial t} = \frac{1+\nu}{1-\nu} \alpha\kappa\nabla^2 T$$

We may therefore take

$$\frac{\partial \psi}{\partial t} = \frac{1+\nu}{1-\nu} \alpha\kappa T$$

The integral of this which is appropriate for a temperature that approaches zero as time goes on is

$$\psi = -\frac{1+\nu}{1-\nu} \alpha\kappa \int_t^\infty T\, dt \tag{o}$$

as may be verified by substitution in Eq. (*d*), making use of Eq. (*n*).

Consider for instance a long circular cylinder (plane strain) which is cooling or being heated toward a steady state of heat conduction. The temperature is not symmetrical about the axis but is independent of the axial coordinate z. The temperature is then representable by a series of terms of the form

$$T_{sn} = e^{-\kappa s^2 t} J_n(sr) e^{in\theta} \tag{p}$$

where the real or imaginary parts of $e^{in\theta}$ may be taken to obtain $\cos n\theta$ or $\sin n\theta$. From Eq. (*o*) the function ψ corresponding to this temperature term will be

$$\psi_{sn} = -\frac{1+\nu}{1-\nu} \alpha\kappa \frac{1}{s^2} T_{sn} \tag{q}$$

A series of such terms, corresponding to the series for T, will represent a particular solution of the general equations (*b*). The displacements may be calculated according to Eqs. (*a*), or their polar equivalents,

$$u = \frac{\partial \psi}{\partial r} \qquad v = \frac{1}{r}\frac{\partial \psi}{\partial \theta}$$

u and v here being the radial and tangential components. The axial component w is zero in plane strain.

The strain components follow from the results of Art. 30. The stress components can then be found from the plane strain formulas (*a*) and (*b*) of Art. 151, together with the last of Eqs. (51) for the shear stress $\tau_{r\theta}$.

When such a solution has been obtained it will, in general, be found that it gives nonzero boundary forces (σ_r, $\tau_{r\theta}$) on the curved surface of the cylinder. The effects of removing these are found by solving an ordinary plane strain problem, using the general stress function in polar coordinates given in Art. 43.[1]

More generally, we may include internal generation of heat at a time rate q per unit volume. Then a term $q/c\rho$ is added to the right-hand side of Eq. (*n*), c being the specific heat and ρ density. The form

$$\psi = \frac{1+\nu}{1-\nu} \alpha \int_{t_1}^{t} (\kappa T + Q)\, dt + f(x,y,z)$$

[1] This problem is worked out for a hollow cylinder, with temperature corresponding to Eq. (*p*), in the paper by J. N. Goodier cited above.

satisfies Eq. (*d*) provided that

$$\nabla^2 Q = \frac{q}{c\rho} \qquad \text{and} \qquad \nabla^2 f = \frac{1+\nu}{1-\nu}\alpha T_1$$

Here Q is in general a function of t as well as x, y, z, and T_1 is T at time t_1.

163 | The General Two-dimensional Problem for Circular Regions

In Arts. 160 and 162 we have made use, in different ways, of the differential equation of heat conduction. Other methods are required if we wish to deal with a completely arbitrary temperature distribution as, for instance, some given initial distribution throughout the body. Here we consider such a method for plane strain or plane stress in polar coordinates.

Using a thermoelastic displacement potential ψ, giving displacements as in Eq. (*k*) of Art. 162, we may write Eq. (*f*) for plane strain or Eq. (*k*) for plane stress in the polar form

$$\frac{\partial^2 \psi}{\partial r^2} + \frac{1}{r}\frac{\partial \psi}{\partial r} + \frac{1}{r^2}\frac{\partial^2 \psi}{\partial \theta^2} = \beta T \tag{a}$$

where $\beta = (1+\nu)/(1-\nu)\alpha$ for plane strain and $(1+\nu)\alpha$ for plane stress.

The *polar* components of displacement are

$$u = \frac{\partial \psi}{\partial r} \qquad v = \frac{1}{r}\frac{\partial \psi}{\partial \theta} \tag{b}$$

The temperature T, a function of r and θ, is taken in the Fourier series form

$$T = \sum_{n=0}^{\infty} T_n(r)\cos n\theta + \sum_{n=1}^{\infty} T_n'(r)\sin n\theta \tag{c}$$

However, we shall limit the discussion to the cosine series, the treatment of the sine series being similar. Correspondingly, we take

$$\psi = \sum_{n=0}^{\infty} \psi_n(r)\cos n\theta \tag{d}$$

Then Eq. (*a*) requires

$$\frac{d^2\psi_n}{dr^2} + \frac{1}{r}\frac{d\psi_n}{dr} - \frac{n^2}{r^2}\psi_n = \beta T_n(r) \tag{e}$$

A particular solution of this, obtainable by the method of variation of parameters, is

$$\psi_n = -\frac{\beta}{2n}\left[r^n \int_r^b T_n(\rho)\rho^{1-n}\,d\rho + r^{-n}\int_a^r T_n(\rho)\rho^{1+n}\,d\rho\right] \tag{f}$$

for $n = 1, 2, 3 \ldots$. For $n = 0$, we find directly from (*e*)

$$\psi_0 = \beta\left[-\log r \int_a^r \rho T_0(\rho)\,d\rho + \int_r^b \rho T_0(\rho)\log\rho\,d\rho\right] \tag{g}$$

Here a and b are the inner and outer radii of the circular region and ρ is merely the dummy variable of integration. The functions given by (*f*) and (*g*) are introduced in (*d*). The displacements then follow from (*b*), and from them the strain components ϵ_r, ϵ_θ, $\gamma_{r\theta}$ are obtained by the formulas (48), (49), and (50). These lead to the stress components through Eqs. (*b*) and (*c*) of Art. 150 for plane stress, or Eqs. (*b*) of Art. 151 for plane strain. The relation between shear stress and shear strain is simply $\tau_{r\theta} = G\gamma_{r\theta}$.

The state of stress and deformation so found from a displacement potential as a particular solution of the differential equations will not itself satisfy prescribed boundary conditions on the circles $r = a$, $r = b$. It will require for its maintenance certain boundary tractions, which can of course be determined from the solution outlined above by evaluating σ_r and $\tau_{r\theta}$ at $r = a$ and $r = b$. But the problem of satisfying prescribed conditions, for instance, $\sigma_r = 0$ and $\tau_{r\theta} = 0$ on the bounding circles, can now be completed by superposing an isothermal Fourier solution as given in Art. 43.

It will be recognized that ψ_0 in Eq. (*g*) leads to the solutions already given in Arts. 150 and 151 for a temperature independent of θ. It is then apparent that the ψ_n given by Eq. (*f*) lead to a generalization[1] of those solutions for temperature depending on θ as well as r.

164 | The General Two-dimensional Problem in Complex Potentials

We have seen that any displacement potential Ψ, corresponding to a given temperature distribution T and yielding continuous displacement, reduces the problem to one of boundary loading only. We may therefore use complex potentials $\psi(z)$, $\chi(z)$, as in Chap. 6, for plane strain or plane stress, as soon as an appropriate displacement potential is found.

Writing Ψ for such a potential, a function of x and y only, we shall have in plane strain

$$u + iv = \frac{\partial \Psi}{\partial x} + i\frac{\partial \Psi}{\partial y} \qquad w = 0 \tag{a}$$

and Ψ must satisfy Eq. (*f*) of Art. 162, i.e.,

$$\left(\frac{\partial^2}{\partial x^2} + \frac{\partial^2}{\partial y^2}\right)\Psi = \frac{1+\nu}{1-\nu}\alpha T \tag{b}$$

The temperature T is, of course, to be a function of x and y only, and as in Chap. 6 we write here $z = x + iy$, and $\bar{z} = x - iy$. Also

$$x = \frac{1}{2}(z + \bar{z}) \qquad y = \frac{1}{2i}(z - \bar{z}) \tag{c}$$

When these are put into $T(x,y)$, we have a function $t(z,\bar{z})$. Thus

$$T(x,y) = t(z,\bar{z}) \tag{d}$$

We can take the partial derivatives

$$\frac{\partial}{\partial z}t(z,\bar{z}) \qquad \frac{\partial}{\partial \bar{z}}t(z,\bar{z}) \tag{e}$$

[1] Such a generalization was given by N. N. Lebedev in the Russian monograph "Temperature Stresses in the Theory of Elasticity," *ONTI*, Leningrad-Moscow, 1937. The derivation above is from C. E. Wallace, Thermoelastic Stress in Plates—Problems in Curvilinear Coordinates, doctoral dissertation, Stanford University, 1958. This also deals with elliptic and bipolar coordinates, and with problems of discontinuous temperature, such as a disk half cold and half hot, which are not well suited for Fourier analysis on account of slow convergence.

formally (in spite of the fact that we cannot change z as a point in the xy plane without at the same time changing $\bar{z}$). We can also take *indefinite* integrals such as

$$\int t(z,\bar{z})\,dz \qquad \int t(z,\bar{z})\,d\bar{z} \tag{f}$$

and also

$$\int\int t(z,\bar{z})\,dz\,d\bar{z} \tag{g}$$

As will be shown immediately, an appropriate form of displacement potential is

$$\Psi(x,y) = f(z,\bar{z}) = \frac{1}{4}\frac{1+\nu}{1-\nu}\alpha\int\int t(z,\bar{z})\,dz\,d\bar{z} \tag{h}$$

It provides displacement and stress in the forms[1]

$$2G(u+iv) = \frac{E\alpha}{2(1-\nu)}\int t(z,\bar{z})\,dz \qquad \sigma_x + \sigma_y = -\frac{E\alpha}{1-\nu}t(z,\bar{z})$$
$$\sigma_y - \sigma_x + 2i\tau_{xy} = \frac{E\alpha}{1-\nu}\int \frac{\partial}{\partial\bar{z}}t(z,\bar{z})\,dz \tag{i}$$

This state is maintained by boundary loading (which can be calculated from these formulas) in combination with the temperature distribution T. The problem of equal and opposite boundary loading on the curved, or lateral, surfaces can then be solved by means of complex potentials $\psi(z)$ and $\chi(z)$ for plane strain without body force as explained in Chap. 6.

To derive Eqs. (h) and (i), we first observe that from Eq. (d) we have

$$\frac{\partial T}{\partial x} = \frac{\partial t}{\partial z}\frac{\partial z}{\partial x} + \frac{\partial t}{\partial\bar{z}}\frac{\partial\bar{z}}{\partial x} = \frac{\partial t}{\partial z} + \frac{\partial t}{\partial\bar{z}} \qquad i\frac{\partial T}{\partial y} = \frac{\partial t}{\partial\bar{z}} - \frac{\partial t}{\partial z} \tag{j}$$

Therefore,

$$2\frac{\partial t}{\partial z} = \frac{\partial T}{\partial x} - i\frac{\partial T}{\partial y} \qquad 2\frac{\partial t}{\partial\bar{z}} = \frac{\partial T}{\partial x} + i\frac{\partial T}{\partial y} \tag{k}$$

and, applying the operation $2\partial/\partial z$ to the last equation,

$$4\frac{\partial^2 t}{\partial z\,\partial\bar{z}} = \left(\frac{\partial}{\partial x} - i\frac{\partial}{\partial y}\right)\left(\frac{\partial T}{\partial x} + i\frac{\partial T}{\partial y}\right) = \frac{\partial^2 T}{\partial x^2} + \frac{\partial^2 T}{\partial y^2}$$

Here T stands for any function of x and y. Thus in Eq. (b) we can use the last result for the left-hand side and use Eq. (d) for the right-hand side to obtain

$$4\frac{\partial^2}{\partial z\,\partial\bar{z}}f(z,\bar{z}) = 4\beta\alpha t(z,\bar{z}) \tag{l}$$

where

$$4\beta = \frac{1+\nu}{1-\nu} \tag{m}$$

Integrating Eq. (l) with respect to z, treating $\bar{z}$ as fixed, we have

$$\frac{\partial}{\partial\bar{z}}f(z,\bar{z}) = \beta\alpha\int t(z,\bar{z})\,dz \tag{n}$$

[1] N. N. Lebedev, "Temperature Stresses in the Theory of Elasticity," *ONTI*, Leningrad-Moscow, 1937, pp. 55, 56. The third of Eqs. (i) above differs from Lebedev's form, which introduces the conjugate of $t(z,\bar{z})$. A similar development is given by B. E. Gatewood, *Phil. Mag.*, vol. 32, pp. 282–301, 1941.

From the first of Eqs. (*a*) and the second of Eqs. (*k*) we now find

$$u + iv = 2\frac{\partial}{\partial \bar{z}} f(z,\bar{z}) = \beta\alpha \int t(z,\bar{z})\, dz \tag{o}$$

from which the first of Eqs. (*i*) follows.

The (indefinite) integral on the right-hand side of Eq. (*n*) is again a function of z and $\bar{z}$. Integration with respect to $\bar{z}$, treating z as fixed, provides

$$f(z,\bar{z}) = \beta\alpha \int [\int t(z,\bar{z})\, dz]\, d\bar{z} \tag{p}$$

which is equivalent to Eq. (*h*). No arbitrary functions of integration are specified because we require only the simplest available solution of Eq. (*b*) or (*l*). To validate the second of Eqs. (*i*), we turn to the stress-strain relations for plane strain. These can be taken from the three-dimensional forms given in Art. 153. But for plane strain we have $\epsilon_z = 0$ (as well as $\gamma_{yz} = \gamma_{xz} = 0$). The first two of the three equations (*c*) of Art. 153 are

$$\sigma_x = \lambda e + 2G\epsilon_x - \frac{\alpha ET}{1-2\nu} \qquad \sigma_y = \lambda e + 2G\epsilon_y - \frac{\alpha ET}{1-2\nu} \tag{q}$$

Thus, since now $e = \epsilon_x + \epsilon_y$,

$$\sigma_x + \sigma_y = 2(\lambda + G)e - 2\frac{\alpha ET}{1-2\nu} \tag{r}$$

But Eqs. (*a*) and (*b*) of the present article show that in the state derived from the displacement potential Ψ we shall have

$$\epsilon_x + \epsilon_y = \frac{1+\nu}{1-\nu}\alpha T \tag{s}$$

With this, and the relations between the elastic constants given as Eqs. (5) and (10), Eq. (*r*) above becomes

$$\sigma_x + \sigma_y = -\frac{E\alpha}{1-\nu} T$$

By Eq. (*d*), this is the same as the second of Eqs. (*i*).

To validate the third of Eqs. (*i*) we start with the first and apply the operation $2\partial/\partial\bar{z}$ to the right-hand side and the equivalent operation $\partial/\partial x + i\partial/\partial y$ to the left-hand side. Then we have

$$2G\left[\frac{\partial u}{\partial x} - \frac{\partial v}{\partial y} + i\left(\frac{\partial v}{\partial x} + \frac{\partial u}{\partial y}\right)\right] = \frac{E\alpha}{1-\nu}\frac{\partial}{\partial \bar{z}}\int t(z,\bar{z})\, dz \tag{t}$$

But from Eqs. (*q*),

$$\sigma_x - \sigma_y = 2G(\epsilon_x - \epsilon_y) = 2G\left(\frac{\partial u}{\partial x} - \frac{\partial v}{\partial y}\right)$$

On the left of Eq. (*t*) the expression in parentheses is the same as γ_{xy}. On the right we may, with acceptable restrictions on $t(z,\bar{z})$, apply the differentiation to $t(z,\bar{z})$. Then (*t*) becomes

$$\sigma_x - \sigma_y + 2i\tau_{xy} = \frac{E\alpha}{1-\nu}\int \frac{\partial}{\partial \bar{z}} t(z,\bar{z})\, dz$$

and this is the same as the third of Eqs. (*i*).

chapter 14

The Propagation of Waves in Elastic Solid Media

165 | Introduction

The preceding chapters have dealt with problems of *elastostatics*. The elastic body was at rest under the action of unchanging loads. Or if changes were contemplated, they were sufficiently gradual to justify the assumption of a statical state at each instant (e.g., the Hertz collision theory, page 420)—the *quasi-statical* problem.

Sudden loading as from an explosion or sudden displacement as in slip at a seismic fault in the earth causing an earthquake presents essentially dynamic problems. Equations of equilibrium must be replaced by equations of motion. When a force is first applied, its action is not transmitted instantaneously to all parts of the body. Waves of stress and deformation radiate from the loaded region with finite velocities of propagation. As in the familiar case of sound in air, there is no disturbance at a point until the wave has time to reach it. But in an elastic solid there is more than one kind of wave and more than one characteristic wave velocity.

We shall begin with the general equations for three-dimensional problems, in rectangular coordinates, and the simplest solutions representing the simplest types of waves.[1] Approximate representations of wave motions in special cases, such as tensile waves in bars, are introduced later when the general theory is available to clarify the nature of the assumptions involved.

[1] Other forms of motion, e.g., vibrations, are not discussed here. For vibrations of bars, rings, and plates see, for instance, S. Timoshenko, "Vibration Problems in Engineering," chap. 5, McGraw-Hill Book Company, New York, 1955.

166 | Waves of Dilatation and Waves of Distortion in Isotropic Elastic Media

In discussing the propagation of waves in an elastic medium, it is of advantage to use differential equations in terms of displacements [Eqs. (127)]. To obtain the equations of small motion from these equations of equilibrium, it is only necessary to add the inertia forces. Then, assuming that there are no body forces, the equations of motion are

$$
\begin{aligned}
(\lambda + G)\frac{\partial e}{\partial x} + G\,\nabla^2 u - \rho\frac{\partial^2 u}{\partial t^2} &= 0 \\
(\lambda + G)\frac{\partial e}{\partial y} + G\,\nabla^2 v - \rho\frac{\partial^2 v}{\partial t^2} &= 0 \\
(\lambda + G)\frac{\partial e}{\partial z} + G\,\nabla^2 w - \rho\frac{\partial^2 w}{\partial t^2} &= 0
\end{aligned}
\tag{269}
$$

in which e is the volume expansion and the symbol ∇^2 represents the operation

$$
\nabla^2 = \frac{\partial^2}{\partial x^2} + \frac{\partial^2}{\partial y^2} + \frac{\partial^2}{\partial z^2}
$$

Assume first that the deformation produced by the waves is such that the volume expansion is zero, the deformation consisting of shearing distortion and rotation only. Then Eqs. (269) become

$$
\begin{gathered}
G\,\nabla^2 u - \rho\frac{\partial^2 u}{\partial t^2} = 0 \\
\cdots\cdots\cdots\cdots
\end{gathered}
\tag{270}
$$

These are equations for waves called equivoluminal waves, or *waves of distortion.*

Consider now the case when the deformation produced by the waves is not accompanied by rotation. The rotation of an element is (see Art. 83)

$$
\omega_x = \frac{1}{2}\left(\frac{\partial w}{\partial y} - \frac{\partial v}{\partial z}\right) \qquad \omega_y = \frac{1}{2}\left(\frac{\partial u}{\partial z} - \frac{\partial w}{\partial x}\right) \qquad \omega_z = \frac{1}{2}\left(\frac{\partial v}{\partial x} - \frac{\partial u}{\partial y}\right) \tag{a}
$$

The conditions that the deformation is *irrotational* can therefore be represented in the form

$$
\frac{\partial v}{\partial x} - \frac{\partial u}{\partial y} = 0 \qquad \frac{\partial w}{\partial y} - \frac{\partial v}{\partial z} = 0 \qquad \frac{\partial u}{\partial z} - \frac{\partial w}{\partial x} = 0 \tag{b}
$$

When these equations are satisfied, the displacements u, v, w are derivable from a single function ϕ as follows:

$$
u = \frac{\partial \phi}{\partial x} \qquad v = \frac{\partial \phi}{\partial y} \qquad w = \frac{\partial \phi}{\partial z} \tag{c}
$$

Then

$$e = \nabla^2\phi \qquad \frac{\partial e}{\partial x} = \frac{\partial}{\partial x}\nabla^2\phi = \nabla^2 u$$

Substituting these in Eqs. (269), we find that

$$(\lambda + 2G)\,\nabla^2 u - \rho\frac{\partial^2 u}{\partial t^2} = 0$$

$$\cdots\cdots\cdots\cdots\cdots\cdots \tag{271}$$

These are equations for *irrotational waves*, or *waves of dilatation.*[1]

The general case of propagation of waves in an elastic medium is obtained by superposing waves of distortion and waves of dilatation.[2] For both kinds of waves the equations of motion have the common form

$$\frac{\partial^2\psi}{\partial t^2} = a^2\,\nabla^2\psi \tag{272}$$

in which

$$a = c_1 = \sqrt{\frac{\lambda + 2G}{\rho}} \tag{273}$$

for the case of waves of dilatation, and

$$a = c_2 = \sqrt{\frac{G}{\rho}} \tag{274}$$

for the case of waves of distortion. We shall now show that c_1 and c_2 are velocities of propagation of plane waves of dilatation and of distortion.

167 | Plane Waves

If a disturbance is produced at a point of an elastic medium, waves radiate from this point in all directions. At a great distance from the center of disturbance, however, such waves can be considered as *plane waves,* and it may be assumed that all particles are moving parallel to the direction of wave propagation (longitudinal waves) or perpendicular to this direction (transverse waves). In the first case we have *waves of dilatation;* in the second, *waves of distortion.*

Considering *longitudinal waves,* if we take the x axis in the direction of wave propagation, then $v = w = 0$ and u is a function of x only. Equations (271) then give

$$\frac{\partial^2 u}{\partial t^2} = c_1{}^2\frac{\partial^2 u}{\partial x^2} \tag{275}$$

[1] But the dilatation is in general accompanied by shear strain.

[2] Concerning the generality of this combination, and the connection with elastostatics, see E. Sternberg, *Arch. Rational Mech. and Anal.*, vol. 6, pp. 34–50, 1960.

It can be shown by substitution that any function $f(x + c_1t)$ is a solution of Eq. (275). Any function $f_1(x - c_1t)$ is also a solution, and the general solution of Eq. (275) can be represented in the form

$$u = f(x + c_1t) + f_1(x - c_1t) \tag{276}$$

This solution has a very simple physical interpretation, which can easily be explained in the following manner. Consider the second term on the right side of Eq. (276). For a definite instant t, this term is a function of x only and can be represented by a certain curve such as mnp (Fig. 236a), the shape of which depends on the function f_1. After an interval of time Δt, the argument of the function f_1 becomes $x - c_1(t + \Delta t)$. The function f_1 will remain unchanged provided that simultaneously with the increase of t by Δt the abscissas are increased by an amount Δx equal to $c_1 \Delta t$. This means that the curve mnp, constructed for the moment t, can also be used for the instant $t + \Delta t$, if it is displaced in the x direction by the distance $\Delta x = c_1 \Delta t$, as shown by the dotted line in the figure. From this consideration it can be seen that the second term of the solution (276) represents a wave traveling in the direction of the x axis with a constant speed c_1. In the same manner it can be shown that the first term of the solution (276) represents a wave traveling in the opposite direction. Thus, the general solution (276) represents two waves traveling along the x axis in two opposite directions with the constant velocity c_1 given by Eq. (273). This velocity can be expressed in terms of E, ν, and ρ by substituting in Eq. (273) the equivalents of λ and G given as Eqs. (10) and Eq. (5). Then

$$c_1 = \sqrt{\frac{E(1 - \nu)}{(1 + \nu)(1 - 2\nu)\rho}} \tag{277}$$

For steel c_1 may be taken as 19,550 fps.

Considering a "forward" wave motion represented by the function

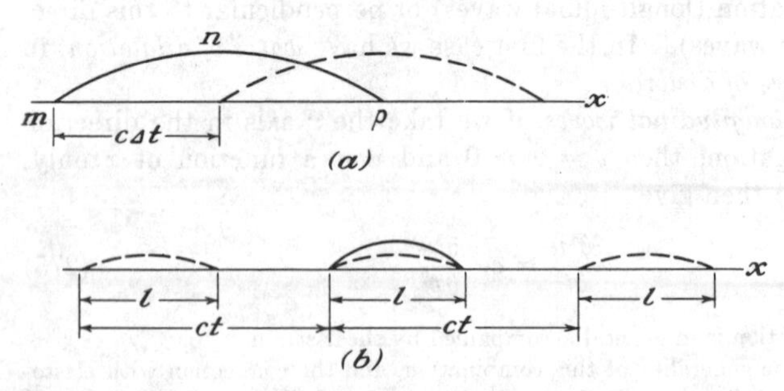

Fig. 236

$f_1(x - c_1t)$ alone in Eq. (276), we have for the *particle velocity* $\dot{u}$

$$\dot{u} = \frac{\partial u}{\partial t} = -c_1 f_1'(\xi) \qquad \text{where } \xi = x - c_1 t \tag{a}$$

and the prime means differentiation of $f_1(\xi)$ with respect to ξ. The kinetic energy of an element $dx\,dy\,dz$ is therefore obtained as

$$\frac{1}{2}\rho\,dx\,dy\,dz\left(\frac{\partial u}{\partial t}\right)^2 = \frac{1}{2}\rho\,dx\,dy\,dz \cdot c_1^2[f_1'(\xi)]^2 \tag{b}$$

The potential energy is the same as the strain energy. The strain components are

$$\epsilon_x = \frac{\partial u}{\partial x} = f_1'(\xi) \qquad \epsilon_y = \epsilon_z = 0 \tag{c}$$

Then by Eq. (132), the strain energy of the element is

$$V_0\,dx\,dy\,dz = \frac{1}{2}(\lambda + 2G)[f_1'(\xi)]^2\,dx\,dy\,dz \tag{d}$$

Comparing Eqs. (*b*) and (*d*), recalling Eq. (273), it is evident that the kinetic and potential energies are equal at any instant.

For the stress we have

$$\sigma_x = (\lambda + 2G)\epsilon_x \qquad \sigma_y = \sigma_z = \lambda\epsilon_x \tag{e}$$

and therefore

$$\frac{\sigma_y}{\sigma_x} = \frac{\sigma_z}{\sigma_x} = \frac{\lambda}{\lambda + 2G} = \frac{\nu}{1 - \nu} \tag{f}$$

These components σ_y, σ_z are as required to maintain $\epsilon_y = \epsilon_z = 0$. Comparing σ_x in Eqs. (*e*) with $\dot{u}$ in Eq. (*a*), using $\epsilon_x = f_1'(\xi)$ from Eqs. (*c*), we find

$$\sigma_x = -\rho c_1 \dot{u} \tag{g}$$

If we had considered the "backward" wave motion represented by the function $f(x + c_1t)$ alone in Eq. (276), the minus signs in Eq. (*g*), and in Eq. (*a*), would be replaced by plus signs.

The functions f and f_1 should be determined in each particular case from the initial conditions at the instant $t = 0$. For this instant we have, from Eq. (276),

$$\begin{aligned} (u)_{t=0} &= f(x) + f_1(x) \\ \left(\frac{\partial u}{\partial t}\right)_{t=0} &= c[f'(x) - f_1'(x)] \end{aligned} \tag{h}$$

Assume, for instance, that the initial velocity is zero and there is an initial displacement given by the equation

$$(u)_{t=0} = F(x)$$

Conditions (h) are satisfied by taking

$$f(x) = f_1(x) = \tfrac{1}{2}F(x)$$

Thus, in this case the initial displacement will be split into halves which will be propagated as waves in two opposite directions (Fig. 236b).

Consider now *transverse waves.* With the x axis in the direction of wave propagation, and the y axis in the direction of transverse displacement, we find that the displacements u and w are zero and the displacement v is a function of x and t. Then, from Eqs. (270),

$$\frac{\partial^2 v}{\partial t^2} = c_2{}^2 \frac{\partial^2 v}{\partial x^2} \tag{i}$$

This equation has the same form as Eq. (275), and we can conclude that waves of distortion propagate along the x axis with the velocity

$$c_2 = \sqrt{\frac{G}{\rho}}$$

or, by (277),

$$c_2 = c_1 \sqrt{\frac{1 - 2\nu}{2(1 - \nu)}}$$

For $\nu = 0.25$, the above equation gives

$$c_2 = \frac{c_1}{\sqrt{3}}$$

Any function

$$f(x - c_2 t) \tag{j}$$

is a solution of Eq. (i) and represents a wave traveling in the x direction with the velocity c_2. Take, for example, solution (j) in the form

$$v = v_0 \sin \frac{2\pi}{l} (x - c_2 t) \tag{k}$$

The wave has in this case a sinusoidal form. The length of the wave is l and the amplitude v_0. The velocity of transverse motion is

$$\frac{\partial v}{\partial t} = -\frac{2\pi c_2}{l} v_0 \cos \frac{2\pi}{l} (x - c_2 t) \tag{l}$$

It is zero when the displacement (k) is a maximum and has its largest value when the displacement is zero. The shearing strain produced by the wave is

$$\gamma_{xy} = \frac{\partial v}{\partial x} = \frac{2\pi v_0}{l} \cos \frac{2\pi}{l} (x - c_2 t) \tag{m}$$

It will be seen that the maximum distortion (m) and the maximum of the absolute value of the velocity (l) occur at a given point simultaneously.

We can represent this kind of wave propagation as follows: Let *mn* (Fig. 237) be a thin fiber of an elastic medium. When a sinusoidal wave (*k*) is being propagated along the x axis, an element A undergoes displacements and distortions, the consecutive values of which are indicated by the shaded elements 1, 2, 3, 4, 5 . . . . At the instant $t = 0$, the element A has a position as indicated by 1. At this moment its distortion and its velocity are zero. Then it acquires a positive velocity and after an interval of time equal to $l/4c_2$ its distortion is as indicated by 2. At this instant the displacement of the element is zero and its velocity is a maximum. After an interval of time equal to $l/4c_2$ the conditions are as indicated by 3, and so on.

The cross-sectional area of the fiber being *dy dz*, the kinetic energy of the element A is

$$\frac{1}{2}\rho\,dx\,dy\,dz\left(\frac{\partial v}{\partial t}\right)^2 = \frac{1}{2}\rho\,dx\,dy\,dz\,\frac{4\pi^2c_2^2}{l^2}v_0^2\cos^2\frac{2\pi}{l}(x - c_2t)$$

and its strain energy is

$$\frac{1}{2}G\gamma_{xy}^2\,dx\,dy\,dz = \frac{1}{2}G\frac{4\pi^2v_0^2}{l^2}\cos^2\frac{2\pi}{l}(x - c_2t)\,dx\,dy\,dz$$

Remembering that $c_2^2 = G/\rho$, it can be concluded that the kinetic and the potential energies of the element at any instant are equal.

In the case of an earthquake both kinds of waves, those of dilatation and those of distortion, are propagated through the earth with velocities c_1 and c_2. They can be recorded by a *seismograph*, and the interval of time between the arrival of these two kinds of waves gives some indication regarding the distance of the recording station from the center of disturbance.

Plane waves of sinusoidal and other forms can be combined in various ways so as to satisfy the physical conditions at a free plane surface or at an interface between two different media. When the directions of propagation are not parallel to the surface, results corresponding to reflection at a free surface, or reflection and refraction at an interface, can be

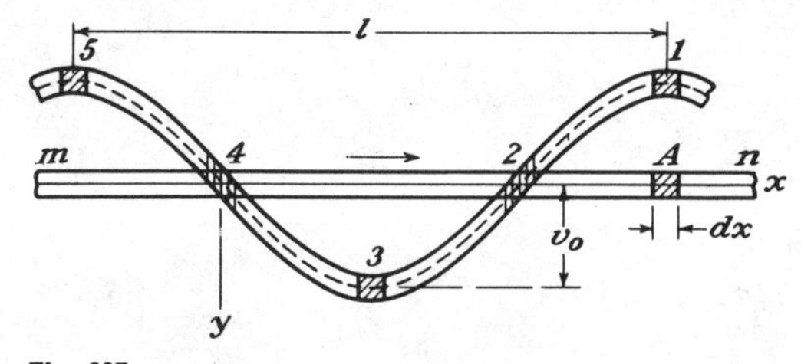

Fig. 237

obtained.[1] A wave motion propagating parallel to a free plane surface at a velocity different from both c_1 and c_2 (Rayleigh surface wave) is considered later in Art. 170.

168 | Longitudinal Waves in Uniform Bars. Elementary Theory

The simple plane longitudinal waves considered in Art. 167 could exist in a bar of rectangular cross section only if the stress components σ_y, σ_z given by Eqs. (*f*) were maintained on the lateral faces of the bar. For a bar of any cross section, corresponding tractions are required on the lateral surface.

When the lateral surface is *free*, it is far more difficult to find appropriate solutions of the complete equations of motion[2] (269). However, there are many practical cases for which a much simpler approximate theory is adequate. In this elementary theory each slice of the bar is regarded as in simple tension corresponding to the axial strain $\partial u/\partial x$, u being a function of x and t only. Then

$$\sigma_x = E\frac{\partial u}{\partial x} \tag{a}$$

The other stress components are taken as negligible. Considering an element originally between cross sections at x and $x + dx$, Fig. 238, the equation of motion is simply (after canceling the cross-sectional area)

$$\frac{\partial \sigma_x}{\partial x}\,dx = \rho\,dx\,\frac{\partial^2 u}{\partial t^2}$$

[1] See, for instance, H. Kolsky, "Stress Waves in Solids," Oxford University Press, Fair Lawn, N.J., 1953; reissued by Dover Publications, New York.

[2] Numerical results for special cases have been obtained by digital computer techniques. See, for instance, L. D. Bertholf, *J. Appl. Mech.*, vol. 34, pp. 725–734, 1967. For general surveys of the bar problem, as well as other major problems of stress wave propagation, with extensive bibliographies, see the articles by J. Miklowitz, Elastic Wave Propagation, pp. 809–839, and by R. M. Davies, Stress Waves in Solids, pp. 803–807, in H. N. Abramson, H. Liebowitz, J. M. Crowley, and S. Juhasz (eds.), "Applied Mechanics Surveys," Spartan Books, Washington, D.C., 1966.

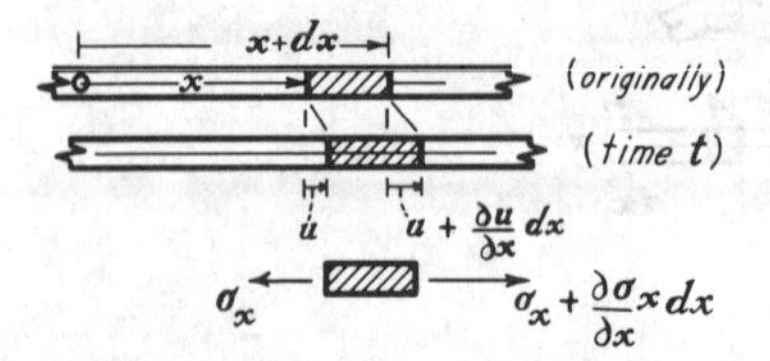

Fig. 238

or

$$\frac{\partial^2 u}{\partial t^2} = c^2 \frac{\partial^2 u}{\partial x^2} \tag{b}$$

where

$$c = \sqrt{\frac{E}{\rho}} \tag{278}$$

Equation (*b*) has the same form as Eq. (275) of Art. 167, and its general solution is

$$u = f(x + ct) + f_1(x - ct) \tag{c}$$

The interpretation follows that already given for Eq. (276). Here, however, the wave velocity[1] is c, as given by Eq. (278). It is lower than the wave velocity c_1 in Eq. (277). The ratio is

$$\frac{c_1}{c} = \sqrt{\frac{1 - \nu}{(1 + \nu)(1 - 2\nu)}}$$

For $\nu = 0.30$ this is 1.16. For steel we may take $c = 16{,}850$ fps.

When only the function f_1 is retained in Eq. (*c*) (forward wave propagation), we have from this equation and Eq. (*a*)

$$\sigma_x = -\rho c \dot{u} \tag{d}$$

whereas for f alone (backward propagation) we have

$$\sigma_x = \rho c \dot{u} \tag{e}$$

The results in Eqs. (278) and (*d*) can be derived without recourse to differential equations. Consider a uniformly distributed compressive stress suddenly applied to the left end of a bar (Fig. 239). It will produce at the first instant a uniform compression of an infinitely thin layer at the end of the bar. This compression will be transmitted to the adjacent layer, and so on. A wave of compression begins to travel along the bar with a certain velocity c, and, after a time interval t, a portion of the bar of length ct will be compressed and the remaining portion will be at rest in an unstressed condition.

The velocity of wave propagation c should be distinguished from the velocity v, given to the particles in the compressed zone of the bar by the compressive forces. The velocity of the particles v can be found by taking into account the fact that the compressed zone (shaded in the figure) shortens due to compressive stress σ by the amount $(\sigma/E)ct$. Hence, the velocity of the left end of the bar, equal to the velocity of particles in the compressed zone, is

$$v = \frac{c\sigma}{E} \tag{f}$$

[1] Often referred to as the "bar velocity."

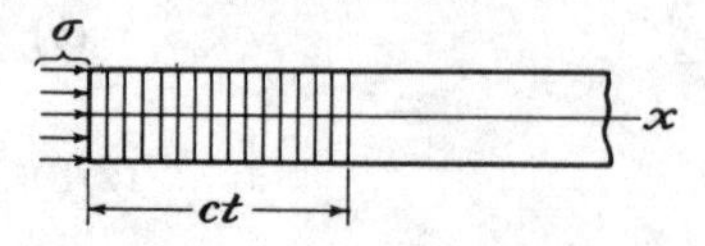

Fig. 239

The velocity c of wave propagation can be found by applying the equation of momentum. At the beginning the shaded portion of the bar was at rest. After the elapse of time t it has velocity v and momentum $Act\rho v$. Putting this equal to the impulse of the compressive force, we find

$$A\sigma t = Act\rho v \tag{g}$$

Using Eq. (*f*), we find for c the value given by Eq. (278)[1] and for the velocity of particles we find

$$v = \frac{\sigma}{\sqrt{E\rho}} \tag{279}$$

This corresponds to Eq. (*d*) in which $\dot{u}$ denoted particle velocity. It will be seen that whereas c is independent of the compressive force, the velocity v of particles is proportional to the stress σ.

If instead of compression a tensile force is suddenly applied at the end of the bar, a tension is propagated along the bar with the velocity c. The velocity of particles again is given by Eq. (279). But the direction of this velocity will be opposite to the direction of the x axis. Thus, in a compressive wave the velocity v of particles is in the same direction as the velocity of wave propagation, but in a tension wave the velocity v is in the opposite direction from that of the wave.

From Eqs. (278) and (279) we have

$$\sigma = E\frac{v}{c} \tag{280}$$

The stress in the wave is thus determined by the ratio of the two velocities and by the modulus E of the material. If an absolutely rigid body, moving with a velocity v, strikes longitudinally the left-hand end of the bar, the compressive stress on the surface of contact at the first instant is given by Eq. (280).[2] If the velocity v of the body is above a certain limit, depending on the mechanical properties of the material of the bar, a

[1] This elementary derivation of the formula for the velocity of wave propagation is due to Babinet; see Clebsch, "Théorie de l'Élasticité des Corps Solides," Saint-Venant (trans.), p. 480*d*, 1883.

[2] This conclusion is due to Thomas Young; see his "Course of Lectures on Natural Philosophy . . . ," vol. 1, pp. 135 and 144, 1807.

permanent set will be produced in the bar although the mass of the striking body may be very small.[1]

Consider now the energy of the wave shown shaded in Fig. 239. This energy consists of two parts: strain energy of deformation equal to

$$\frac{Act\sigma^2}{2E}$$

and kinetic energy equal to

$$\frac{Act\rho v^2}{2} = \frac{Act\sigma^2}{2E}$$

It will be seen that the total energy of the wave, equal to the work done by the compressive force $A\sigma$ acting over the distance $(\sigma/E)\ ct$, is half potential and half kinetic.

Equation (*b*), governing the wave propagation, is linear, so that if we have two solutions of the equation, their sum will also be a solution of this equation. From this it follows that in discussing waves traveling along a bar, we may use the method of superposition. If two waves traveling in opposite directions (Fig. 240) come together, the resulting stress and the resulting velocity of particles are obtained by superposition. If both waves are, for instance, compressive waves, the resultant compression is obtained by simple addition, as shown in Fig. 240*b*, and the resultant velocity of particles by subtraction. After passing, the waves return to their initial shape, as shown in Fig. 240*c*.

Consider a compression wave moving along the bar in the x direction and a tension wave of the same length and with the same magnitude of stress moving in the opposite direction (Fig. 241). When the waves come together, tension and compression annul each other, and in the portion of the bar in which the two waves are superposed we have zero stress. At

[1] It is assumed that contact occurs simultaneously at all points of the end section of the bar.

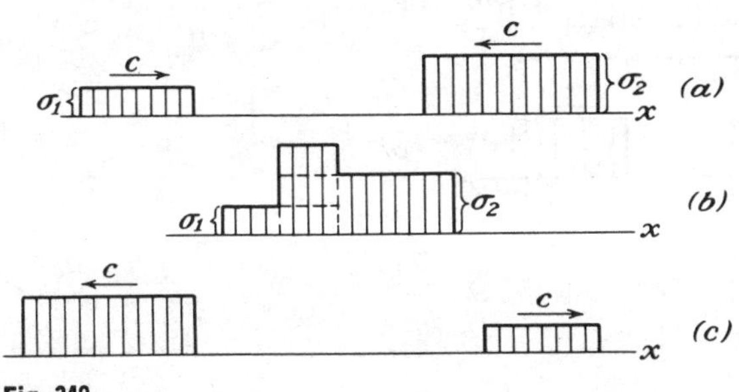

Fig. 240

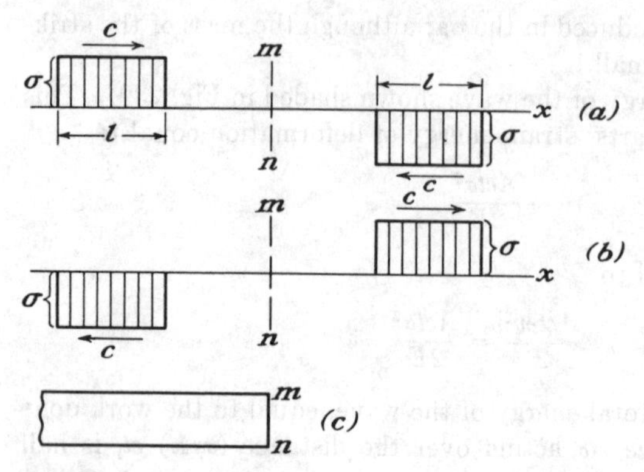

Fig. 241

the same time the velocity of particles in this portion of the bar is doubled and equal to $2v$. After passing, the waves return to their initial shape, as shown in Fig. 241*b*. At the middle cross section *mn* there will be at all times zero stress and we may consider it as a *free end* of a bar (Fig. 241*c*). Then comparing Figs. 241*a* and *b* it can be concluded that in the case of a free end a compressive wave is reflected as a similar tension wave, and vice versa.

If two identical waves, moving toward one another (Fig. 242*a*), come together, there will be doubled stress and zero velocity in the portion of the bar in which the waves are superposed. At the middle cross section *mn* we always have zero velocity. This section remains immovable during passage of the waves and we may consider it as a fixed end of the bar (Fig. 242*c*). Then, from comparison of Figs. 242*a* and *b*, it can be concluded that a wave is reflected from a fixed end entirely unchanged.

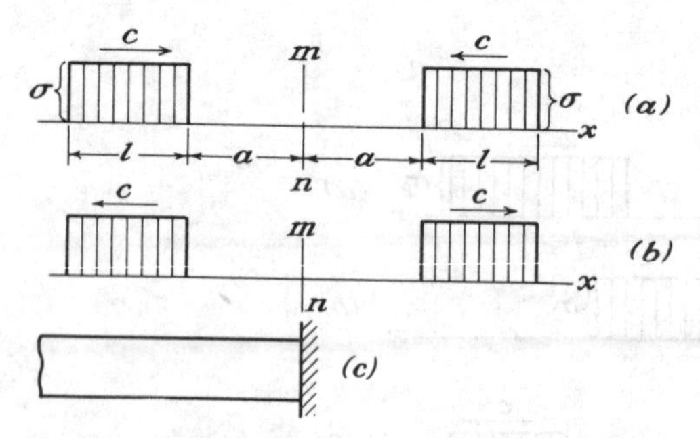

Fig. 242

169 | Longitudinal Impact of Bars

If two equal rods of the same material strike each other longitudinally with the same velocity v (Fig. 243a), the plane of contact mn will not move during the impact[1] and two identical compression waves start to travel along both bars with equal velocities c. The velocities of particles in the waves, superposed on the initial velocities of the bars, bring the zones of waves to rest, and at the instant when the waves reach the free ends of the bars ($t = l/c$), both bars will be uniformly compressed and at rest. Then the compression waves will be reflected from the free ends as tension waves which will travel back toward the cross section of contact mn. In these waves the velocities of particles, equal to v, will now be in the direction away from mn, and when the waves reach the plane of contact the bars separate with a velocity equal to their initial velocity v. The duration of impact in this case is evidently equal to $2l/c$ and the compressive stress, from Eq. (279), is equal to $v\sqrt{E\rho}$.

Consider now a more general case when the bars 1 and 2 (Fig. 243b) are moving[2] with the velocities v_1 and $v_2 (v_1 > v_2)$. At the instant of impact two identical compression waves start to travel along both bars. The corresponding velocities of particles relative to the unstressed portions of the moving bars are equal and are directed in each bar away from the surface of contact. The magnitude of these velocities must be equal to $(v_1 - v_2)/2$ in order to have the absolute velocities of particles of the two bars at the surface of contact equal. After an interval of time equal to l/c, the compression waves arrive at the free ends of the bars. Both bars are at this instant in a state of uniform compression, and the absolute velocities of all particles of the bars are

$$v_1 - \frac{v_1 - v_2}{2} = v_2 + \frac{v_1 - v_2}{2} = \frac{v_1 + v_2}{2}$$

[1] It is assumed that contact takes place at the same instant over the whole surface of the ends of the rods.

[2] Velocities are considered positive if they are in the direction of the x axis.

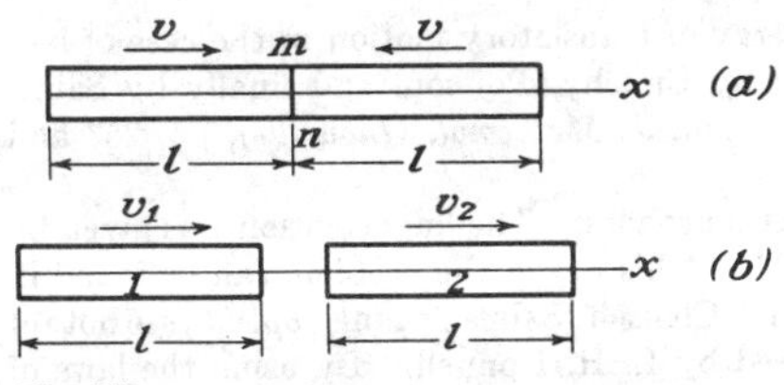

Fig. 243

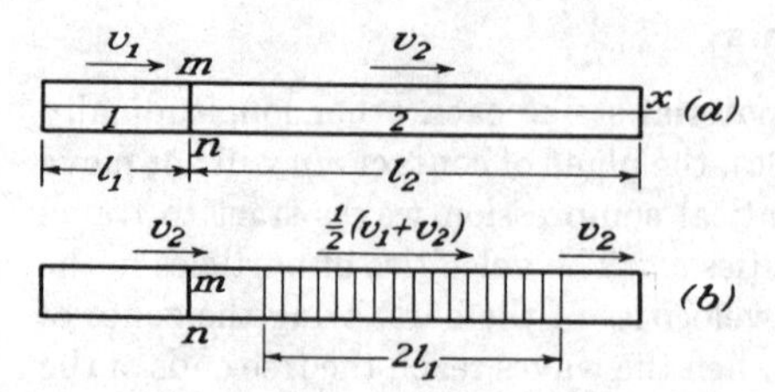

Fig. 244

The compression waves will then be reflected from the free ends as tension waves and at the instant $t = 2l/c$, when these waves arrive at the surface of contact of the two bars, the velocities of bars 1 and 2 become

$$\frac{v_1 + v_2}{2} - \frac{v_1 - v_2}{2} = v_2 \qquad \frac{v_1 + v_2}{2} + \frac{v_1 - v_2}{2} = v_1$$

Thus the bars, during impact, exchange their velocities.

If the above bars have different lengths, l_1 and l_2 (Fig. 244*a*), the conditions of impact at first will be the same as in the previous case. But after a time interval $2l_1/c$, when the reflected wave of the shorter bar 1 arrives at the surface of contact mn, it is propagated through the surface of contact along the longer bar and the conditions will be as shown in Fig. 244*b*. The tension wave of the bar l_1 annuls the pressure between the bars, but they remain in contact until the compression wave in the longer bar (shaded in the figure) returns, after reflection, to the surface of contact (at $t = 2l_2/c$).

In the case of two bars of equal length, each of them, after rebounding, has the same velocity in all points and moves as a rigid body. The total energy is the energy of translatory motion. In the case of the bars of different lengths, the longer bar, after rebounding, has a traveling wave in it, and in calculating the total energy of the bar the energy of this wave must be considered.[1]

Consider now a more complicated problem of a bar with a fixed end struck by a moving mass at the other end[2] (Fig. 245). Let M be the mass

[1] The question of the lost kinetic energy of translatory motion in the case of longitudinal impact of bars was discussed by Cauchy, Poisson, and finally by Saint-Venant; see *Compt. Rend.*, p. 1108, 1866, and *J. Mathémat.* (*Liouville*), pp. 257 and 376, 1867.

[2] This problem was discussed by several authors. The final solution was given by J. Boussinesq, *Compt. Rend.*, p. 154, 1883. A history of the problem can be found in "Théorie de l'Élasticité des Corps Solides," Clebsch (Saint-Venant), *op. cit.*, see note of par. 60. The problem was also discussed by L. H. Donnell. By using the laws of wave propagation he simplified the solution and extended it to the case of a conical bar. See *Trans. ASME*, Applied Mechanics Division, 1930.

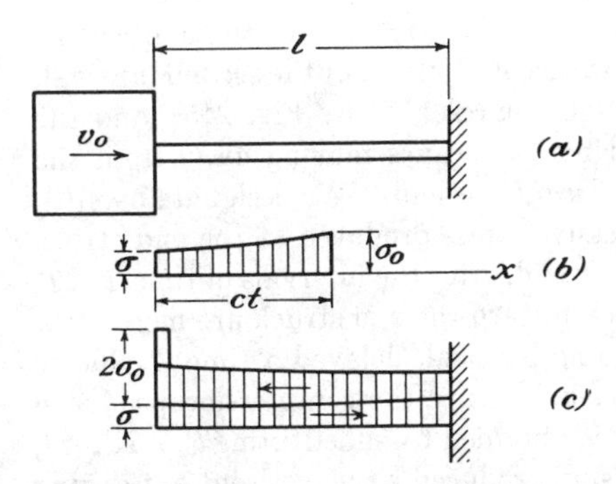

Fig. 245

of the moving body per unit area of the cross section of the bar and v_0 the initial velocity of this body. Considering the body as absolutely rigid, the velocity of particles at the end of the bar at the instant of impact ($t = 0$) is v_0, and the initial compressive stress, from Eq. (279), is

$$\sigma_0 = v_0 \sqrt{E\rho} \tag{a}$$

Owing to the resistance of the bar, the velocity of the moving body and hence the pressure on the bar will gradually decrease, and we obtain a compression wave with a decreasing compressive stress traveling along the length of the bar (Fig. 245*b*). The change in compression with the time can easily be found from the equation of motion of the body. Denoting by σ the variable compressive stress at the end of the bar and by v the variable velocity of the body, we find

$$M \frac{dv}{dt} + \sigma = 0 \tag{b}$$

or, substituting for v its expression from Eq. (279),

$$\frac{M}{\sqrt{E\rho}} \frac{d\sigma}{dt} + \sigma = 0$$

from which

$$\sigma = \sigma_0 e^{-(t\sqrt{E\rho}/M)} \tag{c}$$

This equation can be used so long as $t < 2l/c$. When $t = 2l/c$, the compressive wave with the front pressure σ_0 returns to the end of the bar that is in contact with the moving body. The velocity of the body cannot change suddenly, and hence the wave will be reflected as from a fixed end and the compressive stress at the surface of contact suddenly increases by $2\sigma_0$, as is shown in Fig. 245*c*. Such a sudden increase of pressure occurs during impact at the end of every interval of time $T = 2l/c$, and we must obtain a separate expression for σ for each one of these intervals. For the first interval, $0 < t < T$, we use Eq. (c). For the second interval, $T < t < 2T$, we have the conditions represented by Fig. 245c, and the compressive stress σ is produced by two waves moving *away* from the end struck and one wave moving *toward* this end. We designate by $s_1(t)$, $s_2(t)$, $s_3(t)$, . . . the total compressive stress produced at the end struck by all waves moving away from this end, after the intervals of time T, $2T$, $3T$. . . . The waves coming back toward the end struck are merely the waves sent out during the preceding interval, delayed a time T, due to their travel across the bar and back. Hence the compression produced by these waves at the end struck is obtained by substituting $t - T$, for t, in the expression for the compression produced by waves sent out during the preceding interval. The general expression for the total compressive stress during any interval $nT < t < (n + 1)T$ is therefore

$$\sigma = s_n(t) + s_{n-1}(t - T) \tag{d}$$

The velocity of particles at the end struck is obtained as the difference between the velocity due to the pressure $s_n(t)$ of the waves going away, and the velocity due to the pressure $s_{n-1}(t - T)$ of the waves going toward the end. Then, from Eq. (279),

$$v = \frac{1}{\sqrt{E\rho}}[s_n(t) - s_{n-1}(t - T)] \tag{e}$$

The relation between $s_n(t)$ and $s_{n-1}(t - T)$ will now be obtained by using the equation of motion (b) of the striking body. Denoting by α the ratio of the mass of the bar to the mass of the striking body, we have

$$\alpha = \frac{l\rho}{M} \qquad \frac{\sqrt{E\rho}}{M} = \frac{cl\rho}{Ml} = \frac{2\alpha}{T} \tag{f}$$

Using this, with (d) and (e), Eq. (b) becomes

$$\frac{d}{dt}[s_n(t) - s_{n-1}(t - T)] + \frac{2\alpha}{T}[s_n(t) + s_{n-1}(t - T)] = 0$$

Multiplying by $e^{2\alpha t/T}$,

$$e^{2\alpha t/T}\frac{ds_n(t)}{dt} + \frac{2\alpha}{T}e^{2\alpha t/T}s_n(t) = e^{2\alpha t/T}\frac{ds_{n-1}(t - T)}{dt} + \frac{2\alpha}{T}e^{2\alpha t/T}s_{n-1}(t - T) - \frac{4\alpha}{T}e^{2\alpha t/T}s_{n-1}(t - T)$$

or

$$\frac{d}{dt}[e^{2\alpha t/T}s_n(t)] = \frac{d}{dt}[e^{2\alpha t/T}s_{n-1}(t - T)] - \frac{4\alpha}{T}e^{2\alpha t/T}s_{n-1}(t - T)$$

from which

$$s_n(t) = s_{n-1}(t - T) - \frac{4\alpha}{T} e^{-2\alpha t/T} \left[\int e^{2\alpha t/T} s_{n-1}(t - T) dt + C \right] \tag{g}$$

in which C is a constant of integration. This equation will now be used for deriving expressions for the consecutive values $s_1, s_2, \ldots$. During the first interval $0 < t < T$, the compressive stress is given by Eq. (c), and we can put

$$s_0 = \sigma_0 e^{-(2\alpha t/T)} \tag{h}$$

Substituting this for s_{n-1} in Eq. (g),

$$\begin{aligned} s_1(t) &= \sigma_0 e^{-2\alpha[(t/T)-1]} - \frac{4\alpha}{T} e^{-(2\alpha t/T)} \left(\int \sigma_0 e^{2\alpha}\, dt + C \right) \\ &= \sigma_0 e^{-2\alpha[(t/T)-1]} \left(1 - \frac{4\alpha t}{T} \right) - C \frac{4\alpha}{T} e^{-2\alpha t/T} \end{aligned} \tag{k}$$

The constant of integration C is found from the condition that at the instant $t = T$ the compressive stress at the end struck increases suddenly by $2\sigma_0$ (Fig. 245c). Hence, using Eq. (d),

$$\left[\sigma_0 e^{-(2\alpha t/T)} \right]_{t=T} + 2\sigma_0 = \left\{ \sigma_0 e^{-2\alpha[(t/T)-1]} + \sigma_0 e^{-2\alpha[(t/T)-1]} \left(1 - \frac{4\alpha t}{T} \right) - C \frac{4\alpha}{T} e^{-(2\alpha t/T)} \right\}_{t=T}$$

from which

$$C = -\frac{\sigma_0 T}{4\alpha} (1 + 4\alpha e^{2\alpha})$$

Substituting in Eq. (k),

$$s_1 = s_0 + \sigma_0 e^{-2\alpha[(t/T)-1]} \left[1 + 4\alpha \left(1 - \frac{t}{T} \right) \right] \tag{l}$$

Proceeding further in the same manner and substituting s_1, instead of s_{n-1}, into Eq. (g), we find

$$s_2 = s_1 + \sigma_0 e^{-2\alpha[(t/T)-2]} \left[1 + 2(4\alpha) \left(2 - \frac{t}{T} \right) + 2(4\alpha^2) \left(2 - \frac{t}{T} \right)^2 \right] \tag{m}$$

Continuing in the same way,

$$\begin{aligned} s_3 = s_2 + \sigma_0 e^{-2\alpha[(t/T)-3]} \bigg[& 1 + 2(6\alpha) \left(3 - \frac{t}{T} \right) \\ & + 2(3)4\alpha^2 \left(3 - \frac{t}{T} \right)^2 + \frac{2(2)3}{3(3)} 8\alpha^3 \left(3 - \frac{t}{T} \right)^3 \bigg] \end{aligned} \tag{n}$$

and so on.[1] In Fig. 246 the functions $s_0, s_1, s_2, \ldots$ are represented

[1] The effects of successive reflections illustrated by this problem can be concisely derived by the method of Laplace transforms. See, for instance, W. T. Thomson, "Laplace Transformation," p. 123, Prentice-Hall, Inc., Englewood Cliffs, N. J., 1950.

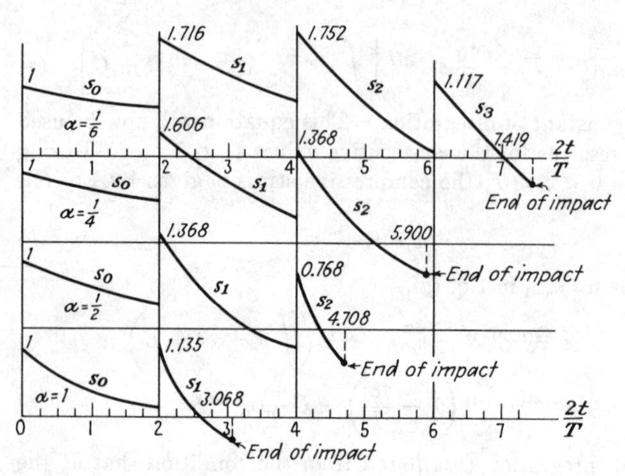

Fig. 246

graphically for $\sigma_0 = 1$ and for four different ratios,[1] $\alpha = \frac{1}{6}, \frac{1}{4}, \frac{1}{2}, 1$. By using these curves, the compressive stress σ at the end struck can easily be calculated from Eq. (*d*). In Fig. 247 this stress is represented graphically for $\sigma_0 = 1$ and for $\alpha = \frac{1}{4}, \frac{1}{2}, 1$. It changes at intervals T, $2T$, . . . by jumps. The maximum value of this stress depends on the ratio α. For $\alpha = \frac{1}{2}$ and $\alpha = 1$ the stress has its maximum value at $t = T$. In the case of $\alpha = \frac{1}{4}$, the maximum stress occurs at $t = 2T$. The instant when σ becomes equal to zero indicates the end of the impact. It will be seen that the duration of the impact increases when α decreases. Calculations of Saint-Venant give the following values for this duration:

$\alpha =$	$\frac{1}{6}$	$\frac{1}{4}$	$\frac{1}{2}$	1
$\frac{2t}{T} =$	7.419	5.900	4.708	3.068

[1] These curves were calculated by Saint-Venant and Flamant. See *Compt. Rend.*, pp. 127, 214, 281, and 353, 1883.

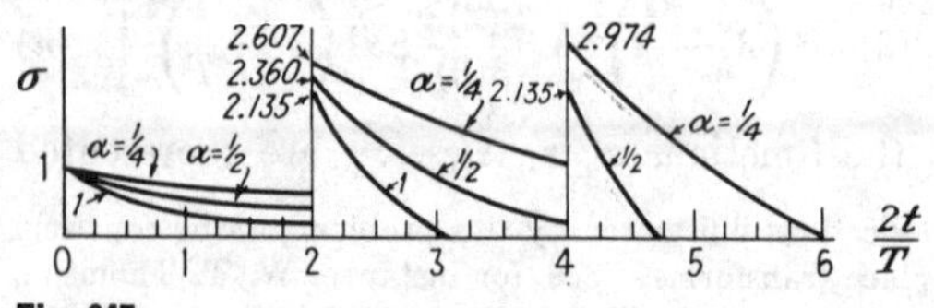

Fig. 247

For a very small α the time of contact can be calculated from the elementary formula

$$t = \frac{\pi l}{c} \sqrt{\frac{1}{\alpha}} \tag{p}$$

which is obtained by neglecting the mass of the rod entirely and assuming that the duration of the impact is equal to half the period of simple harmonic oscillation of the body attached to the rod.

Functions $s_1, s_2, s_3, \ldots$ calculated above can also be used for determining the stresses in any other cross section of the bar. The total stress is always the sum of two values of s [Eq. (d)], one value in the resultant wave going toward the fixed end and one in the resultant wave going in the opposite direction. When the portion of the wave corresponding to the maximum value of s (the highest peak of one of the curves in Fig. 246) arrives at the fixed end and is reflected there, both of the waves mentioned above will have this maximum value: the total compressive stress at this point and at this instant is as great as can occur during the impact. From this we see that the maximum stress during impact occurs at the fixed end and is equal to twice the maximum value of s. From Fig. 246 it can be concluded at once that for $\alpha = \frac{1}{6}, \frac{1}{4}, \frac{1}{2}, 1$, the maximum compressive stresses are $2 \times 1.752\sigma_0$, $2 \times 1.606\sigma_0$, $2 \times 1.368\sigma_0$, and $2 \times 1.135\sigma_0$, respectively. In Fig. 248 the values of σ_{max}/σ_0 for various values of the ratio $\alpha = \rho l/M$ are given.[1] For comparison there is also

[1] See papers by Saint-Venant and Flamant, fn. 1, p. 502.

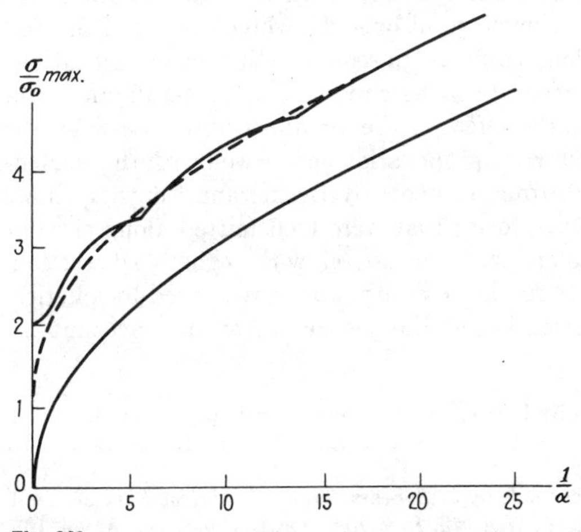

Fig. 248

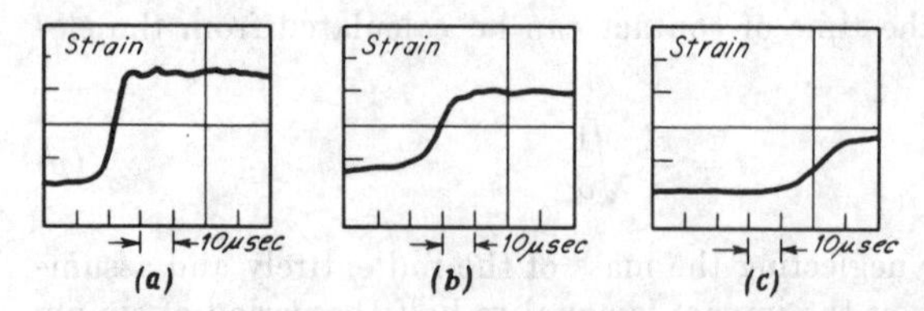

Fig. 249 Oscillograms of strain-gauge signals showing dependence of wake formation on impact velocity. Rod diameter ½ in., strain gauges 30 in. from impact; hammer velocities at impact (*a*) 6 in. $\sec^{-1}$, (*b*) 4 in. $\sec^{-1}$, and (*c*) 3 in. $\sec^{-1}$.

shown the lower parabolic curve calculated from the equation

$$\sigma = \sigma_0 \sqrt{\frac{M}{\rho l}} = \frac{\sigma_0}{\sqrt{\alpha}} \tag{q}$$

which can be obtained at once in an elementary way by neglecting the mass of the rod entirely and equating the strain energy of the rod to the kinetic energy of the striking body. The dotted line shown is a parabolic curve[1] defined by the equation

$$\sigma = \sigma_0 \left(\sqrt{\frac{M}{\rho l}} + 1 \right) \tag{r}$$

It will be seen that for large values of $1/\alpha$, it always gives a very good approximation.

The theory of impact developed above is based on the assumption that contact takes place at the same instant over the whole surface of the end of the rod. This condition is difficult to realize. Elaborate precautions are necessary to ensure accurate plane ends and accurate alignment of the rods and to minimize the effect of the air film trapped between the ends. Then the observed wave propagation agrees well with the elementary theory. Figure 249, from a paper[2] by Becker and Conway, shows oscillograph records of wave forms that were transmitted along circular rods and reflected from plane ends, in case (*c*), with negligible distortion. In earlier experimental work[3] the colliding ends were curved to spherical form, and the local deformation at the contact taken into account by means of the Hertz theory.

[1] This curve was proposed by Boussinesq; see *Compt. Rend.*, p. 154, 1883.

[2] E. C. H. Becker and H. D. Conway, *Brit. J. Appl. Phys.*, vol. 15, pp. 1225–1231, 1964.

[3] Such an investigation was made by J. E. Sears, *Trans. Cambridge Phil. Soc.*, vol. 21, p. 49, 1908. See also J. E. P. Wagstaff, *Proc. Roy. Soc. (London)*, ser. A, vol. 105, p. 544, 1924; and W. A. Prowse, *Phil. Mag.*, vol. 22, p. 209, 1936.

170 | Rayleigh Surface Waves

In Arts. 166 and 167 the propagation of disturbances in the isotropic homogeneous medium obeying Hooke's law was represented as a superposition of irrotational waves at the velocity c_1 and equivoluminal waves at the velocity c_2. Even when there are discontinuities of particle velocity and stress at the wave fronts, c_1 and c_2 are the only possible wave velocities in the infinite medium when the initial disturbance is confined to a finite internal region.[1]

When there are free boundaries (or interfaces between two media), other velocities of propagation are possible. "Surface waves" can appear, involving motion in essentially only a thin surface layer. They are similar to the ripples on smooth water caused by a stone thrown in, and are also closely related to the "skin effect" in conductors carrying high-frequency alternating current. Lord Rayleigh,[2] who first pointed out the existence of surface-wave solutions of the general equations, remarked

> It is not improbable that the surface waves here investigated play an important part in earthquakes, and in the collision of elastic solids. Diverging in two dimensions only, they must acquire at a great distance from the source a continually increasing preponderance.

The study of records of seismic waves supports Rayleigh's expectation.

At a great distance from the source, the deformation produced by these waves may be considered as a two-dimensional one. We assume that the body is bounded by the plane $y = 0$ and take the positive sense of the y axis in the direction toward the interior of the body and the positive direction of the x axis in the direction of wave propagation. Expressions for the displacements are obtained by combining dilatation waves [Eqs. (271)] and distortion waves [Eqs. (270)]. Assuming in both cases that $w = 0$, the solution of Eqs. (271) representing waves of dilatation can be taken in the form

$$u_1 = se^{-ry} \sin (pt - sx) \qquad v_1 = -re^{-ry} \cos (pt - sx) \tag{a}$$

in which p, r, and s are constants. The exponential factor in these expressions indicates that for real positive values of r, the amplitude of waves rapidly diminishes with increase of the depth y. The argument $pt - sx$ of the trigonometrical functions shows that the waves are propagated in the x direction with the velocity

$$c_3 = \frac{p}{s} \tag{281}$$

[1] A. E. H. Love, "Mathematical Theory of Elasticity," 4th ed., pp. 295–297, Cambridge University Press, New York, 1927.

[2] *Proc. London Math. Soc.*, vol. 17, pp. 4–11, 1885; or "Scientific Papers," vol. 2, pp. 441–447, Cambridge University Press, New York, 1900.

Substituting expressions (*a*) into Eqs. (271), we find that these equations are satisfied if

$$r^2 = s^2 - \frac{\rho p^2}{\lambda + 2G}$$

or, by using the notation

$$\frac{\rho p^2}{\lambda + 2G} = \frac{p^2}{c_1^2} = h^2 \tag{b}$$

we have

$$r^2 = s^2 - h^2 \tag{c}$$

We take solutions of Eqs. (270), representing waves of distortion, in the form

$$u_2 = Abe^{-by} \sin (pt - sx) \qquad v_2 = -Ase^{-by} \cos (pt - sx) \tag{d}$$

in which A is a constant and b a positive number. It can be shown that the volume expansion corresponding to the displacements (*d*) is zero and that Eqs. (270) are satisfied if

$$b^2 = s^2 - \frac{\rho p^2}{G}$$

or, by using the notation

$$\frac{\rho p^2}{G} = \frac{p^2}{c_2^2} = k^2 \tag{e}$$

we obtain

$$b^2 = s^2 - k^2 \tag{f}$$

Combining solutions (*a*) and (*d*) and taking $u = u_1 + u_2$, $v = v_1 + v_2$, we now determine the constants A, b, p, r, s, so as to satisfy the boundary conditions. The boundary of the body is free from external forces; hence, for $y = 0$, $\bar{X} = 0$ and $\bar{Y} = 0$. Substituting this in Eqs. (130) and taking $l = n = 0$, $m = -1$, we obtain

$$\begin{aligned} \frac{\partial u}{\partial y} + \frac{\partial v}{\partial x} &= 0 \\ \lambda e + 2G \frac{\partial v}{\partial y} &= 0 \end{aligned} \tag{g}$$

The first of these equations indicates that the shearing stresses, and the second that the normal stresses on the surface of the body, are zero. Substituting the above expressions for u and v in these equations, we find that

$$\begin{aligned} 2rs + A(b^2 + s^2) &= 0 \\ \left(\frac{k^2}{h^2} - 2\right)(r^2 - s^2) + 2(r^2 + Abs) &= 0 \end{aligned} \tag{h}$$

where
$$\frac{k^2}{h^2} - 2 = \frac{\lambda}{G}$$
from (*b*) and (*e*).

Eliminating the constant A from Eqs. (*h*) and using (*c*) and (*f*), we obtain
$$(2s^2 - k^2)^2 = 4brs^2 \tag{k}$$
or, by (*c*) and (*f*),
$$\left(\frac{k^2}{s^2} - 2\right)^4 = 16\left(1 - \frac{h^2}{s^2}\right)\left(1 - \frac{k^2}{s^2}\right)$$
By using Eqs. (*b*), (*e*), and (281), all the quantities of this equation can be expressed by the velocities c_1 of waves of dilatation, c_2 of waves of distortion, and c_3 of surface waves, and we obtain
$$\left(\frac{c_3^2}{c_2^2} - 2\right)^4 = 16\left(1 - \frac{c_3^2}{c_1^2}\right)\left(1 - \frac{c_3^2}{c_2^2}\right) \tag{l}$$
Using the notation
$$\frac{c_3}{c_2} = \alpha$$
and remembering that
$$\frac{c_2^2}{c_1^2} = \frac{1 - 2\nu}{2(1 - \nu)}$$
Eq. (*l*) becomes
$$\alpha^6 - 8\alpha^4 + 8\left(3 - \frac{1 - 2\nu}{1 - \nu}\right)\alpha^2 - 16\left[1 - \frac{1 - 2\nu}{2(1 - \nu)}\right] = 0 \tag{m}$$
Taking, for example, $\nu = 0.25$, we obtain
$$3\alpha^6 - 24\alpha^4 + 56\alpha^2 - 32 = 0$$
or
$$(\alpha^2 - 4)(3\alpha^4 - 12\alpha^2 + 8) = 0$$
The three roots of this equation are
$$\alpha^2 = 4 \qquad \alpha^2 = 2 + \frac{2}{\sqrt{3}} \qquad \alpha^2 = 2 - \frac{2}{\sqrt{3}}$$
Of these three roots only the last one satisfies the conditions that the quantities r^2 and b^2, given by Eqs. (*c*) and (*f*), are positive numbers. Hence,
$$c_3 = \alpha c_2 = 0.9194\sqrt{\frac{G}{\rho}}$$
Taking, as an extreme case, $\nu = \frac{1}{2}$, Eq. (*m*) becomes
$$\alpha^6 - 8\alpha^4 + 24\alpha^2 - 16 = 0$$

and we find

$$c_3 = 0.9553\sqrt{\frac{G}{\rho}}$$

In both cases the velocity of surface waves is slightly less than the velocity of waves of distortion propagated through the body. Having α, the ratio between the amplitudes of the horizontal and vertical displacements at the surface of the body can easily be calculated. For $\nu = \frac{1}{4}$, this ratio is 0.681. The above velocity of propagation of surface waves can also be obtained by a consideration of the vibrations of a body bounded by two parallel planes.[1]

171 | Spherically Symmetric Waves in the Infinite Medium

A disturbance such as a symmetrical explosion inside a spherical cavity sends out a wave or pulse that is also spherically symmetrical. The displacement is purely radial. It is a function, u, of the spherical radial coordinate[2] r and the time t. The deformation is *irrotational* by symmetry, and therefore only the propagation velocity c_1, Eq. (273) or (277), will be involved.

The differential equation for u is easily found by consideration of the typical volume element defined by four radii as indicated in Fig. 250, a small "spherical square" with radial thickness dr. The dynamical equa-

[1] See H. Lamb, *Proc. Roy. Soc. (London)*, ser. A, vol. 93, p. 114, 1917. See also S. Timoshenko, *Phil. Mag.*, vol. 43, p. 125, 1922.

[2] In Chap. 12 this coordinate was denoted by R, r being there the radial coordinate in the cylindrical system.

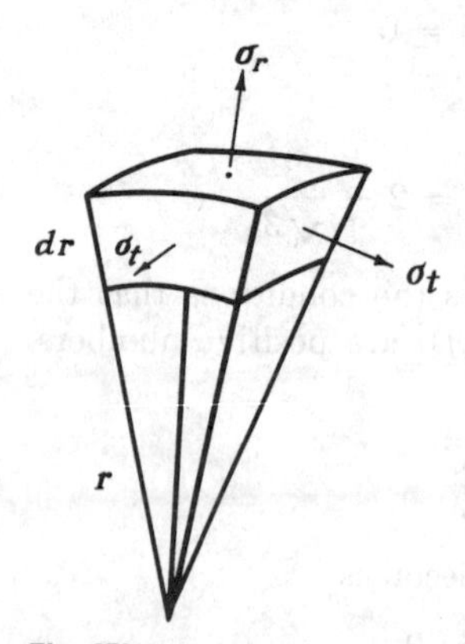

Fig. 250

tion for the radial motion is

$$\frac{\partial \sigma_r}{\partial r} + \frac{2}{r}(\sigma_r - \sigma_t) = \rho \frac{\partial^2 u}{\partial t^2} \tag{a}$$

The strain components are

$$\epsilon_r = \frac{\partial u}{\partial r} \qquad \epsilon_t = \frac{u}{r} \tag{b}$$

and the Hooke's law relations yield

$$\sigma_r = \frac{E}{(1+\nu)(1-2\nu)}\left[(1-\nu)\frac{\partial u}{\partial r} + 2\nu \frac{u}{r}\right]$$

$$\sigma_t = \frac{E}{(1+\nu)(1-2\nu)}\left(\frac{u}{r} + \nu \frac{\partial u}{\partial r}\right) \tag{c}$$

Introducing these in (*a*), we find

$$\frac{\partial^2 u}{\partial r^2} + \frac{2}{r}\frac{\partial u}{\partial r} - \frac{2u}{r^2} = \frac{1}{c_1{}^2}\frac{\partial^2 u}{\partial t^2} \tag{d}$$

Corresponding to the use of the function ϕ in Eqs. (*c*) in Art. 166, we may write

$$u = \frac{\partial \phi}{\partial r} \tag{e}$$

Then, as may readily be verified by carrying out the differentiations on the left-hand side, (*d*) is equivalent to

$$\frac{\partial}{\partial r}\left[\frac{1}{r}\frac{\partial^2}{\partial r^2}(r\phi)\right] = \frac{1}{c_1{}^2}\frac{\partial}{\partial r}\frac{\partial^2 \phi}{\partial t^2} \tag{f}$$

This implies that

$$\frac{1}{r}\frac{\partial^2}{\partial r^2}(r\phi) - \frac{1}{c_1{}^2}\frac{\partial^2 \phi}{\partial t^2} = F(t) \tag{g}$$

where $F(t)$ is an arbitrary function. If it is not zero, we can find a particular solution of (*g*) that is also a function of t only, $\phi(t)$. But this would contribute nothing to the displacement (*e*). Thus $F(t)$ may be discarded. Then in (*g*) we multiply by r to obtain

$$\frac{\partial^2}{\partial r^2}(r\phi) = \frac{1}{c_1{}^2}\frac{\partial^2}{\partial t^2}(r\phi) \tag{282}$$

Comparison with Eq. (275) and its solution (276) shows that the general solution of (*f*) is given by

$$r\phi = f(r - c_1 t) + g(r + c_1 t) \tag{283}$$

The interpretation is like that following Eq. (276). The function

$f(r - c_1t)$ represents an outgoing wave, and the function $g(r + c_1t)$ an ingoing wave. The former is suitable for the problem of *explosion*. The latter is suitable for the problem of *implosion* exemplified by a wave converging toward the center of a finite solid sphere, following sudden application of a pressure to the whole outer surface.

172 | Explosive Pressure in a Spherical Cavity

Discarding the function g in Eq. (283), the problem is reduced to the determination of the single function f to satisfy the *boundary conditions* and also the *initial conditions*.

The initial conditions are that at $t = 0$ the infinite medium with the spherical cavity has zero displacement and zero velocity everywhere. For $t > 0$, a pressure $p(t)$, any given function of t, acts on the cavity surface $r = a$. This is one of the boundary conditions. The other is that the material at infinity remains undisturbed.

Since we have a boundary condition at $r = a$, it is convenient to take in place of (283) the form

$$\phi = \frac{1}{r} f(\tau) \qquad \text{where } \tau = t - \frac{1}{c_1}(r - a) \tag{a}$$

Then at $r = a$ we have $\tau = t$. Also τ measures time, at radius r greater than a, after the arrival of a signal sent out from radius a at time $t = 0$. Writing

$$f' = \frac{d}{d\tau} f(\tau)$$

we find from (*e*) and (*c*) of Art. 171

$$u = -\frac{1}{c_1}\frac{1}{r} f' - \frac{1}{r^2} f \tag{b}$$

$$\frac{1}{\rho c_1{}^2}(1 - \nu)\sigma_r = (1 - \nu)\frac{1}{c_1{}^2}\frac{1}{r} f'' + 2(1 - 2\nu)\left(\frac{1}{c_1}\frac{1}{r^2} f' + \frac{1}{r^3} f\right) \tag{c}$$

$$\frac{1}{\rho c_1{}^2}(1 - \nu)\sigma_t = \nu\frac{1}{c_1{}^2}\frac{1}{r} f'' - (1 - 2\nu)\left(\frac{1}{c_1}\frac{1}{r^2} f' + \frac{1}{r^3} f\right) \tag{d}$$

The cavity boundary condition is $\sigma_r = -p(t)$ on $r = a$. We substitute this value of σ_r on the left of (*c*) and take $r = a$ on the right, implying $\tau = t$. The boundary condition thus requires

$$f''(t) + 2\gamma f'(t) + 2\gamma\frac{c_1}{a} f(t) = -\frac{a}{\rho} p(t) \tag{e}$$

where the primes now can be taken as differentiations with respect to t, and

$$\gamma = \frac{1 - 2\nu}{1 - \nu} \frac{c_1}{a} \tag{f}$$

The ordinary differential equation (e) is of the form

$$x''(t) + a_1 x'(t) + a_0 x(t) = F(t) \tag{g}$$

in which a_1, a_0 are constants. This form is well known in dynamics through the problem of the general forced motion of a simple mass-spring oscillator with viscous damping. Its general solution can be expressed as

$$x(t) = \int_0^t F(\xi) g_1(t - \xi)\, d\xi + C_1 e^{\alpha t} + C_2 e^{\beta t} \tag{h}$$

Here C_1, C_2 are the arbitrary constants of the complementary function (the general solution of the homogeneous equation), in which α and β are the two roots of the quadratic in z,

$$z^2 + a_1 z + a_0 = 0 \tag{i}$$

In the integral, which is a particular solution of (g), the function $g_1(t - \xi)$ is obtained from the function

$$g_1(t) = \frac{1}{\alpha - \beta} (e^{\alpha t} - e^{\beta t}) \tag{j}$$

which is simply the complementary function with C_1 and C_2 chosen so that

$$g_1(0) = 0 \qquad g_1'(0) = 1 \tag{k}$$

The particular solution of Eq. (e), corresponding to the integral in (h), is

$$f(t) = -\frac{1}{\alpha - \beta} \frac{a}{\rho} \int_0^t p(\xi)[e^{\alpha(t-\xi)} - e^{\beta(t-\xi)}]\, d\xi \tag{l}$$

where now

$$\begin{matrix}\alpha \\ \beta\end{matrix} = \gamma(-1 \pm is) \qquad \text{and} \qquad s = \sqrt{\frac{1}{1 - 2\nu}} \tag{m}$$

and s, γ [given by (f) above] are real positive numbers. Although α and β here are complex numbers, the right-hand side of Eq. (l) is real.

We can now show that the particular solution (l) is all that is required for the explosion problem. The initial condition of zero displacement requires, from Eq. (b) with $t = 0$,

$$-\frac{1}{c_1}\frac{1}{r} f'\left(-\frac{r - a}{c_1}\right) - \frac{1}{r^2} f\left(\frac{r - a}{c_1}\right) = 0 \qquad \text{for } r \geq a \tag{n}$$

The initial condition of zero velocity is expressed by first differentiating Eq. (*b*) with respect to t to give

$$\frac{\partial u}{\partial t} = -\frac{1}{c_1}\frac{1}{r} f''(\tau) - \frac{1}{r^2} f'(\tau)$$

and then setting $t = 0$ in τ. The condition is thus

$$-\frac{1}{c_1}\frac{1}{r} f''\left(-\frac{r-a}{c_1}\right) - \frac{1}{r^2} f'\left(-\frac{r-a}{c_1}\right) = 0 \qquad (o)$$

So far we have contemplated only positive values of t as the argument of the function $f(t)$. But in the initial conditions (*n*) and (*o*), the argument is $-(r-a)/c_1$, which is negative for the region $r > a$ we are concerned with. Evidently it is essential to define $f(\eta)$, using η for any argument we may have, for negative as well as positive real values.

Consider the following definition: $f(\eta)$ is given by (*l*) above, with η in place of t, for η positive; $f(\eta)$ is zero when η is negative.

Then when η is negative, the derivatives $f'(\eta)$, $f''(\eta)$ are also zero, and the initial conditions (*n*) and (*o*) are satisfied. Moreover, it follows from (*l*) that for τ positive

$$\lim_{\tau \to 0} f(\tau) = 0 \qquad \lim_{\tau \to 0} f'(\tau) = 0 \qquad (p)$$

Hence, recalling Eq. (*b*) above, the displacement at radius r remains zero until $t = (r-a)/c_1$ (that is, until $\tau = 0$), and then nonzero values develop, without discontinuity. This implies further that the material at infinity remains undisturbed. Also, if we consider the whole range of r, at any instant, there is no discontinuity in the displacement, as the physical conditions require. Evidently the definition given for $f(\eta)$ satisfies all the conditions of the problem.[1]

Cavity Pressure Suddenly Applied and Maintained In this case we may take $p(t) = p_0$, a constant,[2] for $t > 0$. Then in Eq. (*l*) we have $p(\xi) = p_0$, and the integral is easily evaluated. The result is, after replacing t by τ,

$$f(\tau) = -\frac{p_0 a^2}{2\rho\gamma c_1}\left[1 - e^{-\gamma\tau}\left(\cos \gamma s\tau + \frac{1}{s}\sin \gamma s\tau\right)\right] \qquad (a)$$

[1] Several solutions by transform methods which have appeared since 1935 are listed in the article by H. G. Hopkins, Dynamic Expansion of Spherical Cavities in Metals, in "Progress in Solid Mechanics," vol. 1, pp. 84–164, 1960, where further developments for the elastic-plastic medium, and for large deformation, are surveyed.

[2] For references to earlier solutions, and to related problems such as impulsive pressure, see p. 43 in J. N. Goodier and P. G. Hodge, "Elasticity and Plasticity," John Wiley & Sons, Inc., New York, 1958.

The displacement and stress are now obtained by using this in Eqs. (*b*), (*c*), and (*d*). S. C. Hunter (see footnote 1, page 512) has evaluated the stress-difference ratio $(\sigma_t - \sigma_r)/p_0$ at the cavity, as a function of the dimensionless time $\bar{t} = c_1 t/a$. At $\bar{t} = 0$, when the pressure is suddenly applied, the ratio rises suddenly to 0.592. It then rises further to 1.75 at $\bar{t} = 2.19$, then falls asymptotically to 1.5, which is the value corresponding to the statical problem.

Appendix

The Application of Finite-difference Equations in Elasticity

1 | Derivation of Finite-difference Equations

We have seen that the problems of elasticity usually require solution of certain partial differential equations with given boundary conditions. Only in the case of simple boundaries can these equations be treated in a rigorous manner.

Very often we cannot obtain a rigorous solution and must resort to approximate methods. As one of such methods we will discuss here the numerical method, based on the replacement of differential equations by the corresponding finite-difference equations.[1]

If a smooth function $y(x)$ is given by a series of equidistant values

[1] It seems that the first application of finite-difference equations in elasticity is due to C. Runge, who used this method in solving torsional problems. (*Z. Math. Phys.*, vol. 56, p. 225, 1908.) He reduces the problem to the solution of a system of linear algebraic equations. Further progress was made by L. F. Richardson, *Trans. Roy. Soc. (London)*, ser. A, vol. 210, p. 307, 1910, who used for the solution of such algebraic equations a certain iteration process, and so obtained approximate values of the stresses produced in dams by gravity forces and water pressure. Another iteration process and the proof of its convergence was given by H. Liebmann, *Sitzber. Bayer. Akad. Wiss.*, 1918, p. 385. The convergence of this iteration process in the case of harmonic and biharmonic equations was further discussed by F. Wolf, *Z. Angew. Math. Mech.*, vol. 6, p. 118, 1926, and R. Courant, *Z. Angew. Math. Mech.*, vol. 6, p. 322. The finite-difference method was applied successfully in the theory of plates by H. Marcus, *Armierter Beton*, 1919, p. 107; H. Hencky, *Z. Angew. Math. Mech.*, vol. 1, p. 81, 1921, and vol. 2, p. 58, 1922. Subsequently the finite-difference method found very wide application in publications by R. V. Southwell and his pupils. See R. V. Southwell, "Relaxation Methods in Theoretical Physics," Oxford University Press, Fair Lawn, N.J., 1946. Automatic computers, providing numerical solutions to several thousand simultaneous linear algebraic equations in a few minutes, have greatly extended the practical possibilities.

$y_0, y_1, y_2, \ldots$ for $x = 0, x = \delta, x = 2\delta, \ldots$, we can, by subtraction, calculate the *first differences* $(\Delta_1 y)_{x=0} = y_1 - y_0$, $(\Delta_1 y)_{x=\delta} = y_2 - y_1$, $(\Delta_1 y)_{x=2\delta} = y_3 - y_2, \ldots$. Dividing them by the value δ of the interval, we obtain approximate values for the first derivatives of $y(x)$ at the corresponding points:

$$\left(\frac{dy}{dx}\right)_{x=0} \approx \frac{y_1 - y_0}{\delta} \qquad \left(\frac{dy}{dx}\right)_{x=\delta} \approx \frac{y_2 - y_1}{\delta}, \ldots \tag{1}$$

Using the first differences, we calculate the second differences as follows:

$$(\Delta_2 y)_{x=\delta} = (\Delta_1 y)_{x=\delta} - (\Delta_1 y)_{x=0} = y_2 - 2y_1 + y_0$$

With second differences we obtain the approximate values of second derivatives such as

$$\left(\frac{d^2y}{dx^2}\right)_{x=\delta} \approx \frac{(\Delta_2 y)_{x=\delta}}{\delta^2} = \frac{y_2 - 2y_1 + y_0}{\delta^2} \tag{2}$$

If we have a smooth function $w(x,y)$ of two variables, we can use for approximate calculations of partial derivatives equations similar to Eqs. (1) and (2). Suppose, for example, that we are dealing with a rectangular boundary, Fig. 1, and that the numerical values of a function w at the nodal points of a regular *square net* with *mesh side* δ are known to us. Then we can use as approximate values of the partial derivatives of w at a point O the following expressions

$$\begin{aligned} \frac{\partial w}{\partial x} &\approx \frac{w_1 - w_0}{\delta} & \frac{\partial w}{\partial y} &\approx \frac{w_2 - w_0}{\delta} \\ \frac{\partial^2 w}{\partial x^2} &\approx \frac{w_1 - 2w_0 + w_3}{\delta^2} & \frac{\partial^2 w}{\partial y^2} &\approx \frac{w_2 - 2w_0 + w_4}{\delta^2} \end{aligned} \tag{3}$$

In a similar manner we can derive also the approximate expressions for partial derivatives of higher order. Having such expressions we can transform partial differential equations into equations of finite differences.

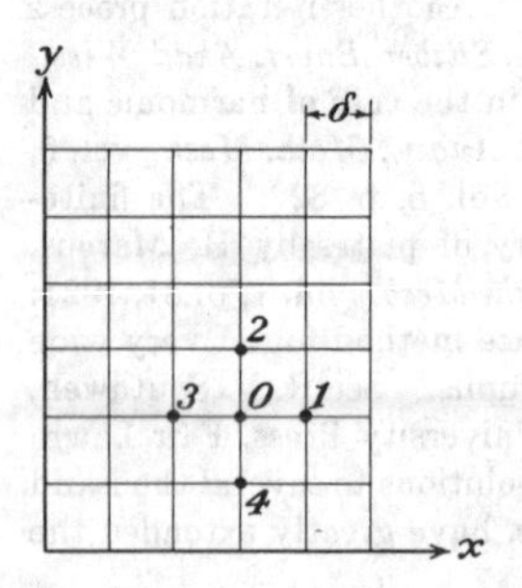

Fig. 1

Take as a first example the torsion of prismatical bars. The problem can be reduced, as we have seen,[1] to the integration of the partial differential equation

$$\frac{\partial^2 \phi}{\partial x^2} + \frac{\partial^2 \phi}{\partial y^2} = -2G\theta \tag{4}$$

in which ϕ is the stress function, which must be constant along the boundary of the cross section, θ is the angle of twist per unit length of the bar, and G is the modulus of shear. Using formulas (3), we can transform the above equation into the finite-difference equation

$$\frac{1}{\delta^2}(\phi_1 + \phi_2 + \phi_3 + \phi_4 - 4\phi_0) = -2G\theta \tag{5}$$

In this way every torsional problem reduces to the finding of a set of numerical values for the stress function ϕ which satisfy Eq. (5) at every nodal point within the boundary of the cross section and become constant along the boundary.

As the simplest example, let us consider a bar of a square cross section $a \times a$, Fig. 2, and use a square net with mesh side $\delta = \frac{1}{4}a$. From symmetry we conclude that in this case it is sufficient to consider only one-eighth of the cross section, shaded in the figure. If we determine the values α, β, γ of the function ϕ at the three points shown in Fig. 2, we shall know ϕ at all nodal points of the net within the boundary. Along the boundary we can assume it equal to zero. Thus the problem reduces to the calculation of three quantities α, β, γ, for which we can write three equations of the form (5). Observing the conditions of symmetry we obtain

$$\begin{aligned} 2\beta - 4\alpha &= -2G\theta\delta^2 \\ 2\alpha + \gamma - 4\beta &= -2G\theta\delta^2 \\ 4\beta - 4\gamma &= -2G\theta\delta^2 \end{aligned}$$

[1] See Eq. (150), p. 295.

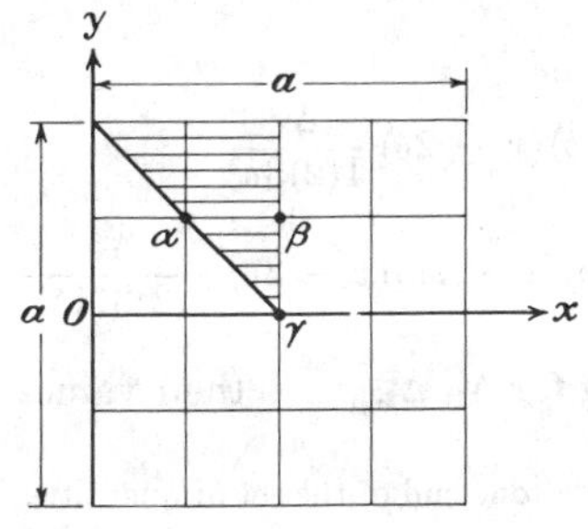

Fig. 2

Solving these equations, we find

$$\alpha = 1.375G\theta\delta^2 \qquad \beta = 1.750G\theta\delta^2 \qquad \gamma = 2.250G\theta\delta^2$$

The required stress function is thus determined by the above numerical values at all nodal points within the boundary and by zero values at the boundary.

To calculate partial derivatives of the stress function, we imagine a smooth surface having as ordinates at the nodal points the calculated numerical values. The slope of this surface at any point will then give us the corresponding approximate value of the torsional stress. Maximum stress occurs at the middle of the sides of the boundary square. To get some idea of the accuracy that can be obtained with the assumed small number of nodal points of the net, let us calculate torsional stress at point O, Fig. 2. To get the necessary slope we take a smooth curve having at the nodal points of the x axis the calculated ordinates β, γ, β. These values, divided by $\frac{1}{4}G\theta\delta^2$, are given in the second line of the table below. The remaining lines of the table give the values of the consecutive finite differences.[1] The required smooth curve is then given by Newton's interpolation formula:

$x =$	0	δ	2δ	3δ	4δ
$\phi =$	0	7	9	7	0
$\Delta_1 =$	7	2	−2	−7	
$\Delta_2 =$	−5	−4	−5		
$\Delta_3 =$	1	−1			
$\Delta_4 =$	−2				

$$\phi = \phi_0 + x\frac{\Delta_1}{\delta} + x(x - \delta)\frac{\Delta_2}{1(2\delta^2)} + x(x - \delta)(x - 2\delta)\frac{\Delta_3}{1(2)3\delta^3}$$
$$+ x(x - \delta)(x - 2\delta)(x - 3\delta)\frac{\Delta_3}{1(2)3(4\delta^4)}$$

Taking the derivative of ϕ and substituting for $\Delta_1, \Delta_2, \ldots$ their values

[1] We consider here the differences as all existing at one end of the set of quantities and use them in Newton's formula.

from the table multiplied by $G\theta\delta^2/4$ we obtain, for $x = 0$,

$$\left(\frac{\partial\phi}{\partial x}\right)_{x=0} = \frac{124}{48}\,G\theta\delta = 0.646Ga\theta$$

Comparing this result with the correct value given on page 312, we see that the error in this case is about 4.3 percent. To get better accuracy we have to use a *finer net.* Taking, for example, $\delta = a/6$, Fig. 3, we have to solve six equations and we obtain

$$\alpha = 0.952(2G\theta\delta^2) \qquad \beta = 1.404(2G\theta\delta^2) \qquad \gamma = 1.539(2G\theta\delta^2)$$
$$\alpha_1 = 2.125(2G\theta\delta^2) \qquad \beta_1 = 2.348(2G\theta\delta^2) \qquad \gamma_1 = 2.598(2G\theta\delta^2)$$

Using now seven ordinates along the x axis and calculating[1] the slope at point O, we obtain the maximum shearing stress

$$\left(\frac{\partial\phi}{\partial x}\right)_{x=0} = 0.661G\theta a$$

The error of this result is about 2 percent. Having the results for $\delta = \frac{1}{4}a$ and $\delta = \frac{1}{6}a$ a better approximation can be obtained by extrapolation.[1] It can be shown[2] that the error of the derivative of the stress function ϕ, due to the use of finite difference rather than differential equations, is proportional to the square of the mesh side, when this is small. If the error in maximum stress for $\delta = \frac{1}{4}a$ is denoted by Δ, then for $\delta = \frac{1}{6}a$ it can be assumed equal to $\Delta(\frac{2}{3})^2$. Using the values of maximum stress calculated above, we obtain Δ from the equation

$$\Delta - \Delta(\tfrac{2}{3})^2 = 0.015G\theta a$$

[1] The calculation of derivatives of an interpolation curve is greatly simplified by using the tables calculated by W. G. Bickley. These tables are given in the book by Southwell, *op. cit.*

[2] See Richardson, *loc. cit.*

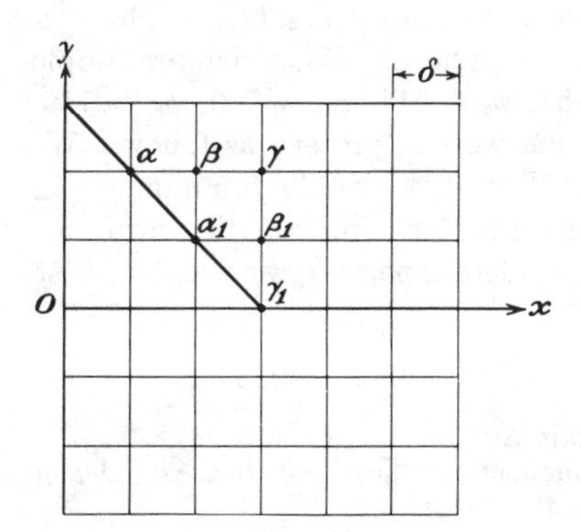

Fig. 3

from which

$$\Delta = 0.027G\theta a$$

The more accurate value of the stress is then

$$0.646G\theta a + 0.027G\theta a = 0.673G\theta a$$

which differs from the exact value $0.675G\theta a$ by less than ⅓ percent.

2 | Methods of Successive Approximation

From the simple example of the preceding article it is seen that to increase the accuracy of the finite-difference method, we must go to finer and finer nets. But then the number of equations that must be solved becomes larger and larger.[1] The solution of the equations can be greatly simplified by using a method of successive approximations. To illustrate this let us consider the equation[2]

$$\frac{\partial^2 \phi}{\partial x^2} + \frac{\partial^2 \phi}{\partial y^2} = 0 \tag{6}$$

The corresponding finite-difference equation, from Eq. (5), will be

$$\phi_0 = \tfrac{1}{4}(\phi_1 + \phi_2 + \phi_3 + \phi_4) \tag{7}$$

It shows that the true value of the function ϕ at any nodal point O of the square net is equal to the average value of the function at the four adjacent nodal points. This fact will now be utilized for calculation of the values of ϕ by successive approximations. Let us take again, as the simplest example, the case of a square boundary, Fig. 4, and assume that the boundary values of ϕ are such as shown in the figure. From the symmetry of these values with respect to the vertical central axis, we conclude that ϕ also will be symmetrical with respect to the same axis. Thus we have to calculate only six nodal values, a, b, a_1, b_1, a_2, b_2, of ϕ. This can easily be done by writing and solving six equations (7), which are simple in this case and which give $\phi_a = 854$, $\phi_b = 914$, $\phi_{a_1} = 700$, $\phi_{b_1} = 750$, $\phi_{a_2} = 597$, $\phi_{b_2} = 686$.[3] Instead of this we can proceed as follows: We assume some values of ϕ, for instance those given by the top numbers in each column written in Fig. 4. To get a better approximation for ϕ, we use Eq. (7)for each nodal point. Considering point a, we take as a first approximation the value

$$\phi_a' = \tfrac{1}{4}(800 + 1{,}000 + 1{,}000 + 900) = 925$$

[1] Use of a digital computer is considered in Art. 10.

[2] It was shown on p. 299 that torsional problems can be reduced to the solution of this equation with prescribed values of ϕ at the boundary.

[3] We make the calculation with three figures only and neglect decimals.

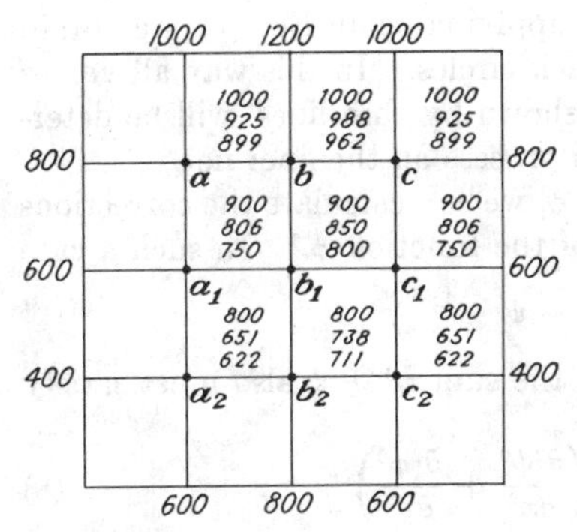

Fig. 4

In calculating a first approximation for point b, we use the calculated value ϕ_a' and also the condition of symmetry, which requires that $\phi_c' = \phi_a'$. Equation (7) then gives

$$\phi_b' = \tfrac{1}{4}(925 + 1{,}200 + 925 + 900) = 988$$

Making similar calculations for all inner nodal points, we obtain the first approximations given by the second (from the top) numbers in each column. Using these numbers we can calculate the second approximations such as

$$\phi_a'' = \tfrac{1}{4}(800 + 1{,}000 + 988 + 806) = 899$$

$$\phi_b'' = \tfrac{1}{4}(899 + 1{,}200 + 899 + 850) = 962$$

. . . .

The second approximations are also written in Fig. 4 and we can see how the successive approximations gradually approach the correct values given above. After repeating such calculations 10 times, we obtain in this case results differing from the true values by no more than one unit in the last figure and we can accept this approximation.

Generally, the number of repetitions of the calculation necessary to get a satisfactory approximation depends very much on the selection of the initial values of the function ϕ. The better the starting set of values, the less will be the labor of subsequent corrections.

It is advantageous to begin with a *coarse net* having only few internal nodal points. The values of ϕ at these points can be obtained by direct solution of Eqs. (7) or by the iteration process described above. After this we can advance to a finer net, as illustrated in Fig. 5, in which heavy lines represent the coarser net. Having the values of ϕ for the nodal points, shown by small circles, and applying Eq. (7), we calculate the values for the points marked by crosses. Using now both sets of values marked by circles and by crosses and applying again Eq. (7), we obtain values for points marked by small black circles. In this way all values for the nodal points of the finer net, shown by thin lines, will be determined, and we can begin the iteration process on the finer net.

Instead of calculating the values of ϕ, we can calculate the corrections ψ to the initially assumed values ϕ^0 of the function ϕ.[1] In such a case

$$\phi = \phi^0 + \psi$$

Since the function ϕ satisfies Eq. (6), the sum $\phi^0 + \psi$ also must satisfy it, and we obtain

$$\frac{\partial^2 \psi}{\partial x^2} + \frac{\partial^2 \psi}{\partial y^2} = -\left(\frac{\partial^2 \phi^0}{\partial x^2} + \frac{\partial^2 \phi^0}{\partial y^2}\right) \tag{8}$$

At the boundary the values of ϕ are given to us, which means that there the corrections ψ are zero. Thus the problem is now to find a function ψ satisfying Eq. (8) at each internal point and vanishing at the boundary. Replacing Eq. (8) by the corresponding finite-difference equation, we obtain for any point O of a square net (Fig. 1)

$$\psi_1 + \psi_2 + \psi_3 + \psi_4 - 4\psi_0 = -(\phi_1{}^0 + \phi_2{}^0 + \phi_3{}^0 + \phi_4{}^0 - 4\phi_0{}^0) \tag{8'}$$

The right-hand side of this equation can be evaluated for each internal nodal point by using the assumed values ϕ^0 of the function ϕ. Thus the problem of calculating the corrections ψ reduces to the solution of a system of equations similar to Eqs. (5) of the preceding article, and these can be treated by the iteration method.

3 | Relaxation Method

A method for treating difference equations, such as Eqs. (8′) in the preceding article, was developed by R. V. Southwell and was called by him

[1] This method simplifies the calculations since we will have to deal with comparatively small numbers.

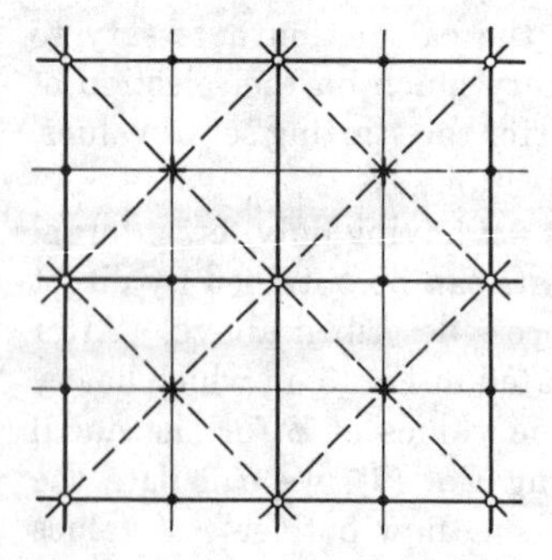

Fig. 5

the *relaxation method*. Southwell begins with Prandtl's membrane analogy,[1] which is based on the fact that the differential equation (4) for torsional problems has the same form as the equation

$$\frac{\partial^2 w}{\partial x^2} + \frac{\partial^2 w}{\partial y^2} = -\frac{q}{S} \tag{9}$$

for the deflection of a uniformly stretched and laterally loaded membrane. In this equation w denotes deflection from the initially horizontal plane surface of the membrane, q is the intensity of the distributed load, and S is the constant tensile force per unit length of the boundary of the membrane. The problem is to find the deflection w as a function of x and y which satisfies Eq. (9) at every point of the membrane and which vanishes at the boundary.

Let us derive now the corresponding finite-difference equation. For this purpose we replace the membrane by the square net of uniformly stretched strings, Fig. 1. Considering point O and denoting by $S\delta$ the tensile force in the strings, we see that the strings $O1$ and $O3$ exert on the node O, Fig. 6, a force in the upward direction, equal to[2]

$$S\delta\left(\frac{w_0 - w_1}{\delta} + \frac{w_0 - w_3}{\delta}\right) \tag{10}$$

A similar expression can be written for the force exerted by the two other strings, $O2$ and $O4$. Replacing the continuous load acting on the membrane by concentrated forces $q\delta^2$ applied at nodal points, we can now write the equation of equilibrium of a nodal point O as follows:

$$q\delta^2 + S(w_1 + w_2 + w_3 + w_4 - 4w_0) = 0 \tag{11}$$

This is the finite-difference equation, corresponding to the differential equation (9). To solve the problem, we have to find such a set of values of the deflections w that Eq. (11) will be satisfied at every nodal point.

We begin with some starting values $w_0^0, w_1^0, w_2^0, w_3^0, w_4^0, \ldots$ of the deflection. Substituting them into Eq. (11), we shall usually find that the conditions of equilibrium are not satisfied and that, to maintain the

[1] See p. 303.

[2] We consider the deflections as very small.

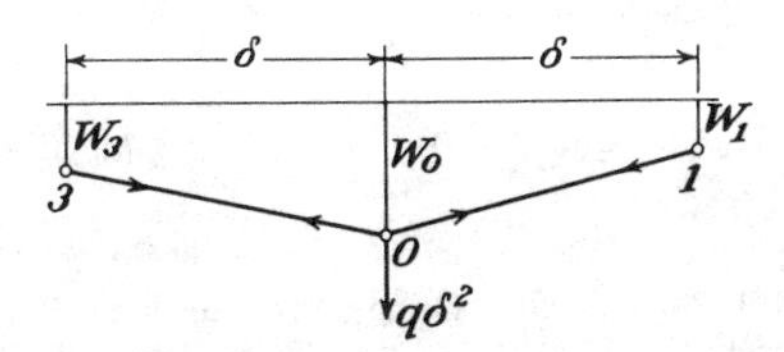

Fig. 6

assumed deflections of the net, we need to introduce supports at the nodal points. The quantities such as

$$R_0 = q\delta^2 + S(w_1{}^0 + w_2{}^0 + w_3{}^0 + w_4{}^0 - 4w_0{}^0) \tag{12}$$

will then represent the portions of the load transmitted to the supports. We call these forces *residual forces*, or residuals. Imagine now that the supports are of the *screw-jack* type, so that a controlled displacement may be imposed at any desired nodal point. Then by proper displacements of the supports we can ultimately make all residual forces (12) vanish. Such displacements will then represent the corrections that must be added to the initially assumed deflections $w_0{}^0$, $w_1{}^0$, $w_2{}^0$, . . . to get the true values of w.

The procedure that Southwell follows in manipulating the displacements of the supports is similar to that developed by Calisev[1] in handling highly statically indeterminate frames. We first displace one of the supports, say support O, Fig. 6, keeping the other supports fixed. From such equations as (11) we can see that to a downward displacement w_0' will correspond a vertical force $-4Sw_0'$ acting on the nodal point O. The minus sign indicates that the force acts upward. Adjusting the displacement so that

$$R_0 - 4Sw_0' = 0 \qquad \text{that is, } w_0' = \frac{R_0}{4S} \tag{13}$$

we make the residual force (12) vanish and there will no longer be a pressure transmitted to the support O, but at the same time pressures Sw_0' will be transferred to the adjacent supports and their residual forces will be increased by this amount. Proceeding in the same way with all the other supports and repeating the procedure several times, we can reduce all residual forces to small quantities, which can be neglected. The total displacements of the supports, accumulated in this procedure, represent then the corrections that must be added with the proper signs to the starting values $w_0{}^0$, $w_1{}^0$, $w_2{}^0$, . . . in order to obtain the true deflections of the stretched square net.

To simplify the calculations required by the procedure described, we first put Eq. (11) in nondimensional form by substituting

$$w = \frac{q\delta^2}{S}\psi \tag{14}$$

In this way we obtain

$$1 + (\psi_1 + \psi_2 + \psi_3 + \psi_4 - 4\psi_0) = 0 \tag{15}$$

where ψ_0, ψ_1, . . . are pure numbers.

[1] K. A. Calisev, Tehnicki List, 1922 and 1923, Zagreb. German translation in *Publ. Intern. Assoc. Bridge Structural Eng.*, vol. 4, p. 199, 1936. A similar method was developed in this country by Hardy Cross, *Trans. ASCE*, vol. 96, pp. 1–10, 1932.

The problem then reduces to finding such a set of values of ψ that Eq. (15) will be satisfied at all inner points of the net. At the boundary ψ is zero. To get the solution we proceed in the manner described above and take some starting values $\psi_0^0, \psi_1^0, \psi_2^0, \ldots$. They will not satisfy the equilibrium equations (15) and we shall have residuals

$$r_0 = 1 + (\psi_1^0 + \psi_2^0 + \psi_3^0 + \psi_4^0 - 4\psi_0^0) \tag{16}$$

which in this case are pure numbers.

Our problem now is to add to the assumed values $\psi_0^0, \psi_1^0, \psi_2^0, \ldots$ such corrections as to annul all residuals. Adding to ψ_0^0 a correction ψ_0' we add to the residual r_0 the quantity $-4\psi_0'$, and to the residuals of the adjacent nodal points the quantities ψ_0'. Taking $\psi_0' = r_0/4$ we shall annul the residual at the nodal point O and shall somewhat change the residuals at the adjacent nodal points. Proceeding in the same way with all nodal points and repeating the procedure many times, we shall in due course reduce the residual forces to negligible values and so obtain the values of ψ with sufficient accuracy. The corresponding values of w will then be obtained from Eq. (14).

To illustrate the procedure, let us consider the problem of torsion of a square bar, already discussed in Art. 1. In this case we have the differential equation (4). To bring it to nondimensional form let us put

$$\phi = \frac{2G\theta\delta^2}{1{,}000}\psi \tag{17}$$

The finite-difference equation (5) then becomes

$$1{,}000 + (\psi_1 + \psi_2 + \psi_3 + \psi_4 - 4\psi_0) = 0 \tag{18}$$

The denominator 1,000 is introduced into Eq. (17) for the purpose of making the ψ's such large numbers that half a unit of the last figure can be neglected. Thus we have to deal with integer numbers only. To make our example as simple as possible we will start with the coarse net of Fig. 2. Then we have to find values of ψ only for three internal points for which we already have the correct answers (see page 518). We make our square net to a large scale to have enough space to put on the sketch the results of all intermediate calculations (Fig. 7). The calculation starts with assumed initial values of ψ, which we write to the left above each nodal point. The values 700, 900, and 1,100 are intentionally taken somewhat different from the previously calculated correct values. Substituting these values together with the zero values at the boundary into the left-hand side of Eq. (18), we calculate the residual forces for all nodal points. These forces are written above each nodal point to the right. The largest residual force, equal to 200, occurs at the center of the net, and we start our relaxation process from this point. Adding to the assumed

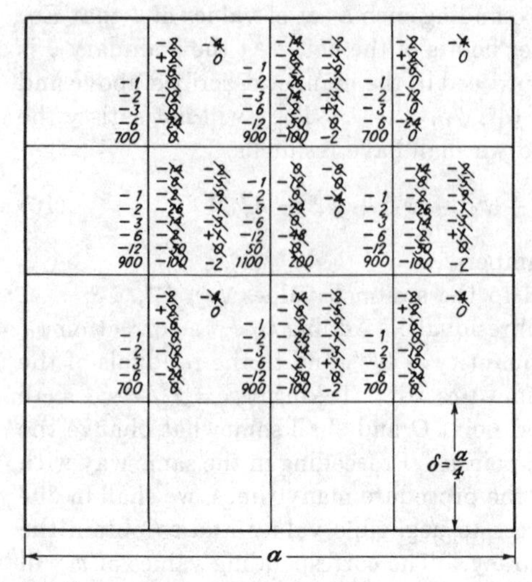

Fig. 7

value 1,100 a correction 50, which is written in the sketch above the number 1,100, we eliminate entirely the residual force at the center. Thus we cross out the number 200 in the sketch and put zero instead. Now we have to change the residuals in the adjacent nodes. We add 50 to each of those residuals and write the new values -50 of the residuals above the original values as shown in the figure. This finishes the operation with the central point of the net. We have now four symmetrically located nodal points with residuals -50 and it is of advantage to make corrections to all of them simultaneously. Let us take for all these points the same correction, equal to -12.[1] These corrections are written in the sketch above the initial values, 900. With these corrections the values $12(4) = 48$ must be added to the previous residuals, equal to -50, and we will obtain residuals equal to -2, as shown in the sketch. At the same time the forces -12 will be added to the residuals in the adjacent points. Thus, as it is easy to see, $-12(4) = -48$ must be added to the residual at the center and $-12(2) = -24$ must be added at the points closest to the corners of the figure. This finishes the first round of our calculations. The second round we again begin with the point at the center and make the correction -12, which eliminates the residual at this point and adds -12 to the residuals of the adjacent points. Taking now the points near the corners and introducing corrections -6, we eliminate

[1] We take the correction -12, instead of $-{}^{50}\!/\!_{4} = -12.5$, because it is preferable to work with integer numbers.

the residuals at these points and make the residuals equal to -26 at the four symmetrically located points. To finish the second round we introduce corrections -6 at these points. The sketch shows further corrections at all points, bringing the residuals to zero at the center and at the four points closest to the corners. The residuals at the remaining four symmetrically located points are each -2, and at these points therefore, instead of satisfying Eq. (18) exactly, we have

$$\psi_1 + \psi_2 + \psi_3 + \psi_4 - 4\psi_0 = -1{,}000 - 2$$

The residual -2 on the right is to be compared with the $-1{,}000$. Evidently it corresponds to a residual force that is acceptably small. To find the values of ψ, we add to the starting values all the corrections introduced, obtaining

$$700 - 6 - 3 - 2 - 1 = 688 \qquad 900 - 12 - 6 - 3 - 2 - 1 = 876$$
$$1{,}100 + 50 - 12 - 6 - 3 - 2 - 1 = 1{,}126$$

Equation (17) then gives for ϕ the values

$$\frac{688}{500} G\theta\delta^2 = 1.376G\theta\delta^2 = 0.0860G\theta a^2$$

$$\frac{876}{500} G\theta\delta^2 = 1.752G\theta\delta^2 = 0.1095G\theta a^2$$

$$\frac{1{,}126}{500} G\theta\delta^2 = 2.252G\theta\delta^2 = 0.1408G\theta a^2$$

which are in very good agreement with the results previously obtained (see page 518).

It is seen that Southwell's method gives us a physical picture of the iteration process of solving Eqs. (15) that may be helpful in selecting the proper order in which the nodes of the net should be manipulated.

To get a better approximation, we must advance to a finer net. Using the method illustrated in Fig. 5, we get starting values of ψ for a square net with mesh side $\delta = \frac{1}{8}a$. Applying to these values the standard relaxation process, the values of ψ for the finer net can be obtained and a more accurate value of the maximum stress can be calculated. With the two values of maximum stress found for $\delta = \frac{1}{4}a$ and $\delta = \frac{1}{8}a$ a better approximation can be found by extrapolation, as explained in Art. 1.

4 | Triangular and Hexagonal Nets

In our previous discussion a square net was used, but sometimes it is preferable to use a triangular or hexagonal net, Figs. 8*a* and *b*. Considering the triangular net, Fig. 8*a*, we see that the distributed load within the hexagon shown by dotted lines will be transferred to the nodal point O.

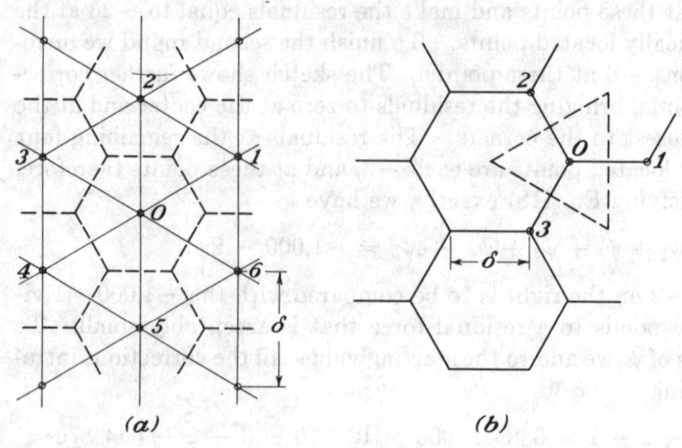

Fig. 8

If δ denotes the mesh side, the side of the above hexagon will be equal to $\delta/\sqrt{3}$ and the area of the hexagon is $\sqrt{3}\,\delta^2/2$, so that the load transferred to each nodal point will be $\sqrt{3}\,\delta^2 q/2$. This load will be balanced by forces in the strings $O1, O2, \ldots, O6$. To make the string net correspond to the uniformly stretched membrane, the tensile force in each string must be equal to the tensile force in the membrane transmitted through one side of the hexagon, i.e., equal to $S\delta/\sqrt{3}$. Proceeding now as in the preceding article, we obtain for the nodal point O the following equation of equilibrium:

$$\frac{w_1 + w_2 + \cdots + w_6 - 6w_0}{\delta}\frac{S\delta}{\sqrt{3}} + \frac{\sqrt{3}\,q\delta^2}{2} = 0$$

or

$$w_1 + w_2 + \cdots + w_6 - 6w_0 + \frac{3}{2}\frac{q\delta^2}{S} = 0 \tag{19}$$

We introduce a nondimensional function ψ defined by the equation

$$w = \frac{3}{2}\frac{q\delta^2}{S}\psi \tag{20}$$

and the finite-difference equation becomes

$$\psi_1 + \psi_2 + \cdots + \psi_6 - 6\psi_0 + 1 = 0 \tag{21}$$

Such an equation can be written for each internal nodal point, and for the solution of these equations we can, as before, use iteration or relaxation methods.

In the case of a hexagonal net, Fig. 8*b*, the load distributed over the equilateral triangle, shown in the figure by dotted lines, will be transferred to the nodal point O. Denoting by δ the length of the mesh side, we see that the side of the triangle will be $\delta\sqrt{3}$ and its area $3\sqrt{3}\,\delta^2/4$. The corresponding load is $3\sqrt{3}\,q\delta^2/4$. This load will be balanced by the tensile forces in the three strings, $O1$, $O2$, $O3$. To make the string net correspond to the stretched membrane, we take the tensile forces in the strings equal to $S\delta\sqrt{3}$. The equation of equilibrium will then be

$$\frac{w_1 + w_2 + w_3 - 3w_0}{\delta} S\delta\sqrt{3} + \frac{3\sqrt{3}\,q\delta^2}{4} = 0$$

or

$$w_1 + w_2 + w_3 - 3w_0 + \frac{3}{4}\frac{q}{S}\delta^2 = 0 \tag{22}$$

To get the finite-difference equations for torsional problems, we have to substitute in Eqs. (19) and (22) $2G\theta$ instead of q/S.

As an example let us consider torsion of a bar the cross section of which is an equilateral triangle,[1] Fig. 9. The rigorous solution for this case is given on page 300.

Using the relaxation method, it is natural to select for this case a triangular net. Starting with a coarse net, we take the mesh side δ equal to one-third of the length a of the side of the triangle. Then there will be only one internal point O of the net, and the values of the required stress function ϕ are zero at all adjacent nodal points 1, 2, . . . , 6 since these points are on the boundary. The finite-difference equation for point O is then obtained from Eq. (19) by substituting ϕ_0 for w_0 and $2G\theta$ for q/S, which gives

$$6\phi_0 = 3G\theta\delta^2 = \frac{G\theta a^2}{3}$$

and

$$\phi_0 = \frac{G\theta a^2}{18} \tag{23}$$

[1] This example is discussed in detail in Southwell's book, referred to above.

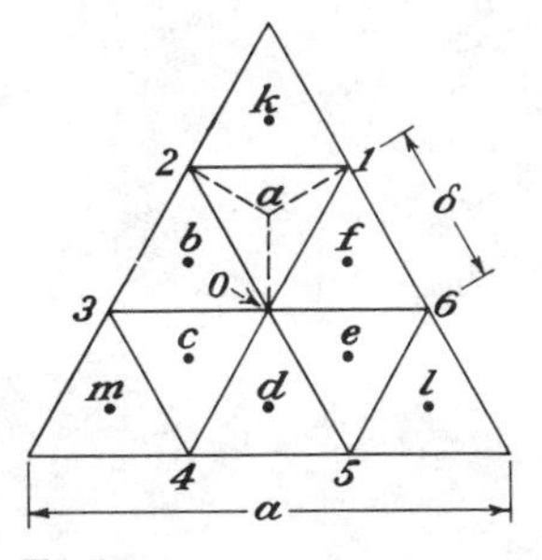

Fig. 9

Let us now advance to a finer net. To get some starting values for such a net let us consider point a—the centroid of the triangle $12O$. Assume that this point is connected to the nodal points 0, 1, and 2 by the three strings aO, $a1$, $a2$ of length $\delta/\sqrt{3}$. Considering the point a as a nodal point of a hexagonal net, Fig. 8b, substituting into Eq. (22) $\delta/\sqrt{3}$ for δ, $2G\theta$ for q/S, and taking $w_1 = w_2 = 0$, $w_3 = \phi_0$, $w_0 = \phi_a$, we obtain

$$\phi_a = \frac{1}{3}\left(\phi_0 + \frac{G\theta\delta^2}{2}\right) = \frac{G\theta a^2}{27} \tag{24}$$

The same values of the stress function may be taken also for the points $b, c, d, e,$ and f in Fig. 9. To get the values of the stress function at points $k, l, m,$ we use again Eq. (22) and, observing that in this case

$$w_1 = w_2 = w_3 = 0$$

we find

$$\phi_k = \phi_m = \phi_l = \frac{G\theta a^2}{54} \tag{25}$$

In this way we find the values of ϕ at all nodal points marked by small circles in Fig. 10. It is seen that at each of the nodal points a, c, and e there are six strings as required in a triangular net, Fig. 8a. But at the remaining nodes the number of strings is smaller than six. To satisfy the conditions of a triangular net at all internal points we proceed as indicated by dotted lines in the upper portion of Fig. 10. In this way the cross section will be divided into equilateral triangles with sides $\delta = a/9$. From symmetry we conclude that it is sufficient to consider only one-sixth of the cross section, which is shown in Fig. 11a. The values of ϕ at the nodal points O, a, b, and k are already determined. The values at the points 1, 2, and 3 will now be determined, as before, by using Eq. (22) and the values of ϕ at three adjacent points. For point 1, for example, we

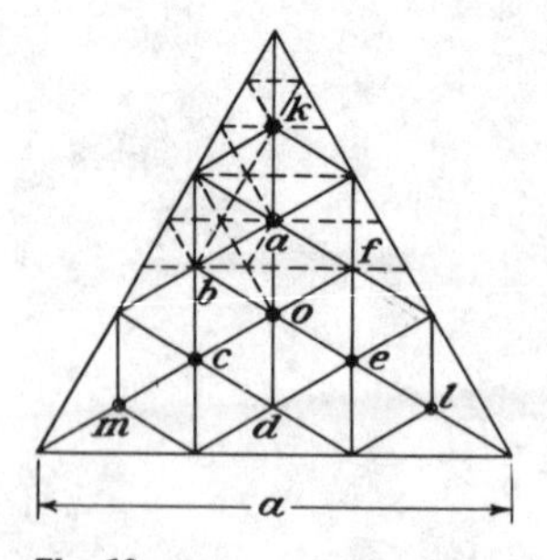

Fig. 10

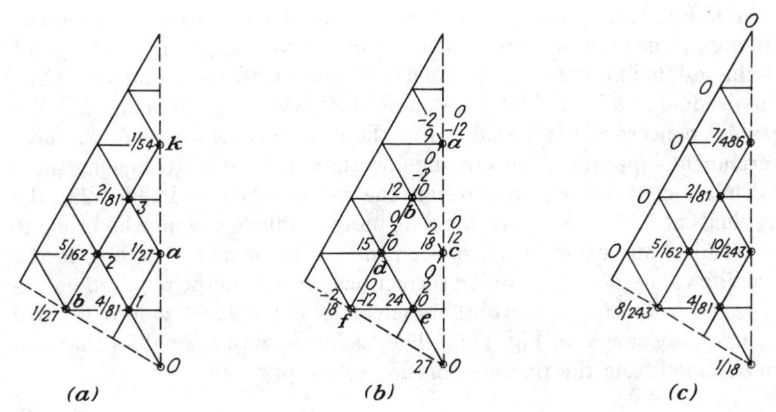

Fig. 11

will get

$$\phi_0 + \phi_b + \phi_a - 3\phi_1 + \frac{3}{4}2G\theta\left(\frac{a}{9}\right)^2 = 0$$

and substituting for ϕ_a, ϕ_b, ϕ_0 the previously calculated values, we obtain

$$\phi_1 = \tfrac{4}{81}G\theta a^2 \tag{26}$$

In a similar way ϕ_2 and ϕ_3 are calculated. All these values are written down to the left of the corresponding nodal points in Fig. 11*a*.[1] They will now be taken as starting values in the relaxation process.

In the case of torsion, Eq. (19) will be replaced by the equation

$$\phi_1 + \phi_2 + \cdots + \phi_6 - 6\phi_0 + 3G\theta\frac{a^2}{81} = 0$$

To bring it into purely numerical form we introduce the notation[2]

$$\phi = \frac{G\theta a^2\psi}{486} \qquad \text{or} \qquad \psi = \frac{486\phi}{G\theta a^2} \tag{27}$$

and obtain

$$\psi_1 + \psi_2 + \cdots + \psi_6 - 6\psi_0 + 18 = 0 \tag{28}$$

The starting values of ψ, calculated from Eq. (27), are written to the left of the nodal points in Fig. 11*b*. Substituting these values into the left-hand side of Eq. (28), we find the corresponding residuals

$$R_0 = \psi_1{}^0 + \psi_2{}^0 + \cdots + \psi_6{}^0 - 6\psi_0{}^0 + 18 \tag{29}$$

The residuals, calculated in this way, are written to the right of each nodal

[1] The constant factor $G\theta a^2$ is omitted in the figure.

[2] The number 486 is introduced so that we may work with integers only.

point in Fig. 11*b*. The liquidation of these residuals is begun with point *a*. Giving to this point a displacement $\psi_a' = -2$ we add [see Eq. (29)] $+12$ to the residual at *a* and -2 to the residuals at all adjacent points. Thus the residual at *a* is liquidated and a residual -2 appears at point *b*. We are not concerned with residuals at the boundary, since there we have permanent supports. Considering now the point *c* and introducing there a displacement $+2$, we bring to zero the residual there and add $+2$ to the residuals at *b*, *d*, and *e*. All the remaining residuals will now be brought to zero by imposition of a displacement -2 at point *f*. Adding to the starting values of ψ all recorded corrections, we obtain the required values of ψ, and from Eq. (27) we obtain the values of ϕ. These values, divided by $G\theta a^2$, are shown in Fig. 11*c*. They coincide with the values that can be obtained from the rigorous solution (*g*) on page 300.

5 | Block and Group Relaxation

The operation used up to now in liquidating the residuals consisted in manipulation of single nodal points, considering the rest of the points as fixed. Sometimes it is better to move a group of nodal points simultaneously. Assume, for example, that Fig. 12 represents a portion of a square net and that we give to all points within the shaded area a displacement equal to unity while the rest of the nodal points remain fixed. We can imagine that all nodal points of the shaded area are attached to an absolutely rigid weightless plate and that this plate is given a unit displacement, perpendicular to the plate. From considerations of equilibrium (Fig. 6), we conclude that the displacement described will produce changes of residuals at the end points of the strings attaching the shaded plate to the remaining portion of the net. If O and 1 denote the nodes at the ends of one string, the contributions to the residuals due to displacement

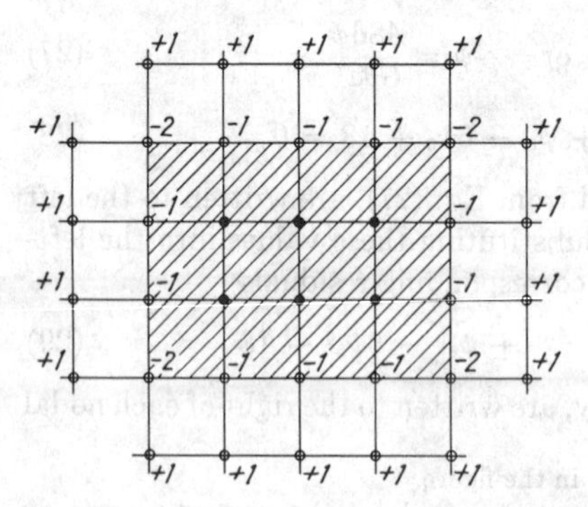

Fig. 12

w_0 and w_1 are

$$R_0 = -S\delta\frac{w_0 - w_1}{\delta} \quad \text{and} \quad R_1 = S\delta\frac{w_0 - w_1}{\delta}$$

If now we keep point 1 fixed and give to point O an additional displacement Δw_0, we get the increments of the residual forces

$$\Delta R_0 = -S\,\Delta w_0 \qquad \Delta R_1 = S\,\Delta w_0$$

Introducing dimensionless quantities according to our previous notation,

$$\frac{R}{q\delta^2} = r \qquad w = \frac{q\delta^2}{S}\psi$$

we find
$$\Delta r_0 = -\Delta\psi_0 \qquad \Delta r_1 = \Delta\psi_0$$

We see that unit increment in ψ_0 produces changes in the residuals equal to

$$\Delta r_0 = -1 \qquad \Delta r_1 = +1$$

These changes are shown in the figure. The residuals of the rest of the nodal points of the net remain unchanged. If n denotes the number of strings attaching the shaded plate to the rest of the net, the unit displacement of the plate results in diminishing by n the resultant of the residual forces of the shaded portion of the net. Choosing the displacement so that the resultant vanishes, we get residual forces that are self-equilibrating and as such lend themselves more readily to liquidation by subsequent point relaxation of the normal kind. In practical applications it is advantageous to alternate sequences of block displacement with sequences of point relaxation. Assume, for example, that the shaded area in Fig. 13 represents a portion of the triangular net. The number n of strings attaching this portion to the rest of the net is 16 and the resultant of the residuals shown in the figure is 8.8. Consequently an appropriate block displacement in this case will be $8.8/16 = 0.55$. After such a displacement the resultant of the residual forces, acting on the shaded portion of the net, vanishes and the liquidation of the residuals by subsequent point relaxation will proceed more rapidly.

Instead of giving the fictitious plate a displacement perpendicular to the plate and constant for all the nodal points attached to the plate, we

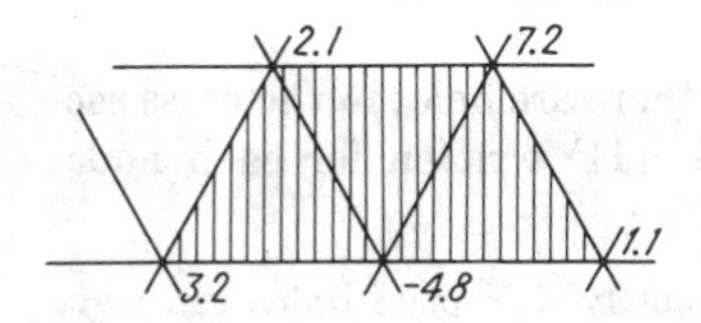

Fig. 13

can rotate the plate about an axis lying in its plane. The corresponding displacements of nodal points and changes of residuals can be readily calculated. So we can liquidate not only the resultant residual force sustained by the fictitious plate but also the resultant moment about any axis chosen in the plane of the plate.

We can also discard the notion of the fictitious plate and assign to a group of points arbitrary selected displacements. If we have some idea of the shape of the deflection surface of the net, we can select group displacements that may result in acceleration of the liquidation process.

6 | Torsion of Bars with Multiply-connected Cross Sections

It was shown[1] that in the case of bars with multiply-connected cross sections the stress function ϕ must not only satisfy Eq. (4), but along the boundary of each hole we must have

$$-\int \frac{\partial \phi}{\partial n}\, ds = 2G\theta A \tag{30}$$

where A denotes the area of the hole.

In using the membrane analogy, the corresponding equation is

$$-S\int \frac{\partial w}{\partial n}\, ds = qA \tag{31}$$

which means that the load uniformly distributed over the area of the hole[2] is balanced by the tensile forces in the membrane. Now applying finite-difference equations and considering a square net, we put $S\delta$ for the tension in the strings, w_0 for the deflection of the boundary of the hole, and w_i for the deflection of a nodal point i adjacent to the hole. Instead of Eq. (31) we then have

$$S\delta \sum \frac{(w_i - w_0)}{\delta} + qA = 0$$

or

$$S\left(\sum_{i=1}^{n} w_i - nw_0\right) + qA = 0 \tag{32}$$

where n is the number of strings attaching the area of the hole to the rest of the net. The equilibrium equation (11) is only a particular case of Eq. (32) in which $n = 4$.

We can write as many equations (32) as there are holes in the cross section. These equations together with Eqs. (11) written for each nodal

[1] See p. 330.

[2] The hole is represented by a weightless absolutely rigid plate which can move perpendicularly to the initial plane of the stretched membrane.

point of the square net are sufficient for determining the deflections of all nodal points of the net, and of all the boundaries of the holes.

Consider as an example the case of a square tube, the cross section of which is represented in Fig. 14. Taking the coarse square net, shown in the figure, and considering the conditions of symmetry, we observe that it is necessary in this case to calculate only five values, a, b, c, d, and e, of the stress function. The necessary equations will be obtained by using Eq. (32) and the four Eqs. (11) written for the nodal points a, b, c, d. Substituting $2G\theta$ for q/S and observing that $n = 20$ and $A = 16\delta^2$ we write these equations as follows:

$$
\begin{aligned}
20e - 8b - 8c - 4d &= 16(2G\theta\delta^2) \\
2b - 4a &= -2G\theta\delta^2 \\
a - 4b + c + e &= -2G\theta\delta^2 \\
b - 4c + d + e &= -2G\theta\delta^2 \\
2c - 4d + e &= -2G\theta\delta_2
\end{aligned}
$$

These equations can be readily solved and in this way we obtain

$$e = \frac{1{,}170}{488} 2G\theta\delta^2$$

and also the values a, b, c, and d.

These values, obtained with a coarse net, do not give us the stresses with sufficient accuracy, and an advance to a finer net is necessary. The results of such finer calculations, made by the relaxation method, can be found in Southwell's book.[1]

[1] R. V. Southwell, *op. cit.*, p. 60.

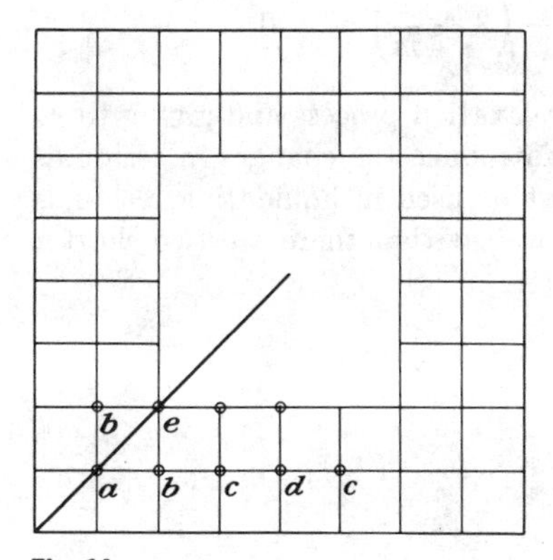

Fig. 14

7 | Points Near the Boundary

In our previous examples the nodal points of the net fall exactly on the boundary and the same standard relaxation procedure is used for all the points. Very often the points close to the boundary are connected with it by shorter strings. Due to difference in lengths of strings some changes in equilibrium equations (11) and (19) should be introduced. The necessary changes will now be discussed in connection with the example shown in Fig. 15. A flat specimen with semicircular grooves is submitted to the action of tensile forces uniformly distributed over the ends. Suppose that the difference of principal stresses at each point has been determined by the photoelastic method, as explained in Chap. 5, and that we have to determine the sum of the principal stresses, which, as we have seen (page 30), must satisfy the differential equation (6). For the points at the boundary one of the two principal stresses is known, and using the results of the photoelastic tests, the second principal stress can be calculated, so that the sum of these two stresses along the boundary is known. Thus we have to solve the differential equation (6), the values of ϕ along the boundary being known. In using the finite-difference method and taking a square net, we conclude, from symmetry, that only one-quarter of the specimen should be considered. This portion with the boundary values of ϕ is shown in Fig. 16. Considering point A of this figure, we see that three strings at that point have standard length δ while the fourth is shorter, say of length $m\delta$ ($m \approx 0.4$ in this case). This must be taken into consideration in the derivation of the equation of equilibrium of point A. This equation must be written as follows:

$$S\delta\left(\frac{\phi_a - \phi_1}{\delta} + \frac{\phi_a - \phi_2}{\delta} + \frac{\phi_a - \phi_3}{\delta} + \frac{\phi_a - \phi_4}{m\delta}\right) = 0$$

or

$$\phi_1 + \phi_2 + \phi_3 + \frac{1}{m}\phi_4 - \left(3 + \frac{1}{m}\right)\phi_a = 0$$

In applying to point A the standard relaxation process and giving to ϕ_a an increment equal to unity, we will introduce the changes in residuals shown in Fig. 17*a*. This pattern must be used in liquidating residuals at point A. In considering point B, we see that there are two shorter

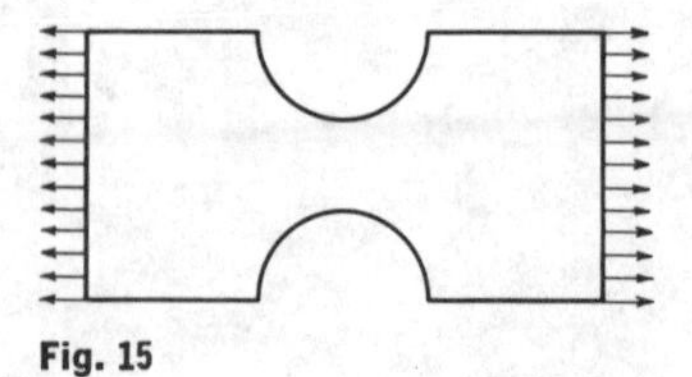

Fig. 15

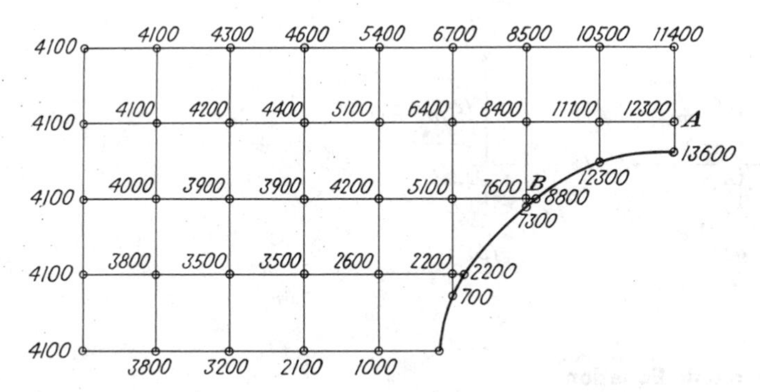

Fig. 16

strings. Denoting their lengths by $m\delta$ and $n\delta$, we find that in the liquidation of residuals at B the pattern shown in Fig. 17*b* should be used. Introducing these changes at the points near the boundary and using the standard relaxation process at all other points, the values of ϕ, shown in Fig. 16, will be obtained.[1]

In a more general case, when we are dealing with Eq. (9) and there is external load at the nodal points, we denote by $m\delta$, $n\delta$, $r\delta$, $s\delta$ the lengths of the strings at an irregular point O of a square net and take the load at O corresponding to pressure q as $\frac{1}{4}q\delta^2(m + n + r + s)$. The equation of equilibrium then will be[2]

$$\frac{q\delta^2}{4}(m + n + r + s) + S\left[\frac{w_1}{m} + \frac{w_2}{n} + \frac{w_3}{r} + \frac{w_4}{s} - w_0\left(\frac{1}{m} + \frac{1}{n} + \frac{1}{r} + \frac{1}{s}\right)\right] = 0 \quad (33)$$

For $m = n = r = s = 1$ this equation coincides with our previous Eq. (11) derived for a regular point. Using Eq. (33), the proper pattern, similar to those shown in Fig. 17, can be developed in each particular case.

With the changes discussed in this article, the relaxation process is extended to cases in which we have irregular points near the boundary.

[1] This example, using a different approximation for B, was treated by R. Weller and G. H. Shortley, *J. Appl. Mech.*, vol. 6, pp. A-71–78, 1939. A derivation of their equation may be seen in the book by S. H. Crandall, "Engineering Analysis," p. 263, McGraw-Hill Book Company, New York, 1956.

Numerical results for a complete stress problem of this kind are given, and compared with earlier results by Southwell, in D. S. Griffin and R. S. Varga, *J. Soc. Ind. Appl. Math.*, vol. 2, pp. 1047–1062, 1963.

[2] Southwell (see n. 1, p. 515) obtains an equation for a nodal point with N unequal strings at equal angles. Equation (33) is the special case $N = 4$.

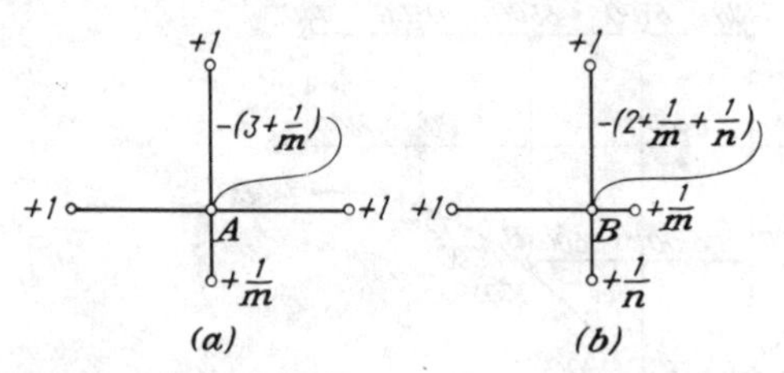

Fig. 17

8 | Biharmonic Equation

We have seen (page 32) that in the case of two-dimensional problems of elasticity, in the absence of volume forces and with given forces at the boundary, the stresses are defined by a stress function ϕ, which satisfies the biharmonic equation

$$\frac{\partial^4\phi}{\partial x^4} + 2\frac{\partial^4\phi}{\partial x^2\,\partial y^2} + \frac{\partial^4\phi}{\partial y^4} = 0 \tag{34}$$

and the boundary conditions (20), which, in this case, become

$$\begin{aligned} l\frac{\partial^2\phi}{\partial y^2} - m\frac{\partial^2\phi}{\partial x\,\partial y} &= \bar{X} \\ m\frac{\partial^2\phi}{\partial x^2} - l\frac{\partial^2\phi}{\partial x\,\partial y} &= \bar{Y} \end{aligned} \tag{35}$$

Knowing the forces distributed along the boundary, we may calculate ϕ at the boundary by integration[1] of Eqs. (35). Then the problem is reduced to that of finding a function ϕ which satisfies Eq. (34) at every point within the boundary and at the boundary has, together with its first derivatives, the prescribed values. Using the finite-difference method, let us take a square net (Fig. 18) and transform Eq. (34) to a finite-difference equation. Knowing the second derivatives,

$$\left(\frac{\partial^2\phi}{\partial x^2}\right)_0 \approx \frac{1}{\delta^2}(\phi_1 - 2\phi_0 + \phi_3)$$

$$\left(\frac{\partial^2\phi}{\partial x^2}\right)_1 \approx \frac{1}{\delta^2}(\phi_5 - 2\phi_1 + \phi_0)$$

$$\left(\frac{\partial^2\phi}{\partial x^2}\right)_3 \approx \frac{1}{\delta^2}(\phi_0 - 2\phi_3 + \phi_9)$$

[1] We consider here only simply connected regions.

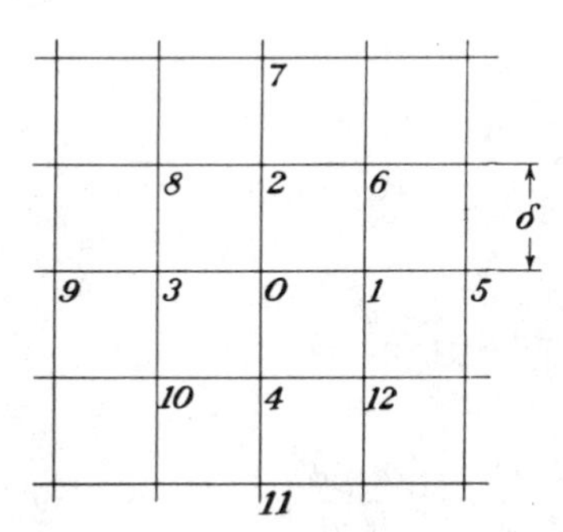

Fig. 18

we conclude that

$$\left(\frac{\partial^4\phi}{\partial x^4}\right)_0 = \frac{\partial^2}{\partial x^2}\left(\frac{\partial^2\phi}{\partial x^2}\right) \approx \frac{1}{\delta^2}\left[\left(\frac{\partial^2\phi}{\partial x^2}\right)_1 - 2\left(\frac{\partial^2\phi}{\partial x^2}\right)_0 + \left(\frac{\partial^2\phi}{\partial x^2}\right)_3\right]$$
$$\approx \frac{1}{\delta^4}(6\phi_0 - 4\phi_1 - 4\phi_3 + \phi_5 + \phi_9)$$

Similarly we find

$$\frac{\partial^4\phi}{\partial y^4} \approx \frac{1}{\delta^4}(6\phi_0 - 4\phi_2 - 4\phi_4 + \phi_7 + \phi_{11})$$

$$\frac{\partial^4\phi}{\partial x^2\,\partial y^2} \approx \frac{1}{\delta^4}4[\phi_0 - 2(\phi_1 + \phi_2 + \phi_3 + \phi_4) + \phi_6 + \phi_8 + \phi_{10} + \phi_{12}]$$

Substituting into Eq. (34), we obtain the required finite-difference equation

$$20\phi_0 - 8(\phi_1 + \phi_2 + \phi_3 + \phi_4) + 2(\phi_6 + \phi_8 + \phi_{10} + \phi_{12}) + \phi_5 + \phi_7 + \phi_9 + \phi_{11} = 0 \quad (36)$$

This equation must be satisfied at every nodal point of the net within the boundary of the plate. To find the boundary values of the stress function ϕ we integrate Eqs. (35). Observing (Fig. 20) that

$$l = \cos\alpha = \frac{dy}{ds} \qquad \text{and} \qquad m = \sin\alpha = -\frac{dx}{ds}$$

we write Eqs. (35) in the following form:

$$\begin{aligned}\frac{dy}{ds}\frac{\partial^2\phi}{\partial y^2} + \frac{dx}{ds}\frac{\partial^2\phi}{\partial x\,\partial y} &= \frac{d}{ds}\left(\frac{\partial\phi}{\partial y}\right) = \bar{X}\\ -\frac{dx}{ds}\frac{\partial^2\phi}{\partial x^2} - \frac{dy}{ds}\frac{\partial^2\phi}{\partial x\,\partial y} &= -\frac{d}{ds}\left(\frac{\partial\phi}{\partial x}\right) = \bar{Y}\end{aligned} \quad (37)$$

and by integration we obtain[1]

$$-\frac{\partial \phi}{\partial x} = \int \bar{Y}\, ds \qquad \frac{\partial \phi}{\partial y} = \int \bar{X}\, ds \tag{38}$$

To find ϕ, we use the equation

$$\frac{\partial \phi}{\partial s} = \frac{\partial \phi}{\partial x}\frac{dx}{ds} + \frac{\partial \phi}{\partial y}\frac{dy}{ds}$$

which, after integration by parts, gives[1]

$$\phi = x\frac{\partial \phi}{\partial x} + y\frac{\partial \phi}{\partial y} - \int \left(x\frac{d}{ds}\frac{\partial \phi}{\partial x} + y\frac{d}{ds}\frac{\partial \phi}{\partial y}\right) ds \tag{39}$$

Substituting in this equation the values of the derivatives given by Eqs. (37) and (38), we can calculate the boundary values of ϕ. It should be noted that in calculating the first derivatives (38), two constants of integration, say A and B, will appear and the integration in Eq. (39) will introduce a third constant, say C, so that the final expression for ϕ will contain a linear function $Ax + By + C$. Since the stress components are represented by the second derivatives of ϕ, this linear function will not affect the stress distribution and the constants A, B, C can be taken arbitrarily.

From the boundary values of ϕ and its first derivatives we can calculate the approximate values of ϕ at the nodal points of the net adjacent to the boundary, such as points A, C, E in Fig. 19. Having, for example, at point B the values ϕ_B and $(\partial\phi/\partial x)_B$, we obtain

$$\phi_C = \phi_B + \left(\frac{\partial \phi}{\partial x}\right)_B \delta \qquad \phi_A = \phi_B - \left(\frac{\partial \phi}{\partial x}\right)_B \delta$$

Similar formulas can be written also for point E. A better approximation for these quantities can be obtained later when, by further calculation, the shape of the surface representing the stress function ϕ becomes approximately known.

Having found the approximate values of ϕ at the nodal points adjacent

[1] Equivalent forms were obtained in Art. 59 as Eqs. (*d*) and (*e*).

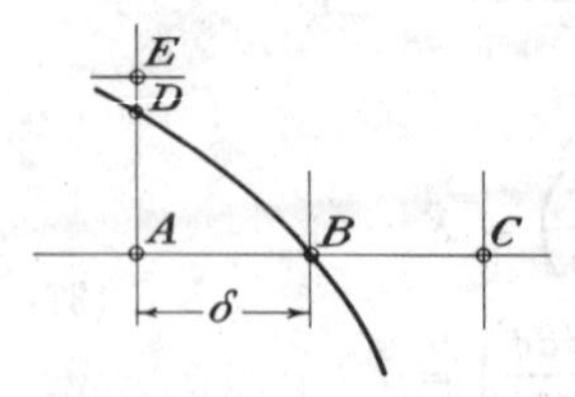

Fig. 19

to the boundary, and writing for the remaining nodal point within the boundary the equations of the form (36), we shall have a system of linear equations sufficient for calculating all the nodal values of ϕ. The second differences of ϕ can then be used for approximate calculation of stresses.

The system of Eqs. (36) may be solved directly, or we can find an approximate solution by one of the processes already described. The various methods of solution will now be illustrated by the simple example of a square plate loaded as shown[1] in Fig. 20.

Taking coordinate axes as shown in the figure,[2] we calculate the boundary values of ϕ starting from the origin. From $x = 0$ to $x = 0.4a$ we have no forces applied to the boundary; hence

$$\frac{\partial^2 \phi}{\partial x^2} = \frac{\partial^2 \phi}{\partial x \, \partial y} = 0$$

[1] Numerical solutions for many such "deep-beam" cases have been given. The set given by L. Chow, H. D. Conway, and G. Winter, *Proc. Am. Soc. Civil Eng.*, vol. 78, 1952, includes some by H. Bay (1931). The procedure in the text follows P. M. Varvak, "Collection of Papers in Structural Mechanics," vol. 3, p. 143, Kiev Structural Institute, 1936 (Russian).

[2] The system is obtained by rotating clockwise by π the axes used in Fig. 20.

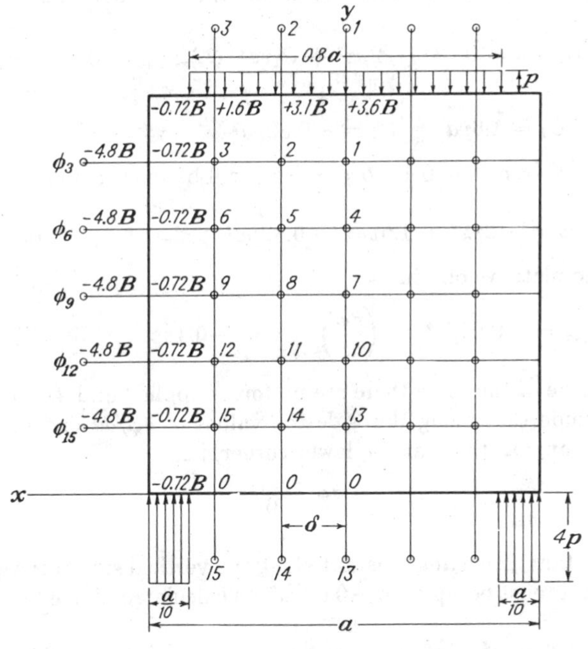

Fig. 20

and integration gives

$$\frac{\partial \phi}{\partial x} = A \qquad \phi = Ax + B \qquad \frac{\partial \phi}{\partial y} = C$$

Here A, B, C are constants along the x axis, which, as mentioned before, can be chosen arbitrarily. We assume $A = B = C = 0$. Then ϕ vanishes along the unloaded portion of the bottom side of the plate, which ensures the symmetry of ϕ with respect to the y axis. From $x = 0.4a$ to $x = 0.5a$ there acts a uniformly distributed load of intensity $4p$ and Eqs. (38) give

$$\frac{\partial \phi}{\partial x} = -\int 4p\,dx = -4px + C_1$$

$$\frac{\partial \phi}{\partial y} = 0$$

The second integration gives

$$\phi = -2px^2 + C_1x + C_2$$

The constants of integration will be calculated from the conditions that for the point $x = 0.4a$, the common point of the two parts of the boundary, the values of ϕ and $\partial\phi/\partial x$ must have the same values for both parts. Hence

$$(-4px + C_1)_{x=0.4a} = 0 \qquad (-2px^2 + C_1x + C_2)_{x=0.4a} = 0$$

and we find

$$C_1 = 1.6pa \qquad C_2 = -0.32pa^2$$

The stress function ϕ, from $x = 0.4a$ to $x = 0.5a$, will be represented by the parabola

$$\phi = -2px^2 + 1.6pax - 0.32pa^2 \tag{a}$$

At the corner of the plate we obtain

$$(\phi)_{x=0.5a} = -0.02pa^2 \qquad \left(\frac{\partial \phi}{\partial x}\right)_{x=0.5a} = -0.4pa \tag{b}$$

Along the vertical side of the plate there are no forces applied and, from Eqs. (38), we conclude that along this side the values of $\partial\phi/dx$ and of $\partial\phi/dy$ must be the same as those at the lower corner, i.e.,

$$\frac{\partial \phi}{\partial x} = -0.4pa \qquad \frac{\partial \phi}{\partial y} = 0 \tag{c}$$

From this it follows that ϕ remains constant along the vertical side of the plate. This constant must be equal to $-0.02pa^2$, as calculated above for the lower corner.

Along the unloaded portion of the upper side of the plate the first derivatives of ϕ remain constant and will have the same values (c) as calculated for the upper corner. Thus the stress function will be

$$\phi = -0.4pax + C$$

Since at the upper corner to the left ϕ must have the previously calculated value, equal to $-0.02pa^2$, we conclude that $C = 0.18pa^2$ and the stress function is

$$\phi = -0.4pax + 0.18pa^2 \tag{d}$$

Taking now the loaded portion of the upper side of the plate and observing that for this portion $ds = -dx$ and $\bar{Y} = -p$, $\bar{X} = 0$, we obtain, from Eqs. (38),

$$\frac{\partial \phi}{\partial x} = -px + C_1 \qquad \frac{\partial \phi}{\partial y} = C_2$$

For $x = 0.4a$ these values must coincide with the values (c). Hence, $C_1 = C_2 = 0$, and the stress function must have the form

$$\phi = -\frac{px^2}{2} + C$$

For $x = 0.4a$ it must have a value equal to that obtained from Eq. (d). We conclude that $C = 0.1pa^2$ and

$$\phi = -\frac{px^2}{2} + 0.1pa^2 \tag{e}$$

The stress function is represented by a parabola symmetrical with respect to the y axis. This finishes the calculation of the boundary values of ϕ and its first derivatives, since for the right-hand portion of the boundary all these values are obtained from symmetry.

With the notation

$$\frac{pa^2}{36} = B$$

we can now write all the calculated boundary values of ϕ as shown in Fig. 20.

Next, by extrapolation, we calculate the values of ϕ for the nodal points taken outside the boundary. Starting again with the bottom side of the plate and observing that $\partial\phi/\partial y$ vanishes along this side, we can take for the outside points the same values ϕ_{13}, ϕ_{14}, ϕ_{15} as for the inside points adjacent to the boundary.[1] We proceed similarly along the upper side of the plate. Along the vertical side of the plate we have the slope

$$\left(\frac{\partial \phi}{\partial x}\right)_{x=0.5a} = -0.4pa$$

[1] This manner of extrapolation, used in Varvak's paper, is different from that described on p. 540.

and we can, as an approximation, obtain the values for the outside points by subtracting the quantity

$$0.4pa(2\delta) = \frac{0.4pa^2}{3} = 4.8B$$

from the inside points adjacent to the boundary, as shown in Fig. 20.

Now we can start the calculation of ϕ values for the inside nodal points of the net. Using the method of direct solution of the difference equations, we have to write in this symmetrical case the Eqs. (36) for the 15 points shown in Fig. 20. The solution of these equations gives for ϕ the values shown in the table below.

	1	2	3	4	5	6	7	8	9	10	11	12	13	14	15
ϕ/B	3.356	2.885	1.482	2.906	2.512	1.311	2.306	2.024	1.097	1.531	1.381	0.800	0.634	0.608	0.396

Let us calculate the normal stress σ_x along the y axis. The values of this stress are given by the second derivative $\partial^2\phi/\partial y^2$. Using finite differences, we obtain for the upper point $(y = a)$

$$(\sigma_x)_{y=a} \approx \frac{[3.356 - (2)3.600 + 3.356]B}{\delta^2} = -\frac{0.488pa^2}{36\delta^2} = -0.488p$$

For the lower point $(y = 0)$ we find

$$(\sigma_x)_{y=0} = \frac{(0.634 - 0 + 0.634)B}{\delta^2} = 1.268p$$

If we consider the plate as a beam on two supports and assume a linear distribution of σ_x over the middle cross section $(x = 0)$, we find $(\sigma_x)_{\max} = 0.60p$. We can see that for a plate of such proportions the usual beam formula gives a very unsatisfactory result.

To solve the finite-difference equations (36) by iteration, we assume some starting values $\phi_1, \quad \phi_2, \; . \; . \; . \; , \phi_{15}$ for the stress function. Substituting these into Eqs. (36), we obtain residual forces for all internal nodal points that can be liquidated by a relaxation process. The proper pattern, as obtained from Eqs. (36), is shown in Fig. 21, in which the changes in residuals due to unit change of ϕ_0 are given. In applying this method to the square plate discussed above, it must be observed that the ϕ values along the boundary are restricted by the boundary conditions, which means that the residual forces at points on the boundary need not be liquidated.

We can next advance to a finer net, obtaining starting values of ϕ from the results of the calculation on the coarse net.

In the case of a nonsymmetrical loading such as shown in Fig. 22a,

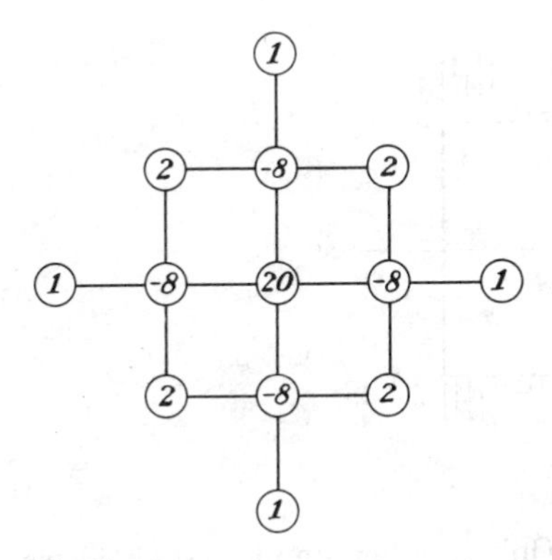

Fig. 21

we can split the load as shown in Figs. 22*b* and *c* into symmetrical and antisymmetrical loadings. In both latter cases we have to consider one-half of the plate, since $\phi(x,y) = \phi(-x,y)$ for the symmetrical case and $\phi(x,y) = -\phi(-x,y)$ for the antisymmetrical loading.

The work can be further reduced by considering also the horizontal axis of symmetry of the rectangular plate. The load shown in Fig. 20 can be resolved into symmetrical and antisymmetrical cases as shown in Fig. 23. For each of these cases only one-quarter of the plate should be considered in calculating numerical values of the stress function.

9 | Torsion of Circular Shafts of Variable Diameter

In this case, as we have seen (page 343), it is necessary to find a stress function that satisfies the differential equation

$$\frac{\partial^2 \phi}{\partial r^2} - \frac{3}{r}\frac{\partial \phi}{\partial r} + \frac{\partial^2 \phi}{\partial z^2} = 0 \tag{40}$$

at every point of the axial section of the shaft, Fig. 24, and is constant

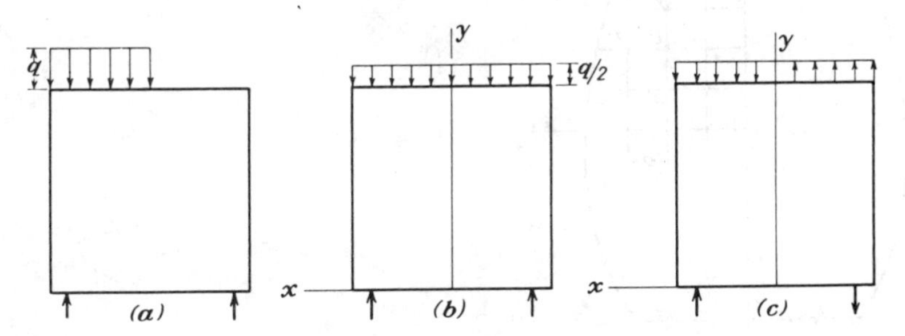

Fig. 22

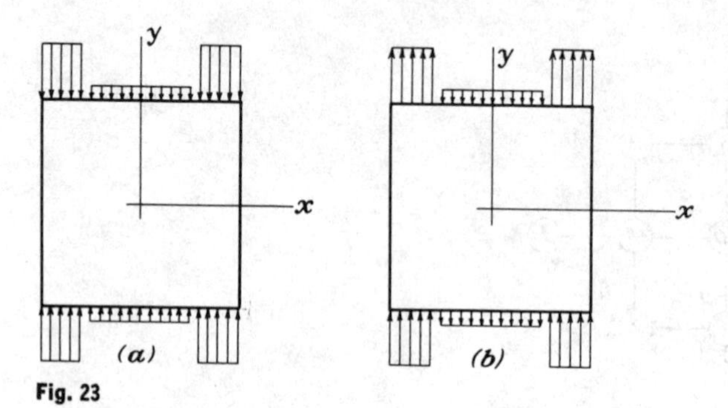

Fig. 23

along the boundary of that section. Only in a few simple cases have we a rigorous solution of the problem and in practical cases we must usually resort to approximate methods.

Using the finite-difference method, we shall take a square net. Considering a nodal point O, Fig. 24, we can treat the second derivatives in Eq. (40) as before. For the first derivative we can take

$$\left(\frac{\partial \phi}{\partial r}\right)_{r=r_0} \approx \frac{1}{2}\left(\frac{\phi_1 - \phi_0}{\delta} + \frac{\phi_0 - \phi_3}{\delta}\right) = \frac{\phi_1 - \phi_3}{2\delta}$$

Then the finite-difference equation, corresponding to Eq. (40), is

$$\phi_1 + \phi_2 + \phi_3 + \phi_4 - 4\phi_0 - \frac{3\delta}{2r_0}(\phi_1 - \phi_3) = 0 \qquad (41)$$

The problem then is to find such a set of values of ϕ that Eqs. (41) will be satisfied at every nodal point of the net and ϕ will be equal to the assumed constant value at the boundary. This problem can be treated either by direct solution of Eqs. (41) or by one of the iteration methods.

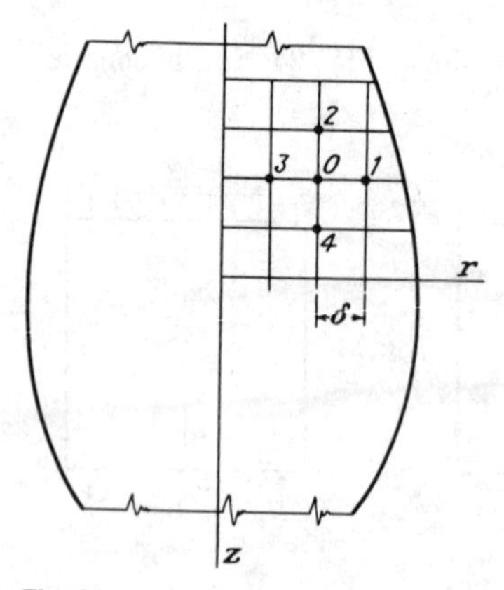

Fig. 24

As an example, let us consider the case shown in Fig. 25. In the region of rapid change in the diameter there will be a complicated stress distribution, but at sufficient distances from the fillets a simple Coulomb solution will hold with sufficient accuracy, and the stress distribution will be independent of z. Equation (40) for such points becomes

$$\frac{d^2\phi}{dr^2} - \frac{3}{r}\frac{d\phi}{dr} = 0 \tag{42}$$

The general solution of this equation is

$$\phi = Ar^4 + B \tag{43}$$

and the corresponding stresses are (see page 343)

$$\tau_{z\theta} = \frac{1}{r^2}\frac{d\phi}{dr} = 4Ar \qquad \tau_{r\theta} = 0$$

Comparing this result with the Coulomb solution, we find

$$4A = \frac{M_t}{I_p}$$

where M_t is the applied torque and I_p is the polar moment of inertia of the shaft. Omitting the constant B in the general solution (43) as having no effect on the stress distribution, we find for the stress function at sufficient distances from the fillet the expressions

$$\phi_a = \frac{M_t}{2\pi a^4}r^4 \qquad \phi_b = \frac{M_t}{2\pi b^4}r^4 \tag{44}$$

These expressions vanish at the axis of the shaft and assume at the bound-

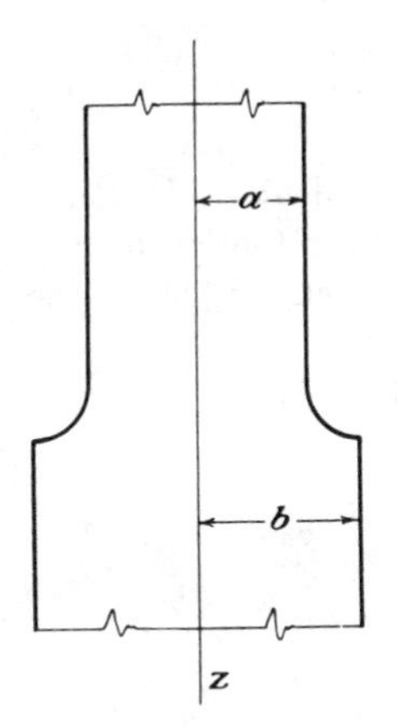

Fig. 25

ary a common value $M_t/2\pi$. Since ϕ is constant along the boundary, the value $M_t/2\pi$ holds also for the fillets. Thus, selection of the constant at the boundary in solving Eqs. (41) is equivalent to assuming a definite value for the torque.

In solving Eqs. (41), we can again apply the membrane analogy. We begin with points where Eq. (42) holds. The corresponding finite-differences equation is

$$\phi_1 + \phi_3 - 2\phi_0 - \frac{3\delta}{2r_0}(\phi_1 - \phi_3) = 0 \tag{45}$$

This equation is of the same form as that for deflections to a cylindrical form of a membrane with tension varying inversely as r^3. To show this let us consider three consecutive points of the net, Fig. 26. The corresponding deflections we denote by w_3, w_0, w_1.

The tension at the middle of the strings $3O$ and $O1$ will be

$$\frac{S\delta}{(r_0 - \delta/2)^3} \approx \frac{S\delta}{r_0^3}\left(1 + \frac{3\delta}{2r_0}\right)$$

and

$$\frac{S\delta}{(r_0 + \delta/2)^3} \approx \frac{S\delta}{r_0^3}\left(1 - \frac{3\delta}{2r_0}\right)$$

The equation of equilibrium for the point O is then

$$\frac{S\delta}{r_0^3}\left(1 - \frac{3\delta}{2r_0}\right)\frac{w_1 - w_0}{\delta} + \frac{S\delta}{r_0^3}\left(1 + \frac{3\delta}{2r_0}\right)\frac{w_3 - w_0}{\delta} = 0$$

or

$$w_1 - 2w_0 + w_3 - \frac{3\delta}{2r_0}(w_1 - w_3) = 0$$

This is the same as Eq. (45).

Similarly, in the general case, observing that the tension in the membrane does not depend on z, we obtain the equation

$$w_1 + w_2 + w_3 + w_4 - 4w_0 - \frac{3\delta}{2r_0}(w_1 - w_3) = 0 \tag{46}$$

which agrees with Eqs. (41). It is seen that we can calculate the stress function as the deflection of a membrane with nonuniform tension having constant deflection $M_t/2\pi$ along the boundary and deflections (44) at the

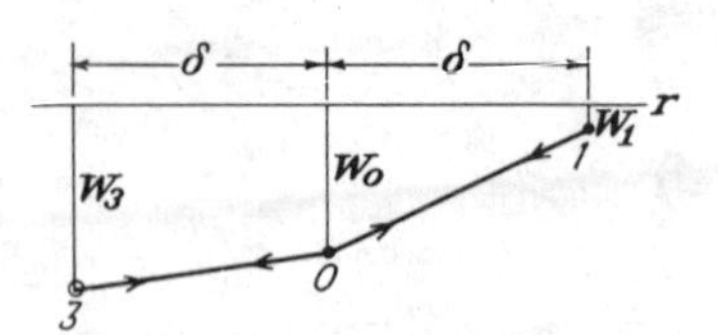

Fig. 26

points at large distances from the fillets. We assume some starting values for w at the nodal points, substitute them into the left-hand sides of Eqs. (46), and calculate the residuals. Now the problem is to liquidate all these residuals by the relaxation process. From Fig. 26 we see that by giving to point O a displacement unity we add to the residuals at points 1 and 3 the quantities

$$\frac{S}{r_0^3}\left(1 - \frac{3\delta}{2r_0}\right) \quad \text{and} \quad \frac{S}{r_0^3}\left(1 + \frac{3\delta}{2r_0}\right)$$

which indicates that the pattern for the relaxation process is as shown in Fig. 27. It varies from point to point with variation of the radial distance r_0. Calculations of this kind were carried out by R. V. Southwell and D. N. de G. Allen.[1]

10 | Solutions by Digital Computer[2]

In plane and axisymmetric problems involving boundary shapes and loading conditions more complex than the simple cases we have already considered, the number of finite-difference equations required for practical accuracy soon becomes far too great for desk calculation. It is then appropriate to program the solution for a high-speed automatic digital computer.

The program must put into effect one or another of the basic methods available for the solution of such systems of equations. The relaxation technique is not readily adaptable to automatic computation. Direct

[1] *Proc. Roy. Soc. (London)*, ser. A, vol. 183, pp. 125–134. See also Southwell's book, *op. cit.*, p. 152.

[2] This article is based on the paper by D. S. Griffin and R. B. Kellogg, A Numerical Solution for Axially Symmetrical and Plane Elasticity Problems, *Intern. J. Solids Struc.*, vol. 3, pp. 781–794, 1967, and Figs. 28 and 29 are reproduced by permission from this paper. Computer procedures for problems beyond linear elasticity are illustrated in B. Alder, S. Gernbach, and M. Rötenberg (eds.), "Methods in Computational Physics," Academic Press Inc., New York, 1964, in particular, in the article in vol. 3 by M. L. Wilkins, "Calculation of Elastic-Plastic Flow." See also the references in n. 1, p. 551.

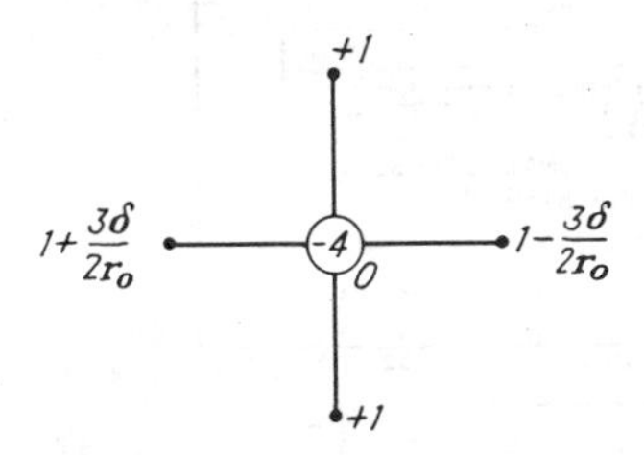

Fig. 27

methods such as gaussian elimination (or Cramer's rule) can be used, but the number of equations is still unduly limited. Iterative schemes,[1] however, permit efficient solution for several thousand unknowns if the matrix of coefficients in the equations has suitable properties. This requirement makes it appropriate to work with displacements rather than stress functions.

Results obtained for a nonuniform mesh having 525 interior and boundary nodal points are shown in Fig. 28. The physical problem is to find the stress in an internally pressurized cylinder, the wall thickness changing through a fillet as indicated by the axial section in Fig. 29. As an axisymmetric problem with two displacement components at each point, there are some 1,050 unknowns to be found. The surface stress values passing round the fillet (angular coordinate α indicated in Fig. 28) are shown by the curves. The circles and squares show photoelastic results[2] for comparison.

The finite-difference equations to be solved by the computer can be derived in various ways. Article 1 of this Appendix illustrates the mathematical conversion from the partial differential equations of a continuum.

[1] See, for instance, (1) G. E. Forsythe and W. R. Wasow, "Finite Difference Equations for Partial Differential Equations," John Wiley & Sons, Inc., New York, 1960; (2) R. S. Varga, "Matrix Iterative Analysis," Prentice-Hall, Inc., Englewood Cliffs, N.J., 1962.

[2] By M. M. Leven.

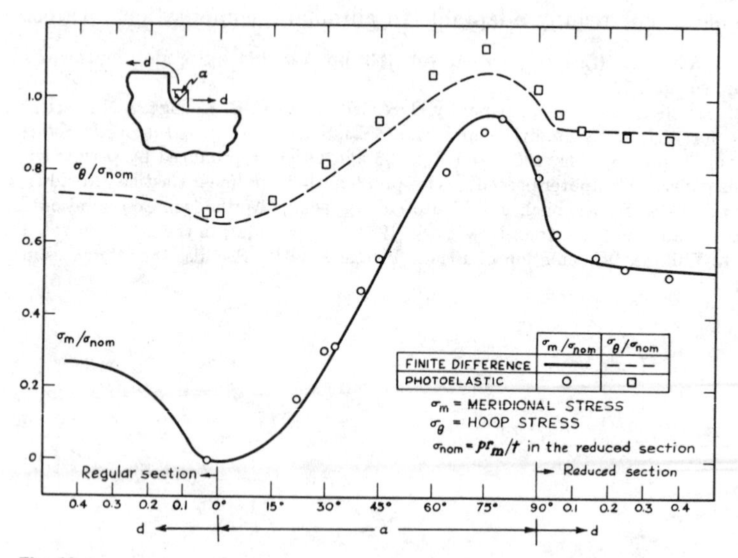

Fig. 28

Variational methods can also be used. For instance, in the problem in Fig. 29 the potential energy was expressed as a sum involving the nodal point displacements, then minimized. Article 3 of this Appendix illustrates the "physical" conversion from the continuum (membrane) to a net of uniformly stretched strings. The finite-difference equations are then derived as the physical equations for a finite element of the net. Similar procedures for more elaborate problems are included in what is now known as the *finite-element method.*[1]

[1] See, for instance, (1) R. W. Clough, The Finite Element Method in Structural Mechanics, in O. C. Zienkiewicz and G. S. Hollister (eds.), "Stress Analysis," John Wiley & Sons, Inc., New York, pp. 85–119, 1965; (2) several technical notes by J. H. Argyris in *J. Roy. Aeron. Soc.*, vols. 69 and 70, 1965 and 1966.

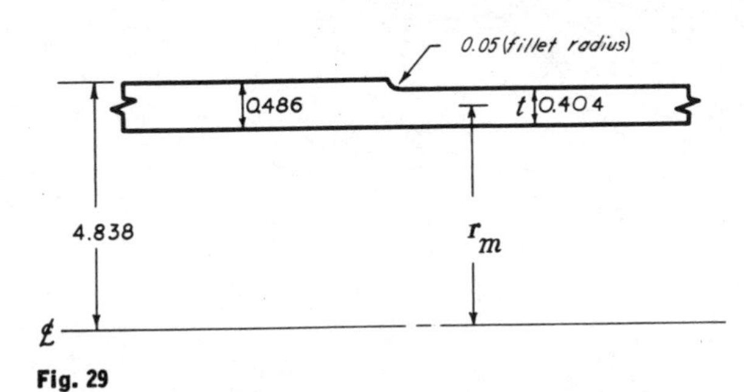

Fig. 29

Name Index

Subject Index